Advances in Thick Section Composite and Sandwich Structures

Sung W. Lee
Editor

Advances in Thick Section Composite and Sandwich Structures

An Anthology of ONR-sponsored Research

Springer

Editor
Sung W. Lee
Department of Aerospace Engineering
University of Maryland, College Park
College Park, MD, USA

ISBN 978-3-030-31064-6 ISBN 978-3-030-31065-3 (eBook)
https://doi.org/10.1007/978-3-030-31065-3

This Springer imprint is published by the registered company Springer Nature Switzerland AG
The registered company address is: Gewerbestrasse 11, 6330 Cham, Switzerland

Preface

This book contains a collection of contributed papers based on the research achievements in the area of composites and sandwich structures sponsored by the US Office of Naval Research (ONR) Solid Mechanics Program over the last decade or so, with Dr. Yapa D.S. Rajapakse as program manager. Composite materials exhibit higher specific stiffness and strength relative to conventional metals as well as higher resistance to fatigue and corrosion. Sandwich structures with composite facesheets and various types of core can provide lightweight solutions to naval structures. The facesheets are the primary load-carrying members providing in-plane strength and stiffness, while the core maintains the distance between the facesheets and carries any out-of-plane shear and compressive loads. The naval composite and sandwich structures tend to be thick with a large number of plies through thickness. They are also subject to highly dynamic, multiaxial loading and severe environmental conditions, including exposure to sea water, moisture, extreme temperatures such as in the Arctic regions, hydrodynamic pressure, and fluid-structure interactions.

The ONR Solid Mechanics Program, under the leadership of Dr. Rajapakse, has supported research projects in understanding the fundamentals of mechanics associated with nonlinear response and integrity of composite and sandwich structures subjected to severe and extreme loading and environmental conditions. This book with 22 chapters is an embodiment of tremendous advances made by ONR contractors and recognized experts and may serve as a reference and guide for future research. The various topics covered in the book are described as follows:

"Dynamic Response of Composite Structures in Extreme Loading Environments," by A. Shukla, C. Salazar, S. Kishore, and H. Matos, describes the research on the response of composite and sandwich structures to various extreme in-air and underwater loading conditions investigated using 3D digital image correlation (DIC), along with various mitigation techniques. A shock tube facility was used to generate in-air shock loading, and underwater experiments were conducted in a large, semispherical pressure vessel to investigate response of composite and sandwich structures under extreme hydrostatic loading. For composite and sandwich structures subjected to in-air blast loading, stepwise foam core gradation and strategic placement of polyurea coatings were shown to successfully mitigate structural damage. For underwater implosion of composite cylinders, it was shown that polyurea coatings reduced the energy emitted by implosion pulses when located on the inside of the implodable structure and increased the energy emitted when located on the outside of the structure.

"The Response of Composite Materials Subjected to Underwater Explosive Loading: Experimental and Computational Studies," by J. LeBlanc, E. Gauch, C. Javier, and A. Shukla, presents a summary of research investigations focused on the response of composite materials subject to highly transient loading conditions, primarily underwater explosions. Included in the results are the influence of material aging, elastomeric coatings, and structural geometry on the dynamic response and material damage characteristics. Each study incorporates advanced experimental approaches coupled with detailed computational simulations to present a full breadth of the available techniques for the study of such problems.

"Blast Performance of Composite Sandwich Panels," by J.P. Dear, summarizes the work done by the author's group, including Emily Rolfe, Mark Kelly, Hari Arora, and others, at Imperial College London. The results of their work showed the benefits of hybrid CFRP/GFRP sandwich composites when subjected to air and underwater blast loading. They also showed that graded cores in sandwich materials could much improve blast resistance, as first shown by Shukla, LeBlanc, and co-researchers at URI. The high-speed photography was applied in conjunction with digital image correlation (DIC) to monitor the deformation and strain of composite sandwich materials under full-scale explosive blast loading.

"Explosive Blast Response of Marine Sandwich Composites," by A. Gargano, R. Das, and A.P. Mouritz, examines the effect of water absorption on the deformation and damage caused to sandwich composite materials with fiber-polymer laminate facesheets when subjected to explosive blasts. Water absorbed by composites from the environment can weaken the material and consequently lower the blast resistance. The effect of increasing amounts of absorbed water on the blast response of several types of sandwich composites used in naval vessels and commercial ships is investigated. The deterioration of the blast performance is dependent on the materials used for the laminate facesheets and the core.

"Dynamic Response of Polymers Subjected to Underwater Shock Loading or Direct Impact," by V. Eliasson and R. Chavez, summarizes experimental findings in two areas: underwater shock wave focusing with emphasis on fluid structure interaction and dynamic fracture in polymeric specimens. Underwater shock focusing was generated by means of a gas gun projectile that impacted onto a specimen containing a convergent water-filled region, and results were obtained using high-speed visualization techniques. The surrounding materials ranged from bulk specimens made of polycarbonate to thin specimens made of carbon fiber vinyl ester composites. The dynamic fracture experiments were performed on both dry samples and samples conditioned to contain water. Evaluation was done using high-speed visualization techniques and strain gages.

"Recent Developments on Ballistic Performance of Composite Materials of Naval Relevance," by V. Sánchez-Gálvez, D.A. Cendón, R. Sancho, and F. Gálvez, describes the work on ballistic performance of composites which is very important to assess the safety of naval ships. Analytical model developed is a useful tool to assess the performance of composite panels against different threats. The chapter includes experimental results of ballistic performance of carbon, glass, and hybrid reinforced laminates at different temperatures and environmental conditions as well as analytical results, showing excellent agreement between experimental and analytical results.

"Fluid-Structure Interaction of Composite Structures," by Y.W. Kwon, presents various topics associated with fluid-structure interaction of composite structures. These include underwater impact, structural coupling by fluid, fluid-solid particle-structure interaction, etc. The problems were investigated experimentally, numerically, or using both. The study suggests that the effects of fluid-structure interaction are important for the dynamic response of marine composite structures.

"Low-Velocity Impact of Marine Composites: Experiments and Theory," by V. Lopresto, I. Papa, V. Pagliarulo, P. Russo, and M. Porfiri, seeks to provide an overview on the dynamic behavior of Navy-relevant carbon fiber composite materials to low-velocity impact at room and extreme temperatures. The study focuses on laminates toward establishing a compelling body of empirical results in order to support numerical and semi-empirical models for understanding of the fluid-structure interaction and investigate the role of geometric and physical parameters on the dynamic response of water-backed panels. The main novelty of the research lies in the experimental scheme, based on a modified drop-weight machine, which was realized to dynamically load panels, across a range of temperatures in the presence of the water to simulate realistic operating conditions of marine vessels. Nondestructive techniques were adopted to investigate the internal damages which are difficult to detect.

"Inferring Impulsive Hydrodynamic Loading During Hull Slamming from Water Velocity Measurements," by M. Porfiri, reviews technical progress toward quantification of the hydrodynamic loading experienced during hull slamming. Building upon established techniques in experimental fluid mechanics, he develops a viable methodology to achieve spatially distributed, temporally resolved inference of impulsive hydrodynamic loadings from velocity measurements. The methodology is demonstrated through a series of exemplary problems, illuminating the critical role of physical and geometric factors on hull slamming.

"Response of Sandwich Structures to Blast Loads," by R.C. Batra, A.P. Chattopadhyay, and P.H. Shah, summarizes the author's their group's work on studying transient deformations of doubly curved sandwich shells by using a third-order shear and normal deformable theory, optimizing their designs for the maximum first failure load, fluid-structure interaction related to water slamming, and high-speed viscous flows interacting with rigid solids.

"The Extended High-Order Sandwich Panel Theory for the Static and Dynamic Analysis of Sandwich Structures," by G.A. Kardomateas, presents a summary of research investigations focused on the development of a theory for sandwich structures that includes the transverse shear and in-plane stiffness and transverse compressibility of the core. The accuracy of his theory is validated by considering the cases of static distributed loading, dynamic blast loading, and wrinkling behavior of sandwich beams, in comparison with the elasticity solutions.

"Mechanics Based Modeling of Composite and Sandwich Structures in the Naval Environment: Elastic Behavior, Fracture, and Damage Evolution," by R. Massabò, summarizes selected works on mechanics-based modeling of composite and sandwich structures for marine applications. Elastic response, wave propagation, and progressive delamination fracture in layered plates subject to thermo-mechanical loadings are studied using the theory of elasticity, matrix methods, and a novel homogenized structural model. The evolution and interaction of multiple damage modes in sandwich beams with compressive yielding cores are investigated. Interface fracture mechanics solutions are derived, which are useful for the characterization of sandwich composites, and account for the effects of shear on energy release rate and mode mixity.

"On Characterizing Multiaxial Polymer Foam Properties in Sandwich Structures," by M.S. Hoo Fatt, C. Zhong, and X. Tong, describes the work to develop accurate material models to address multiaxial, crushable elastic-plastic behavior of polymer foams. Such a material model is needed to determine the behavior of the polymer foam core of externally loaded curved sandwich panels and sandwich panels subjected to underwater explosions. Novel pressure chamber experiments were performed to obtain triaxial foam material properties, and a new elastic-plastic, viscoelastic-damage constitutive equation was developed to predict the response of the foam.

"3D Printing of Syntactic Foams for Marine Applications," by N. Gupta and M. Doddamani, describes the method of developing filament of syntactic foams for material extrusion-type 3D printers. The filaments are tested for density and mechanical properties. These filaments are used for a standard desktop 3D printer to print tensile test specimens. The results show that the properties of 3D printed parts are better than the injection molded syntactic foam specimens of the same composition. Furthermore, recyclability of syntactic foams is studied by cutting the drawn filaments and redrawing them two and three times. Results show that syntactic foams can be recycled without significant change in their properties.

"Damage Tolerance Assessment of Naval Sandwich Structures with Face-Core Debonds," by C. Berggreen and B. Hayman, describes research carried out at the Technical University of Denmark (DTU) during the past 15–20 years and related work at collaborating universities and research institutions in Europe and the USA. Much of this research has been supported by the ONR Solid Mechanics Program. Modeling and physical testing of sandwich structures with face-core debonds, at both the small characterization test specimen scale and the larger ship structural component scale, is addressed. Previously developed schemes for assessing damage experienced by sandwich structures in naval ships and their influence on residual structural strength are reviewed, and possibilities for devising similar procedures for assessing residual fatigue life are explored.

"Modeling Nonlinear and Time-Dependent Behaviors of Polymeric Sandwich Composites at Various Environmental Conditions," by B. Davoodi, A. Gomez, B. Pinto, A. Muliana, and V. La Saponara, notes that when exposed to mechanical loading and hostile environments, the constituents of the sandwich structures experience different time-dependent behavior and degradation, leading to complex failure mechanisms in sandwich composites. Accordingly, their study aims at predicting the performance of sandwich composites subjected to various mechanical loadings and environmental conditions, by taking into account the influence of environments on the viscoelasticity of the constituents in the sandwich composites.

"Toward More Representative Accelerated Aging of Marine Composites," by P. Davies, describes the widely used approach to accelerated testing of wet aging by raising water temperature. The advantages and difficulties of this method are discussed, and alternative strategies are proposed. Unique data-sets from aging tests lasting up to 8 years are used to examine the validity of extrapolating long-term predictions from short-term data.

"Statistical Long-term Creep Failure Time of Unidirectional CFRP," by Y. Miyano and M. Nakada, proposes a formulation for the statistical time- and temperature-dependent static and creep strengths for CFRP based on the matrix resin viscoelasticity. The validity of formulation is clarified experimentally for tension loading along the longitudinal direction of unidirectional CFRP using the CFRP strand system they have developed. Results demonstrate that the long-term creep failure time of unidirectional CFRP under tension loading can be predicted statistically and easily using static tests for CFRP strands at various temperatures.

"Effect of Seawater on Carbon Fiber Composite Facings and Sandwich Structures with Polymeric Foam Core," by D. Penumadu, notes that exposure to sea environment and coupled effects with temperature change to ship sandwich structures over extended period of time often leads to degradation of their mechanical properties. Accordingly, this work provides a comprehensive treatise on static, fatigue, and fracture response of carbon fiber vinyl ester-based laminates with H100 PVC foam core, considering three-dimensional micro- and meso-structure effects.

"Failure Mechanics of Low-Velocity Dynamic Impact on Woven Polymeric Composites in Arctic Conditions," by P. Prabhakar, investigates low-velocity single and repeated impact behavior of carbon fiber/vinyl ester composites in arctic temperature ($-50°C$). Impact responses, such as the contact force, displacement, and absorbed energy, at four impact energies of 20J, 25J, 30J, and 35J under single impact loading and repeated impact loading until perforation are determined at $-50°C$ and compared against those at room temperature ($25°C$). For the repeated impact cases, the number of impacts required for perforation, the rate of reduction in impact force, the degree of damage, and the failure mechanisms change significantly with varying impact energies and in situ ambient temperatures and are elucidated in detail in this chapter.

"Behavior of Composite Materials and Structures in Low-Temperature Arctic Conditions," by K.T. Tan, presents his work on the behavior of composite materials and composite sandwich structures in low-temperature Arctic conditions. Impact and postimpact compressive and flexural response of foam core sandwich composites are analyzed. Postmortem inspection using X-ray microcomputed tomography revealed complex failure mechanisms due to the influence of low temperature. The findings in this chapter help provide fundamental understanding in guiding future naval composite structures for application in low-temperature environment.

"Mapping Interior Strain Fields in Thick Composites and Sandwich Plates with Digital Volumetric Speckle Photography Technique," by L. Mao and F. Chiang, presents a new 3D whole-field strain analysis technique called digital volumetric speckle photography which is capable of probing the interior deformation of opaque materials. They have successfully applied it to mapping the interior deformation fields of a woven composite beam under bending and a sandwich beam with a foam core under bending. They are able to show the interior damage of the beams before it is visible from outside, thus foretelling the eventual failure of the specimens.

The editor would like to thank the authors for their excellent contributions to this book through individual chapters and summaries of their work as described above. The help provided by Dr. Ananth Virakthi for editing this book is also greatly appreciated.

College Park, MD, USA Sung W. Lee

Acknowledgment

Under the leadership of Dr. Yapa D.S. Rajapakse, the ONR Solid Mechanics Program has made fundamental and groundbreaking contributions in understanding the mechanics of composite materials. As head of the ONR Solid Mechanics Program, Dr. Rajapakse has repeatedly exemplified his role as a leader in research areas of solid mechanics including the effect of temperature extremes such as in Arctic conditions on marine composites and sandwich structures, failure theories for 3D composite materials, response and integrity of naval composite and sandwich structures to highly dynamic loading, the implosion of sandwich and composite structures, and innovative techniques for mitigating the adverse effects of shock, blast, implosion events in air and underwater, and a 3D volumetric speckle technique that can map the interior deformation of opaque materials such as woven composite and sandwich plate with foam core. The completion of these studies supported by Dr. Rajapakse has resulted in a plethora of new knowledge and discoveries which will help designers and engineers adopt composite and sandwich structures for marine and naval application with greater confidence than ever before.

In his research program, Dr. Rajapakse has taken a balanced approach in which experimental work and theoretical work are complementary to each other, emphasizing the need of theories based on physics. In addition, he has advanced professional development of younger faculty by supporting their research. He has enthusiastically promoted Navy's research interest to international researchers by attracting them to his program and has played a key role in the effort to establish an Indo-US collaborative program on blast mitigation in composites.

Dr. Rajapakse's dedication to the fields of solid mechanics is embodied by his active leadership roles in organizations such as the American Society of Mechanical Engineers (ASME), the American Society for Composites (ASC), and the Society of Engineering Science (SES) and on the editorial boards for journals including *Composite Science and Technology* and *Journal of Sandwich Structures and Materials*. In particular, Dr. Rajapakse has served as president, vice president, and member of Board of Directors of SES. He has also served as chair of the Composite Materials Committee of the Applied Mechanics Division of the ASME and the

Polymer-Matrix Composites Division of the ASC. He has organized ONR Symposia at several past ICCMs including ICCM16, ICCM17, ICCM18, and ICCM19. He has also edited/coedited numerous books, including *Major Accomplishments in Composite Materials and Sandwich Structures: An Anthology of ONR Sponsored Research*, *Durability of Composites in a Marine Environment*, *Durability of Composites in a Marine Environment 2*, *Dynamic Failure of Composite and Sandwich Structures*, and *Blast Mitigation Strategies in Marine Composite and Sandwich Structures*.

It is to be noted that as ONR Solid Mechanics Program manager, Dr. Rajapakse has been dedicated to providing high-quality, Navy-relevant, basic research in solid mechanics, in accordance with the traditions established by several generations of outstanding leaders in mechanics, such as Ludwig Prandtl (University of Goettingen) and Stephen Timoshenko and J. Norman Goodier (both from Stanford University). Goodier's research on fracture mechanics was supported by the ONR Solid/Structural Mechanics Program, and Dr. Rajapakse's doctoral research was directed by Goodier. We who have had the privilege to be associated with Dr. Rajapakse over an extended period are grateful that, through his continued leadership and dedication to the Navy and its scientific interests, he has been able to identify and support key topics of research which will continue to advance the field and provide opportunities for students and the mechanics community to grow and prosper.

September 2019

Arun Shukla
Fu-Pen Chiang
Romesh Batra
Sung W. Lee

Contents

Contributors

Romesh C. Batra Department of Biomedical Engineering and Mechanics, M/C 0219, Virginia Polytechnic Institute and State University, Blacksburg, VA, USA

Christian Berggreen Department of Mechanical Engineering, Technical University of Denmark, Kongens Lyngby, Denmark

Brian Pinto Department of Mechanical and Aerospace Engineering, University of California, Davis, CA, USA

David A. Cendón Materials Science Department, (ETSI Caminos, Canales y Puertos), Universidad Politécnica de Madrid, Madrid, Spain

Arka P. Chattopadhyay Department of Biomedical Engineering and Mechanics, M/C 0219, Virginia Polytechnic Institute and State University, Blacksburg, VA, USA

Rodrigo Chavez University of California San Diego, La Jolla, CA, USA

Fu-Pen Chiang Laboratory for Experimental Mechanics Research, Department of Mechanical Engineering, Stony Brook University, Stony Brook, NY, USA

Raj Das School of Engineering, RMIT University, Melbourne, Victoria, Australia

Peter Davies Marine Structures Laboratory, IFREMER Centre Bretagne, Plouzané, France

Bentolhoda Davoodi Department of Mechanical Engineering, Texas A&M University, College town, TX, USA

John P. Dear Department of Mechanical Engineering, Imperial College London, London, UK

Mrityunjay Doddamani Advanced Manufacturing Laboratory, Department of Mechanical Engineering, National Institute of Technology Karnataka, Surathkal, India

Veronica Eliasson University of California San Diego, La Jolla, CA, USA

Francisco Gálvez Materials Science Department, (ETSI Caminos, Canales y Puertos), Universidad Politécnica de Madrid, Madrid, Spain

Alexander Gargano School of Engineering, RMIT University, Melbourne, Victoria, Australia

Erin Gauch Naval Undersea Warfare Center, Division Newport, Newport, RI, USA

Antonio Gomez Department of Mechanical and Aerospace Engineering, University of California, Davis, CA, USA

Nikhil Gupta Composite Materials and Mechanics Laboratory, Mechanical and Aerospace Engineering Department, New York University, Tandon School of Engineering, Brooklyn, NY, USA

Brian Hayman Department of Mechanical Engineering, Technical University of Denmark, Kongens Lyngby, Denmark

Michelle S. Hoo Fatt Department of Mechanical Engineering, The University of Akron, Akron, OH, USA

Carlos Javier Department of Mechanical, Industrial and Systems Engineering, University of Rhode Island, Kingston, RI, USA

George A. Kardomateas School of Aerospace Engineering, Georgia Institute of Technology, Atlanta, GA, USA

Shyamal Kishore Department of Mechanical, Industrial and Systems Engineering, University of Rhode Island, Kingston, RI, USA

Young W. Kwon Department of Mechanical & Aerospace Engineering, Naval Postgraduate School, Monterey, CA, USA

James LeBlanc Naval Undersea Warfare Center, Division Newport, Newport, RI, USA

Valentina Lopresto Department of Chemical, Materials and Production Engineering, University of Naples Federico II, Naples, Italy

Lingtao Mao State Key Laboratory of Coal Resources and Safe Mining, China University of Mining & Technology, Beijing, China

Roberta Massabò Department of Civil, Chemical and Environmental Engineering, University of Genova, Genova, Italy

Helio Matos Department of Mechanical, Industrial and Systems Engineering, University of Rhode Island, Kingston, RI, USA

Yasushi Miyano Materials System Research Laboratory, Kanazawa Institute of Technology, Hakusan, Japan

Adrian P. Mouritz School of Engineering, RMIT University, Melbourne, Victoria, Australia

Anastasia Muliana Department of Mechanical Engineering, Texas A&M University, College town, TX, USA

Masayuki Nakada Materials System Research Laboratory, Kanazawa Institute of Technology, Hakusan, Japan

Vito Pagliarulo CNR National Research Council, ICIB, Institute of Cybernetics, Pozzuoli, NA, Italy

Ilaria Papa Department of Chemical, Materials and Production Engineering, University of Naples Federico II, Naples, Italy

Dayakar Penumadu Department of Civil and Environmental Engineering, 227 Tickle College of Engineering, University of Tennessee, Knoxville, TN, USA

Maurizio Porfiri Department of Mechanical and Aerospace Engineering and Department of Biomedical Engineering, New York University Tandon School of Engineering, Brooklyn, NY, USA

Pavana Prabhakar Civil & Environmental Engineering, Engineering Physics (Affiliate), University of Wisconsin – Madison, Madison, WI, USA

Pietro Russo CNR National Research Council, Institute for Polymers, Composites and Biomaterials, Pozzuoli, NA, Italy

Christopher Salazar Department of Mechanical, Industrial and Systems Engineering, University of Rhode Island, Kingston, RI, USA

Vicente Sánchez-Gálvez Materials Science Department, (ETSI Caminos, Canales y Puertos), Universidad Politécnica de Madrid, Madrid, Spain

Rafael Sancho Materials Science Department, (ETSI Caminos, Canales y Puertos), Universidad Politécnica de Madrid, Madrid, Spain

Valeria La Saponara Department of Mechanical and Aerospace Engineering, University of California, Davis, CA, USA

Priyal H. Shah Department of Biomedical Engineering and Mechanics, Virginia Polytechnic Institute and State University, Blacksburg, VA, USA

Arun Shukla Department of Mechanical, Industrial and Systems Engineering, University of Rhode Island, Kingston, RI, USA

K. T. Tan Department of Mechanical Engineering, The University of Akron, Akron, OH, USA

Xiaolong Tong Department of Mechanical Engineering, The University of Akron, Akron, OH, USA

Chong Zhong Department of Mechanical Engineering, The University of Akron, Akron, OH, USA

Dynamic Response of Composite Structures in Extreme Loading Environments

Arun Shukla, Christopher Salazar, Shyamal Kishore, and Helio Matos

1 Introduction

As the ever-advancing stage of modern warfare threatens civilian and defense structures with extreme loadings under complex conditions [1, 2], the necessity has arisen for new, more blast-resistant structural materials to be researched and understood to ensure the ongoing safety of civilians and military personnel. This has led to the development of composite and sandwich structures, which have important applications in the naval and aerospace industry. Composite materials offer increased strength/weight ratios, energy absorption capabilities, improved corrosion resistance, and low thermal and magnetic signatures. For these reasons, composites are currently used in several military applications, including armored Army vehicles and Navy sonar domes and hull sheathings [3, 4]. However, the principal hindrance to the widespread application of composite and sandwich structures is a lack of complete understanding and simple design rules for their use, especially under extreme loading conditions [4]. For this reason, the current chapter presents the findings of various experimental studies conducted over the past 10 years by researchers at the University of Rhode Island's Dynamic Photo-Mechanics Laboratory which examine the behavior of composite and sandwich structures under extreme in-air and underwater loading environments.

Sandwich structures consist of two thin, stiff facesheets, usually the same thickness, separated by a lightweight, thicker core. The facesheets carry almost all of the bending and in-plane loads, while the core helps to stabilize the facesheets and defines the flexural stiffness and out-of-plane shear and compressive behavior. When sandwich structures are subjected to high-intensity impulse loadings, such as air blasts, the core materials play a crucial role in the dynamic behavior and overall

A. Shukla (✉) · C. Salazar · S. Kishore · H. Matos
Department of Mechanical, Industrial and Systems Engineering, University of Rhode Island, Kingston, RI, USA
e-mail: shuklaa@uri.edu; chris_salazar@uri.edu; shyamal_kishore@uri.edu

S. W. Lee (ed.), *Advances in Thick Section Composite and Sandwich Structures*,
https://doi.org/10.1007/978-3-030-31065-3_1

structural response. Their properties assist in dispersing the mechanical impulse that is transmitted into the structure and thus protects anything located behind it [5–7].

In recent years, with its ability to improve structural performance and damage resistance of structures, as well as effectively dissipate blast energy, the application of polyurea to sandwich structures has become a new area of interest. Although the behavior of polyurea has been investigated [8–11], there have been no studies regarding the dynamic behavior of sandwich structures with a functionally graded core and polyurea interlayer. Tekalur et al. [12] experimentally studied the blast resistance and response of polyurea-based layered composite materials subjected to blast loading. Results indicated that sandwich materials prepared by sandwiching the polyurea between two composite skins had the best blast resistance compared to the EVE composite and polyurea layered plates, verifying the material's potential for blast-mitigation.

All underwater loading events presented in this chapter deal with the problem of implosion. An implosion event occurs when a closed, hollow structure of lower internal pressure is submerged in a medium of higher pressure such that the pressure differential results in instability in the structure walls, causing it to collapse inwards on itself. Implosion can also occur below this critical buckling pressure (P_c) if the structure is subjected to dynamic deformations exceeding a certain limit [13]. When this occurs underwater, the sudden increase and subsequent arrest in fluid motion emits a pressure pulse into the fluid which can be damaging to and even initiate the implosion of nearby structures [14, 15]. The problem of implosion itself has been a topic of study for many decades, with the first equations for the critical buckling pressure of a hydrostatically loaded cylindrical shell having been derived by von Mises in the early 1900's [16, 17]. The implosion of aluminum tubes has been widely investigated, with Turner and Ambrico having identified the key stages of the implosion process in a free-field environment with respect to the local pressure about the collapsing volume [14]. The harmful effect of these pressure waves on surrounding structures was gained public attention in 2001, at Japan's Super-Kamiokande neutrino observatory, when the implosion of a single photomultiplier tube released a pressure wave powerful enough to trigger a chain reaction of implosion. This accident resulted in the implosion of nearly 7000 nearby tubes, causing \$20–\$30 million in damage [18]. The danger posed to surrounding structures by implodable volumes has thus sparked investigations from researchers in the naval community, with the implosion of composite and sandwich structures at the forefront of most recent investigations.

The experimental investigations detailed in the current chapter are divided into two sections: the response of composite and sandwich structures to air-blast loading and the response of composite and sandwich shells to extreme underwater loadings. In these sections, composite shells, sandwich structures, and structures with the inclusion of polyurea coatings are subjected to extreme loadings, with special attention paid to failure mechanisms and mitigation effects. Note that in the context of air blast loading, mitigation refers to a structure's ability to resist deformation under blast loading. In the context of underwater loading, mitigation refers to a structure's ability to minimize the strength of the pressure pulse resulting from implosion. In many experimental studies, the capture of high-speed photos coupled with 3D digital image correlation (DIC) is used to quantify the deformation behavior of the structure. Dynamic pressure sensors are used to quantify important pressure data, namely incident and reflected pulses in air blast experiments, and pressures emitted into the surrounding fluid due to UNDEX events, implosion, or both, for underwater experiments. Careful analysis of this data provides key insights into the response of composite and sandwich structures to extreme loading environments, which will aid in the development of optimally designed structures necessary to ensure the ongoing safety of civilians and military personnel.

2 Response of Composite and Sandwich Structures to Air-Blast Loading

This section details various experimental studies conducted to provide insight as to the performance and dynamic behavior of composite and sandwich structures, including those with PU coatings, subject to air blast loading. The section begins by providing a brief description of the physics and operation of the shock tube facility used to create air blast loadings, followed by in-depth theoretical considerations in which the Rankine-Hugoniot relations are utilized to derive expressions which predict reflected pressure during an air blast loading event. The applicability of the expressions to sandwich composite structures is then demonstrated. This section proceeds to detail the effect of foam core gradation in composite sandwich structures and then ends with an overview of two studies conducted to determine the effect of polyurea coatings on the response of composite and sandwich structures to air blast loading. The first of these studies deals with the location of PU coatings on a single composite plate, while the second deals with the location of PU interlayers within composite sandwich structures with sandwich cores.

2.1 Experimental Methods: The Shock Tube Facility

The shock tube apparatus used to obtain controlled dynamic loading is shown in Fig. 1a. Shock tubes offer the advantages of plane wavefronts, wave parameters that are easily controllable and repeatable, and uniform loading over the shock tube muzzle diameter [19]. A complete description of the shock tube and its calibration is given by LeBlanc et al. [20].

In principle, the shock tube consists of a long rigid cylinder, divided into a high-pressure driver section and a low pressure driven section, which are separated by a diaphragm. By pressurizing the high-pressure driver section, a pressure difference across the diaphragm is created. When this pressure differential reaches a critical value, the diaphragm ruptures. The subsequent rapid release of gas creates a shock

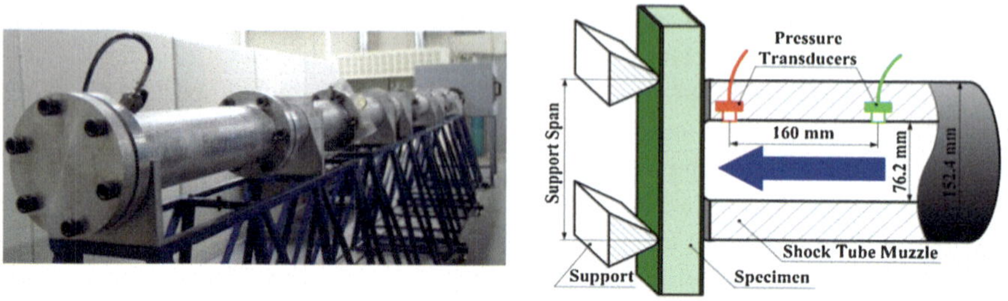

(a) Shock tube

(b) Detailed dimensions of the muzzle

Fig. 1 Shock tube apparatus

wave, which travels down the shock tube to impart shock loading on the specimen at the muzzle end.

When the shock wave impacts the specimen located at the end of the muzzle, the wave is reflected at a higher pressure than that of the incident shock pressure. The theoretical details on the equations for shock tubes have been previously established in the literature [21]. There are four basic theoretical assumptions which are used to describe the gas flow in shock tube:

1. The gas flow is one-dimensional.
2. The gas is ideal and has constant specific heats.
3. Heat transfer and viscosity effects are neglected.
4. Diaphragm rupture is instantaneous and does not disturb the subsequent gas flow.

The shock tube apparatus utilized in the present study has an overall length of 8 m, consisting of a driver, driven, converging, and muzzle section. The diameter of the driver and the driven section is 0.15 m. Figure 1b shows detailed dimensions and locations of the muzzle, specimen, supports and the pressure sensors (PCB102A). The sensors are mounted at the end of the muzzle section to measure the incident pressure and the reflected pressure during the experiment. The final muzzle diameter is 0.076 m. The distance between the two sensors is 0.160 m and the distance between the second sensor and the end of the muzzle is ~ 0.020 m.

2.2 *Theoretical Considerations*

In this section, a comprehensive fluid-structure interaction model describing the incident and reflected pressure pulses impinging on a flat structure is presented. Unlike previous models [6, 22–25], this model, derived and validated by Wang et al. [26], considers compressibility of the fluid by utilizing the Rankine-Hugoniot relation of the fluid. The model was then used alongside experimental results to show that the reflected pressure profile can be predicted in sandwich composites by assuming that they are monolithic plates with the same areal density.

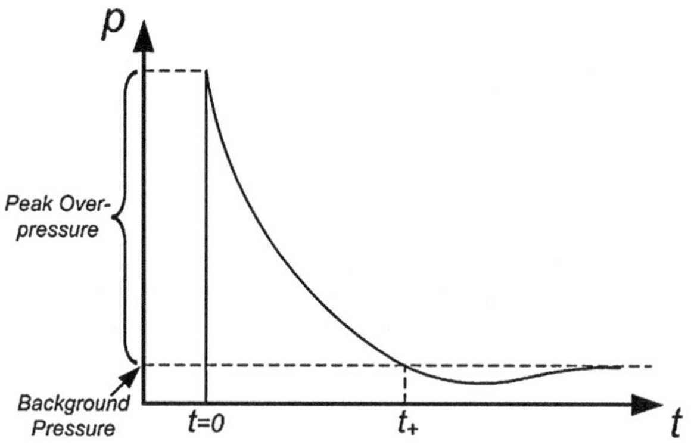

Fig. 2 Typical shock pressure profile

2.2.1 Preliminary Considerations

Consider a uniform, planar blast wave propagating with a constant velocity, U, in a fluid of density ρ_f, and impinging normally upon a free-standing flat target with areal density ρ_s. An approximate blast pressure profile is shown in Fig. 2. It has a very sharp jump at time $t = 0$ with a very high overpressure in relation to the background pressure (normally the atmospheric pressure). After the sharp jump, the pressure decays very quickly and even drops to a pressure level lower than the background pressure. The time at which the blast pressure is equal to the background pressure is defined as positive time period, t_+. The most common approximation of this pulse is expressed with an exponential decay profile [27–29]:

$$p(t) = p_{peak} e^{-\frac{t}{\theta}}, \quad 0 \le t \le \infty \tag{1}$$

where, p_{peak} is the peak pressure. θ is the time constant and can be expressed by the positive time period t_+ and the peak pressure p_{peak}:

$$\theta = \frac{t_+}{\ln\left(p_{peak}\right) - \ln\left(p_0\right)} \tag{2}$$

where p_0 is the background pressure or base pressure.

Note, for the presented model, the traveling direction of the wave is defined as the positive x direction, $x = 0$ is chosen as the original position of the plate, and the out-of-plane displacement of the plate is defined as $w(t)$.

2.2.2 Model by Wang et al

In the model by Wang et al. [26], compressibility is considered by utilizing the Rankine-Hugoniot relation of the fluid. The incident pressure (represented by the subscript 1 in Eq. (3a)) and reflected pressure (represented by the subscript 2 in Eq. (3b)) profiles are all assumed to have a similar expression as that of Eq. (1):

$$p_1(t) = p_{1\_peak} e^{-\frac{t}{\theta_1}}, \quad 0 \le t \le \infty \tag{3a}$$

$$p_2(t) = p_{2\_peak} e^{-\frac{t}{\theta_2}}, \quad 0 \le t \le \infty \tag{3b}$$

where p_{1_peak} and p_{2_peak} are the peak pressures of the incident and reflected pressure pulses, respectively. θ_1 and θ_2 are the time constants for the incident and reflected pressure. Furthermore, assume t_{1+} and t_{2+} are the positive time periods for the incident and reflected pressures which can be related to θ_1 and θ_2 through Eq. (2).

The reflected pressure profile, p_2, can thus be calculated from the known incident pressure profile, p_1, through two steps. The first step is to calculate the peak pressure, p_{2_peak}, of the reflected pressure profile by assuming that the incident shock wave is impinging upon a rigid wall:

$$p_{2\_peak} = p_{1\_peak} \left\{ \left[\frac{u_{1\_peak}}{(1-\mu^2)c_{1\_peak}} + \sqrt{\frac{u_{1\_peak}^2}{(1-\mu^2)^2 c_{1\_peak}^2} + 4} \right]^2 \frac{1+\mu^2}{4} - \mu^2 \right\} \tag{4}$$

where u_{1_peak} and c_{1_peak} are the peak particle velocity and the peak sound velocity of the incident fluid, respectively. They can be calculated using the incident pressure profile. Note, $\mu^2 = \frac{\gamma-1}{\gamma+1}$, where γ is the adiabatic coefficient of the fluid.

The second step is to utilize the conservation of momentum of the plate at the end of the fluid-structure interaction stage:

$$\int_0^{t_+} [p_2(t) - p_0] dt = \rho_s u_2 |_{t=t_+} \tag{5}$$

The reflected target particle velocity, u_2, can be calculated from the incident and reflected pressure profiles p_1 and p_2. Since p_{2_peak} has been determined in the first step, there is only one unknown parameter in Eq. (5). Therefore, the value of the unknown parameter (θ_2 or t_+) can be obtained by solving Eq. (5). Consequently, all of the parameters, such as impulse, can be generated using the calculated reflected pressure profile, p_2, and the known incident pressure profile p_1. Further details regarding the steps and generation of parameters can be found in [26].

2.2.3 Application of the Model by Wang et al. to Sandwich Composite Structures of Varying Areal Density

The fluid-structure interaction model derived in the previous sections of Sect. 2.2 was developed by Wang et al. and thus will be referred to as the Wang model. Other models have been previously derived to describe the magnitude of reflected pressure resulting from shock loading generated by a shock tube, namely the Taylor model [23] and the Kambouchev model [24, 25]. These models, however, do not consider the compressibility of gas that the Wang model considers, and thus result in a large deviation from experimental results for shock loading in air [30]. To demonstrate the applicability of the Wang model to air blast loading scenarios, a series of experiments were conducted in which composite structures and a monolithic Al plate under simply supported conditions were subjected to air blast loading in air, and the reflected pressure profiles compared to values predicted using the Wang model.

The structures subjected to blast loading were an EVE composite plate, a sandwich composite with 25.4 mm monolithic foam core, a sandwich composite with 38.1 mm monolithic foam core, and a monolithic Al alloy 6061-T6 plate. The weight and areal density of the structures used are found in Table 1. The shock tube facility discussed in Sect. 2.1 was used to provide the air blast loading with a peak incident pressure of 1 MPa. Further experimental details can be found in [30].

The predicted reflected pressure profiles, as obtained using Taylor's model and the Wang model are shown in Fig. 3. Since the two models can only predict the behavior of a free-standing specimen, the simply-supported boundary conditions are completely ignored in the calculation. Therefore, the results are only valid during the characteristic fluid-structure interaction time, in which the simply supported boundary conditions have not affected the dynamic behaviors of the structure. The results in Fig. 3 validate the above assumption. It can be seen clearly that during the characteristic fluid-structure interaction time (~250 μs), the predictions from the Wang model agrees well with experimental results. The predictions even agree with the experimental results through t = 500 μs. After this time, the predictions drift off the experimental data, which shows the boundary condition effects on the results. Unlike the Wang model, the Taylor model does not consider the compressibility of the gas and thus the theoretical predictions exhibit a large deviation from the experimental results. This reveals that the Taylor model cannot be used in a highly compressible fluid such as air.

Table 1 Weight and areal density of shock-loaded structures

Specimen name	Weight (g)	Areal density (kg/m^3)
EVE composite	177	6.8
Sandwich composite with 25.4 mm foam core	446	17.2
Sandwich composite with 38.1 mm foam core	600	23.3
Al alloy 6061-T6	660	25.7

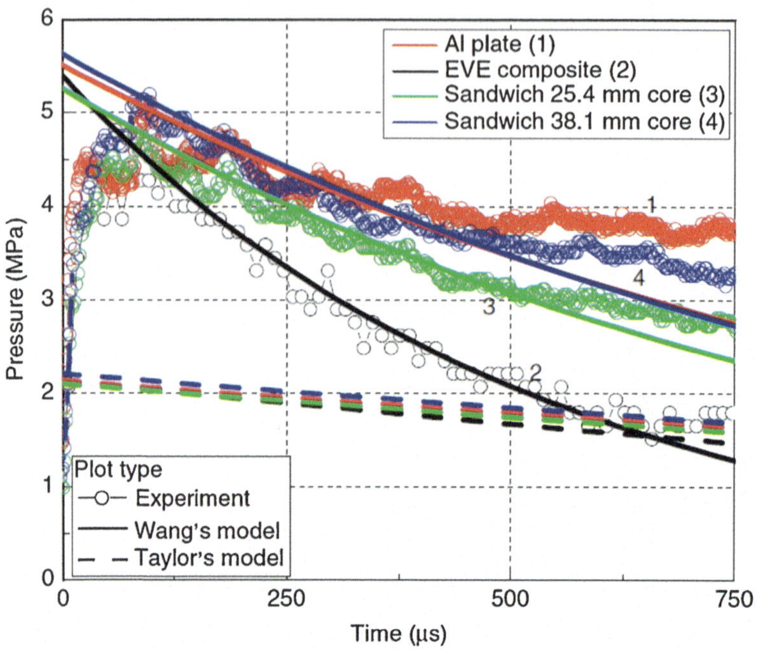

Fig. 3 Comparison of theoretical and experimental reflected pressure profiles

It has been assumed in previous studies [6, 22] that when a sandwich composite structure is subject to blast loading, fluid-structure interaction only occurs between the fluid and the facesheet of the structure. That is, it was assumed that core material and core thickness do not play a role. Numerical studies [31] have shown that the fluid-structure interaction is indeed affected by the core material. The results obtained by Wang et al. provide a much clearer and more comprehensive understanding of the fluid-structure interaction in sandwich composites. The experimental results and theoretical predictions of the sandwich composites and the facesheets (EVE composites) are plotted in Fig. 3. For the theoretical prediction, the sandwich composite is considered to be an equivalent monolithic plate. It can be clearly seen that the fluid-structure interaction between the fluid and the sandwich composites is completely different than that between the fluid and the EVE facesheet. Using the prediction of the facesheet will induce significant error. Furthermore, sandwich composites which have a core of identical material but different thicknesses exhibit different responses from the very beginning. This indicates that the core material affects the fluid-structure interaction. These results indicate that using the Wang model and assuming the sandwich composite to be an equivalent monolithic plate with same areal density allows for a close prediction of the reflected pressure profiles during the characteristic fluid-structure interaction time.

2.3 *Air Blast Response of Composite Sandwich Structures*

2.3.1 Effect of Functional Foam Core Gradation

This section details a series of experiments conducted by Gardner et al. [32] which consisted of subjecting simply supported composite sandwich structures of various core gradation configurations to air blast loading with 1 MPa incident pressure and 5 MPa reflected pressure. The shock tube apparatus used to generate air blast loading is detailed in Sect. 2.1 of this chapter, and the sandwich structures being loaded were fabricated in the configurations given by Table 2.

All sandwich structures were fabricated using the Vacuum Assisted Resin Transfer Molding (VARTM) process with 4 mm thick EVE facesheets, and were 254 mm long with a width of 102 mm. Note that areal densities for all configurations are very similar. Finally, it should be noted that the core materials, read from left to right, indicate the gradation sequence, with the leftmost core adjacent to the front (loaded) faceplate, and the rightmost core adjacent to the back faceplate. Properties of the materials used can be found in [33, 34].

In a study conducted by Wang et al. [35], it was shown that sandwich structures with graded foam cores of increasing nominal density showed least back-face deflection, as this allowed for more core compression. That is, sandwich structures with graded cores arranged with the softest nearest the shock and the hardest furthest from the shock resulted in more effective mitigation. Thus, the sequence of the foam core material for this study was designed based on increasing the one-dimensional acoustic wave impedance of the layers. That is, for the A-series foam, the A300 foam has the lowest nominal density (ρ) and compressive modulus (E) of the four foams, followed by the A400, A500, and A800 foams respectively. Since both the nominal density and the compressive modulus are increasing from the A300 foam to the A800 foam, the one-dimensional acoustic wave impedance (Z) is also increasing, and shows the following relationship:

$$Z = \rho C = \rho\sqrt{E/\rho} \tag{6}$$

$$Z_{A300} < Z_{400} < Z_{A500} < Z_{A800} \tag{7}$$

Note, the cell structures of the four A-series foams are very similar and the only difference appears in the cell wall thickness and node sizes, which accounts for the

Table 2 Material system configurations for sandwich structures with functionally graded foam cores

Core materials	Core thickness	Areal density (kg/m^3)
A500	38 mm	19.0
A300, A800	19 mm each	18.5
A300, A500, A800	12.7 mm each	19.0
A300, A400, A500, A800	9.5 mm each	20.0

Fig. 4 Cell microstructure of A-series foam core layers

different densities of the foams. The SEM images of the cell microstructures can be seen in Fig. 4.

It was shown that the number of core layers has an influence on the dynamic response of structures under blast loading. More specifically, by increasing the number of monotonically graded layers, the acoustic wave impedance mismatch between successive layers is reduced. Therefore, the strength of the initial shock wave (stress wave) can be weakened by the time it reaches the back facesheet, resulting in lower back face deflection, in-plane strain, and velocity. More importantly, the overall damage imparted on the structure can be reduced and structural integrity can be maintained. Also, increasing the number of monotonically graded foam core layers, thus introducing more material interfaces, allows for blast wave (stress wave) attenuation through the following mechanisms: (1) stepwise compression of the core (energy dissipation mechanism) and (2) scattering/dispersion of the wave through interface variations. Combining these mechanisms results in lengthened timescales for pressure rises across the samples, allowing for a time-delay of the peak stress arrival, and thus delaying the time of damage initiation. The following sections elucidate the effects a functionally graded styrene acrylonitrile (SAN) foam core on the blast response of sandwich composites.

2.3.1.1 Deflection

The mid-point deflections of each graded sandwich panel and all of its constituents were obtained from the high-speed images and a typical response can be seen in Fig. 5. The difference between the deflection of the front face and deflection of the back face signifies the total amount of compression observed in the core.

It should be noted that for all of the configurations studied, the core layers were graded monotonically by increasing the acoustic wave impedance, therefore allowing for a stepwise compression of the core. This stepwise compression is more evident in the three and four-layer core configurations, i.e. Fig. 5c and 5d.

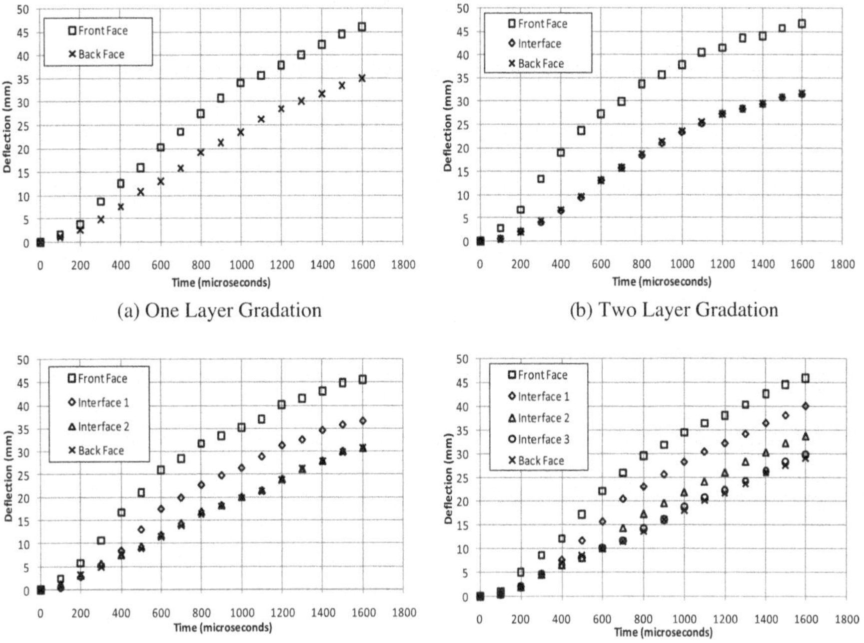

(a) One Layer Gradation

(b) Two Layer Gradation

(c) Three Layer Gradation

(d) Four Layer Gradation

Fig. 5 Mid-point deflection of all configurations

2.3.1.2 DIC Analysis

Utilizing the DIC technique, the full-field deflection, in-plane strain and particle velocity of the back facesheet of each configuration were generated. Figures 6a, 6b, 6c and 6d, show the full-field results for the back facesheet of all core gradations respectively. Figure 6 shows the full-field out-of-plane deflection (W) during the initial fluid-structure interaction ($t \leq \sim 250$ μs), with an emphasis on the shape of the loading, as indicated by the localized areas of larger deflection. Note that the scale used to represent each core gradation is different in order to highlight these areas. For one-layer core gradation, as shown in Fig. 6a, by $\sim t = 150$ μs, the loading can be observed as a circular region in the center of the back facesheet. For two, three and four-layer core gradation, as shown in Figs. 6b, 6c, and 6d respectively, by $t = 150$ μs the loading is more dispersed across the back facesheet, resulting in two and even three areas of localized deflection (loading). Therefore, it can be concluded that utilizing multiple layers of core gradation, and thus introducing more material interfaces, aids in dispersing the initial loading on the structure, resulting in up to three areas of localized deflection (loading) on the back facesheets.

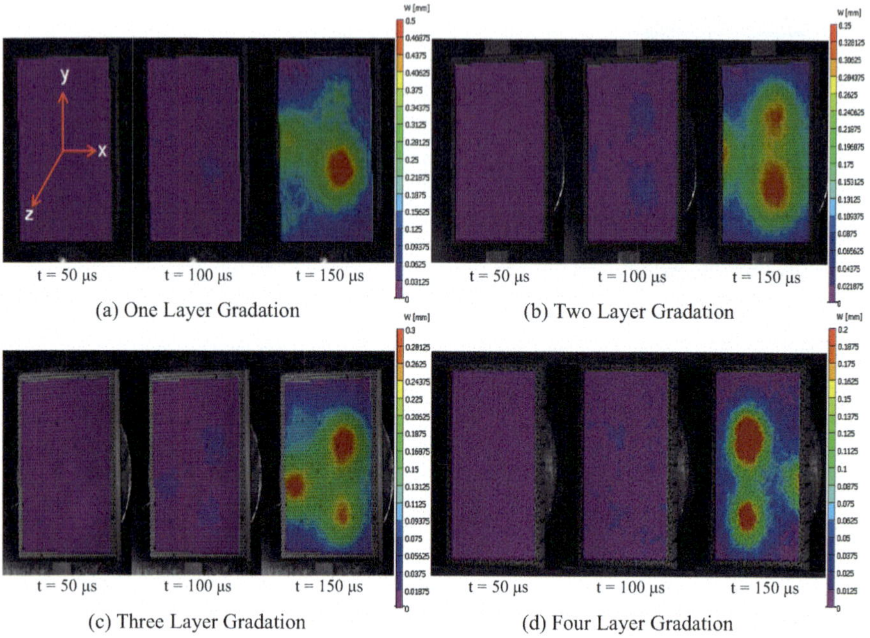

Fig. 6 Localized areas of larger deflections (loading) during fluid-structure interaction

2.3.1.3 Damage Mechanisms

After the blast loading event occurred, the damage patterns of the sandwich composites with four different core layer arrangements were visually examined and recorded using a high-resolution digital camera and are shown in Fig. 7. When the sandwich composite with one-layer core gradation was subjected to highly-transient loading, as shown in Fig. 7a, the damage was confined to the areas where the supports were located in the shock tube and core cracking is visible in these two areas. The core cracks propagated completely through the foam core. Core delamination is visible between the two core layers of A500 foam. Some core compression is also visible in the first core layer of A500 foam.

For the sandwich composite with two layers of core gradation, the damage patterns after being subjected to the shock loading are shown in Fig. 7b. For this core configuration, the damage was again confined to the areas where the supports were located in the shock tube and core cracking is evident. The core cracks propagated completely through the foam core. Skin delamination is obvious between the front facesheet and the foam core, as well as back skin delamination between the back facesheet and the foam core. Core delamination between the first and second core layers of foam, A300, and A800 respectively, is also evident, along with core compression in the first core layer of foam (A300).

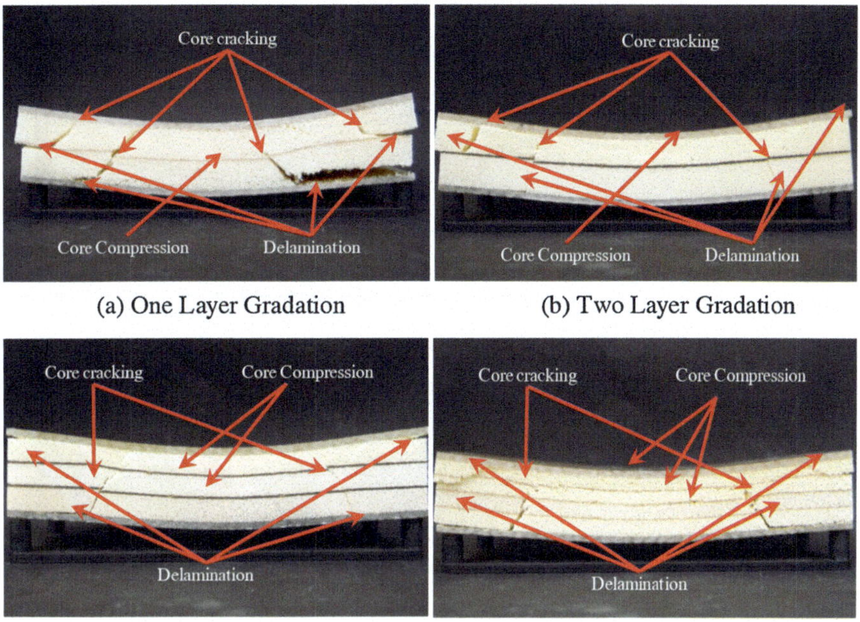

(a) One Layer Gradation (b) Two Layer Gradation

(c) Three Layer Gradation (d) Four Layer Gradation

Fig. 7 Post-mortem examination of all core configurations

Figure 7c shows the damage patterns of the sandwich composite with three layers of core gradation after the blast loading event occurred. Again, the damage to this core configuration was confined to the areas where the supports were located in the shock tube and core cracking is visible in these two areas. These core cracks propagated completely through the foam core. Also, skin delamination is visible between the front facesheet and the foam core, as well as back skin delamination located between the back facesheet and the foam core. Core compression is also evident in both the first and second core layers of foam, A300, and A500 respectively.

When the sandwich composite with four layers of core gradation was subjected to the shock loading, as shown in Fig. 7d, the damage was again confined to the areas where the supports from the shock tube were located and core cracking is evident in these two areas. Unlike the previous three configurations, the core cracks did not propagate completely through the foam core. Core delamination is obvious between the first and second core layers of foam, A300, and A400, as well as back skin delamination between the back facesheet and the foam core. Core compression is very obvious in this configuration. The first, second and third layers of foam, A300, A400, and A500 respectively, all exhibit various amounts of core compression.

The permanent deflection (deformation) for each graded core configuration was measured after the shock loading experiment. A schematic of the specimen and how the measurements were taken can be seen in Fig. 8. The distance between the top

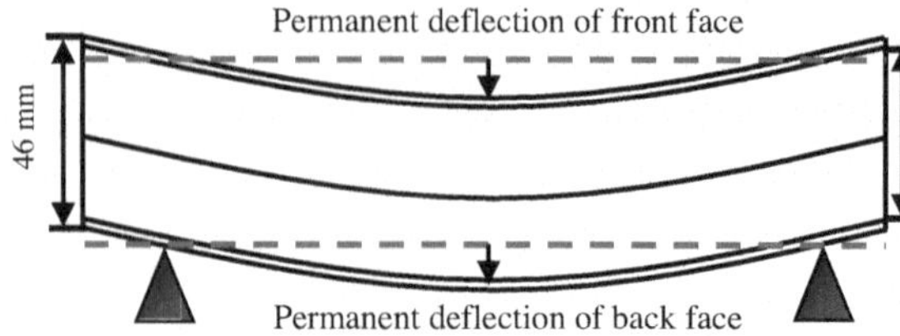

Fig. 8 Specimen schematic for permanent deflection measurements (between simple-supports)

Table 3 Permanent deflection and core compression

	1 Layer	2 Layer	3 Layer	4 Layer
Permanent front face deflection (mm)	11.1	14.1	12.9	12.4
Permanent back face deflection (mm)	8.5	9.2	7.3	6.1
Final core thickness (mm)	35.4	33.1	32.4	31.7
Permanent core compression (mm/%)	2.6 (7%)	4.9 (13%)	5.6 (15%)	6.3 (17%)
Real-time core compression (mm/%)	11 (30%)	13 (35%)	15 (40%)	17 (45%)
Recovered (mm/%)	8.4 (23%)	8.1 (22%)	9.4 (25%)	10.7 (28%)

dotted line (red) and the surface of the front facesheet is defined as the permanent deflection of the front face. Similarly, the distance between the bottom dotted line and the top surface of the back facesheet is defined as the permanent deflection of the back face. Subtracting the total permanent deflection of the back face from the front face, the final core thickness and thus total core compression (permanent) can be obtained. These values are shown in Table 3.

2.4 *Air Blast Response of Composite Structures With Polyurea Coatings*

With its ability to improve structural performance and damage resistance of structures, as well as effectively dissipate blast energy, the application of polyurea to composite structures has become a new area of interest. The following sections reveal the importance of polyurea and the effect of its location on the blast response of composite plate and sandwich structures.

2.4.1 Effect of Polyurea on Composite Plates

To determine the deformation and damage mechanisms at play in the air blast loading of EVE composite panels with PU coatings, a series of shock tube experiments were conducted which varied the location of PU on EVE composite panels [12]. Results were compared against plain-woven composite panels without PU coating. The shock tube shown in Fig. 1 was used to provide the loading on simply supported specimens, and their response was recorded using the high-speed digital imaging system detailed in [12]. The panels of plain-woven composite were 6 mm (0.25″) nominal thickness and panels of layered composite were 12 mm (0.5″) nominal thickness that includes 6 mm (0.25″) of plain-woven composite and 6 mm (0.25″) of polyurea coating. The layered composites were tested in two different ways, namely:

1. PU side facing the shock blast (henceforth referred to as PU/EVE)
2. EVE side facing the shock blast (henceforth referred to as EVE/PU)

For EVE plates subjected to simply supported boundary conditions, the lamina facing the oncoming shock front is largely subjected to compression, while the lamina on the back of the plate is largely in tension. This is important because, in EVE composite plates, the tensile strength is much greater than the compressive strength as tensile strength is dominated by the fibers and compressive strength is dominated by the vinyl-ester matrix. Due to this fact, it was found in this study that PU/EVE plates responded much better to air blast loading than EVE/PU plates. This is because when polyurea is applied to the composite lamina that largely undergoes compression, the increase in compressive strength is large compared to that of the composite. When applied to the back face, however, the additional strengthening is relatively little. This fact plays a large role in the center point deflections and damage mechanisms of EVE plates subjected to air blast loading.

2.4.1.1 Center Point Deflection

The center point deflection of the plate was calculated from the high-speed images and normalized to unit thickness to provide a good comparison of structure response. Figure 9 provides a normalized plot of deflections per unit thickness of the plain composite and layered materials under the same or comparable input blast loadings. Also shown in Fig. 9 are the deformed EVE, PU/EVE, and EVE/PU specimens showing clear variations in damage. The input pressure is stated in the legend for each material. These plots reveal that the deflections observed in the layered constructions were lower than those observed in the plain composite plates, as

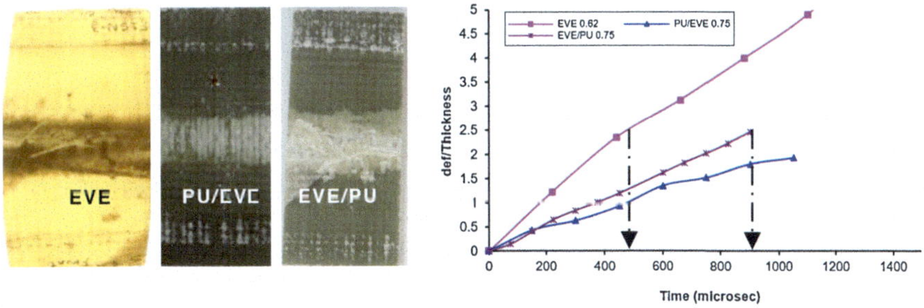

Fig. 9 Damaged EVE, PU/EVE, and EVE/PU structures shown alongside respective deformation plots normalized to unit thickness, under comparable input blast loading

expected. The quantitative estimate of the reduction in deflections can be observed from these plots. Also to be noted is that the input blast pressure is much lower for the plain composite compared to the layered constructions.

In case of the plain composite materials, the "failure" point (deflections equaling 2.5 times the thickness) is produced at an earlier time compared to the layered system under comparable input blast loadings. Also, in the case of polyurea facing the blast, the failure point is not observed at all. The macroscopic damage in the plate also corroborates with the observed real-time trend, vis-à-vis, the PU/EVE configuration showing lower damage area compared to the EVE/PU configuration.

Under the same input blast loadings, the sandwich configuration showed normalized deflections less than one, which is well within the elastic limits of the plate. Here again, it was observed that the PU/EVE configuration reached the failure point at a later time stage compared to the EVE/PU configuration. The delay in the attainment of this failure point between the layered configurations can be attributed to the internal strengthening mechanisms that are present in the PU/EVE system.

2.4.1.2 Failure Mechanisms

Microscopic analysis of undamaged and damaged specimens was conducted using a Nikon SMZ Microscope to provide insight as to the impact of polyurea on damage in the structure. This analysis was conducted for EVE, PU/EVE, and EVE/PU configurations. Detailed microscopic used for this analysis are given by Tekalur et al. [12].

(a) EVE composite

The damage modes observed in blast loaded plain-woven composite included fiber breakage and interface failure. The tensile properties of these composites are superior to the compressive properties. This explains the initiation and the mode of damage on the impact side, which is predominantly under compression. Crushing and cleavage of the longitudinal fiber are observed in these panels. The straight cleavage of a longitudinal fiber bundle near the midsection observed in suggests that the fiber bundle had been crushed compressively during the initial phase of loading and subsequently pulled in tension due to the reflection of the waves from the rear surface, leading to an interface failure between the longitudinal and transverse fiber regions.

(b) PU/EVE layered structure

Addition of a PU layer provides additional modes of damage and hence added energy dissipation mechanisms. In addition to the interface between the transverse and longitudinal fiber directions, the layered materials also have a PU interface with these directions. When studied under the microscope, the predominant damage modes observed in PU/EVE layered materials were:

1. Fiber direction tensile failure
2. Matrix direction failure (observed as voids created through tensile separation)

3. PU-EVE Interface failure (with both the directions)

As noted, the damage modes observed in PU/EVE were predominantly tensile failure patterns. It is also observed that the bonding between the Polyurea and the transverse layer is weaker than the bonding between Polyurea and longitudinal fiber directions when a layered composite plate is subjected to blast.

(c) EVE/PU layered composite

When the loading direction was reversed (i.e., EVE on the impact side), compression dominated failure mode (fiber crushing) was observed. Again, the interface between polyurea and the transverse fiber direction was observed to be weaker than the interface between polyurea and longitudinal direction. In these layered materials, the microscopic failure mode was dominated by compressive failure patterns like crushing of the transverse and longitudinal fibers. The macroscopic failure patterns also correspond to compressive and shear failure on the strike face of composite plates.

The strengthening or the enhanced blast performance in the layered composite can be attributed to the following factors:

1. Energy dissipation due to the nonlinear and highly rate dependent properties of the polyurea layer
2. Energy dissipation in the failure of polyurea-composite interfaces.

However, the explanation for the observance of better performance of a specific orientation (PU/EVE) needs a thorough understanding. It was seen during dynamic material characterization [12] that when the loading is uni-axial, the stress response is not significantly different whether the polyurea faces the impact or composite faces the impact. But, in case of blast loading of a layered plate, wherein the loading induces multi-dimensional stress fields, an additional strengthening mechanism is involved. When polyurea is on the strike face, the composite lamina that is in direct contact with the polyurea is provided with strengthening against compressive and shear failure. So the damage initiation in this lamina will require additional energy from the blast. This will not be true when the composite lamina faces the initial blast directly. The impact face, wherein the first lamina is exposed to a severe compressive zone, begins to fail and thus, the overall strength of the structure reduces progressively as the blast loading progresses. Since the reinforcement of polyurea was on the tensile zone and not the compressive zone, the enhancement in blast performance of these layered composites was comparatively lower than when the polyurea faced the blast loading.

2.4.2 Effect of Polyurea Location Within Composite Sandwich Structures

Thus far, Sects. 2.3.1 and 2.4.1 have discussed the effect of foam core gradation in sandwich structures and the effect of PU coatings on EVE plates, respectively.

Table 4 Material system configurations for sandwich structures with graded foam cores and PU interlayer

Core materials	Core thickness (mm)	Areal density
PU, A300, A500, A800	6.35, 12.7, 12.7, 6.35	26.0
A300, A500, A800, PU	12.7, 12.7, 6.35, 6.35	26.0

However, the question remains as to the effect of PU interlayers within sandwich structures with graded foam cores. To address this question, this section presents the results of a series of air blast experiments conducted by Gardner et al. [36] on A300/A500/A800 graded sandwich structures (identical to those of Figs. 5c, 6c, and 7c, of Sect. 2.3.1) with the inclusion of a PU interlayer. The air blast loading with peak incident and reflective pressures of 1 MPa and 5 MPa respectively was produced using the shock tube apparatus discussed in Sect. 2.1. Graded sandwich structures with PU interlayers were simply supported and fabricated using the VARTM process in the configurations shown in Table 4.

All sandwich structures were fabricated with 4 mm thick EVE facesheets, and were 254 mm long with a width of 102 mm. It should be noted that the core materials, read from left to right, indicate the gradation sequence, with the leftmost material adjacent to the front (loaded) faceplate, and the rightmost material adjacent to the back faceplate. Dragonshield-HT polyurea was used as the PU interlayer. Properties of the materials, including EVE facesheet used, can be found in [33, 34, 37].

When the polyurea interlayer is located behind the graded foam core (configuration 1), and in front of the back face (configuration 2), the core layer arrangement allows for a stepwise compression of the core. Larger compression was visible in the A300 and A500 foam core layers of configuration 2 than configuration 1. This compression lowers the strength of the initial shock wave by the time it reaches the back facesheet and thus the overall deflection, in-plane strain, and velocity were reduced in comparison to the sandwich composite with the polyurea interlayer located behind the front facesheet and in front of the foam core (configuration 1). Therefore, it can be concluded that placing the polyurea interlayer behind the foam core and in front of the back facesheet (configuration 2) improves the blast resistance of the sandwich composite and better maintains structural integrity. It was observed that the location of the polyurea layer has a significant positive effect on the response of composite sandwich panels to shock wave loading, both in terms of failure mitigation and energy absorption, if it is placed opposite the blast-receiving side (configuration 2). On the contrary, the presence of polyurea on the blast-receiving side (configuration 1), amplifies the destructive effect of the blast, promoting (rather than mitigating) the failure of the composite sandwich panels. The following sections detail the experimental results of subjecting functionally graded sandwich structures with PU interlayers to blast loading.

2.4.2.1 Interface Deflection

The mid-point deflections of the constituents of sandwich composites with different core configuration were obtained from the high-speed side-view images and shown in Fig. 10. For configuration 1, the mid-point deflection of the front face (front skin), interface 1 (between first and second core layer), interface 2 (between second and third core layer), interface 3 (between third and fourth core layer), and back face (back skin) were plotted and are shown in Fig. 10a. Since the front face and interface 1 deflect in the same manner to the same value (43 mm), it signifies that the polyurea interlayer, which is located between the front face and interface 1, exhibits no compression. Since interface 2, interface 3, and the back face all deflected in a similar manner to the same value of approximately 34 mm, it can be concluded that the A500 foam layer (located between interface 2 and interface 3) and the A800 foam layer (located between interface 3 and the back face) showed no compression. Therefore, the core layer arrangement of configuration 1 allows for compression only in the A300 layer of foam and has a front face and back face deflection of approximately 43 mm and 34 mm respectively.

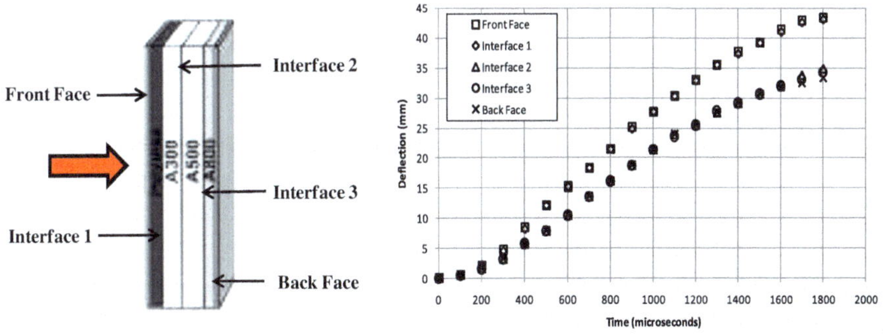

(a) Configuration 1 (PU/A300/A500/A800)

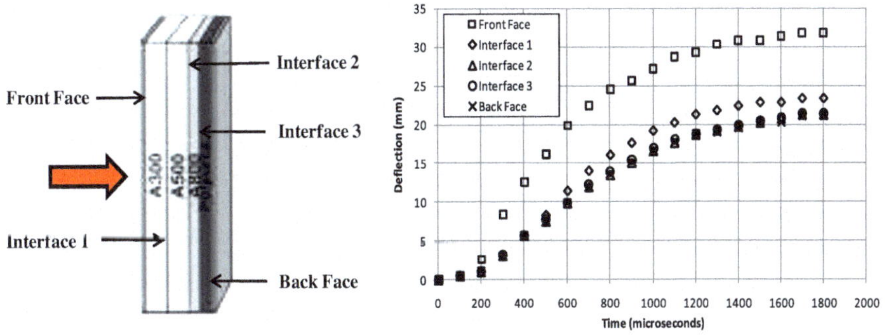

(b) Configuration 2 (A300/A500/A800/PU)

Fig. 10 Mid-point deflection of both configurations

For configuration 2, the mid-point deflection of the front face (front skin), interface 1 (between first and second core layer), interface 2 (between second and third core layer), interface 3 (between third and fourth core layer), and back face (back skin) were plotted and are shown in Fig. 10b. It should be noted that interface 2, interface 3 and the back face all deflected in a similar manner to the same value of approximately 21 mm, and thus there was no compression in the third and fourth core layer (A500 foam layer and the polyurea interlayer). Therefore, the core arrangement of configuration 2 allows for a stepwise compression through the core and the front face and back face deflect to approximately 33 mm and 21 mm respectively.

2.4.2.2 DIC Analysis

Utilizing the DIC technique, the in-plane deflection contours of the back facesheet for each configuration were generated. Figure. 11 shows the full-field out-of-plane deflection (W) with a scale of 0 mm (purple) to 32 mm (red). It is evident from the figure that for configuration 1, the back face exhibits very little out-of-plane deflection until approximately t = 400 μs. Between t = 400 μs and t = 1800 μs, the panel continues to show deflection.

By t = 1800 μs, it can be observed that the central region of the panel has deflected approximately 32 mm. For configuration 2, as shown in Fig. 11b, the back

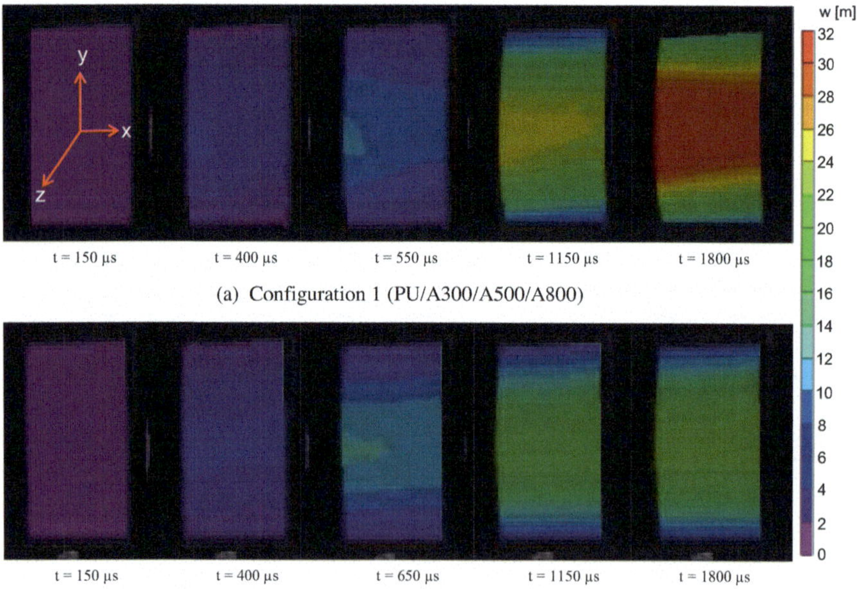

(a) Configuration 1 (PU/A300/A500/A800)

(b) Configuration 2 (A300/A500/A800/PU)

Fig. 11 Full-field out-of-plane deflection (W) of both configurations

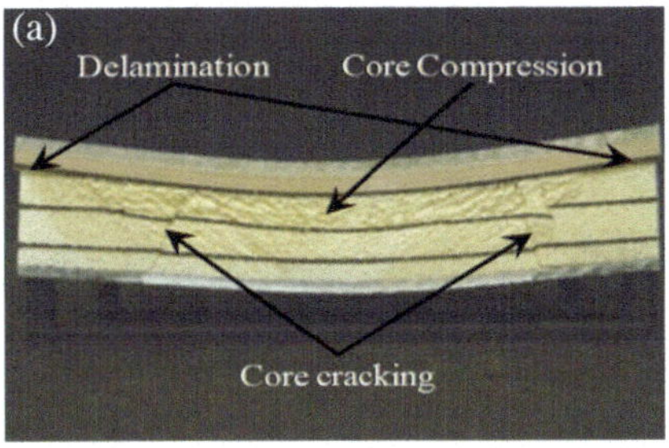

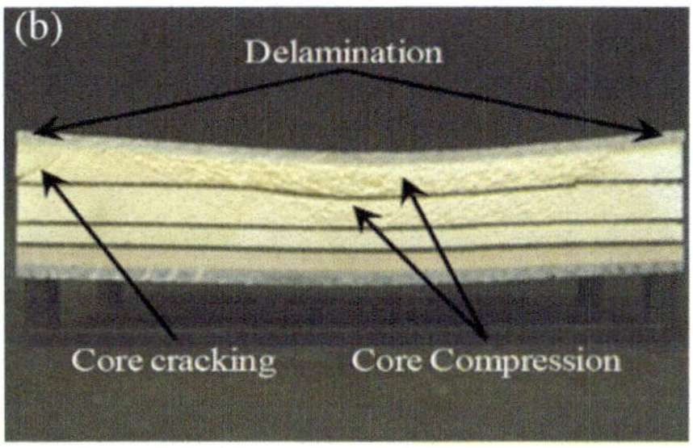

Fig. 12 Post-mortem examination of (a) configuration 1 and (b) configuration 2

face shows very little out-of-plane deflection until t = 400 μs. Between t = 400 μs and t = 1800 μs, the panel continues to exhibit deflection. By t = 1800 μs, the central region of the panel has deflected approximately 22 mm. Therefore, configuration 2 deflects approximately 35% less than configuration 1.

2.4.2.3 Failure Mechanisms

After the blast loading event occurred, the damage patterns of both configuration 1 and configuration 2 were visually examined and recorded using a high-resolution digital camera and are shown in Fig. 12. When configuration 1 was subjected to transient shock wave loading, as shown in Fig. 12a, the damage was confined to the areas where the supports were located in the shock tube and core cracking is visible in these two areas. The core cracks propagated completely through the foam core to the polyurea interlayer. Core delamination is visible between the polyurea interlayer, and the first layer of the foam core (A300). Core compression is visible in the first core layer of A300 foam.

When configuration 2 was subjected to transient shock wave loading, the damage patterns can be seen in Fig. 12b. For this configuration, very little core damage was observed. Core delamination between the first two layers of the foam core (A300 and A500) led to a crack that propagated through the first foam core layer (A300) to the front facesheet. Skin delamination was evident between the front face and the first foam core layer (A300). Also core compression can be observed in the first two layers of the foam core (A300 and A500).

3 Response of Composite and Sandwich Structures to Extreme Underwater Loading Environments

The studies presented in this section cover the advances in the underwater implosion of both thin and sandwich composite shell structures. In Sects. 3.2 and 3.5 the mechanics of the hydrostatic and shock initiated collapse, as well as the emitted pressure pulses released, are characterized for the two structures respectively in free-field. Finally, Sect. 3.3 mitigation explores strategies to reduce the strength of the implosion pulse resulting from the collapse of thin shell composites.

3.1 Experimental Methods: Free-Field Implosion

To simulate free-field conditions experiments were conducted in a large spherical (2.1 m dia.) pressure vessel with a maximum pressure rating of 6.89 MPa to provide constant hydrostatic pressure throughout the collapse event (Fig. 13). Several high-pressure windows mounted about the midspan of the vessel allow the specimens to be viewed by high-speed cameras (Photron SA1, Photron USA, Inc.), and adequately lit by high-intensity light sources. The specimens were securely suspended horizontally in the center of the pressure vessel using several steel cables attached to the inner walls of the vessel. To measure the changes in local pressure during the collapse event, several dynamic pressure transducers (PCB 138A05, PCB Piezotronics, Inc., Depew, NY) were mounted at different locations about the specimens either axially or circumferentially or both depending on the study. A few locations of pressure sensors with a standoff distance R_s are illustrated in Fig. 13 for illustration. The vessel is then flooded with filtered water for maximum optical clarity, leaving a small air pocket at the top. Once the vessel is filled, nitrogen gas is introduced into the air pocket to pressurize the vessel. The pressure inside the vessel is increased at a gradual rate (0.083 MPa/min) until the specimen collapsed triggering the cameras and the pressure sensors to record data. 3-D Digital image correlation (DIC) a well-known experimental tool is used to determine real-time, full-field displacements across the viewable surface of the specimen throughout the implosion event [38]. This technique is calibrated for underwater testing based on previous work [39] for good confidence in the accuracy of measured displacements and

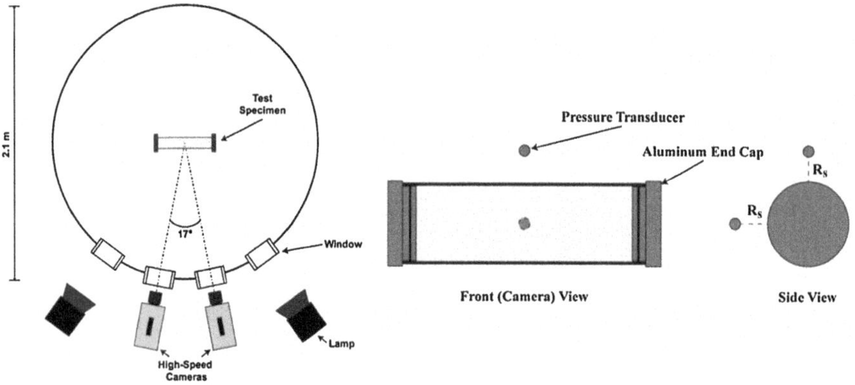

Fig. 13 Experimental apparatus to simulate free-field conditions

velocities. A generalized description of this pressure chamber facility is given by Gupta et al. [39], however, specific details of experimental setups of the studies discussed in this chapter are given in those studies' respective published journal articles.

3.2 Hydrostatic and Shock Initiated Implosion of Thin Cylindrical Composite Shells in Free-Field Environment

A review of free-field implosion related investigations of thin cylindrical composite shells are presented in several following subsections:

1. In Sect. 3.2.1 an experimental study of the hydrostatic implosion of carbon fiber reinforced epoxy composite tubes to examine the failure and damage mechanisms of collapse is presented.
2. In Sect. 3.2.2 the mechanisms and pressure fields associated with the hydrostatic implosion of glass-fiber reinforced polymer (GFRP) tubes with varying reinforcement are presented.
3. In Sect. 3.2.3 a comprehensive investigation of the implosion of thin composite cylinders subjected to a nearby explosion is presented.

3.2.1 Hydrostatic Implosion of Thin Cylindrical Carbon Composite Shells

This section details experiments conducted by Pinto et al. [40] on the hydrostatic implosion of carbon fiber reinforced epoxy composite tubes to examine the failure and damage mechanisms of collapse. Experiments were performed in the pressure vessel shown in Sect. 3.1. The implodable volumes used in this study were laminated carbon/epoxy composite tubes. Three different reinforcing fabric architectures are studied: filament-wound, roll-wrapped, and braided carbon fabric. The filament-wound tubes consist of seven layers of unidirectional carbon fabric reinforcement arranged in a [±15/90/±45/±15] layup. Tubes of two different inner diameters were studied, 76.2 mm (notated as CT) and 60.3 mm (notated as CTRD). The roll-wrapped tubes contain 11 layers of unidirectional tape, arranged in a [0_3/±45/90/±45/0_3] layup. The braided composite tubes were constructed of two layers of braided carbon fabric reinforcement sandwiching one unidirectional layer to obtain a [±45/90/±45] layup. Table 5 gives the details of the specimens used in this study. The dimensions were selected to provide specimens with a relatively low expected collapse pressure, and a high radius-to-thickness ratio (>18) so that thin-wall assumptions may be utilized. It was found that the quantitative characteristics of the released pressure pulse during the implosion of the specimens as well as the nature of their collapse itself are significantly affected by the architecture of reinforcing fibers and geometry of the tube. In particular, the composites with

Table 5 Geometric and material properties of studied specimens

Specimen	Reinforcement	Inner diameter (mm)	Wall thickness (mm)	L/D	Effective modulus (GPa)	Collapse pressure, P_{cr} (MPa)
CT	Filament-wound	76.4	1.39	3.7	74.2	0.73[±0.01]
CTRD	Filament-wound	60.4	1.58	4.6	78.4	1.61[±0.09]
RC	Roll-wrapped	76.3	1.63	3.7	91.1	0.75[±0.06]
BC	Braided	75.8	2.12	3.7	41.0	1.83[±0.10]

braided fabric reinforcements have more damage potential to adjacent structures than those containing unidirectional reinforcements, as they release pressure waves with significantly greater impulse. The results of this study have been succinctly discussed in the following subsections.

3.2.1.1 Deformation and Post-buckling Analysis

The deformation and the post-buckling response of the different specimens are compared and contrasted below. For the sake of brevity, only the first case will be examined in detail and the other cases will be discussed highlighting the differences to previous cases.

(a) Large Diameter Filament-Wound Specimens (CT)

The larger diameter CT specimens failed violently at approximately 0.73 MPa. These tubes buckled in a mode 3 shape, collapsing completely during the failure process then rebounding to a roughly cylindrical shape following the release of pressure. The averaged local dynamic pressure trace about the midspan of the specimen at a standoff distance of 48.3 mm is shown in Fig. 14, and high-speed images, as well as DIC, generated displacement and velocity contours corresponding to key points on the trace are seen in Fig. 15.

Before the start of the rapid and unstable collapse of the structure, a significant amount of slow, quasi-static deformation is measured by DIC. For CT tubes, this initial ovalling amounts to −4.31 mm (11% of the tube radius), a very significant amount of deformation which may serve as an early warning of implosion risk. As the collapse initiates, the local pressure about the midspan decays somewhat smoothly for about 4.5 ms, from time A to time B. By examining the correlated images for this period, it is seen that this pressure decays corresponding with the buckling of the tube in a mode 3 shape, without visible damage. At time B, a longitudinal crack is observed on the bottom visible lobe of the structure, pursued immediately by a small amount of cavitation on that surface. This is followed by an abrupt drop in local pressure, caused by the increased acceleration due to the loss of structural stability in that region brought on by cracking. The local pressure continues to fall until it reaches a minimum at time C. At this instant, it is observed that surface cavitation has intensified, reflecting this drop in pressure. A sudden increase

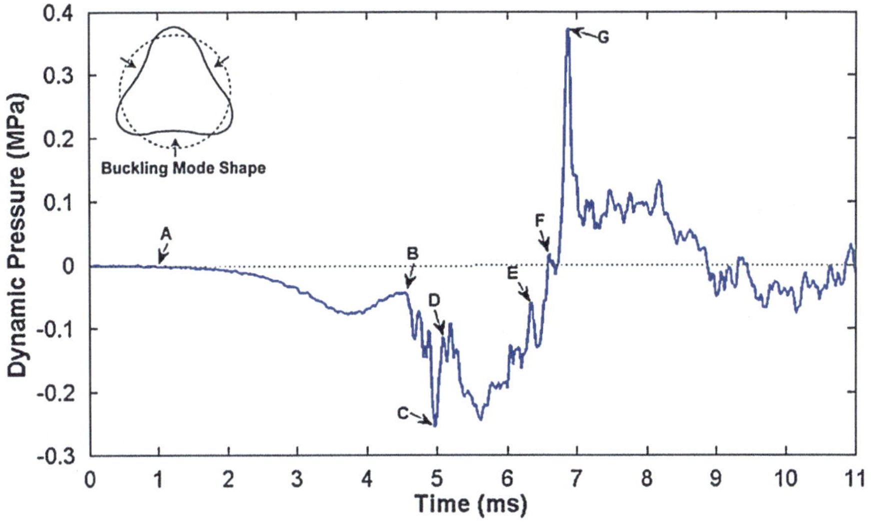

Fig. 14 Local pressure history and buckling mode shape about midspan of CT specimens

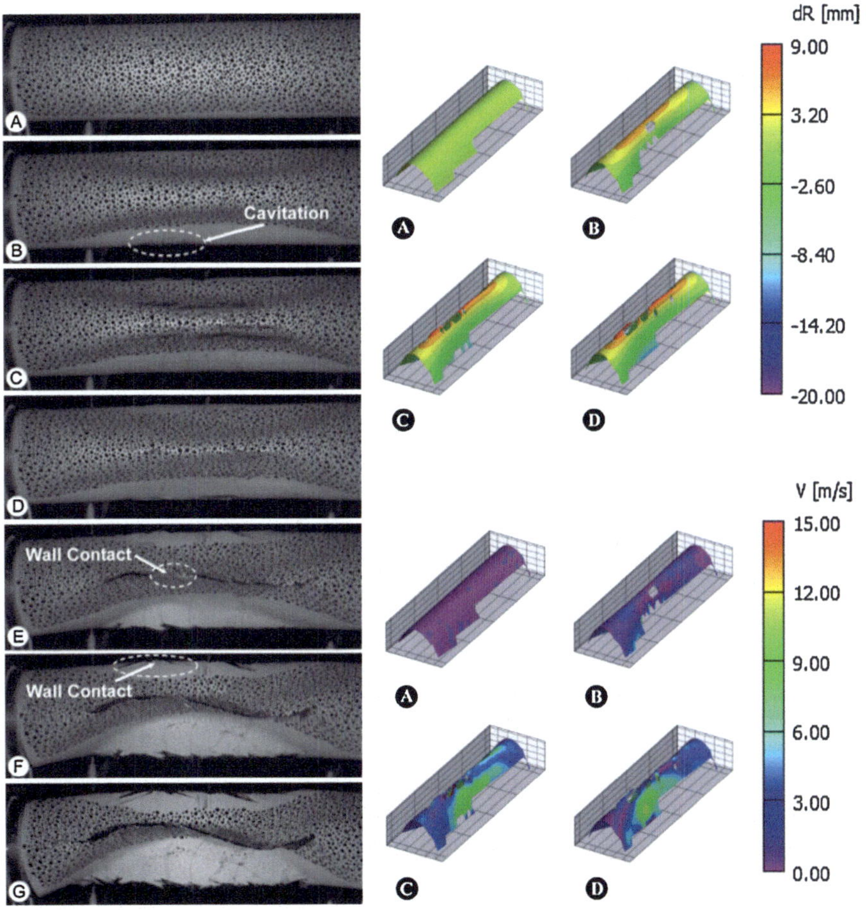

Fig. 15 High-speed photographs (left) and DIC contours of radial displacement (top-right) and velocity (bottom-right) for CT specimens

in velocity magnitude is also seen at time C, increasing from ~4 m/s at time B to ~10 m/s over a period of 0.5 ms, demonstrating the sudden loss of stability brought on by longitudinal cracking. Following the minimum pressure, a sudden spike in pressure is seen at time D. This spike is relatively large, ~0.15 MPa, and has not been observed in previous implosion studies on homogeneous materials. The cause of this peak can be found by comparing images at C and D. In image C, as previously discussed, the bubbles formed by surface cavitation are at their maximum size. In image D, all visible bubbles have vanished. Between these two images, the surface cavitation bubbles collapse inward upon themselves and the surface on which they reside. The collapse of such bubbles has been well documented as resulting in large pressure pulses [41] and is, therefore, the cause of the observed spike at time D in the local pressure trace. Near the end of the under-pressure region of the pressure history, two small yet distinct peaks can be seen at times E and F. These peaks are analogous to those seen in the implosion of aluminum tubes [14, 42] and are indicative of wall-to-wall contacts at the center of the structure. After the walls contact, the collapse of the structure is abruptly halted. This causes the surrounding fluid, which accelerated inward with the surface of the structure, to rapidly decelerate. This drastic change in momentum causes the release of a high magnitude pressure pulse, observed in the pressure history at time G. Following this pulse is a broad region of overpressure as the buckling propagates through the length of the tube. Here it is seen that the tube has closed completely in the circumferential direction assuming a mode 3 buckling shape. After this point, longitudinal buckling waves develop from the applied axial compression imparted by hydrostatic pressure in conjunction with the loss in structural integrity from accumulated damage.

(b) Reduced Diameter Filament-Wound Specimens (CTRD)

The smaller diameter filament-wound CTRD specimens collapsed at 1.60 MPa. Interestingly, both of these specimens appeared to initiate a mode 3 collapse shape, then reduce to a mode 2 shape as the implosion progresses and the tube flattens. As with the larger specimens, the tubes return to a roughly cylindrical shape following depressurization. CTRD specimens also experience some initial quasi-static deformation, though it is significantly less than CT specimen, amounting to only −1.02 mm (3% of tube radius). The reason for this difference is the increased structural rigidity of these specimens brought on by reduced diameter and increased thickness-to-radius ratio. The collapse and resulting local pressure history for CTRD specimens are quite similar to that of the larger diameter filament-wound tubes previously discussed.

(c) Roll-Wrapped Specimens (RC)

The RC specimens collapsed in a mode 3 buckling shape at 0.75 MPa. RC tubes experience quasi-static ovalling of −2.61 mm (7% of tube radius) prior to the dynamic implosion event. This is significantly less than seen in filament-wound tubes of the same diameter, mainly due to the enhanced directional stiffness of these specimens as significantly more fibers are oriented along the major axes of the tube.

The collapse of these tubes shows many similarities to the filament-wound specimens discussed previously.

(d) Braided Specimens (BC)

In both tests of BC tubes, the specimens fail very rapidly at 1.81 MPa, and in both cases the collapse is so violent that the tube tears free of the sealing end caps and falls from the support structure. In these cases, the collapse mode is an imperfect mode 2 shape, with one wall drawn in toward the other, analogous to an indentation. Braided specimens show the least quasi-static ovalling of all 76.2 mm diameter specimens, amounting to −1.39 mm (2.6% of tube radius). This shows that braided tubes give the least "warning" before undergoing catastrophic failure. This is a result of the deformation behavior characteristic of braided composite tubes. When braided tubes are loaded in tension, interlaced fiber tows attempt to "scissor" and straighten out. The tows then reach a point where the matrix material prohibits scissoring, causing them to "lock-up" and result in significant stiffening and ultimately brittle and catastrophic failure [43].

3.2.2 Hydrostatic Implosion of Thin Cylindrical Glass Fiber Composite Shells

This section further expands on the knowledge of composite shell implosion by presenting a study on the implosion of thin cylindrical glass-fiber reinforced polymer (GFRP) shells [44]. Among different composite systems, GFRPs are of particular interest to the naval industry due to their low cost and excellent resistance to both corrosion and water absorption. Specimens of varying geometry and reinforcement architecture are studied to examine the effects of these parameters on the damage and failure progression of the structure as well as the emitted pressure pulse. Two different material systems were studied, filament-wound glass fiber/polyester (PE) and braided glass fiber/epoxy. Filament-wound glass/PE tubes consist of three unidirectional glass fiber reinforcement plies arranged in an [±55/90] orientation. The material properties of the specimens studied are listed in Table 6. The dimensions were selected to provide specimens with a low expected collapse pressure, and a high radius-to-thickness ratio (>14) so that thin-wall assumptions may be utilized.

Table 6 Geometric and material properties of studied specimens and summary of experimental results

Specimen	Reinforcement	Inner diameter (mm)	Wall thickness (mm)	L/D	Effective modulus (GPa)	P_{cr} (MPa)
RT	Filament-wound	57.2	2.08	6.4	35.4	2.05[±0.01]
GT	Filament-wound	101.8	1.91	3.7	20.2	0.68[±0.03]
BG	Braided	60.1	2.03	6.3	14.7	1.94[±0.08]

Filament-wound tubes are shown to release a significantly less damaging shockwave than similar braided tubes, as more energy intensive damage mechanisms dominate the failure process. Differences in the dominating failure mechanisms present in each material produce significant differences in the pressure pulse released. Some key results are discussed below.

3.2.2.1 Deformation and Post-buckling Analysis

The deformation and post-buckling response of the different specimens are compared and contrasted below. For the sake of brevity, only the first case will be examined in detail and the other cases will be discussed highlighting the differences to previous cases.

(a) 57.2 mm Filament-Wound Tube (RT)

The smaller diameter filament-wound specimens fail at 2.05 MPa in a mode 2 buckling shape, forming two lobes. The tubes flattened completely during the dynamic failure event but were seen to rebound to a nearly cylindrical shape following the release of pressure. The local dynamic pressure history measured about the midspan of these specimens is seen in Fig. 16. High-speed photographs corresponding to marked events on that trace are included in Fig. 17.

Qualitatively, the pressure trace obtained at first appears very much like those measured for aluminum tubes by previous researchers [14, 41, 45], with a gradual decay in pressure followed by a pressure spike and a broad overpressure region. Prior to the dynamic collapse event, these specimens begin to oval significantly in a

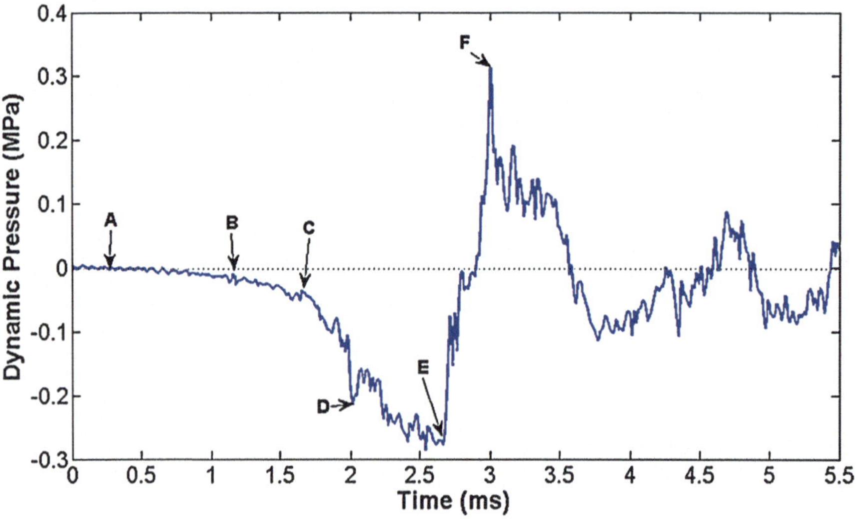

Fig. 16 Average dynamic pressure measure about the midspan of RT specimens

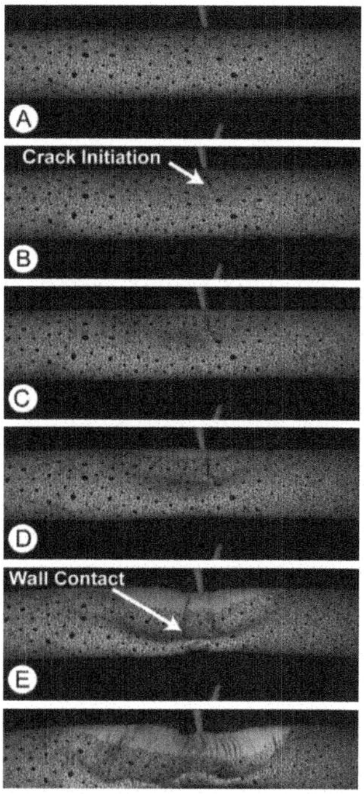

Fig. 17 High-speed photographs of key events for RT specimens

quasi-static manner. The DIC displacement data recorded a maximum radial deflection of −3.9 mm (13.6% of specimen radius) at time A, indicating that this initial slow ovalling is quite significant for these tubes and this continues until time B when an unstable collapse appears to initiate. Examining the corresponding high-speed photograph, it is observed that damage also initiates at this time as indicated by the white arrow. This damage comes in the form of inter-fibrillar cracking/ debonding, which will come to dominate the damage behavior of these materials. Pressure decays steadily until time C when the collapse intensifies and the matched image in Fig. 17 shows the appearance of a mode 2 dimple near the initiation of collapse. Recorded DIC data revealed that the area outside the dimple is largely unchanged from previous key point suggesting that initially, the instability is highly local.

The new rate of pressure decay is maintained fairly constant until time D where the pressure is seen to drop quite suddenly. The photograph at this time shows that the dimple formed previously has grown significantly, and that additional fiber/ matrix debonding is apparent on the specimen surface. This accumulated damage effectively reduces the stiffness and structural integrity of the tube, allowing for more rapid deformation and as a result, steeper pressure drops. After this time, pressure continues to decay until a minimum value is reached. Shortly thereafter, a distinct, fairly small peak is observed at time E as pressure steeply rises. This peak is consistent with observations on the implosion of aluminum tubes [14, 42, 45] and represents initial contact of the walls of the structure. This is confirmed using high-speed photography, as the tube is seen to make wall contact at that time in the location indicated by the white arrow. Pressure continues to rise until a maximum is reached at time F. The image at this time shows the propagation of the wall contacts both circumferentially as well as axially. This peak is followed by a broad overpressure region as the collapse continues to propagate through the length of the specimen, after which the pressure returns to oscillate about the hydrostatic value.

(b) 101.8 mm Filament-Wound Tube (GT)

The larger diameter filament-wound specimens fail at 0.68 MPa in a mode 2 buckling shape. As with the smaller specimens, these tubes regain much of their initial circularity following the collapse, and also recover significant structural stability. As is the case with the smaller RT tubes, the larger diameter GT specimens experience a significant amount of gradual ovalling prior to dynamic collapse. DIC data recorded a maximum radial deflection of −4.1 mm (8.1% of specimen radius) at the start of the collapse process, similar in magnitude to that seen in the smaller specimens. Overall, the extent of damage in these specimens is remarkably less than for the smaller diameter RT specimens of the same make. This is due to the larger size of the GT specimens, coupled with a much lower collapse pressure. Therefore, the potential hydrostatic energy at collapse may be dissipated in the deformation of the structure without resulting in catastrophic damage.

(c) Braided Tube (BG)

Braided glass specimens fail at 1.94 MPa in a mode 2 buckling shape, flattening to a similar degree as aluminum specimens [14, 42, 45]. Like other specimens, circularity and structural integrity are regained following the release of pressure though to an even larger degree. The average local pressure history measured about the midspan of a typical specimen is seen in Fig. 18. The pressure trace for these specimens appears quite different from filament-wound glass/PE tubes discussed previously. Because dimensions and collapse pressures are very similar, this can be attributed to the change in reinforcement architecture. An amount of initial ovalling did occur in these specimens, although it is considerably less than either filament-wound specimen. DIC records a maximum deflection of −1.3 mm (4.3% of specimen radius) at time A. The dynamic collapse begins with a relatively slow decay in pressure from time A to time B, lasting ~2.25 ms. This is much longer than the initial gradual decay seen in RT tubes, which lasted ~1 ms before intensifying. At time B, a slight recovery in pressure is seen followed by a very sharp drop to a minimum at time C which indicates initiation of major longitudinal cracking. The under-pressure region continues until time E where a distinct peak is observed followed by a rapid increase in pressure, indicating wall contact. After wall contact is made the high magnitude pressure pulse typical of the implosion arrives at time F.

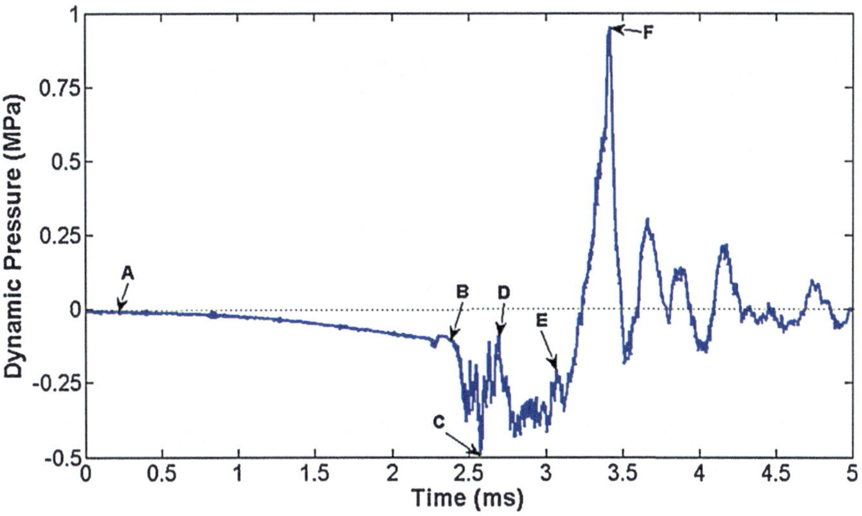

Fig. 18 Average dynamic pressure measure about the midspan of BG specimens

3.2.3 Shock-Initiated Buckling of Carbon/Epoxy Composite Tubes at Sub-Critical Pressures

An implosion caused by quasi-static pressurization to this critical pressure is known as a natural implosion. However, implosions can also be initiated at sub-critical pressures with an underwater explosive (UNDEX) loading. If a structure experiences a high magnitude shock –wave loading such as that generated by a near-field UNDEX, it may collapse at a relatively low hydrostatic pressure [46]. The study discussed here [47] focuses on this extreme condition for composite materials. The implodable volumes in this study were carbon/epoxy filament wound composite tubes, consisting of seven layers of unidirectional carbon fabric reinforcement arranged in a [±15/0/±45/±15] layup with a 60.3 mm inner diameter and 381 mm unsupported length with a nominal wall thickness of 1.63 mm. The dimensions are selected to provide specimens with a relatively low expected collapse pressure. The critical natural collapse pressure of these tubes was found to be 0.90 MPa. To generate the explosive load, an exploding-bridgewire detonator of 80 mg PETN and 1031 mg RDX (RP-85, Teledyne RISI, Inc., Tracy, CA) is mounted directly behind the specimen at a controlled standoff distance of either 102 mm (near-field), 203 mm (mid-field) or 305 mm (far-field).

The model proposed by Shin [48] is fitted to recorded peak pressure data at several standoff distances, to obtain a relationship by which the peak pressure could be predicted as a function of any standoff distance. The fit model was used to determine maximum incident pressures listed in Table 7 for the afore-mentioned standoff distances.

Table 7 Summary of explosive-initiated implosion experiments

Case	Hydrostatic pressure (MPa)	Explosive standoff (mm)	Peak predicted incident pressure (MPa)	Time before implosion (ms)
1	0.71	102	45.9	0.93
2	0.71	203	20.2	6.83
3	0.71	305	12.4	15.22

When a sealed composite tube is subjected to explosive loading, the pressure waves that travel through the walls of the structure are not solely compressive. If this tube is filled with a low impedance material such as air, tensile waves of significant magnitude will be caused in the tube wall by the impedance mismatch at the interior interface. This tensile loading has the potential to degrade the stability of the structure, as it may cause damage in the matrix material or at the fiber-matrix interface which is highly sensitive to out-of-plane tension [49]. When an underwater explosive is detonated, the decomposed gases of the charge create an expanding superheated gas bubble. After reaching its maximum critical size, the surrounding hydrostatic pressure of water causes the bubble to collapse, and upon completion of this collapse, a relatively strong and fairly broad pressure pulse known as the "bubble pulse"is emitted [50]. The magnitude of the bubble pulse is typically much smaller than that of the initial shock wave, however, due to its long duration, the impulse is of comparable strength and it interacts with the deforming shell to potentially cause its implosion. Based on the stand-off distance, the following behaviors were noted:

(a) Near-Field Explosion

In this case, the tube implodes in mode 2 shortly after experiencing the initial shock from the explosive. The average local pressure history measured about the midspan of a typical specimen is seen in Fig. 19. The initiation of the instability occurs at approximately 0.93 ms, well before the arrival of reflections from the tank wall, so it may be concluded that the initial shock wave alone possesses sufficient energy to trigger the collapse at this sub-critical static pressure. Wall contact occurs 5.00 ms after the initiation of the implosion, roughly consistent with results from natural implosion experiments. This implies that though the initiation mechanism of the collapse is different, the subsequent mechanics of the collapse are the same as for a natural implosion. In this case, no implosion pulse is detectable in the pressure history following wall contact due to the noisy nature of the pressure signal at this time caused by the superposition of reflections of the initial shock and bubble pulses from the tank walls.

(b) Mid-Field Explosion

Implosion in mode 2 again occurs in this case, though it is much later in the time than the 102 mm standoff case, initiating 6.83 ms after the detonation of the charge. This delayed collapse shows that neither the initial shock wave nor the reflections of that

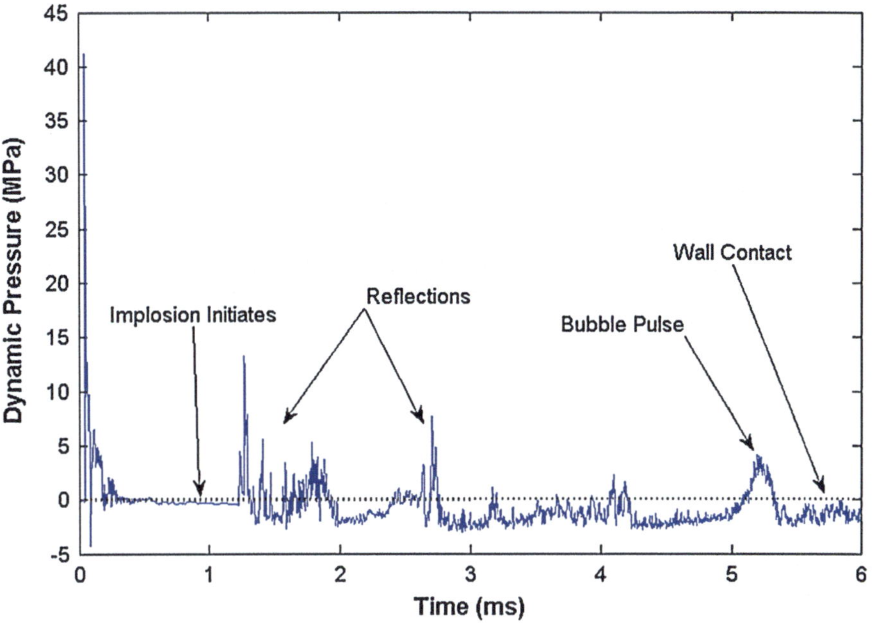

Fig. 19 Pressure history for explosive experiments with 102 mm (near-field) standoff distance

shock from the tank walls are sufficient to cause the collapse at this charge distance. The instability is triggered after the first bubble pulse showing that for this standoff distance, the bubble pulse is the cause of the implosion. The tube continues to collapse until wall contact is made 2.43 ms after initiation, which is significantly shorter than the collapse time of the natural implosion (4.7 ms). The reason behind this reduction in collapse time, as well as the reason for the timing of the collapse, is related to the wave interaction discussed in the reference publication by Pinto and Shukla [47].

(c) Far-field Explosion

Similar to the previous two cases, these tubes implode in a mode 2 shape, however, they collapse very late in the event, initiating 15.22 ms after the detonation of the charge. This is long after both the initial shock and the first bubble pulse, showing that neither of these events triggers the collapse on their own. Wall contact occurs at 1.83 ms following the initiation of the implosion, even shorter than for the 203 mm standoff case. Like the 203 mm standoff case, the cause for the delayed collapse as well as the duration of the collapse is due to the accumulation of damage caused by each successive loading event from the bubble pulses. This damage continually degrades the structural stiffness, until the vibrations caused by the impulse coupled with hydrostatic pressure can drive the implosion.

3.3 *Mitigation of Pressure Pulses from Implosion of Thin Composite Shells*

This section details a recent work done on mitigation of dangerous pressure released from the implosion of composite shells by addition of polymeric layers [51]. It was seen in Sect. 2.4 that polyurea (PU) coatings can serve as a useful addition to mitigate the blast response of composite materials. Layers of controlled thicknesses were applied to carbon/epoxy tubes on both the interior and exterior of the tube to determine the effect of coating thickness on the composite structure.

The implodable volumes used in this study were the same geometry and material as the ones used in Sect. 2.2.3.

Specimens were coated with a two-part PU which has a low stiffness (~10 MPa), but very high elongation of approximately 500%. For this reason, polyurea is very attractive for energy absorption applications. A simple fixture was built to coat both exterior and interior of the composite tubes with a controlled thickness of the mixed polymer. In this study, two coating thicknesses were used: a "thin" coating having equal volume to the tube walls, and a "thick" coating, with double the thickness of the "thin" coating. It should be noted that the interior and exterior coatings have slightly different thicknesses due to their locations on the tube, but their volumes, and therefore masses, are approximately the same (as shown in Table 8). Results show that thick interior coatings significantly reduce the energy released in the pressure pulse, at the cost of an increase in the overall mass of the structure. Surprisingly, thick exterior coatings increase the energy released, thus making the collapse more dangerous. These findings provide useful guidelines in the design of marine composite structures susceptible to implosion. The flow energy obtained directly as a function of pressure, is a useful tool for comparing the total energy emitted during the implosion process. By making an assumption of spherical symmetry of the emitted pressure pulse, the energy flux through a surface at distance, *R*, up to an arbitrary time, *t*, is defined simply as [52]:

$$E_F = \frac{1}{2\rho_0 R_s}\left[\int_0^t \Delta p dt\right]^2 \tag{9}$$

Table 8 Summary of experimental results

Specimen	Collapse pressure, P_{cr} (MPa)	P_{min}/P_{cr}	P_{max}/P_{cr}	Maximum collapse velocity (m/s)	Pre-buckling displacement (mm)
Aluminum	1.68 [±0.01]	0.69 [±0.04]	1.70 [±0.17]	20.0 [±2.0]	2.5 [±0.2]
Carbon/ epoxy	0.97 [±0.03]	0.51 [±0.05]	2.57 [±0.56]	16.5 [±2.3]	1.7 [±0.5]
Glass/PE	2.05 [±0.01]	0.84 [±0.02]	1.24 [±0.06]	27.2 [±1.4]	3.9 [±0.2]

where Δp is the dynamic pressure at time t, R_s is the standoff distance, and ρ_0 is the density of the fluid. To find the total flow energy, a spherical wave is assumed and the flux is multiplied by the area of an imaginary sphere with radiaus equal to the standoff distance of the pressure sensor as follows:

$$E_A = 4\pi R_s^2 E_F \tag{10}$$

To normalize this measure for differences in the collapse pressure of different materials, the total flow energy is divided by the potential hydrostatic energy available just prior to the collapse, defined as:

$$E_H = P_{cr} V_{tube} \tag{11}$$

where P_{cr} is the hydrostatic collapse pressure and V_{tube} is the internal volume of the tube. In this way, the flow energy released in the pressure pulse may be presented as a percentage of the available hydrostatic energy at the collapse pressure. Due to the knowledge of the shape of the pulse, the signal of a single sensor may be used to calculate the flow energy as described above. Some of the results are discussed below:

(a) Over-pressure comparison

For exterior coatings, thin coatings have little effect on the peak pressure, while thick coatings produce a significantly higher magnitude pressure peak. The maximum pressure for the thick exterior coated case is approximately 80% greater than the uncoated case which gives more evidence that thick exterior coatings actually intensify the collapse of the structure. This appears to be due to a containment of the damage in the structure and is discussed further in later sections.

The interior coatings show even more interesting changes to the overpressure. Not only is the magnitude of the pressure pulse altered, but so is the timing of the pulse. As the thickness of the coating increases, the magnitude of the pressure pulse decreases, and the arrival of the pulse is delayed. The mitigation of the peak pressure again gives evidence that these coatings slow the collapse of these tubes, thereby lessening the severity of the pressure pulse. The change in timing of the peak suggests these coatings extend the contact event itself, meaning that it takes longer to arrest the structure and the deceleration of the tube walls is therefore less.

(b) Flow energy

The flow energy as a percentage of the total available hydrostatic potential energy is dependent on the different material cases. Starting with thin coatings, it is seen that both interior and exterior coatings show little effect on the flow energy developed. Though there is some change in mean value, any change from the control uncoated case lies within error bars. Coatings of this thickness had little effect on the severity of the collapse. Thick coatings, on the other hand, do have a significant effect on the flow energy released. Thick interior coatings show a significant (30%) reduction in flow energy developed in the collapse. This implies that this coating drastically reduces the severity of the collapse, and this is supported by the analyses previously discussed. Tubes with thick interior coatings showed less of a pressure drop, reduced collapse velocity, and lower peak acceleration. All of these factors contribute to lessening the intensity of the failure event, and in turn reducing the flow energy released. Conversely, thick exterior coatings show a significant (14%) increase in flow energy, however, much higher peak pressure is observed for these cases.

3.4 Hydrostatic and Shock-Initiated Instabilities in Double-Hull Composite Cylinders

This section presents a study on the stability and dynamic behavior of concentric composite cylinders, with and without foam cores, during underwater an implosion [53]. Both hydrostatic and shock-initiated implosions of these double-hull cylinders are considered. These structures provide increased stability through bending strength, as well as pulse mitigation through foam core crushing. The specimens used in this study consist of commercially available filament wound sanded carbon/epoxy carbon fiber cylinders placed concentrically and bonded with PVC foam cores filling up the gap. Both the inner and outer shells have a filament winding architecture of [±15°/+87°/±45°/±15°]. The outer shell has a 60.3 mm ID and the inner shell a 38.1 mm ID, with both shells having a 279.4 mm unsupported length. The measured average wall thicknesses of the outer and inner shells used in this study are 1.69 mm and 1.68 mm, respectively. The closed cell PVC foam cores used in the specimens are cut from Divinycell H series foam-sheets provided by DIAB, Inc. (DeSoto, TX). Cores are made by cutting rings of 41.2 mm ID and 9.6 mm radial thickness. Rings are then stacked concentrically between the inner and outer shells. Foam cores of different densities are used: experiments are performed with no foam core, Divinycell H35, H60, H80, and H100. Results show a substantial increase in structural stability when the foam core is added, with critical collapse pressure increasing linearly with core crushing strength under hydrostatic conditions, and collapse delayed substantially or prevented under dynamic pulse loading.

3.4.1 Collapse Pressure

It is important to be able to predict the buckling behavior of a double hull structure as a function of material and geometrical properties. Figure 20 shows that the ΔP_{cr} values (defined as the increase in collapse pressure of the outer shell due to the sandwich structure [54]) measured previously increase linearly with core crushing strength, intersecting the origin and having a non-dimensional slope of 2.6. This implies that for the given core geometry, it would be possible to find a higher

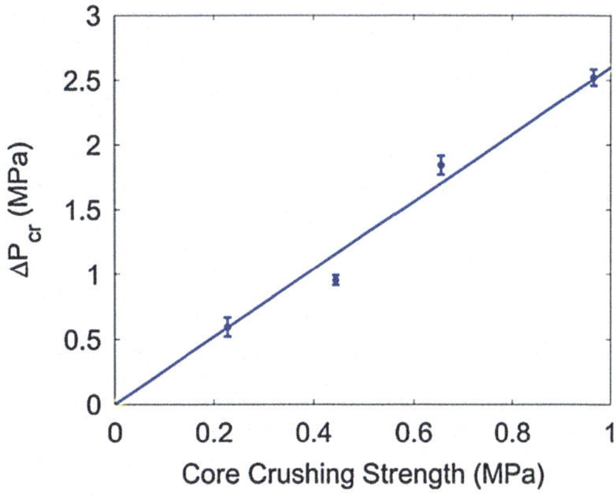

Fig. 20 Increase in collapse pressure as a function of foam core crushing strength

strength material than those studied here and that using such a core would likely increase buckling strength along the line in Fig. 20.

3.4.2 Implosion Under Hydrostatic Pressure: Observed Collapse Behaviors

Three overarching behaviors were observed in the collapse of the double hull specimens. In the first case, the outer cylinder collapsed and crushed the foam core but the inner cylinder did not collapse. In the second case, the outer cylinder collapsed and crushed the foam core, after which there was a brief dwell time before the inner cylinder collapsed. The last case shows that the outer cylinder collapsed, crushed the foam core, and collapsed the inner cylinder in a singular motion, with no dwell in between. Only the one case of collapse with dwell is discussed for illustration.

3.4.2.1 Complete Collapse With Dwell

Figure 21 shows the pressure pulse recorded in cases where the outer cylinder collapses, a brief dwell occurs, and then the inner cylinder collapses. Here, buckling initiation of the outer cylinder occurs at (A), followed by a smooth under-pressure region that exists until the onset of matrix cracking and the associated drop in fluid pressure at (B). The under-pressure region then continues until (C), where the pressure trace crosses the zero dynamic pressure line and an initial over-pressure corresponding to the collapse of the outer cylinder against the inner cylinder is recorded at (D). Following that, the inner cylinder begins to buckle from the hydrostatic pressure. What follows is a secondary implosion pulse corresponding to the collapse of the inner cylinder, with a new under-pressure region in the pressure signal being created at (E). At (F), there is a spike in pressure which corresponds to the development of damage in the inner cylinder as it collapses. The under-pressure

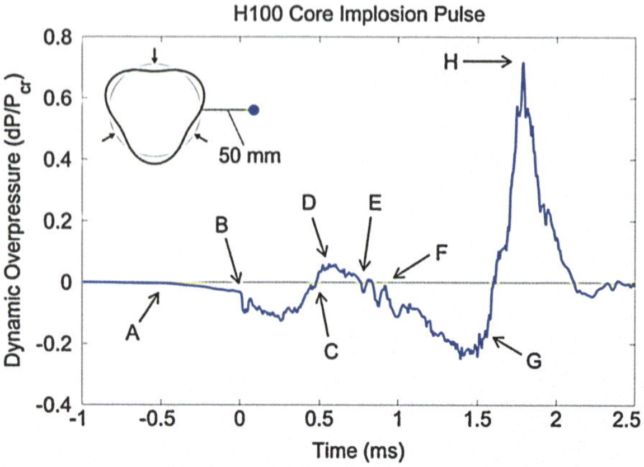

Fig. 21 Normalized pressure pulses from cases where the outer cylinder collapsed, dwelled briefly, then the inner cylinder collapsed; H100 foam core case, measured at midsection; outer cylinder buckling mode shape and sensor location

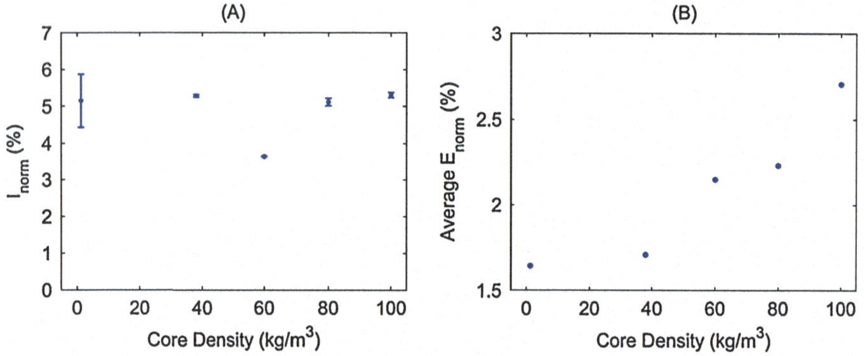

Fig. 22 Implosion emissions as a function of core density showing (**A**) normalized under-pressure impulse; (**B**) normalized average waveform energy

region continues until (G), where the inner cylinder makes wall contact with itself, and a secondary pressure peak at H is created which exceeds that of the first in amplitude.

3.4.3 Implosion Under Hydrostatic Pressure: Impulse and Energy

It is useful to look at normalized impulse, I_{norm}, defined as the area of the under-pressure region in the dynamic pressure versus time plot normalized by the product of collapse pressure and total under-pressure duration. It is plotted for various implosion cases in Fig. 22A with foam core density on the horizontal axis. The almost constant value of the impulse with foam core density implies that the impulse is mainly a function of collapse pressure, and is not directly influenced by core density or other properties. A plot of normalized average energy values ($E_{norm} = E_A/E_H$) which present the energy emitted by the implosion event as a function of available hydrostatic energy, is given in Fig. 22B.

The exception to this trend in the data is the case of the H60 core, which has a lower magnitude normalized impulse because of the uniquely long time delay between the collapse of the inner and outer cylinder. There, normalized average energy released increases with core density. The experiments showed a more violent collapse phenomenon with the H60, −80, and − 100 cases. These experiments also showed increased variation in energy released because the inner cylinder also collapses in the aforementioned cases, and introduces new damage mechanisms. The increase in normalized energy with core density is also significant, as there is a 57% increase in average normalized energy released between H35 and H100 core specimens. This highlights the diminishing returns on energy absorption for higher density cores. Increasing core density increases both hydrostatic collapse pressure and the energy released from the implosion. However, the increase in energy released from implosion is greater than the increase in potential energy from collapse pressure ($dE_A/dE_H > 0$) as core density increases. Thus, increasing core density is not an effective means of mitigating energy release from the natural implosion of double hull composite cylinders.

3.4.4 Implosion Under Combined Hydrostatic and Shock Initiated Loadings: Pressure History

For shock-initiated implosion experiments, an exploding-bridgewire detonator of 80 mg PETN and 1031 mg RDX (RP-85, Teledyne RISI, Inc., Tracy, CA) is mounted directly behind the specimen at a controlled standoff distance of either 102 mm or 203 mm. After filling, the vessel is pressurized to 80% of the measured natural buckling pressure of the specimen and the charge is detonated causing the specimen to experience shock loading. The superposition of the explosive pulse on top of the hydrostatic pressure causes complex loading on the underwater structure. The structures without foam cores collapsed completely in all experiments. Specimens collapsed in the same stages, regardless of whether they were subject to 102 mm or 203 mm explosive standoff. There, collapse initiation of the outer cylinder is seen, followed by outer-inner cylinder impact. After that, there is a period of time wherein the two cylinders vibrate and deform inward radially before the collapse of the inner cylinder begins with inner-inner wall contact occurring. Following this, the collapsed area propagates along the length of the structure, until it reaches the final collapse state. The specimens with foam core either implode quickly or undergo long-term vibrations. In the latter case, specimens may either survive completely, or implode after an extended period of time as a result of damage development during vibration. These vibrations provide an additional mechanism of energy absorption. The time period of vibration of the cylinders is influenced by both the added mass of water and the initial hydrostatic pressure [13].

Acknowledgments The authors acknowledge the financial support provided by Dr. Yapa D.S. Rajapakse, Solid Mechanics Program Manager ONR under grant numbers N00014-01-1-1033, N00014-04-1-0248, N00014-10-1-0662, and N00014-15-1-2046. The authors also acknowledge the support provided by the Department of Homeland Security (DHS) under Cooperative Agreement No. 2008-ST-061-ED0002. The authors would like to thank TPI Composites for providing the facilities used to create some of the composite structures discussed in this chapter. Finally, the help of graduate and undergraduate students working in the Dynamic Photo Mechanics Lab during this period is gratefully acknowledged.

References

1. Perl R, O'Rourke R (2001) Terrorist attack on USS Cole: background and issues for congress. Emerging technologies: recommendations for counter-terrorism. Institute for Security Technology Studies, Dartmouth College
2. Perl R (1998) Terrorism: US responses to bombings in Kenya and Tanzania: a new policy direction? Congressional report. Congressional Research Service, The Library of Congress
3. Thomas GP (2013) Composites in combat: composites for military vehicles. AZO Materials. https://www.azom.com/article.aspx?ArticleID=8166>. Accessed 12 Oct 2018
4. Mouritz AP, Gellert E, Burchill P, Challis K (2001) Review of advanced composite structures for naval ships and submarines. Compos Struct 53(1):21–42
5. Xue Z, Hutchinson JW (2003) Preliminary assessment of sandwich plates subject to blast loads. Int J Mech Sci 45(4):687–705
6. Fleck NA, Deshpande VS (2004) The resistance of clamped sandwich beams to shock loading. J Appl Mech 71(3):386–401
7. Dharmasena KP, Wadley HNG, Xue Z, Hutchinson JW (2008) Mechanical response of metallic honeycomb sandwich panel structures to high-intensity dynamic loading. Int J Impact Eng 35 (9):1063–1074
8. Yi J, Boyce MC, Lee GF, Balizer E (2005) Large deformation rate-dependent stress-strain behavior of polyurea and polyurethanes. Polymer 47(1):319–329
9. Amirkhizi AV, Isaacs J, McGee J, Nemat-Nasser S (2006) An experimentally-based constitutive model for polyurea, including pressure and temperature effects. Philos Mag 86 (36):5847–5866
10. Hoo Fatt MS, Ouyang X, Dinan RJ (2004) Blast response of walls retrofitted with elastomer coatings. Struct Mater 15:129–138
11. Roland CM, Twigg JN, Vu Y, Mott PH (2006) High strain rate mechanical behavior of polyurea. Polymer 48(2):574–578
12. Tekalur SA, Shukla A, Shivakumar K (2008) Blast resistance of polyurea based layered composite materials. Compos Struct 84(3):271–281
13. Gupta S, Matos H, LeBlanc J, Shukla A (2016) Shock initiated instabilities in underwater cylindrical structures. J Mech Phys Solids 95:188–212
14. Turner SE, Ambrico JM (2012) Underwater implosion of cylindrical metal tubes. J Appl Mech 80(1):1–11
15. Gupta S, LeBlanc J, Shukla A (2015) Sympathetic underwater implosion in a confining environment. Extreme Mech Lett 3:123–129
16. von Mises R (1914) The critical external pressure of cylindrical tubes Zeitschrift des Vereines Dtsch. Ingenieurs 58(19):750–767
17. von Mises R (1929) The critical external pressure of cylindrical tubes under uniform radial and axial load. Stodola's Festschrift, Zurich, pp 418–430
18. Cartlidge E (2001) Accident grounds neutrino lab. IOP Publishing Physicsworld. https://physicsworld.com/a/accident-gro
19. Kumar P, LeBlanc J, Stargel D, Shukla A (2012) Effect of plate curvature on blast response of aluminum panels. Int J Impact Eng 46:74–85

20. LeBlanc J, Shukla A, Rousseau C, Bogdanovich A (2007) Shock loading of three-dimensional woven composite materials. Compos Struct 79(3):344–355
21. Wright J (1961) Shock tubes. Wiley, New York
22. Xue Z, Hutchinson JW (2004) A comparative study of impulse-resistant metal sandwich plates. Int J Impact Eng 30(10):1283–1305
23. Taylor GI (1963) Pressure and impulse of submarine explosion waves on plates. In: Batchelor GK (ed) The scientific papers of Sir Geoffrey Ingram Taylor, aerodynamics and the mechanics of projectiles and explosions, vol 2. Cambridge university press, Cambridge, pp 287–303
24. Kambouychev N (2007) Ph. D dissertation, Massachusetts Institute of Technology
25. Kambouchev N, Noels L, Radovitzky R (2006) Nonlinear compressibility effects in fluid-structure interaction and their implications on the air-blast loading of structures. J Appl Phys 100:063519
26. Wang E, Jefferson W, Shukla A (2011) Analytical and experimental study on the fluid structure interaction during air blast loading. J Appl Phys 110:114901
27. Baker WE, Cox PA, Westine PS, Kulesz JJ, Strehlow RA (1983) Explosion hazards and evaluation. Elsevier Publishing Company, New York
28. Glasstone S, Dolan PJ (1964) The effects of nuclear weapons. Third edition. United States, p 1977. https://doi.org/10.2172/6852629
29. Smith PD, Hetherington JG (1994) Blast and ballistic loading of structures. Butterwirth-Heinmann, Elsevier Science Ltd, Oxford
30. Wang E, Gardner N, Gupta S, Shukla A (2012) Fluid-structure interaction and its effect on the performance of composite structures under air-blast loading. Int J Multiphys 6(3):219–239
31. Li R, Kardomateas GA, Simitses GJ (2009) Point-wise impulse (blast) response of a composite sandwich plate including core compressibility effects. Int J Solids Struct 46(10):2216–2223
32. Gardner N, Wang E, Shukla A (2012) Performance of functionally graded sandwich composite beams under shock wave loading. Compos Struct 94(5):1755–1770
33. http://www.gurit.com
34. Gardner N (2012) Novel composite materials and sandwich structures for blast mitigation. Ph. D dissertation, University of Rhode Island
35. Wang E, Gardner N, Shukla A (2009) The blast resistance of sandwich composites with stepwise graded cores. Int J Solids Struct 46(18–19):3492–3502
36. Gardner N, Kumar P, Wang E, Shukla A (2012) Blast mitigation in a sandwich composite using graded core and polyurea interlayer. Exp Mech 52(2):119–133
37. http://specialty-products.com
38. Schreier H, Orteu JJ, Sutton MA (2009) Image correlation for shape, motion and deformation measurements: basic concepts, theory and applications. Springer, New York
39. Gupta S, Parameswaran V, Sutton MA, Shukla A (2014) Study of dynamic underwater implosion mechanics using digital image correlation. Proc R Soc A Math Phys Eng Sci 470:20140576–20140576. https://doi.org/10.1098/rspa.2014.0576
40. Pinto M, Gupta S, Shukla A (2015) Study of implosion of carbon/epoxy composite hollow cylinders using 3-D digital image correlation. Compos Struct 119:272–286. https://doi.org/10.1016/J.COMPSTRUCT.2014.08.040
41. Brennen CE (Christopher E (2014) Cavitation and bubble dynamics. Cambridge University Press, New York
42. Ikeda CM, Wilkerling J, Duncan JH (2013) The implosion of cylindrical shell structures in a high-pressure water environment. Proc R Soc A Math Phys Eng Sci 469:20130443–20130443. https://doi.org/10.1098/rspa.2013.0443
43. Harte A-M, Fleck NA (2000) On the mechanics of braided composites in tension. Eur J Mech A Solids 19:259–275. https://doi.org/10.1016/S0997-7538(99)00164-3
44. Pinto M, Gupta S, Shukla A (2015) Hydrostatic implosion of GFRP composite tubes studied by digital image correlation. J Press Vessel Technol 137:051302. https://doi.org/10.1115/1.4029657

45. Farhat C, Wang KG, Main A et al (2013) Dynamic implosion of underwater cylindrical shells: experiments and computations. Int J Solids Struct 50:2943–2961. https://doi.org/10.1016/J.IJSOLSTR.2013.05.006
46. Lindberg HE, Florence AL (1987) Dynamic pulse buckling. Springer Netherlands, Dordrecht
47. Pinto M, Shukla A (2016) Shock-initiated buckling of carbon/epoxy composite tubes at sub-critical pressures. Exp Mech 56:583–594. https://doi.org/10.1007/s11340-015-0033-1
48. Shin YS (2004) Ship shock modeling and simulation for far-field underwater explosion. Comput Struct 82:2211–2219. https://doi.org/10.1016/J.COMPSTRUC.2004.03.075
49. Sridharan S (2008) Delamination behaviour of composites. CRC Press, Boca Raton
50. Cole RH (1948) Underwater explosions. Princeton University Press, Princeton
51. Pinto M, Shukla A Mitigation of pressure pulses from implosion of hollow composite cylinders. https://doi.org/10.1177/0021998315624254
52. Arons AB, Yennie DR (1948) Energy partition in underwater explosion phenomena. Rev Mod Phys 20:519–536. https://doi.org/10.1103/RevModPhys.20.519
53. DeNardo N, Pinto M, Shukla A (2018) Hydrostatic and shock-initiated instabilities in double-hull composite cylinders. J Mech Phys Solids 120:96–116. https://doi.org/10.1016/j.jmps.2017.10.020
54. Glasstone, S, and Dolan, P J. The Effects of Nuclear Weapons. Third edition. United States. 1977. https://doi.org/10.2172/6852629

The Response of Composite Materials Subjected to Underwater Explosive Loading: Experimental and Computational Studies

James LeBlanc, Erin Gauch, Carlos Javier, and Arun Shukla

1 Introduction

Within many marine industries there is a significant and increased interest in the use of composite materials to support advanced structural requirements as well as reduce maintenance concerns. These advanced materials provide many advantages over traditional structural materials, including high stiffness/strength to weight ratios, superior resistance to corrosion, reduced maintenance costs, and near net geometry part manufacturing capabilities. Correspondingly, there are also considerations that must be taken into account when considering the use of such materials, including reduced impact damage tolerances, jointed connections, and repair methods should damage be sustained. Furthermore, in certain applications there is an increased risk of severe loading and environmental conditions such as impacts, collisions, shock loading, and long term seawater exposure. While the static response of such materials is well understood, there is less of an understanding in terms of what happens to the same composite material when subjected to high loading rates. Therefore, there is a fundamental need to understand the behavior of these materials not only at static load levels but also at loading rates many orders of magnitude higher. Correspondingly, the ability to predict the load carrying capability of these materials after a damaging or shock loading event is critical to provide assurance that a component retains structural integrity and functionality. Furthermore, the degrading effects of long term environmental exposure on the mechanical properties of the materials should be quantified and understood in order to provide designs that will function

J. LeBlanc (✉) · E. Gauch
Naval Undersea Warfare Center, Division Newport, Newport, RI, USA
e-mail: James.M.LeBlanc@Navy.Mil; Erin.Gauch@Navy.Mil

C. Javier · A. Shukla
Department of Mechanical, Industrial and Systems Engineering, University of Rhode Island, Kingston, RI, USA
e-mail: Carlos_Javier@my.uri.edu; ShuklaA@uri.edu

S. W. Lee (ed.), *Advances in Thick Section Composite and Sandwich Structures*,
https://doi.org/10.1007/978-3-030-31065-3_2

throughout the expected service life of a specific structure. Due in large part to these uncertainties in the knowledge base related to these material systems there is a tendency to introduce large safety factors into structural designs which can have the effect of designing out the weight savings afforded by the materials.

When a submerged structure is exposed to an underwater explosion, it undergoes a complex and highly transient loading condition including high peak pressures and spherical wave fronts. When explosions occur at sufficiently large standoff distances from a structure, the shock fronts are nearly planar and act over the entire structure in a nearly uniform manner. This loading results in structural responses consisting primarily of flexure with large center-point deflections. However, there tends to be low levels of material damage (primarily inter-laminar delamination) and plate perforations or ruptures are minimal. In the absence of plate rupture, the shock wave is almost fully reflected away from the structure, shielding any occupants and/or internal equipment from the effects of the high pressure waves. Conversely, when an explosion occurs directly on, or very close to, the surface of a structure, the loading area is limited to the vicinity of the detonation itself. This loading is generally characterized by a spherical shock front impinging upon the structure as well as interaction of the UNDEX bubble and the target structure. The result is highly localized pressure loadings and significant damage, oftentimes including plate penetration or complete rupture. Upon rupture of the plate the pressure waves enter the structure, subsequently exposing any occupants to the adverse effects of high pressure gases as well as any shrapnel which may become dislodged from the blast area.

In recent years, there have been a multitude of research investigations which have studied the response of composite materials when subjected to highly transient and severe loading conditions. These investigations have explored the effects of blast and shock conditions through the use of a variety of advanced experimental techniques including the direct use of explosive charges, shock tubes, and underwater shock simulators. The work has also focused on a diverse range of structural configurations of interest ranging from flat solid laminates, to curved panels, cylinders, and sandwich constructions. Nurick et al. [1, 2] have studied the effects of boundary conditions on plates subjected to air blast loading and identified distinct failure modes depending on the magnitude of the impulse and standoff. Tekalur et al. [3] investigated the effects of shock loading on both E-Glass and Carbon based laminates. Mouritz [4] studied the effectiveness of adding a light weight, through thickness stitching material to increase the damage resistance of composites. The same author has also studied the effect of shock loading on the flexural [5] and fatigue [6] properties of composite laminates when subjected to underwater shock loading. These studies have shown that under relatively low impulsive loading the material sustains little damage (primarily matrix cracking) and the mechanical properties remain the same as for undamaged laminates. However, once a critical loading threshold is exceeded then the panels experience fiber breakage and the material strengths are significantly degraded. Dear and Brown [7] have conducted a detailed study on the damage mechanisms and energy absorption in composite plates when subjected to impact loading. Franz et al. [8] and LeBlanc et al. [9] have studied the effects of shock loading on three-dimensional woven composite materials. Studies on the response of composites subjected to UNDEX have generally focused on far field loading in which the encroaching shock front is nearly planar and there is no interaction between the UNDEX bubble and the structure. LeBlanc and Shukla [10, 11] have studied the response of both flat and curved E-glass/epoxy composite plates to far field loading. Mouritz [4], conducted a study of the development of damage in a glass reinforced composite subjected to underwater explosive loading at increasing pressures. Both air backed and water backed conditions were evaluated. In the case of the water backed laminates no damage or degradation in strength was noted. In the air backed laminates delamination and matrix cracking led to a degradation of the residual strength of the composite. Recently, there has been an increased interest in the study of the effect of shock loading on sandwich structures. These studies include the effects of shock and impact loading conditions (Jackson et al. [12], Schubel et al. [13], Arora et al. [14]). Avachat and Zhou [15] studied the effects of underwater shock loading on filament wound and sandwich composite cylinder and found that while both constructions exhibited similar damage mechanisms, including delamination, fiber failure and matrix cracking, the sandwich structure had overall better performance than a monolithic cylinder with similar mass. The same authors [16] also utilized an Underwater Shock Loading Simulator combined with digital image correlation to show that for sandwich constructions lower density cores yield higher blast performance than high density cores due to their larger core compression capability. Work by Latourte et al. [17] utilized a scaled fluid structure method [18] to study the failure modes and damage mechanisms in both monolithic and sandwich plates subjected to underwater impulsive loads.

Analytical damage models for composites have been widely developed and are continually being refined and updated. These models typically assign an internal damage variable to each of the types of damage of interest (ie. matrix cracking, fiber rupture) which, in simple form, are ratios of the stress state to a failure criteria (Matzenmiller et al. [19], Zako et al. [20], Dyka et al. [21]). Based upon the expression representing each damage variable, the effective elastic properties can be degraded when the variable reaches a critical value. As the mechanical properties must be continually updated to account for the damage degradation this methodology lends itself well to implementation in finite element codes. The finite element modeling of damage in composites has been performed primarily on models simulating strain rates up to those representing drop test experiments with some work performed at the high strain rate regimes expected in shock loading. Material models are continually being implemented and refined in existing commercial finite element codes (O'Daniel et al. [22], McGregor et al. [23]). Recent publications involving computational modeling of damage progression in composites have utilized Ls-Dyna and the Mat_162 (Mat_Composite_OPTION) material model which simulates fiber breakage, matrix cracking and delamination damage. This material model combines the progressive failure theory of Hashin and the damage mechanics approach of Matzenmiller et al. [19]. Gama et al. [24] have published results from quasi-static punch shear loading experiments which correlate well with simulations utilizing the Mat_162 material model. Simulations of low velocity impact experiments have been documented in the work by Donadon et al. [25], Hosseinzadeh et al. [26], and Tagarielli et al. [27]. Furthermore, Batra and Hassan [28] studied the response of composites to UNDEX loading through numerical simulations; however, there are no comparisons to experimental results. LeBlanc et al. [10] have presented a modeling methodology which simulates composite plates subjected to underwater explosive loading with comparisons to both the transient strain response as well as post mortem damage.

During the service life of marine structures there is typically an extended/prolonged exposure to the wetted environment, specifically aggressive salt water. As a result there is an ongoing body of research into the adverse effects of these conditions on component materials to the continuous exposure [29]. Additionally, these structures may be deployed into operations settings in which impact and shock loading is a concern. With these loading conditions in mind, failure to consider degradation of mechanical properties could be catastrophic. A significant cause of mechanical degradation in composites in a marine environment is the diffusion of water into the matrix material [29]. The diffusion process is relatively well established and can be described by a diffusion coefficient that is a function of parameters such as temperature, type of resin and curing agent, surrounding medium composition, fillers, void content, and so on. The value for diffusion coefficient and the theoretical models used to describe the diffusion varies in previous studies of diffusion in composites [30–37]. A standard and well-accepted model for epoxy resins, in terms of mass diffusions, is a Fickian model which uses Fick's second law to predict how the concentration of a diffusive substance changes over time within a material [38, 39]. Previous works used the Fickian model to study the properties changes during low strain rate loading of diffused composites. Current research on the high strain rate response of weathered composites is limited. Recently, there has been a study that analyzes the shock response of weathered composites plates within an air medium [40].

2 Far Field UNDEX Loading of Curved Plates

Two primary, underwater applications of composite materials are vehicle hull forms (ROV's, UUV's) and underwater pipelines. These uses generally result in curved structural geometries in which the curvature is oriented into the fluid domain (convex as viewed from the wetted surface). Thus there is an ongoing need to understand the complex loading, both temporally and spatially, that results when a highly transient shock pressure load is applied to a curved plate geometry. The transient structural response of the plates themselves is also of key interest. To this end, a detailed experimental and numerical investigation of the response of curved plates when subjected to far field shock loading was conducted. The far field pressure loading is characterized by a shock wave front that is nearly planar upon arrival at the structure of interest (i.e. shallow curvature with respect to the structural geometry of interest). The primary objective was to develop a controlled experimental approach to subject curved composite plates to far field underwater explosive loading while capturing the transient response, along with development of a computational modeling method to accurately simulate the experiments.

2.1 *Conical Shock Tube Facility and Experimental Method*

A conical shock tube (CST) facility was utilized to impart shock loading to the composite plates in the study. The shock tube provides an effective and repeatable means for the conduct of underwater explosive testing representative of a far field detonation, generally characterized by planar, or very shallow wave front curvature. The shock tube is a horizontally mounted, water filled tube with a conical internal shape as illustrated in Fig. 1. The geometry of the tube itself is intended to represent an angular segment of the pressure field that results from the detonation of an explosive charge in a free field underwater environment; generally, a spherically spreading wave front. In the CST, the rigid tube walls confine the expansion of the pressure wave in a manner that simulates a conical sector of the pressure field. An amplification factor which relates the amount of explosive charge used in the CST to an equivalent spherical charge in a free field is defined by Poche and Zalesak

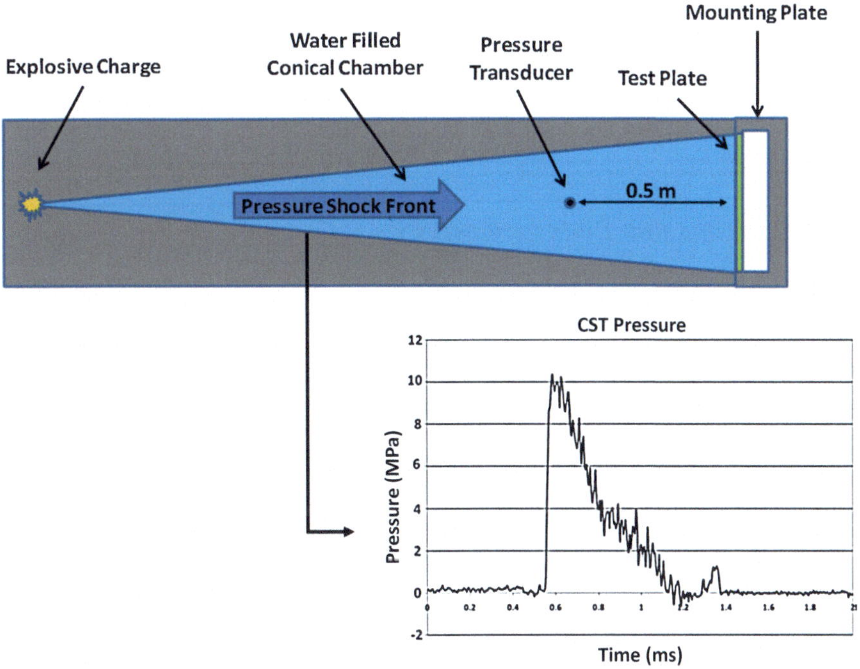

Fig. 1 Conical shock tube schematic (not to scale)

[41]. This is taken to be the ratio between the weight of a spherical charge, W, required to produce the same peak pressure at a given standoff distance as that produced in the shock tube by a segment of charge weight, w. Further discussion on the development and history of the water filled conical shock tube is provided by [42, 43]. The tube utilized in the current study has an internal cone angle of 2.6 degrees, is 5.25 m (207 in.) in length, and when flooded contains 98.4 L (26 gal.) of water at atmospheric pressure. The pressure shock wave is initiated by the detonation of an explosive charge at the breech end of the tube. A typical pressure profile obtained from the use of the tube is shown in Fig. 1 and is highlighted by the rapid rise time of the pressure followed by the exponential decay of the wave. The length of the tube is sufficient so that plane wave conditions are nearly established at the test specimen.

During testing, a mounting fixture which holds the composite plates with fixed edges and allows for air backing, was utilized as shown in Fig. 1. The plates are mounted with the convex face oriented towards the incoming shock front so as to represent a submerged underwater structure. The study has employed the use of high speed photography coupled with digital image correlation to capture the transient response of the plates during loading and subsequent deformation in real time. The advantage of such a non-contact, optical method in the experiments is full-field measurement as well as the elimination of strain gages and their inherent practice of debonding from the specimens at high shock levels and large plate flexures. The use of two cameras in a stereo configuration allowed for the three-dimensional, out of plane response to be captured. A framing rate of 20,000 frames/second was used with an inter-frame time of 50 μs. The explosive charge used in the study is an M6 blasting cap which yields peak pressures of 10.3 MPa at the pressure sensor location, .508 m (1.67 ft).

2.2 *Materials and Plate Geometry*

The composite plates utilized in the study are comprised of a 0°–90°, balanced construction biaxial, E-Glass / Vinyl laminate. The fabric consists of 0/90 layers that are stitched (non-woven) and have a dry areal weight of 0.406 kg/m^2. Each plate has a finished part thickness of 1.37 mm (0.054 in.), a fiber content of 62% by weight, has 3 plies of the fabric, and has mechanical properties provided in Table 1. The plates contain a curved midsection with the convex face having a radius of curvature

Table 1 E-Glass /Vinyl ester biaxial laminate -mechanical properties (ASTM 638)

	MPa (lb/in^2)
Tensile modulus (0°)	15.8e3 (2.3e6)
Tensile modulus (90°)	15.8e3 (2.3e6)
Tensile strength (0°)	324 (47,000)
Tensile strength (90°)	324 (47,000)

of 18.28 cm (7.2 in.) and an outer diameter of 26.54 cm (10.45 in.) with a 22.86 cm (9 in.) unsupported middle section.

2.3 Computational (Finite Element) Model Overview

The conical shock tube experiments have been simulated through the use of the LS-DYNA finite element code. The objective of the numerical modeling is to demonstrate an approach which is able to capture both the complex fluid structure interaction that occurs between the shock front and the curved plate surface as well as the transient deformation mechanisms that the plate undergoes as a result of the loading. The complete finite element model of the CST test setup utilized in the study is shown in Fig. 2. The model consists of the internal fluid contained within the shock tube itself as well as the composite plate. The fluid within the tube is considered in the simulation so as to accurately represent the fluid structure interaction (FSI) at the interface of the fluid and test plate. The simulations show that during the interaction of the pressure wave and the curved surface of the plate the pressure loading on the plate is not uniform across its face. Thus the inclusion of the fluid is a critical aspect of the modeling approach. In the model, the pressure profile that is

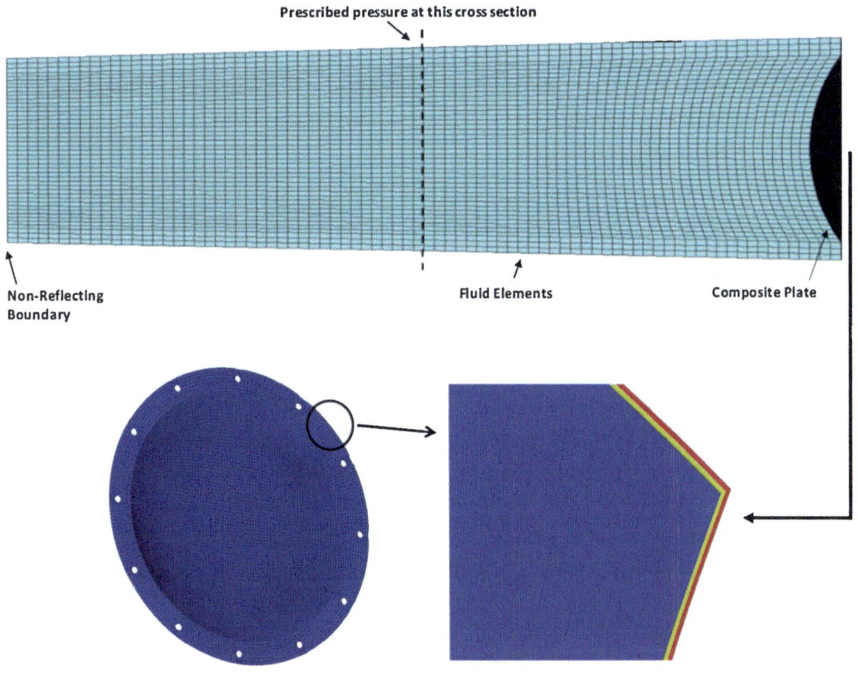

Fig. 2 CST and Composite plate model

recorded by the pressure sensor is mapped into the fluid domain at a tube cross section corresponding to the sensor location and the wave is propagated numerically from that point towards the plate surface. The fluid is modeled with solid elements and a null material definition which allows for the material to be defined with an equation of state (EOS). The linear polynomial EOS is utilized in the models and allows for the bulk modulus and density of the water to be defined. This allows for an accurate propagation of the pressure wave in the water in a computationally efficient manner. Fluid–structure interaction between the water and the composite plate is numerically coupled through the use of a tied-surface-to-surface contact definition.

The composite plate in the simulations is modeled using shell elements with the plate model consisting of 3 layers of shell elements, each layer representing a 0° and 90° combined ply, Fig. 2. The mid-surface of each ply is meshed and the individual shell layers are offset by the ply thickness. The material model utilized in this work is Mat_Composite_Failure_*Option*_Model which is an orthotropic material definition capable of modeling the progressive failure of the material due to any of several failure criterion including tension / compression in the longitudinal and transverse directions, compression in the through thickness direction, and through thickness shear. It is important to note that failure in one direction does not cause the element to be deleted. An element is only deleted from the analysis after it has failed in all directions and can no longer carry any load. Delamination damage is considered and is taken into account through the use of a surface-to-surface tiebreak contact definition. The tie break definition initially ties the nodes between plies together to inhibit sliding motion. The force at each node is monitored by the software and the corresponding normal and shear stresses are computed. If the current stress state at any node in the contact definition exceeds the failure criteria then the tie definition for that node is deleted and the node is free to slide, thus individual plies can separate but not pass thorough one another.

2.4 Simulation Results and Correlation to Experiments

The complex fluid structure interaction between the incident shock front and the curved composite plates was visualized through the use of the computational models. Due to the solid walls of the CST only the pointwise pressure history at the pressure sensor was obtained during the experiments. The pressure wave propagation and corresponding interaction with the curved surface of the plate is detailed in Fig. 3. Correspondingly, the transient plate response is shown in the right side of Fig. 3. The results presented identify several significant observations. The first is that while the pressure wave in the tube is nearly planar during its propagation towards the panel, upon its interaction with the curved surface the pressure loading becomes complex and not uniform. Noticeably, there is a low pressure area that develops in the center of the plate while the clamped edge sustains a higher pressure magnitude. Due to the air-backed nature of the plates and the relative thinness, the middle of the plate has a low stiffness as compared to the clamped edge of the plate. A second key

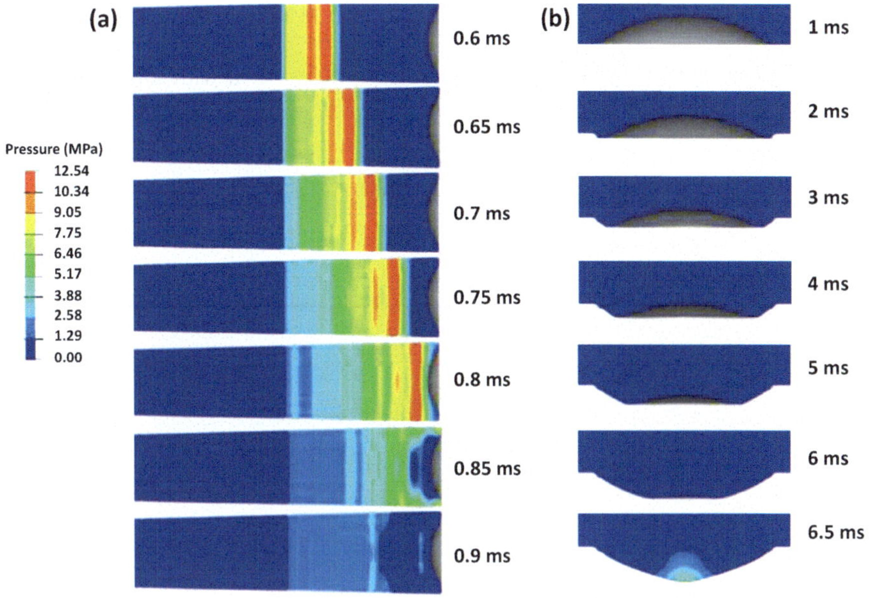

Fig. 3 (**a**) Fluid structure interaction, (**b**) Plate deformation progression

observation is that there are two distinct time regimes in the problem of interest as it relates to the plate mechanics. The first is a very short duration of time, 0.2 ms, in which the pressure wave loads the plate surface followed by a longer period of time, ~5 ms, in which the plate undergoes the physical deformation. The mechanics of the plate deformation consist of a full inversion process, with three key mechanisms. Initially, a hinge forms at the outer edge of the plate at the clamped boundary, which subsequently propagates towards the center of the plate, and finally is arrested upon reaching full plate inversion at 6.5 ms. Corresponding to the arrest of the inversion process there is a high pressure region that develops at the apex of the inverted shape caused by the fluid that was following the plate surface coming to a sudden stop.

The displacement and velocity data that was captured during the experiments is used to demonstrate correlation between the results and the simulation data. The back face, transient data as recorded through DIC is the primary means for correlation, namely the center-point time history displacement and the corresponding full field deformation profiles. The agreement between the simulation and experiment time histories is shown in Fig. 4 and the full field correspondence is highlighted in Fig. 5. The center-point displacement comparison shows that the experiment and simulation results agree nearly exactly early in the event and then the displacement in the experiment occurs slightly faster than the simulation. The full field comparison of the deformation evolution exhibits good correlation in terms of the displacement evolution and the full inversion of the panels as a result of the shock pressure loading. There is some level of non-symmetric behavior observed in the experimental results whereas in the simulations the deformation is symmetric as expected. In

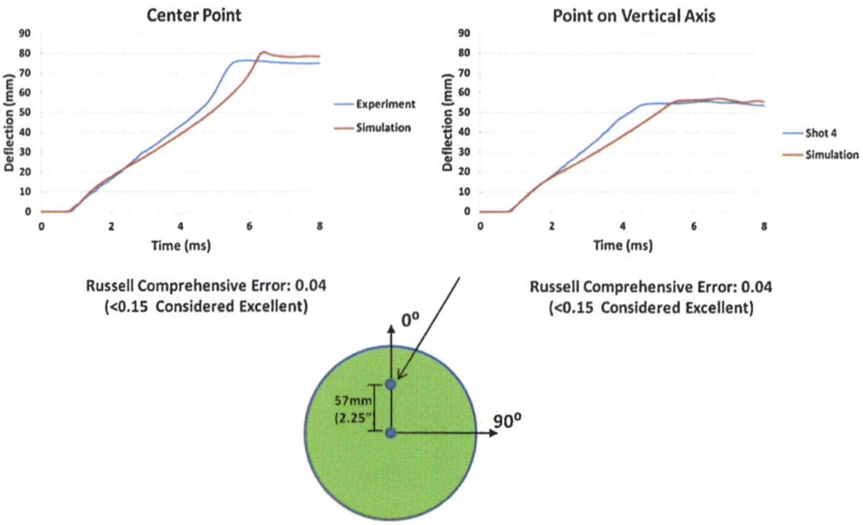

Fig. 4 Time history deformation comparison of experiment and simulation

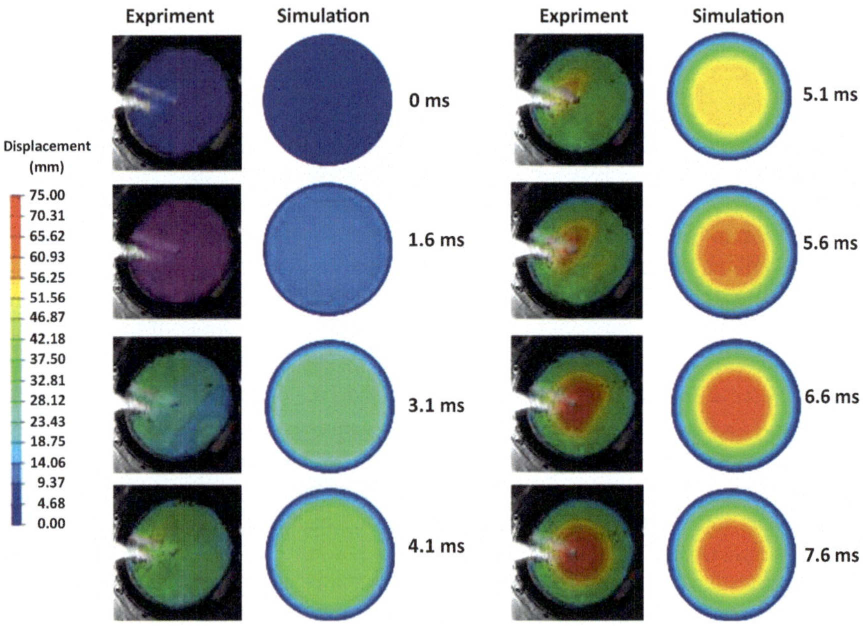

Fig. 5 Full field deformation comparison of experiment and simulation

addition to the transient correlations, the relationship between the material damage observed during the test and the damage level predicted by the simulations is correlated. In the experiment the main damage mechanism that was observed is delamination between the plies with minimal fiber rupture or matrix cracking. The

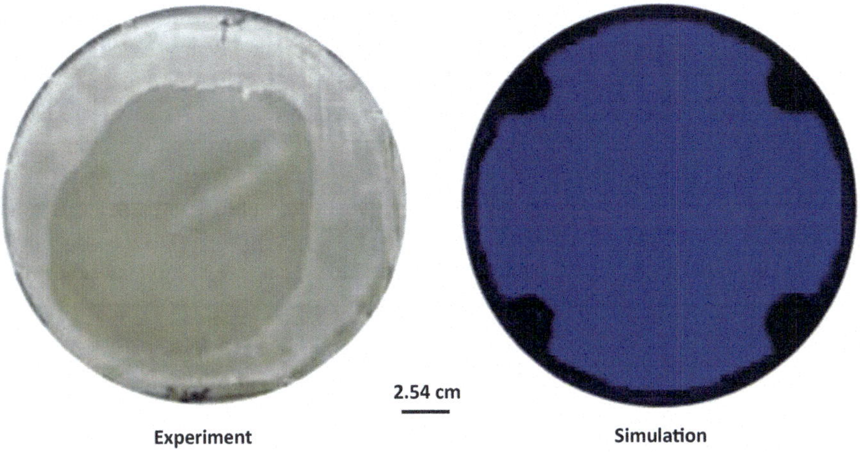

Fig. 6 (**a**) Material damage during test, (**b**) Material damage from simulation

final damage state from the shock test and corresponding simulation is shown in Fig. 6 where in the experimental image the damage is indicated by the lighter material and in the simulation the delamination area is highlighted by the black area. Although the amount of delamination is somewhat larger in the experiment than is observed in the computational model, it is encouraging that the model is able to predict the onset of the delamination itself and propagate it to a comparable distance. In the models the delamination criterion was taken to be 36 MPa (5250lb/in^2) for both tensile and shear stresses which corresponds to approximately one-half of the tensile strength of the pure vinyl ester. The degradation by ½ of the tensile strength accounts for voids, and interfacial defects / flaws between the layers of fibers during the manufacturing of the material. This observation is provided to aid in the development of delamination modeling best practices but is not meant to be definitive, rather further work in this area is required.

2.5 Key Findings

The investigation described in this section has studied the response of curved composite panels to far field underwater explosive loading conditions through both novel experiments and advanced computational models. The experiments were performed through the use of a conical shock tube facility which provides a controlled and repeatable means for the conduct of underwater shock testing representative of much larger charges in a laboratory setting. The use of computational models to supplement the experimental work allows for additional insight into the physics of the experiments which would otherwise be difficult to ascertain, namely the fluid structure interaction between the pressure wave and the plate surface. The specific material evaluated in the study is an E-Glass / Vinyl Ester bi-axial laminate with fibers balanced in the 0 and 90 degree directions. The plates are convex in geometry and oriented with the curved face towards the incident shock front. The correlation between the experimental and simulation results demonstrated high agreement at both the pointwise and full field deformation levels. Furthermore, the simulations were able to highlight that the loading of the plates by the pressure and the resulting deformation evolution could be separated into two distinct time regimes, with the pressure loading occurring over a fraction of a millisecond and the deformation occurring over several ms. The significant finding of the work is that computational tools can serve to support experimental test results and show promise for use as an alternative to testing to support structural designs utilizing composite materials.

3 Near Field UNDEX Flat Plates

Subsequent to the far field UNDEX experiments which were described in the preceding section, an experimental and numerical study was undertaken to investigate the response of flat composite plates when subjected to near field loading conditions. The near field blast loading condition is encountered when an explosion occurs directly on, or very close to, the surface of a structure. The result is highly localized pressure loadings and the structure sustains higher amounts of damage, oftentimes including plate penetration or complete rupture. The work presented in the following section discusses the effects of the near field blast loading on not only baseline composite plates, but also investigates the effects of structural plate thickness and the application of polyurea coatings on both the transient response as well as damage levels sustained in the material. The investigation consisted of experiments performed in an underwater blast tank including the use of Digital Image Correlation (DIC) to capture the transient response of the plates along with corresponding computational simulations performed with the commercial finite element code LS-DYNA.

3.1 Materials and Plate Configurations

The composite material utilized in the study consists of a commercially available, bi-axial E-Glass/Epoxy laminate, specifically Cyply® 1002. The panels are of non-woven construction with continuous, parallel fiber orientations. The cross ply construction has alternating 0 and 90° plies of which each is 0.254 mm thick. The laminates have an areal weight of 0.46 kg/m^2 (0.85 lb./yd^2) per ply and a resin content of ~36%. To determine the relative influence of panel thickness on the transient and damage performance of the plates, two thicknesses are considered, 0.762 mm and 1.524 mm. Furthermore, for the 0.762 mm plates a polyurea coating of thickness 0.762 mm has been applied to the back surface of the plates, resulting in

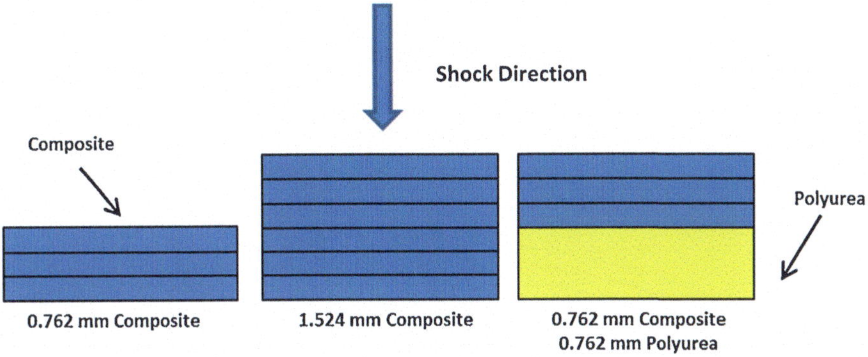

Fig. 7 Composite plate construction – Schematic (Not to scale)

an overall coated plate thickness of 1.524 mm. The plate configurations that have been studied are: (1) 0.762 mm thick uncoated plate, (2) 1.524 mm thick uncoated plate, and (3) 0.762 mm thick plate with a 0.762 mm polyurea coating on the back face. The schematic of the panel configurations investigated are provided in Fig. 7. The polyurea was sprayed onto the plates and then post cured for 48 h at a temperature of 160 °F. The specific polyurea material used is Dragonshield-BC and is a 2 part, spray cast material. Mechanical characterization of the coatings was performed in both tension and compression from strain rates from 0.01 s^{-1} to 2000 s^{-1}. Characterization up to 100 s^{-1} was performed using standard material testing machine whereas a split Hopkinson pressure bar was used to characterize the response of the material at 2000 s^{-1}. The full material characterization including loading direction and strain rate dependence is shown in Fig. 8. Moreover, the material exhibits strong strain rate dependence and becomes stiffer with increasing loading rate.

3.2 UNDEX Facility and Experimental Method

One of the primary focuses of the current study is the implementation of an experimental approach for the subject of flat plates to near field UNDEX conditions in a controlled laboratory environment. The conduct of such testing allows for highly controlled and repeatable experiments to be performed while also employing measurement techniques that would be difficult to utilize in a larger scale or operation test theatre. To that end, the study described in this section makes use of a water filled blast tank coupled with high speed photography and Digital Image Correlation to impart UNDEX loading to the fully clamped plates while capturing the full-field transient response. The experiments in this study were conducted in a water filled tank with internal dimensions of 1.21 m × 1.21 m × 1.21 m (3.94 ft. × 3.94 ft. × 3.94 ft) as shown in Fig. 9. Four window ports allow for the lighting and high speed photography of the UNDEX event and plate motion. Mounted to the inner surface of

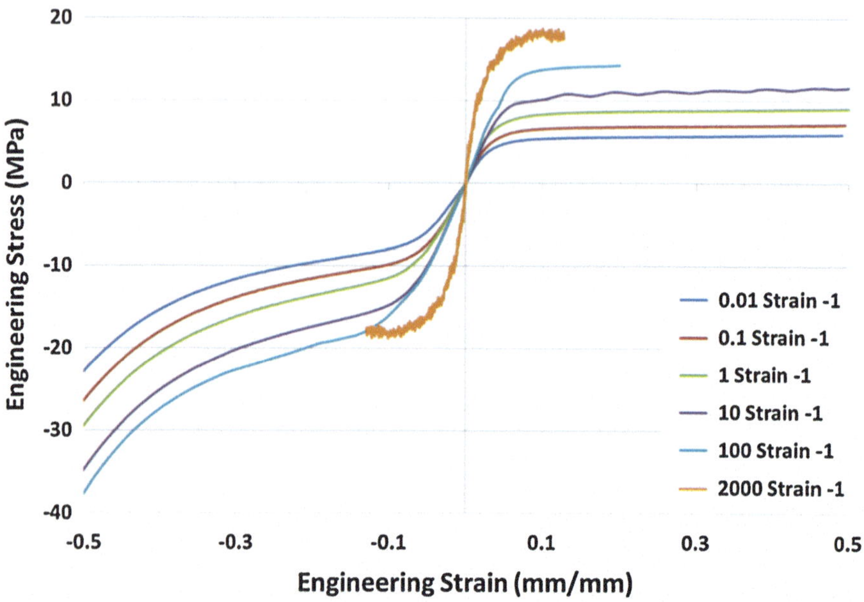

Fig. 8 Dragon shield BC polyurea stress-strain behavior

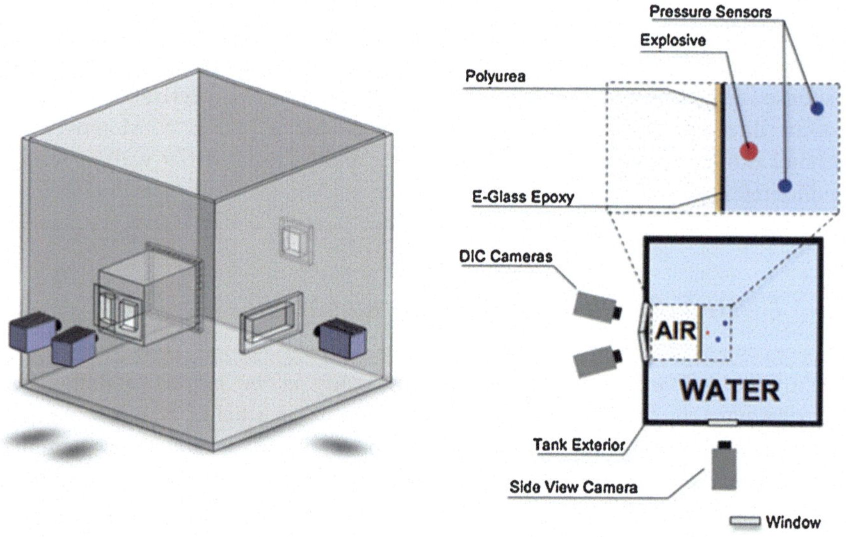

Fig. 9 UNDEX test tank

one wall is a 304.8 mm × 304.8 mm (1.0 ft. × 1.0 ft), rectangular tunnel with a wall thickness of 12.7 mm (0.5in), which serves as the base for the mounting of the composite plates. The use of the tunnel and a water tight mounting fixture allow for the plates to be air backed. The explosive used in the near field blast experiments is an RP-503 charge manufactured by Teledyne RISI. The charge is comprised of 454 mg RDX and 167 mg PETN contained within an outer plastic sleeve. It is suspended by its detonation wire into the tank and placed 50.8 mm (2.0in) from the center of the composite plate. To ensure consistent charge standoff distances for each experiment a 3.18 mm (0.125in) diameter foam spacer is placed between the charge and plate. The foam is secured to both the charge and plate by a fast setting epoxy. Free field dynamic pressure transducers located within the fluid field capture the pressure wave which results from the explosive detonation. High speed photography, coupled with three dimensional Digital Image Correlation (DIC) was used to capture the full-field deformation of the back-face (side opposite of the explosive) of the plates during the UNDEX loading. Two high speed cameras were positioned 330 mm (12.9in) behind the tank walls perpendicular to the viewing windows to avoid any distortion effects from the windows themselves. A third high speed camera was positioned at the side of the tank to view the detonation of the explosive, resulting bubble growth, and interaction of the bubble with the composite plate. The cameras used during experimentation were Photron FastCam SA1 and a frame rate of 27,000 fps was utilized for an inter-frame time of 37 μs.

3.3 Computational Model

A fully coupled fluid-structure interaction modeling approach for the simulation of the near field explosive experiments was developed through the use of the commercial LS-DYNA code. Specifically, the Lagrange-Eulerian formulation of the code has been adopted in the investigation as it allows for accurate representation of the detonation of the explosive charge as well as the fluid structure interaction between the fluid and the composite plate. A schematic of the full simulation model is shown in Fig. 10, and is seen to consist of the composite plates (uncoated and coated), the

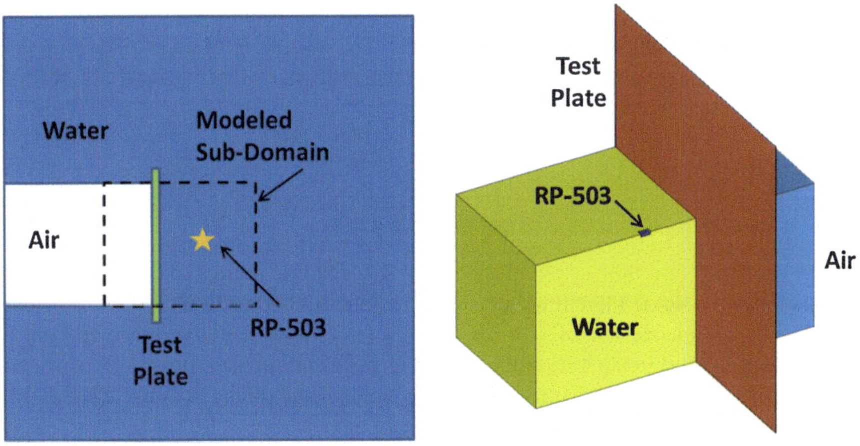

Fig. 10 Finite element model of UNDEX experiment (3 Quadrants of fluid domain hidden)

water within the tank, the air behind the plates and, the RP-503 charge. For efficiency only a subdomain of the full tank facility is explicitly modeled in the simulations consisting of the unsupported section of the composite plate, 120 mm (4.72in) of air extending behind the plate, and 200 mm (7.87in) of water extending from the plate surface towards the charge. This subdomain approach is deemed appropriate as the loading of the plate and subsequent response occurs sufficiently fast that reflections from the tank walls do not affect the overall transient response of the plate. Furthermore, the outer surface of the fluid sub-domain is prescribed a non-reflecting boundary condition which allows the associated pressure waves to freely leave the domain rather than reflect off of the free numerical surface. In the model the water, air, and explosive charge are modeled using an ALE multi-material element formulation and is fully described through the use of a material definition in combination with an equation of state (EOS). The water and air utilize the *Mat_Null material definition with the density of the water and air given as 1 g/cm^3 and 0.0013 g/cm^3 respectively. The Gruneisen EOS is used for the definition of the water with the speed of sound taken to be 149,000 cm/s. A Linear Polynomial EOS describes the air domain in the model by defining C_0, C_1, C_2, C_3, and C_6 equal to zero, and C_4, and C_5 equal to γ-1. The RP-503 charge is modeled with the *Mat_High_Explosive_Burn material model combined with the JWL EOS. The structural components of the model, namely the composite plate and polyurea coating, are represented by shell and solid elements, respectively. The plate in the simulations is modeled using a single layer of shell elements with the laminate schedule prescribed in the shell section definition including the angle of each respective ply and the orthotropic properties of each layer. The composite mechanical properties are defined through an orthotropic material definition (Mat_Composite_Damage) capable of modeling the progressive failure of the material due to any of several failure criterions including in-plane shear, tension in the longitudinal/transverse directions, and compression transverse direction. The polyurea material is modeled as solid elements and are assumed to be perfectly bonded to the composite plate. The material model for the polyurea coating is a viscoelastic material definition which captures both the strain and strain-rate effects through the use of a family of load curves. The model reproduces the uniaxial tension and compression behavior as obtained through material testing at discreet strain rates.

3.4 Experimental and Numerical Results

The response of the composite plates in this study is characterized by the transient center-point displacement of the back-face of the plate, deformation evolution mechanisms during the displacement, and full-field DIC observations.

The pressure profiles resulting from the detonation of the RP-503 charge, as measured by the two free field pressure sensors at 100 mm (3.93in) and 175 mm (2.95in) standoff distances from the charge, are shown in Fig. 11. The pressure

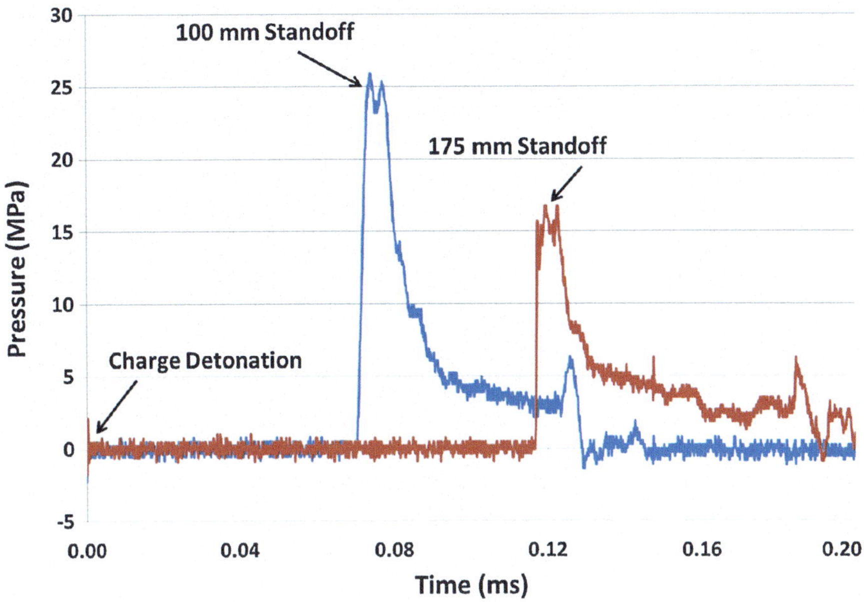

Fig. 11 UNDEX pressure profiles (Time zero corresponds to charge detonation)

profiles display the characteristic components of an UNDEX, namely: a rapid pressure increase associated with the shock front, followed by an exponential decay and a reduction in peak pressure with increasing radial standoff from the charge center. The behavior of the bubble resulting from the detonation and its associated interaction with the composite plate is shown in Fig. 12. The sequence of images shows the clear formation of the bubble at 80 μs and its subsequent growth in size due to the combustion of the explosive products. Due to the high pressure of these gaseous products the bubble expands, reaching a diameter of ~50 mm (1.97in) at 320 μs at which point it reaches and interacts with the surface of the composite plate. As a result of this interaction with the plate it is prevented from further expansion in the direction of the plate but continues a spherical expansion in the remaining directions.

As a result of the initial pressure loading from the detonation of the charges the composite plates deform primarily in the form of outward flexure (towards the air backing). The DIC data is used to capture both the pointwise (center) as well as the full field deformation evolution. Center-point time histories for each of the respective plate configurations is provided in Fig. 13 from which several key observations are made. The first observation that can be made is drawn from the comparison of the overall peak displacements of the plate. It is evident that, as compared to the baseline 0.762 mm (.03in) plate, increasing the plate thickness or including a polyurea coating reduces the peak overall deflection for a given level of loading. The center-point deflection comparison between the 1.524 mm (0.06in) uncoated plate and the 0.762 mm (0.03in) plate with a 0.762 mm (0.03in) coating of polyurea

Fig. 12 UNDEX gas bubble behavior

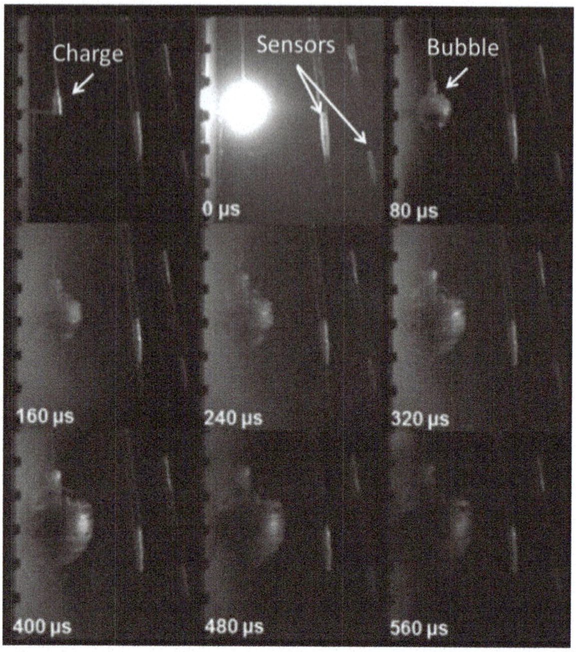

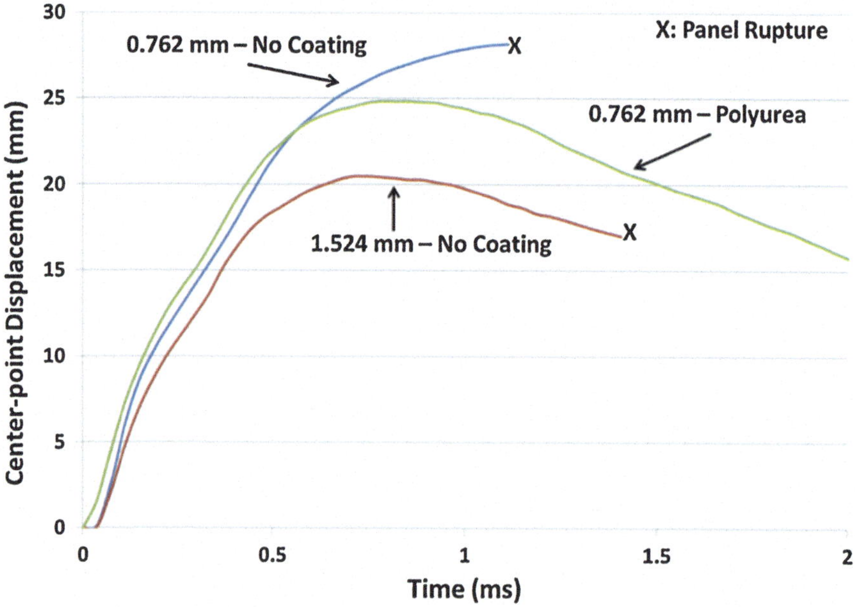

Fig. 13 Plate center-point deflections

indicate that for a given plate thickness it is more advantageous to utilize additional structural plies rather than an elastomeric coating. However, when a structure has previously been designed and further thickening of the structural shape is not possible, the application of a polyurea coating can improve the transient response to shock loading. The second primary observation from the time histories is related to the material damage initiation. Both the uncoated 0.762 mm (0.03in) and 1.524 mm (0.06in) specimens experienced significant through-thickness tearing at the plate boundaries at approximately 1.1 and 1.4 ms respectively. Furthermore, it is observed that although the 0.762 mm (0.03in) plate with the polyurea coating did experience larger deflections than the 1.524 mm (0.06in) uncoated plate, there was no edge tearing of the plate itself. Thus in terms of reducing material damage itself, the polyurea coatings offer an advantage over a thicker uncoated plate. The full field deformation evolution for each of the respective panels is shown in Fig. 14. The contours of out of plane displacement show that the deformation is initially dominated by localized deflections at the center with minimal deflection near the boundaries. As the plate responds to the pressure loading, it gradually transitions to an overall plate flexure mode as shown by the cross sectional shape at 0.74 and 1.11 ms. The significant observation is that the initial plate deformation is governed by the

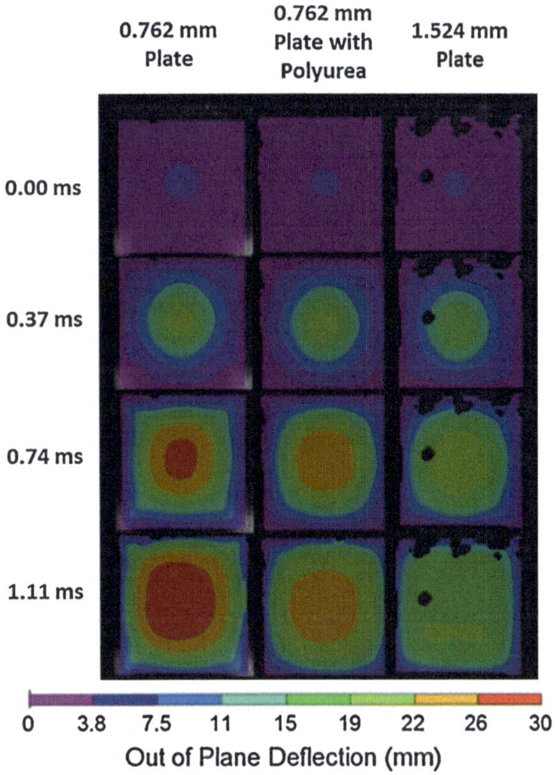

Fig. 14 Full-field deflection contours

highly localized pressure loading and then subsequently shifts to a mode I flexure deformation profile later in time.

The primary means for demonstrating the accuracy of the computational models in the investigation is through correlation of the center-point time histories and onset of material damage. The quality of the correlation between the test data and numerical results in this study is shown in the transient time histories shown in Fig. 15 for each of the panel configurations considered. Observation of the trends show that there is a high level of correlation between the experiment and simulations, both in terms of timing and overall peak deflections. The comparisons exhibit consistent results in the early time frame of the event (0–0.4 ms) in terms of displacement with minor discrepancy observed beyond this point. Furthermore, in terms of material damage onset, for both of the uncoated plate configurations (0.762 mm (0.03in) and 1.524 mm(0.06in)), it is seen that the onset of edge tearing occurs slightly later (0.1 ms) in time as compared to the experimental results. The timing differences in the onset of damage is expected as the model assumes a uniform plate in terms of material properties and does not account for manufacturing variability or minor internal defects which can contribute to the onset of damage or slightly weaker/stronger areas of the plates as compared to the gross material strengths.

3.5 *Significant Findings*

With the objective of investigating the characteristics of composite plates (including the effects of polyurea coatings) subjected to near field underwater explosions a detailed experimental and computational study was conducted. The primary focus of the work was on determining how the response of the plates was effected by increased plate thickness or through the application of a rate-sensitive, elastomeric coating to the baseline plate. The experiments were conducted in a water filled blast tank which allowed for laboratory scale, controlled testing to be completed while also allowing for the use of Digital Image Correlation to capture the full-field, transient response of the back (dry) surface of the plates. The experiments were complemented by corresponding fully coupled fluid structure interaction simulations. The response of three unique plate configurations was evaluated: (1) 0.762 mm (0.03in) baseline plate, (2) 1.524 mm (0.06in) plate, and (3) 0.762 mm (0.03in) plate with a 0.762 mm (0,03in) polyurea coating applied to the back-face. The significant findings of the study are:

1. The transient response of the plates is influenced by both overall plate thickness as well as the application of the polyurea coating. In terms of net effectiveness, the additional structural thickness reduced the transient deflections more than the inclusion of the coatings.

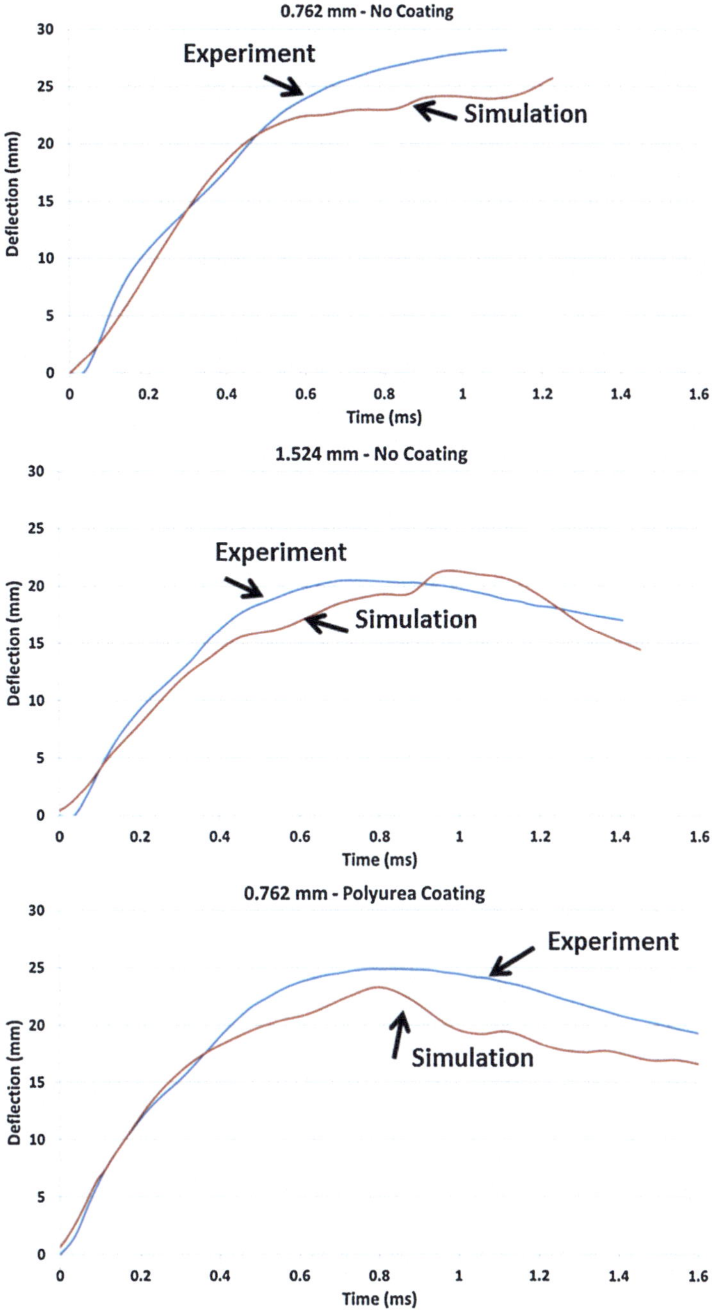

Fig. 15 Center-point displacement model correlation

2. As it relates to material damage sustained by the plates, the polyurea coating was more effective in reducing material damage as compared to both the baseline and thicker uncoated plates.
3. The computational modeling approach employed in the study is able to accurately capture the transient deformation of the plates undergoing shock loading as well as capture the onset of material damage consistent with the experimental results.
4. When considering a plate design, the desired performance metric of the plate response should be considered. A thicker plate of structural material (composite) is preferable to reduce center-point deflection, while the use of polyurea coating are effective in reducing overall damage. However, in the case of an existing design the use of polyurea coatings can be an effective retrofitting application to improve the blast resistance of a structure while reducing overall material damage.

4 Near Field UNDEX of Cylinders

Many undersea vehicles take the form of cylindrical, or cylindrically based, hull forms and therefore there exists a need to understand the physics of underwater blast waves with such geometric shapes. Thus, an experimental and computational study was conducted to investigate the transient response and material damage characteristics of composite cylinders when subjected to highly localized pressure loading conditions resulting from an underwater explosive (UNDEX) detonation. The key aspects of the study include evaluating the transient response of the cylinders due to the complex loading due to the detonation, the resulting material damage levels, as well as evaluating the effects of the elastomeric coatings in terms of damage reduction and energy levels.

4.1 Cylinder Specimens

The investigation included E-glass/epoxy, roll wrapped composite cylinders with a bi-axial laminate structure. The cylinders were tested in an uncoated base configuration as well as a configuration in which the polymeric coatings were applied to the outer surface in two thicknesses as shown in Fig. 16. The cylinders themselves have a wall thickness of 1.14 mm with 4 plies through the thickness and a laminate schedule of [0/45/45/0]. The laminates had a resin content of ~38% by weight and an areal weight of 0.49 kg/m2 per ply. The outside diameter of the cylinder is 7.44 cm with an unsupported length of 38.1 cm. The cylinder is fitted with an aluminum endcap protruding 12.7 mm into the length of the cylinder which seals against water intrusion via a rubber o-ring. Several cylinders were coated with polyurea, specifically Dragonshield-BC, through a spray-cast process to the outer surface. Thicknesses of 100% and 200% of the composite wall thickness were applied to evaluate

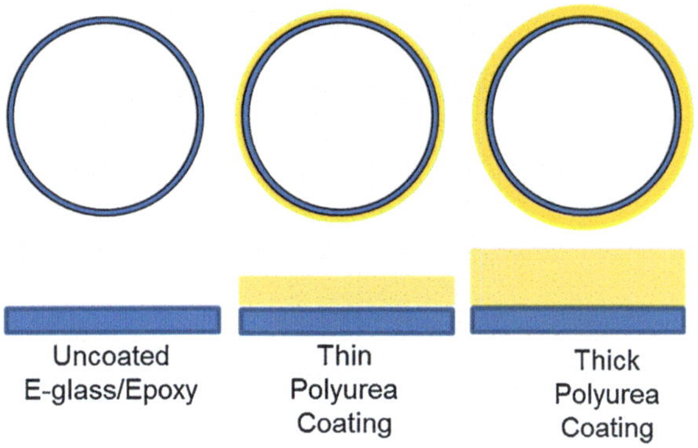

Fig. 16 Cylinder construction

the effect of the coating thickness parameter. This configuration is intended to be representative of a post-design (retrofit) coating application. The polyurea material properties are identical to those previously used during the flat plate experiments.

4.2 *Experimental Method*

The following is an overview of the experimental methodology that was employed in the study. A large diameter pressure vessel has been utilized for the conduct of all experiments in the study. The vessel has an internal diameter and height of 2.1 m with an array of windows along the horizontal axis of the test tank to allow for high speed photography to be conducted during the experiments. Each cylinder is mounted and held in the center of the tank with tensioned cables to minimize rigid body motion of the test article during transient loading. Dynamic pressure sensors located around the cylinders capture the pressure resulting from the explosive detonation as well as the subsequent bubble loading phases. Each cylinder configuration (base composite, thick coating, and thin coating) was tested at two charge stand-offs, 2.54 cm and 5.08 cm. All experiments were conducted at ambient pressure within the flooded tank. The test facility and experimental setup is detailed in Fig. 17. In each respective experiment, an RP-503 explosive charge containing 454 mg of RDX and 167 mg of PETN was utilized. Three high speed video cameras, FastCam SA1, were used to capture video during experiments with one camera mounted to align with the longitudinal axis of the cylinder, providing a side view of the UNDEX event and two cameras were arranged to provide a stereoscopic view of the cylinder on the opposite side of the explosive. Images were captured at a frame rate of 36,000 fps.

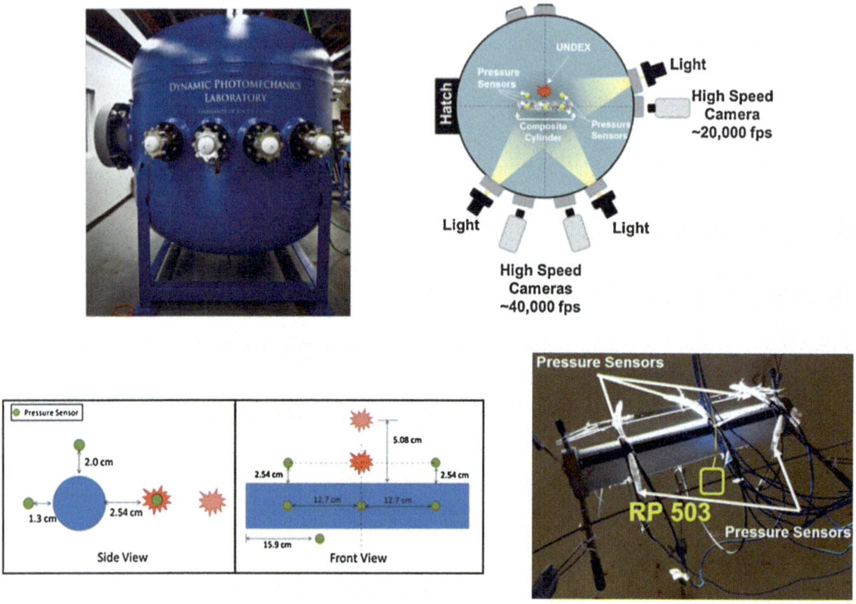

Fig. 17 Test configuration and sensor layout

4.3 *Key Experimental Results and Findings*

4.3.1 Bubble-Cylinder Interaction and Local Pressures

The near field nature of experiments resulted in a complex interaction between the UNDEX bubble and the cylinders, highlighted by a unique splitting of the bubble around the cylinder. Subsequent to the formation of the initial UNDEX bubble, a split occurred in which one portion of the bubble wrapped around onto the non-charge side with the bulk remaining on the charge side. As the initial pressure wave from the detonation passes over the cylinder, small cavitation bubbles form on the surface of the cylinder, which then subsequently coalesce and collapse against the surface after about 1 msec. Figure 18 provides images of key developments observed during the bubble-structure interaction during an experiment conducted at a charge standoff of 2.54 cm on a cylinder with a thick coating applied. No significant differences were noted in the bubble interaction between uncoated and coated cylinders. The large bubble which forms on the non-charge side of the cylinder, due to the splitting phenomenon, collapses upon itself at approximately 12.7 msec and serves to cause a secondary loading of the cylinder.

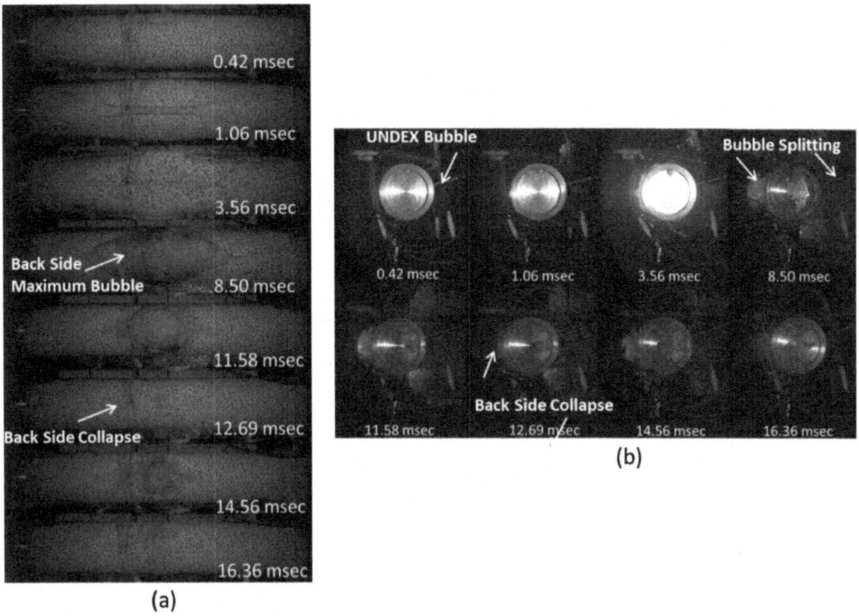

Fig. 18 Bubble growth and interaction (**a**) Front view, (**b**) Side view

4.3.2 Transient Cylinder Response

The physical response of the cylinders to the near field UNDEX loading is characterized primarily by flexure of the cylinder and quantified by the radial displacement of the center point on the non-charge side of each cylinder. Comparisons between the coated and uncoated specimens are limited to the time period for which DIC results are available due to cavitation effects which preclude DIC correlation later in time. The results for both the 2.54 and 5.08 cm charge standoff are similar in nature in overall deformation mechanics and thus only the 5.08 cm results are presented for brevity as this charge standoff allows for a longer time period prior to correlation loss of the DIC data.

The radial displacement of the cylinders exposed to the UNDEX is characterized by an initial global deformation in the positive radial direction (away from the charge) followed by an inflection and dimpling in the center of the cylinder toward the charge location as the cylinder recovers. The upper image in Fig. 19, depicts the radial displacement of line segments along the cylinder centers for all three cylinder configurations (uncoated, thin coated and thick coated) over time. At 0.5 ms the center point displacement for the coated cylinders is 2.5 mm in the positive direction with the uncoated cylinder lagging with a center point displacement of 1.9 mm in the positive direction. Full field displacement contours over the initial 2.75 ms of the experiments can be seen in the lower portion of Fig. 19. The full field contours confirm the general shape suggested by the center line displacements.

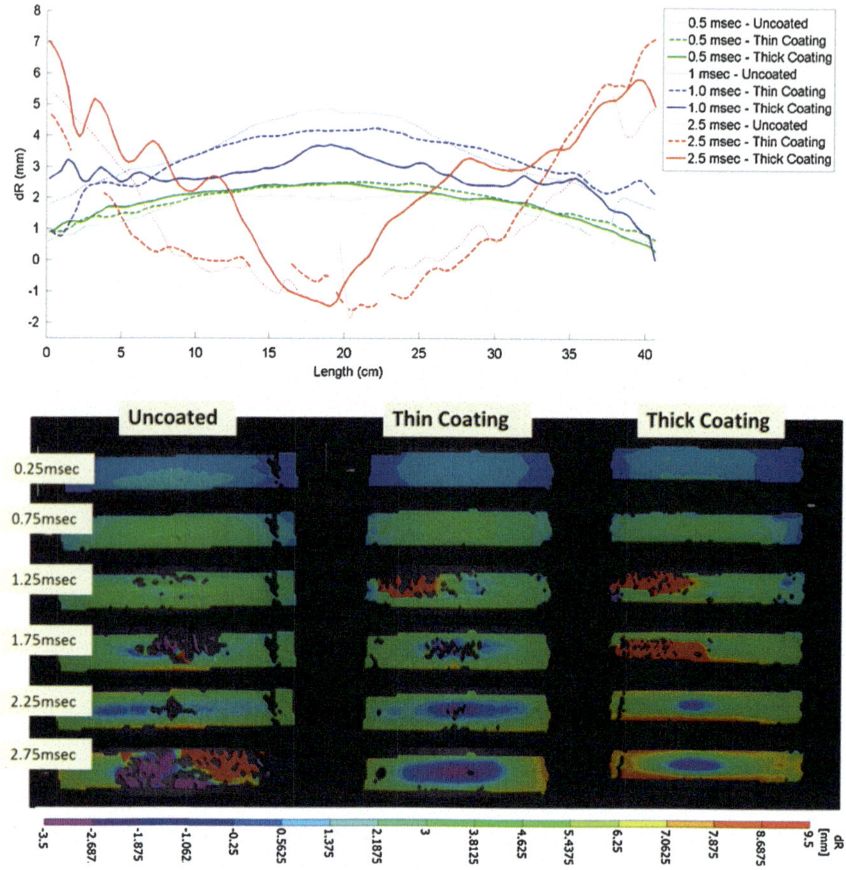

Fig. 19 Centerline and full field displacements for 5.08 cm standoff

4.3.3 Material Damage

While the application of the polyurea had minimal effect on the transient response of the cylinders, there was a significant effect of the polyurea coatings on the material damage sustained by the cylinders during the shock event. For both charge standoffs considered in the study there was a significant reduction in damage sustained with increasing coating thickness. The damage in the cylinders from the interior and exterior view are provided in Fig. 20 for the charge standoff of 2.54 cm. It is observed that the uncoated cylinder damage is dominated by large cracks and missing sections of material. Furthermore, nearest the charge location, sections of delamination can be seen along the edges of the missing portions of the cylinder. Additionally, curving cracks, suggestive of an ellipsoid indenting of the cylinder, at approximately ±90° from the cylinder centroid can be seen. The damage in the thinly coated cylinders is dominated by large circumferential and longitudinal cracks emanating from the point closest to the charge location. At the nexus of the

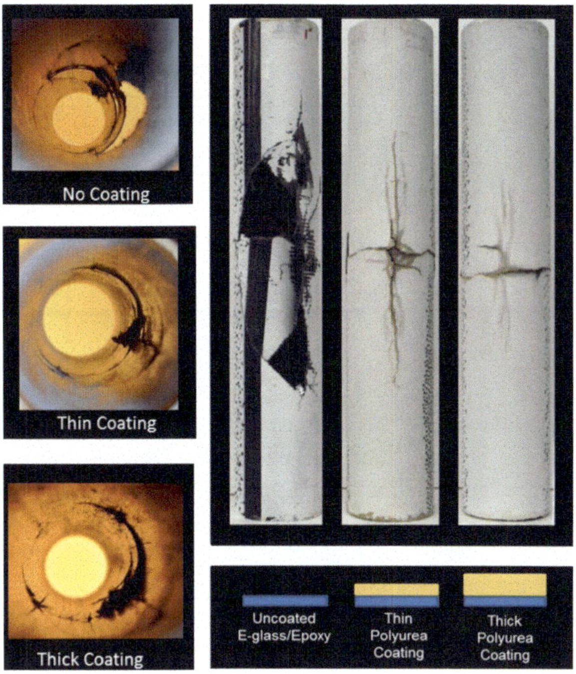

Fig. 20 Material damage for 2.54 cm charge standoff

longitudinal and circumferential cracks the damage extends through the thickness of both the composite and coating. The cylinder with the thick coating contains damage that is similar in character to that observed in the thinly coated cylinder, although at a reduced magnitude.

4.4 Computational Model Overview

The experiments which have been previously discussed, have been simulated utilizing the LS-DYNA finite element code. A fully coupled fluid structure Lagrange-Eulerian formulation is utilized due to the nature of the problem: namely highly curved wave fronts and dependence of the decay of the pressure wave during propagation through the fluid domain. This approach allows for accurate representation of the detonation of the explosive charge, resulting pressure wave propagation into the fluid, and the transient fluid structure interaction between the pressure wave and the cylinder.

The finite element model of the experiments consists of the cylinder body and endcap, polyurea coating, surrounding tank water, internal air, and the RP-503 as shown in Fig. 21. Included in the model is the ¼ representation of the system with the fluid domain extending out to a distance of 12.93 cm from the outer surface of the cylinder. The maximum charge standoff considered in the experiments was 5.08 cm,

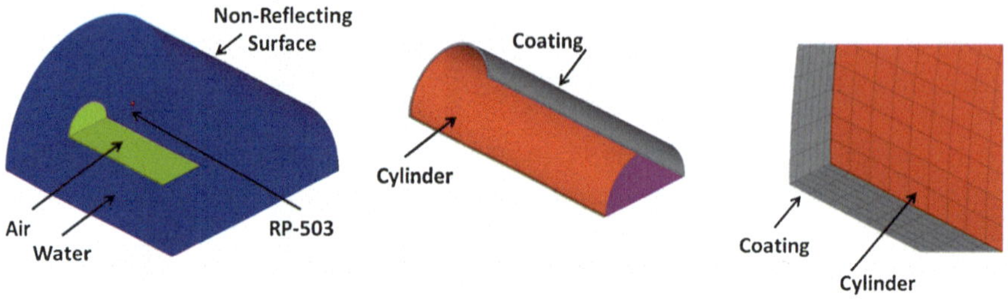

Fig. 21 Computational model detail

and thus the inclusion of the domain to a distance larger than twice the standoff value ensures boundary effects do not influence the fluid structure interaction. The Eulerian components of the coupled model consist of the water, air, and explosive charge, and are modeled with solid elements utilizing the ALE multi-material element formulation. The fluid and explosive components of the model are fully defined through the combined use of a material model and an equation of state (EOS). The structural aspect of the coupled model consists of the composite cylinder and the polyurea coating. The composite cylinder in the simulations is modeled using a single layer of shell elements with the appropriate prescribed lamina properties. Each ply is represented as having two through thickness integration points so as to capture the correct bending behavior on a per ply basis. The polyurea material is represented in the model by solid elements and are assumed to be perfectly bonded to the cylinders.

4.5 *Significant Computational Results and Findings*

4.5.1 Material Energy Comparisons

The internal energy of the cylinders (including coatings) during the explosive loading event is provided in Fig. 22 for the 5.08 cm charge standoff cases. The time history trends are similar in nature for the 2.54 cm charge standoff scenarios. The internal energy histories are broken down individually by the cylinder and coating, while the kinetic energies are presented as the net sum of the system. The results are presented in this manner so as to differentiate the internal energy distribution between the individual components during deformation, whereas the kinetic energy is a measure of the net motion of the system as a whole. There are several key aspects related to the energy characteristics that are illustrated by the results. It is evident that in terms of the energy experienced by the cylinder itself (internal), there is an increase as a function of coating thickness. Furthermore, for a given coating thickness it is evident that the cylinders themselves comprise ~90% of the total internal energy (cylinder plus coating) sustained with the coatings comprising 10% of the net peak energies occurring at 0.1 ms. This result is anticipated as the

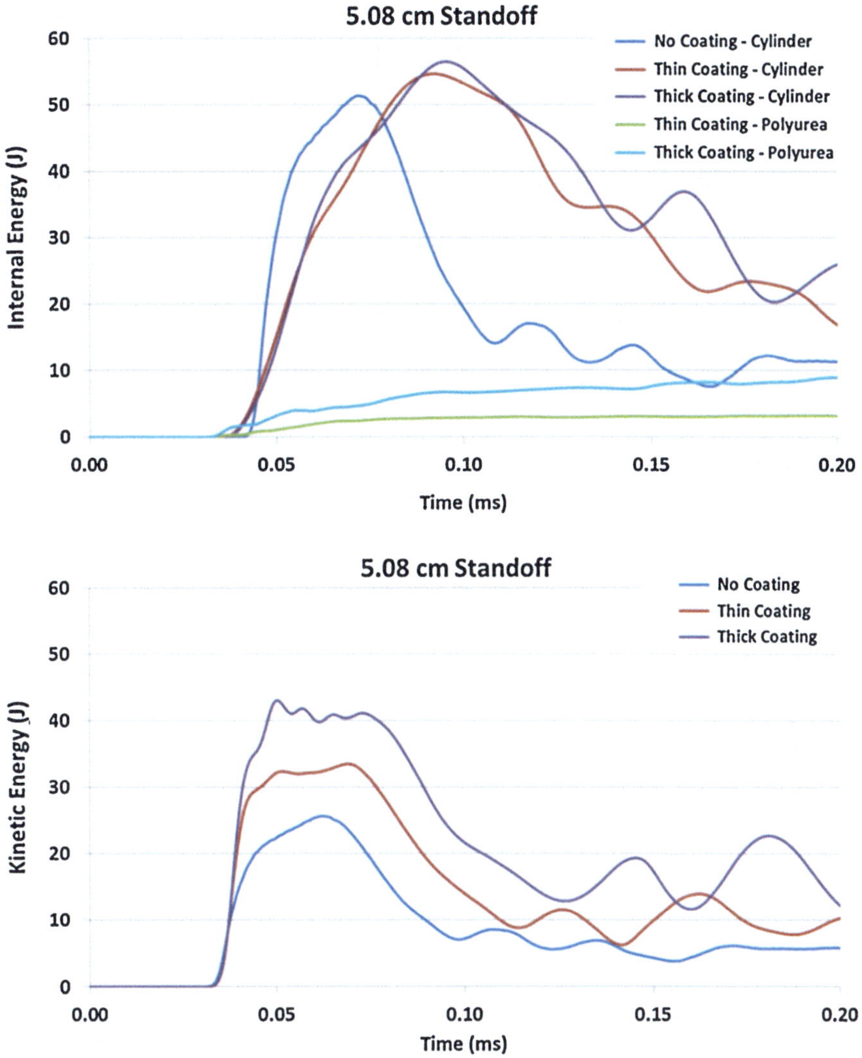

Fig. 22 Energy time histories, 5.08 cm standoff

composite material is significantly stiffer than the coating and thus for a given deformation would represent the primary load carrying mechanism. Finally, it is noted that as the coating thickness is increased there is a corresponding increase in the amount of internal energy that can be absorbed by the system (cylinder plus coating) as a whole. For the case of an uncoated cylinder, the sole mechanisms for energy absorption/dissipation are strain energy in the composite and fracture energy corresponding to the evolution of damage through fiber and matrix failure. In the presence of the coatings, there is the additional energy absorption/dissipation reservoir of the coating itself. Thus, whereas the uncoated cylinders sustain damage, the coated cylinders can dissipate that energy into the coating itself and reduce the overall composite material loading. Hence, the coated cylinders experience higher levels of loading, but also a corresponding decrease in material damage. In terms of the internal energy observations, through comparison of the energies of the coatings themselves, the thicker coating does experience a higher level of internal energy as compared to the thin coating.

4.5.2 Strain Comparison

The strain time histories, radial and longitudinal, for the back surface of the cylinder are presented in Fig. 23 as measured on the surface of the composite cylinder itself. It is shown that the overall trends in the strain histories are consistent with those observed in the internal energy comparisons. Specifically, there is an increase in overall strain level with increasing coating thickness in both the radial and longitudinal directions. It is further noted that in observing the temporal evolution of the strains, the time to reach the peak strains is longer as the coating thickness in increased for all cases. The increase in strain as a function of increasing thickness can be attributed to the additional mass that the coatings contribute to the overall structure, while providing limited additional stiffness to the system. For the case of

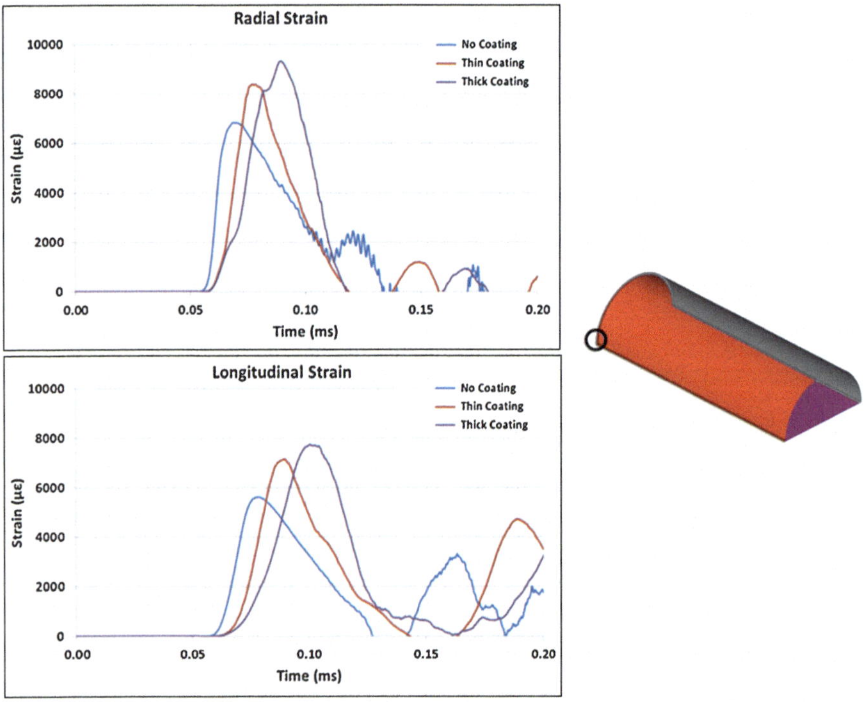

Fig. 23 Back surface strain time history, 5.08 cm charge standoff

the cylinders with a thick coating, there is an overall doubling of the structural mass of the composite/coating system. As the cylinders are accelerated and undergo deformation due to the UNDEX pressure loading, the composite cylinder is the primary load carrying mechanism due to its overall higher stiffness as compared to the coating. During this initial response the coating is adding additional mass to the system which must be arrested primarily by the composite through additional deformation which leads to resulting increases in strain. Additionally, in a similar manner as was observed with the internal energies of the system, the increase in the rear face surface strains with increasing thickness can be partially attributed to the coatings reducing or preventing the onset of material damage. The uncoated cylinders sustain significant material damage on the charge side which has the effect of dissipating a certain level of energy. By reducing the damage levels, the coating has the effect of allowing the cylinders to undergo larger overall deformations, and corresponding strains, as the energy is distributed through the system as a whole.

4.6 Results and Findings

The effects of polyurea coatings on the response and damage of submerged composite cylinders when subjected to near field UNDEX loading was investigated through experiments with corresponding computational simulations. Three unique cylinder configurations were evaluated to determine the effects of coating application and thickness The computational modeling of the experiments has been conducted with the LS-DYNA finite element code and specifically utilizes the ALE methodology so as to develop fully coupled fluid structure interaction models. The primary parameters of interest in the study are deformations, damage extents, energy levels, and material strains. The significant results of the investigation were:

1. For both the charge standoffs investigated there is a splitting of the UNDEX bubble upon interaction with the cylinders. The bubble on the non-charge side of the cylinder collapsed in close proximity to the surface of the cylinder and produced localized pressure loading on the non-charge side of the cylinder.
2. The inclusion of the polyurea coating serves to reduce the material damage levels compared to an uncoated cylinder and furthermore, damage was reduced as a function of increasing coating thickness.
3. During the transient loading of the cylinders, both the internal material energy and the overall system kinetic energy increased with increasing coating thickness. Furthermore, the composite material experiences ~90% of the overall internal energy with the coatings carrying the remaining 10%. The radial and longitudinal surface strains during the early time response of the cylinders increases with increasing coating thickness.
4. Polyurea coatings can affect structures subjected to shock loading in both beneficial and adverse means. There is an observed increase in deformation and strains with increasing coating thickness during the early time deformation, whereas there is an overall reduction in material damage and failure due to the presence of the coating over the entire time duration.

5 Weathering and Ageing Effects

Within many marine industries the use of composites is expanding to include applications which require long term and continual submergence or exposure to seawater. The deployment in such marine applications include the structures being subjected to aggressive and harsh environmental conditions, including high salinity seawater and/or salt spray that can significantly degrade their structural performance over time. These effects are of particular concern when composite vehicles are deployed in a setting where impact, shock, and blast are a concern over the period of the service life. As a result of these growing considerations there has been a body of both ongoing and recently established research investigating the manner in which composite materials with exposure to marine environments respond to dynamic loading events and how that response is adversely effected by the long term exposure. Work conducted by the authors has recently aimed at quantifying the prolonged exposure to aggressive conditions effects on the dynamic and shock response of composite materials. Accelerated ageing techniques have been developed to simulate the ageing process in a controlled laboratory environment and have been coupled with novel underwater blast loading experiments to identify the fundamental effects of ageing on structural performance.

5.1 Accelerated Ageing Method

An Accelerated Life Testing (ALT) method to simulate the long term exposure to marine environments has been implemented to investigate the effect of long term seawater submersion on the mechanical properties and shock response of a carbon fiber material. The approach utilized in the study for simulating the long term seawater immersion is to subject the material to a seawater bath at an elevated temperature. The use of the elevated temperature bath increases the rate of diffusion of the seawater into the material as compared to the absorption rate at the typical operating temperatures while also accelerating the degradation of the material. The determination of an acceleration factor, the relationship between the exposure time in the accelerated life test and an equivalent time in a typical operating environment is established. The dominant factor contributing to material degradation during prolonged submersion is fluid absorption in the matrix. In order to mathematically relate the experimental submergence of the composite plate to actual service submergence time, a water diffusion study of the matrix material has been conducted. The relationship between the accelerated life test and service time immersion is governed by the Arrhenius equation which describes the temperature dependence of the rate of reaction for a given process. The diffusion process is relatively well established and can be described by a diffusion coefficient that is a function of parameters such as temperature, type of resin and curing agent, surrounding medium composition, fillers, void content, and so on. A standard and well-accepted model for epoxy resins, in terms of mass diffusions, is a Fickian model [44] which uses Fick's second law to predict how the concentration of a diffusive substance changes over time within a material.

The facility which has been developed to age the composite specimens is shown in Fig. 24 and allows the specimens to be submerged in a 3.5% NaCl solution (representative of sea water conditions) while also being held at an elevated temperature. The elevated temperature in the study is chosen so as to remain below the glass transition temperature of the material. Temperatures above this point would result in changes to the mechanical properties independent of the seawater ageing. The diffusion acceleration factor was determined with respect to a given temperature by conducting a water absorption evaluation performed at various temperatures; specifically: 5, 25, 45, 65, and 85 °C. Figure 25 (a) and (b) show the mass diffusion for different temperatures and the logarithmic relationship between D and E_a respectively. The diffusion coefficient was calculated from a point that is within the initial linear portion of the mass diffusion curve ($\leq$ 50% mass saturation). The diffusion coefficient was related to E_a by using Arrhenius' equation. After obtaining the activation energy for the composite material, the acceleration factor (AF) can be found as the ratio of diffusions at different temperatures as shown in Eq. (1) [45]. The submersion specimens were kept at a constant temperature (T_1 = 338 K), but

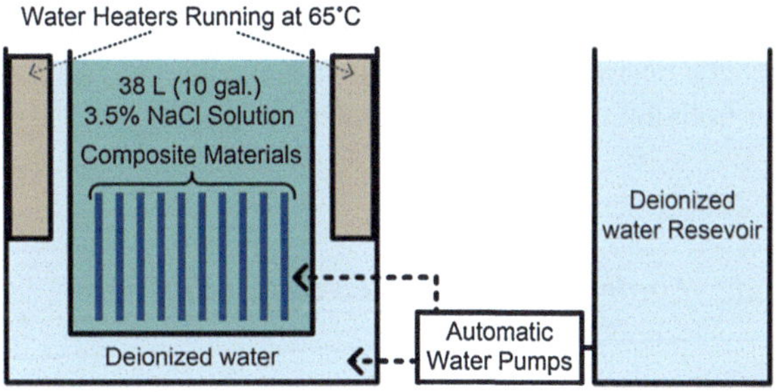

Fig. 24 Accelerated weathering facility setup

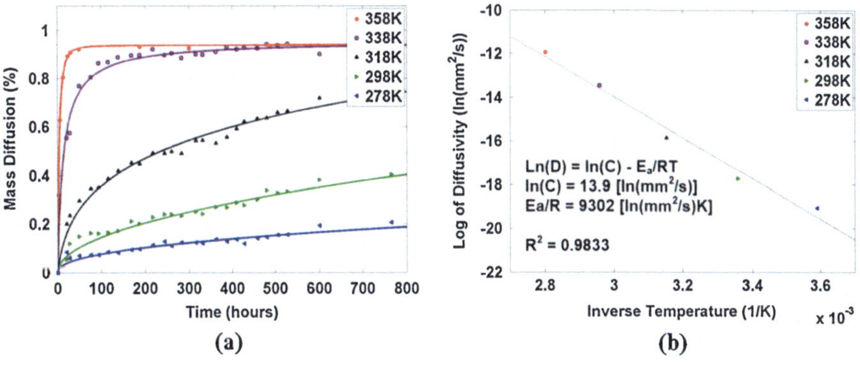

Fig. 25 (**a**) Mass diffusion and (**b**) Logarithmic relationship for diffusivity

the service temperature (T_2) can vary depending on the application. Hence, AF is application dependent. For reference, if the average ocean temperature (17 °C) is assumed to be the operational temperature, then 35 and 70 days of submersion approximates to 10 and 20 years of service respectively.

$$AF = \frac{Ce^{-\frac{E}{RT_2}}}{Ce^{-\frac{E_a}{RT_1}}} = e^{\left(\frac{E_a}{R}\right)\left(\frac{T_2 - T_1}{T_1 T_2}\right)} \quad (1)$$

5.2 *Material Summary*

The composite materials used in the study consist of four unidirectional carbon fiber layers with [0, 90]s and [45, −45]s layups. The laminates are manufactured from two layers of +/− 45° biaxial carbon fabric, Tenax HTS40 F13 24 K 1600tex carbon fibers with an epoxy resin matrix. Vacuum infusion was utilized in the manufacturing of all laminates plates and post curing at 70 °C for 10 h was conducted. The plates were 1.26 mm (0.050 in) in thickness. The mechanical properties as determined through ASTM standards tests is provided in Table 2 and contains both the baseline and aged properties. From the material summary it is seen that there is a general decrease in material properties with increasing submergence/exposure time, for both the moduli and strengths.

5.3 *Underwater Blast Experiments*

An advanced underwater blast facility is used to conduct the explosive experiments. The facility is identical to that described in Sect. 3.2 and shown in Fig. 9. The unsupported area of the panels is 254 × 254 mm^2 (10×10 in^2). An RP-503 explosive comprised of 454 mg RDX and 167 mg PETN was used as the explosive in all experiments conducted. The charge is submerged in the center of the specimens, and is placed at a 152 mm (6 in) standoff distance from the plate. The real-time response of the panels during loading is captured through the use of high speed photography

Table 2 Composite's effective mechanical properties

Weathering time (Days)	0	35	70
E_x, E_y (GPa)	78.4 +/− 1.8	78.0 +/− 2.1	74.9 +/− 2.6
V_{xy}	0.039 +/− 0.014	0.040 +/− 0.010	0.042 +/− 0.009
G_{xy} (GPa)	7.38 +/− 0.19	5.32 +/− 0.24	4.92 +/− 0.22
Yield shear stress (kPa)	36.1 +/− 1.1	25.3 +/− 1.0	21.7 +/− 0.6
Failure shear stress (kPa)	45.3 +/− 1.2	41.3 +/− 1.9	38.7 +/− 2.6

Table 3 Experimental details

Cases	Layup	Standoff distance, mm (in)	Weathering time, days (simulated years)
E45-0wd	[45,-45]s	152 (6)	0
E45-0wd-2	[45,-45]s	114 (4.5)	0
E45-0wd-3	[45,-45]s	76 (3)	0
E45-35wd	[45,-45]s	152 (6)	35 (10)
E45-70wd	[45,-45]s	152 (6)	70 (20)
E90-0wd	[0,90]s	152 (6)	0
E90-70wd	[0,90]s	152 (6)	70 (20)

coupled with digital image correlation. The two Photron SA1 high-speed cameras are used to record the shock event at 10,000 frames per second. A side view camera is also employed to capture the explosive bubble-to-structure interaction. The details of the experimental cases are summarized in Table 3.

5.4 *Experimental Results*

The underwater blast experiments on the composite plates are characterized by a rapid detonation of the RP-503 explosive charge, a high pressure/short duration pressure loading, and a subsequent longer duration growth of the UNDEX bubble. These characteristics are highlighted in the high speed side-view images in Fig. 26. From the images it is also seen that due to the relatively small standoff of the charge to the plate as compared to the maximum bubble size there is a physical interaction between the bubble and the plate itself. The high pressures characteristic of an underwater explosion are also shown in Fig. 26 as recorded at varying standoffs from the charge location. The shock from the explosive is distinguished by an immediate rise in pressure followed by exponential decay. The amplitude of the explosive pressure decreases spherically by 1/R from the explosive location. From the pressure time histories it is observed that the pressures from the explosive detonation are very large, 20–30 MPa, but also very short in duration, less than 0.25 ms from peak to full decay.

The center-point out of plane deformation of the 45/−45 and 0/90 panels with a charge standoff of 152 mm are provided in Fig. 27. The deformation evolution is highlighted by the following trend: (1) A flexure towards the air side during the initial pressure loading phase upto a maximum displacement value, (2) a rebound phase during the surface cavitation (at vacuum pressure) with a rapid flexure towards the water-side of the plate to a magnitude beyond its initial displacement, and finally, (3) an abrupt increase in displacement towards the air side of the plate due to the cavition bubble collapse. Furthermore, the ageing of the plates led to an increase in maximum displacements for the same loading condition as highlighted in the time histories. After weathering the [45,-45]s composite for 35 days, the maximum center point displacements increase by an average of 20%. An additional 5% increase in

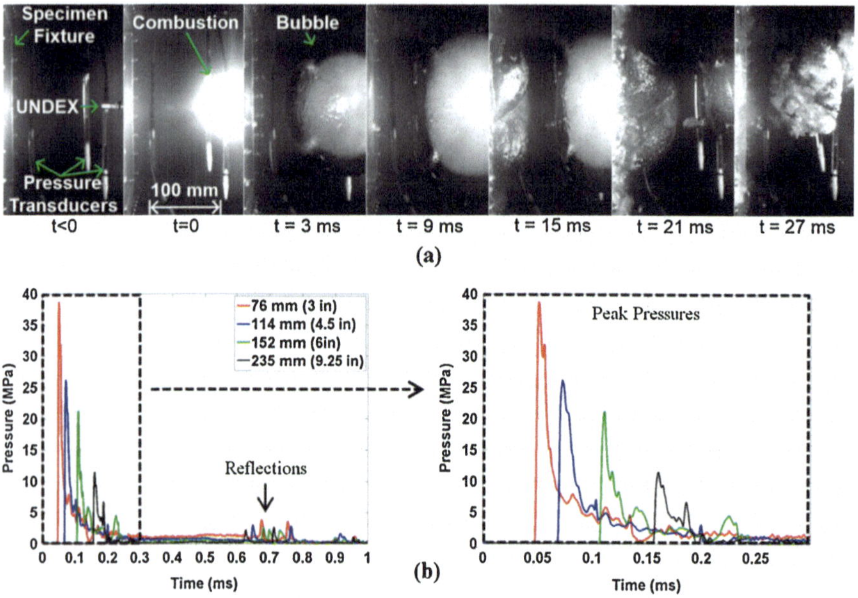

Fig. 26 (**a**) Fluid-structure interaction images, (**b**) Pressure history from the explosive

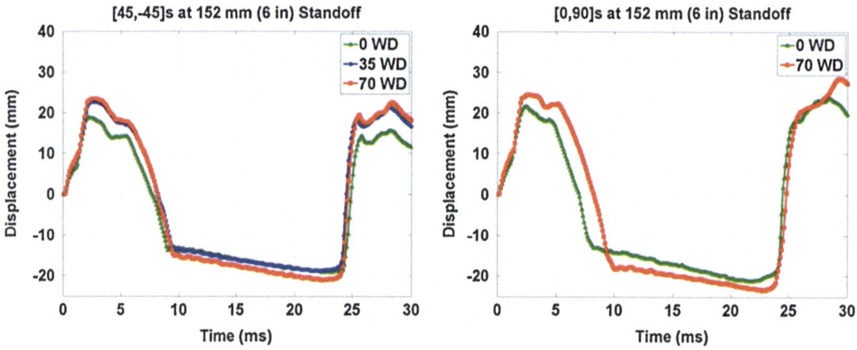

Fig. 27 Center point displacements for [45,-45]s and [0,90]s weathered composites

displacement is seen for the 70 weathering days (WD) cases indicating further material degradation. Furthermore, a post mortem evaluation of the panels to investigate the material damage characteristics was performed. The analysis revealed that damage increased with weathering time. Damage observed in the [45,-45]s cases consisted primarily of through-thickness damage near the plate's corners as illustrated in Fig. 28. Additionally, in terms of the 35 WD and 70 WD, there is a notable increase in average crack length. This increase in crack length suggests further material degradation from fiber/matrix debonding after saturation. A similar trend in increasing damage with ageing time was seen in the [0,90]s laminate cases,

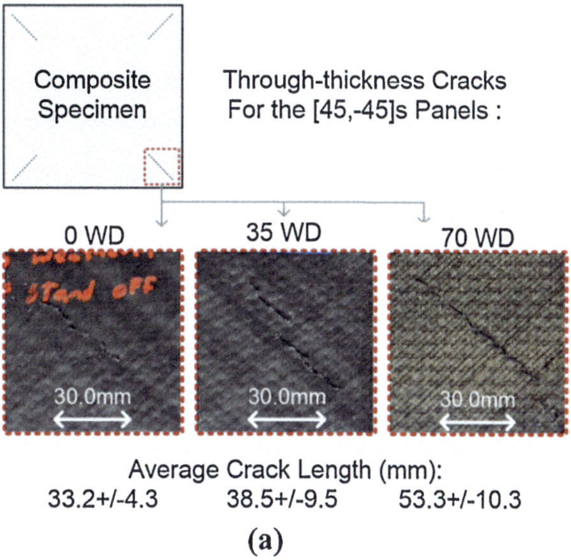

Fig. 28 Interfibrillar and through-thickness cracking for the [45,-45]s cases

although the damage was predominately seen in the form of delamination near its corners rather than cracking. The difference in damage between [45,-45]s and [0,90] s arises from how the boundary interacts with the fiber orientations.

5.5 *Significant Findings*

A detailed experimental investigation into the effects of long term seawater immersion on the mechanical and shock response of carbon fiber laminates was conducted. The significant findings of the study are:

1. There is a measurable degradation of the mechanical properties of the laminates as a result of the seawater submersion. The shear properties, which are primarily governed by the epoxy matrix exhibited the largest degradation.
2. The peak center point displacements during shock loading increased with ageing time, namely for the 35 day weathering period of both the laminate constructions considered. The additional weathering time of 70 days had only a minor additional effect.
3. The length of weathering time in seawater is also related to the amount of accumulated damage sustained by the plates during loading. In the 45/−45 plates there was an increase in through-thickness cracking whereas in the 0/90 laminates the delamination levels increased.

6 Conclusion

Composite material systems are increasingly being considered for the design and construction of advanced structures and vehicles. The materials are able to meet the competing design interests of increased stiffness and strength with decreased overall weight as well as ease of manufacture and decreased maintenance requirements. As a result of the performance increases afforded by these materials they are being increasingly employed in applications pertaining to aerospace, marine, infrastructure, and defense. As these materials find wider application consideration must also be given to the extreme loading conditions which are imparted to them in the expected operating theatres, including shock, blast, and environmental exposure. It is with these considerations in mind that the research program discussed in the preceding sections has been undertaken and continues. The early part of the program investigated the response of composite plates to far-field underwater explosive conditions and subsequently transitioned to the study of near field blast loading. More recently the complex interactions between near field pressure waves and cylindrical geometries was considered, including bubble interaction phenomena. Furthermore, the degrading effects of long term submersion in a salt water environment on mechanical performance was investigated. Each of the components of the program have developed and utilized both advanced experimental techniques with detailed numerical modeling approaches. The significant findings from the research are:

1. The fluid structure interactions resulting from the impingement of an UNDEX pressure wave (near or far field) on a nearby composite structure are complex, both temporally and spatially. That complexity translates to the bubble-structure interaction as well for near-field explosions.
2. The application of rate-dependent, elastomeric coatings, namely polyurea, can be beneficial during loading of a structure in terms of both transient deflection reduction and, more importantly, as a damage mitigation and/or reduction method.
3. Long term seawater submersion has a degrading effect on the mechanical performance of the materials considered with reductions in both modulus and strength observed. Additionally, increased damage levels with increased ageing were noted.
4. Numerical/computational models can be employed to accurately simulate the complex fluid structure interactions, transient responses, and damage levels as compared to the corresponding experiments.

Acknowledgments The work was supported by the Office of Naval Research under the Solid Mechanics Program managed by Dr. Y.D.S. Rajapakse and by the Naval Undersea Warfare Center Division Newport In-House Laboratory Independent Research Program (Chief Technology Office) and Internal Investment Program (Strategic Investment Office).

References

1. Nurick G, Olson M, Fagnan J, Levin A (1995) Deformation and tearing of blast loaded Stiffened Square plates. Int J Impact Eng 16:273–291
2. Nurick G, Shave G (1996) The deformation and tearing of Thin Square plates subjected to impulsive loads–an experimental study. Int J Impact Eng 18:99–116
3. Tekalur AS, Shivakumar K, Shukla A (2008) Mechanical behavior and damage evolution in E-glass vinyl Ester and Carbon composites subjected to static and blast loads. Compos Part B 39:57–65
4. Mouritz AP (2001) Ballistic impact and explosive blast resistance of stitched composites. Compos Part B 32:431–439
5. Mourtiz AP (1996) The effect of underwater explosion shock loading on the flexural properties of GRP laminates. Int J Impact Eng 18:129–139
6. Mouritz AP (1995) The effect of underwater explosion shock loading on the fatigue behavior of GRP laminates. Compos 26:3–9
7. Dear J, Brown S (2003) Impact damage processes in reinforced polymeric materials. Compos A: Appl Sci Manuf 34:411–420
8. Franz T, Nurick G, Perry M (2002) Experimental investigation into the response of chopped-strand mat glassfibre laminates to blast loading. Int J Impact Load 27:639–667
9. LeBlanc J, Shukla A, Rousseau C, Bogdanovich A (2007) Shock loading of three-dimensional woven composite materials. Compos Struct 79:344–355
10. LeBlanc J, Shukla A (2010) Dynamic response and damage evolution in composite materials subjected to underwater explosive loading: an experimental and computational study. Compos Struct 92:2421–2430
11. LeBlanc J, Shukla A (2011) Dynamic response of curved composite plates to underwater explosive loading: experimental and computational comparisons. Compos Struct 93:3072–3081
12. Jackson M, Shukla A (2010) Performance of Sandwich composites subjected to sequential impact and air blast loading. Compos Part B 42:155–166. https://doi.org/10.1016/j.compositesb.2010.09.005
13. Schubel PM, Luo J, Daniel I (2007) Impact and post impact behavior of composite Sandwich panels. Compos Part A 38:1051–1057
14. Arora H, Hooper P, Dear JP (2010) Impact and blast resistance of glass fibre reinforced Sandwich composite materials. In: Proceedings of IMPLAST 2010
15. Avachat S, Zhou M (2014) Response of cylindrical composite structures to underwater impulsive loading. Procedia Eng 88:69–76
16. Avachat S, Zhou M (2015) High-speed digital imaging and computational modeling of dynamic failure in composite structures subjected to underwater impulsive loads. Int J Impact Eng 77:147–165
17. Latourte F, Gregoire D, Zenkert D, Wei X, Espinosa H (2011) Failure mechanisms in composite plates subjected to underwater impulsive loads. J Mech Phys Solids 59:1623–1646
18. Espinosa H, Lee S, Moldovan N (2006) A novel fluid structure interaction experiment to investigate deformation of structural elements subjected to impulsive loading. Exp Mech 46 (6):805–824
19. Matzenmiller A, Lubliner J, Taylor RL (1995) A constitutive model for anisotropic damage in fiber-composites. Mech Mater 20:125–152
20. Zako M, Uetsuji Y, Kurashiki T (2003) Finite element analysis of damaged woven fabric composite materials. Compos Sci Technol 63:507–516
21. Dyka CT, Badaliance R (1998) Damage in marine composites caused by shock loading. Compos Sci Technol 58:1433–1442
22. O'Daniel JL, Koudela KL, Krauthammer T (2005) Numerical simulation and validation of distributed impact events. Int J Impact Eng 31:1013–1038

23. McGregor CJ, Vaziri R, Poursartip A, Xiao X (2007) Simulation of progressive damage development in braided composite tubes under axial compression. Compos Part A 38:2247–2259
24. Gama B, Xiao J, Haque M, Yen C, Gillespie J (2004) Experimental and numerical investigations on damage and delamination in thick plain weave S-2 glass composites under quasi-static punch shear loading. Center for Composite Materials, University of Delaware
25. Donadon MV, Iannucci L, Falzon BG, Hodgkinson JM, de Almeida SFM (2008) A progressive failure model for composite laminates subjected to low velocity impact damage. Comput Struct 86:1232–1252
26. Hosseinzadeh R, Shokrieh MM, Lessard L (2006) Damage behavior of Fiber reinforced composite plates subjected to drop weight impacts. Compos Sci Technol 66:61–68
27. Tagarielli VL, Deshpande VS, Fleck NA (2010) Prediction of the dynamic response of composite Sandwich beams under shock loading. Int J Impact Eng 37:854–864
28. Batra RC, Hassan NM (2007) Response of Fiber reinforced composites to underwater explosive loads. Compos Part B 38:448–468
29. Davies P (2016) Environmental degradation of composites for marine structures: new materials and new applications. Phil Trans Math Phys Eng Sci 374(2071):20150272. https://doi.org/10.1098/rsta.2015.0272
30. Shirrell C, Halpin J (1977) Moisture absorption and desorption in epoxy composite laminates. Compos mater: testing and design (Fourth Conference). https://doi.org/10.1520/stp26963s
31. Browning C, Husman G, Whitney J (1977) Moisture effects in epoxy matrix composites. In: Davis J (ed) Composite materials: testing and design (Fourth Conference), STP26961S. ASTM International, West Conshohocken, pp 481–496. https://doi.org/10.1520/STP26961S
32. Blikstad M, Sjoblom PO, Johannesson TR (1984) Long-term moisture absorption in graphite/epoxy angle-ply laminates. J Compos Mater 18(1):32–46. https://doi.org/10.1177/002199838401800103
33. Neumann S, Marom G (1987) Prediction of moisture diffusion parameters in composite materials under stress. J Compos Mater 21(1):68–80. https://doi.org/10.1177/002199838702100105
34. Choqueuse D, Davies P (2008) Aging of composites in underwater applications. Ageing of Composites. https://doi.org/10.1201/9781439832493.ch18
35. Sar B, Fréour S, Davies P, Jacquemin F (2012) Coupling moisture diffusion and internal mechanical states in polymers – a thermodynamical approach. Eur J Mech A Solid 36:38–43. https://doi.org/10.1016/j.euromechsol.2012.02.009
36. Fichera M, Totten K, Carlsson LA (2015) Seawater effects on transverse tensile strength of carbon/vinyl ester as determined from single-fiber and macroscopic specimens. J Mater Sci 50 (22):7248–7261. https://doi.org/10.1007/s10853-015-9279-3
37. Choqueuse D, Davies P, Mazéas F, Baizeau R (1997) Aging of composites in water: comparison of five materials in terms of absorption kinetics and evolution of mechanical properties. In: High temperature and environmental effects on polymeric composites, vol 2. https://doi.org/10.1520/stp11369s
38. Davies P, Rajapakse Y (2014) Durability of composites in a marine environment. Springer, Dordrecht
39. Crank J (1975) The mathematics of diffusion, 2nd edn. Oxford University Press, London
40. Shillings C, Javier C, LeBlanc J, Tilton C, Corverse L, Shukla A (Submitted 2017) Experimental and computational investigation of blast response of Carbon-Epoxy weathered composite materials. Composites B
41. Poche L, Zalesak J (1992) Development of a water-filled conical shock tube for shock testing of small sonar transducers by simulation of the test conditions for the heavyweight MIL-S-901D (Navy). NRL Memorandum Report 7109
42. Coombs A, Thornhill CK (1967) An underwater explosive shock gun. J Fluid Mech 29:373–383

43. Filler WS (1964) Propagation of shock waves in a hydrodynamic conical shock tube. Phys Fluids 7:664–667
44. Popineau S, Rondeau-Mouro C, Sulpice-Gaillet C, Shanahan ME (2005) Free/bound water absorption in an epoxy adhesive. Polymer 46(24):10733–10740. https://doi.org/10.1016/j.polymer.2005.09.008
45. Rice M (2011) Activation energy calculation for the diffusion of water into PR-1590 and Pellethane 2103-80AW polyurethanes. NUWC-NPT Technical Memo 11–062

Blast Performance of Composite Sandwich Panels

John P. Dear

1 Introduction

There has been increasing demand for composite materials in the naval industry since the 1960s when Swedish defense institutes began to develop composite sandwich panels for naval applications [1, 2]. Mouritz et al. highlighted that this demand has been due to the goals of reducing acquisition and maintenance costs whilst improving the operational performance of the vessels [3]. Along with structural performance requirements, naval vessels are required to withstand a range of demanding loads due to the environment in which they operate. This can include wave slamming, high strain rate loading, impact loading and blast loading. These loading phenomena are very complex and can result in devastating consequences. The performance of composite sandwich panels against such loads must be investigated.

Due to the high costs, there is relatively limited literature available on the air blast performance of composite structures using explosives. Comtois et al. [4] placed small explosive charges on the surface of glass-fiber and carbon-fiber laminates. The authors found that support conditions strongly influenced laminate damage and that damage was a function of both peak pressure and impulse. Furthermore, the carbon-fiber laminates suffered less fiber breakage compared to the glass-fiber laminates. Gargano et al. [5, 6] have subjected composite laminates to small scale charges to investigate the effect of glass-fiber versus carbon-fiber and weave types. The authors found that glass-fiber woven laminates with vinyl ester resin demonstrated the best blast resilience. Non-crimp continuous fiber fabrics with z-binder yarn were found to outperform woven fabrics with the same fibers and resin system.

J. P. Dear (✉)
Department of Mechanical Engineering, Imperial College London, London, UK
e-mail: j.dear@imperial.ac.uk

S. W. Lee (ed.), *Advances in Thick Section Composite and Sandwich Structures*,
https://doi.org/10.1007/978-3-030-31065-3_3

Shock tubes have been widely used by researchers to evaluate materials under shock loading. Shukla and co-workers have investigated the effects of using a stitched and unstitched foam core along with graded core composite sandwich panels [7, 8]. Composite sandwich panels offer advantages over composite laminates due to their increased second moment of area and presence of an energy absorbing core. The authors concluded that a core configuration of low/medium/high density demonstrated superior blast performance compared to a panel with core configuration medium/low/high density. The same research group has looked into the addition of poly-urea (PU) interlayers into the skins of composite sandwich panels [9]. The location of the PU layer was varied and found to affect deflection and strain of the sandwich panels. Placing the interlayer behind the front skin or behind the core was found to reduce back skin deflection, velocity and in-plane strain. This improves the overall blast performance of the panel.

Shock tubes have also been developed to simulate underwater blast loading. LeBlanc et al. [10, 11] have used a conical shock tube (CST) to determine the effects of underwater blast loading on composite sandwich panels. The initial experiment relied on strain gauge data to record panel response [10], however, the experimental setup was developed and transient plate response was recorded using 3-dimensional digital image correlation (DIC) [11]. The second experiment investigated the effect that PU coatings have on composite sandwich panels with GFRP skins. The authors found that for a given PU thickness, the panel responded best when it was applied to the back skin.

Composite structures absorb blast energy through deformation and damage mechanisms. The presence of damage in a composite panel can lead to reduced post-blast or residual, strength. This can be problematic if a vessel is required to remain in service following a blast event. Understanding the effect of damage on composite strength is, therefore, important. Zenkert et al. [12] have sought to establish the size and extent of damage that a structural ship component can have without the risk of growth or failure. This highlights the importance of documenting and comparing the extent of damage sustained by a panel during blast loading.

Experimental blast tests generate valuable data, however, numerical models can provide additional insight into material response that may be challenging to measure. Furthermore, simulations that have been validated against reliable experimental data can be used during the design process. This will reduce the number of experiments required. To support their experimental investigations, Langdon, Cantwell, Nurick and co-workers have extensively developed finite element (FE) simulations in ABAQUS/Explicit. A FE model successfully predicted the post-blast deformed shapes of the composite sandwich panels with polyvinyl chloride (PVC) foam cores and aluminum alloy skins [13]. The foam core was found to absorb over 50% of the total energy. LS-DYNA FE simulations have been developed by Bahei-El-Din et al. [14, 15] to model the blast response of composite sandwich beams with stiff PU rubber and compliant polyethylene foam interlayers between the top carbon-fiber reinforced polymer (CFRP) skin and PVC foam core. A variety of layups were investigated and the inclusion of both PU and polyethylene was found to be most effective. The presence of the interlayers was shown to reduce core crushing by over 50% and reduce panel displacement by approximately 15%.

The literature highlighted in this introduction has provided motivation for a number of the investigations detailed in this chapter. The nature and quality of scaling from small scale to large scale blast response is still in development. Therefore, large-scale data and novel approaches applied to data acquisition are an important contribution to this field of research. Furthermore, it is important to perform experiments on representative materials under realistic scenarios to gather results relevant to the final application. Simulations can be a powerful tool during a design process and can minimize the requirement for expensive experiments. However, experimental data remains vital to validate and further develop these simulations.

2 Air Blast Loading of Composite Sandwich Panels

Full scale air blast experiments have been performed on a range of composite sandwich panels to investigate multiple mechanical, geometrical and material properties. The aim of the air blast experiments has been to capture displacement-time histories of the composite sandwich panels and understand how damage develops during a blast event. The experiments provide full-scale data to validate analytical and numerical models of structures. The experimental series has been performed over several years hence the experimental setup has been adapted and improved. This section details a generic air blast experimental design along with the key aims of each investigation.

2.1 Experimental Design

The experiments have been carried out at two specialized test facilities, located at RAF Spadeadam and Radnor Range both in the UK. The composite sandwich panels were bolted to the front of test cubicles with a large unsupported target area appropriate for each panel size. The cubicle was positioned at a measured stand-off distance from the center of the charge. The stand-off distance has been varied to alter the pressure and impulse delivered to the panels by the explosive charge. In each case a spherical charge was raised off the floor to the mid-height of the target panels using polystyrene foam blocks and positioned at the center of the test pad. A large steel plate was placed beneath the charge to prevent cratering during blast. Figure 1 shows the generic experimental setup featuring all external instrumentation that was used during any of the air blast experiments (not all instrumentation was present for each test).

There were in total three different test fixtures used. One was a steel test cubicle at RAF Spadeadam which was 3.0 m in length. The second cubicle was a larger

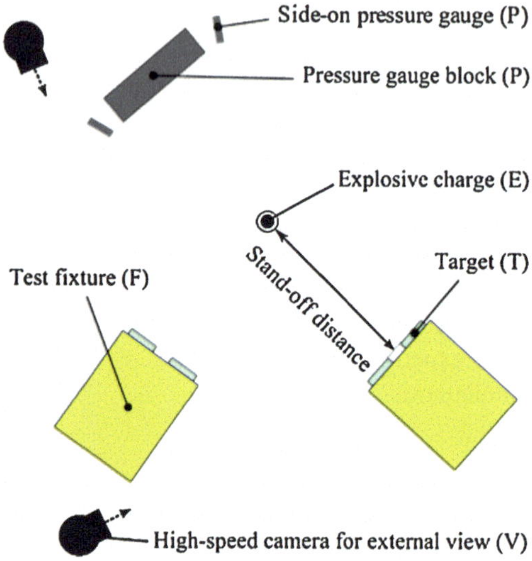

Fig. 1 Schematic diagram showing the blast experiment setup

concrete cubicle with a steel front at RAF Spadeadam which was 8.0 m in length. The final cubicle was a steel cubicle at Radnor Range which was also 3.0 m in length.

To capture the response of the composite sandwich panels under blast loading, high-speed 3-dimensional digital image correlation (3D DIC) was employed. A pair of high-speed Photron cameras were positioned within the test cubicle behind each panel. Placing the cameras within the test cubicle ensured they were protected from the blast wave. To achieve the desired 25° angle between the cameras, appropriate lenses were used. This was dictated by the cubicle length and hence achievable perpendicular distance between the cameras and panel surface. The cameras were mounted on heavy duty camera stands which were weighed down with sand bags during blast testing. Halogen lamps or LED light banks were used to illuminate the panels during blast testing. During some initial experiments, point displacement measurements were taken with a laser gauge arrangement to verify the results obtained through DIC. During many tests, additional cameras were setup to record: the front view of the panels, the progression of the blast wave front and to ensure no debris impacted the front of the panels.

During all experiments, the blast pressure was recorded using either a reflected pressure gauge and/or a side-on pressure gauge. The reflected pressure gauge was positioned within a concrete pressure gauge block or mounted onto the front of the test cubicle. The reflected pressure sensor was situated normal to the blast wave direction, measuring the pressure as the air particles came to rest against the surface. When a side-on pressure gauge was adopted, it was situated at the same stand-off distance as the test cubicle. The side-on pressure gauge was positioned parallel to the blast wave and measured the static pressure as the blast wave moved over the sensor.

Various investigations have been performed to understand material or geometrical properties along with the influence of experimental variables. The following sections detail investigations that have been carried out over many years by the research group.

2.1.1 Effect of Core Thickness

During this investigation, the performance of four composite sandwich panels with different core thicknesses were compared. Each panel was constructed from two plies of E-glass quadriaxial skins (Gurit QE1200) either side of a styrene acrylonitrile (SAN) foam core with density 155 kg m^{-3} (Gurit P800). The panels were fabricated using vacuum infusion with epoxy resin and hardener (Ampreg 22). The composite sandwich panels were 1.7 m × 1.5 m in size. Panels with a 40 mm thick core and a 30 mm thick core were subjected to a 30 kg C4 charge at 14 m stand-off distance. In addition, panels with a 40 mm thick core and a 50 mm thick core were subjected to a 30 kg C4 charge at 8 m stand-off. Table 1 lists the panels investigated during this series of experiments. A proportion of this work has been previously published in reference [16].

2.1.2 Effect of Core Material and Core Density

Four composite sandwich panels with varying core configurations were subjected to an identical blast load. Once again, each panel was constructed from two plies of E-glass quadriaxial skins (QE1200) either side of a foam core. The composite sandwich panels were 1.7 m × 1.5 m in size. The panels were infused with epoxy resin and hardener (SR8500 /ST94). The four foam cores adopted were:

- 40 mm 100 kg m^{-3} SAN core
- 40 mm 100 kg m^{-3} polyvinyl chloride (PVC) core
- 40 mm 110 kg m^{-3} polymethacrylimid (PMI) core
- 30 mm graded SAN core: 10 mm 100 kg m^{-3}, 10 mm 140 kg m^{-3}, 10 mm 200 kg m^{-3}

All four panels were subjected to a 100 kg nitromethane charge at 15 m stand-off distance and are detailed in Table 2. A proportion of this work has been previously detailed in reference [17].

Table 1 Details of the panels with varying core thicknesses investigated

Panel code	Skin material	Core material	Core thickness (mm)	Charge size (kg)	Stand-off distance (m)
G30	GFRP	SAN	30	30	14
G40a	GFRP	SAN	40	30	14
G40b	GFRP	SAN	40	30	8
G50	GFRP	SAN	50	30	8

Table 2 Details of the panels with varying core materials investigated

Skin material	Core material	Core thickness (mm)
GFRP	SAN	40
GFRP	PVC	40
GFRP	PMI	40
GFRP	Graded SAN	30

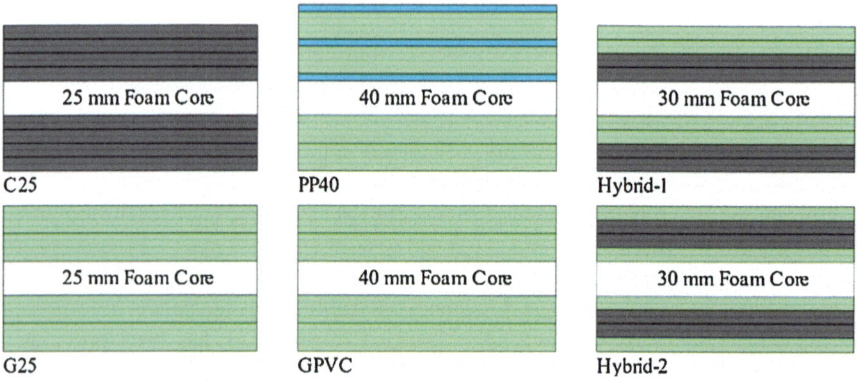

Fig. 2 Schematic diagram showing the layup order of the hybrid composite sandwich panels

2.1.3 Effect of Skin Configuration

A range of composite sandwich panels with varying skin configurations have been subjected to full-scale air blast loading. The performance of composite sandwich panels with E-glass fiber skins either side of a 25 mm thick 140 kg m^{-3} SAN foam core has been compared to panels with carbon-fiber skins and an identical core. These panels were subjected to a 100 kg nitromethane charge at 14 m stand-off distance and were 1.7 m $\times$ 1.5 m in size. This investigation was carried out as these are the two most common fiber types used within the composites industry. This work has been previously detailed in references [18, 19].

The effect of polypropylene (PP) interlayers within E-glass fiber skins has been investigated. The panel had quadriaxial E-glass skins with four layers of woven PP included in the front skin and a 40 mm 100 kg m^{-3} PVC core. The panel was directly compared to an identical panel without the PP interlayers. This panel was simultaneously used in Sect. 2.1.2 and this investigation. Composite sandwich panels with hybrid glass-fiber and carbon-fiber skins have also been studied. The layup order of the glass-fiber and carbon-fibers was varied to understand the effect this has on blast resilience. These panels were subjected to a 100 kg nitromethane charge at 15 m stand-off distance and were 1.7 m $\times$ 1.5 m in size. Diagrams illustrating the layup order of these composite sandwich panels are shown in Fig. 2. The panels in this investigation are detailed in Table 3. Aspects of this work can be found in references [17, 20].

Table 3 Details of the panels with varying skin configurations investigated

Panel code	Skin material	Core material	Core thickness (mm)	Charge size (kg)	Stand-off distance (m)
G25	GFRP	SAN	25	100	14
C25	CFRP	SAN	25	100	14
PP40	GFRP/PP	PVC	40	100	15
GPVC	GFRP	PVC	40	100	15
Hybrid-1	GFRP/CFRP	PVC	30	100	15
Hybrid-2	GFRP/CFRP	PVC	30	100	15

Table 4 Details of the panels subjected to varying blast loads

Panel code	Skin material	Core material	Core thickness (mm)	Panel size (m × m)	Charge size (kg)	Stand-off distance (m)
G40a	GFRP	SAN	40	1.7 × 1.5	30	14
G40b	GFRP	SAN	40	1.7 × 1.5	30	8
G40c	GFRP	SAN	40	1.7 × 1.5	30	16
G40d	GFRP	SAN	40	1.7 × 1.5	30	16, 12, 8
GS40	GFRP	PVC	40	0.8 × 0.8	8	8, 8
CS40	CFRP	PVC	40	0.8 × 0.8	8	8, 8

2.1.4 Effect of Progressive Loading and Multiple Blast

Air blast tests were conducted on three identical composite sandwich panels with quadriaxial E-glass skins either side of a 40 mm thick 155 kg m^{-3} (Gurit P800) SAN foam core. A panel size of 1.7 m × 1.5 m was used. The panels were subjected to separate 30 kg C4 explosions at 16 m, 14 m and 8 m to compare the difference in displacement and damage caused by reducing stand-off and hence increasing blast severity. The results from two of the panels were used in the investigation detailed in Sect. 2.1.1 as well as this section. In addition, an identical panel was subjected to three 30 kg C4 charges in succession at distances of 16 m, 12 m and 8 m. This work has been previously detailed in [21].

In a separate experimental series, two composite sandwich panels were subjected to two 8 kg PE4 charges both at 8 m. This was carried out to understand the effect of existing damage on panel blast response. These composite sandwich panels had either quadriaxial E-glass fiber skins or quadriaxial carbon-fiber skins and 40 mm thick 130 kg m^{-3} (Divinycell H100) PVC foam core. The panels were 0.8 m × 0.8 m in size.

These investigations were designed to represent a real combat scenario, where a structure may be subjected to a number of different blast conditions and multiple strikes. Table 4 details the panel types subjected to varying blast loads.

2.2 *Results*

2.2.1 Effect of Core Thickness

Two thickness comparisons were made in this investigation. Figures 3 and 4 show data gathered through DIC for the G30 and G40a panel, respectively. It can be seen that increasing the core thickness lowers the amplitude of oscillations. This is due to the increased stiffness of the panel resulting from the increased core thickness. The G40a panel had a maximum displacement of 61 mm whilst the G30 panel had a max displacement of 78 mm. The DIC deflection measurements agreed well with the laser gauge measurements for the central point data, hence the laser gauge was dropped in subsequent experiments. For the G40a panel the major principal strain peaked at 1% whilst it peaked at 1.2% for the G30 panel. Upon visual inspection, the skins of both panels appeared undamaged following blast loading.

The second blast, comparing the thicker panels, results in more severe loading of the panel due to the reduced stand-off distance. The response of the G40b panel to the more severe loading is shown in Fig. 5. In addition to the increase in displacement magnitude, the difference in smoothness of the deformation path taken during the first period of oscillation is noticeable. A flattening is observed as the displacement curve reaches its maximum which coincides with the time at which a crack is observed to initiate. The major principal strain peaked at approximately 3% prior to the crack developing. Upon post inspection, the front skin sustained interlaminar skin failure and front ply fiber breakage whilst the core suffered from skin to skin

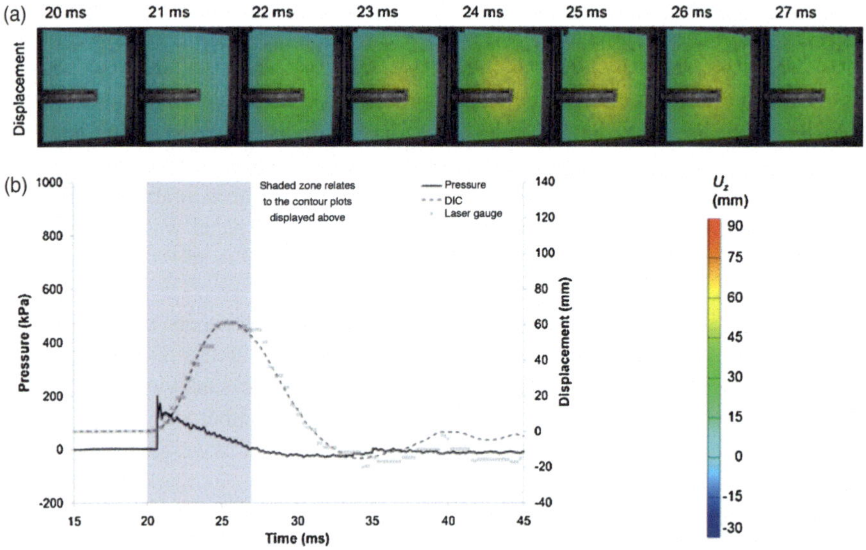

Fig. 3 Blast summary for 30 kg C4 at 14 m stand-off from G40a panel showing: (**a**) DIC analysis and (**b**) plot of pressure-time and displacement-time both using DIC and laser gauge center point measurements

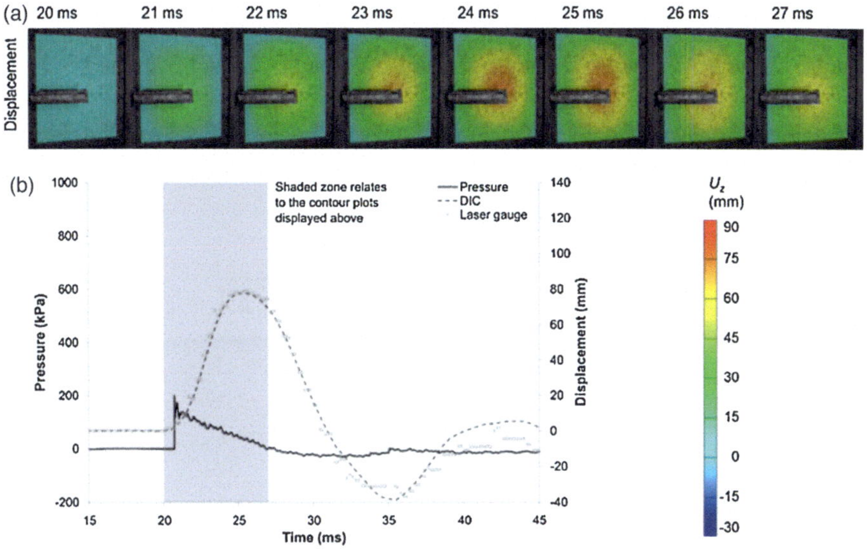

Fig. 4 Blast summary for 30 kg C4 at 14 m stand-off from G30 panel showing: (**a**) DIC analysis and (**b**) plot of pressure-time and displacement-time both using DIC and laser gauge center point measurements

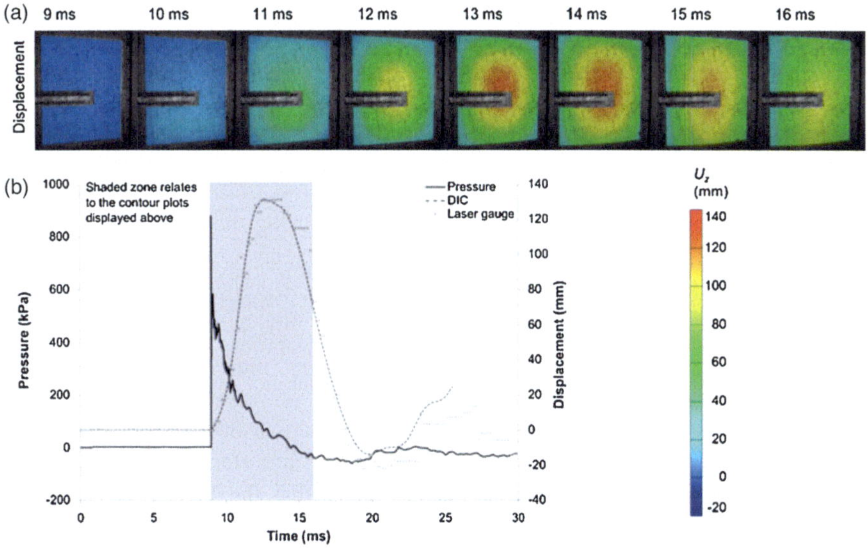

Fig. 5 Blast summary for 30 kg C4 at 8 m stand-off from G40b panel showing: (**a**) DIC analysis and (**b**) plot of pressure-time and displacement-time both using DIC and laser gauge center point measurements

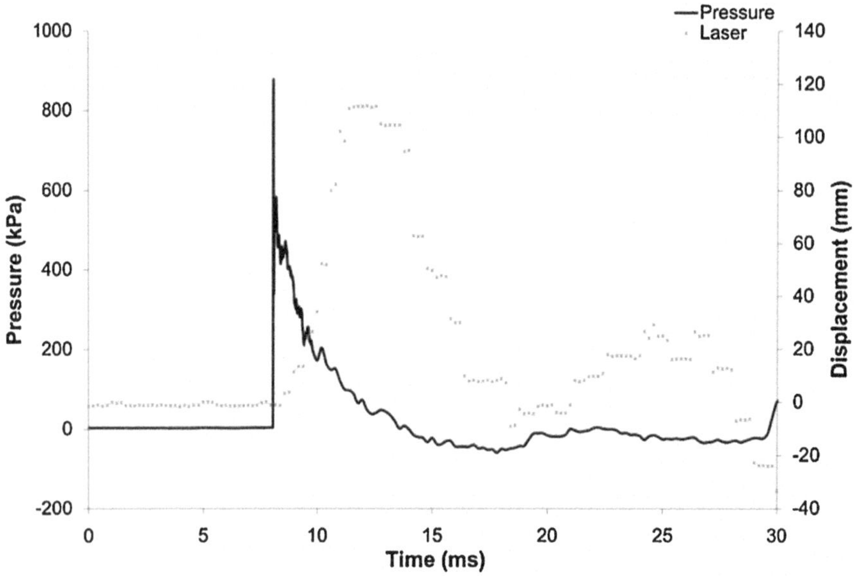

Fig. 6 Blast summary for 30 kg C4 at 8 m stand-off from G50 panel showing plot of pressure-time and displacement-time using laser gauge center point measurements

shear failure. Comparing the results to the G50 panel, it can be seen in Fig. 6 that there is a saving of around 20 mm in maximum displacement for the same blast. The G50 panel deflected to a maximum of 111 mm compared to the G40b panel deflecting to 134 mm. This result is expected given the higher section modulus of the panel created by the thicker core.

2.2.2 Effect of Core Material and Core Density

The deflections of the central points of all the panels in this study are shown in Fig. 7. It is clear that there is little difference between the initial displacements of the panels. This is due to the initial deflection being predominantly elastic with damage occurring late in the deflection. Around the maximum peaks, the PMI, PVC and SAN curves demonstrate a flattening and distortion of the return stroke. This is due to damage initiating within the panels. The PMI panel demonstrates the most significant flattening and return stroke distortion indicating it suffers from the greatest damage. The rebound displacements clearly show that the PMI panel deflects towards the blast the most, followed by the PVC panel and then the SAN panel. This is due to the differing levels of panel damage coming into effect. Although the graded density core does not exhibit the lowest deflection, it is 10 mm thinner than the other panels. Based on previous experiments, a decrease in core thickness of 10 mm results in an increase in deflection of approximately 25%. Therefore, if the graded SAN panel was 40 mm thick, a deflection of 80 mm would be expected which is an improvement on the single density SAN panel. This highlights that the graded

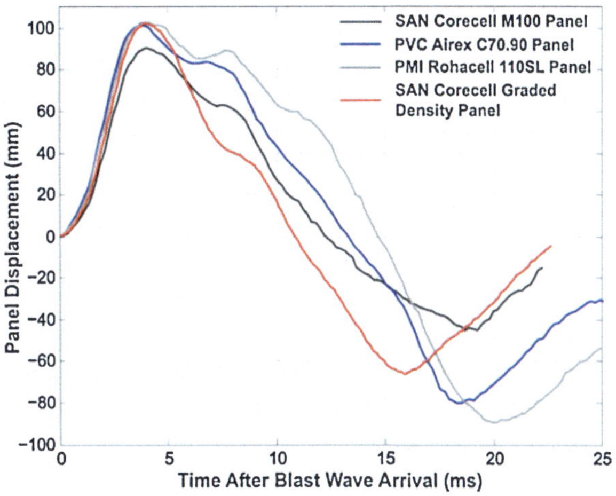

Fig. 7 Displacement-time using DIC center point measurements for the four different composite sandwich panels subjected to 100 kg TNT at 15 m stand-off

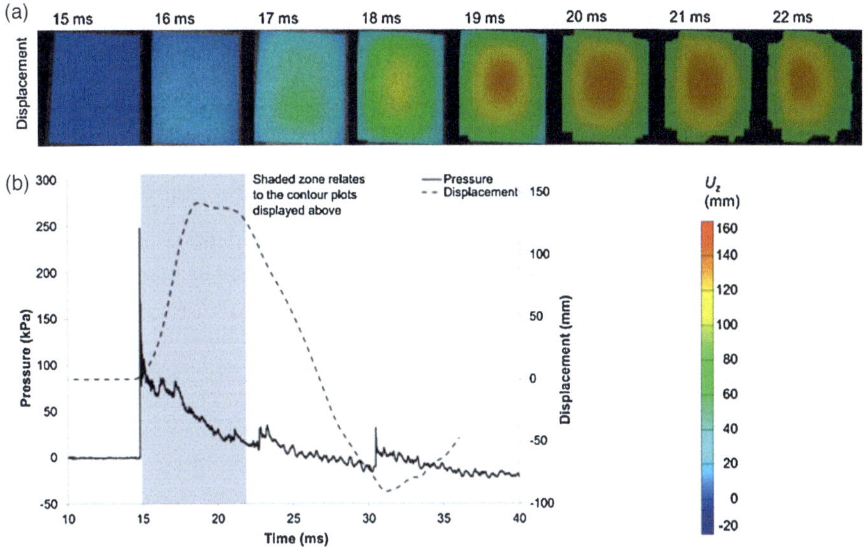

Fig. 8 Blast summary for 100 kg TNT equivalent at 14 m stand-off from G25 panel showing: (**a**) DIC analysis and (**b**) plot of pressure-time and displacement-time using DIC center point measurements

core reduces deflection compared to a single density core. In addition, the deflection profile of the rear skin of the graded panel is far smoother than the deflection profiles of the other panels.

2.2.3 Effect of Skin Configuration

The G25 and C25 panels were tested side-by-side to directly compare the panel responses. Both panels had a similar mass per unit area and thickness. Figure 8

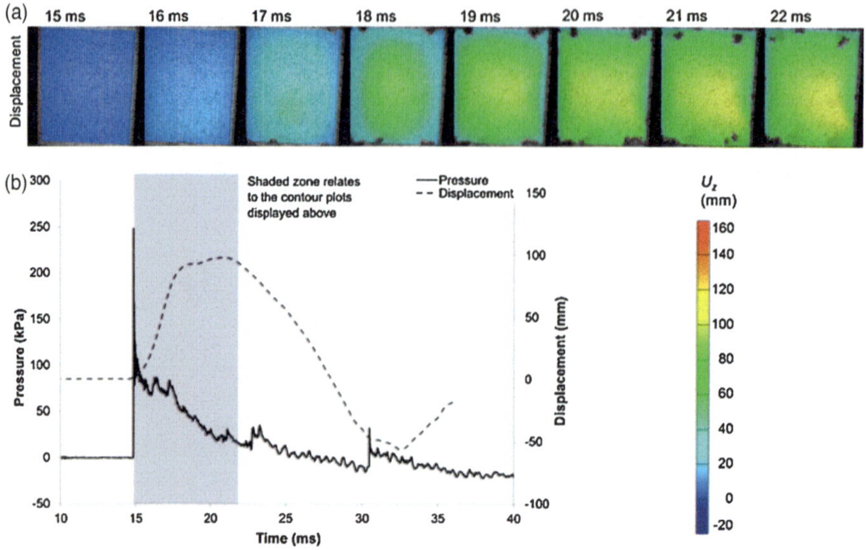

Fig. 9 Blast summary for 100 kg TNT equivalent at 14 m stand-off from C25 panel showing: (**a**) DIC analysis and (**b**) plot of pressure-time and displacement-time using DIC center point measurements

shows the data recorded using DIC during the blast event for the G25 panel. There is a clear distortion of the return stroke of the panel due to damage initiation and propagation. In addition, the curve has a flattened section around its maximum which coincides with when a crack was observed to initiate and propagate. Maximum deflection was found to be 140 mm and strain peaked in the region of 1.6%. Figure 9 shows the data for the C25 panel. This panel is observed to deflect noticeably less than the glass-fiber panel. The C25 panel has a larger displaced central region. The C25 panel exhibited a maximum deflection of 107 mm and maximum strain of approximately 0.8%. Upon visual inspection post-blast, the front face-sheet of the G25 panel suffered from a severe crack. The G25 panel was more severely damaged than the carbon-fiber panel as shown in Fig. 10.

The central point out-of-plane displacement for the panel with compliant PP plies (PP40) compared to a panel without the PP plies is shown in Fig. 11. It can be seen that the panel with PP plies deflects less, which was expected due to thicker front face-sheets. Following post-blast visual inspection, no damage to the front skin was identified in the panel with PP plies unlike the panel with solely glass-fiber skins.

The deflection of the center points for the two hybrid panels are shown in Fig. 12. The response of the panels are very similar. This is because the panels are responding elastically as the blast was not severe enough to cause significant damage within the skins. In addition, the mass per unit area of the panels was equal and the stiffness of the panels was very similar. The maximum central displacement of Hybrid-1 was 74.6 mm and maximum central strain in the region of 0.7%. The maximum central displacement of Hybrid-2 was 72.8 mm and strain 0.7%.

Fig. 10 Images taken of the front view of the G25 and C25 panels before and after blast loading of 100 kg TNT equivalent at 14 m stand-off

Fig. 11 Displacement-time using DIC center point measurements for GPVC and PP40 panels subjected to 100 kg TNT at 15 m stand-off

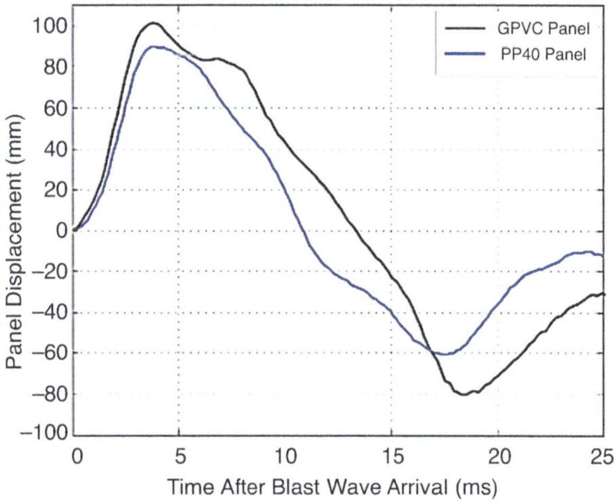

2.2.4 Effect of Progressive Loading and Multiple Blast

Figure 13 shows the response of the identical panels due to blast at different stand-off distances. The plot highlights that reduced stand-off leads to larger deflections. For the panel that underwent multiple blasts at reducing stand-off distances the deflection of the central point measured using a laser gauge is shown in Fig. 14. The first blast at 16 m stand-off resulted in no detectable damage being sustained. The target deforms uniformly and elastically. This can be stated due to the smoothness and symmetry of both the forward and rebound stroke of the panel response. The

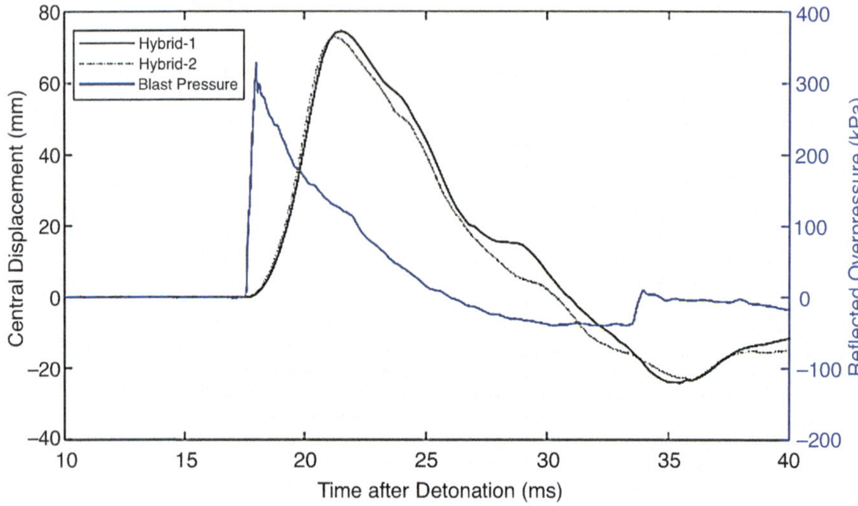

Fig. 12 Displacement-time using DIC center point measurements for the hybrid composite sandwich panels subjected to 100 kg TNT at 15 m stand-off

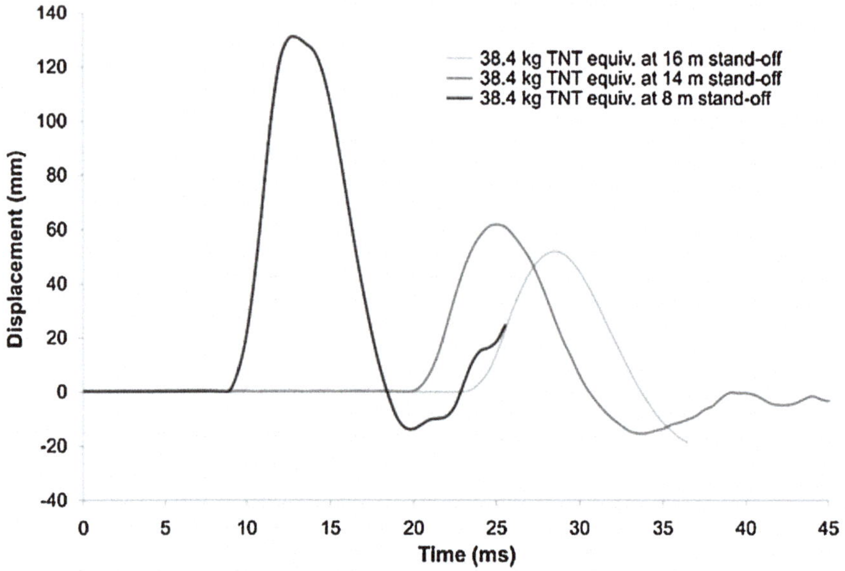

Fig. 13 Displacement-time using DIC center point measurements for G40a, G40b and G40c panels which were subjected to blast loads at varying stand-off distances [21]

magnitude of major principal strain remained well below the failure strains and experience from previous blast support these conclusions. The subsequent two blasts at reduced stand-off distances of 12 m and 8 m respectively, result in progressive increase in panel deflection. The period of panel response remains relatively constant

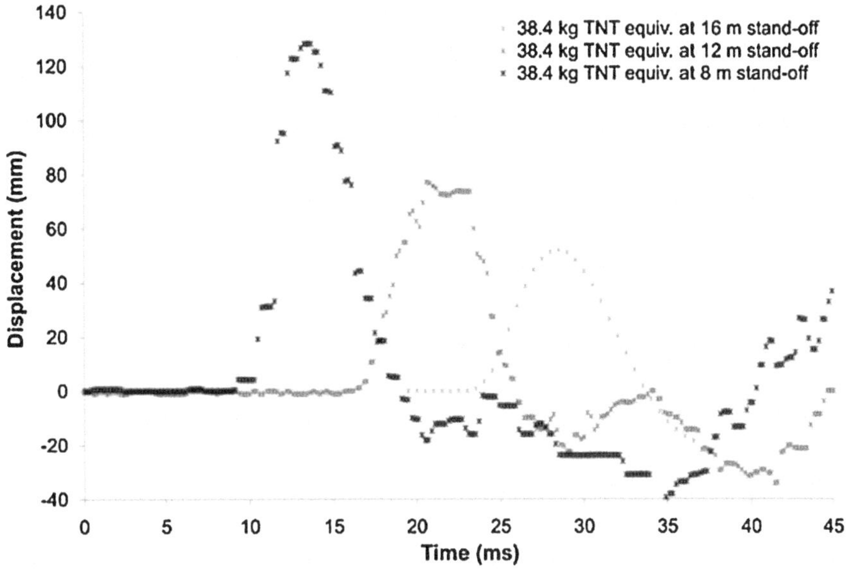

Fig. 14 Displacement-time using laser gauge center point measurements for G40d panel subjected to repeated blast loads and reducing stand-off distances [21]

during the three blast experiments since the panel properties and geometry did not change. The dominant failure mode in the composite sandwich panel was core shear failure. In addition, the panel suffered from skin cracking and skin delamination during the final blast load.

The deflection of the panels that were subjected to a repeated 8 kg PE4 charge are shown in Figs. 15 and 16. Despite the second blast loading being identical to the first, the panel deflections increase between first and second blast. This is due to the damage sustained by the panel being exacerbated by the second load. The CFRP panel experiences central deflection of 4.4 mm and 5.1 mm during the first and second blast loads respectively. The deflections for the GFRP panel are 5.1 mm and 6.0 mm for the first and second blast load. The GFRP panel deflects further under both blast loads due to its reduced stiffness compared to the CFRP panel. Additionally, the GFRP undergoes a greater relative increase in deflection indicating it has suffered from greater damage or has an inferior blast resilience.

2.3 *Discussion*

2.3.1 Effect of Core Thickness

These panels were designed to withstand a peak shock pressure of up to 200 kPa without resulting in catastrophic failure, and the panels attained this performance.

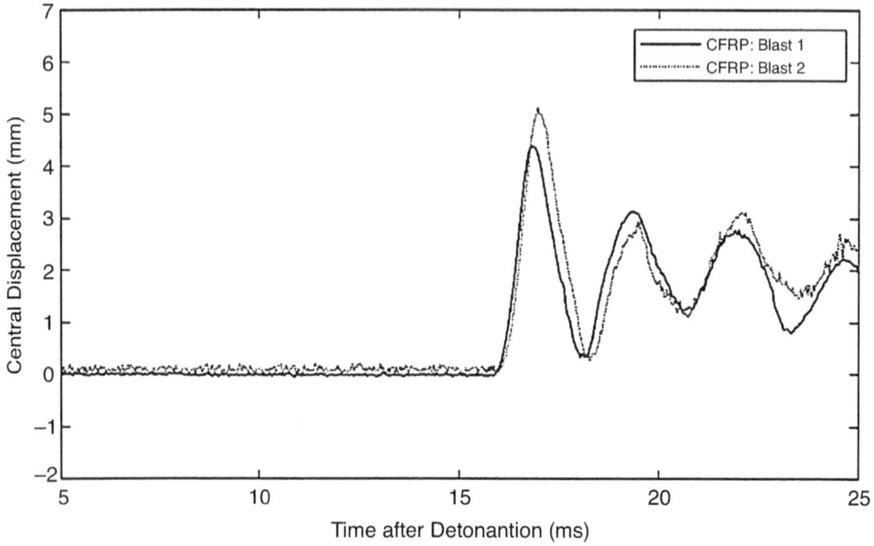

Fig. 15 Displacement-time using DIC center point measurements for CS40 panel subjected to repeated 8 kg PE4 blast loads at 8 m stand-off

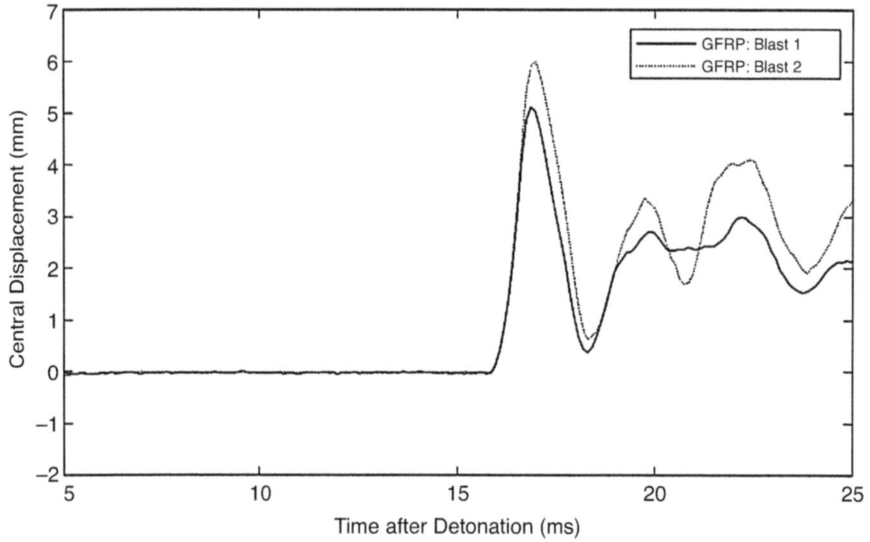

Fig. 16 Displacement-time using DIC center point measurements for GS40 panel subjected to repeated 8 kg PE4 blast loads at 8 m stand-off

When the charge weight and stand-off distances were altered to increase the impulse exerted, the front face sheet and core sustained considerable damage (core cracking, formations of delaminations and skin breakage). However, the back-face sheet remained intact after the blast and this was consistent with each test sample.

Two pairs of panels were tested to investigate the effect of core thickness. A 40 mm thick core (G40a) and a 30 mm thick core (G30a) GFRP sandwich panel were subject to a 30 kg C4 charge at 14 m stand-off. A second 40 mm thick core (G40b) and a 50 mm thick core (G50a) GFRP sandwich panel were subject to a 30 kg charge C4 at 8 m stand-off. When comparing the response of G40a and G30a, it was shown that increasing the core thickness lowered the amplitude of oscillations. Increasing the core thickness increases the second moment of area and the equivalent flexural rigidity which is proportional to the square of core thickness. Therefore, increasing the core thickness increased the stiffness of the panel and this resulted in smaller peak amplitude of displacement oscillation. It is clear, when the charge is positioned closer to the same type of panel, that an exaggerated response of the same nature is expected. When the second 40 mm core panel (G40b) was subject to a more intense blast, a skin crack formed on the front face of the target.

2.3.2 Effect of Core Material and Core Density

Four composite sandwich panels with varying core configurations were subjected to an identical blast load. SAN, PVC, PMI and a graded SAN polymeric foam core were employed in these panels. Failure of the sandwich panels was clearly visible in the form of sharp changes in displacement gradient where the foam cores and face-sheets have failed, caused by a local drop in bending stiffness. The M100 SAN core was found to deflect the least. The C70.90 PVC core deflected less than the 110SL PMI foam core, and these results correlated with the damage suffered in each case. The amount the sandwich panel pulls back out of the test cubicle, as shown by the negative displacement in Fig. 7, is another useful indicator of the damage suffered. The pull-out was caused by the momentum of the sandwich panel returning from the maximum displacement, as well as the negative blast pressure. At the point of pull-out, the sandwich panel had suffered all of the damage it was likely to suffer. Therefore, the extent of pull-out is largely dictated by the residual bending stiffness in the sandwich panel. The pull-out of the M100 SAN case was the least in the polymer comparisons, followed by the C70.90 PVC sandwich panel, and then the 110SL PMI sandwich panel.

The graded density sandwich panel suffered a maximum out-of-plane displacement of 102 mm and a maximum pull-out of 66 mm. This was much greater than the M100 SAN case to which it was compared. This was expected due to the reduced thickness of the graded foam core. However, the graded SAN core had a very similar out-of-plane displacement to the C70.90 PVC and the 110SL PMI cases, and suffered less pull-out. Furthermore, the displacement shape of the graded sandwich panel was much smoother than the other sandwich panels, which indicates less damage suffered near the rear face-sheet during blast.

2.3.3 Effect of Skin Configuration

Three pairs of panels were tested to investigate the effect of skin material and configuration. An E-glass fiber panel (G25) was tested alongside a carbon-fiber panel (C25) using a 100 kg charge at 14 m stand-off. The GFRP panel experiences significantly more displacement than the CFRP panel. The GFRP structure provides less resistance to the shock wave impact and so the severity of the damage increases faster than in the CFRP structure.

The effect of compliant PP plies within the front face sheet of an E-glass fiber skinned structure was investigated. A direct comparison between the two panels shows the structure containing the PP plies deflected less, which was expected due to the thicker face-sheet. It can also be seen that the damage implied by the defection profiles is similar, and this was also found in the post-blast damage inspection. However, despite similar amounts of core damage present, the PP ply sandwich panel suffered no front face-sheet damage due to the presence of the PP plies, unlike the solely GFRP face-sheet comparison.

Two composite sandwich panels with hybrid glass-fiber and carbon-fiber laminate skins were compared to reveal whether the layup order of the materials affects blast response. Under large-scale blast loading, where the pressure load is approaching uniformity across the panel, the position of the materials does not appear to affect the panel deflection or strain. This is partially due to the blast load causing a largely elastic response as the blast wave was not severe enough to damage the skins. Furthermore, the mass per unit area of the panels was equal and the stiffness of the panels was very similar. A more significant difference in mass per unit area or stiffness caused by layup variation would be required to reveal differences in performance.

2.3.4 Effect of Progressive Loading and Multiple Blast

The experiment investigating stand-off distance shows that reducing stand-off leads to larger deflections, as expected. One panel (G40d) was subjected to multiple loads at a reducing stand-off distance. The first blast at the greatest stand-off distance, 16 m, resulted in no detectable damage being sustained by the panel. The target deforms uniformly and elastically. The following two blasts at reduced stand-off distances result in an increase in panel deflection along with panel damage. This can be seen due to a lack of smoothness and symmetry of the forward and rebound stroke of the panel response.

The multiple blast loading performance of a CFRP and GFRP panel were compared. The panel deflections increase between first and second blast for both panel types. This is due to the damage sustained by the panel reducing the panel stiffness hence enabling a greater second deflection. Furthermore, damage is likely to be increased during the second blast load. The GFRP panel experiences a greater deflection under both blast loads due to its reduced initial stiffness compared to the CFRP panel. The greater relative increase in deflection highlights that the GFRP panel has a lower blast resilience.

3 Underwater Blast Loading of Composite Sandwich Panels

Full scale underwater blast experiments have been performed on composite sandwich panels to investigate multiple variables. The underwater blast experiments have been used to understand the strain experienced by composite sandwich panels during the blast event and how damage has developed during a blast event. Once again, the experimental series has been performed over several years hence the experimental setup has been adapted and improved. This section details the underwater blast experimental design along with the key aims of each investigation.

3.1 Experimental Design

During underwater blast testing, the composite sandwich panels are mounted within a welded steel channel box. This allows a composite sandwich panel to be sealed and bolted onto the front leaving an enclosure behind the panel. This enclosure can be sealed with air or filled with water. Two different sizes of welded steel box have been used. The welded steel box structure was lowered into a specialized test pond either from a gantry or from a crane. Figure 17 shows a diagram of a welded steel box and photographs of the structure being lowered into the test pond. The explosive charge was either lowered into position from an adjacent gantry or was attached using a lightweight wooden frame to the structure itself.

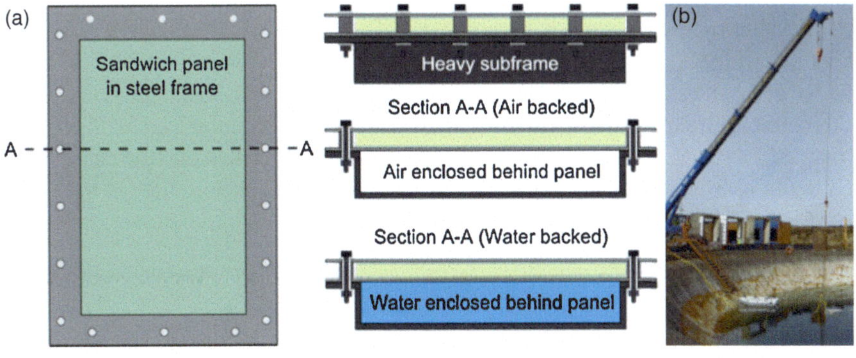

Fig. 17 Underwater blast experiment setup showing: (**a**) schematic diagram of welded steel box and (**b**) photograph of structure being lowered into test pond from crane

To capture the response of the composite sandwich panels, single axis strain gauges were adhered to the front and rear skins of the panels. Since the panels were symmetric, only one quarter of each panel had strain gauges attached. The gauges were made waterproof using a liberal coating of wax.

To record the pressure traces during blast, two pressure gauges were used in each test. The gauges measured the hydrostatic pressure in the water as the blast wave passed over them. The side-on gauge measured the unimpeded blast wave pressure as it passed over. This meant the water particles did not lose momentum as they travelled past the gauge. The side-on pressure gauge was either suspended from an adjacent gantry or attached to a thin steel arm out to the side of steel welded box. The reflected pressure gauge was mounted on the steel frame and measured the blast wave pressure as the particles came to rest against the surface of the test structure. Therefore, measuring the actual loading on the structure.

3.1.1 Effect of Backing Medium

Two composite sandwich panels with quadriaxial E-glass fiber skins were fabricated. One panel had a 30 mm thick 155 kg m^{-3} (Gurit P800) foam core and the second had a 15 mm foam core. The 30 mm thick panel had air enclosed within the welded steel box and was subjected to 1 kg C4 charge at 1 m stand-off at a depth of 6 m. The thinner panel had water enclosed within the welded steel box and was subjected to 1 kg C4 charge at 1.4 m stand-off at a depth of 6 m. Despite the difference in stand-off, these two experiments were performed to observe the effect of backing medium on sandwich response as well as allowing other comparisons of stand-off and core thickness to be made. The panels investigated in this test series were 0.5 m × 0.4 m in size and are detailed in Table 5. This investigation has been previously detailed in reference [16].

3.1.2 Effect of Graded Core

A composite sandwich panel with quadriaxial E-glass fiber skins and a graded core constructed from three layers of SAN foam with densities of 100 kg m^{-3}, 140 kg m^{-3} and 200 kg m^{-3} was compared to a panel with identical skins and a 30 mm thick 140 kg m^{-3} SAN foam core. These panels were subjected to a 1 kg PE4 charge at 1 m stand-off distance. The structure was lowered such that the charge was at a depth of 3 m. The panels both had air sealed in the enclosure behind the panel

Table 5 Details of the panels with varying core thickness and backing mediums investigated

Skin material	Core thickness (mm)	Charge size (kg)	Stand-off distance (m)	Charge depth (m)	Backing medium
GFRP	30	1	1	6	Air
GFRP	15	1	1.4	6	Water

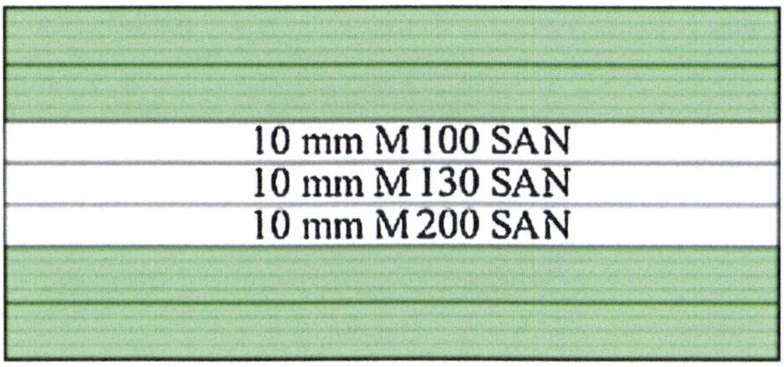

Fig. 18 Schematic diagram showing the layup of the graded core composite sandwich panel

Table 6 Details of the panels with varying core configurations investigated

Skin material	Core material	Core thickness (mm)	Charge size (kg)	Stand-off distance (m)	Charge depth (m)
GFRP	SAN	30	1	1	3
GFRP	Graded SAN	30	1	1	3

and were 0.8 m × 0.8 m in size. Figure 18 shows a diagram of the composite sandwich panel with graded density foam core and Table 6 lists the panels investigated in this experiment. Aspects of this investigation can be found in reference [22].

3.1.3 Effect of Skin Configuration

A range of composite sandwich panels with varying skin configurations have been subjected to full-scale underwater blast loading. The performance of composite sandwich panels with E-glass fiber skins either side of a 30 mm thick 140 kg m^{-3} SAN foam core has been compared to panels with carbon-fiber skins and an identical core. In addition, a panel with a graded SAN core and carbon-fiber skins was fabricated. This panel can be compared to the graded core panel detailed in Sect. 3.1.2. These panels were subjected to a 1 kg PE4 charge at 1 m stand-off distance and were 0.8 m × 0.8 m in size. Air was sealed within the enclosure behind the panel.

The effect of a combination of E-glass fiber and Kevlar fiber within skins has been investigated. The panel had one ply of quadriaxial E-glass and four plies of quadriaxial aramid either side of a 30 mm and a 15 mm 140 kg m^{-3} SAN core. Once again, the panel can be compared to the panels detailed in Sect. 3.1.1. Both panels were air backed. The thicker 30 mm panel was subjected to 1 kg C4 charge at 1 m stand-off whilst the thinner panel had a 1.4 m stand-off distance. Both panels were submerged to a charge depth of 6 m. Table 7 details the panels investigated in this series of experiments. Reference [22] details some aspects of this investigation.

Table 7 Details of the panels with varying skin configurations investigated

Skin material	Core material	Core thickness (mm)	Charge size (kg)	Stand-off distance (m)	Charge depth (m)
GFRP	SAN	30	1	1	3
GFRP	Graded SAN	30	1	1	3
CFRP	SAN	30	1	1	3
CFRP	Graded SAN	30	1	1	3
GFRP/Kevlar	SAN	30	1	1	6
GFRP/Kevlar	SAN	15	1	1.4	6

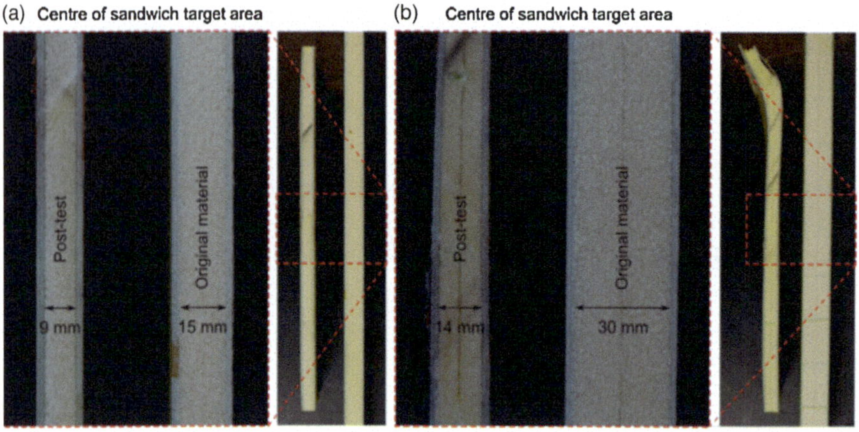

Fig. 19 Images taken of the panels before and after underwater blast loading showing: (**a**) the 15 mm thick water-backed panel and (**b**) the 30 mm thick air-backed panel

3.2 *Results*

3.2.1 Effect of Backing Medium

The air backed panel demonstrated typical plate response. 3% front skin compressive strains were recorded on the top edge of the panel indicating the panel bent and deformed around its edge. Damage was also observed at this location on the panel after blast, as shown in Fig. 19. The panel crushed to under half the original thickness, there was a 16 mm core thickness reduction. The panel remained permanently deformed after the blast. In contrast the water supporting the rear face of the second panel was observed to dampen the overall response of the panel during blast. The recorded strains were not as large. Target crushing was caused by the pressure wave pushing the panel against a backing mass of water which was encapsulated within the steel frame with no means to vent out. Therefore, the backing mass gained in pressure and supported the panel against large out-of-plane motion. The panel crushed to nearly half its original thickness. The maximum front skin strain recorded on the second panel was ±0.6%. Figure 19 shows the thickness of the two panels before and after testing.

3.2.2 Effect of Graded Core

By linearly interpolating the strains measured across the panel sections, it is possible to get an approximation of the central point displacement of each panel. The calculation assumes a simplification that no crushing occurs in the panels during the blast, this is clearly a simplification considering extensive core crushing occurred in the investigation detailed previously. Nevertheless, the values can be used as an indicator to compare the performance of the different panels. These displacements are for the centerline of the sandwich panels, as they are calculated using an average of the two face sheet strains. The deflection is relative to the edge of the sandwich panel. The peak displacement at failure for the panels are detailed in Table 8. The graded density panel had a lower displacement at failure than the equivalent single density panel.

3.2.3 Effect of Skin Configuration

It was expected that in a comparison between sandwich panels with GFRP and CFRP face sheets, the CFRP sandwich panel would suffer catastrophic boundary failure due to the higher stiffness causing greater bending moments at the panel edges. This was found to be the case for the graded density CFRP panel, where very high strains were recorded at the boundaries. Table 9 details damage sustained by the panels following evaluation using X-ray CT scanning. Overall, the CFRP panels suffer from greater damage and the graded core is largely beneficial. The

Table 8 Details of panel displacement at failure as a result of underwater blast testing [22]

Skin material	Core material	Displacement at failure (mm)
GFRP	SAN	48
GFRP	Graded SAN	34
CFRP	SAN	50
CFRP	Graded SAN	13

Table 9 Details of panel damage as a result of underwater blast testing [22]

Skin material	Core material	Fraction of panel damage (%)	Fraction of front debond (%)	Fraction of rear debond (%)
GFRP	SAN	7.2	26.9	18.2
GFRP	Graded SAN	4.4	32.5	9.3
CFRP	SAN	20.6	76.0	15.2
CFRP	Graded SAN	10.3	58.1	31.8

displacement at failure for the graded density and single density CFRP panels are included in Table 8 alongside the values for the GFRP panels. These results support the statement made in Sect. 3.2.2 regarding the reduced displacement caused by the graded core.

The panels with GFRP/Kevlar skins were subjected to the same blast conditions as the panels detailed in Sect. 3.1.1. Both panel types sustained a degree of visible damage. The GFRP panels sustained skin wrinkling, whereas the GFRP/Kevlar panels sustain more severe skin damage. The 30 mm GFRP/Kevlar panel demonstrated a maximum front skin compressive strain of 2.8% which is comparable to the strain recorded for the solely GFRP panel under the same conditions. However, the GFRP/Kevlar panels sustained more severe skin damage in the form of fiber breakage and skin tearing. The GFRP/Kevlar skins did not absorb or redistribute sufficient blast energy. This lead towards a more instant formation of a highly deforming plate. Hence, the back skin damage for the GFRP/Kevlar panels was significantly more severe. For the thinner Kevlar panel under more severe loading, however, the maximum front skin compressive strain was 3.9%. Although this panel exhibited greater strain, the greater charge stand-off distance delivered a lower impulse to the panel. Furthermore, the GFRP/Kevlar panels suffered from greater core crushing.

The core thickness influences the behavior of both the GFRP and GFRP/Kevlar panels. The thinner cores enable greater deformation of the skins prior to failure. Whereas the thicker core introduces high shear stresses, leading to core cracking and hence separation of skin from core. This meant that failure stress was attained quicker in the skin given it was unsupported. The thinner cores may have sustained a degree of cracking to a lesser extent and generally maintained support for the skins. This allows the skins to deform further prior to any fiber breakage. The higher section modulus arising from a thicker core means there would be a greater resistance to flexure and so lower strains are expected for equivalent loads in thicker panels compared to thinner panels.

3.3 Discussion

Adopting water backing damped the overall response of the panel and reduced the strains recorded in comparison to the air backed panel. Significant target crushing occurred for both the air-backed and water-backed panel. The air-backed panel experienced significant deformation around the edges and remained permanently deformed post blast. Both panels suffered significant damage, however, water-backing appears to be advantageous due to the damping and support it offers the panel which reduced deflection and permanent deformation.

An approximate method for calculating central point displacement showed that both the panels with graded density foam cores had a lower displacement at failure compared to the panels with single density foam cores. This displacement calculation was based on the strain gauge reading, therefore, despite incorrectly assuming that no core crushing occurs, the graded core clearly offers advantages. Damage assessment of the panels has revealed that the lower density foams are able to undergo crushing and protect the high density foam at the rear. The GFRP panel with graded density core demonstrated the best performance.

The investigation into panels with GFRP versus CFRP skins revealed that the panels with CFRP skins suffered from significantly greater damage and skin fracture at the boundaries. This indicates that the high stiffness and brittle behavior of CFRP skins prevents energy absorption and leads to catastrophic panel damage.

The GFRP/Kevlar hybrid panels sustained more severe skin damage in the form of fiber breakage and skin tearing compared to the panels with solely GFRP skins. Due to the lower strength of Kevlar, the GFRP/Kevlar panels were unable to absorb sufficient blast energy and are likely to have experienced greater deflection resulting in this damage. Additionally, the hybrid panels suffered from greater core crushing. The hybrid and GFRP panels both experienced comparable strain levels indicating that the hybrid panel has a lower strain to failure. These results show that the replacement of glass-fiber by Kevlar lessened the skin properties rather than enhancing them. This indicates that this combination is not an optimized hybrid.

Furthermore, the core thickness affects the behavior of the GFRP and GFRP/Kevlar panels. A thinner core allows greater deflection prior to fiber breakage. The thicker cores introduce shear stresses which leads to core cracking and separation from the skins. This results in unsupported skins hence attainment of failure stress more quickly.

4 Finite Element Analysis

Finite Element (FE) models have been produced and used during the course of the air blast studies primarily as an evaluative tool. The models have been validated against experimental data such that they can be used to develop and extend the experimental investigations performed. To some extent, the models have been used as a predictive tool. They have been used to establish suitable sample geometries and blast parameters to achieve the desired damage effects. This section details the main models that have been developed and the comparison of numerical data to experimental data, hence the validation of the models.

4.1 Air Blast Modeling Method

Finite element models have been generated using ABAQUS/Explicit. The models have aimed to capture the elastic behavior of the sandwich panels.

4.1.1 The Effect of Support Conditions

Despite using purpose built blast test cubicles, there is naturally a degree of elasticity present. This has been observed during the experiments. Three models were developed to capture the most appropriate boundary conditions. 2D shell elements were used to model the composite sandwich panel as through thickness stresses, damage and failure were neglected. The continuum shell elements were sectioned into the various layers of the composite sandwich panel with appropriate material properties and thicknesses. The first model (Model 1) included the front sheet of the steel test cubicle, which was simplified to a homogenous section of continuum shell elements. The effective stiffness of the various I-beams and ribbed plates were taken into account by equivalent thickness of elements for the test cubicle. Boundary conditions were fully built-in around the edges of the cubicle front. The boundary conditions at the sandwich panel edge were effectively built-in to the cubicle structure. The experimentally recorded reflected pressure was used as the load. A diagram of this FE model setup is shown in Fig. 20. Aspects of this work have been previously published in reference [23].

Although this model captures the experimental setup, it does not provide useful insight to a broader audience. Therefore, a second model (Model 2) was developed which investigated the panel alone with simple boundary conditions. These conditions were: (Case 1) fully built-in, where there is zero translational and zero rotational degrees of freedom all around the panel edges; (Case 2) pinned, where there is zero translational degrees of freedom all around the panel edges; and (Case 3) built-in on three edges and pinned on the right hand edge. This takes into account the geometrical effect arising from the non-uniform support from the cubicle as the panels are mounted off-center.

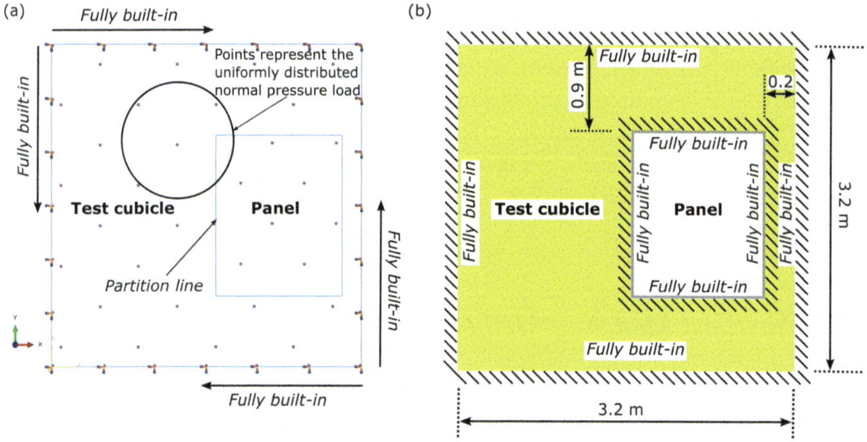

Fig. 20 Boundary and load conditions: (**a**) an annotated ABAQUS visualization of the model; and (**b**) a schematic representation of the same model highlighting the effective boundary conditions [23]

As detailed in Sect. 2.1 some air blast experiments were performed using a concrete cubicle. The third FE model (Model 3) was produced to capture the effect of this setup. The cubicle front face area was slightly larger and more substantial supporting beams were present.

4.1.2 The Effect of Core Thickness

Model 1 was used to investigate the effect of core thickness. The core thickness was varied by altering the section assigned to the sandwich panel. The thickness of the core was varied from 30–50 mm. The two different blast scenarios tested (30 kg C4 at 14 m and 8 m stand-off distances), as detailed in Sect. 2.1.1, were run in this simulation to give a comparison between experimental and numerical data. These simulations were intended to show how well the experimental data correlates to the numerical solution.

4.1.3 The Effect of Skin Configuration

Using Model 3 the skin configuration was changed by editing the section assignments of the sandwich panel from a GFRP to a CFRP panel. The skins were changed to have the appropriate corresponding material properties. The simulations were run to validate the model against the experimental data detailed in Sect. 2.1.3. A proportion of this work has been previously detailed in reference [18].

A fourth model (Model 4) was subsequently created to investigate whether the elastic response of a hybrid glass-fiber and carbon-fiber panel under 100 kg TNT equivalent charge could be captured. The layup of Hybrid-1 was selected for this model trial. ABAQUS/Explicit was used to create a model of the panel and front face of the steel support cubicle, similar to Model 1. C3D8R elements were used to model the sandwich panel. The average values of the carbon/carbon and glass/glass interface properties were used for the glass/carbon interface. An estimated preload of 9 kN was exerted on the front steel frame to represent the pretension of the bolts. As stated, this model was developed as an example to determine whether the experimental results can be reliably reproduced numerically. This work has been previously detailed in reference [20].

4.2 Air Blast Modeling Results

4.2.1 The Effect of Support Conditions

The results from the initial model highlight the obvious effect of the support structure on the panel response, given it formed an asymmetrically deforming surface during the simulation. Figure 21 shows the contour plots of the FE analysis

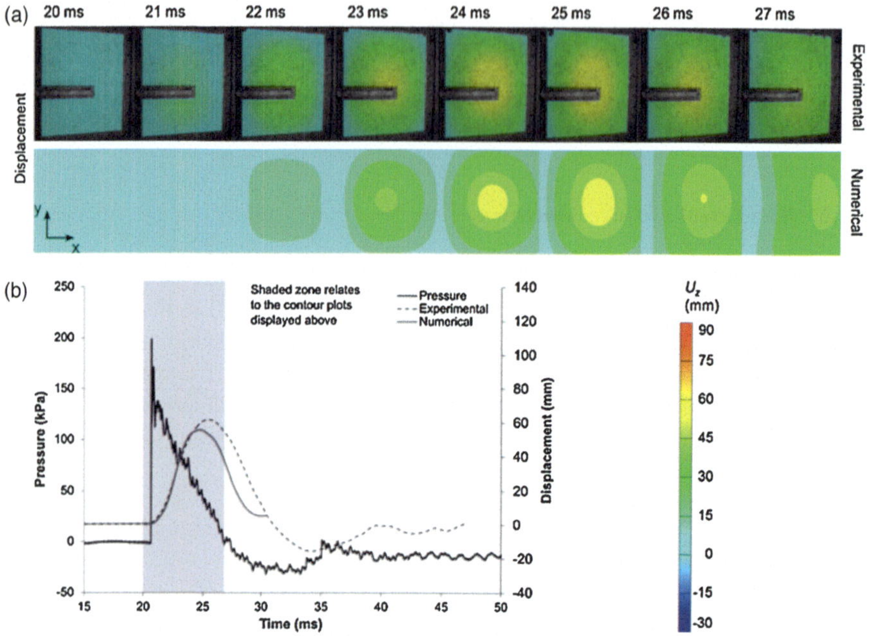

Fig. 21 Blast summary for 30 kg C4 at 14 m stand-off from G40a, displaying both experimental and finite element analyses: (**a**) contour plots of out-of-plane displacement, (**b**) graphical plot of load-time and displacement-time for both numerical and experimental results

displaying the transient out-of-plane displacement response along with the experimental data. These results are for the 30 kg C4 charge at 14 m stand-off against target G40a. The contour plots displayed are just for the panel alone and the deformation of the front of the cubicle is omitted from view. These images are shown alongside a graphical plot of central out-of-plane displacement against time for the numerical and experimental analyses. The magnitude of the two peaks as well as the time period of oscillation show close but not exact agreement. The deviations are due to simplifications in the stiffness of the steel cubicle front as well as the omission of damping. The contour plots clearly shows that after the point of maximum displacement there is a bias towards the right hand edge. The magnitude and characteristics of these contour plots correlate well with the observed experimental data. As the blast load was increased to 30 kg C4 at 8 m, these effects are exaggerated. A limitation of the model, however, is caused by the lack of an implemented damage model. This causes deviation of the model from experimental data in the 8 m stand-off analysis.

The second model investigated the most representative boundary conditions applied to the panel edges. The experimentally observed result was found to lie between Case 2 (pinned) and Case 1 (built-in) as shown in Fig. 22. Cases 3 and 4 show that the relative degrees of freedom from one edge to the other can cause an asymmetrically deformed shape.

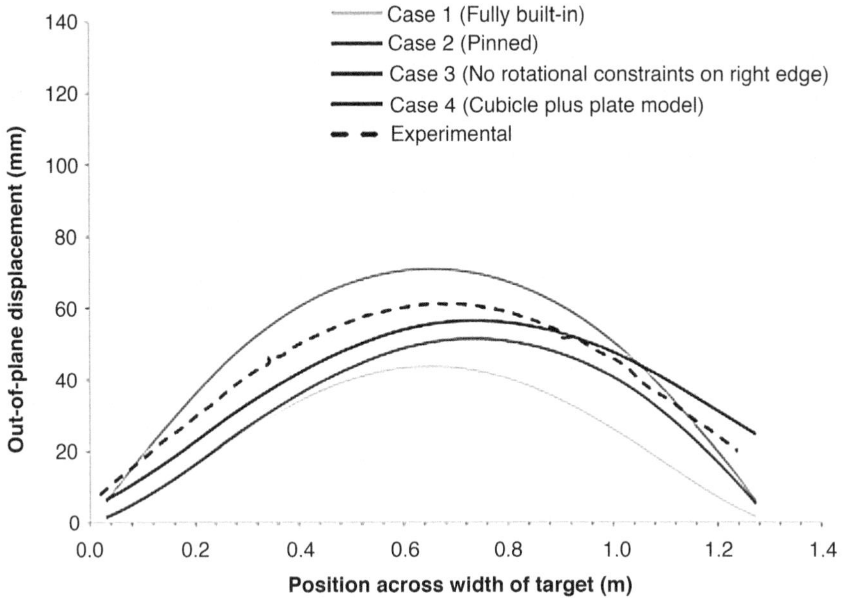

Fig. 22 Out-of-plane displacement plots from numerical modelling showing the central horizontal cross-section of the panel when subject to 30 kg C4 at 14 m stand-off [23]

4.2.2 The Effect of Core Thickness

Model 1 was used to investigate the effect of core thickness and whether the experimental results could be reproduced through FE analyses. Figure 23 shows that the predictions of the simulation fall in line with the experimental data. The simulation observed a comparable central out-of-plane displacement-time history to the experiment, which reinforces the reliability of the model. Table 10 details the central point deflections recorded experimentally and predicted by the FE simulation.

4.2.3 The Effect of Skin Configuration

The results confirm the experimentally observed superior performance of the CFRP skinned sandwich panel compared to the GFRP panel. Figure 24 shows the central displacement recorded using DIC with the FE data overlaid. The FE model predicts that G25 and C25 deflect to 150 mm and 110 mm respectively. The experimental results detailed earlier show that these panels experienced deflections of 140 mm and 107 mm. The nature of the curve deviates after the arrival of the blast wave to the peak. This is due to a combination of target failure and boundary deformations.

The central deflection versus time curves obtained from the experiment and simulation for Hybrid-1 are shown in Fig. 25. The model is an elastic model

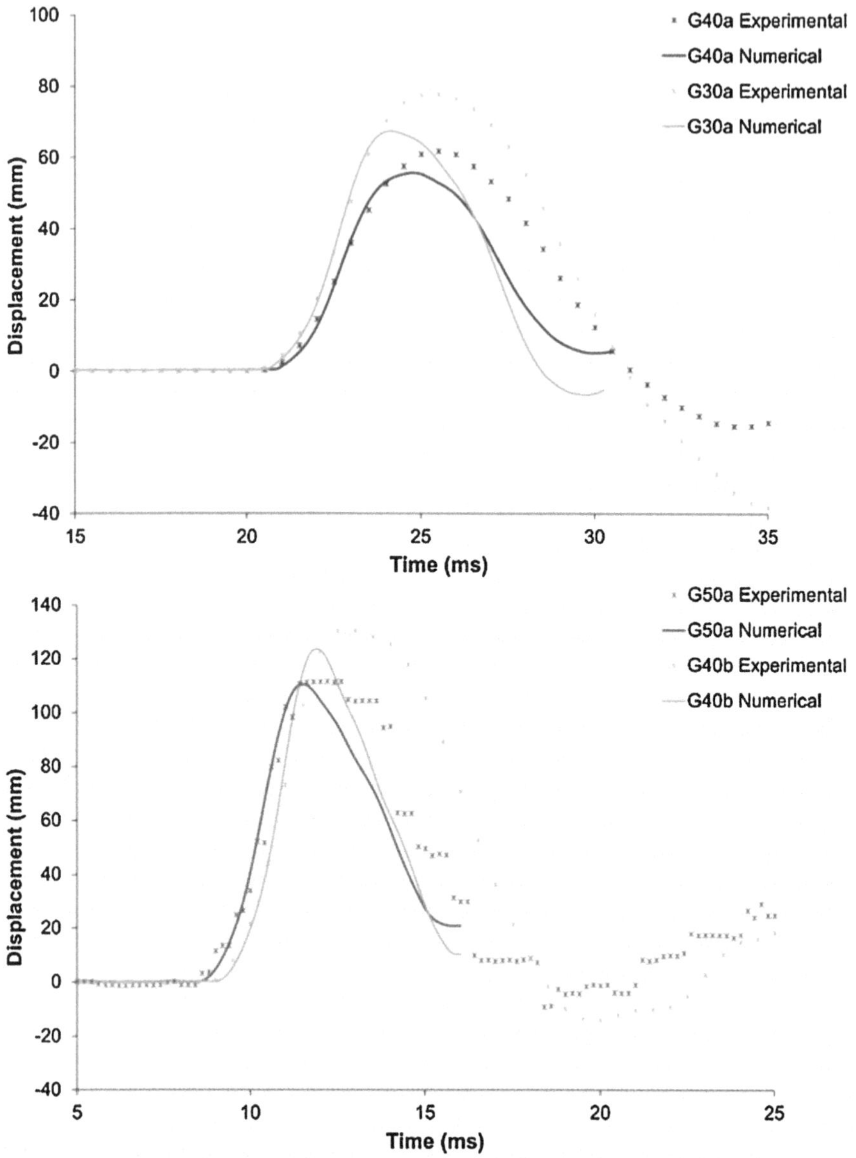

Fig. 23 A pair of plots showing central displacement-time history from numerical modelling and experimental studies on the effect of core thickness on panel response. The blast parameters were 30 kg C4 at 14 m stand-off (top) and 8 m stand-off (bottom)

which has focused on capturing the elastic response of the sandwich panel only. Due to this, the curves deviate beyond the peak deflection. The experimental and numerical maximum central deflections are 74.6 mm and 70.6 mm. The experimentally measured major strain and the computational major strain are 0.70% and

Table 10 Details of panel central displacement recorded experimentally and predicted numerically

Panel code	Core thickness (mm)	Charge size (kg)	Stand-off distance (m)	Numerical displacement (mm)	Experimental displacement (mm)
G30	30	30	14	67	78
G40a	40	30	14	55	61
G40b	40	30	8	126	134
G50	50	30	8	110	111

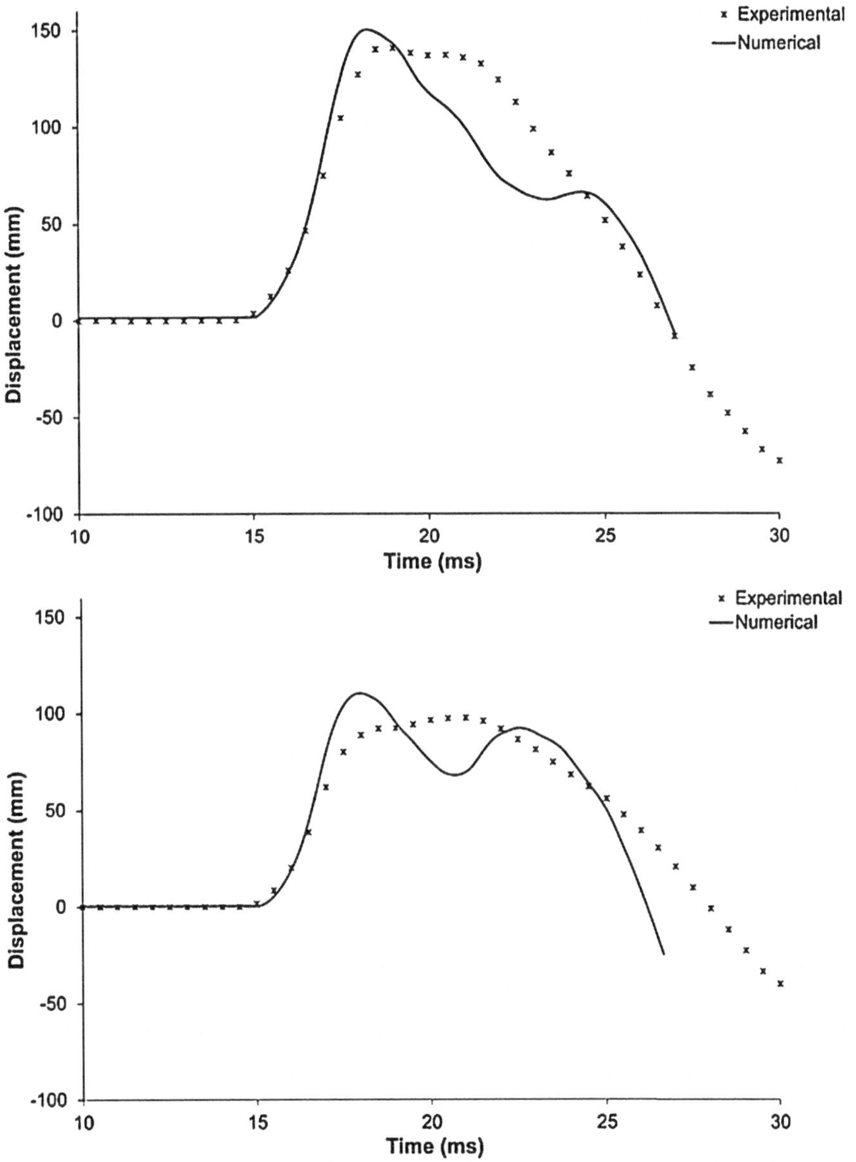

Fig. 24 A pair of plots showing central displacement-time history from numerical modelling and experimental studies on the effect of skin configuration on panel response: (top) G25 and (bottom) C25. The blast parameters were 100 kg nitromethane at 14 m stand-off distance [18]

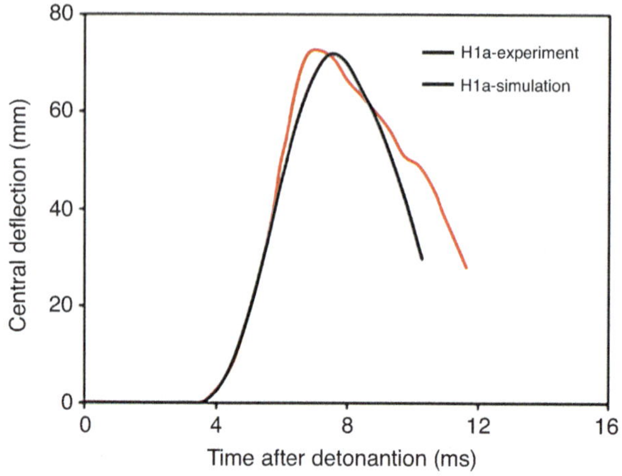

Fig. 25 Out-of-plane displacement-time plot from numerical modelling and experimental study on Hybrid-1 under 100 kg nitromethane charge at 15 m stand-off distance [20]

0.62%. These results confirm that the proposed numerical model has the capability to predict the elastic behavior of the sandwich panel under air blast loading.

4.3 Discussion

The advantage of having full-field experimental data allows for direct comparison of FE simulations.

Models were generated to simulate the effects of the boundary conditions on the target response. However rigid the support structure is, a degree of elastic deformation is always present. A degree of elasticity can aid the blast resistance of a structure due to the distribution and dissipation of energy. Model 1 effectively captured the biased response of the panel due to its support conditions. Model 2 revealed that the experimental result lay within a range of finite element simulation results. More specifically between the pinned and fully built-in cases. These degrees of freedom can be fitted to achieve an exact agreement and transient boundary conditions can be created using experimental data. However, this is not useful information to any user outside of the test facility. The design of the support conditions could be used to influence panel response and even allow failure to occur in favorable locations.

The two blasts simulated with varying core thicknesses showed good agreement with the experimental data available. These results imply that a suitable model of the support structure is established such that when a parameter, such as core thickness, is changed, the predictions of the simulation fall in line with the experimental results. Once again, when the stand-off distance was changed, general trends between the numerical and experimental data were observed. This reinforces that a suitable support structure model has been established. However, damage mechanisms neglected from the model in the core, such as cracking, compression, buckling and crushing, limit the amount of damping present in the model and perhaps limit the energy absorbing mechanisms present in the model. This is an area for further development.

The comparison of CFRP and GFRP panels used Model 3 which had a more rigid support structure. Correlation was shown to be very good. The implementation of material and damage models would further enhance the model. The increased damping would have slowed or inhibited the rebound stroke of the panels, not resulting in the double peak of C25. In C25 the spread of core cracking was significantly higher than in G25, this can account for why the lack of damage model was more significant on the results for C25 than for G25. Carbon-fiber skins were shown to outperform glass-fiber skins numerically and experimentally. However, as detailed in Sect. 3.2.3, this is not always the case. Therefore, composite sandwich panels with hybrid glass-fiber and carbon-fiber skins were investigated. The results from Model 4 showed that the experimental and numerical results differed by ca. 5%. The deflection profile and major strain profile obtained from experimental data and numerical simulation agreed well.

Overall, more advanced and complex material models and damage functions, along with more detailed representations of the support structures and fluid structure interaction would provide a more rigorous and thorough simulation of the blast event.

5 Conclusion

A significant volume of large scale air and underwater blast testing has been performed to capture displacement-time histories of the composite sandwich panels and understand how damage develops during a blast event.

Increasing the panel core thickness reduces panel deflection due to the increased second moment of area. However, this can lead to increased shear stresses and shear cracks which leave the skin unsupported. The skins are susceptible to failure in these unsupported locations. An intermediate core thickness is, therefore, optimal.

From the four structural foam cores tested, SAN demonstrated the best blast performance. The panel with a SAN core underwent the lowest deflection and rebound stroke. The benefits of a graded core were illustrated both in air and underwater blast. The graded cores experience less damage and lower deflection for constant core thickness.

In the air blast experiment, CFRP skins were beneficial due to their increased stiffness which led to lower deflection and less damage. However, during underwater blast, the higher stiffness results in greater damage as energy absorption through deflection is prevented. Hybrid skins have been investigated to determine whether a combination of materials can alleviate this problem. The GFRP/Kevlar panel was not an optimal hybrid, the low strength and stiffness of the Kevlar fibers did not enhance the glass-fiber properties. Hybrid glass-fiber and carbon-fiber composite sandwich panels were tested under air blast loading. The two layups investigated did not affect the response of the panels. Finally, the addition PP interlayers was beneficial. The presence of interlayers reduced deflection and damage to the front skin.

Reducing stand-off distance increases panel displacement, as expected, yet the period of panel response remains largely the same. Progressive and repeated loading leads to the build-up of damage. The GFRP panel underwent a greater relative increase in displacement between the first and second blast load compared to a CFRP panel indicating the superior blast resilience of CFRP. Finally, adopting water-backing for a panel subjected to an underwater blast was advantageous. The panel response was damped, which resulted in lower surface strains and less damage. Comparable core crushing to an air-backed panel was still sustained.

Numerical simulations have been developed to support experimental investigations. The experimental data has been used to validate these simulations such that they can be used during the design process. The importance of boundary conditions has been highlighted by the simulations. It would be advantageous to design the support conditions such that they influence panel response and even dictate allowable failure locations. Further development of the simulations to capture damage and fluid structure interaction is recommended.

Overall, these results illustrate the inherent blast resilience of composite sandwich panels. The various investigations have revealed how this performance can be improved. These studies have used materials that are currently widely available, hence demonstrating the advantages that can arise from simply adopting certain geometries or materials.

Acknowledgments The authors would like to thank Dr. Yapa Rajapakse of the Office of Naval Research for the strong support received by Dr. Emily Rolfe, Dr. Mark Kelly and Dr. Hari Arora during their PhDs. This work was supported by ONR grants N62909-15-1-2004, N00014-08-1151 and N00014-12-1-0403. The authors would also like to thank EPSRC for supporting Emily Rolfe during her PhD, CPNI, DNV GL and Radnor Range Ltd. for testing opportunities along with the assistance from GOM UK, LaVision and Slowmo Ltd.

References

1. Gullberg O, Olsson KA (1990) Design and construction of GRP sandwich ship hulls. Mar Struct 3:93–109
2. Hayman B (2018) Underwater explosion response of sandwich structures with compliant cores. In: Blast mitigation strategies marine composite sandwich sturctures. Springer, Singapore, pp 23–52
3. Mouritz AP, Gellert E, Burchill P, Challis K (2001) Review of advanced composite structures for naval ships and submarines. Compos Struct 53:21–42
4. Comtois JLR, Edwards MR, Oakes MC (1999) The effect of explosives on polymer matrix composite laminates. Compos Part A Appl Sci Manuf 30:181–190
5. Gargano A, Pingkarawat K, Blacklock M, Pickerd V, Mouritz AP (2017) Comparative assessment of the explosive blast performance of carbon and glass fibre-polymer composites used in naval ship structures. Compos Struct 171:306–316
6. Kerber A, Gargano A, Pingkarawat K, Mouritz AP (2017) Explosive blast damage resistance of three-dimensional textile composites. Compos Part A Appl Sci Manuf 100:170–182

7. Tekalur SA, Bogdanovich AE, Shukla A (2009) Shock loading response of sandwich panels with 3-D woven E-glass composite skins and stitched foam core. Dyn Fail Durab 69:736–753
8. Wang E, Gardner N, Shukla A (2009) The blast resistance of sandwich composites with stepwise graded cores. Int J Solids Struct 46:3492–3502
9. Gardner N, Wang E, Kumar P, Shukla A (2012) Blast mitigation in a sandwich composite using graded core and polyuria interlayer. Exp Mech 52:119–133
10. LeBlanc J, Shukla A (2010) Dynamic response and damage evolution in composite materials subjected to underwater explosive loading: an experimental and computation study. Compos Struct 92:2421–2430
11. LeBlanc J, Gardner N, Shukla A (2013) Effect of polyuria coatings on the response of curved E-glass/vinyl ester composite panels to underwater explosive loading. Compos Part B Eng 44:565–574
12. Zenkert D, Shipsha A, Bull P, Hayman B (2005) Damage tolerance assessment of composite sandwich panels with localized damage. Compos Sci Technol 65:2597–2611
13. Hassan MZ, Guan ZW, Cantwell WJ, Langdon GS, Nurick GN (2012) The influence of core density on the blast resistance of foam-based sandwich structures. Int J Impact Eng 50:9–16
14. Bahei-El-Din YA, Dvorak GJ, Fredricksen OJ (2006) A blast-tolerant sandwich plate design with a polyuria interlayer. Int J Solid Struct 43:7644–7658
15. Bahei-El-Din YA, Dvorak GJ (2008) Enhancement of blast resistance of sandwich plates. Compos Part B Eng 39:120–127
16. Arora H, Hooper PA, Dear JP (2012) The effects of air and underwater blast on composite sandwich panels and tubular laminate structures. Exp Mech 52:59–81
17. Kelly M, Arora H, Worley A, Kaye M, Del Linz P, Hooper PA, Dear JP (2015) Sandwich panel cores for blast applications: materials and graded density. Exp Mech 56(4):1–22
18. Arora H, Hooper PA, Del Linz P, Yang H, Chen S, Dear JP (2012) Modelling the behavior of composite sandwich structures when subject to air-blast loading. Int J Multiphys 6:199–218
19. Arora H, Kelly M, Worley A, Del Linz P, Fergusson A, Hooper PA, Dear JP (2014) Compressive strength after blast of sandwich composite materials. Philos Trans R Soc London A Math Phys Eng Sci 372
20. Rolfe E, Quinn R, Sancho A, Kaboglu C, Johnson A, Liu H, Hooper PA, Dear JP, Arora H (2018) Blast resilience of composite sandwich panels with hybrid glass-fibre and carbon-fibre skins. Multiscale Multidiscip Model Exp Des 1:197–210
21. Arora H, Del Linz P, Dear JP (2017) Damage and deformation in composite sandwich panels exposed to multiple and single explosive blasts. Int J Impact Eng 104:95–106
22. Rolfe E, Kelly M, Arora H, Hooper PA, Dear JP (2017) Failure analysis using X-ray computed tomography of composite sandwich panels subjected to full-scale blast loading. Compos Part B Eng 129:26–40
23. Arora H, Hooper PA, Dear JP (2011) Dynamic response of full-scale sandwich composite structures subject to air-blast loading. Compos Part A Appl Sci Manuf 42:1651–1662

Explosive Blast Response of Marine Sandwich Composites

Alexander Gargano, Raj Das, and Adrian P. Mouritz

1 Introduction

Fiber-reinforced sandwich composites are used in a wide variety of light-weight military assets at risk of attack using explosive munitions. These assets include armoured vehicles, tanks, ships, submarines, aircraft, helicopters and drones. The types of sandwich composite materials used in naval ship structures can consist of either fiberglass or carbon fibre-polymer laminate facesheets covering a thick core of polymer foam or balsa wood. A concern with sandwich composites instead of metals (e.g. steels, aluminium alloys) in naval vessels is their brittle-like response under the high strain rate loading caused by the shock wave. Metals often undergo large plastic strain deformation which absorbs a large amount of shock wave energy before catastrophic failure. The laminate facesheets do not usually plastically deform which results in brittle-type damage such as matrix cracking, delamination and fiber fracture. The core material to sandwich composites may plastically deform and crush under the impulse load exerted by a shock wave.

The deformation and damage response of sandwich composites due to explosive blast loading has been studied in detail [1–17]. Researchers have identified that the explosive blast response of sandwich composites is largely governed by their geometry (facesheet-core thickness ratio etc.) [18–20], density of the core material [11, 14, 15, 22], and properties of the individual material constituents used in the laminate facesheets and core [11, 12, 14, 15, 21, 22].

It is known that sandwich composites fail under several damage modes when subjected to an explosive blast. Damage initiation due to blast is most commonly in the form of facesheet delamination [2, 3, 7–9, 14, 26], facesheet-core debonding

A. Gargano · R. Das · A. P. Mouritz (✉)
School of Engineering, RMIT University, Melbourne, Victoria, Australia
e-mail: alexander.gargano@rmit.edu.au; raj.das@rmit.edu.au; adrian.mouritz@rmit.edu.au

S. W. Lee (ed.), *Advances in Thick Section Composite and Sandwich Structures*,
https://doi.org/10.1007/978-3-030-31065-3_4

[4, 5, 10, 15, 22, 23], core crushing [1, 5, 6, 10, 15, 21, 24] and core cracking [4, 5, 8–11, 22, 25, 26]. Failure via core fragmentation [7, 14], core disintegration [1–3, 24] and tearing failure at the supports [11] have also been reported.

Majority of the studies have tested the explosive blast response of sandwich composites with glass fiber facesheets [2, 3, 7, 14] and polymer foam core [4, 5, 8–10, 26]. Little published research is available on the explosive blast resistant properties of sandwich composites containing carbon fiber laminate facesheets or end-grain balsa wood core – which are both commonly used materials in naval structures.

The effect of the laminate facesheet material and the core material on the explosive blast resistance of sandwich composites commonly used in naval ship structures is investigated. The sandwich composites used have facesheets made of woven E-glass or woven carbon fiber reinforced vinyl ester laminate. The core materials were closed-cell PVC foam and end-grain balsa wood. The facesheet (2 mm thick) and core thickness (6 mm) was the same to provide a direct comparison of the blast response of the different types of sandwich composites. Air blast tests were performed on the sandwich composites under three shock wave conditions by varying the mass or stand-off distance of the explosive charge. The dynamic deformation, amount and types of damage, and post-blast mechanical properties are compared for the sandwich materials.

2 Sandwich Composite Materials and Experimental Methodology

2.1 Fabrication of Sandwich Composite Materials

The response to explosive blast loading of four types of sandwich composite commonly used in ship structures was evaluated:

- woven carbon fiber-end grain balsa core composite
- woven carbon fiber-PVC foam core composite
- woven glass fiber-end grain balsa core composite
- woven glass fiber-PVC foam core composite.

Top and side-views of these sandwich materials are shown in Fig. 1. The core material was 6 mm thick and was either end-grain balsa (Baltek SB) with a bulk density of 150 kg/m^3 or Divinylcell PVC foam (Diab Group) with a density of 130 kg/m^3. Both core materials were gently sanded and cleaned prior to manufacturing to ensure effective bonding to the laminate facesheets. The carbon and glass used in the facesheets were both single ply plain woven fabric (areal density of 600 g/m^2). The carbon and glass fabrics were stacked on both sides of the sandwich core, with the warp tows aligned in the same direction, giving a cross-ply fiber [0/90] pattern for the facesheets. The dry fabric to the facesheets were impregnated with liquid

Fig. 1 Top and side views of the sandwich composite panels for blast testing

vinyl ester resin at room temperature using vacuum bag resin infusion (VBRI) in a one-step process (i.e. there was no secondary process to bond the laminate facesheets to the core). The vinyl ester was catalysed using 0.8 wt% MEKP solution (40 wt% MEKP in dimethyl phthalate) (Norox from Nuplex Composites). Due to the closed cell structure of both the PVC and balsa wood, the vinyl ester did not penetrate the core during the VBRI process.

Following the VBRI process, the sandwich composites were allowed to gel and partially cure at 23 °C for 1 day and were then post-cured at 80 °C for 1 h. The glass and carbon fiber facesheets were about 2 mm thick each, which resulted in an overall thickness of about 10 mm for the sandwich composites.

2.2 *Explosive Blast Testing of Sandwich Composite Materials*

The blast response of the sandwich composites was performed using an experimental facility operated by the Defence Science and Technology Group (Australia) that consists of an enclosed steel plate-lined concrete chamber fitted with viewing windows to observe the explosion and the dynamic deformation of the targets (Fig. 2). The sandwich composite targets were flat 275 mm × 275 mm square panels held within a steel window frame having a 250 mm × 250 mm aperture. The frame was lined with soft rubber which allowed the target sandwich plate to bend freely under the pressure exerted by the shock wave without causing edge clamping damage.

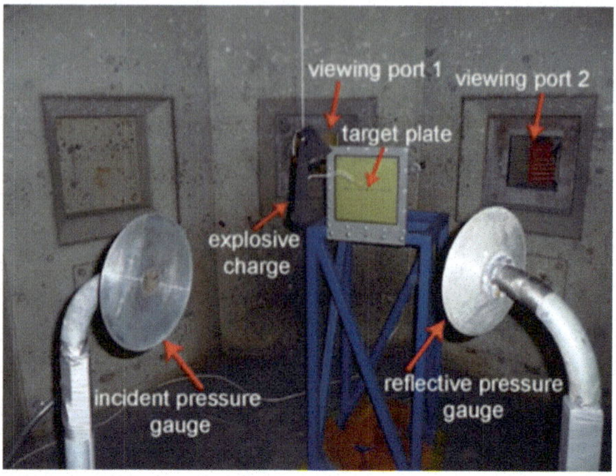

Fig. 2 Explosive blast test chamber

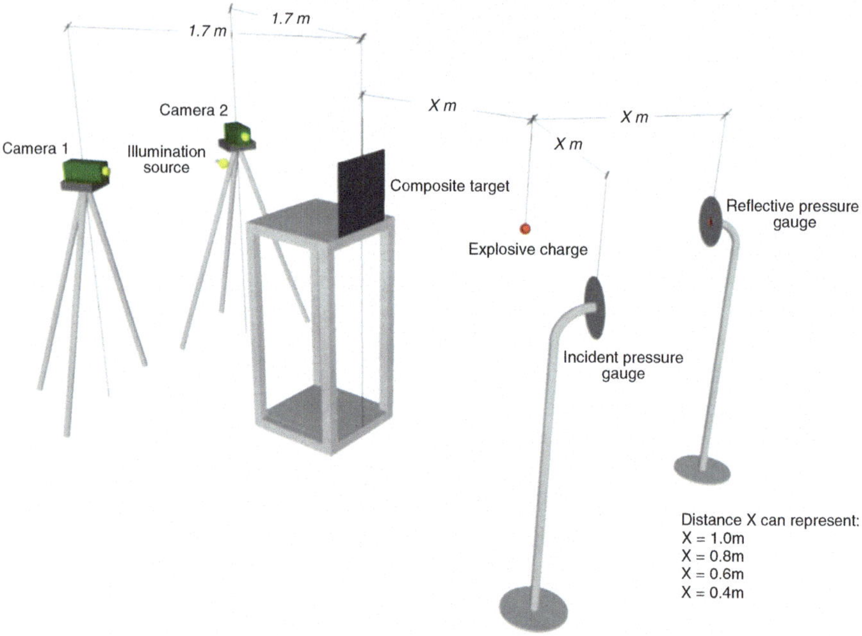

Fig. 3 Schematic representation of the explosive blast test setup

A schematic of the explosive blast experimental setup is shown in Fig. 3. The window frame holding the target panel was bolted to a 1.075 m high steel stand that was placed in the center of the chamber. The out-of-plane deformation of the sandwich panels was captured using high-speed photography and measured using 3D DIC software ARAMIS. Two Photron SA5 cameras were positioned 1.7 m behind the target, at an angle of 22.5° from the center of the panel. 32 mm thick blast and fragment resistant flexi-glass was used in the viewing ports to protect the cameras from damage. A PF300 slow peak flashbulb was used to illuminate the rear of the target panel to capture the deformation. During the explosive blast testing, the SA5 cameras were operated at a frame rate of 7000 per second capturing the out-of-plane deformation of the target. A third Photron SAZ camera was positioned 1.7 m at an angle of 67.5° from the center of the plate. The SAZ camera was operated at frame rates ranging between 12,000–30,000 per second, depending on the proximity of the charge to the target. Near-field blasts required higher camera frame rate to adequately capture the explosive-panel interactions. Time-lapse images of the explosive detonation, fireball and shock wave interaction with the sandwich plate for far- and near-field tests are shown in Figs. 4 and 5, respectively.

The blast was generated using a spherical plastic explosive Type 4 (PE4) charge made of RDX (cyclotrimethylenetrinitramine). The explosive was fired using a 3.8 g RP-80 EBW electric detonator. The peak overpressure and impulse of the shock wave was controlled by varying the explosive charge mass and stand-off distance from the sandwich panel. The overpressure-time response of the incident and reflected shock waves were measured using two free-field pressure transducers (Kulite XTL-190). The

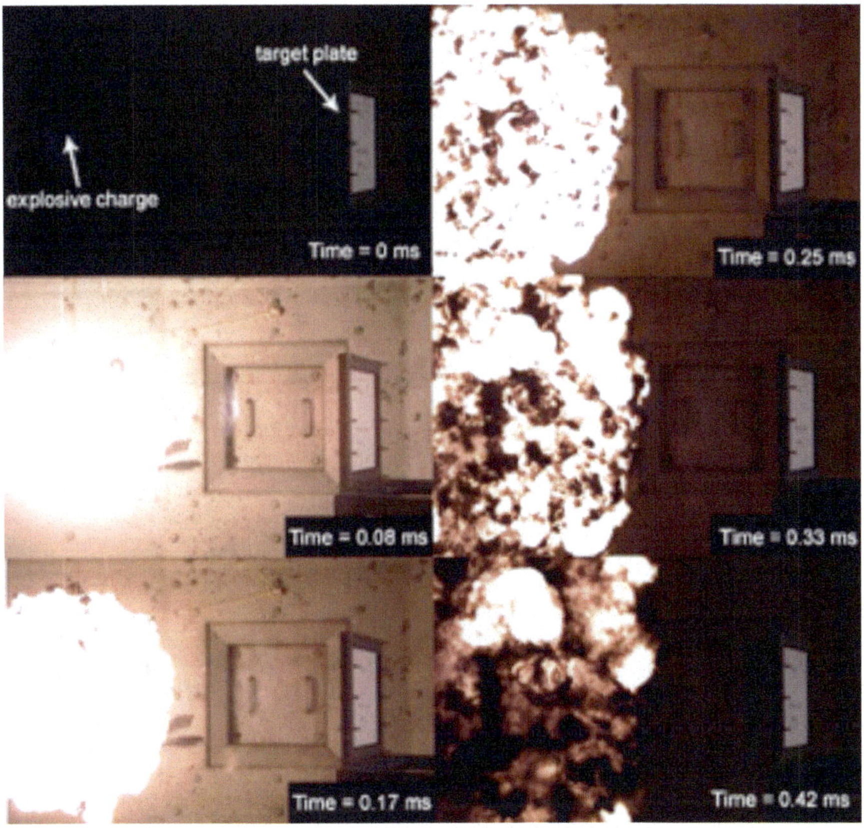

Fig. 4 Time-lapse photographs of a far-field explosive blast event. The times after detonation are given

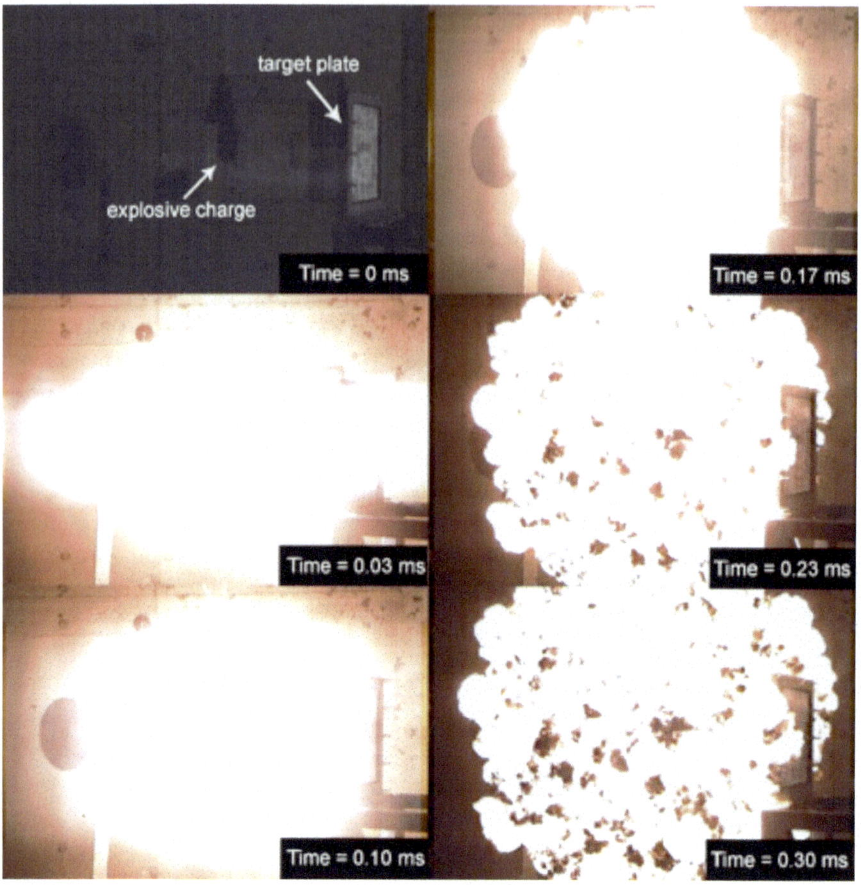

Fig. 5 Time-lapse photographs of a near-field explosive blast event. The times after detonation are given

near-field blast pressures and impulses were predicted using a two-dimensional, axially symmetric coupled Euler-Lagrangian model in ANSYS AutoDyn 16.2. The explosive blast overpressures and impulses used in this study are given in Table 1. One sandwich composite sample was tested at each blast test condition.

2.3 Mechanical Property Testing of Sandwich Composite Materials

The flexural properties were determined for the laminate facesheet materials. The stress-strain curves are shown in Fig. 6, and indicate that the response of both facesheet materials was linearly elastic followed by progressive failure at approximately 450 MPa and 650 MPa for the carbon and glass fiber laminates, respectively.

Table 1 Conditions used for explosive blast testing of the sandwich composites. The symbol * indicates the blast conditions when the peak pressure and impulse were calculated because they could not be precisely measured using pressure transducers

Explosive weight (g)	Stand-off distance (m)	Peak shock wave pressure (MPa)	Shock wave impulse (Pa.s)	Field condition
100	0.6	3.36	219	Far
100*	0.4	10.9	348	Near
160*	0.4	16.0	472	Near
200*	0.4	18.0	500	Near

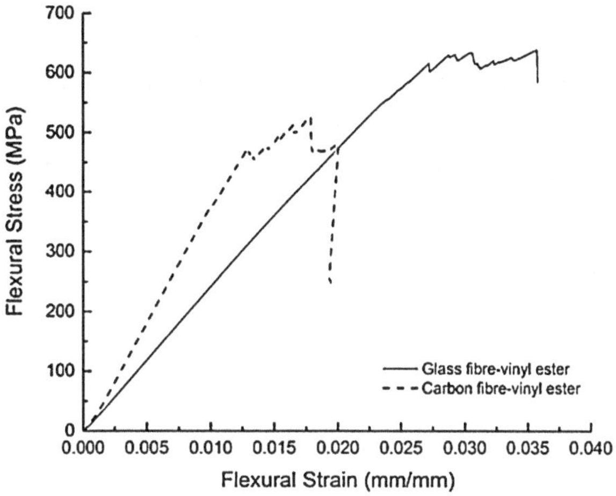

Fig. 6 Flexural stress-strain curves for the carbon- and glass-vinyl ester laminate facesheets

The flexural modulus of the carbon fiber facesheet was higher than the glass fiber facesheet, although the flexural failure stress, failure strain and strain energy density (area under the stress-strain curve) are substantially lower.

Three-point flexure tests were also performed on long, slender beams of the sandwich composites according to ASTM C393 [27] and D7249 [28] specifications. Bending tests were performed on 20 mm wide rectangular coupons of the sandwich materials. A span-to-thickness ratio of 16:1 was used, and therefore testing involved a support span of 160 mm and a total beam length of 200 mm. The coupons were loaded at a displacement rate of 5 mm/min until failure. The relation for the calculation of facesheet failure stress in three-point bending is given by:

$$\sigma_{facing} = \frac{P_{max}S}{4bt(d-t)} \tag{1}$$

where P_{max} is the maximum force, S is the support span length, t is the facesheet thickness, d is the sandwich thickness, and b is the specimen width. The core shear failure stress under three-point bending was calculated using:

$$\tau = \frac{P_{max}}{(d + c)b} \quad (2)$$

Flexural stress-strain curves measured for the sandwich materials are shown in Fig. 7. The balsa core sandwich composites exhibited linear elastic behavour up to a peak stress of ~120 MPa and ~170 MPa for the glass fiber- and carbon fiber-balsa sandwich composites, respectively. Damage initiated as core shear failure at the supports, resulting in a gradual reduction of the applied load. The shear cracking in the core propagated to the bond line, where the balsa core composite failed abruptly due to facesheet-core debonding. The PVC core sandwich composites initially displayed linear elastic behavour, followed by a large non-linear region due to progressive compression failure of the core. The PVC sandwich beams ultimately failed in compression of the facesheet under the central loading point. The test results in Table 2 show that the carbon-balsa sandwich composite had the highest face sheet

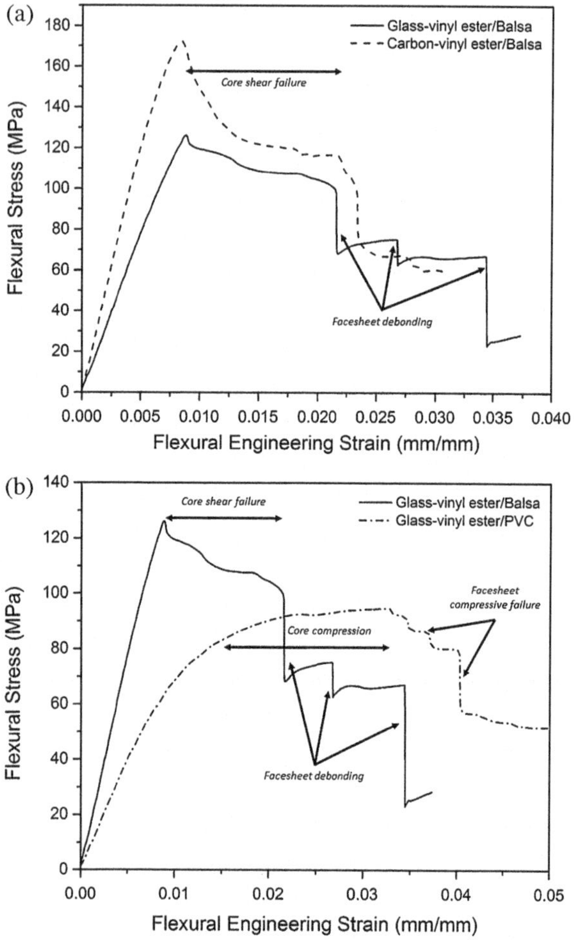

Fig. 7 Flexural stress-displacement curves for the (**a**) balsa-core and (**b**) glass fiber sandwich composites

and core shear strengths. In general, the flexural facing and core shear strength properties of the PVC foam sandwich composites were lower than the balsa core sandwich composites. Although, the highest flexural strain energy density (area under stress-strain curve) was measured for the glass fiber/PVC sandwich composite.

The mechanical properties measured for the sandwich composites are given in Table 2.

3 Results and Discussion

3.1 Blast-Induced Deformation of Sandwich Composite Materials

The dynamic deformation response of the sandwich composites when impulsively loaded by the shock wave and detonation products (in the case of near-field blast tests) was measured using the high-speed DIC technique. Typical DIC generated images showing the displacement and surface strain maps of the back surface of a target panel are shown in Fig. 8. In this case, the panel was tested using a 100 g PE4 explosive charge at 0.4 m stand-off distance, which generated a blast impulse of ~350 Pa.s. The images are shown directly after the explosive was detonated (0.1 ms), and at the point of maximum center-point deflection (2.1 ms). The impulse from the explosive detonation caused the panel to deform in the direction of the shock wave. As expected, the point of maximum deflection was in the center of the plate.

The center-point deflection versus time curves measured for the glass fiber- and carbon fiber-balsa sandwich composites are compared in Fig. 9. When subjected to a ~350 Pa.s impulse, the sandwich composites rapidly deformed to reach the maximum center-point deflection. The maximum positive displacements were similar for both materials, however the glass fiber composite deformed ~10% less than the carbon. Also, the maximum negative deformation was also less for the glass fiber composite. This was due to the development of front facesheet damage in the carbon panel, whereas no facesheet damage occurred in the glass panel. At a higher blast impulse (~470 Pa.s), the maximum center-point deflection of the carbon fiber-balsa sandwich composite was much higher (~50%) than the glass fiber-balsa composite. Also, the residual deformation of the carbon fiber-balsa sandwich composite was higher, which was due to the more severe damage experienced under blast loading (as described later). The same result was found for the glass fiber- and carbon fiber-PVC sandwich composites subjected to ~470 Pa.s shock wave impulse, as shown in Fig. 10.

The center-point deflection versus time curves of the carbon fiber-PVC and balsa core sandwich composites are compared in Fig. 11. When subjected to a ~350 Pa.s blast impulse, both carbon sandwich panels reached maximum out-of-plane displacement at approximately the same time (~1 ms post detonation). However, the

Table 2 Flexural properties of sandwich composites

Material	Facesheet flexural modulus (GPa)	Facesheet flexural strength (MPa)	Facesheet failure strain (%)	Sandwich bending strength (MPa)	Sandwich core shear strength (MPa)	Strain energy density (MJ/mm^3)
Glass-fiber/balsa	23.1 (±1.83)	647 (±32.9)	3.4 (±0.18)	104 (±12.1)	3.0 (±0.35)	0.46 (±0.09)
Glass-fiber/PVC	23.1 (±1.83)	647 (±32.9)	3.4 (±0.18)	96 (±2.9)	2.6 (±0.03)	2.26 (±0.08)
Carbon-fiber/balsa	42.5 (±2.05)	450 (±16.0)	1.5 (±0.04)	151 (±15.3)	3.5 (±0.38)	0.64 (±0.19)
Carbon-fiber/PVC	42.5 (±2.05)	450 (±16.0)	1.5 (±0.04)	124 (±4.7)	2.7 (±0.07)	1.32 (±0.15)

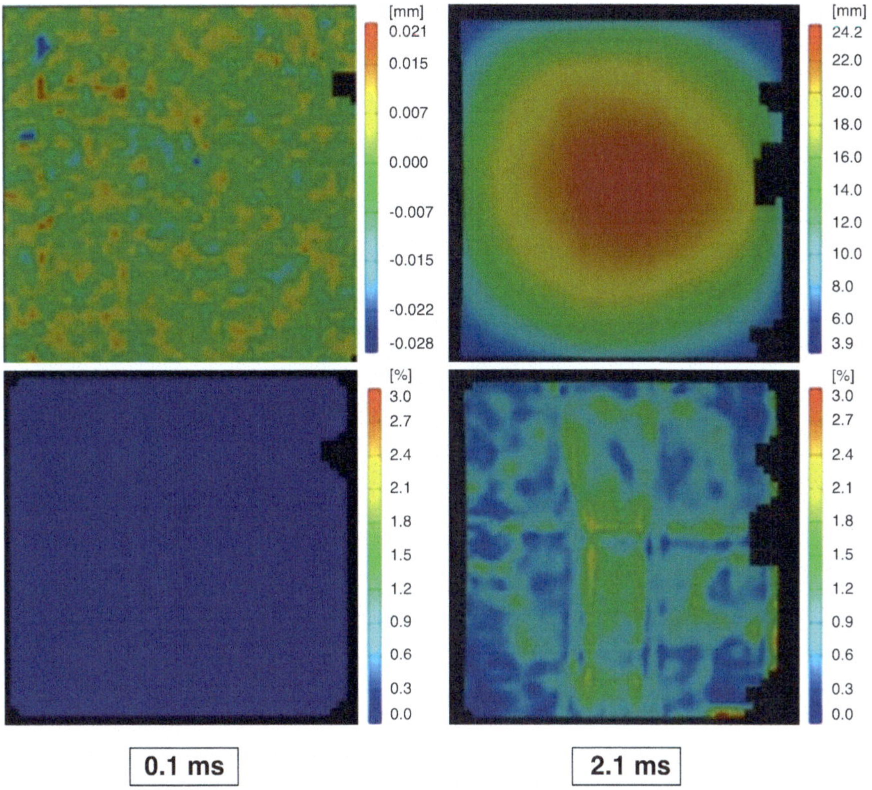

Fig. 8 DIC generated images for the carbon-balsa sandwich composite tested at a 353 Pa.s blast impulse. The times following detonation are given. The upper and lower images show the out-of-plane displacement and strain, respectively

maximum center-point displacement of the carbon fiber-balsa composite was lower than the carbon fiber-PVC composite. This is expected given the higher stiffness of the balsa core composite. Both panels suffered from front facesheet tearing at the edges, however this damage was more extensive in the PVC core composite. The PVC foam compressed under the force exerted by the shock wave, which limited the load transferred to the back facesheet. Similar behaviour was observed in the PVC foam composite under three-point bend loading (substantial core compression resulting in failure of the top facesheet). At a higher blast impulse (~470 Pa.s), the deflection response and maximum center-point displacement of the carbon balsa and PVC sandwich composites were similar. Despite the higher flexural stiffness of the balsa core panel, the amount of damage sustained by the core reduced the panels rigidity during deformation.

The center-point deflection versus time curves of the glass fiber-PVC and balsa core sandwich composites are compared in Fig. 12. When subjected to a 472 Pa.s blast impulse, both sandwich composites reached their maximum out-of-plane

Fig. 9 Comparison of the center-point deflection versus time histories for the glass fiber- and carbon fiber-balsa sandwich composites subjected to (**a**) 348 Pa.s and (**b**) 472 Pa.s blast impulse

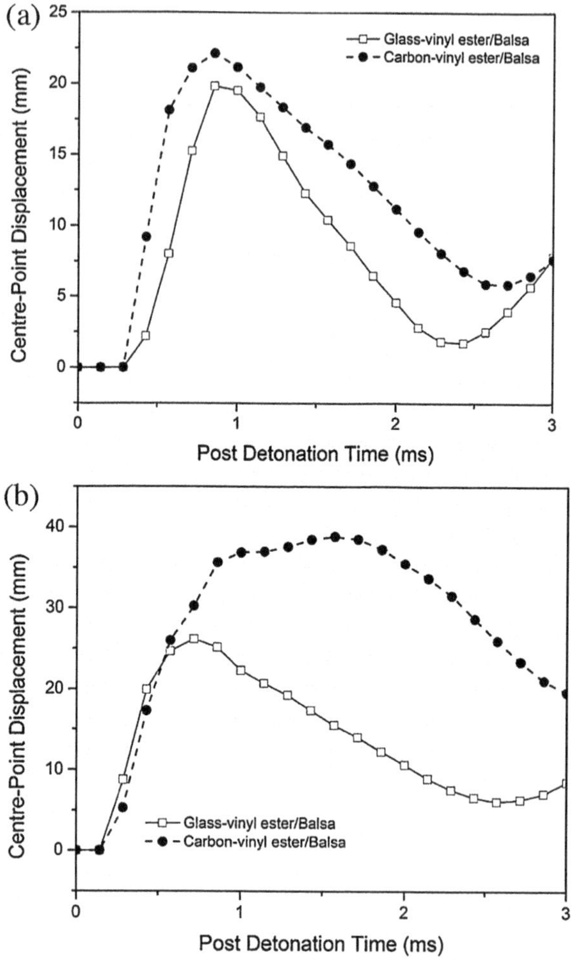

Fig. 10 Comparison of the center-point deflection versus time histories for the glass fiber- and carbon fiber-PVC composites subjected to 472 Pa.s blast impulse

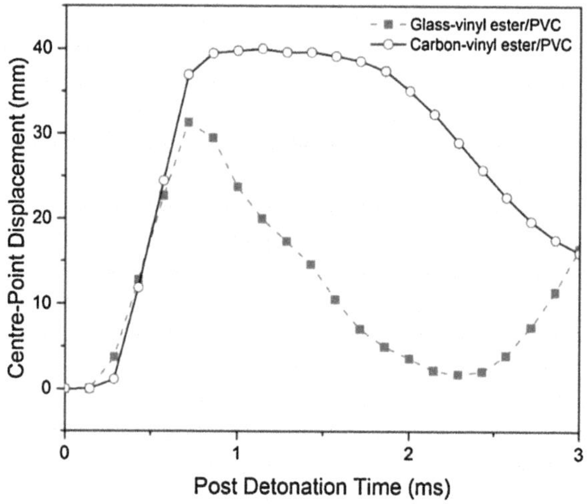

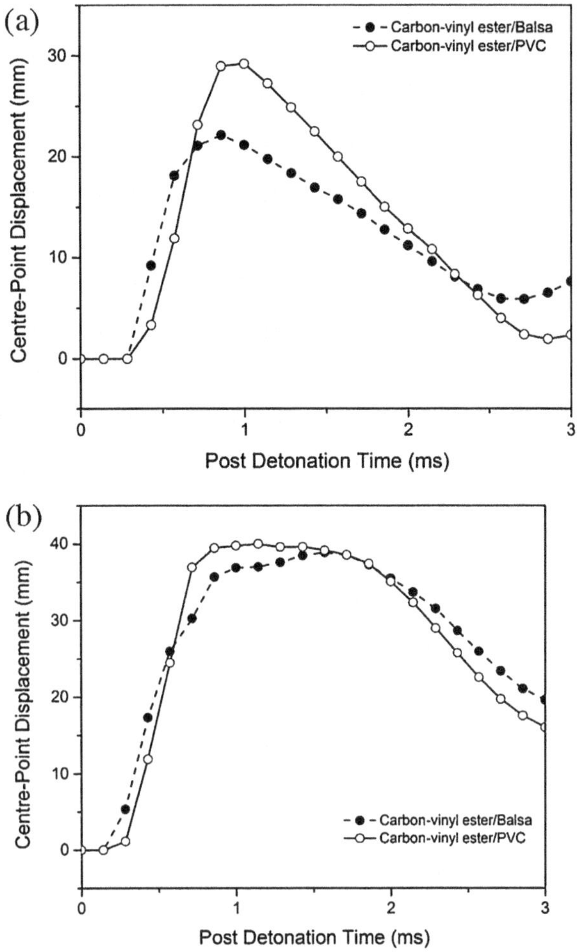

Fig. 11 Comparison of the center-point deflection versus time histories for the carbon fiber-balsa and PVC sandwich composites subjected to (**a**) 348 Pa.s and (**b**) 472 Pa.s blast impulse

deflection at approximately the same time (~0.7 ms post detonation). However, similar to that found for the carbon composites, the balsa-core sandwich composite had a lower maximum displacement than the PVC-core composite owing to its higher bending stiffness. Comparatively, at the highest shock wave impulse (500 Pa.s), both sandwich materials reached similar maximum out-of-plane deflections (~40 mm). The glass-balsa composite reached maximum out-of-plane deflection 1.7 ms post detonation, which was later than the glass-PVC composite (~0.86 ms). Also, the glass-balsa composite failed to return to its initial starting position, both of which were representative of extensive damage and permanent deformation.

The effect of increasing blast impulse loading on the maximum center-point displacements for the different sandwich panels is shown in Fig. 13. The maximum center point deflection of the carbon fiber sandwich composites was greater than

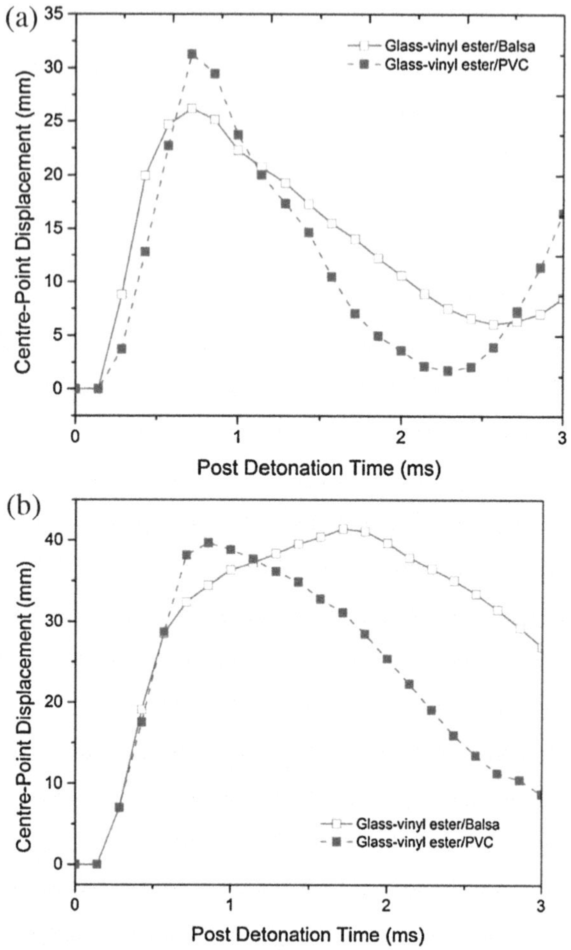

Fig. 12 Comparison of the center-point deflection versus time histories for the glass fiber-balsa and PVC sandwich composites subjected to (**a**) 472 Pa.s and (**b**) 500 Pa.s blast impulse

the glass fiber composites. Also, up to the blast impulses of 353 Pa.s and 472 Pa.s for the carbon and glass, respectively, the maximum out-of-plane displacement of the PVC core composites was higher than the balsa core composites. When subjected to greater respective blast impulses, the maximum displacements of the balsa and PVC core composites were similar due to the extensive damage to the balsa core.

3.2 *Blast-Induced Damage to Sandwich Composite Materials*

Following blast testing, the sandwich composites were examined using visual inspection, through-transmission ultrasonics and X-ray computed tomography to

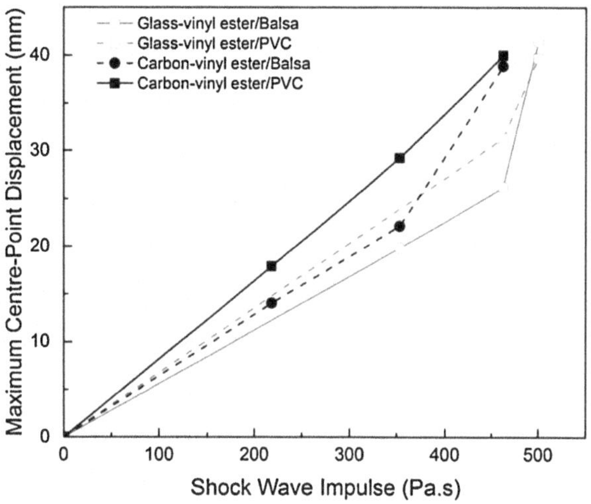

Fig. 13 Effect of blast impulse on the maximum center-point displacement of the sandwich composites

determine the amount and types of damage. A front-view (blast side) comparison of the sandwich panels for all blast test conditions is shown in Fig. 14. The regions highlighted with red denote the location of through-thickness rupture of the front facesheet. Front facesheet rupture in the carbon sandwich composites initiated at a lower blast impulse compared to the glass sandwich panels. This was consistent with the recent study performed by Gargano et al. [29] which revealed that fiberglass laminates are more blast resistant than carbon fiber laminates. Therefore, for the sandwich composites, the higher flexural failure strain and flexural strain energy density of the glass fiber facesheet (given in Table 2) resulted in the fiberglass laminate facesheets withstanding larger out-of-plane displacements without damaging. The results in Fig. 14 also show that facesheet damage was influenced by core type, with the PVC-core composites having greater amounts of front facesheet cracking than the balsa-core composites for a given blast impulse.

Figure 15 presents C-scan ultrasound images of the sandwich composites after blast testing. The images show the facesheet-core interface on the side that was exposed to the explosive blast. The dark-blue 'sparkle' region in the C-scan images for the balsa core represent regions where the facesheet has debonded, which is a common failure mode of sandwich composites exposed to shock wave loading [7–9, 14, 22]. The effect of increasing blast impulse loading on the amount of front facesheet-core debonding damage sustained by the balsa-core composites is summarised in Table 3. The percentage debonded area was measured from the C-scan ultrasound images shown in Fig. 15, and was defined as a percentage of the total surface area of the sandwich target. The C-scan images were analysed via a colour contrast technique using Adobe Photoshop CC software to determine the facesheet-core debonding area. After testing at the 219 Pa.s blast impulse, the majority of the front facesheet-core interface of the carbon fiber-balsa composite was undamaged. At a blast impulse of ~350 Pa.s, both the glass- and carbon-balsa composites exhibited substantial front facesheet-core debonding, however it was more extensive in the carbon-balsa composite than in the glass-balsa composite.

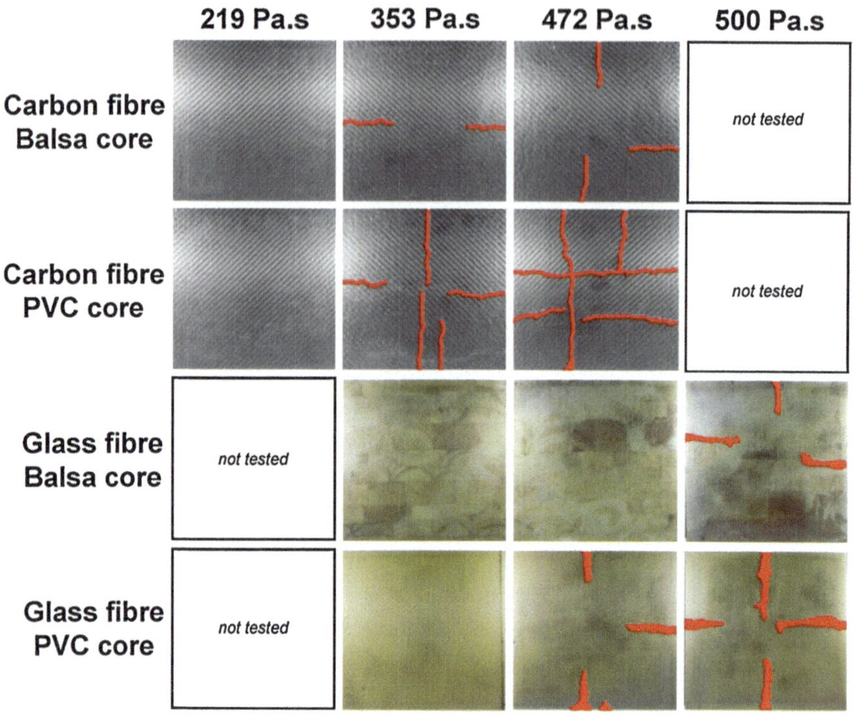

Fig. 14 Comparison of blast-induced surface damage to the sandwich composites. Note the red markings represent front facesheet cracking. The images show the front facesheet of the sandwich panels

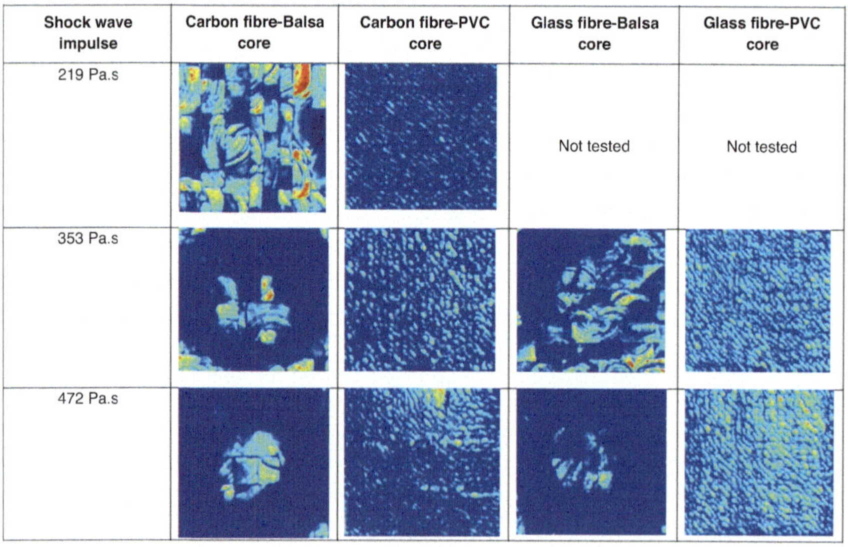

Fig. 15 Effect of increasing blast impulse on ultrasonic C-scan images for the sandwich composites. The images show the front skin-core interface

Table 3 Effect of increasing blast impulse loading on the debonding damage area to the balsa-core sandwich composites

Shock wave impulse	Carbon fiber – balsa core	Glass fiber – balsa core
219 Pa.s	3.4%	Not tested
353 Pa.s	73.6%	51.6%
472 Pa.s	81.7%	83.9%
500 Pa.s	Not tested	75.5%

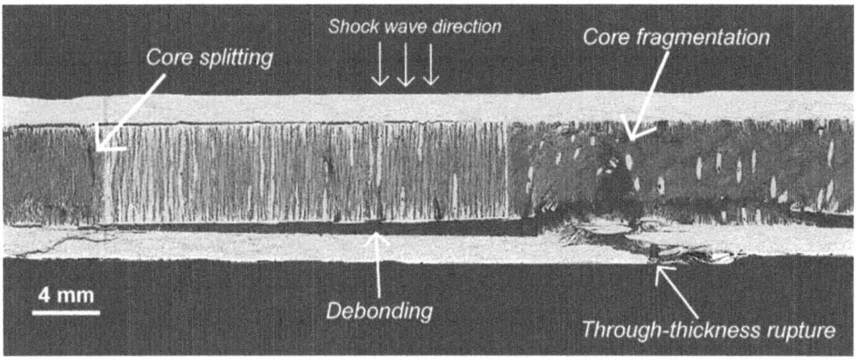

(a)

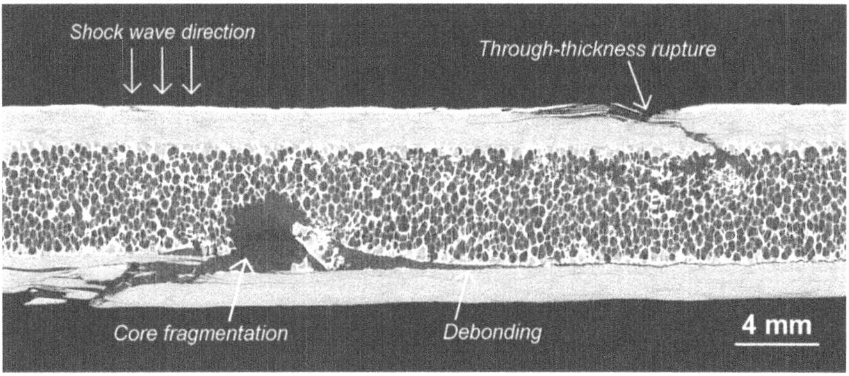

(b)

Fig. 16 Side-on X-ray computed tomography images of the damage to the (**a**) carbon-balsa and (**b**) carbon-PVC sandwich composites subjected to 472 Pa.s blast impulse

Damage types such as front and rear facesheet fiber failure, longitudinal shear failure of the balsa core and cracking and splitting of the balsa were found in both the carbon- and glass-balsa sandwich composites (as shown for example in Fig. 16(a)). However, the magnitude of the blast impulse needed to cause these damage types in the glass-balsa composite was higher than the carbon-balsa composite. Similar observations were made by Rolfe et al. [10], who found that a carbon fiber sandwich panel had substantially greater amounts of debonding damage compared to the glass fiber sandwich panel when subjected to the same explosive blast load.

For the carbon-balsa composite subjected to a 353 Pa.s blast impulse, the center region remained bonded to the front facesheet. This was believed to be due to compaction of the front facesheet and core when impulse loaded by the shock wave. Similar damage patterns have been observed in carbon and glass fiber sandwich panels using X-ray computed tomography; extensive debonding around the edges with core crushing and compaction in the center [10]. Andrews et al. [30] calculated failure mode maps for sandwich panels with PVC and balsa wood cores, whereby based on the core type, core density and facesheet thickness-length ratio the expected damage modes were calculated (either face sheet wrinkling, face sheet failure or core shear failure). For a core density of 130–150 kg/m^3 and a facesheet thickness-length ratio of 0.007, it would be expected that balsa core composites fail by core shear at the supports when subjected to air blast loading. Core shear failure would then propagate to facesheet-core debonding once the cracks reached the bondline, as found previously from the three-point bend flexural tests. This failure map confirms the debonding pattern observed around the edges of the blast loaded balsa core panels.

The same damage pattern (debonding around the boundaries with a center region intact) was observed for the carbon- and glass-balsa composites subjected to a 472 Pa.s blast impulse. Also, at this shock wave impulse, the amount of facesheet-core debonding was similar for both the glass- and carbon-balsa sandwich composites. Unexpectedly, at the highest blast impulse (500 Pa.s), the debonding area in the glass-balsa composite panel was less than that of glass-balsa panel tested at 472 Pa.s blast impulse. It was possible that the region of compaction could increase with larger charge sizes due to the larger shock wave radius generated from a bigger charge. However, this is difficult to confirm considering only one panel was tested per explosive blast test condition. Further experimental tests would be needed to confirm this finding.

From the ultrasound images shown in Fig. 15, damage to the facesheet-core interface from explosive blast loading could not be determined for the PVC composites. The cells of the PVC foam are air-filled, which has a similar acoustic impedance value as debonding damage thereby making it difficult to isolate the debonding damage from the air in the PVC cells. Damage such as front and rear facesheet fiber failure, facesheet delamination, longitudinal shear and fragmentation failure of the core were instead observed using X-ray computed tomography, as shown in Fig. 16. Similar damage modes were observed by Langdon et al. [7, 14], who subjected glass fiber-PVC sandwich composites to localised air blast loading. The damage types occurred regardless of the type of fiber reinforcement (glass or carbon). However, the blast impulse to initiate damage in the carbon-PVC composite was lower than the glass-PVC, similar to the results for the balsa-core composites. This was due to the glass fibers having greater energy absorption capacity than the carbon fibers, as mentioned previously.

Figure 17 shows X-ray computed tomography images taken at different through-thickness locations of the carbon-balsa and carbon-PVC sandwich composite plates after loaded by a 472 Pa.s shock wave impulse. The dimensions below each image represent the through-thickness distance with respect to the top (blast side) surface (i.e. 0 mm is the top surface, 5 mm is directly in the middle and 10 mm is the rear

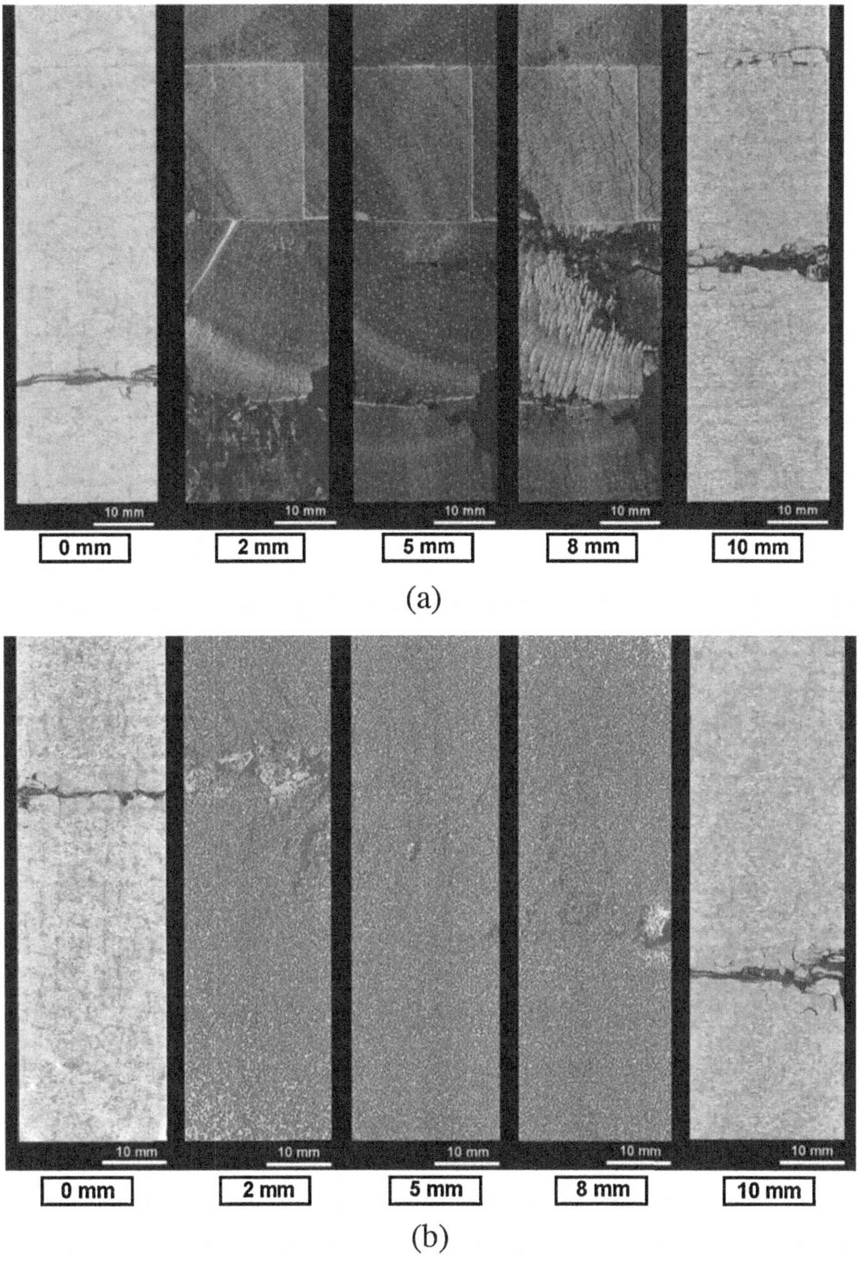

Fig. 17 Through thickness X-ray computed tomography images of (**a**) carbon-balsa and (**b**) carbon-PVC sandwich composites subjected to 472 Pa.s blast impulse. The numbers below each image represent the distance from the top surface with 0 mm, 5 mm and 10 mm for the top, middle and bottom of the panel

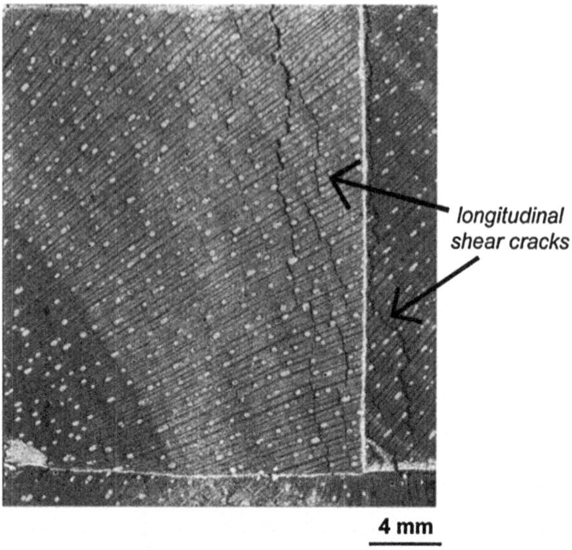

Fig. 18 X-ray computed tomography image showing longitudinal shear cracking between the balsa wood grains. This image is taken for carbon fiber-balsa sandwich composite after subjected to a 472 Pa.s blast impulse

surface). Both materials experienced fiber rupture on the front and rear facesheets. However, damage sustained by the balsa core was significantly greater than the PVC core. Large portions of the balsa core were fragmented throughout the core thickness, and thin cracks had propagated perpendicular to the grain direction, as shown in Fig. 18. Under the same blast impulse, the PVC-core in the carbon-PVC sandwich composite sustained localised core fragmentation. The greater damage in the balsa core composites was believed to be due to the balsa core's lower shear ductility compared to the PVC, which has been previously observed to be a weakness in balsa sandwich composite design when subjected to blast loads [11]. Another consideration is that the greater amount of damage in the balsa core composites was due to the weaker face/core interfacial bond strength. Truxel et al. [31] compared the face/core fracture resistance of sandwich panels with glass fiber-vinyl ester facesheets with either end-grain balsa or PVC foam cores using various surface preparation techniques. They found that when specimens were prepared with no infusion aid (i.e. no grooves or interface mat) the glass fiber-PVC face/core interface was ~50% higher than the glass fiber-balsa face/core bond strength.

Following post-blast damage assessment, the residual flexural properties of the sandwich composites were measured, and the results are shown in Figs. 19 and 20. Five flexural coupons were cut from each blast panel. The coupons were cut from the same locations across all blast panels to ensure the strength retention comparisons between materials was consistent. When subjected to the lowest intensity blast (219 Pa.s), the flexural ultimate facing stress decreased by 13% and 10% for the carbon-balsa and carbon-PVC composites, respectively. Similar reductions to the core shear strength were also recorded. The reduction in strength was expected to be due to fiber-matrix interfacial cracking, transverse inter-tow cracking and matrix

Fig. 19 Post-blast (**a**) facing strength and (**b**) core shear strength of the carbon fiber-balsa and carbon fiber-PVC sandwich composites. The error bars represent one standard deviation

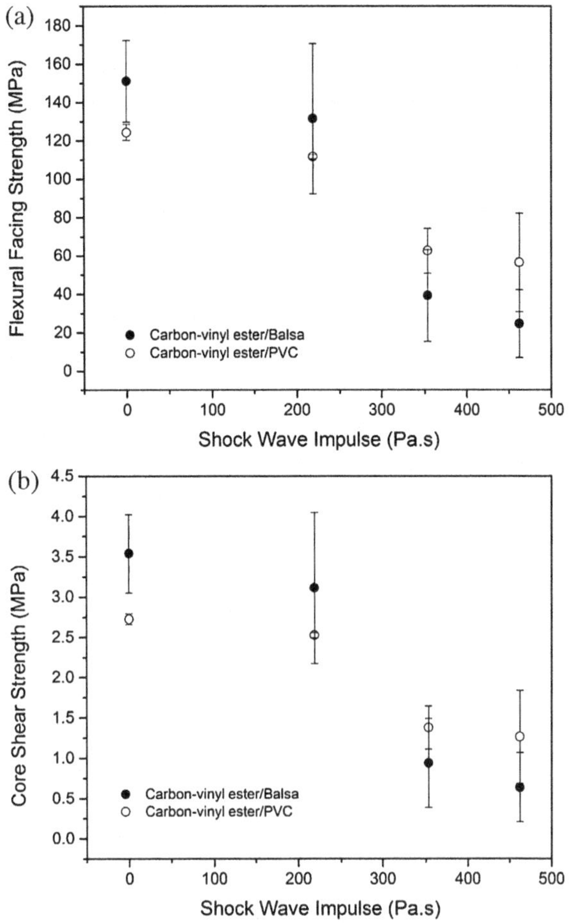

cracking in the carbon facesheets, as previously found to occur in carbon laminates subjected to 219 Pa.s blast impulse load [29]. The residual flexural properties of the carbon sandwich composites were significantly reduced after loaded by a 353 Pa.s shock wave impulse. The facing strength and core shear strength were both reduced, on average, by 75% and 50% for the carbon-balsa and carbon-PVC, respectively. However, the magnitude of the strength properties of both carbon sandwich composites is similar, and statistically there is no difference between in the structural integrity of PVC and balsa core carbon composites when subjected to any shock wave impulse tested in this study. At the highest blast impulse tested on the carbon composite (472 Pa.s), the flexural properties were similar to those after the 353 Pa.s blast impulse.

For the lowest blast impulse tested on the glass fiber sandwich composites (353 Pa.s), the ultimate facing stress decreased by 63% for the glass-balsa composite, whereas there was no change for the glass-PVC composite. Similar results were

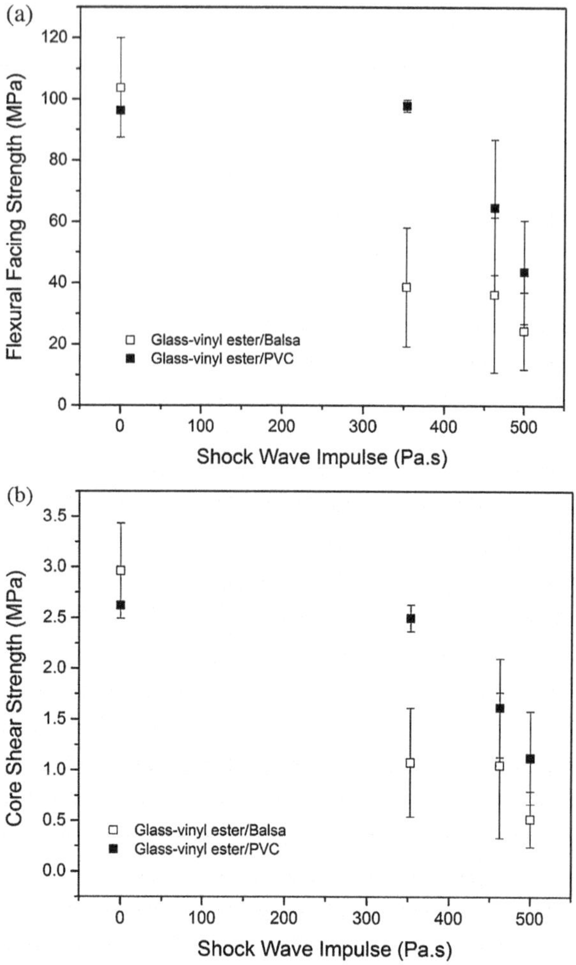

Fig. 20 Post-blast (**a**) facing strength and (**b**) core shear strength of the glass fiber-balsa and glass fiber-PVC sandwich composites. The error bars represent one standard deviation

observed for the ultimate core shear strength, with a 64% and 5% decrease for the glass-balsa and glass-PVC composites, respectively. The reduction in ultimate facing stress and core shear strength for the glass-balsa composite was due to significant damage sustained to the balsa core as mentioned previously. Similar observations have been reported by Tagarielli et al. [11], in that the sandwich beams with end-grain balsa wood cores failed at a lower blast impulse than the PVC core composite, despite being the better performing sandwich construction at low projectile impulses. The balsa-core composites suffered less front facesheet compressive failure then the PVC core composites, yet much more extensive core damage. The post-blast flexural results show that the core damage in the balsa-core composites was much more detrimental to the structural integrity of the sandwich construction than the front facesheet damage. However, when the shock wave impulse is high enough (>450 Pa.s), there is no significant difference in the post-blast flexural strength properties of the glass-fiber PVC and balsa core composites.

4 Conclusion

The explosive blast resistance of sandwich composites used in naval ship structures is influenced by both the type of fiber reinforcement used in the laminate facesheets and the type of core material. For the explosive blast test conditions used in this work, the resistance to out-of-plane deformation was higher for the glass fiber sandwich panels than the carbon fiber sandwich panels, and the balsa core composites were superior to the PVC core composites at low blast impulses. The blast impulse required to initiate front facesheet failure in the PVC composites was lower than the balsa composites. However, above a threshold impulse of ~350 Pa.s, the resistance of the balsa core sandwich composites to out-of-plane deformation was heavily reduced due to extensive core cracking and splitting, which developed into facesheet-core debonding over the majority of the panel. This resulted in large reductions to the flexural facing and core shear strengths of the balsa core composites. Also, the reduction in flexural properties was similar between glass fiber and carbon fiber balsa core sandwich composites as the threshold impulse to cause extensive core damage and facesheet-core debonding was the same (353 Pa.s). However, the impulse required to initiate damage in the carbon sandwich composites was lower than the glass sandwich composites, which was due to the lower energy absorption capacity of the former.

Acknowledgments The research was supported by the United States Office of Naval Research (ONR) and Office of Naval Research Global (ONRG) under the direction of Dr. Y.S. Rajapakse (N62909-15-1-2000). The support of the Defence Science Technology & Group (DSTG), Australia is also gratefully acknowledged, including the technical assistance provided by V. Pickerd, T. Delaney, P. McCarthy, M. Ibrahim, Z. Mathys and W. Reid. The authors also thank P. Tkatchyk and R. Ryan at RMIT University for technical assistance.

References

1. Wang E, Gardner N, Shukla A (2009) The blast resistance of sandwich composites with stepwise graded cores. Int J Solids Struct 46(18–19):3492–3502
2. Tekalur SA, Shukla A. (2010) Failure of polymer-based sandwich composites under shock loading. In: Shukla A, Ravichandran G, Rajapakse YDS (eds) Dynamic failure of materials and structures. Springer, New York, pp 235–268
3. Tekalur SA, Bogdanovich AE, Shukla A (2009) Shock loading response of sandwich panels with 3-D woven E-glass composite skins and stitched foam core. Compos Sci Tech 69 (6):736–753
4. Gardner N, Wang E, Shukla A (2012) Performance of functionally graded sandwich composite beams under shock wave loading. Compos Struct 94(5):1755–1770
5. Gardner N, Gupta S, Wang E, Shukla A (2014) Blast response of sandwich composites: effect of core gradation, pre-loading, and temperature. In: Shukla A, Rajapakse YDS, Ellen M (eds) Blast mitigation: experimental and numerical studies. Springer, New York, pp 279–330
6. Langdon GS, Chi Y, Nurick GN, Haupt P (2009) Response of GLARE© panels to blast loading. Eng Struct 31(12):3116–3120
7. Arora H, Hooper PA, Dear JP (2011) Dynamic response of full-scale sandwich composite structures subject to air-blast loading. Compos A Appl Sci Manuf 42(11):1651–1662

8. Arora H, Hooper PA, Dear JP (2012) The effects of air and underwater blast on composite sandwich panels and tubular laminate structures. Exp Mech 52(1):59–81
9. Rolfe E, Kelly M, Arora H, Hooper PA, Dear JP (2017) Failure analysis using X-ray computed tomography of composite sandwich panels subjected to full-scale blast loading. Compos Part B Eng 129:26–40
10. Tagarielli VL, Deshpande VS, Fleck NA (2007) The dynamic response of composite sandwich beams to transverse impact. Int J Solids Struct 44(7):2442–2457
11. Tagarielli VL, Deshpande VS, Fleck NA (2010) Prediction of the dynamic response of composite sandwich beams under shock loading. Int J Impact Eng 37(7):854–864
12. Karagiozova D, Nurick GN, Langdon GS, Yuen SCK, Chi Y, Bartle S (2009) Response of flexible sandwich-type panels to blast loading. Comp Sci Tech 69(6):754–763
13. Langdon GS, von Klemperer CJ, Rowland BK, Nurick GN (2012) The response of sandwich structures with composite face sheets and polymer foam cores to air-blast loading: preliminary experiments. Eng Struct 36:104–112
14. Hassan MZ, Guan ZW, Cantwell WJ, Langdon GS, Nurick GN (2012) The influence of core density on the blast resistance of foam-based sandwich structures. Int J Impact Eng 50:9–16
15. Fatt MSH, Palla L (2009) Analytical modeling of composite sandwich panels under blast loads. J Sandw Struct Mater 11(4):357–380
16. Karagiozova D, Nurick GN, Langdon GS (2009) Behavour of sandwich panels subject to intense air blasts – Part 2: numerical simulation. Compos Struct 91(4):442–450
17. Hua Y, Akula PK, Gu L (2014) Experimental and numerical investigation of carbon fiber sandwich panels subjected to blast loading. Compos Part B Eng 56:456–463
18. Xue Z, Hutchinson JW (2004) A comparative study of impulse-resistant metal sandwich plates. Int J Impact Eng 30(10):1283–1305
19. Ahmed S, Galal K (2017) Effectiveness of FRP sandwich panels for blast resistance. Compos Struct 163:454–464
20. Ye N, Zhang W, Li D, Huang W, Xie W, Huang X, Jiang X (2017) Dynamic response and failure of sandwich plates with PVC foam core subjected to impulsive loading. Int J Impact Eng 109:121–130
21. Chen A, Kim H, Asaro RJ, Bezares J (2011) Non-explosive simulated blast loading of balsa core sandwich composite beams. Compos Struct 93(11):2768–2784
22. Rolfe E, Kelly M, Arora H, Hooper PA, Dear JP, (2016) X-ray CT analysis after blast of composite sandwich panels. In International symposium on dynamic response and failure of composite materials, Elsevier Ltd. Naples, Italy
23. Wang E, Shukla A (2011) The blast response of sandwich composites with in-plane pre-loading. In SEM annual conference on experimental and applied mechanics. Indianapolis
24. Jackson M, Shukla A (2011) Performance of sandwich composites subjected to sequential impact and air blast loading. Compos Part B 42(2):155–166
25. Arora H, Hooper PA, Dear JP (2013) Blast loading of sandwich structures and composite tubes. In: Abrate S, Castanie B, Rajapakse YDS (eds) Dynamic failure of composite and Sandwich structures. Springer, Dordrecht, pp 47–92
26. Arora H, Del Linz P, Dear JP (2017) Damage and deformation in composite sandwich panels exposed to multiple and single explosive blasts. Int J Impact Eng 104:95–106
27. ASTMC393 (2016) Standard test methof for core shear properties of sandwich constructions by beam flexure. ASTM International, West Conshohocken
28. D7249 (2018) Standard test method for facesheet properties of sandwich constuctions by long beam flexure. ASTM International, West Conshohocken
29. Gargano A, Pingkarawat K, Blacklock M, Pickerd VK, Mouritz AP (2017) Comparative assessment of the explosive blast performance of carbon and glass fibre-polymer composites used in naval ship structures. Compos Struct 171:306–316
30. Andrews EW, Moussa NA (2009) Failure mode maps for composite sandwich panels subjected to air blast loading. Int J Impact Eng 36(3):418–425
31. Truxel A, Aviles F, Carlsson L, Grenestedt J, Millay K (2006) Influence of face/core interface on debond toughness of foam and balsa cored sandwich. J Sandw Struct Mater 8(3):237–258

Dynamic Response of Polymers Subjected to Underwater Shock Loading or Direct Impact

Veronica Eliasson and Rodrigo Chavez

1 Introduction

Polymeric materials are often utilized in naval and aircraft applications as the matrix component of composite materials. This means they are exposed to outdoor environments, in which conditions may vary significantly and material properties could change significantly with time. In order to properly design for and use these materials in such changing environments, it is important to understand how environmental conditions affect their performance under different types of loading conditions — with particular interest of loading rate dependence. Many uses for these materials may involve exposure to high loading rates in liquid environments, such as in seawater for naval purposes, or in varying humidity levels, such as an aircraft traveling through various climates thousands of times during its life cycle. Therefore, because the performance of many practical applications depends on our knowledge of the behavior of polymers under high strain rate loading conditions and hostile environments, it is important to study these scenarios in detail.

The use of composite materials is steadily increasing in all kinds of naval applications. Any changes in subjected loading amplitude and loading rate can affect the properties of the polymer matrix, which could in turn influence the fracture toughness of the composite. Thus, the mechanical properties and rate sensitivity of the polymer matrix directly determine the rate sensitivities of the polymer composite [1]. Furthermore, it is of interest to understand the difference in dynamic response when the load is applied as a transient load resulting in stress waves propagating back and forth in the sample leading to influence of sample geometry and boundary conditions. One might ask if this is relevant to real world naval structures? The answer is yes. Naval vessels have parts with complex geometries, for example

V. Eliasson (✉) · R. Chavez
University of California San Diego, La Jolla, CA, USA
e-mail: eliasson@ucsd.edu; rchavezm@eng.ucsd.edu

S. W. Lee (ed.), *Advances in Thick Section Composite and Sandwich Structures*,
https://doi.org/10.1007/978-3-030-31065-3_5

convergent sections, such as rudder–hull junctions, propeller shafts, and bow thrusters.

Now, assume that a nearby underwater explosion occurs. Sometime later, the resulting shock wave will have reached the convergent section and begins to interact with the structure as it propagates into the convergent section. Depending on the material of the structure, the dynamic response may differ significantly. Further, the pressure increase that the structure will experience is much more severe — the increase is on the order of 10^4 — in an underwater explosion than in an air explosion simply due to the water having a higher density and speed of sound than the air.

Hence, it is important that both the matrix component materials and the composite materials are well understood and characterized for the highly dynamic loading conditions to which they can be subjected.

2 Underwater Explosions and Shock Wave Focusing in Convergent Structures

To study in detail the fundamental nature of shock wave focusing occurring in liquid media and the coupling to wave propagation in the surrounding solid structures, an underwater shock focusing laboratory experiment was developed, see a detailed description in Eliasson et al. [2]. The specimen consisted of a water-filled convergent section that was sandwiched between two windows with a square-shaped piston that was used to seal-off the entrance to the water chamber and provide a point of impact for the gas gun projectile. A schematic representation of the specimen assembly is shown in Fig. 1.

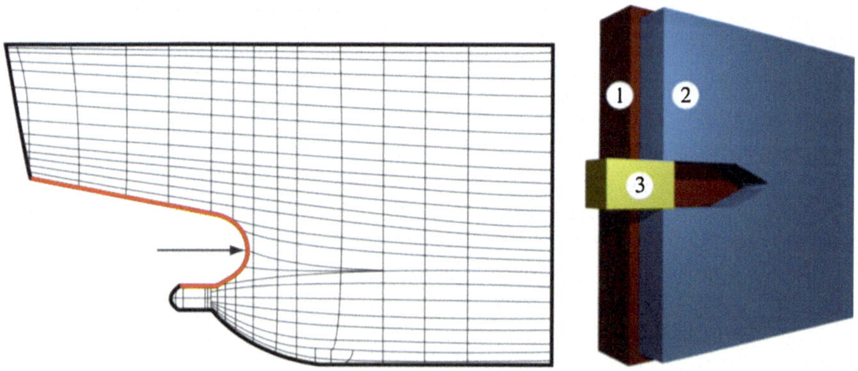

Fig. 1 Example of a convergent section on a naval structure (left). A schematic of the sandwich structure with a water-filled convergent section (right): (1) polycarbonate window, (2) core material with water-filled convergent wedge, and (3) piston that the gas gun projectile impacts upon. From [2]

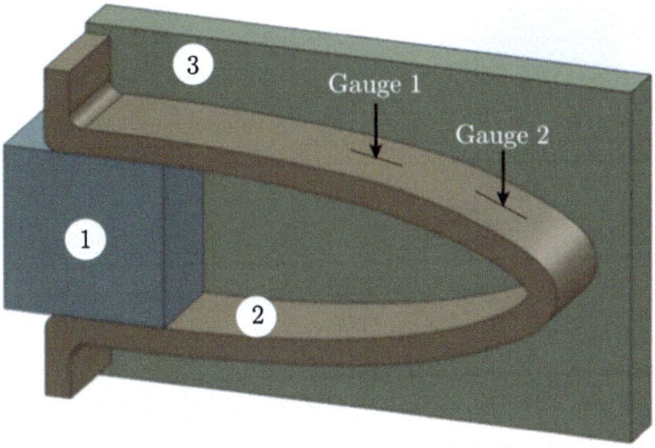

Fig. 2 Sketch showing a thin-walled convergent structure for underwater shock loading. The sample consists of a (1) piston; (2) test specimen; and (3) polycarbonate window. From [3]

In subsequent experiments different types of core materials were used (aluminum, polycarbonate, solithane, carbon fiber vinyl ester composite, and steel) with different shaped convergent sections (wedge and logarithmic spiral). Further, the core material was changed from the large square bulk material (as depicted in Fig. 1) to a much thinner structure (see Fig. 2) that allowed for a large number of internal wave interactions with the outside boundary.

Results showed that the fluid-structure interaction highly depended on the type of material of the convergent structure and its geometry and size.

Experiments using a polycarbonate core material showed simultaneous shock waves in air and water alongside stress waves in the core material, Fig. 3. The impact between the projectile and the piston generated a toroidal shock wave that could be used as a time measure of when the impact happened compared to when the actual schlieren photograph was captured. The impact generated a shock wave in the water-filled region, which in turn produced longitudinal and transverse waves in the polycarbonate core material. Other experiments were performed in which the wave speeds (longitudinal and transverse) of the core materials were either both faster than or slower than the shock wave speed.

By changing the core material from polycarbonate to aluminum, the schlieren photographs of the shock focusing event reveals a strong fluid-structure interaction. The stress waves that propagate faster than the shock wave produce disturbances that feed back into the water-filled region. Thus, the incident shock wave no longer propagates into an undisturbed fluid, but instead it encounters a fluid containing oscillating cavitation bubbles, see Fig. 4. Further details are published in reference [2]. Similar experiments were reported in reference [4] but instead of a wedge-shaped concave water-filled section it was shaped as a logarithmic spiral to avoid shock reflections off the edges of the cavity. The logarithmic spiral shape was used also for thin walled specimens, Fig. 5, and in this case the precursor waves caused by a strong fluid-structure interaction are even more pronounced [3]. Numerical simulations indicated that flexural waves in the thin-walled structure was the source of the precursor waves.

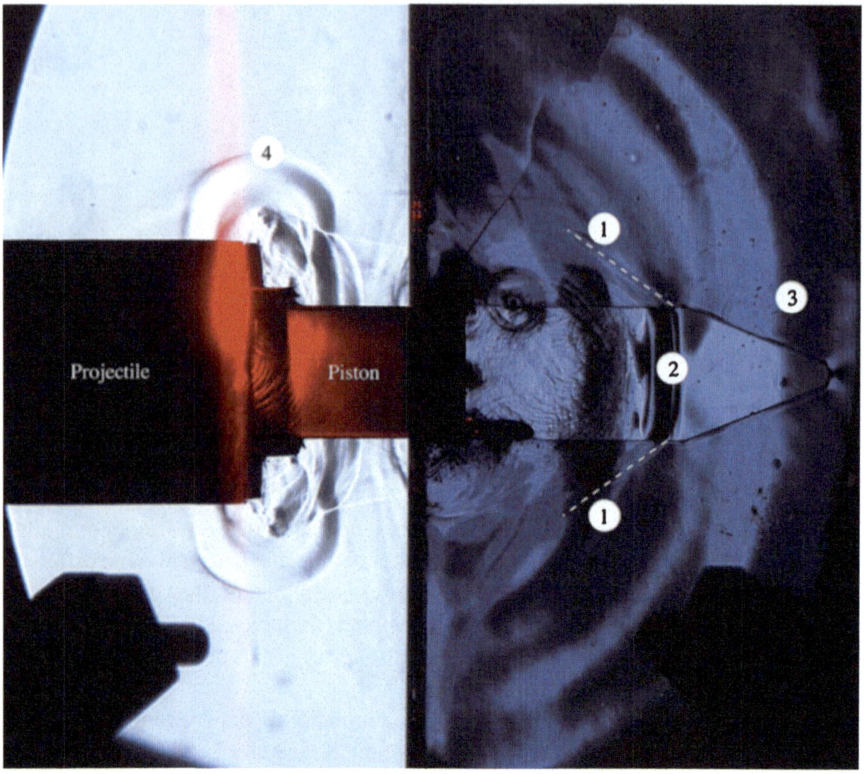

Fig. 3 A schlieren photograph showing the resulting shock waves and stress waves in a polycarbonate core material. The impact velocity of the projectile was 57 m/s. The waves are (1) Oblique shear wave inside the polycarbonate samples, (2) shock wave in the water-filled region, (3) pressure wave inside the polycarbonate sample, and (4) shock waves in air resulting from the impact between the projectile and the piston. From [2]

Fig. 4 Three schlieren photographs showing the shock wave entering a wedge-shaped water filled section made of aluminum. The precursor waves are highlighted with black dots in the left frame. The right frame shows compression waves from the oscillating cavitation bubbles are seen as faint circles ahead of the shock wave. From [2]

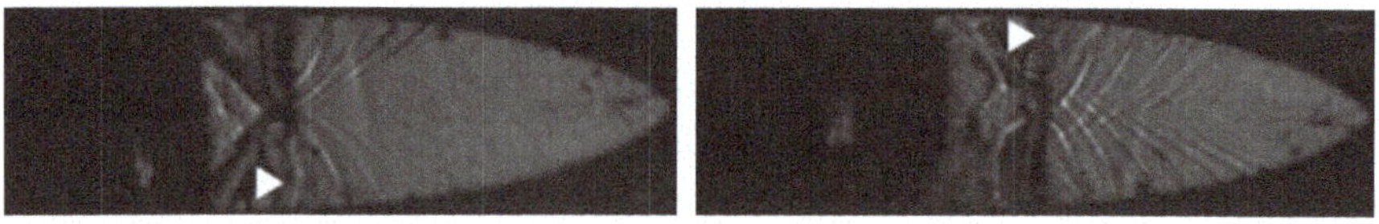

Fig. 5 Two schlieren photographs showing a shock wave propagating into a convergent water-filled section shaped as a logarithmic spiral. The surrounding material is 1.3 mm thick steel, and the precursor waves ahead of the shock wave are visible. The white arrow indicates the location of the shock wave. From [3]

To obtain quantitative data from the underwater schlieren photographs, a background oriented schlieren (BOS) technique was developed and used [5]. BOS had been used by other research groups to obtain the density variation in gaseous fluids. Here, the technique was adapted to record the density variation in the water, and then relate this to a pressure by using a suitable equation of state. Similar to digital image correlation (DIC), BOS uses displacements of a background pattern. The displacements are then correlated to a change in density in the flow. Therefore, one can estimate the pressure increase due to the shock passing through a convergent section filled with water (similar to cases shown in Figs. 3, 4 and 5), see Fig. 6.

Lastly, water-filled thin-walled convergent specimens, made from a carbon fiber vinyl ester composite, were studied. During the shock focusing experiment, the thinner specimens fractured at the apex of the convergent shape, see Fig. 7, while

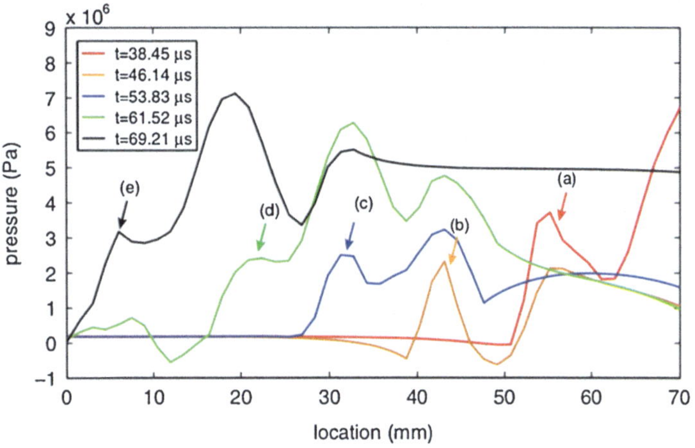

Fig. 6 Successive pressure plots evaluated inside a water-filled convergent specimen under shock loading. The shock is in this case traveling from right to left. From [5]

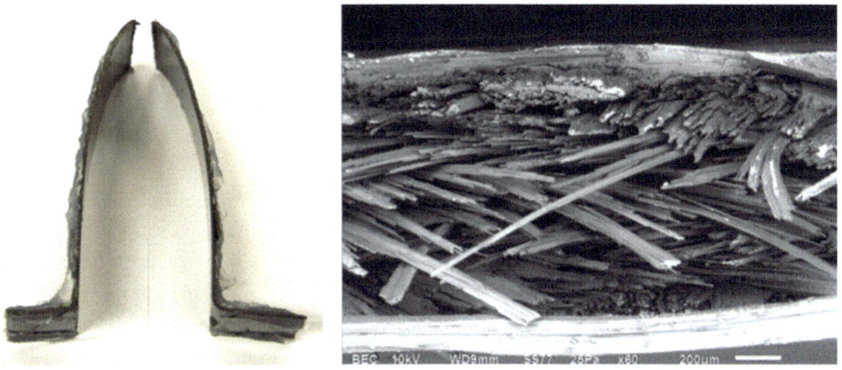

Fig. 7 A fractured CFVE specimen shown to the left, and an SEM image of the fractured edge shown on the right. Right image from [6]

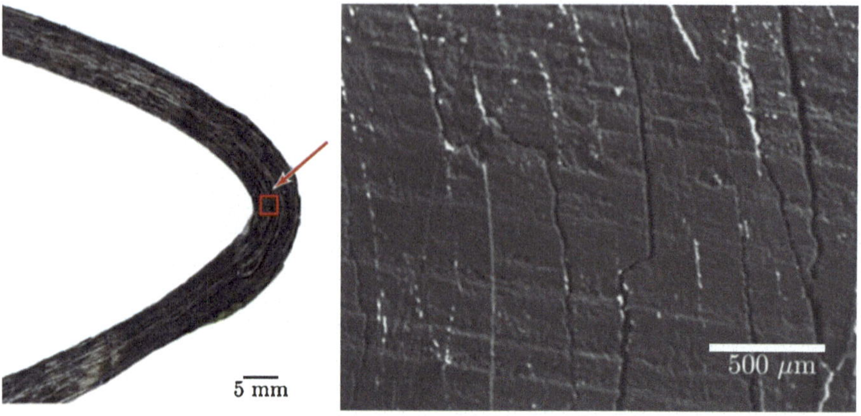

Fig. 8 To the left is a photograph of the 5.8 mm thick CFVE specimen illustrating the delamination damage, and to the right is an enlarged region captured using scanning electron microscopy. From [3]

the thicker specimens suffered delamination but did not fracture, see Fig. 8. Results from high-speed photography and strain gages combined with numerical simulations indicated a complex loading scenario, especially if taking the perturbations caused by the strong fluid-structure interaction into account. Thus, decisions were made to further study the crack initiation process in the composite specimens but using an experimental approach that removed the complex fluid-structure interaction. Hence, new research directions were taken where the specimens were to be planar and subjected to an edge-on impact. This would ensure a highly dynamic but transient load delivered through the specimen, and this is introduced in the next section.

3 Dynamic Fracture Behavior of Polymeric Materials

A modern need with aging structures is to assess how much their survivability has decreased since the moment they were put in service. Currently with a rising trend of integrating fiber reinforced polymers (FRPs) in naval structural elements [7], it is necessary to understand the full extent of how their failure behavior changes across their lifespan.

Fiber reinforced polymers rely on the interaction between fiber and matrix to attain their full strength. Usual construction involves laying plies of a type of fiber and applying a resin in between each ply. The laminate is then vacuum bagged and cured to attain full strength.

Failure of the fiber and matrix plies to fully bond, such as voids can decrease the effectiveness of such laminates. Furthermore, damage that leads to a similar result, such as delamination can lead to premature failure.

Environmental conditions such as water absorption pose the risk of affecting the fiber/matrix bond and decreasing the strength of the composite. This is a serious risk in naval applications due to the constant exposure of water to ships. With the increased use in FRPs in ship construction, it is a necessary to understand how moisture content has an effect on their failure mechanisms.

Previous research has shown that water absorption has an effect on matrix dominated material properties, such as the shear modulus [8]. Most of this research, to the best knowledge of the authors, has focused on static or quasi-static behavior.

The need to understand failure behavior coupled with long term environmental effects, and the lack of dynamic behavior research in this area led to the development of a series of experiments that explore different dynamic failure modes on polymeric materials subject to environmental conditioning.

Doing a study of this type in a scaled structure like a ship hull, would bring to many complexities and the resources and facilities required to perform a controlled experiment would be cost prohibitive. To start this project, it was decided to use 2D geometries that would allow to carefully study mode-I and mode-II fracture modes. A 2D geometry would also allow for the visualization of stress waves and how they interact with cracks.

3.1 Mode-I Dynamic Fracture Behavior of Poly (Methyl Methacrylate)

The study started with experiments on poly (methyl methacrylate) (PMMA). This is a material that has been widely used in fracture experiments and allows for a baseline on how to properly conduct this type of experiment. For this type experiment the same experimental procedure as in the work by Theocaris [9] was used.

The first set of experiments focused on mode-I fracture behavior. PMMA notched samples would be loaded at strain rates of 100/s by launching projectiles from a gas gun. The notches would be loaded in tension from the impact, leading to mode-I fracture. Details and results from this work can be found in [10].

The fracture samples were made from commercially available PMMA. The samples were 3.175 mm thick and have dimensions of 300 mm $\times$ 40 mm. At halfway through the long side they had a thin notch machined parallel to the short end, the notch was 10 mm long and 0.152 mm wide. The geometry can be seen in Fig. 9.

A 10 mm-diameter spherical projectile carried by a 25 mm thick Delrin sabot was launched from a 2 m long gas gun. Before leaving the barrel, a laser velocity sensor would be used to measure the velocity. The projectile and sabot had an average velocity of 38 m/s for all experiments. Upon exiting the barrel, the sabot was stopped by a steel stopper plate that allowed the projectile to go through and impact on the short end of the sample. A schematic of this setup is seen in Fig. 10.

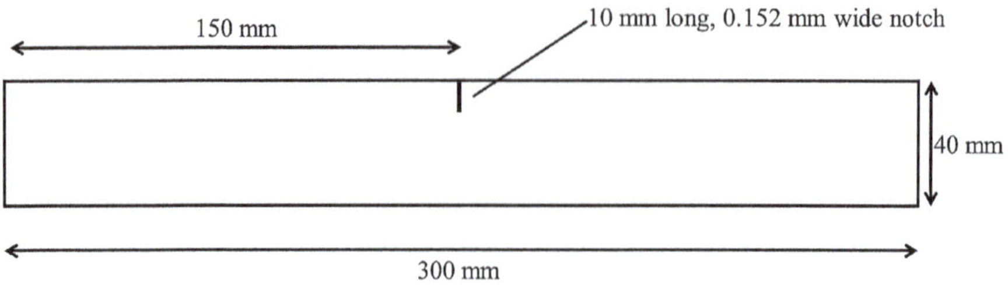

Fig. 9 Mode-I sample geometry. As seen in [10]

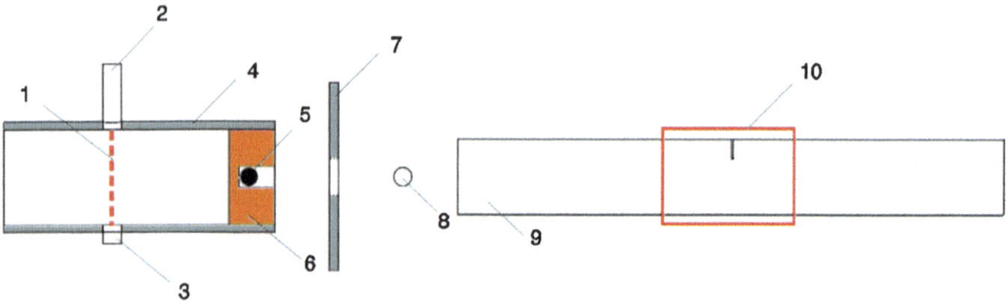

Fig. 10 Schematic of impact setup. Impact occurs from left to right. (1) Laser beam, (2) Laser, (3) Photodiode receiver, (4) Gun barrel, (5) 10 mm diameter steel projectile carried inside sabot (6) Delrin sabot, (7) Metal stopper, (8) 10 mm diameter steel projectile before impact, (9) Sample, (10) Area of interest upon which camera is focused. From [10]

The sample is aligned along the impact axis and free to displace along that axis, but constrained in any other direction, meaning that the motion would be happening along a plane. Great care was put into ensuring plane stress conditions, since out of plane effects would go beyond the scope of this study.

Upon impact a stress wave would travel through the specimen. This wave would be a compressive wave, which would not cause fracture to occur. However, upon reaching the end of the sample, the wave would reflect as a tensile wave, loading the notch in tension.

The samples were conditioned as follows: Submersion in distilled water, submersion in sea water, exposure to 98%, 60% and 11% relative humidity environment. All samples were conditioned in these environments for up to 40 days at 25 °C as seen in Figs. 11 and 12.

All samples had strain gages bonded in order to record the strain response of each experiment, ensuring loading was similar for all samples. The average strain response for all experiments can be seen in Fig. 13. The strain gage was located at the centerline, 50 mm behind the crack. Furthermore, the strain rate was able to be calculated from these experiments.

In order to obtain quantifiable data, the mode-I stress intensity factor at the crack tip needed to be found. The method of caustics was chosen to extract the stress intensity factor from high-speed footage obtained from the experiment. This technique has been used by Theocaris [9] and Kalthoff [11] in similar experiments.

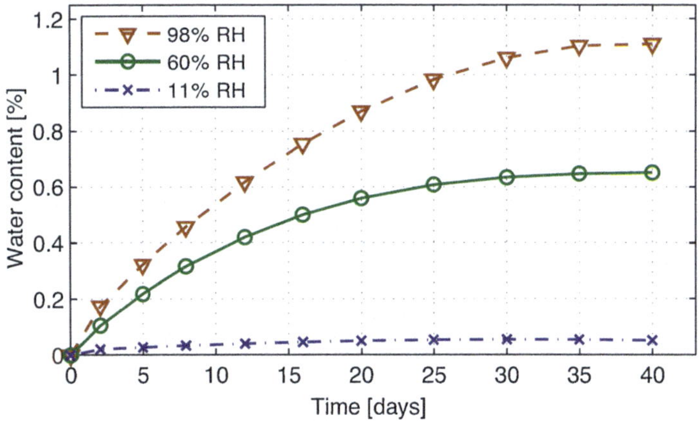

Fig. 11 Water absorption trends for PMMA samples at different RH levels. From [10]

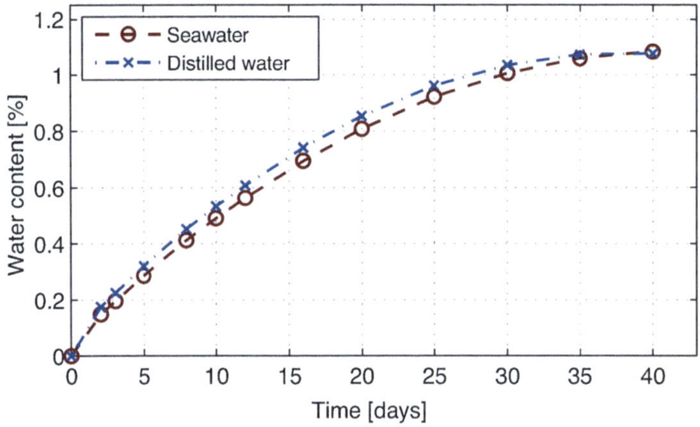

Fig. 12 Water absorption trends for PMMA samples exposed to distilled water and seawater. From [10]

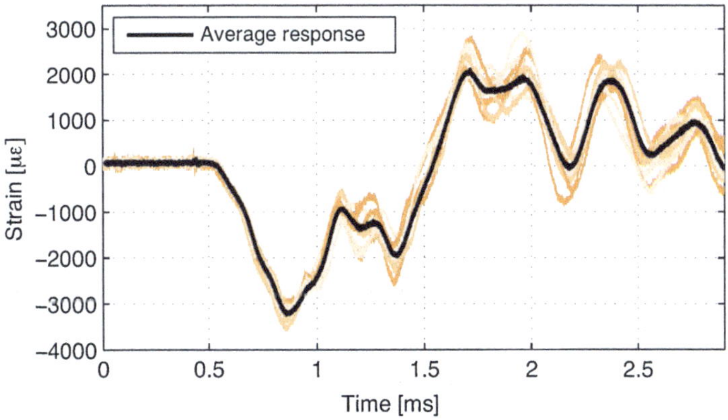

Fig. 13 Strain gage response for mode-I experiments. From [10]

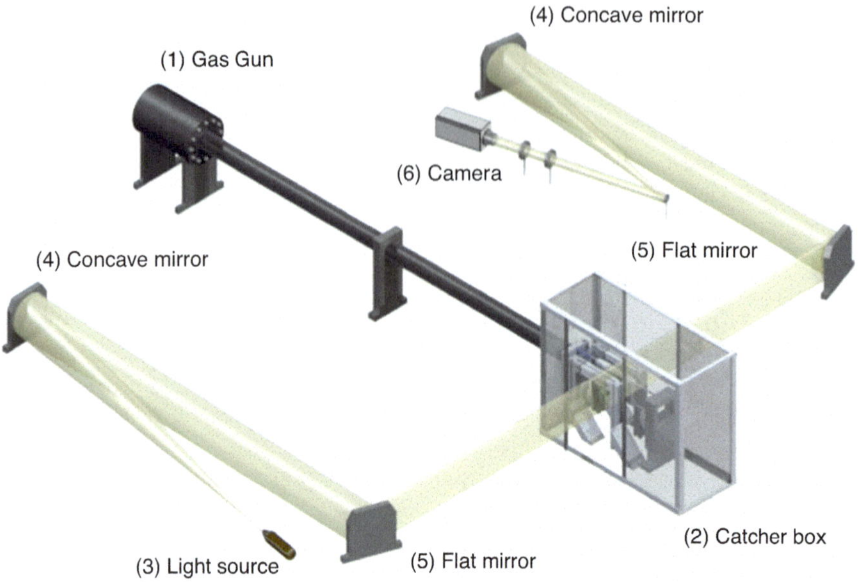

Fig. 14 Experimental setup consisting of an (1) gas gun, a (2) catcher box, a (3–6) visualization system, and a test section placed inside of the catcher box. From [10]

The method of caustics involved using an optical setup shown in Fig. 14. In this setup the transparent sample of PMMA is placed along a path of collimated rays of light. This is achieved by having a small LED light source and a combination of spherical and flat mirrors used to redirect and collimate the light. The light is then recombined using another set of mirrors and lenses into a high-speed camera. For this experiment, the camera used was a Phantom V711.

Upon fracture, the Poisson effect at the crack tip will make the material act as a diverging lens. When the light is projected onto a parallel plane away from the specimen a dark circular shadow is seen around the crack tip, with a light concentration around it called a caustic as observed in Fig. 15.

The size of the caustic is determined by the stress state at the crack tip. Meaning that by measuring the diameter of the caustic it is possible to calculate the stress intensity factor at the crack tip by using Eq. (1):

$$
\begin{aligned}
K_I^d &= \frac{2\sqrt{2\pi}F(v)}{3m^{\frac{3}{2}}ctz_0}\left(\frac{D}{3.17}\right)^{\frac{5}{2}},\\
F(v) &= \frac{4\alpha_1\alpha_2 - \left(1+\alpha_2^2\right)^2}{\left(\alpha_1^2-\alpha_2^2\right)\left(1+\alpha_2^2\right)},\\
\alpha_1 &= \left(1-\frac{v^2}{c_L^2}\right)^{\frac{1}{2}},\\
\alpha_2 &= \left(1-\frac{v^2}{c_T^2}\right)^{\frac{1}{2}}.
\end{aligned}
\tag{1}
$$

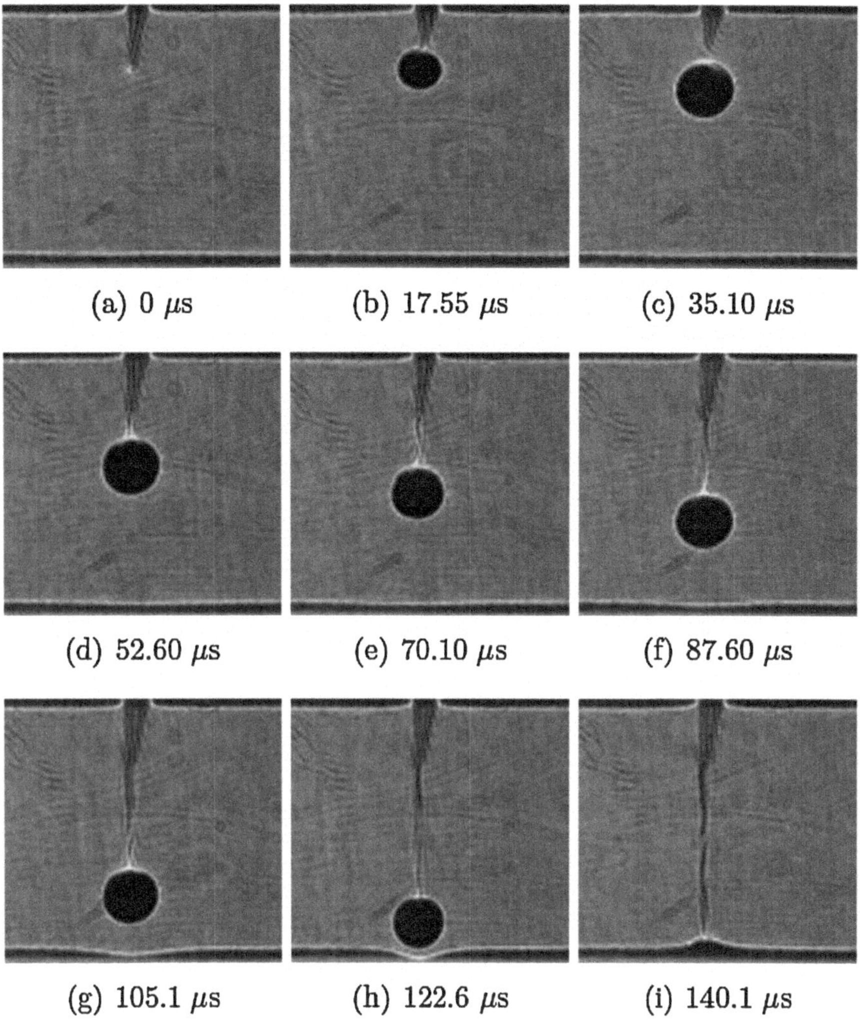

Fig. 15 Series of caustics images progression as crack grows. From [10]

Here D is the caustic diameter, m is the magnification factor, c is the stress optical coefficient, t is the material thickness, z_0 is the distance to the reference plane, v is the crack speed, c_L is the longitudinal wave speed and c_T is the transverse wave speed.

One key variable that will affect the SIF calculation is the stress optical coefficient (SOC) of PMMA. Since the objective is to analyze variations due to moisture content, the authors needed to confirm that the SOC was invariant to moisture contents at this strain rates.

For this reason, a study was performed to analyze how SOC changed in PMMA with different water contents and different strain rates. This work can be found in [12]. It was observed that at strain rates of 5×10^{-4} there was a magnitude change of

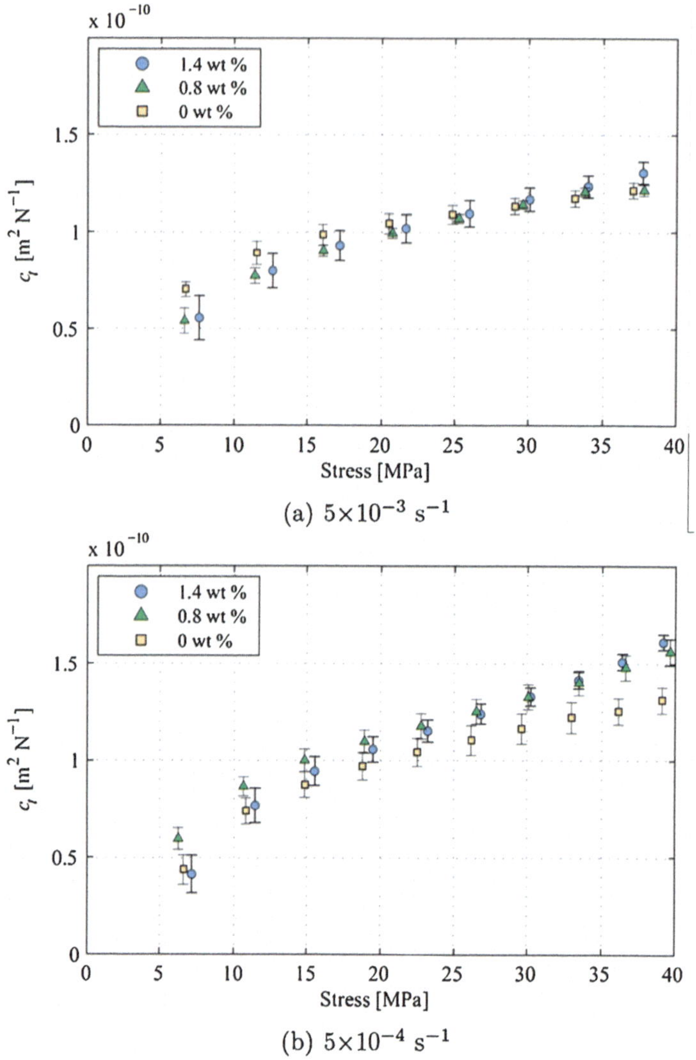

Fig. 16 Stress optical coefficient comparison for PMMA samples subjected to two different with varying water contents. From [12]

11% in the SOC between dry and soaked samples. However, at a higher strain rate of 5×10^{-3} any differences were negligible. The author concluded that as the strain rates became higher, the SOC became independent of moisture content. The results from this study are shown in Fig. 16.

When analyzing each frame of the experiment, as the ones in Fig. 15, not only the SIF at the crack tip can be determined. The critical stress intensity factor is found by finding the SIF at which the crack starts to grow. This time instant can be determined by finding at which frame the caustics starts to translate. The critical SIFs between the different sets of samples can be compared, and any meaningful differences observed.

By performing experiments on all five sets of samples, no significant difference was observed in the SIF, and no significant difference was observed in the fracture behavior between sets. In Fig. 17 it can be observed that at the shown conditions, the crack starts to grow at the same SIF, and as the crack grows the SIF follows similar trends for all conditions.

This was surprising, considering the fact that under static loading conditions, Bokoi et al. [14] showed that there was a change in critical stress intensity factor due to water content. A comparison between both findings is seen in Fig. 18.

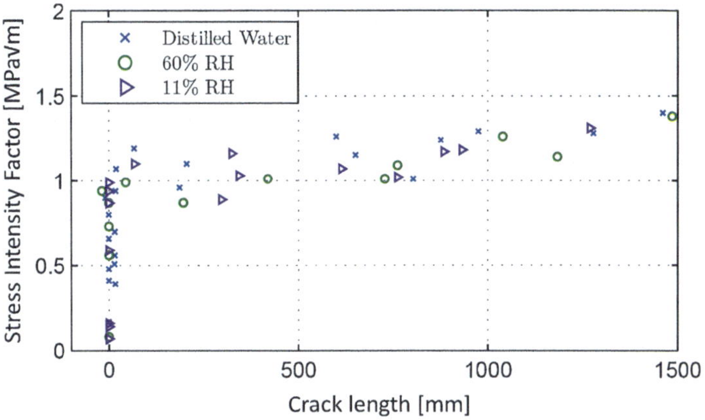

Fig. 17 Mode-I stress intensity factor at different crack lengths for different conditioning scenarios of PMMA. From [13]

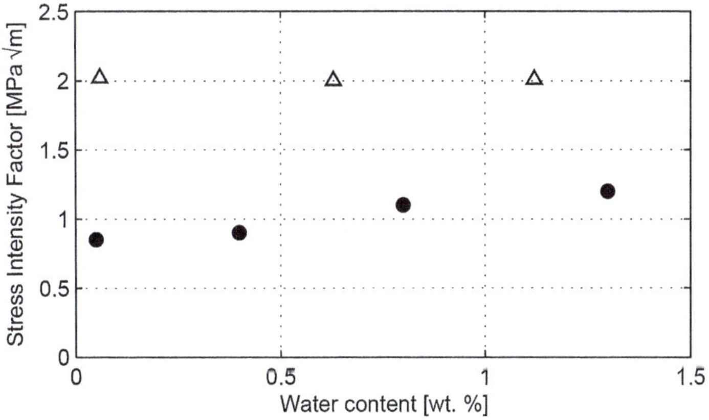

Fig. 18 Comparison of the critical SIF variation with water content: the results obtained for static tensile testing by Bokoi et al. [14] (●) and the results obtained in this study (△) for PMMA. From [13]

Furthermore, it's been shown that at lower strain rates, water content will affect the crack speeds. However, this was not the case at high strain rates. The author concluded that at strain rates of $10^2\ s^{-1}$ moisture content played no significant role in fracture behavior.

3.2 Mode-I Dynamic Fracture Behavior of Carbon Fiber Vinyl Ester

With a good understanding of polymeric materials, and experience with the setup, the authors were able to move to a more complex and expensive material. Carbon fiber vinyl ester (CFVE) was chosen due to its being a popular material in naval applications.

For this work, medium modulus CFVE samples provided by the Naval Sea Warfare Center (NSWC) Carderock Division. The samples were made from unidirectional carbon fiber, with all fibers oriented along the short end. They had dimensions of 300 mm × 40 mm × 3.3 mm and had a notch machined along half length, that was then made into a crack with a razor blade as shown in Fig. 19.

The loading method was the same as that for the PMMA samples. The main difference lied in how the stress intensity factor was obtained. Since CFVE is not transparent, the optical setup used for PMMA was not viable. Therefore, a digital image correlation (DIC) setup was acquired for this experiment.

For this setup, a high-speed camera was placed perpendicular to the sample as show in Fig. 20. This was done to ensure all 2D motions were captured without capturing any out-of-plane displacements. For DIC to be used, the sample had to be painted white, and a black speckle pattern had to be applied with spray paint.

The fracture footage was then post processed using Correlated Solutions VIC 2D DIC software. The DIC algorithm tracks how the different speckles deform when subject to a load. From this analysis vertical and horizontal displacement fields are obtained.

The stress intensity factor for an orthotropic material can then be found by using Eq. (2) [15] shown below. From these equations it can be observed that the vertical and horizontal displacement fields can be expressed as functions of the mode-I and

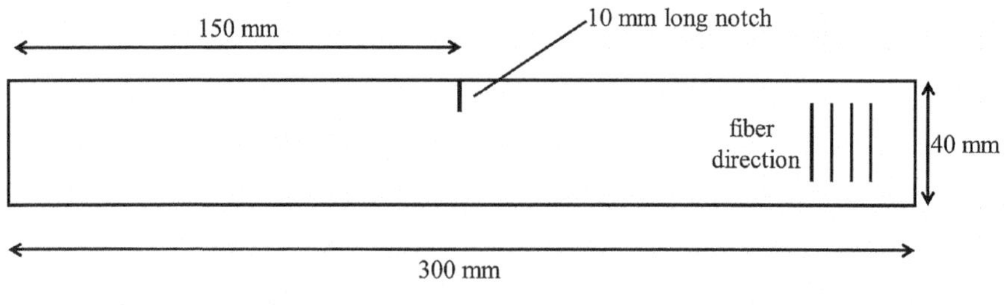

Fig. 19 Geometry of CFVE samples. From [13]

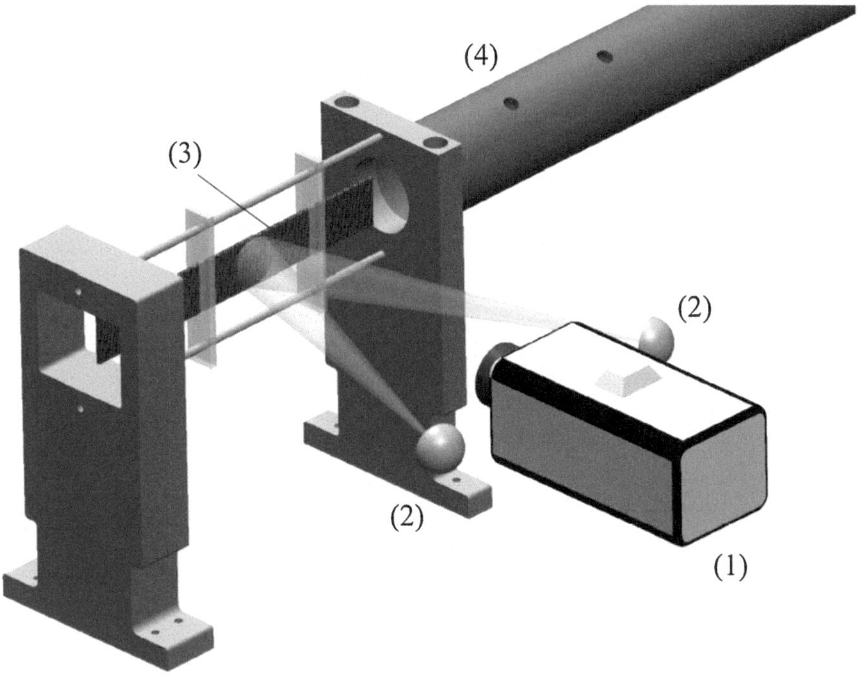

Fig. 20 Experimental DIC setup: (1) Phantom camera, (2) high-intensity LED lights, (3) sample placement, and (4) gun barrel. From [13]

mode-II stress intensity factors, polar coordinates for every displacement point defined from the crack tip and the compliance matrix of the material,

$$\begin{aligned}
u &= K_I\sqrt{\frac{2r}{\pi}}Re\left[\frac{1}{\mu_2-\mu_1}(p_1\mu_2 z_1 - p_2\mu_1 z_2)\right] + K_{\text{II}}\sqrt{\frac{2r}{\pi}}Re\left[\frac{1}{\mu_2-\mu_1}(p_1 z_1 - p_2 z_2)\right],\\
v &= K_I\sqrt{\frac{2r}{\pi}}Re\left[\frac{1}{\mu_2-\mu_1}(q_1\mu_2 z_1 - q_2\mu_1 z_2)\right] + K_{\text{II}}\sqrt{\frac{2r}{\pi}}Re\left[\frac{1}{\mu_2-\mu_1}(q_1 z_1 - q_2 z_2)\right],\\
p_j &= \mu_j^2 S_{11} + S_{12} - \mu_j S_{16},\\
q_j &= \mu_j S_{12} + \frac{S_{22}}{\mu_j} - S_{26},\\
z_j &= \sqrt{\cos\theta + \mu_j \sin\theta}.
\end{aligned} \tag{2}$$

where: $\mu_j(j = 1, 2)$ are the two roots of:

$$S_{11}\mu^4 - 2S_{16}\mu^3 + (2S_{12} + S_{66})\mu^2 - 2S_{26}\mu + S_{22} = 0. \tag{3}$$

Table 1 Material properties of CFVE samples, provided by NSWC Carderock division

Property	Value
E_1	55 GPa
E_2	7 GPa
G_{12}	26 GPa
ν_{12}	0.05

Here, r and θ, are the polar coordinates relative to the crack tip, S_{ij} is the compliance matrix and u and v are the horizontal and vertical displacements respectively.

The material properties of this particular CFVE were provided by NSWC and are shown in Table 1:

Additionally, vertical and horizontal rigid body translation and rotation terms can be added as observed, leading to Eq. (4):

$$u = K_I\sqrt{\frac{2r}{\pi}}Re\left[\frac{1}{\mu_2-\mu_1}(p_1\mu_2 z_1 - p_2\mu_1 z_2)\right] + K_{II}\sqrt{\frac{2r}{\pi}}Re\left[\frac{1}{\mu_2-\mu_1}(p_1 z_1 - p_2 z_2)\right] + T_x - Rr\sin\theta$$
$$v = K_I\sqrt{\frac{2r}{\pi}}Re\left[\frac{1}{\mu_2-\mu_1}(q_1\mu_2 z_1 - q_2\mu_1 z_2)\right] + K_{II}\sqrt{\frac{2r}{\pi}}Re\left[\frac{1}{\mu_2-\mu_1}(q_1 z_1 - q_2 z_2)\right] + T_y + Rr\cos\theta \quad (4)$$

This leads to a system of equations with five unknowns, them being the mode-I and mode-II SIFs and the rigid body translation and rotation terms. Since DIC gives displacement points for every pixel within a defined area, the number of equations in the system will be more than five, leading to an overdetermined system.

A least square regression algorithm was implemented to solve this system. In order to determine which displacement points had to be implemented in the system of equations one has to think of the limitations of the model. The proposed model does not account for edge effects, far field interactions and assumes for an ideal crack. Accordingly, many trials had to be made using different regions around the crack tip.

Since there is only one solution of SIF, a region where results were repeatable was chosen. For this case, a region of $0.5 < r/t < 1.5$ was chosen, where r is the radius from the crack tip and t is the material thickness.

Like the PMMA samples, the CFVE samples were also subject to environmental aging. The two conditions explored were dry samples subject to no conditioning and soaked samples submerged in distilled water for 43 days at 25 °C. The change in weight of the samples was constantly monitored to measure the water content in the samples and is shown in Fig. 21.

The samples were impacted at 40 m/s and the mode-I SIF was extracted using the process described above. From the footage it was observed that the crack grows along the fiber direction, as expected. Additionally, full field strain fields were extracted using DIC as seen in Fig. 22, were the strain concentration at the crack tip is observed. The average critical mode-I SIF for both dry and soaked samples are

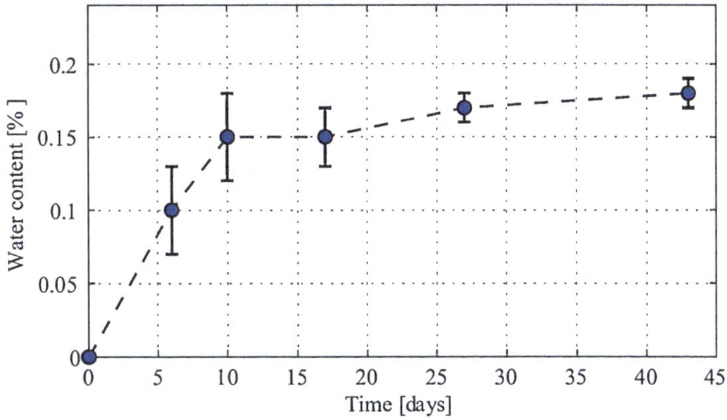

Fig. 21 Water absorption for CFVE samples. From [13]

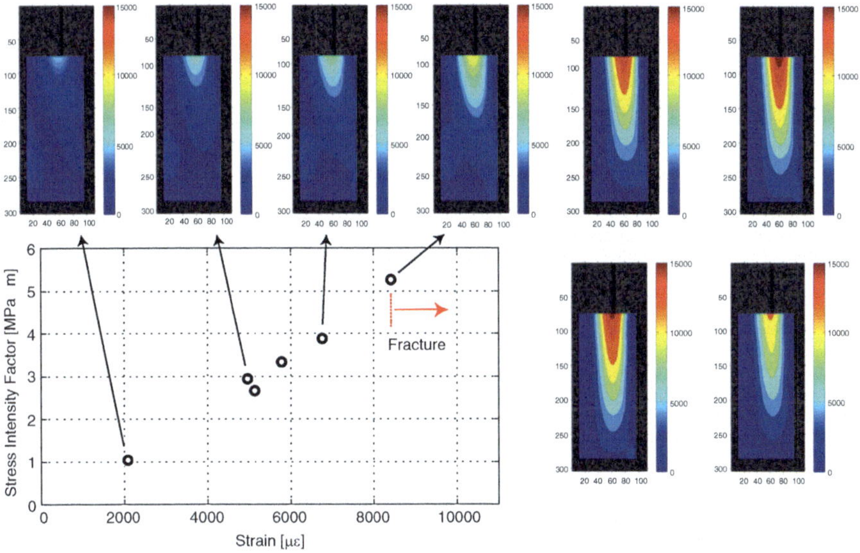

Fig. 22 CFVE SIF measurements and full-field strain maps until fracture initiation. From [13]

Table 2 CFVE critical stress intensity factors

Fracture Type	*CFVE dry*	*CFVE soaked*
K_{Ic} [MPa√m]	4.9	3.1

shown in Table 2. It can be seen that the soaked samples experienced a degradation of 37% of its critical SIF compared to the dry condition.

The authors concluded that a decrease in fracture toughness is present in CFVE. Two processes can be attributed to this degradation: plasticization of the matrix and matrix swelling, both due to water absorption.

3.3 Mode-II Dynamic Fracture Behavior of Carbon Fiber Epoxy

With water degradation observed in CFVE composites, a mode-II experiment28 was conducted. In this case the material chosen was carbon fiber epoxy (CFE). CFE is one of the most prominent FRPs in use and therefore it was deemed a good material to start a mode-II study. Furthermore, all manufacturing and material characterization of the CFE was performed in-house allowing for total quality control.

The mode-II experiment setup was inspired by the experiments performed by Coker and Rosakis [16]. The fracture specimen geometry consisted of rectangles with dimensions 127 mm × 203 mm and thickness of 2 mm. For this laminate all fibers were oriented along short end. Furthermore, a notch was machined halfway through the long end following the fiber direction. The notch was then made into a natural crack by using a blade, with a total length of 25 mm. The specimen geometry is shown in Fig. 23.

It is widely known that for an isotropic material loaded in mode-II conditions, the local loading of the crack tip will transition to mode-I once the crack starts to grow. However, the matrix between the fiber acts as a weak plane, that allows for the crack to be loaded and grow under mode-II conditions.

In order to generate mode-II loading of the crack, a cylindrical projectile was launched from a gas gun. The projectile would impact a steel buffer attached to the bottom of the crack as shown in Fig. 24. The buffer would redistribute the load below the crack, generating asymmetric loading of the specimen, leading to mode-II fracture.

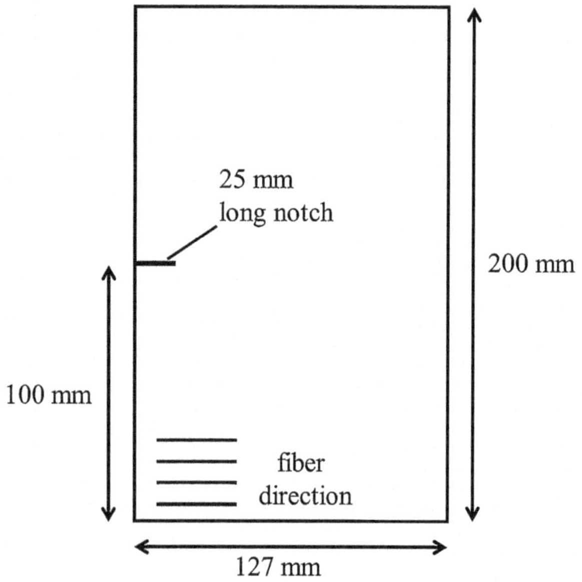

Fig. 23 Geometry for CFE mode-II fracture samples. Similar to sample in [16]

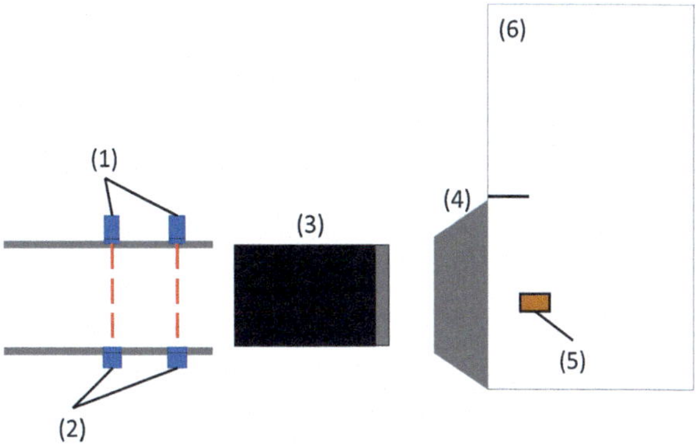

Fig. 24 Schematic of the mode-II impact setup. Impact occurs from left to right. (1) LEDs, (2) photodiode receiver, (3) 75 mm Delrin projectile (4) steel buffer, (5) strain gage, (6) mode-II sample

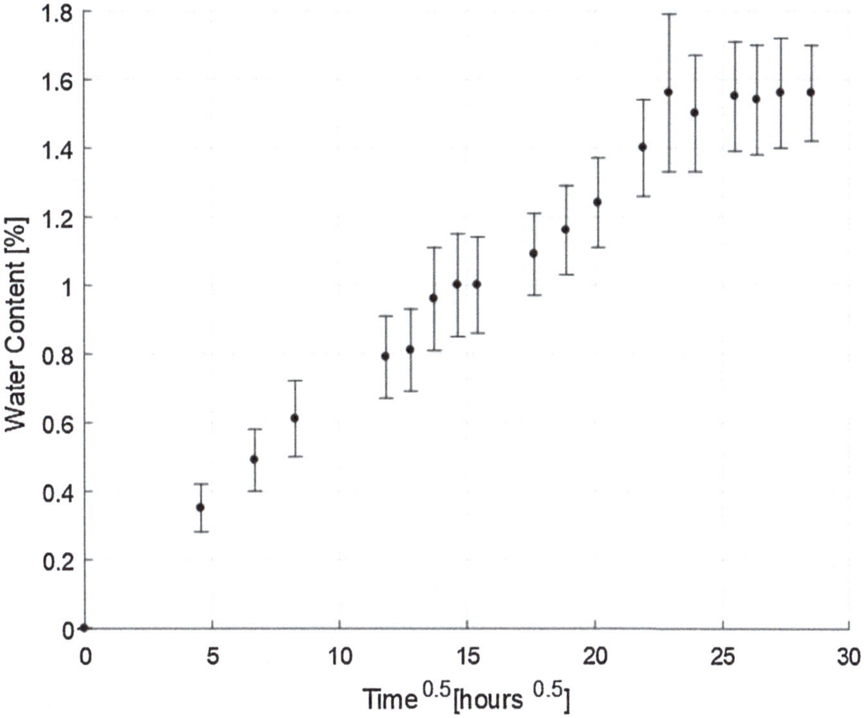

Fig. 25 Water absorption of CFE samples

The samples were divided in two sets. The first set remained dry, while the second set was submerged in distilled water at 65 °C, to allow for the accelerated aging of the samples. The samples remained in water for a total of 48 days until a total saturation water content of 1.56% was attained, as seen in Fig. 25.

The in-plane material properties of the CFE in both dry and soaked conditions were obtained by following ASTM standards D3039/D3039M-17 for the

Table 3 Coupon geometries for material characterization

Coupon	Fiber orientation	Thickness, [mm]	Length, [mm]	Width, [mm]	Tab length, [mm]	Tab thickness, [mm]
E_1	0° unidirectional	1	250	15	56	1.5
E_2	90° unidirectional	2	175	25	25	1.5
G_{12}	$[\pm 45°]_{4S}$	2	229	25	-	-

Table 4 Material properties of CFE laminates in dry and soaked conditions

	Dry	Soaked
E_1	153.5 ± 8 GPa	150.7 ± 7 GPa
E_2	10.2 ± 0.8 GPa	8.6 ± 0.6 GPa
G_{12}	7.6 ± 0.5 GPa	5.8 ± 0.2 GPa
ν_{12}	0.4 ± 0.05	0.35 ± 0.05
ν_{21}	0.022 ± 0.006	0.019 ± 0.003

longitudinal and transverse Young's modulus and Poisson ratio and D3518/D3518M-13 for the in-plane shear modulus. For the soaked conditions coupons that followed the geometries outlined in the ASTM standards where manufactured and subject to the same conditioning as the fracture samples. The coupon geometries used are shown in Table 3.

From Table 4, it can be observed that the longitudinal modulus did not present a significant degradation while the transverse modulus and the shear modulus, which are matrix dominated showed a 16% and 24% degradation respectively.

For this experiment the fracture event was recorded at 1,000,000 frames per second, using an ultra high-speed camera. Since ultra high-speed cameras have a short recording window, a strain gage bonded to the specimens was used to trigger the camera upon detection of a stress wave travelling through the specimen. The average strain response for all experiments is shown in Fig. 26.

DIC was used to obtain quantifiable data from the experiments. The camera setup was the same as for the mode-I CFVE experiments and the same DIC software was used and can be seen in Fig. 27. Furthermore, the SIFs were obtained using the same overdeterministic process.

For selection of the displacement points to use for determination of the stress intensity factor, it was decided that the displacement points in the regions between $0.5 < r/t < 3.5$ and $-2\pi/3 < \theta < 2\pi/3$ should be used, where r and θ are the polar coordinates measured from the crack tip and t is the thickness of the material.

When examining the CFE SIFs, some mode-mixity was observed, however the mode-I SIF was negligible compared to the mode-II SIF. As for the differences of mode-II SIFs between the dry and soaked conditions it was seen that there was a 48% decrease in mode-II fracture toughness, as shown in Fig. 28.

As was also observed in the mode-I experiments, plasticization of the matrix combined with matrix swelling played a role in decreasing the mode-II fracture

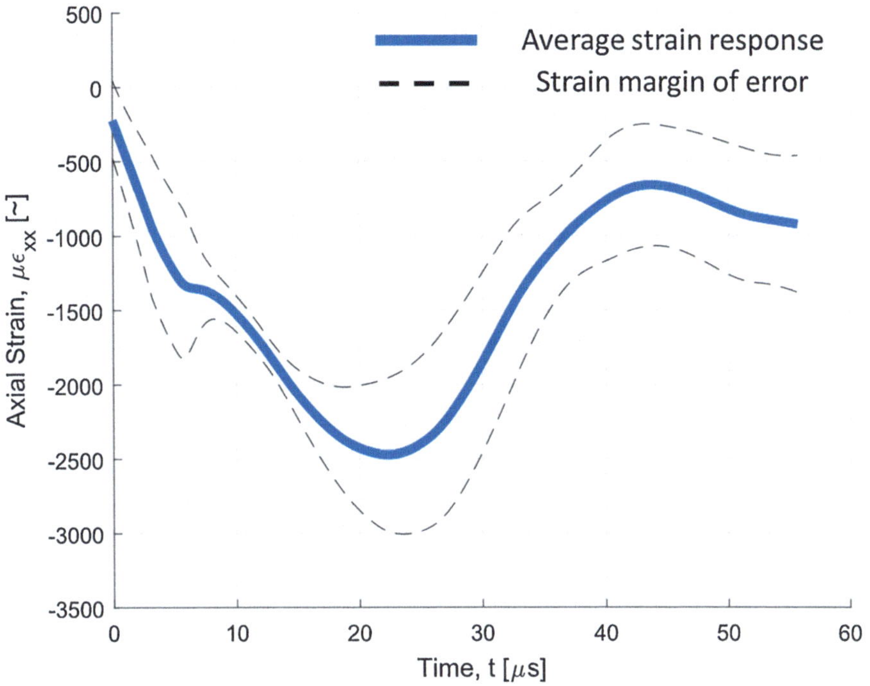

Fig. 26 Strain response for mode-II CFE experiments

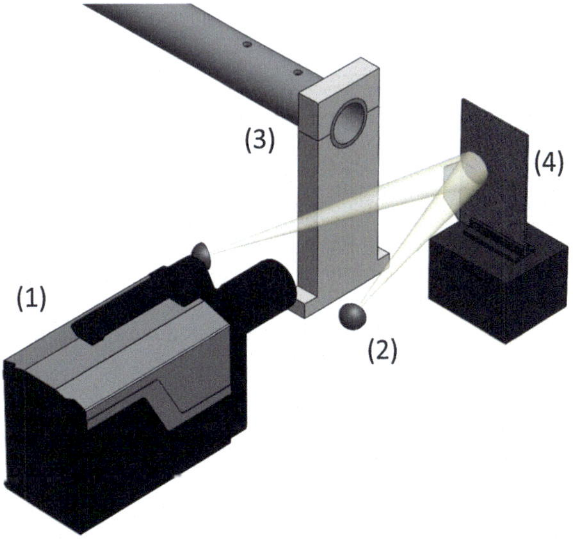

Fig. 27 DIC setup for mode-II experiments. (1) Ultra high-speed camera, (2) high-intensity LED lights, (3) end of gun barrel, (4) mode-II sample

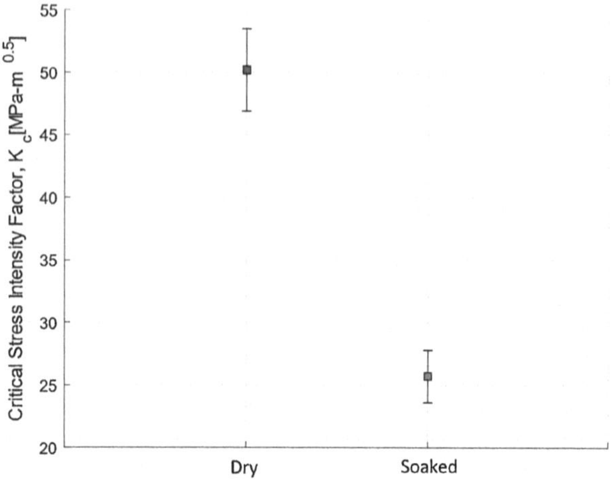

Fig. 28 Comparison of mode-II stress intensity factors between dry and soaked conditions

toughness of these CFE laminates. This indicates the necessity of taking into consideration the potential intake of water when assessing the survivability of ship structures. However further studies need to be performed with more complex laminates and even sandwich structures to properly reflect current constructions techniques.

4 Conclusion

During the last decade, our research group has delved into the effect of high strain rate loading and environmental effects on different composites and polymeric materials. The complexity of solid-fluid interaction has been studied, and different effects due to high strain loading were observed with varying geometries.

When studying the fracture behavior of polymers, it was observed that their response to fracture at high strain rates is invariant to water uptake, in contrast to loading at low strain rates, where the water content plays a role. However, this was not the case for carbon fiber reinforced polymers. It was observed that for both mode-I and mode-II fracture, water uptake reduced the magnitude of the critical stress intensity factors. This indicates the effect of moisture content on fiber/matrix interactions which reduce the fracture toughness of the laminate.

Acknowledgments The work was supported by ONR MURI grant number N00014-06-1-0730, ONR grant number N00014-13-1-0607, ONR Grant Number N00014-16-1-3215 with Dr. Y.D.-S. Rajapakse, Program Manager, and the National Science Foundation through Grant Number CMMI-1332840. Thank you also to our previous students: Dr. Chuanxi Wang, Dr. Orlando Delpino Gonzales, Dr. Shi Qui, and all former and present undergraduate students that have in one way or another helped us.

References

1. Jacob GC, Starbuck JM, Fellers JF, Simunovic S, Boeman RG (2005) The effect of loading rate on the fracture toughness of fiber reinforced polymer composites. J Appl Polym Sci 96 (3):899–904
2. Eliasson V, Mello M, Rosakis AJ, Dimotakis PE (2010) Experimental investigation of converging shocks in water with various confinement materials. Shock Waves 20:395–408
3. Wang C, Grunenfelder LK, Patwardhan R, Qiu S, Eliasson V (2015) Investigation of shock wave focusing in water in a logarithmic spiral duct, Part 2: strong coupling. Ocean Eng 102:185–196
4. Wang C, Qiu S, Eliasson V (2013) Investigation of shock wave focusing in water in a logarithmic spiral duct, Part 1: weak coupling. Ocean Eng 102:174–184
5. Wang C, Qiu S, Eliasson V (2016) Quantitative pressure measurement of shock waves in water using a Schlieren-based visualization technique. Exp Tech 40(1):323–331
6. Wang C, Eliasson V (2012) Shock wave focusing in water inside convergent structures. Int J Multiphys 6(3):267–281
7. Mouritz AP, Gellert E, Burchill P, Challis K (2001) Review of advanced composite structures for naval ships and submarines. Compos Struct 53(1):21–42
8. Browning CE, Husman GE, Whitney JM (1977) Moisture effects in epoxy matrix composites, composite materials: testing and design (Fourth conference), ASTM STP 617. American Society for Testing and Materials, 481–496
9. Theocaris PS, Katsamanis P (1978) Response of cracks to impact by caustics. Eng Fract Mech 10:197–210
10. Delpino Gonzales O, Eliasson V (2015) Influence of water uptake on dynamic fracture behavior of poly (methyl methacrylate). Exp Mech 56(1):59–68
11. Kalthof JF (1990) Transition in the failure behavior of dynamically shear loaded crack. Appl Mech Rev 43(5):247–250
12. Delpino Gonzales O, Nicassio A, Eliasson V (2016) Evaluation of the effect of water content on the stress optical coefficient in PMMA. Polym Test 50:119–124
13. Delpino Gonzales O (2016) On the dynamic fracture behavior of polymeric materials subjected to extreme conditions. Ph.D. Thesis, University of Southern California, Los Angeles
14. Bokoi Y, Ishiyama C, Shimojo M, Shiraishi Y, Higo Y (2000) Effects of sorbed water on crack propagation in poly(methyl methacrylate) under static tensile stress. J Mater Sci 35:5001–5011
15. Sih GC, Paris PC, Irwin GR (1965) On cracks in rectilinearly anisotropic bodies. Int J Fract Mech 1(3):189–203
16. Coker D, Rosakis AJ (2001) Experimental observations of intersonic crack growth in asymmetrically loaded unidirectional composite plates. Philos Mag A 81(3):571–559

Recent Developments on Ballistic Performance of Composite Materials of Naval Relevance

Vicente Sánchez-Gálvez, David A. Cendón, Rafael Sancho, and Francisco Gálvez

1 Introduction

Fiber reinforced polymer matrix composites (FRP) are increasingly used to manufacture components in many industrial sectors. Their excellent values of specific strength and stiffness make them a good solution to produce components in those industrial applications where weight saving of the component is an important factor. This is the case of the transportation industry (aeronautical, naval and automotive).

It is well known however, that fiber reinforced polymer composites are highly susceptible to impulsive loads, being damaged by low velocity impact, their mechanical properties being severely reduced even by a low energy impact such as a falling object. This is an important drawback for the extension of the use of FRPs in the naval sector and it is conceivable that many investigations have been carried out to analyze, predict and detect low velocity impact damage on a great number of fiber reinforced polymer composites at different temperatures and environmental conditions [1–15].

On the other hand, ballistic performance of fiber reinforced polymer composites is also a very important issue to assess the safety of military platforms (aircrafts, helicopters, vehicles, ships) against the attack by machine guns or grenades. Ballistic performance of composites has been widely investigated for those fibers used to produce body armors and helmets, such as aramid (Kevlar) or high molecular weight polyethylene (Spectra) as well as carbon reinforced epoxy matrix composites being used to produce components of aerospace platforms [16–19]. In the naval sector FRPs have substituted traditional materials (timber, steel) to manufacture hulls of small length boats. Their great advantages (weight reduction, corrosion resistance)

V. Sánchez-Gálvez (✉) · D. A. Cendón · R. Sancho · F. Gálvez
Materials Science Department (ETSI Caminos, Canales y Puertos),
Universidad Politécnica de Madrid, Madrid, Spain
e-mail: vicente.sanchez@upm.es; david.cendon.franco@upm.es; rafael.sancho@upm.es; f.galvez@upm.es

S. W. Lee (ed.), *Advances in Thick Section Composite and Sandwich Structures*,
https://doi.org/10.1007/978-3-030-31065-3_6

have led them to be almost the only material used to produce hulls of yachts, fishing ships and sport boats. In the marine sector, FRPs are used to produce hulls of mine hunters and disembark boats. Recently, they are being considered to produce superstructures of long vessels and even to produce hulls of battleships or submarines [20–22].

Glass fiber reinforced polyester or vinylester matrix composites have been almost the only composites used to produce boats hulls, mainly due to their lower cost, although carbon fiber reinforced vinylester is being increasingly considered for applications where its higher mechanical properties (strength, Young's modulus) are more attractive than its higher cost, such as sport boats.

The research on ballistic performance of these composites (glass, carbon reinforced vinylester) is scarce. Therefore, the Office of Naval Research (ONR) Solid Mechanics Program launched a Research Project to analyze the dynamic response and failure of composites of naval relevance under high-speed impact in the extreme conditions of temperature and humidity that may be found in the ocean. This chapter summarizes the recent developments achieved in the first-year research period including experimental results of ballistic tests, development of analytical and numerical tools to simulate high-speed impact phenomena on composite materials, analysis of alternative reinforcements (hybrid reinforcement) and the influence of temperature and humidity on the performance of composites of naval relevance.

2 Materials

Two different fibers have been considered, glass and carbon. With respect to the matrix only vinylester has been considered, being the most widely used one for naval applications.

Experimental tests have been carried out on six different configurations:

- Plain-carbon-reinforced vinylester
- Plain-S2-glass-reinforced vinylester
- Mixed-hybrid-S2-glass/carbon-reinforced vinylester
- Non-mixed-hybrid-S-2-glass-backed by carbon vinylester
- Non-mixed carbon backed by S-2 glass, vinylester
- Mateglass, described below

Plates with approximate dimensions of 400×600 mm were produced by infusion of resin into a vacuum mold method. Except Mateglass, reinforcement was carried out by means of woven fabric fibers. The number of plies was 20 for plain carbon reinforcement, 40 for plain glass and 10 Carbon +20 Glass for hybrid reinforcement, leading to about 6 mm thickness for all plates. On the other hand, Mateglass is a composite material manufactured by stacking several layers of E-glass reinforcement: mat + woven roving + pro core + woven roving + mat. This configuration is duplicated to achieve a thickness of about 6 mm similar to those of the other configurations.

All data of the fabrics (except for Mateglass) required to feed the analytical model are summarized in Tables 1 and 2. It is important to point out that mechanical properties of single fibers (Young's modulus, tensile strength) were determined by FAVIMAT method (see Fig. 1) while other quantities such as fiber cross-section area, surface density, distance between cross-overs, fiber density, etc. were simply measured on the fabric with the help of a profile projector.

Plates were checked by ultrasonic inspection to guarantee the absence of flaws. After that, plates were cut to produce specimens for all kind of tests. All composites were subjected to static tests to determine those mechanical properties that will be used for numerical simulations of the impact process. Tensile tests were performed according with ASTM Standard D3039. Results are summarized in Table 3. In-plane

Table 1 Carbon-fiber data

Number of plies	Surface density per ply	Young's modulus	Cross-over distance	Critical strain	Fiber cross-section	Fiber density
20	0.482 kg/m^2	230 GPa	2.6 mm	1.51%	0.23 mm^2	1770 kg/m^3

Table 2 S2-glass-fibre data

Number of plies	Surface density per ply	Young's modulus	Cross-over distance	Critical strain	Fiber cross-section	Fiber density
40	0.283 kg/m^2	87 GPa	0.28 mm	3.26%	0.012 mm^2	2100 kg/m^3

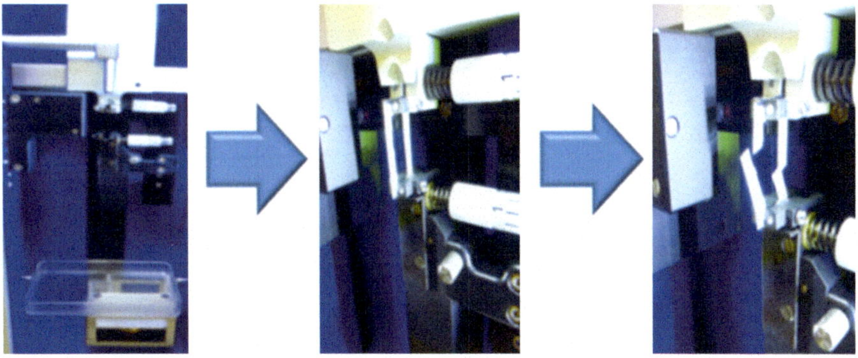

Fig. 1 Determination of the mechanical properties of the fibers using FAVIMAT

Table 3 Mechanical properties of composites

Material	E_1 (GPa)	E_2 (GPa)	UTS_1 (MPa)	UTS_2 (MPa)	G_{12} (GPa)	UTS_{12} (MPa)
CFRC	49	44	715	710	2.6	56
S-2 GFRC	21.3	21	454	417	4.0	107
Mateglass	7.8	7.8	110	110		

shear tests results performed according to Standard ASTM D3518, except for Mateglass are also summarized in Table 3.

3 Experimental Program

The whole experimental program included both static as well as high-speed impact testing of all materials at different temperatures and humidity conditions. Static tests have been described in the preceding section.

3.1 Impact Test Set-up

All impact tests were performed in a compressed gas gun with 7.62 mm caliber barrel (see Fig. 2). Projectiles were 6.49 mm diameter steel balls. Plastic sabots were used to increase impact velocity (see Fig. 3). A video camera with 80.000 fps was used to measure both the impact velocity as well as the residual velocity after perforation of the target. The use of steel balls avoids any problem of yaw and pitch that may be observed when FSP (fragment simulating projectiles) are employed.

Experimental results are displayed as graphics residual velocity vs. impact velocity for all materials, temperatures and humidity conditions. From such graphics, the ballistic limit V_{50} (impact velocity for which there is 50% probability of full perforation) can be derived and used for comparison.

Testing at high (+50 °C) or low (−50 °C) temperatures requires using a polystyrene box to minimize temperature changes. In these cases, the specimen is placed inside the box, previously heated or cooled and a thermocouple is also inserted into

Fig. 2 Compressed-gas gun

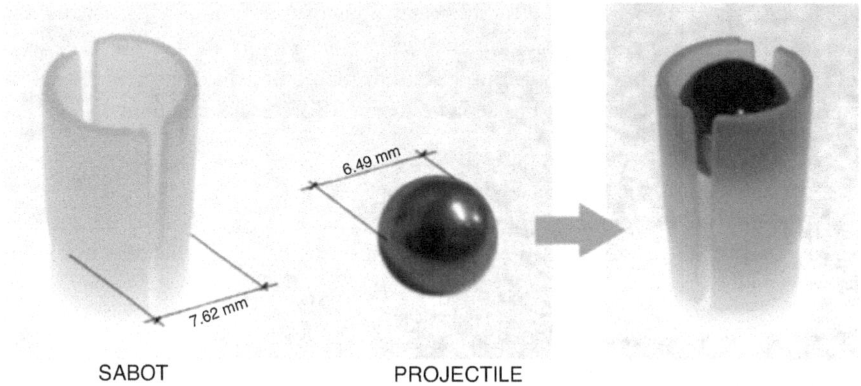

Fig. 3 Sabot and projectile that were used for testing

the box to control the temperature. Test is performed when the temperature measured by the thermocouple is the desired one. Previous tests on an empty box have demonstrated that its influence on the impact velocity is negligible.

3.2 *Material Conditioning*

On the other hand, all tests were repeated on conditioned specimens, to analyze the influence of water intake on high-speed impact performance of composites. The conditioning process was carried out by immersing plates of all materials into a sea water bath kept at 50 °C and measuring the weight increase every week. Saturation was considered when the weight increase in a week was lower than 1%. After conditioning specimens were kept into the bath wrapped with a film to avoid water loss until testing time. This procedure is analogous to that specified in the Standard EN2823AE-CM.

4 Analytical Modelling

In parallel with the experimental program, the analytical model, previously developed by the authors [23–25] has been widely utilized in this research with two objectives: Firstly to check the reliability of the model to simulate accurately high-speed impact performance of composites of naval relevance and secondly to get a robust tool to predict the safety of components produced with composite materials being subjected to ballistic impact (bullets, fragments) in the extreme environments of the ocean.

The analytical model is based on the Mechanics of Continuum Media equations to develop expressions that provide penetration and velocity of the projectile as a function of time during the impact process into the composite laminates. A set of simplifying hypotheses are used to achieve a set of differential equations:

- The projectile is assumed as perfectly rigid.
- Projectile spinning is not considered.
- The composite is assumed to be fully delaminated immediately after the impact.
- The elastic energy stored in the fibers is transmitted to the projectile when the fibers fail.

These assumptions involve that this model may be accurate to simulate the impact of fragments onto thin laminates, but the impact of soft ammunition like for instance lead bullets may give important deviations from the actual performance, because the projectile is highly deformed, and it has spin. Anyway, this situation is such that the model is in the safe side.

The model has been implemented in a MATLAB code to derive penetration and velocity of the projectile and fiber strain step by step until either projectile arrest (when the velocity is zero) or full perforation, when fiber strain reaches a critical value. In case of full perforation, the model gives the residual velocity of the projectile as well as the fiber strain pattern around the impact point.

The analytical model has been extended to simulate high-speed impact process into hybrid reinforced laminates such as those used in this research program (glass/carbon fiber reinforcement). The comparison between experimental and analytical results is carried out by depicting graphically residual velocity against impact velocity for all materials analyzed.

5 Experimental and Analytical Results

Experimental results in the form of residual-velocity vs. impact-velocity plots were obtained for all materials, temperatures and environmental conditions. Also, ballistic limits V_{50} (impact velocity for 50% probability of full perforation of the target) were derived.

On the other hand, a few years ago the authors developed a new analytical model able to simulate accurately high-speed impact response and failure of thin composite materials targets [23]. The model was recently updated to simulate impact onto hybrid-reinforced polymers and to simulate oblique impact on composites [24, 25]. Analytical simulations showed good agreement with experimental results of impact onto Kevlar and carbon/epoxy targets. Now the abovementioned model has been used to simulate the performance of glass, carbon and hybrid-reinforced-vinylester under high-speed impact at different temperatures and environmental conditions.

As previously mentioned, all quantities needed to feed the analytical program, are listed in Tables 1 and 2 for all materials analyzed; except for mateglass which due to

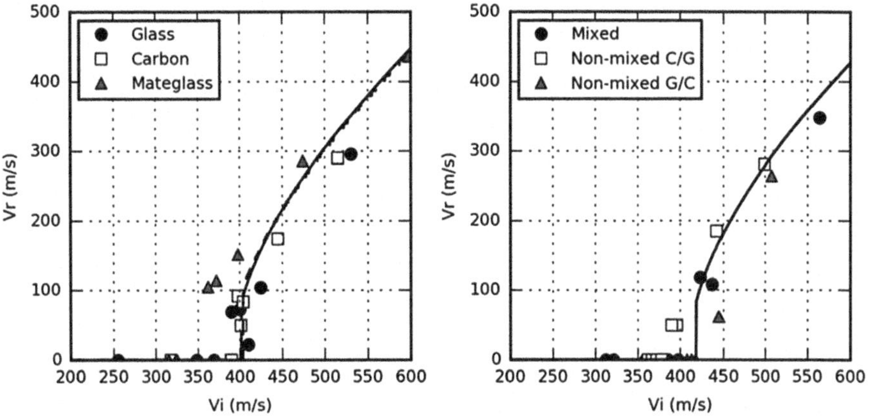

Fig. 4 Experimental data and analytical residual velocity vs. impact-velocity curves at room temperature for non-conditioned specimens. On the left curve, the GFRC (solid line) and CFRC (dashed line) analytical curves are identical. The critical strain was set equal to 3.22% for the S-2 glass fiber and 1.5% for the carbon fiber

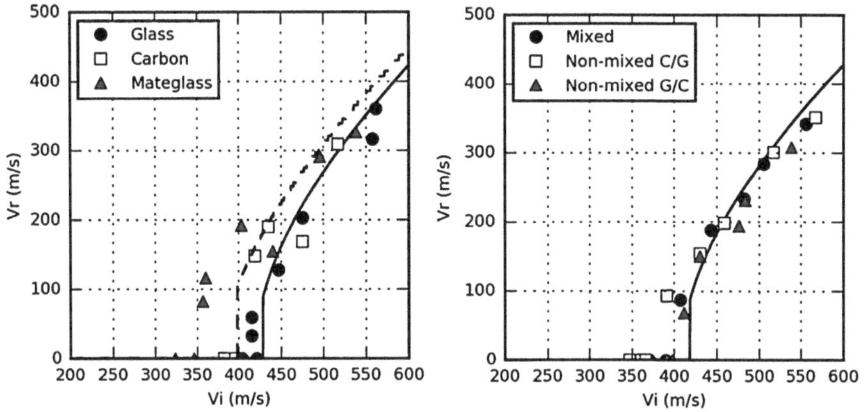

Fig. 5 Experimental data and analytical residual-vs-impact-velocity curves at low temperature (−50 °C) for non-conditioned specimens. The critical strain was set equal to 3.43% for the S-2 glass fiber and 1.5% for the carbon fiber

its complex structure was not simulated with the analytical model but with the commercial hydrocode LS-DYNA. Both experimental and analytical results are illustrated in Figs. 4, 5, 6, 7, 8 and 9. Figures show residual-velocity vs. impact-velocity curves for all materials, temperatures and conditioning, except for mateglass which was not simulated with the analytical model as mentioned. From these curves, the ballistic limit V_{50} is easily obtained for all cases studied. The results, which are very useful to analyze the influence of environment on the performance, are summarized in Figs. 10, 11, 12, 13 and 14.

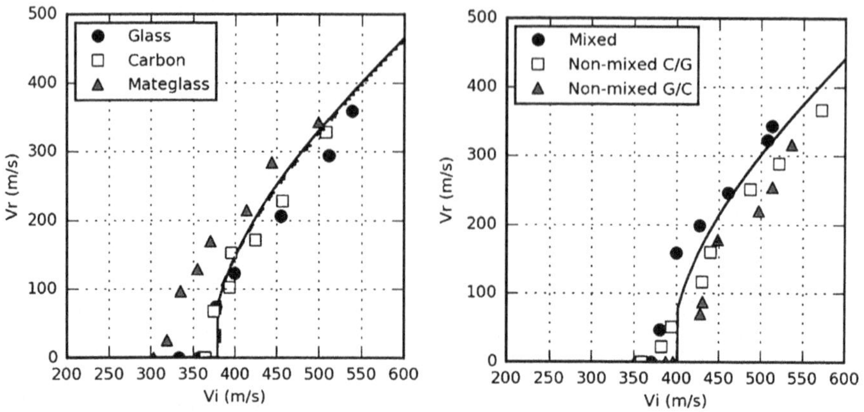

Fig. 6 Experimental data and analytical residual-velocity vs. impact-velocity curves at high temperature (50 °C) for non-conditioned specimens. The critical strain was set equal to 3.02% for the S-2 glass fiber and 1.42% for the carbon fiber

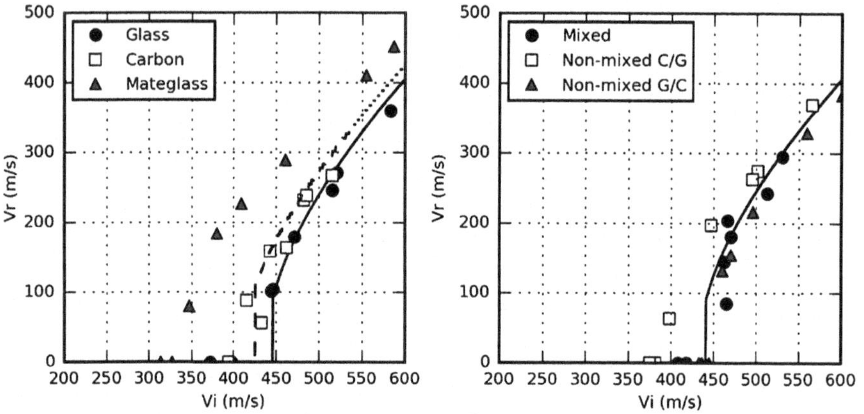

Fig. 7 Experimental data and analytical residual-vs-impact-velocity curves at room temperature for conditioned specimens. The critical strain for the S-2 glass fiber was set equal to 3.55% and for the carbon fiber was set equal to 1.58%

6 Numerical Simulation of Impact on MATEGLASS

As mentioned before, the complex configuration of mateglass involves a great difficulty to be easily modelled by a simple analytical model. Therefore, high-speed impact simulation of mateglass was performed by numerical simulation using the commercial hydrocode LS-DYNA. Due to the abovementioned difficulties associated with this material, only the impact tests performed at room temperature were addressed by these numerical simulations.

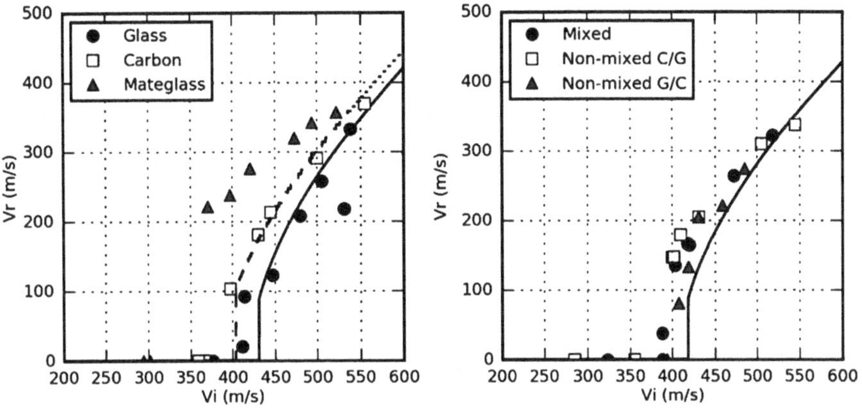

Fig. 8 Experimental data and analytical residual-velocity vs. impact-velocity curves at low temperature (−50 °C) for conditioned specimens. The critical strain for the S-2 glass fiber was set equal to 3.43% and for the carbon fiber was set equal to 1.5%

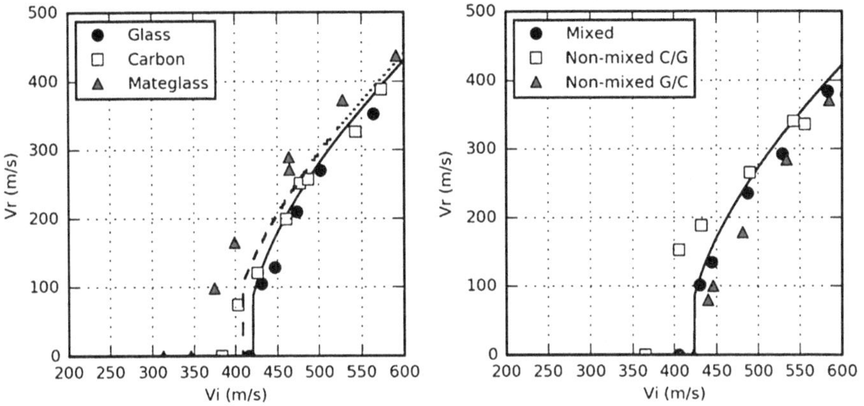

Fig. 9 Experimental data and analytical residual-velocity vs. impact-velocity curves at high temperature (50 °C) for non-conditioned specimens. The critical strain was set equal to 3.35% for the S-2 glass fiber and 1.52% for the carbon fiber

6.1 Model Set-up

Given the good results achieved with the analytical model, in which the fibers are the mainly responsible for decelerating the penetrator, it was decided to decouple the effect of the fibers and the matrix creating therefore a kind of binary model, according to [26]. According to this idea, the contribution of the fibers was lumped in layers of shell elements while the vinylester matrix was modelled through solid elements. Given that mateglass is composed by stacking layers of fibers with different configurations (Mat+woven roving+ProCore+multiaxial +/−45°), each of these layers was modelled as a single layer of shell elements having material

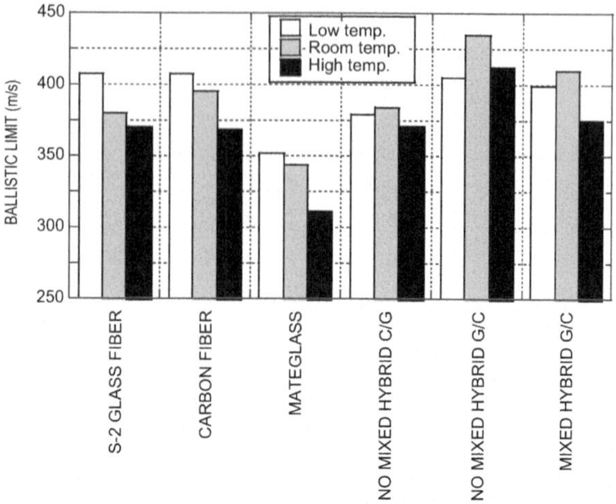

Fig. 10 Graphical representation of the ballistic limit V50 for non-conditioned specimens

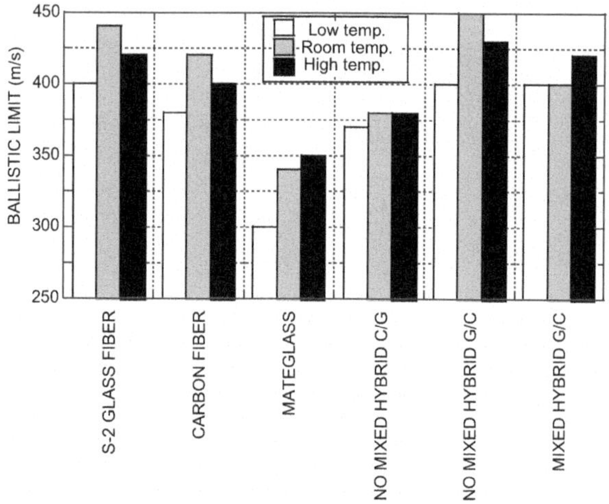

Fig. 11 Graphical representation of the ballistic limit V50 for conditioned specimens

properties in accordance with its fiber content and orientation. Figure 15 shows a sketch of the model used for each plate of mateglass. It must be reminded that the specimens tested were composed of two superimposed plates. For the sake of simplicity, the same mass density averaged for the composite material, which was equal to 1.23 g/cm^3, was used for all constituents of the composite.

As mentioned before, single integration solid elements were used for the mesh of the vinylester matrix, while 0.2 mm thick Belytschko-Tsai shell elements [27] were used for the layers involving glass fibers. The projectile was also modelled through single integration solid elements.

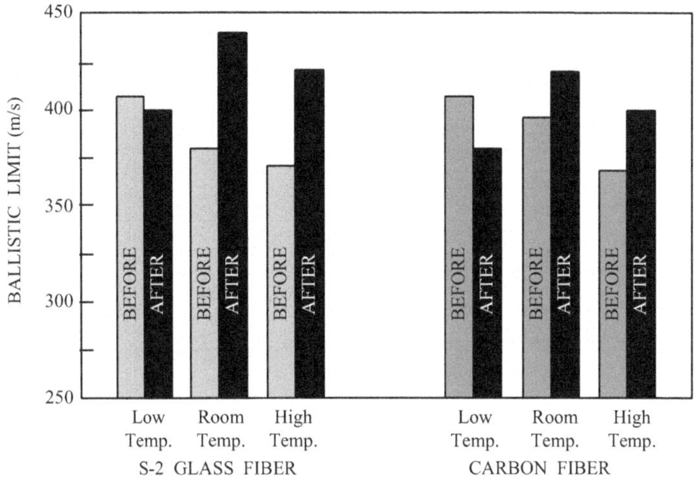

Fig. 12 Graphical representation of the ballistic limit V50 for the S-2 GFRC and CFRC materials at the three different temperatures, comparing the effect of seawater saturation (before and after material conditioning)

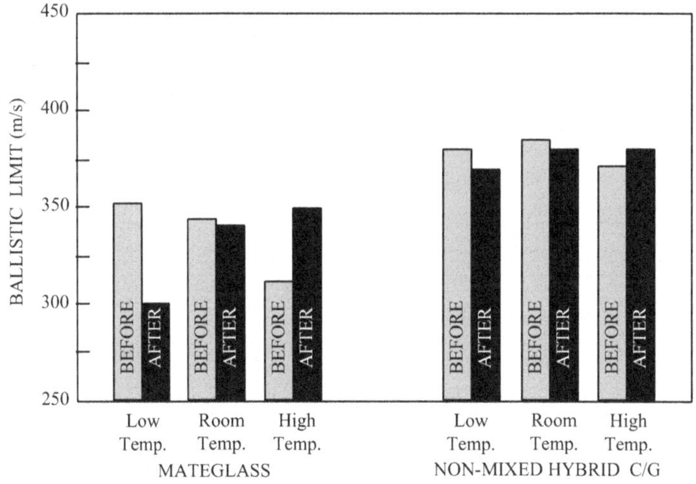

Fig. 13 Graphical representation of the ballistic limit V50 for the Mateglass and non-mixed hybrid carbon/glass composite materials at the three different temperatures, comparing the effect of seawater saturation (before and after material conditioning)

Figure 16 shows the model appearance, including the shell layers used to model the fiber behavior (Fig. 16a). The element size of the shell elements was $0.566 \times 0.566\text{mm}^2$, while the size of the solid elements was $0.566 \times 0.566 \times 0.35\text{mm}^3$, making a total number of 225,000 shell elements plus 462,096 brick elements (including 12,096 elements for the projectile).

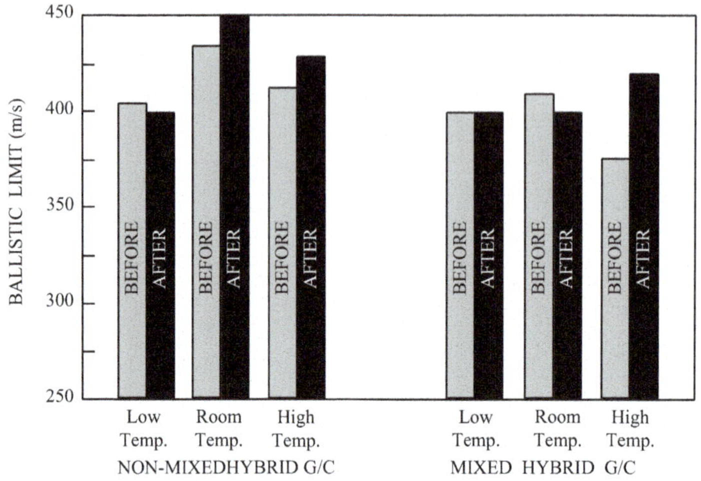

Fig. 14 Graphical representation of the ballistic limit V50 for the mixed hybrid and non-mixed hybrid glass/carbon composite materials at the three different temperatures, comparing the effect of seawater saturation (before and after material conditioning)

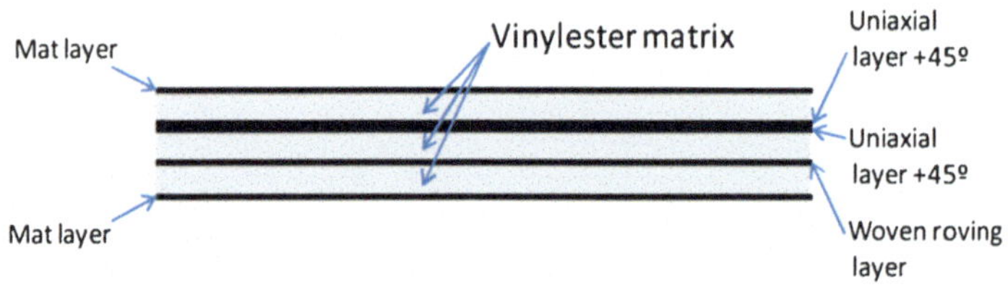

Fig. 15 Sketch showing the structure used in the numerical model for each plate of mateglass. Black lines represent layers of shell elements, while grey color is used for the vinylester matrix involving solid elements

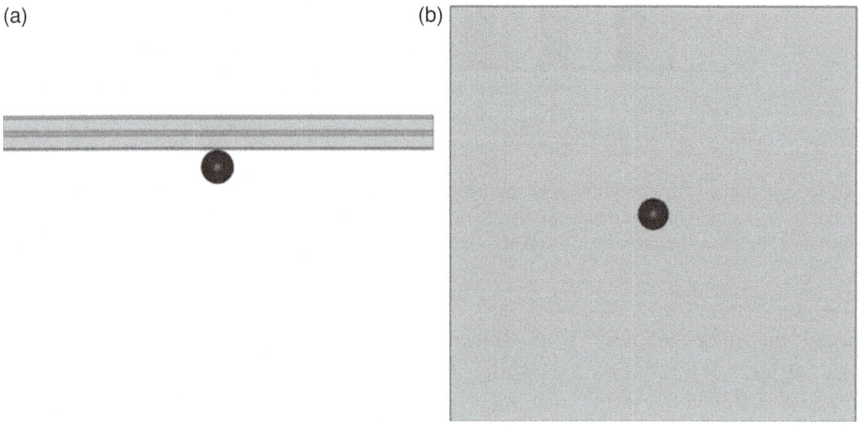

Fig. 16 Upper view (**a**) and front view (**b**) of the model appearance, including the projectile. The finite element mesh is not displayed for the sake of clarity

6.2 Materials Properties

Unfortunately, dynamic properties of each constituent were not available and therefore the material parameters needed to feed the code had to be estimated from experimental data in the literature obtained in static conditions. On the other hand, several studies have shown that the strength of composite materials under high-strain rates may be increased by a factor that may reach values up to about 2 (see for example [28, 29]). For this reason, all material parameters under static conditions were affected by a strain-rate factor which was estimated in a value of 1.5.

The vinylester matrix was modelled as an elastic-plastic material with Von Mises plasticity, plus a maximum-principal-stress-based failure criterion. The material properties estimated are summarized in Table 4. It must be pointed out that in this table, both the yield stress and the tensile strength are already affected by the strain rate enhancing factor of 1.5.

All layers of shell elements used for the contribution of the fibers were modelled by using LS-DYNA's "MAT_LAMINATED_COMPOSITE_FABRIC" material model. This material model is essentially an orthotropic linear-elastic material model with damage, which is introduced progressively as a stiffness degradation until failure conditions are met. Further details about this material model can be found in [30]. As stated above, Mateglass material presents layers of different fiber orientation and content, therefore, each layer required different material properties. Such material properties are shown in Table 5, which again includes the strain rate factor of 1.5 in all strength related parameters, that is to say: tensile strength in the direction of the fibers, X_t, tensile strength in the direction normal to the fibers, Y_t and shear strength, S_c. Besides these strength parameters, the following symbols are used in the table: $E_{//}$ for the Young's modulus in the direction parallel to the fibers, $E_{\perp}$ for the Young's modulus in the direction normal to the fibers, $G_{//\perp}$ for the shear modulus in the plane of the fibers and $\mu_{//\perp}$ for the Poison's ratio in the plane of the fibers.

Table 4 Material parameters used for the vinylester matrix

	Density	Young's modulus	Poisson's ratio	Yield stress	Hardening modulus	Tensile strength
Vinylester	1.23 g/cm^3	4.0 GPa	0.2	0.12 GPa	0.1 GPa	0.15 GPa

Table 5 Material parameters used for the layers involving fibers

Layer	Density (gr/cm^3)	$E_{//}$ (GPa)	$E_{\perp}$ (GPa)	$G_{//\perp}$ (GPa)	$\mu_{//\perp}$ (−)	X_t (GPa)	Y_t (GPa)	S_c (GPa)
Mat	1.23	10.0	10.0	4.54	0.1	0.18	0.18	0.18
Uniaxial +/− 45°	1.23	43.0	4.0	4.0	0.04	1.65	0.0525	0.09
Woven	1.23	20.0	20.0	4.0	0.1	0.90	0.90	0.09

As can be seen in Table 5, inspired by the analytical model the material properties chosen basically focus on the strength of the layer in the direction of the fibers, as the material properties in the directions perpendicular to the fibers are set to lower values.

To allow for modelling interlaminar failure by decohesion, each layer of shell elements was coupled to the rest of the model by using LS-DYNA's "CONTACT_AUTOMATIC_ONE_WAY_SURFACE_TO_SURFACE_TIEBREAK" option, which makes possible to couple nodal pairs until a certain failure criterion is satisfied. In this model, the failure criterion chosen was maximum tensile strength in the direction normal to the layers. The failure tensile strength was estimated in 120 MPa, since this value allowed to reproduce the area of delamination in the rear interface of the targets with a reasonable level of accuracy.

6.3 Results

Figure 17 shows the results in terms of impact-velocity vs. residual-velocity achieved with this numerical model, compared with the experimental results. Generally speaking, good agreement between the numerical results and the experiments is observed.

The experimental result for the impact velocity of 398 m/s was by mistake obtained by shooting at the target from the rear side. Since the numerical model is obviously not symmetric, in accordance with the actual material, the green mark in the plot corresponding to this velocity of 398 m/s represents the residual velocity obtained shooting at the numerical model also from the rear side. It can be observed

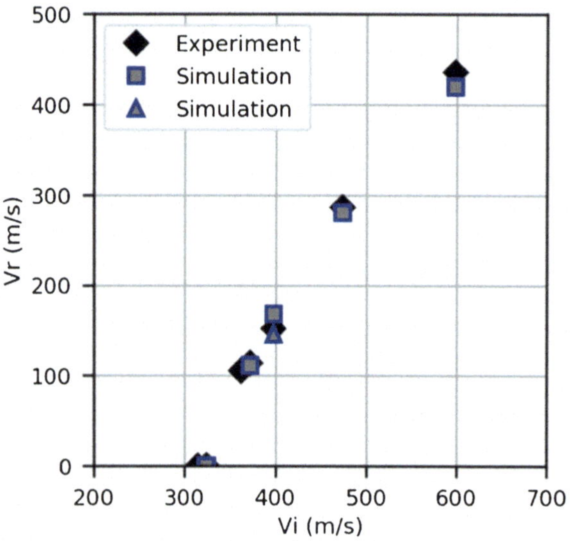

Fig. 17 Impact vs residual velocity curve of mateglass for non-conditioned samples and room temperature

how the differences between the numerical simulation shooting from the front side (square mark) and the back side (triangle mark) are limited but the latter fits better with the experimental result (rhomb). This result gives support to the accuracy of the numerical model presented.

7 Discussion

Observation of figures is interesting to derive some ideas about the performance of all materials in the different environmental conditions. First of all, it is interesting to point out that the ballistic limit is not significantly affected either by temperature or by water intake. The worst performance is observed by mateglass, a result that might be predicted, being this material manufactured with E-glass fiber, cheaper although less resistant than S2 glass fiber reinforcement used in all other materials.

A surprising result achieved is the observation that seawater intake leads to a better impact performance for the majority of environmental conditions. This result may be explained by the assumptions of the analytical model, because the matrix, which could be damaged by the seawater plays not a significant role on the impact resistance of the target and on the other hand, water intake increases the surface density of the target, which finally involves a greater velocity reduction of the penetrator according to the momentum conservation law.

The comparison between the three different configurations for the hybrid reinforcement composites shows clearly that the best performance is achieved with non-mixed configuration, with glass fibers in the front face and carbon fibers in the rear. The highest ballistic limit of the whole investigation has been obtained for this configuration at room temperature and after material conditioning (450 m/s).

8 Conclusions

The experimental program performed allows to conclude that there are no significant differences from one material to another neither for different temperatures nor for conditioned materials.

Nevertheless, even though the behavior of materials is not too much different, it can be pointed out that the worst performance is that of mateglass while the best behavior is that of the no-mixed hybrid glass/carbon (glass fibers in the front face of the target and carbon fibers in the rear one). The influence of low temperature (at least down to -50 °C) is negligible while high temperatures ($+50$ °C) lead to some reduced performance.

Finally, the behavior of materials, saturated with seawater is not significantly different from that of non-conditioned materials. Even, a small increase of ballistic limits can be detected, which could be a surprising result.

With respect to the analytical modelling, it may be pointed out that analytical simulation showed excellent agreement with experimental data for all fabric-reinforced materials for all temperatures studied. Therefore, it may be concluded that the analytical model is a reliable tool for the prediction of the response of fiber-reinforced composites under high-speed impact.

Acknowledgements Authors are indebted to Dr. Yapa Rajapakse and Office of Naval Research (ONR) for financial support of this research trough the Solid Mechanics Program.

References

1. Abrate S (1998) Impact on composite structures. Cambridge University Press, Cambridge
2. Proceedings of Dynamic Response and Failure of Composite Materials and Structures, Ischia 1, 2014 and 2016
3. Gómez del Río T et al (2005) Damage in CFRPs due to low velocity impact at low temperature. Compos Part B 31(1):41–50
4. García-González D et al (2015) Low temperature effect on impact energy absorption capability of PEEK composites. Compos Struct 134:440–449
5. Zhou G (1995) Damage mechanisms in composite laminates impacted by a flat-ended impactor. Compos Sci Technol 54:267–273
6. Ibekwe SI, Mensah PF, Li G, Pang SS (2007) Impact and post impact response of laminated beams at low temperatures. Compos Struct 79:12–17
7. Hawyes VJ, Curtis PT, Soutis C (2001) Effect of impact on the compressive response of composite laminates. Compos Part A Appl Sci Manuf 32:1263–1270
8. Cantwell WJ, Morton J (1991) The impact resistance of composite materials. A review. Composites 22:347–362
9. Reid SR, Zhon G (2000) Impact behavior of fiber reinforced materials and structures. Woodhead Publishing Ltd, Cambridge
10. Barnus SD, Waidva UI (2007) A review. Impact damage of composite materials. J Adv Mater 39(3):3–21
11. Caprino F, Lamanna G, de Luca A, Loprsto V, Riccio A (2015) Numerical investigation of onset and evolution of LVI damages in carbon-epoxy plates. Compos Part B 68:385–391
12. LaMartina B, Li G, Hui D (2003) Blast/impact on engineered (nano) composite materials. Compos Part B 40:413–415
13. Caprino G, Langella A, Lopresto V (2003) Indentation and penetration of carbon fiber reinforced plastic laminates. Compos Part B 34:319–325
14. Panettieri E, Fanteria D, Montemustey M, Froustey C (2016) Low-velocity impact tests on carbon/epoxy composite laminates. A benchmark study. Compos Part B Eng 107:9–21
15. Minak G, Ghelli D (2008) Influence of diameter and boundary conditions on low velocity impact response of CFRP circular laminated plates. Compos Part B 39:962–972
16. Iremonger M, Went AC (1996) Ballistic impact of fiber composite armours by fragment simulating projectiles. Compos Part A 27:575–581
17. Cuniff PM (1996) A semiempirical model for the ballistic impact performance of textile-based personnel armor. Textile Research Journal, 66(1):45–58.
18. Syngellakis S (2014) *Projectile impact. Modelling techniques and target Peformance assessment.* WIT Press Southhampton
19. Sánchez-Gálvez V, Galvez F, Hernandez E, Cendon D (2015) Ballistic performance of carbon/epoxy composites at very low temperatures. In: Proceedings 66 Aeroballistic range association meeting. San Antonio, Texas

20. Department of Defence USA (2000) Composite materials handbook, vol 3
21. Shenoi RA, Dulleu-Barton JM, Quinn S, Blake JIR, Boyd SW (2007) Composite materials for marine applications. Key challenges for the future. School of Engineering Sciences, University of Southampton
22. Grabovac I (2005) Composite reinforcement for naval ships. Concept design, analysis and demonstration. School of Applied Sciences, RMIT University
23. Sanchez Paradela L, Sanchez-Galvez V (2013) Analytical simulation of high-speed impact onto composite materials targets. J Strain Anal Eng Des 48(5):282–290
24. Sánchez-Gálvez V, Galvez F, Sancho R, Cendon D (2017) A new analytical model to simulate high-speed impact onto composite materials targets. Int J Impact Eng 108:322–333
25. Sánchez-Gálvez V, Sánchez Paradela L, Gálvez F (2014) Analytical simulation of high-speed impact onto hybrid glass/carbon epoxy composites targets. Procedia Eng 88:101–108
26. Cox BN, Carter CW, Fleck N (1994) A binary model of textile composites-I. Formulation. Acta Metall Mater 42(10):3463–3479
27. LS-DYNA Theory manual (2006) Livermore software technology corporation, California
28. Arbaoui M, Tarfaoui A, El Malki Alaoui A (2016) Mechanical behavior and damage kinetics of woven E-glass/vinylester laminate composites under high strain rate dynamic compressive loading: experimental and numerical investigation. Int J Impact Eng 87:44–54
29. Barré S, Chotard T, Benzeggagh ML, Townsend, Grabovac I (1996) Comparative study of strain rate effects on mechanical properties of glass fiber-reinforced thermoset matrix composite. Compos Part A 27(12):1169–1181
30. Matzenmiller A, Lubliner J, Taylor RL (1995) A constitutive model for anisotropic damage in fiber-composites. Mech Mater 20(2):125–152

Fluid-Structure Interaction of Composite Structures

Young W. Kwon

1 Introduction

Polymer composite materials have been used for many engineering applications because of their beneficial material properties. Initially, these materials were used for stiffness applications, and later being applied in load-bearing structures. Navy ships are examples of those load-carrying structures. The anti-corrosive nature of polymer composites makes them more attractive for marine applications.

As polymer composites are considered for marine applications, the interaction of the structure and water becomes an important topic due to the density of the water which is very comparable to those of polymer composite materials. As such, the added mass effect on polymer composite structures is critical in changing the behaviors of the composite structures.

This chapter presents the effects of FSI on composite structures subjected to dynamic loading. First, experimental studies were conducted and the behaviors of a composite structure were compared to those of metallic structures in order to demonstrate the importance of the effects of FSI on polymer composite structures. Numerical studies were also undertaken to complement the experimental study. Different contact conditions between a fluid (water in this study) and a structure were considered along with diverse loading conditions such as mechanical impact loading, cyclic loading, and fluid loading. Structures were either open structures or closed structures like containers. The coupling effect of structures separated by a fluid medium was also investigated. The study focused on the effect of FSI on dynamic motions, failure loadings, and failure locations of composite structures. Composite structures were compared to geometrically equivalent metallic structures to show how critical the effect of FSI on polymer composite structures was.

Y. W. Kwon (✉)
Department of Mechanical & Aerospace Engineering, Naval Postgraduate School, Monterey, CA, USA
e-mail: ywkwon@nps.edu

S. W. Lee (ed.), *Advances in Thick Section Composite and Sandwich Structures*,
https://doi.org/10.1007/978-3-030-31065-3_7

2 Experimental Set-Up for FSI with Impact Loading

Marine structures may interact with water in different ways. Some structures are wet only one side and dry on the other side while others may be wet on both sides. External loading can be applied to either the wet or dry side. There are four cases in all as shown in Fig. 1. One of them is a completely dry structure subjected to a dynamic loading, and another is a completely wet structure subjected to a dynamic loading. The other two cases have one side wet and one side dry, and the external loading may be applied to the dry side or the wet side. When the side opposite to the external loading is wet, it is called a water backed structure while the other case is called an air backed structure. All those cases have their own applications for marine structures.

An anechoic water tank, which represents a semi-infinite water medium with a free surface between air and water, was used for the study to minimize the effect of reflected waves. An impact test set-up was designed, fabricated, and installed in the anechoic water tank so that all four cases could be studied. Figure 2 shows the impact tester installed in the anechoic water tank. In the event that a structure was to be tested in the fully dry on both sides case, the water in the tank was drained so that the structure did not come in contact with the water. If a structure was wet on both sides, the tank was filled high enough so that the structure was submerged fully in the tank.

Fig. 1 Four different test cases; (**a**) both sides dry (**b**) both sides wet (**c**) impact side wet while the opposite side dry, and (**d**) impact side dry while the opposite side wet

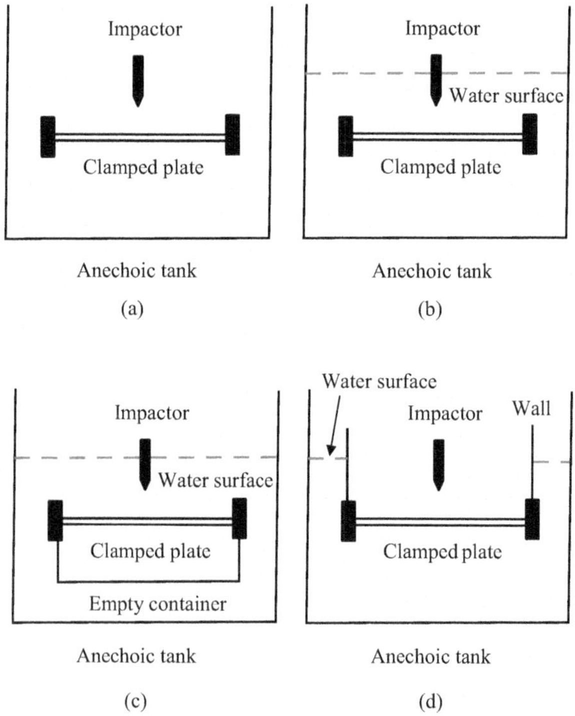

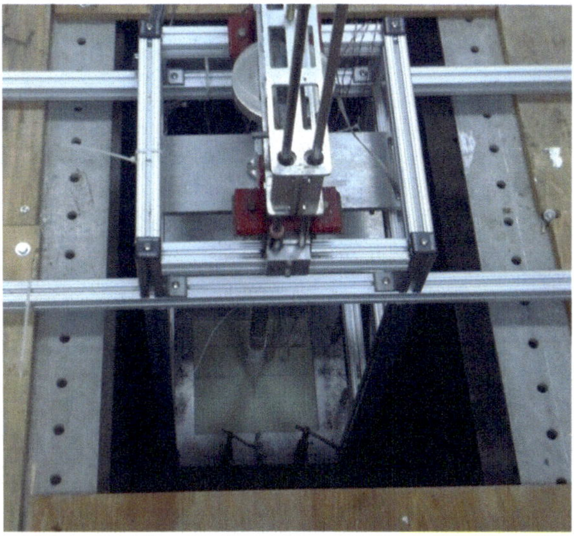

Fig. 2 Impact tester installed in anechoic water tank

If a structure has one side dry and the other side wet, a modification was undertaken to the impact tester.

The two cases, Fig. 1 (a) and (d), have the impactor striking the dry side of the structure while the other two cases, Fig. 1 (b) and (c), have the impactor hitting the wet side of the structure. For the impactor to hit the structure on its wet side, the impactor must travel through water. The resistance of the water is much greater than that of air. As a result, the final velocity of the impactor at the onset of the initial contact with the structure would be different between the dry and wet impact cases even though the specified drop height was the same. This makes any direct comparison between the dry side impact and the wet side impact unfair. In order to address this problem, the impact tester was designed in a special way. The impact mass (i.e. dropping object) does not go into water itself even though the impact is applied to the wet side of a structure. An impact rod was placed between the impact mass and the structure to be impacted. The dropping object hits the top end of the impact rod. The rod is held initially just above the structure so that it moves such a small distance before striking the structure. Because the movement of the impact rod is so minimal, the drag force applied to the impact rod is also minimal. Therefore, the initial impact velocity can be maintained almost the same between the dry side impact and the wet side impact [1–4]. A load cell was attached to the bottom end of the impact rod so that the impact force could be recorded during the impact loading.

Strain gages were attached to composite plates that were clamped along all edges while the impactor hit the center of the plates. In order to investigate the effect of FSI, the results of the completely dry plate were compared to those with FSI effects.

3 Experimental Results of FSI with Impact Loading

First, dynamic responses of composite structures were measured and compared with and without the effect of FSI so as to determine the effect of FSI on the composite structures. To this end, the same composite structures were tested for both cases, with and without FSI. Composite plates constructed using laminated plain woven fabrics were used. Both E-glass and carbon fiber composites were utilized. The plates were clamped along all edges, and the impactor struck the center of the plates. Under the same impact loading conditions, the impact force and strain responses were compared.

Figure 3 compares the impact forces when a composite plate was completely dry or fully submerged in water. The results showed that the impact force was much larger when the structure was in the water for the same impact mass and height. This is believed to be resulting from the added mass effect. When a structure moves in a fluid, there is the added mass effect. In other words, the structure moves as if it is heavier with an additional mass resulting from FSI. This causes the slow-down of the structural motion, and that slow-down yields a larger contact force (i.e. impact force) between the impactor and the structure.

The greater impact force certainly resulted in larger deformation of the structure leading to greater strains. However, the effect of FSI is not uniform over the whole structural surface because the motion of the structure is not uniform. Some surface locations have a larger FSI effect while others have a less effect. Therefore, the effect of FSI on the strain response depends on the location of the structure. For example, Figs. 4 and 5 plot the strain comparison at two different locations of the plate. While strains in Fig. 4 were measured near the center of the plate, strains in Fig. 5 were obtained near the corner of the plate. The difference in the strain response resulting from FSI is much greater near the corner than the center. This illustrates that the effect of FSI varies over the plate because it is not in rigid body translation.

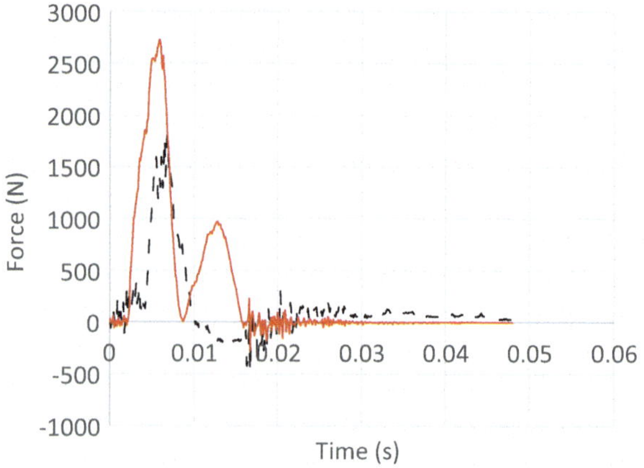

Fig. 3 Comparison of impact forces between the dry impact (broken line) and wet impact (solid line)

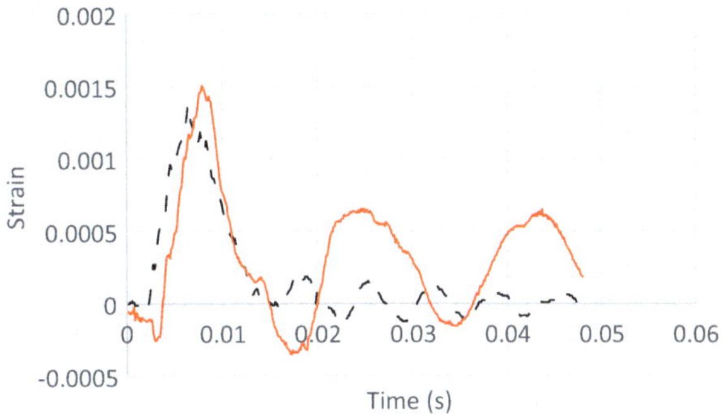

Fig. 4 Comparison of strains near the plate center between the dry impact (broken line) and wet impact (solid line)

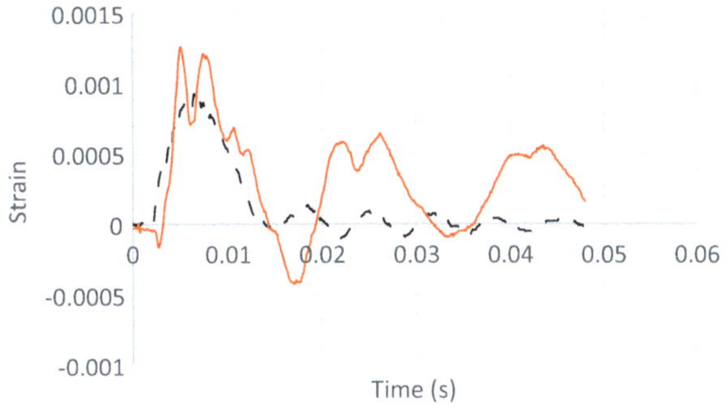

Fig. 5 Comparison of strains near the plate corner between the dry impact (broken line) and wet impact (solid line)

In order to study the non-uniform effect of FSI in further detail, a cantilever composite beam was studied experimentally [5]. A cantilever beam was allowed to oscillate in a state of free vibration while starting from the same initial deflection: in one case submerged in water and in the other case surrounded by air. Its vibrational motion was measured using the Digital Image Correlation (DIC) technique. The vibrational motion in water was very different from that in air. Because the beam was initially deflected with a tip force, its free vibrational motion was very close to the first mode shape of a cantilever beam in air. However, when the beam vibrated in water, its motion was much more complex. In the early time, the cantilever beam in water vibrates like the first mode shape, but later, the vibrational motion contains many higher modes. Figure 6 plots the vibration motion as time progresses. At early times, the initial deflection shape was maintained. As the vibration was damped out gradually, the effect of FSI became more distinctive. When the beam was in air, the vibrational motion was quite consistent until it completely stopped.

Another important observation from the experimental impact study was damage or failures in composite structures. Once a structure sustained damage in one condition, the same structure cannot be tested again in another condition. As a result of this, apple to apple comparisons cannot be made, since the test structures may not be identical. In order to address this problem, tests were conducted for multiple structures constructed using the same technique. The test results showed consistent

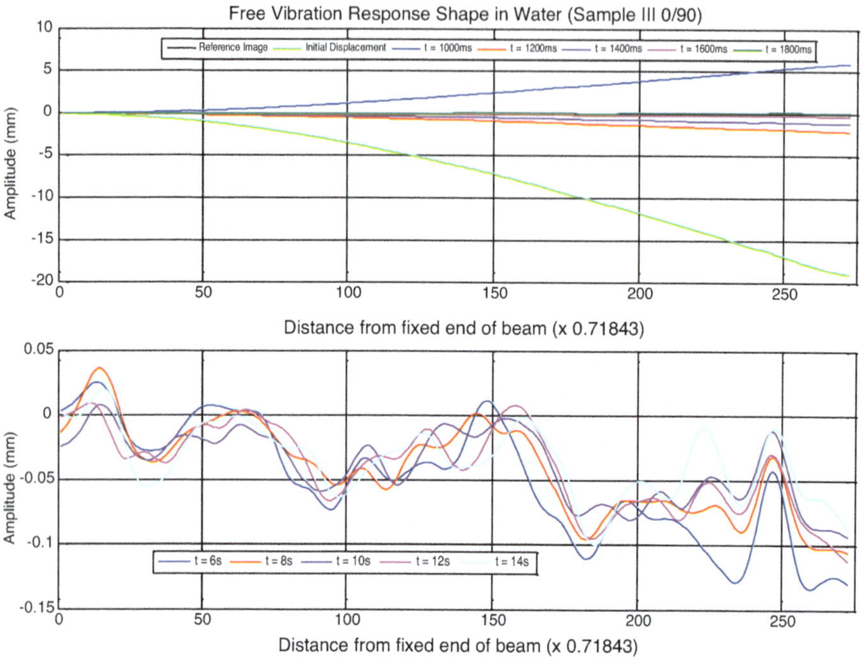

Fig. 6 Plot of free vibrational motion of a cantilever beam in water

findings suggesting they were not dependent on the discrepancy from one structure from another.

The greater strains resulting from FSI produced damage and failures of composite plates and beams at lower impact heights while the mass remained constant. This suggests that the FSI effect can cause premature failure of composite structures unless the effect of FSI was considered in their design and analysis.

Figure 7 shows the comparison of the delamination damage in E-glass skin sandwich plates. The test conditions were the same as described above. In order to minimize the moisture effect on the composite material, the tests were conducted immediately as the plates were submerged in water. The results showed that damage started at a lower impact load and grew faster for the plate in water.

A more interesting point was that the failure location might change because of FSI [2]. Sandwich beams were clamped at both ends, and impact was applied to the center. One set of tests were conducted while the beams were in air and the other sets of tests were undertaken when the beams were in water. The failure locations were different. When beams were in air, the failure occurred at one of their clamped end. On the other hand, the beams in water failed at their center. The tests were repeated about a half dozen times for each case. For the wet beams, five out of seven specimens failed at the mid-span while two failed at clamped ends. Five out of six dry samples failed at the clamped ends. The results were relatively consistent. Such a change in the critical location is also related to the nonuniform FSI effect over a structure.

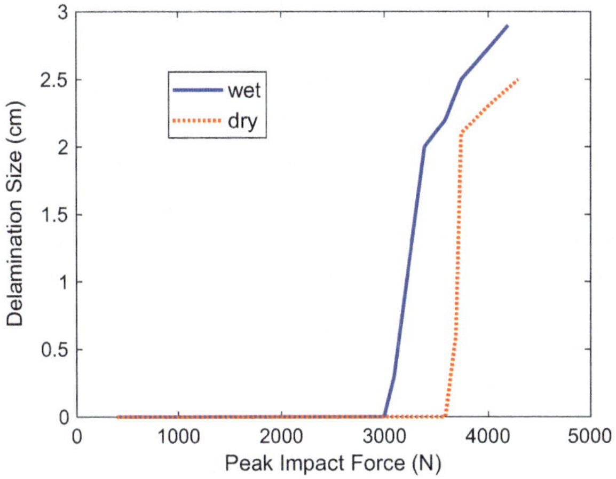

Fig. 7 Comparison of damage in sandwich plate with and without FSI effect

Fig. 8 Cyclic loading maximum and minimum force envelops of composite beam in water and air, respectively

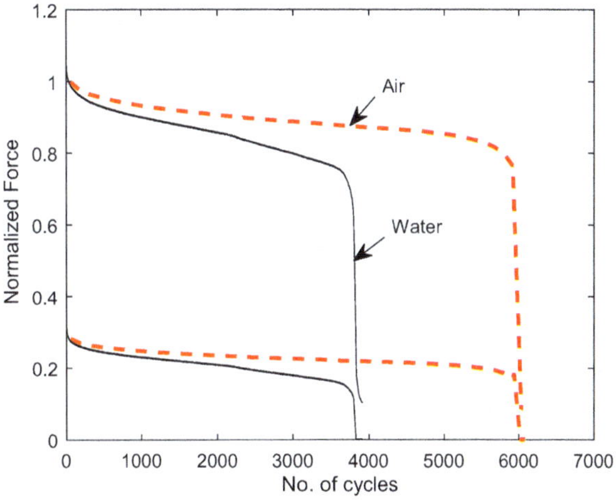

The next study used cyclic loading to composite beams until they failed [6]. The tests were conducted either in water or in air. Then, the numbers of cylices to failure were compared between the two conditions. The beams were subjected to three-point bending tests. The cyclic loads were applied to the center of the beams. The applied loading cycles were very far from their natural frequencies in air or water in order to avoid the resonance. The beams in water failed in lesser cycles resulting from FSI. Both 5 Hz and 10 Hz loadings were used, and the test results were similar. The failure cycles in water were approximately a half of those in air. The period of each test was quite short so that the effect of moisture was not important to influence the test results. Figure 8 shows the maximum and minimum force envelops of the cyclic loading as the beam specimen was controlled by displacements in air and water, respectively.

4 Numerical Studies of FSI Under Dynamic Loading

To complement the experimental study, a series of numerical work was conducted. The composite structures were modelled using the Finite Element Method (FEM) while the water medium was modelled using either the Lattice Boltzmann Method (LBM) or the Cellular Automata (CA) technique. [7–13]

The LBM technique was utilized to solve Navier-Stokes equations for viscous flows. The technique consists of two major steps: collision step and redistribution step. In this technique, a collection of imaginary fluid particles is distributed at every lattice point. Then, because of the collision of particles at the lattices, they are redistributed. This process is repeated to every lattice point and continues as time progresses. The details of the LBM technique are presented Ref. [14] so that its mathematical derivation is omitted here.

The CA technique was used to model the wave equation for the acoustic domain. In this case, a fluid is assumed to be stationary and the pressure wave propagates through the fluid medium. The CA technique is based on a set of rules applied to all lattice points. The major advantage of this technique is its simplicity and efficiency. The computer code is very short compared to other numerical techniques. For example, the 3-D linear wave equation is solved using the following equation:

$$p_c = \frac{1}{3}\left(p_e + p_w + p_n + p_s + p_f + p_b - 3p_c\right) \tag{1}$$

where subscript c denotes the center lattice which is the point for updating the value; and subscripts e, w, n, s, f, and b indicate the neighboring lattice points in the east, west, north, south, front, and back directions. This calculation is conducted at every other lattice points by skipping one. The next cycle updates the skipped lattice points using the same equation. This process continues and the numerical results shows wave propagation in a 3-D space. The CA technique is also very useful to represent various boundary conditions such as non-reflective boundary, free boundary, rigid boundary, etc. The details of the CA technique are also discussed in Ref. [14].

The CA technique was also applied to the dynamic analysis of beam and plate structures [13]. For example, the following rules are applied to every lattice points of a beam.

$$\begin{aligned}
M_i^t &= (EI)_i\left(w_{i+1}^t + w_{i-1}^t - 2w_i^t\right)/(\Delta x)^2 \\
\ddot{w}_i^t &= \left[\left(-M_{i+1}^t - M_{i-1}^t + 2M_i^t\right)/(\Delta x)^2 + p_i^t\right]/m_i \\
\dot{w}_i^{t+\Delta t} &= \dot{w}_i^t + \ddot{w}_i^t(\Delta t) \\
w_i^{t+\Delta t} &= w_i^t + \dot{w}_i^{t+\Delta t}(\Delta t)
\end{aligned} \tag{2}$$

in which M and w are the bending moment and the beam deflection, EI is the beam rigidity, m is the lumped mass of the node, and subscripts denote the lattice points. The over-dot indicates the temporal derivative of the variable, and Δt and Δx are the time and space increment. Similar expressions were also available for the plate bending analysis [13].

Coupling between the structural and fluid domains is conducted at their interface boundary. Unless there is cavitation at the structural boundary, both fluid and structure are considered to move together. Therefore, their velocities should be the same as well as the traction at the boundary should be equal and opposite. Using the developed numerical techniques, FSI problems were analyzed to better understand the experimental findings. The experimental data also served to validate the computational results.

One of the numerical studies was to model an impact load to a water-backed composite plate as discussed above. The top side is dry and the back side of the plate is wet. The plate was clamped along all edges. The plate was modelled using FEM while the water was modelled using CA. The measured impact force-time history was used as the applied force to the numerical model. Then, the strain response was computed numerically and compared to the experimental data. Figure 9 shows a comparison that is quite reasonable.

The next example included delamination damage in an E-glass composite plate with water-back, which was subjected to a sudden concentrated force at the center. The plate with or without inclusion of the delamination damage was compared in Fig. 10. The results show that the delamination reduced the strain on the surface slightly.

Another numerical study investigated the cyclic loading to a water-backed clamped beam or plate. Then, a spring-mass system was attached to the center of the structure. A sinusoidal force was applied to the lumped mass as sketch in Fig. 11. This is a simplistic representation of a structure supporting an equipment containing a rotational component. The frequency of the cyclic loading was varied and the structural deflection was computed as a function of the applied loading frequency. Figure 12 shows the comparison of deflections when the structure was in air or water, respectively. The parameters of the spring-mass system were initially selected such that the first natural frequency matched between the motion with and without

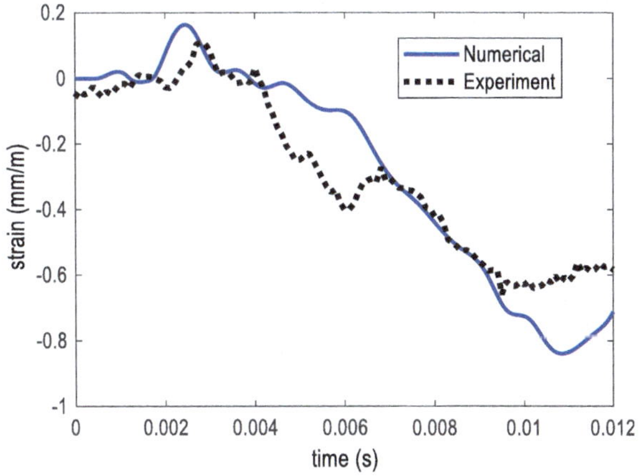

Fig. 9 Comparison of experimental and numerical results of clamped water-backed plate subjected to impact loading

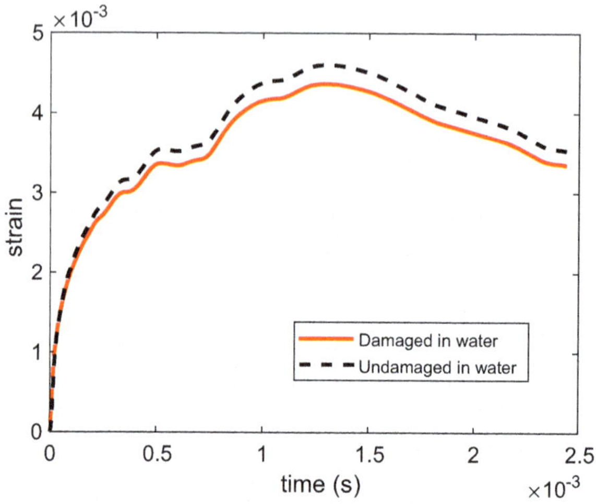

Fig. 10 Comparison of strains in clamped water-backed plate subjected to concentrated load with and without delamination

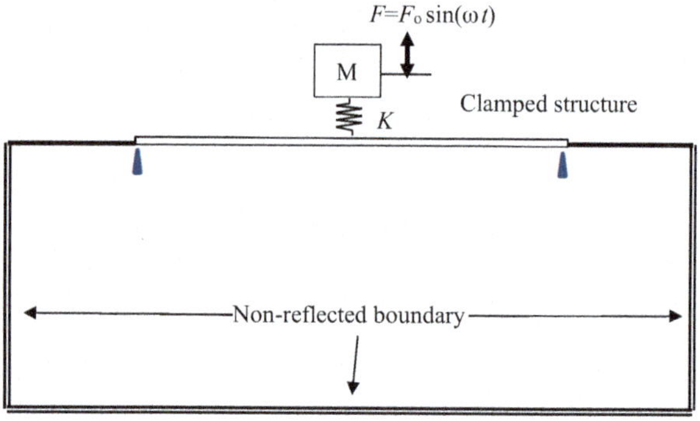

Fig. 11 Water-backed clamped structure subjected to cyclic loading through a spring-mass system

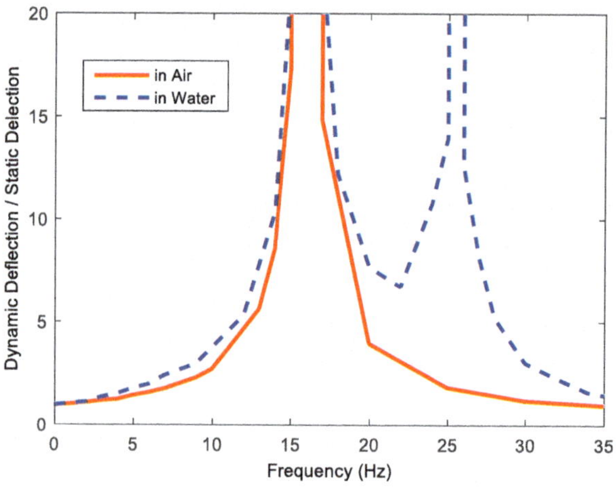

Fig. 12 Water-backed clamped structure subjected to cyclic loading through a spring-mass system

water. The results indicate that the effect of FSI, i.e. in water, yielded a greater dynamic deflection of the structure throughout the variation of the applied loading frequency.

5 Effect on Mode Shapes With FSI

It is well known that FSI changes the vibrational frequencies of a structure because of the added mass effect. The structural frequencies become lower with the effect of FSI. However, its effect on mode shapes and modal curvatures, which are the second derivatives of the mode shapes, have not been well understood. Therefore, a study was conducted to compare the mode shapes between the dry and wet structure. First, an experimental modal analysis was conducted, and a numerical modal analysis was also conducted later [5, 11]

The experimental modal analysis requires sensors to measure the dynamic response of a structure at various locations. Unless non-contact lasers are used, other sensors are attached to the structure to be tested. If accurate mode shapes are to be obtained, many sensors should be attached to the structure at critical locations. Those attached sensors can alter the response of the structure because of the added sensors. In order to avoid this problem, numerical modal analyses were conducted, which does not require any physical sensor and uses as many as degrees of freedom to represent the structure.

Ten small accelerometers were attached to a composite beam for the experimental modal analysis. The beam was freely hanged with a thin fishing rod so that it represented a free-free boundary condition. Then, first three mode shapes associated with non-zero natural frequencies were plotted. One set of tests were conducted with the beam in air while the other set was done with the beam submerged in water. The density of the beam was greater than that of the water so that the beam did not float. Their mode shapes were compared between the two conditions, in air or water, respectively.

The mode shapes were very close each other whether the aluminum beam was in air or water. However, the carbon composite beam showed notable differences in their mode shapes. When the modal curvatures were computed from the mode shapes, they were quite different between the two cases. Because modal curvatures are directly proportional to the bending strain, such a difference in the modal curvatures may explain the different strain responses.

The next study was the numerical modal analysis. An impulse load was applied to a structure to excite it. Then, displacements, velocities and accelerations were determined from the numerical analysis. These computed parameters were used to find the mode shapes and natural frequencies using the same algorithm used for the experimental modal analysis. For a dry structure, a standard eigenvalue analysis could be used to find the natural frequencies and mode shapes. However, this is not applicable for a wet structure with FSI. As a result, the experimental modal analysis technique was adopted for the numerical study.

Numerical modal analysis was conducted for a clamped composite beam. One case was a dry beam and the other was a wet beam submerged in water. Mode shapes were compared between the two cases [11]. There was a small difference in the mode shapes between the dry and wet beams. However, such a small difference resulted in a larger difference in the modal curvatures. Figure 13 compares both mode shapes and modal curvatures of a clamped beam in air and water, respectively.

Another numerical modal analysis was conducted for a clamped composite plate that was in air or water, respectively. Figure 14 compares one of the major mode shapes with and without the FSI effect. There was a striking difference in the two mode shapes around the corners of the plate. This may explain the strain responses from the previous experimental study. The experiment showed that the strain response near the center of the clamped plate was rather similar with or without FSI. However, the strain response was much different near the corners, where the strain with the effect of FSI was much greater than that without FSI. This observation seems to be related to the difference in the mode shapes.

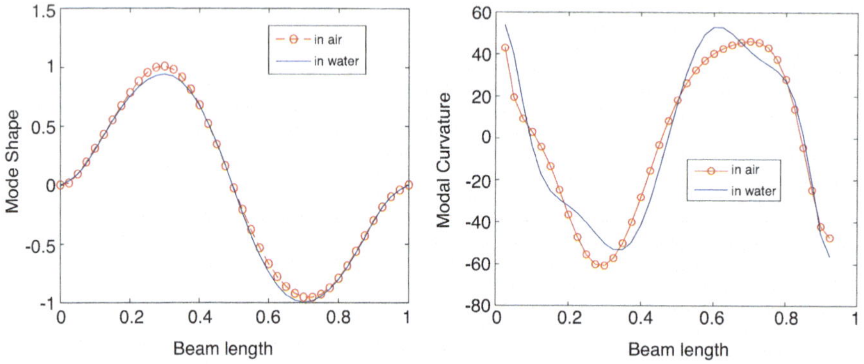

Fig. 13 Comparison of mode shapes and modal curvatures of a clamped composite beam in air and water, respectively

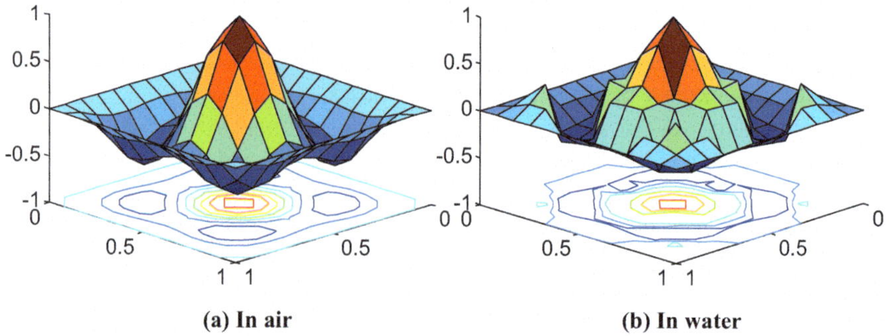

Fig. 14 Comparison of mode shapes of a clamped composite plate

6 FSI Study of Structure Containing Fluid

This section discusses a composite structure containing a fluid inside and subjected to dynamic loading. A thin square box with open top and bottom was fabricated out of E-glass composites as shown in Fig. 15. A very thick aluminum plate was attached to the bottom, and a thick Plexiglas was also attached to the top to enclose the box. Plexiglas was selected so as to be able to observe fluid motion inside the box.

A mechanical impact was applied to the center of the front side of the composite box, and the strain response of the other walls was measured for different water amounts as shown in Fig. 16. A load cell was attached to the tip of the impactor to measure the impact force. The entire set-up was placed on a table which had vibrational isolations so that the test set-up would not be affected by the environmental condition.

Water was filled in the box with different volumes from 0% to 100% of the box volume. Initially the water volume was 0%, 25%, 50%, 75%, and 100% of the square box volume, respectively. If necessary, other volumes were also considered. The impact loading was varied by the drop angle of the impactor. For each case, tests were repeated at least six times to make sure the responses were repeatable. The test results were very consistent as shown in Fig. 17 [15].

Figure 18 compares the impact force-time history for three different water levels; 0%, 50%, and 100%. The impact force increased as the water level increased. When the water level was 25%, the impact force was close to that of the no-water case. Likewise, when the water level was equal to and greater than 75%, the force

Fig. 15 Composite box with open ends at the top and bottom

Fig. 16 Impact test set-up for a composite box

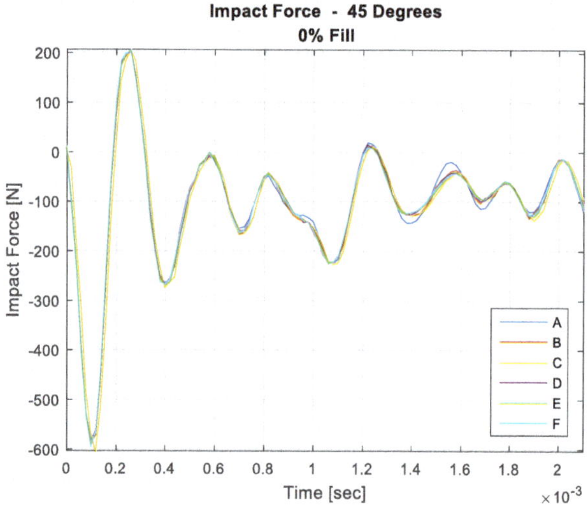

Fig. 17 Impact forces repeated six times at the same condition

response was almost identical. Therefore, the impact force did not increase gradually as a function of the water level. The influence of the water level on the impact force was more significant between 25% and 75% water levels.

Table 1 lists the measured peak impact forces for different water levels at two different drop angles. The peak impact force for 25° was quite consistently 54% of that for the 45° angle. Neglecting any frictional loss, the ratio of the initial impact velocity between 25° and 45° is 0.566. Therefore, the ratio of the peak impact force is very close to the initial impact velocity.

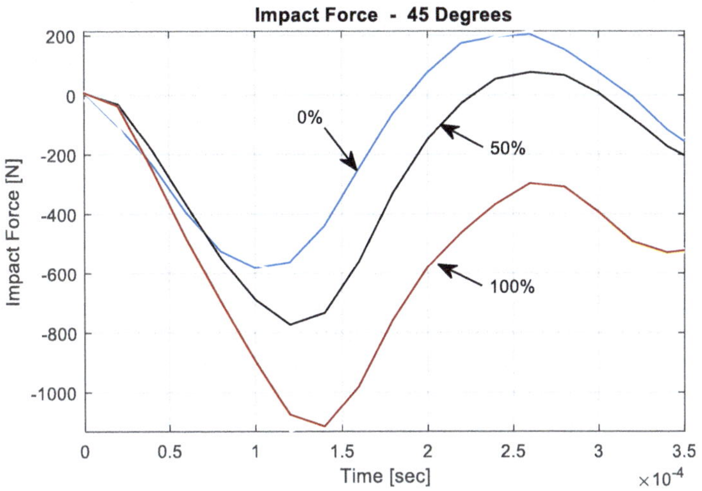

Fig. 18 Comparison of impact forces for three different water levels

Table 1 Measured peak impact force

Fill level	Maximum force (25 Degrees) [N]	Maximum force (45 Degrees) [N]
0%	315.5	590.9
25%	317.6	593.5
50%	417.2	772.5
75%	603.4	1130.5
90%	604.9	1127.2
95%	594.0	1121.2
100%	599.2	1117.6

The water level affected not only the peak impact force but also the contact time between the impactor and the composite structure. As the water level became higher, the contact duration became longer. For example, the contact duration with 100% water was almost twice as long as that with no water. The characteristics of the contact time was very similar to that of the peak impact force. The change in the contact duration was more noticeable between 25% and 75% water levels, and there was an increase as the water level went up from 95% to 100%.

The strain response on the front face was plotted in Fig. 19. Because the impactor struck the center of the front side, the strain was attached slightly off center. It is not surprising to see that the strain was grater with more water because the impact force was greater. Like the comparison for the peak impact force, the peak strain was almost twice larger for the 100% water as compared to the no water case. However, the latter case showed the maximum strain under compression on the front outer face while the former case had the maximum tensile strain at the same location. This suggested that the deformed shape would be very different between the two cases.

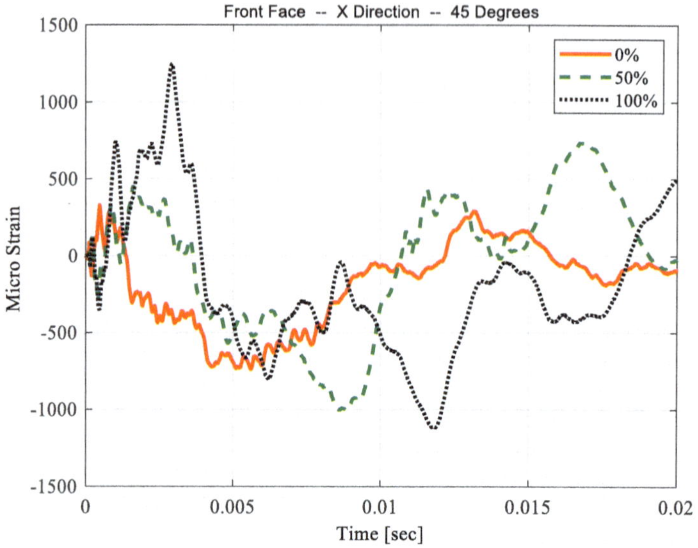

Fig. 19 Comparison of strains at front face for three different water levels

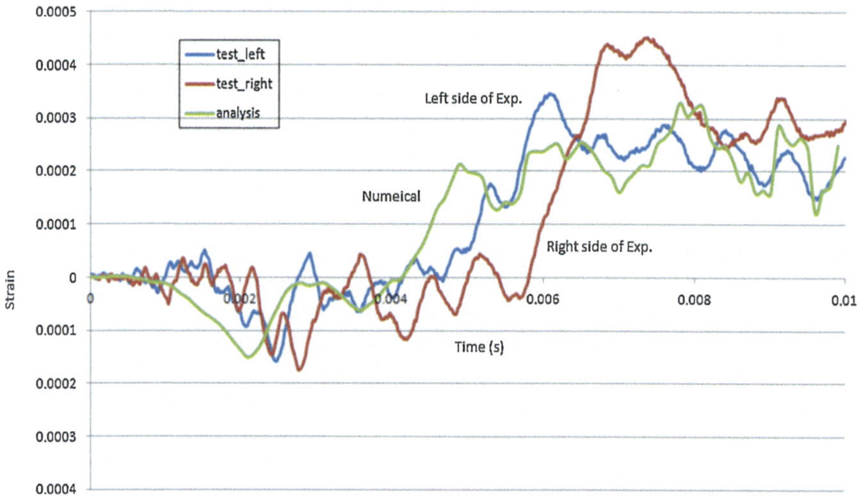

Fig. 20 Comparison of numerical and experimental strains at side faces for 50% water level

Numerical modeling and simulation was conducted for the impact loading on the composite box. Various cases as tested in experiments were simulated. The numerical and experimental results were compared. One of the comparisons is shown in Fig. 20, which is the strain response of the side wall with 50% water. While the fabricated composite box did not have the constant thickness because of the manual fabrication process, the numerical model assumed a constant wall thickness. As a

Table 2 Impactor contact time with structure

Fill level	Contact time [msec]
0%	13
25%	13
50%	18
75%	21
90%	22
95%	22
100%	25

result, the responses of the right wall and the left wall during the experiment were different as shown in Fig. 20. Considering such a difference in in the physical and numerical models, the agreement between the numerical and experimental results is reasonable (Table 2).

Once the numerical results were validated against the experimental data, additional information was obtained from the numerical study. Figure 21 shows the pressure wave plots in water along with time. The modelled water level was 100%. The pressure wave starts from the left side where impacted, and propagates to right side and reflected. The plot was at the mid-plane between two side walls.

In this study, the effect of a baffle was considered inside the water container. This was to study how a baffle interacts with the wave propagation in water and yields the resultant effect on the response of the composite box. The size of the baffle was smaller than the inner space of the composite box such that the baffle did not provide any structural support to the box wall motion. The baffle inserted to the composite box is shown in Fig. 22. Two types of baffles were considered. One was called Baffle 1 and the other called Baffle 2. Both baffles had the same geometry except for the opening to pass fluid motion. Both have the same total cross-sectional area of the opening, but Baffle 1 has a larger opening area with a lesser number of openings as compared to Baffle 2.

The experimental response of the composite box with and without a baffle was compared for different water levels. The effect of the baffle on the impact force was very small. However, its effect on the strain responses was noticeable. Figure 23 plots the strain responses of the front face among no baffle, Baffle 1 and Baffle 2 cases. The baffles reduced the strains on the front surface for every water level. The baffle effect was greater for higher water levels such as at least 75% water. On the other hand, the baffle effect on the back surface was smaller than that on the front side as Fig. 24 is compared to Fig. 23.

The vibration frequency of the box was determined from the strain responses by taking their FFT. The frequency was reduced as the water level increased as expected because of the added mass effect. Figure 25 plotted the frequency variation as a function of the water level. There was an almost linear reduction of the first vibrational frequency as the water level increase to 75% for the no baffle case. Any further increase beyond 75% did not affect the natural frequency. This suggested that the added mass effect was almost the maximum with 75% water

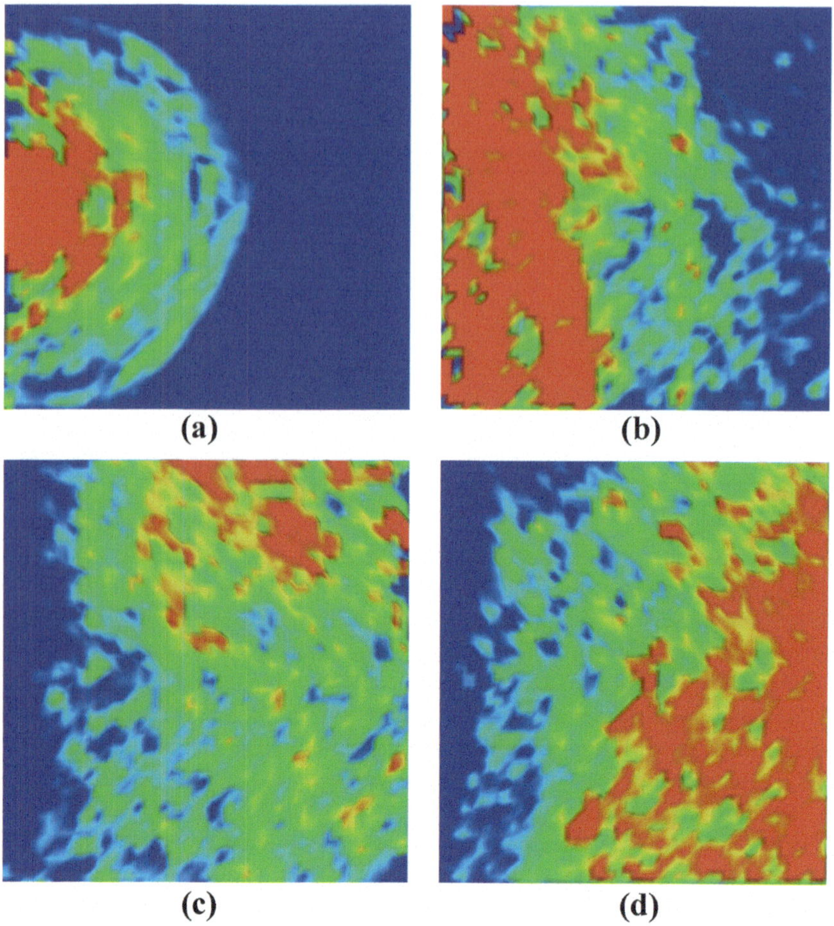

Fig. 21 Pressure wave propagation inside the water for 100% water case as time progresses from (**a**) to (**d**)

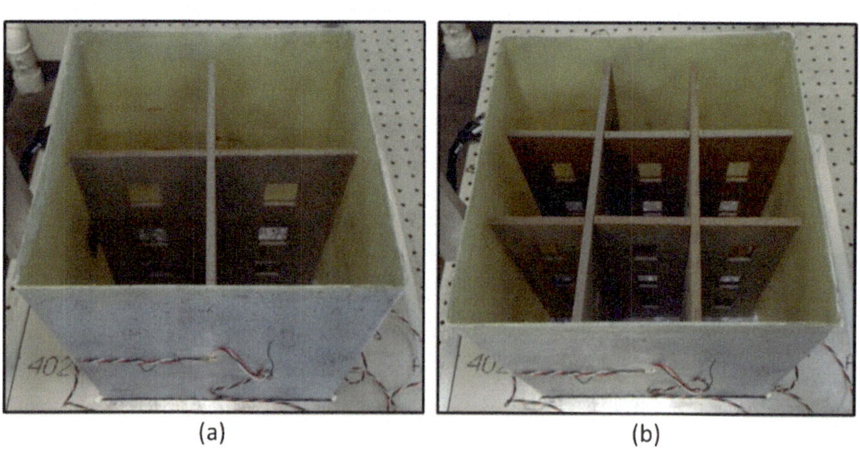

Fig. 22 Composite box containing two different baffles: (**a**) Baffle 1 and (**b**) Baffle 2

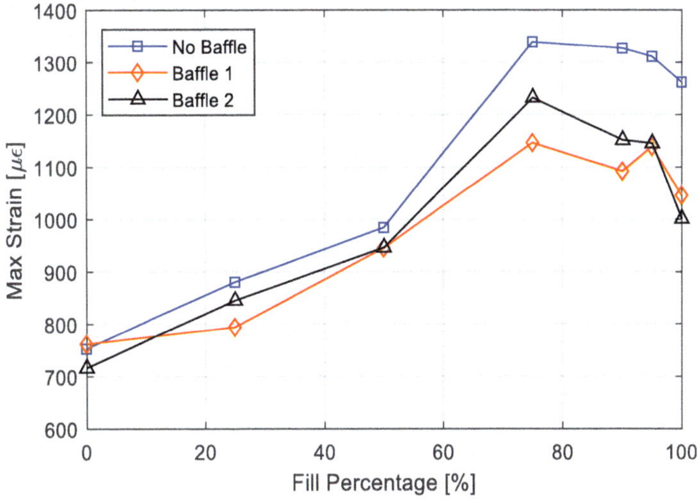

Fig. 23 Comparison of strain at the front surface

Fig. 24 Comparison of strain at the back surface

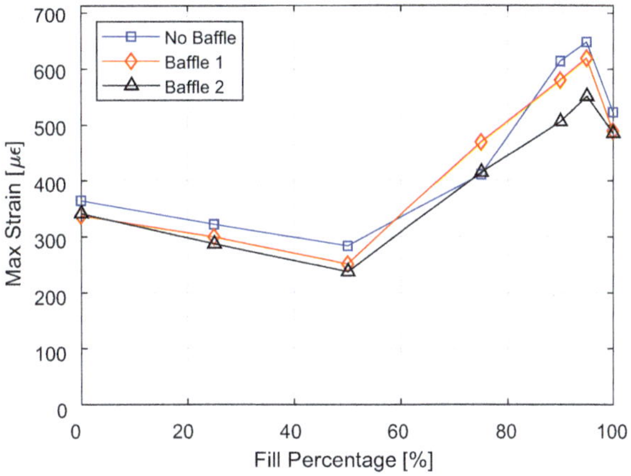

Fig. 25 Plot of the first vibrational frequency

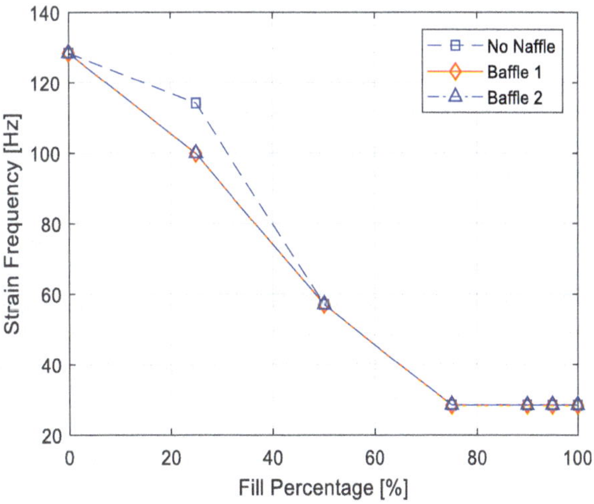

at least for the first mode of vibration. Baffle 1 also followed the same curve as the no baffle case. However, Baffle 2 showed a different value for the 25% water level and the rest of them were the same as other cases.

7 Structural Coupling via FSI

This section considered two separated structures while they were coupled through a fluid medium. For example, two parallel plates or beams were considered while there was water between the two structures. As one structure was impacted, its dynamic response was transferred to the neighboring structure through FSI. Both experimental and numerical studies were conducted [16–18]

Figure 26 shows the sketch of the test set-up. Two or more plates were held by rigid walls along all edges. Water was filled between any two plates. The water level was also varied as discussed in the previous section. Impact loading was applied to the first plate, and strain gages were attached to plates as necessary to measure their responses. The gage locations are shown in Fig. 27.

The impact force was applied using the same pendulum motion as before with a load cell. The impact force was measured for different test conditions. First, two plates were considered, which were 30.5 cm × 30.5 cm with a 12 cm separation distance. The water level was changed. At least ten tests were conducted for each case to make sure the repeatability of the test. The repeated test results confirmed the consistency of the test.

The impact force was plotted as a function of the water level in Fig. 28 for different water levels. Interestingly, the case with no water had the greater peak impact force than those with water. The observation of the impact force may suggest that the strain response would be greater for the no-water case because of the higher impact force. However, the strain results were the opposite.

Figure 29 plots the maximum strain of the front plate as a function of the water level. At all gage locations, the maximum strains with water was larger than that

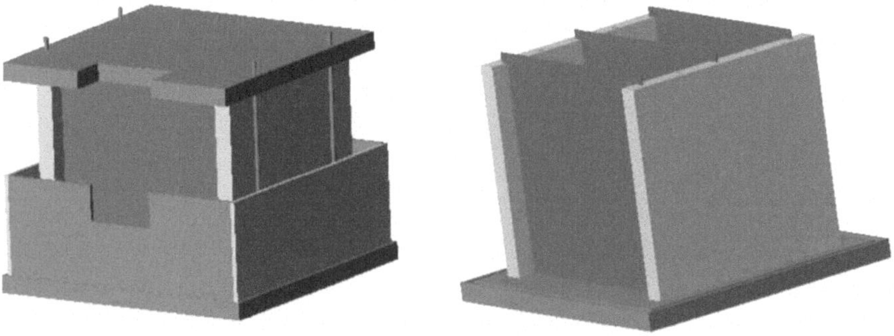

Fig. 26 Sketch of test set-up for structural coupling via fluid

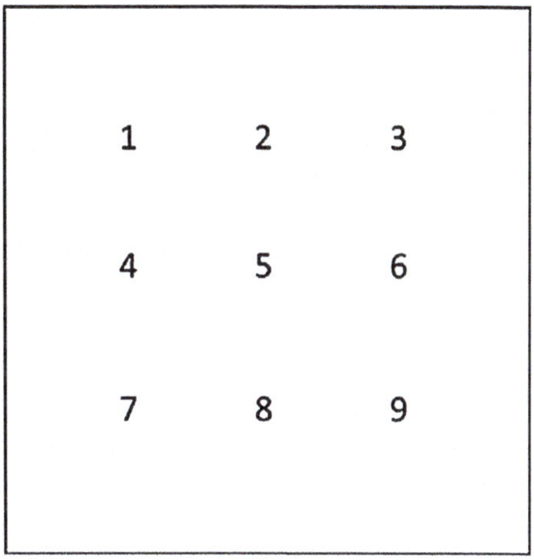

Fig. 27 Equally spaced strain gage locations

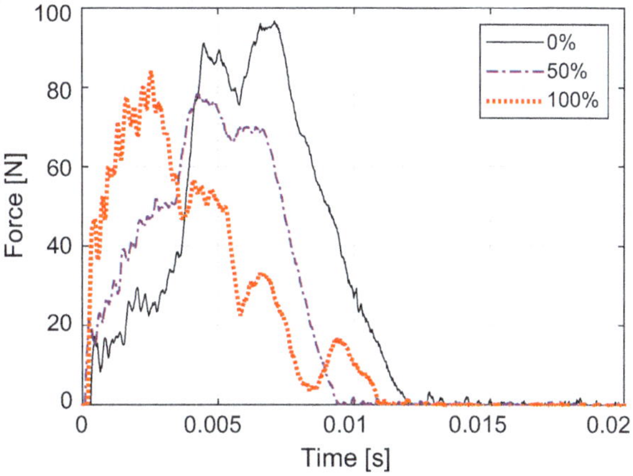

Fig. 28 Comparison of force-time histories with different water levels

without water. This can be explained by different loading rates. Even though the no-water case had a larger impact force, the time-history graph showed a slower loading rate, i.e. the lower slope of the graph in early times. On the other hand, the full water case had a lower impact force but a very fast rise of the force, i.e. higher loading rate. Such different loading rates contributed to the strain response of the plate.

In order to explain the effect of the loading rate, a simple spring-mass system was considered with different loading rates. The loading rate was represented by two linear curves as seen in Fig. 30. One case has a faster slope but a lower peak force, and the other case has the opposite condition. As each loading rate was applied to the spring-mass system, their responses were compared. In this situation, the natural frequency (or period) of the spring-mass system was varied, and the displacement of the higher loading was compared to that of the lower loading rate. That is, the ratio of

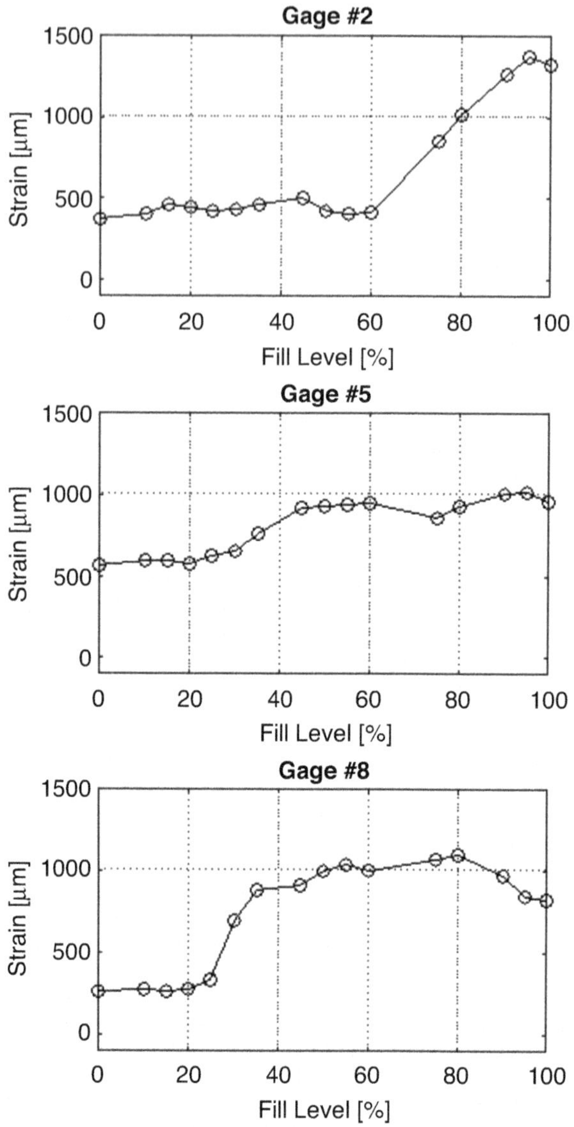

Fig. 29 Plot of maximum strains at different gage locations of the front plate as a function of the water level

the former case to the latter case was plotted as a function of the period of the spring-mass system as seen in Fig. 31. The figure shows that the higher loading rate can result in a larger displacement even though the force magnitude is smaller. This phenomenon seemed to occur for the tests so that the strains were greater with less magnitude of forces with water as compared to that without water.

Different separation distances were also studied to investigate their effects on the structural coupling. The effect of spacing was different depending on the location of each plate. Comparing the strains at gage 5 (i.e. center) for the front and back plates,

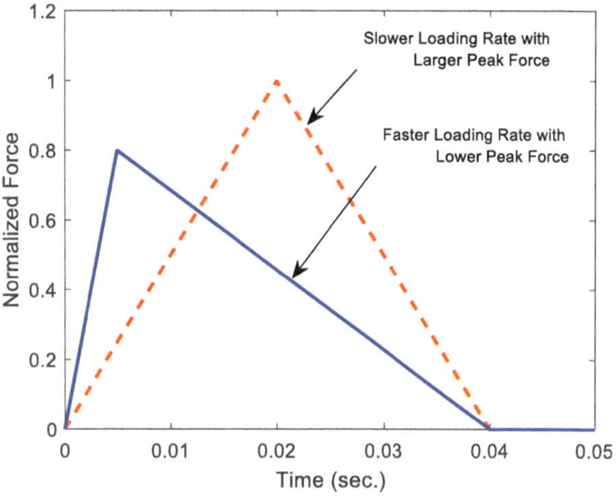

Fig. 30 Two different loading rates

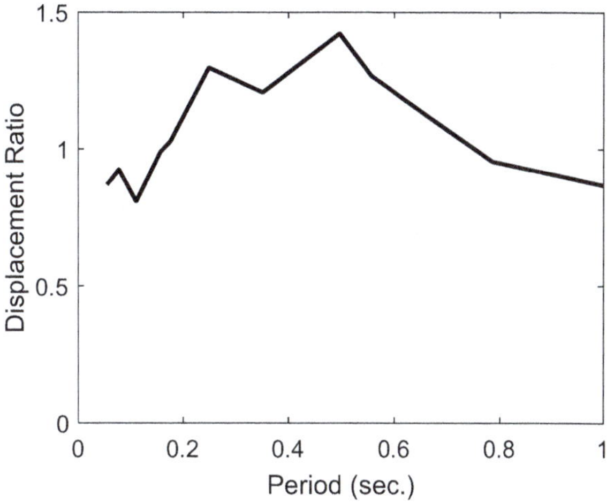

Fig. 31 Displacement ratio vs. period of spring-mass system

the back plate was more influenced by the different spacing as Figs. 32 and 33 are compared. In the figures, Sects. 2 and 4 have the twice and four times distance than Sect. 1, respectively. There was no consistent behavior as a function of the separation distance.

The next study compared the different plate thicknesses. The baseline was the same thickness plates for both front and back as denoted by FT2/BT2. Then, either the front or back plate was replaced by a two times thicker plate. The former cased was called FT4/BT2 and the latter case was called FT2/BT4. Figures 34 and 35 show the results. The figures show that FT2/BT2 yielded the largest strains for the center of the front plate while FT4/BT2 had the largest strain at the center of the back plate.

The impact forces were also compared for different plate thicknesses. Figure 36 compares the impact forces. As expected, a thicker plate resulted in a higher impact

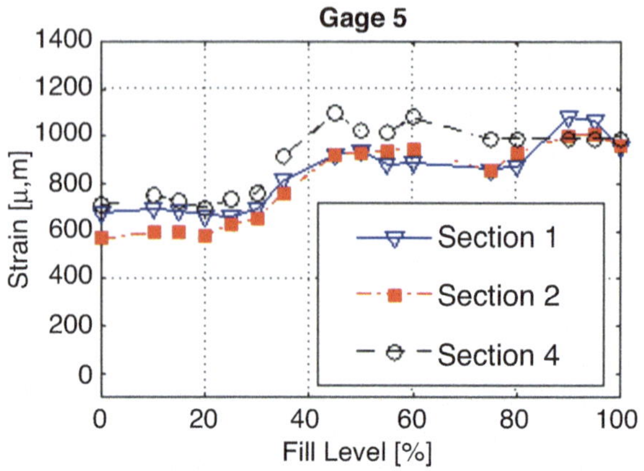

Fig. 32 Plot of maximum strains on the front plate vs. water level for different spacing distances

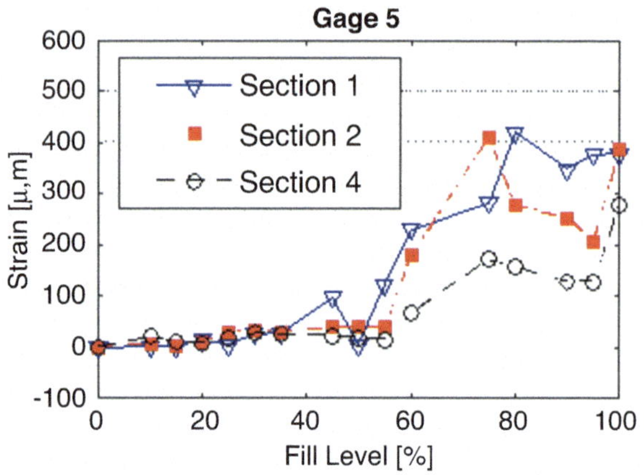

Fig. 33 Plot of maximum strains on the back plate vs. water level for different spacing distances

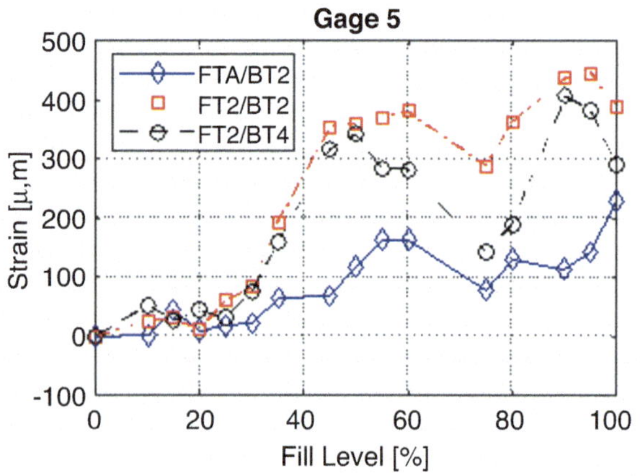

Fig. 34 Plot of maximum strains on the front plate vs. water level for different plate thicknesses

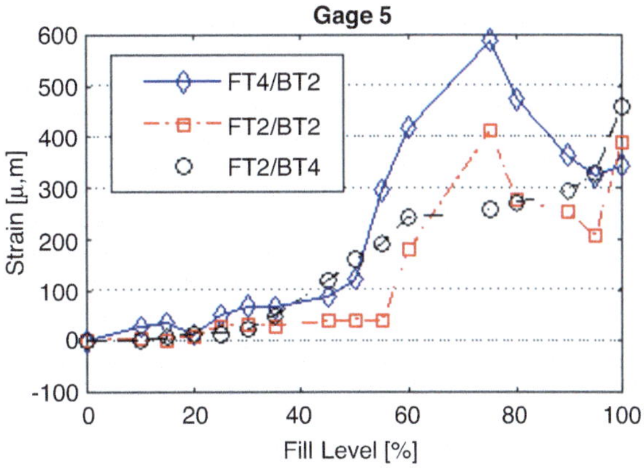

Fig. 35 Plot of maximum strains on the back plate vs. water level for different plate thicknesses

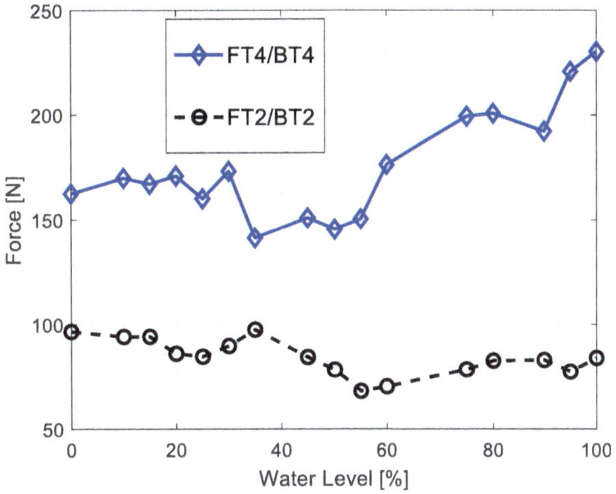

Fig. 36 Plot of peak impact force for two different plate thickness as a function of the water level

force. While the thinner plate showed the peak impact force was smaller with 60% or more water level as compared to that of the no-water case, it was opposite for the thicker plate. Thus, the plate thickness also played an important role with FSI on the impact force.

Numerical simulations were also conducted using the CA technique for the fluid domain. In the numerical study, the loading distribution could be varied. Instead of a local impact loading such as a concentrated force, a uniform pressure load was applied to the structure. The study suggested that the loading distribution also influenced the coupling effect significantly. When a uniform loading was applied, the front and back plates responded more similar each other. On the other hand, a concentrated force resulted in very different responses of the front and back plates. This makes sense. The pressure wave in the fluid has a more unidirectional propagation as a uniform pressure loading is applied to the front plate. This results in a point to point coupling between the front and back plates. However, the concentrated force produces a hemispherical pressure wave, which yields quite different responses between the front and back plates.

8 FSI on Moving Composite Structures

The next study considered a flexible plate structure moving in a fluid medium [19–21]. A composite plate was towed using a carriage in a tow tank as seen in Fig. 37. The towing speed was varied by changing the motor speed. Additionally, the plate orientation could be adjusted to represent the bow angle of different kinds of ships. Figure 38 illustrates the possible plate orientations which are either negative or positive as defined in the figure. The composite plate was held at the top center and two strain gages were attached to the plate. The number of strain gages were limited to the accessible data acquisition system that can transfer the data wirelessly to a computer for storage.

During the tests, three plate orientations were studied: positive, negative and zero (or neutral). The towing speed was also varied. As the speed was changed, both steady-state and transient motions were examined. First of all, an E-glass composite plate and an aluminum plate were compared for their drag forces. Both plates had the same geometric dimensions and the boundary conditions. As the towing velocity increased, the drag force increased as expected. However, there was a difference in the drag forces of both materials at the same towing speed. The composite plate showed greater drag force than the aluminum plate. Figure 39 shows the comparison. As the towing speed increased, the difference in the drag force also increased.

The strain attached to the composite plate was measured for two different steady-state towing speeds as the plate angle was varied. The result is plotted in Fig. 40. While the strain variation was small in terms of the plate angle for the towing speed 1 m/s, the strain was more dependent on the plate angle for the speed 1.5 m/s. This makes some sense since the net surface cross-sectional area becomes smaller as the plate angle increases or decreases. However, the maximum strain did not occur for the 0° plate.

Fig. 37 Tow tank

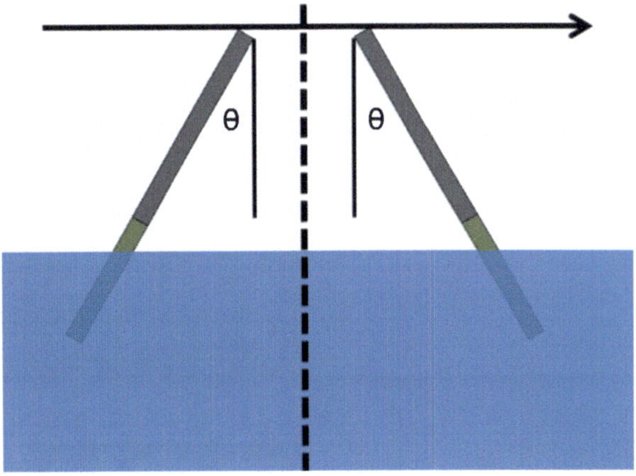

Fig. 38 Orientation of plate (The arrow is the towing direction, the left side angle is negative and the right side angle is positive)

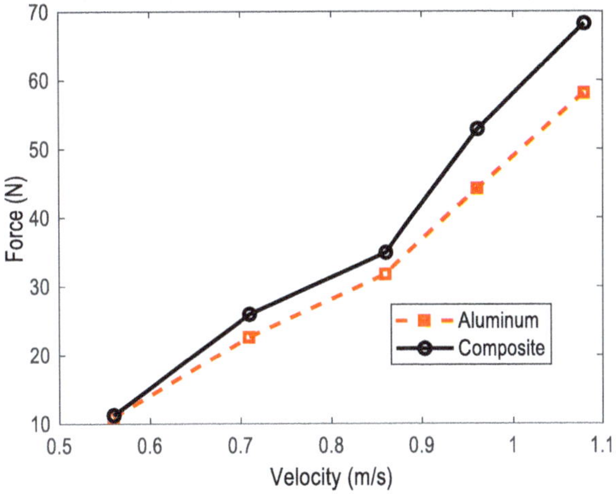

Fig. 39 Comparison of drag forces between composite and aluminum plates

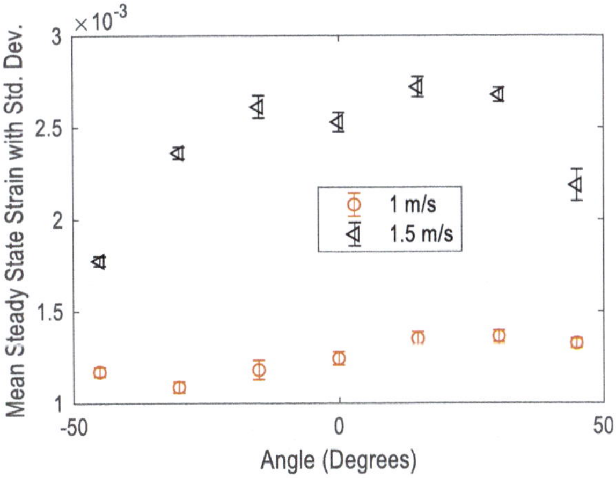

Fig. 40 Plot of strains as a function of the plate angle for two towing speeds

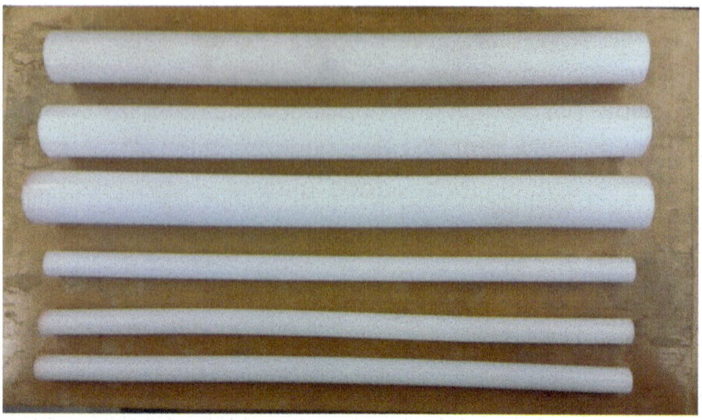

Fig. 41 Low density polyethylene plastic cylinders

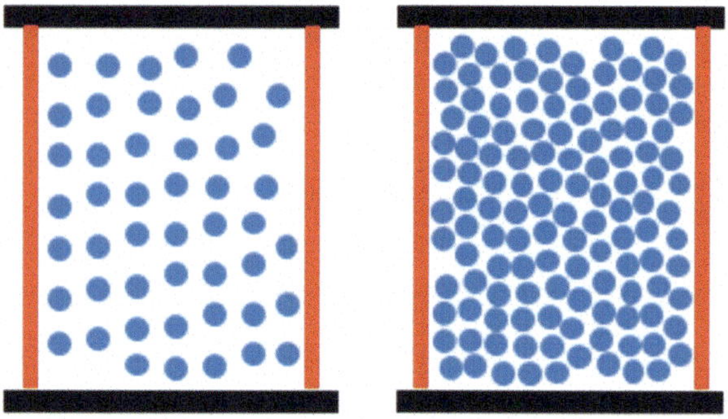

Fig. 42 Two different ice coverage ratios

The next study considered a structure travelling through ice equivalent floating blocks. This was to represent a ship travelling through the arctic region when chuck ice blocks floated in the sea. Because the tow tank laboratory did not have a temperature control capability, actual ice pieces could not been used. Instead, a Low Density Polyethylene (LDPE) plastic material, whose density is very close to that of ice, was used in the experiment. Different sizes of LDPE cylinders, as seen in Fig. 41, were selected and cut along the length directions to make them short cylindrical objects. The cylindrical length was determined such that the cylinder could float with the circular cross-section in the vertical position, i.e. facing up and down.

The circular cross-sectional area of each cylindrical block was measured, and the total number of blocks was decided based on what percentage of the water surface would be covered by the blocks. Figure 42 shows two different ice coverage ratios in the water. The right side had a twice more coverage than the left side. Once the ice equivalent bocks were placed inside the tow tank, the composite plate was towed through them at different plate angles. The towing speed was also varied.

Figure. 43 compares the frag forces with and without ice equivalent blocks in water under a constant towing speed. Three different plate orientations were also considered. The results showed that travelling though the ice equivalent blocks increased the drag force regardless of the plate angle. The drag force was the greatest for the positive angled plate and the least for the negative angled plate.

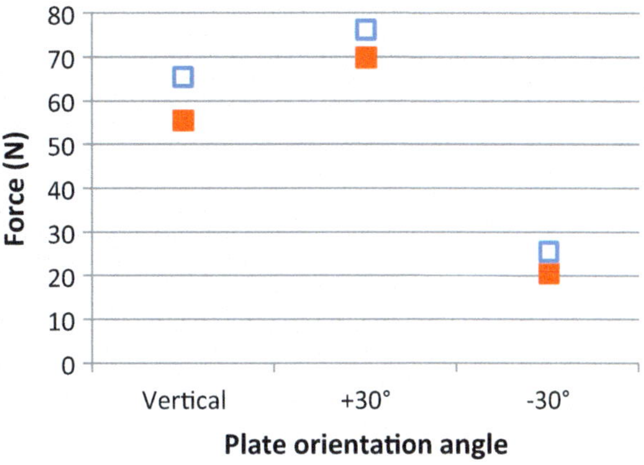

Fig. 43 Plot of drag forces with and without ice equivalent blocks under steady-state towing speed (open square for with ice blocks and solid square for without ice blocks)

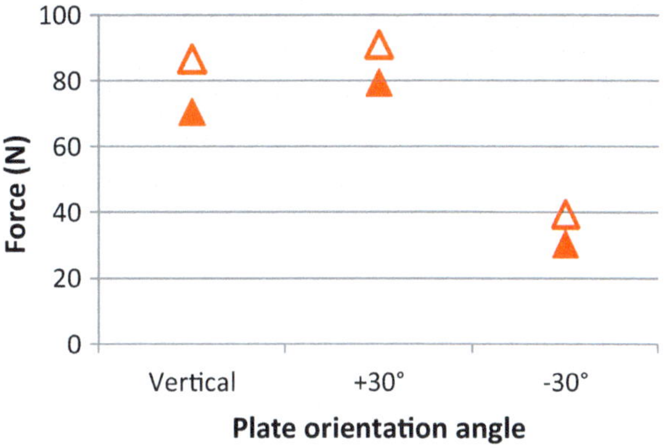

Fig. 44 Plot of drag forces with and without ice equivalent blocks under linearly increasing towing speed (open triangle for with ice blocks and solid triangle for without ice blocks)

A similar graph was plotted in Fig. 44 except that the towing speed was not constant but increasing linearly representing a constant acceleration. The general characteristics of Figs. 43 and 44 are the same. However, the case with acceleration had a greater force than the case of constant speed consistently. The effect of the ice blocks was the greatest for the vertical (i.e. 0°) plate and the smallest for the -30° plate in terms of the net absolute change in the force. In terms of the relative force ratio of the increase with respect to that without ice blocks, the -30° plate was the largest and the 30° plate was the smallest.

The interaction between the plate and ice blocks was recorded. Figures 45 and 46 show the ice blocks and the plate just before their interaction and almost at the end of the interaction, respectively. Their interaction was different depending on the plate orientation. When the plate angle was positive, ice blocks moved up along the plate. On the other hand, the ice blocks moved down along the negative angled plate. This makes sense in terms of their relative motions each other. Such ice motions are also

Fig. 45 Profile view of the ice section as the three plate orientations approach the ice equivalent block

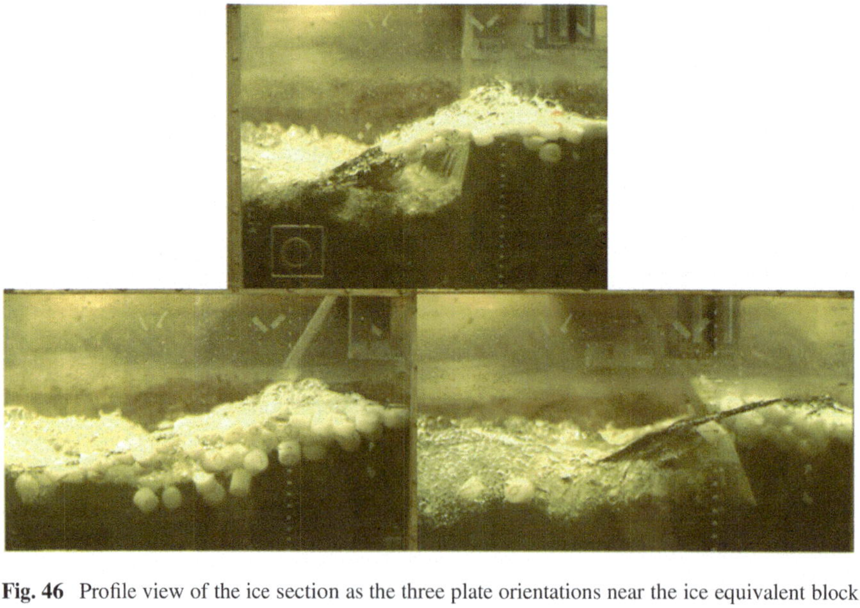

Fig. 46 Profile view of the ice section as the three plate orientations near the ice equivalent block

related to the water pressure distribution just in front of the plate. The negatively angled plate showed the fluid pressure decreased from the free surface toward the downward while the other showed the opposite at least around the free surface zone.

9 Conclusions

This chapter presented FSI problems associated with polymer composite structures. The topics included impact loading, cyclic loading and fluid loading on the composite plates. A composite box containing fluid inside was studied for its dynamic response subjected to impact loading. Structural coupling of two or more independent structures coupled through only fluids was also studied. The last topic was a study of a flexible composite structures travelling through water.

Both experimental and numerical studies were conducted. For the experimental studies, original test set-ups were designed and fabricated for the specific study. For the numerical studies, FEM, LBM, and CA techniques were used for coupling of structures and fluids for computational efficiency.

Some of the major findings are summarized here.

- The effect of FSI resulted in a greater impact force on composite structures as compared to that without FSI under the same external loading condition. This resulted in greater strains and failures at lower loading conditions with FSI. If a structure were deigned without considering FSI, the structure would fail prematurely because of FSI.
- The effect of FSI may also influence the failure location of a structure. Since FSI is not uniform over a structure, some locations can have larger effect than others. This can shift the most critical location from the original locations without FSI to another location with FSI.
- Both experimental and numerical modal analyses were conducted to determine the effect of FSI on frequencies and mode shapes. The FSI influenced not only vibrational frequencies, as well known, but also mode shapes. It also affected the modal curvatures which are the second derivative of mode shapes. The modal curvatures are directly related to bending strains of the structure.
- The nonuniform effect of FSI activated many high frequency modes of structures. As a result, the vibrational motions were much more complicated with FSI as compared to those without FSI.
- The FSI also shortened the number of cycles to failures under fatigue loading. Both experimental and numerical results supported this finding.
- The different water level inside a composite box dictated the FSI effect on the structure.
- There were many parameters that played an important role on structural coupling by fluids. Those parameters were the separation distance, plate thickness, loading type, number of plates, etc.

- The vibrational frequency of the structure decreased as the water level inside or between the structures. However, there was no more reduction in the frequency as the water level reached 75% or beyond.
- The drag force on a flexible composite structure travelling in water was greater than that of the equivalent aluminum structure. The plate orientation angle influenced the drag force as well as interaction with floating ice equivalent blocks.

Acknowledgements The work presented here was conducted by many graduate students. They were Angela Owens, Jarema Didoszak, Eric Priest, Michael Violette, Ryan Conner, Ryan McCrillis, Scott Bolstad, Scott Knutton, Scott Millhouse, Spyridon Plessas, Stephen Arceneaux, Taylor South, Dariush Alaei, Linda Craugh, Stuart Blair, etc. The research was mostly conducted with the financial support by the Office of Naval Research Solid Mechanics Program. Dr. Yapa Rajapakse is the Program Manager of Solid Mechanics program.

References

1. Kwon YW, Owens AC, Kwon AS, Didoszak JM (2010) Experimental study of impact on composite plates with fluid-structure interaction. Int J Multiphys 4(3):259–271
2. Kwon YW, Violette MA, McCrillis RD, Didoszak JM (2012) Transient dynamic response and failure of sandwich composite structures under impact loading with fluid structure interaction. Appl Compos Mater 19(6):921–940
3. Kwon KW, Violette MA (2012) Damage initiation and growth in laminated polymer composite plates with fluid-structure interaction under impact loading. Int J Multiphys 6(1):29–42
4. Kwon YW, Conner RP (2012) Low velocity impact on polymer composite plate in contact with water. Int J Multiphys 6(3):179–197
5. Kwon YW, Priest EM, Gordis JH (2013) Investigation of vibrational characteristics of composite beams with fluid-structure interaction. Compos Struct 105:269–278
6. Kwon YW, Teo HF, Park C (2016) Cyclic loading on composite beams with fluid structure interaction. Exp Mech 56(4):645–652
7. Kwon YW (2011) Study of fluid effects on dynamics of composite structures. ASME J Press Vessel Technol 133:031301–031306
8. Kwon YW (2013) Analysis of laminated and sandwich composite structures using solid-like shell elements. Appl Compos Mater 20(4):355–373
9. Craugh LE, Kwon YW (2013) Coupled finite element and cellular automata methods for analysis of composite structures with fluid-structure interaction. Compos Struct 102:124–137
10. Kwon YW (2014) Dynamic responses of composite structures in contact with water while subjected to harmonic loads. Appl Compos Mater 21(1):227–245
11. Kwon YW, Plessas SD (2014) Numerical modal analysis of composite structures coupled with water. Compos Struct 116:325–335
12. Blair SR, Kwon YW (2015) Modeling of fluid-structure interaction using lattice Boltzmann and finite element methods. ASME J Press Vessel Technol 137:021302
13. Kwon YW (2017) Finite difference based cellular automaton technique for structural and fluid-structure interaction applications. ASME J Press Vessel Tecnol 139:041301
14. Kwon YW (2016) Multiphysics and multiscale modeling: techniques and application. CRC Press, Boca Raton
15. Kwon YW, South T, Yun KJ (2017) Low velocity impact to composite box containing water and baffles, composite structures. ASME J Press Vessel Tecnol 139(3):031304
16. Kwon YW, Bowling JD (2018) Dynamic responses of composite structures coupled through fluid medium. Multiscale Multidiscip Model Exp Des 1(1):69–82. https://doi.org/10.1007/s41939-017-0002-9

17. Bowling JD, Kwon YW (2018) Coupled structural response via fluid medium. Multiscale Multidiscip Model Exp Des 1(3):221–236
18. Kwon YW (2018) Coupling of different computational techniques for fluid-structure interaction analysis. Int J Eng Anal Simul Addit Manuf, Accepted for publication
19. Kwon YW, Bolstad SH, Didoszak JM, Rodriguez JA (2016) Study of composite plate travelling in water containing ice equivalent objects. Compos Struct 135:38–48
20. Kwon YW, Knutton SC (2015) Computational study of effect of transient fluid force on composite structures submerged in water. Int J Multiphys 8(4):367–395
21. Kwon YW, Millhouse SC, Arceneaux S (2015) Study of composite plate in water with transient and steady sate motions. Compos Struct 123:393–400

Low Velocity Impact of Marine Composites: Experiments and Theory

Valentina Lopresto, Ilaria Papa, Vito Pagliarulo, Pietro Russo, and Maurizio Porfiri

1 Introduction

Several elements of naval and aerospace structures are regularly subjected to impact loading at room and low temperatures. Variations in the temperature are responsible for dramatic changes in the onset, architecture, and mechanics of damage [1–8]. Particularly relevant is to elucidate the dynamic response of lightweight composite materials, whose integration in marine vessels and aircraft is steadily growing [9–18]. In these applications, impact loading may trigger complex fluid-structure interactions.

Motivated by limited experimental evidence in the technical literature, this chapter presents a comprehensive database which can be used by high-end practitioners in naval and aerospace engineering, along with experimental and theoretical insight that can support an improved understanding of the dynamic response of composites. Such an improved understanding is critical for designing effective

V. Lopresto (✉) · I. Papa
Department of Chemical, Materials and Production Engineering, University of Naples Federico II, Naples, Italy
e-mail: lopresto@unina.it; ilaria.papa@unina.it

V. Pagliarulo
CNR National Research Council, ICIB, Institute of Cybernetics, Pozzuoli, NA, Italy
e-mail: vito.pagliarulo@cnr.it

P. Russo
CNR National Research Council, Institute for Polymers, Composites and Biomaterials, Pozzuoli, NA, Italy
e-mail: pietro.russo@unina.it

M. Porfiri
Department of Mechanical and Aerospace Engineering and Department of Biomedical Engineering, New York University Tandon School of Engineering, Brooklyn, NY, USA
e-mail: mporfiri@nyu.edu

S. W. Lee (ed.), *Advances in Thick Section Composite and Sandwich Structures*,
https://doi.org/10.1007/978-3-030-31065-3_8

manufacturing techniques to enhance the toughness of the matrix or the inter-ply interface.

Properties of composite laminates with plastic matrices at different temperatures have been extensively investigated, focusing on the onset and mechanics of damage. For example, changes of material properties in response to temperature variations were studied in [1] by Howard and Hollaway. In [2], Shindo et al. investigated the thermal mechanical response of non-metallic woven composites with temperature-dependent properties. A finite element method was used to explore the influence of crack formation, residual thermal stresses, and weave curvature on the mechanical performance of glass–epoxy laminates at low temperatures. The tensile properties of a glass/epoxy composite subjected to thermo-mechanical cyclic loads at low temperatures (up to 10 cycles) from room temperature (r. t.) to −50 °C, to −100 °C, and to −150 °C (c. t.) were studied in [3], using an environmental test chamber. Therein, Kim et al. demonstrated that the tensile stiffness significantly increases with decreasing temperature and the tensile strength decreases with decreasing temperature, while thermo-mechanical cycling plays a secondary role. In [4], Sefrani et al. presented an experimental analysis on the effect of temperature on both the stiffness and the damping of glass fiber laminates. Their experimental results indicate that mechanical properties are nearly constant up to the glass transition temperature, where the damping increased sharply in a rather small temperature interval.

Experimental tests on aramid/epoxy composite samples reinforced with glass fibers were performed in [5] using the Charpy impact method. The effects of the fiber volume fraction and the temperature on the impact toughness were tested. Impact damage was observed using a microscope at a magnification of 100. Experimental results indicate a modest increase of impact toughness of composite samples with increasing temperature in the interval − 40 °C–−10 °C. This is followed by a larger increase of the impact toughness with increasing temperature from −10 °C to room temperature. The aramid/epoxy system was observed to exhibit a higher impact toughness than glass/epoxy for all the tested temperatures [6–8]. Fiber failure was found in the composite sample. Increasing the fiber fraction volume decreases the impact toughness of glass and aramid/epoxy composites.

While many studies have characterized the properties of polymer composite materials at different temperatures, only few of them examined the influence of the temperature on the impact response. Filling this gap in knowledge is one of the key objectives of this chapter, which brings forward the investigation of the impact behavior of composite laminates of different thickness, in air and in the presence of water. Another objective of this chapter is to present a mechanics-based understanding of the fluid-structure interaction occurring during impact in the present of the water. A physically-based understanding of the fluid-structure interaction will help, in fact, to clarify the process of load redistribution during impact, inform the interpretation of empirical observations, and assist in the design of future experiments.

With respect to Navy-relevant research, it is crucial to simulate real conditions for the navigation in arctic environments, which calls for experimentally proxying complex fluid-interactions at extreme temperature. If a panel is supported by water instead of being in air, damage will be triggered by different load distributions [9]. In the case of water-backing, further variations will occur if the loading is exerted from the water or the air side, and variations in the boundary conditions could further add to the complexity of the impact. However, only a few studies [10–13] have examined the behavior of composite laminates subject to impact at low temperature or [14–18] have tackled the problem of the fluid-structure interaction. To the best of our knowledge, the combination of low temperature and fluid-structure interaction is an untapped area of research.

Low velocity impact is one of the most subtle threats to composite materials [19–26]. Due to weak bonds between the plies, even small energies imparted by out-of-plane loads can result in hardly detectable damages, such as matrix cracks, delamination, and fiber breakage. In turn, these damages will cause considerable stiffness and strength losses in tension, and, even more, in compression, thereby severely reducing structural integrity. Hence, it is crucial to understand the mechanisms and the mechanics of impact, in order to predict the residual strength of the damaged structures.

The earliest observable damage affecting a laminate subject to low velocity impact is typically delamination, which is mainly responsible for the compression strength degradation. Intralaminar fractures, the first failure caused by an impact, have a negligible influence on the mechanical behaviour of the laminate, but they cannot be disregarded since delamination starts from them. For this reason, significant research [20, 22–24] has been devoted to studying the mechanisms of delamination initiation and growth. Since residual material properties after impact are of primary concern in applying damage tolerance concepts, many authors have tried to correlate, analytically or experimentally, the residual strength to impact energy with the damage mechanisms. Most of the previous endeavours have focused on understanding whether the structure could continue to function after an impact, which requires the study of how delamination extension influences the residual capability to support structural loads. Toward this end, it is critical to study the onset, extension, and propagation of delamination in composite laminates subject to low velocity impact. Such an analysis could be carried out by varying the thickness of the laminates and modulating the impact energy up to penetration.

Studies on delamination and other damage mechanisms have often been conducted separately [20–24]. A composite laminate could develop more than one delamination along the thickness during impact, depending on the impact energy and the stacking sequence. How do these delaminations start? How do they propagate? Does one delamination play a stronger role than others during subsequent impacts or transverse loading? Are there interactions between different forms of damage? Answering these questions call for new experimental research that could inform the comprehension of these complex phenomena, verify semi-empirical models, and support the refinement of physically-based theories.

This chapter reports low velocity impact tests up to complete penetration for increasing energy levels, carried out in a modular falling weight tower and followed by non-destructive and destructive damage investigations. Through extensive experiments at increasing impact energies, we sought to investigate damage onset and propagation, as well as the interaction between several damages. Semi-empirical models for the prediction of the influence of impact parameters on the dynamic behaviour, from the technical literature, were critically examined and verified in the present research. Impact parameters [25, 26], including penetration and absorbed energy, maximum force, indentation depths and delaminated area, were analyzed and utilized to validate the models toward the final aim of predicting the residual strength after impact.

Experiments were carried out on carbon fiber laminates, subjected to experimental low velocity impact tests at three different temperatures: room (or ambient) temperature, -25 °C, and -50 °C. Compression-after-impact (CAI) tests were performed to characterize the residual strength through a new, simpler, apparatus, which was designed and realized at the University of Naples. Ultra sound (US) technique is used to investigate the delamination, and a new holographic technique, ESPI (electronic speckle pattern interferometry), is proposed as a promising alternative technique to gather real time in situ measurements. Measurements were compared with results from US [17–19].

The rest of the chapter is organized as follows. First, we summarize low velocity impact tests on composite laminates until penetration and for different impact energies. Then, we study the influence of impact parameters and the temperature on the impact behavior. Third, we examine the influence of surrounding water on the impact response. Fourth, we report on CAI tests, carried out using the new equipment designed and realized at University of Naples to estimate and predict the residual strength of the impacted laminates. Fifth, we document damage investigations, comparing among the innovative technique ESPI with and traditional non-destructive testing methods. Then, we turn to semi-empirical and analytical models to predict the dynamic response of composites. Finally, we summarize ongoing theoretical research toward an improved understanding of fluid-structure interactions during impact.

2 Materials and Experiment

2.1 Materials

We studied carbon fiber reinforced plastics laminates (CFRP), obtained by overlapping $n = 7$, 10, and 13 T700 carbon fabric plies (0°/90°) 300 g/m^2, with $t = 2$, 3 and 4 mm nominal thickness. The number of layers was experimentally varied to examine its role on the impact response. Square panels, 600 × 600mm, were fabricated at the Naval Surface Warfare Center (NSWC), by vacuum infusion process using the Ashland Derakane 510A vinyl ester resin, commonly used by the US Navy. The final volumetric fiber percentage, V_f, was 48%. From the panels, we cut rectangular specimens 100 × 150mm with a diamond saw, following the ASTMD7137 Standards for CAI tests.

2.2 Experimental Setup

Experimental tests were carried out in a drop weight machine, Ceast/Instron, where rectangular specimens were impacted at their geometric center by an instrumented impactor of cylindrical shape, with an hemispherical nose 19.8 mm in diameter. The minimum mass of the impactor was 3.6 kg, but additional weights could be added up to 10.6 kg to increase the impact energy. The maximum drop height of the testing machine is 1 m, which corresponds to a maximum impact energy of 35.7 J with the minimum mass of the impactor. Different impact energy values could be obtained using additional weights and/or varying the drop height. A maximum velocity of 20 m/s could be reached by using a preloaded spring. An impact velocity of about 4.0 m/s was measured for the tests reported below.

The drop weight apparatus was equipped with a motorized lifting track. A piezoelectric load cell allowed to record the force–time curve. The acceleration history was calculated by dividing the force by the mass of the impactor. The displacement curve was obtained by double integration of the acceleration, and the force–displacement curves were plotted. At the initial time, the dart coordinate is zero and its initial velocity can be obtained from standard dynamics [24, 25], as follows:

$$\nu_0 = \sqrt{2g\Delta h} \tag{1}$$

where Δh is the height loss of the center of the mass of the dart, with respect to the reference surface of the specimen and g is the gravity [24, 25]. The impact velocity was also measured by an optoelectronic device.

Data were stored after each impact and the impactor was caught by an anti-rebound system to avoid multiple impact on the same specimen. The specimens were loaded until complete penetration; the whole load-displacement curve, up to penetration, is offers important insight into the impact in terms of the relationship between the force and the deflection of the panel. Characteristics points on the rising path of the load curve, such as load drops or local changes in slope, were used to evidence variations in material behavior as a consequence of a damage. Measurements were carried out at increasing energy levels and followed by indentation tests to investigate the damage onset and propagation.

During each test, the complete force-time and force-displacement curves, as well as energy and velocity curves, were recorded by the DAS16000 acquisition program and successively studied to evaluate the impact parameters (maximum force and corresponding displacement, absorbed energy, and failure load). All the parameters obtained in different conditions were compared. We tested three different energy values, 5 J, 10 J and 20 J, to study damage formation and propagation.

Updating the machine with a thermal chamber allowed to perform tests at low temperatures, up to −50 °C. Liquid Nitrogen was used to lower the temperature. From 10 to 15 min were necessary to measure −25 °C and −50 °C in the chamber, before each single test could be conducted. The specimens were stored in a freezer at

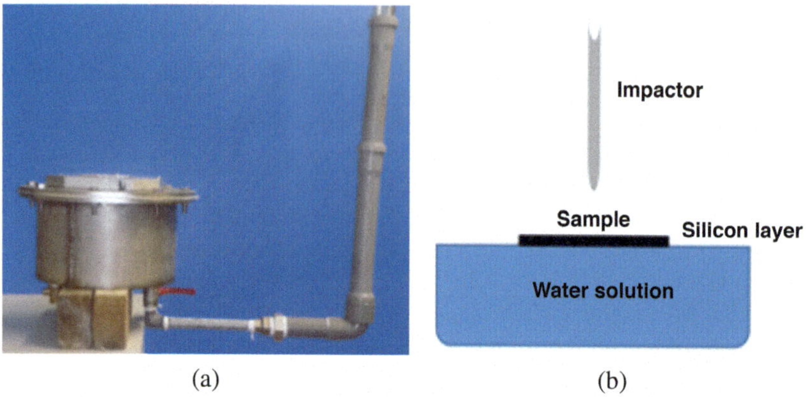

Fig. 1 (**a**) Steel box full of water and (**b**) schematic view of the experiment

−18 °C for 14 h. When the thermal chamber reached the desired temperature for the test, the samples were installed on the support and were kept into the conditioned chamber until they reached the test temperature (−25 °C or −50 °C). The temperature of the specimens was monitored by a thermocouple. Traditional air-backed loading tests were first carried out at room and at the low temperatures of −25 °C and −50 °C. Then, we conducted water-backed loading tests, labeled as "FWB", and tests on samples fully immersed in the water column, labelled as "immersed". Specimens were loaded at the same impact energies and temperatures, reproducing air-backed impact conditions. The aim of this comparative study was to study the influence of the water on the response of the composite due fluid-structure interaction.

In Fig. 1a, the equipment used for FWB tests is shown, while Fig. 1b sketches the proposed experimental scheme. The latter has been proposed by the University of Naples, where the setup is based on a modified drop weight machine, in which an instrumented impactor falls on a clamped specimen which rests upon a water column to obtain a load distribution on the entire surface of the specimen. Specifically, a novel experimental framework was designed to elucidate the impact response of water-backed marine laminates at varying temperatures. The setup is constituted by an instrumented water-filled box used as a support, where panels are first simply supported and then fully clamped and subjected to low velocity impact in a drop tower. During the impact, the deformation of the panel, the motion of the impactor and the pressure field in the box are recorded to ascertain the severity of the impact and quantify fluid-structure interaction.

To avoid water splashing during the impact, a thin silicone deformable bag was used for covering the water solution contained in a cylindrical rigid steel box, 200 mm in diameter (Fig. 2a), and the specimen was positioned on the water pillow. A circular steel plate, 8 mm in thickness, with a rectangular window of 100 × 150mm to house the specimen, was bolted on the box container (Fig. 2b). The specimens were simply supported by the water layer to simulate free boundary

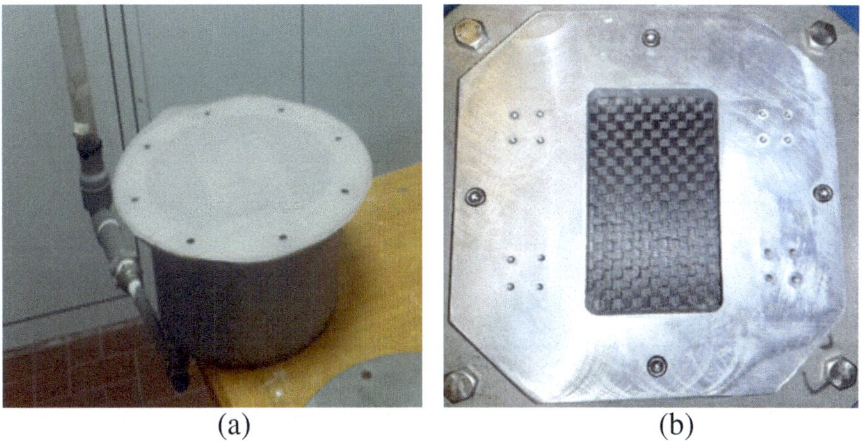

Fig. 2 (**a**) Silicon layer to avoid the leakage of water and (**b**) frame on the steel cover for the specimen positioning

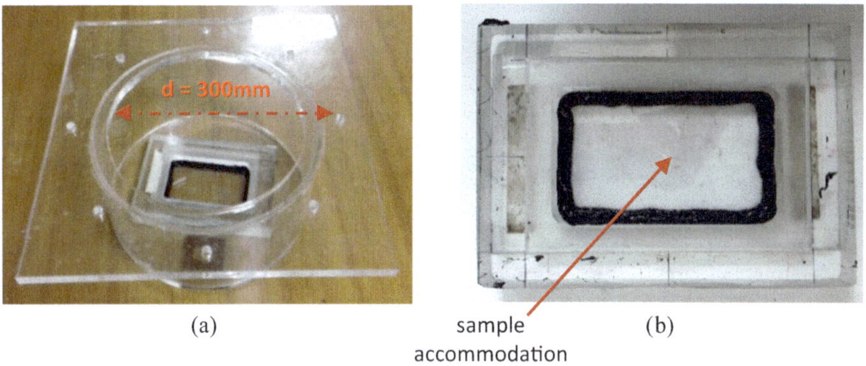

Fig. 3 (**a**) Experimental set up: plexiglass tank containing water (**b**) and detail of the support. Reproduced form [40]

conditions and facilitate a precise assessment of the material response of the panels, in the presence of water backing. The box was filled with a quantity of water greater than the one necessary to reach the edges, to create a water pillow beneath the specimen, minimize any resistance of the sheath during the impact, and allow contact of the entire back surface of the panel. The specimens were impacted on their free side by the cylindrical impactor described above. Specifically, the setup facilitates in diffusing the concentrated loading of the impact to a distributed dynamic loading along the whole wetted surface. To avoid ice formation at $T = -25$ °C and -50 °C, an antifreeze liquid with a density similar to the water, was added to the water.

Tests on specimens immersed in the water (Fig. 3a) were performed in a plexiglass tank, 300 mm in diameter. In these tests, the laminates were impacted

(a)

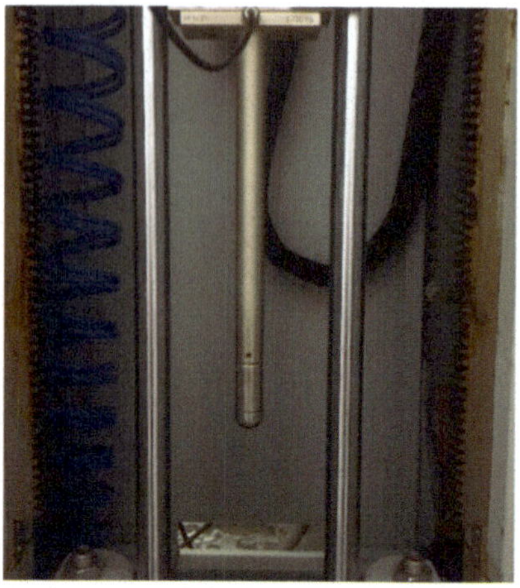

(b)

Fig. 4 (**a**) Impactor modified for the in-water tests (**b**) and impactor suggested by the Standard

from the water side. Figure 3b shows the plexiglass support, which was machined to reproduce the one suggested by the ASTMD 7137 standard with an internal rectangular window 75 $\times$ 150 mm^2 to house a specimen. The support was entirely immersed in the water and fixed on the bottom of the tank. A thin polymeric layer was sealed on the support to avoid water on the back of the specimen.

It was necessary to modify the impactor of the Ceast/Instron impact machine for loading the specimens on the bottom of the tank. Specifically, a longer blunt with the same hemispherical nose (Fig. 4a) was glued and fixed by screws on the first original body (Fig. 4b). Although the longer arm, along with the glue and the screws, caused significant measurement noise in the force signals, the general trend was not affected.

A schematic view of the clamping device suggested by the ASTMD7137 Standard is reported in Fig. 5. To exactly reproduce the same boundary conditions during the water backed tests, a rectangular frame of 75 mm by 125 mm (Fig. 6a) was machined on the steel plate to cover the steel tank containing water shown in Fig. 2. Four black rubbers (Fig. 6b) were used to clamp the specimen at four points, similar to traditional impact tests in air. The latter tests will be labelled as WB to distinguish them from the ones labelled as FWB where the specimens were merely supported by the water pillow.

All the parameters obtained in the different conditions were compared. The use of three impact energies, 5 J, 10 J and 20 J, allowed for the study of the damage onset and propagation.

The indentation depths, *I*, (the plastic deformation produced by the indenter on the material/impactor contact point during the impacts) were measured using a confocal microscope, LEICA DCM3D. This advanced apparatus allows for acquiring three-dimensional images of the sample, thereby extracting salient information about the profile and the depth of the indentation from the three-dimensional shape

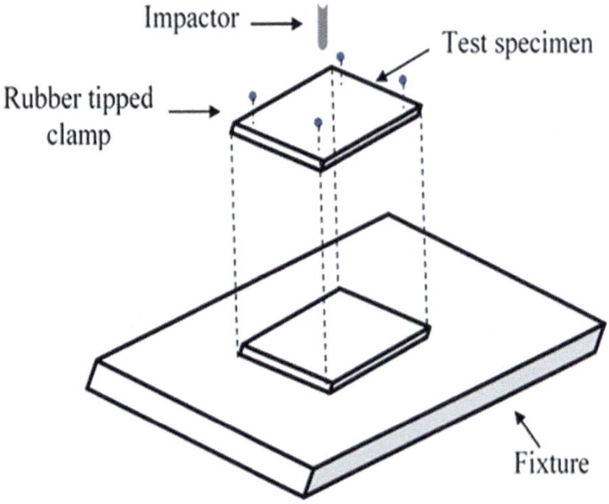

Fig. 5 Clamping device from ASTMD7137 standard. Reproduced form [40]

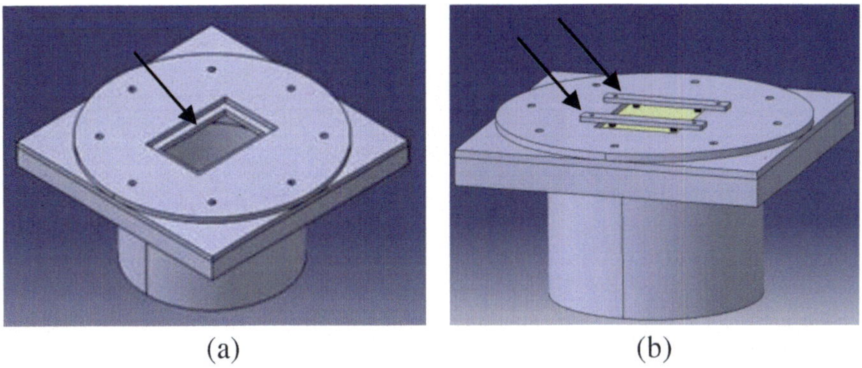

Fig. 6 Rectangular frame 75mm × 125mm (**a**); clamping device (**b**). Reproduced form [40]

of the surface. A dedicated software was used to acquire and manage the indentation data, obtaining the shape and roughness values according to the international standard (ISO 4287).

The microscope is equipped with an x-y table and different magnifications, from 5x up to 150x. In Fig. 7, an example of acquisition of an impacted surface is reported. From the picture, we can appreciate the ability to clearly resolve the indentation (dark blue area) as well as the precision of surface reconstruction.

The ultra sound technique by the Olympus OmniScan® SX Phased Array Probe, 5 MHz Linear Array (automated data acquisition system with a16:64PR phased array unit) was used to characterize the internal damage [27]. The pulse echo technique [28] was adopted to detect the depth of the delaminated area, such that the damage was identified by the echo signals (first echo, back-wall echo and defect echo) captured by the probe. The coupling agent was a layer of water and contact sound testing was performed.

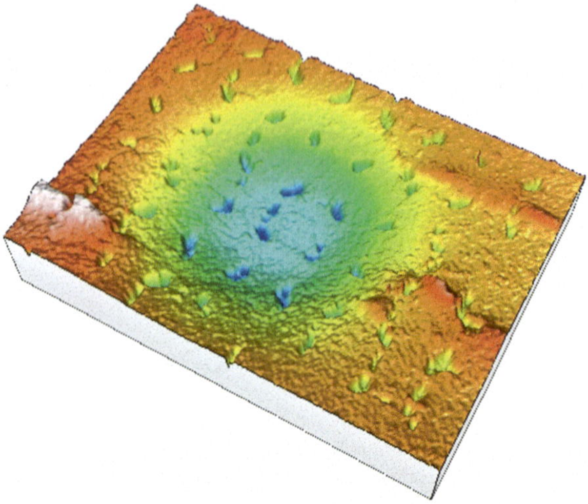

Fig. 7 Example of surface image recorded by the confocal microscope after the impact test at 10 J

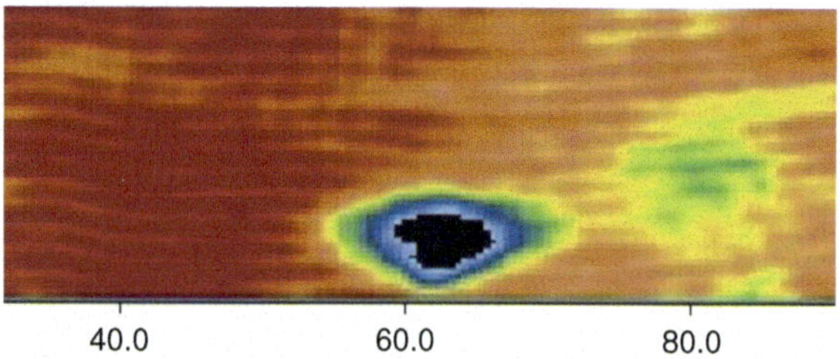

Fig. 8 Example of damage detected by ultra sound

Through C-scan inspection, we obtained a plane view of the damage, from which the software yielded the entire plane image of damage induced by the impact. The acquisition system was calibrated using an undamaged sample. The latter was found to be associated with a uniform color (usually dark red), while damaged areas are usually represented by different colours (Fig. 8) as a function of the severity of the damage. In this way, the damaged area was easily identified and measured. Longitudinal and transverse damage dimensions were recorded using M2M software, whereas each delaminated area was imported in a CAD software, bordered and measured. The acquisition system was calibrated using an undamaged sample by following a procedure standard practice in the aeronautical field.

Internal damage was also investigated via a speckle technique (ESPI). This technique allows us to measure the out-of-plane displacements of a stressed specimen, for identification of strains, along with cracks on rough surfaces with high sensitivity and in real time, without any physical contact [27, 29]. ESPI belongs to the family of coherent light interferometry techniques that includes holographic interferometry and shearography [26]. The detection of invisible and barely visible micro-deformations and micro-cracks is possible by scanning the surface by a laser, at wavelength of 532 nm, and the measurement accuracy is related to the wavelength.

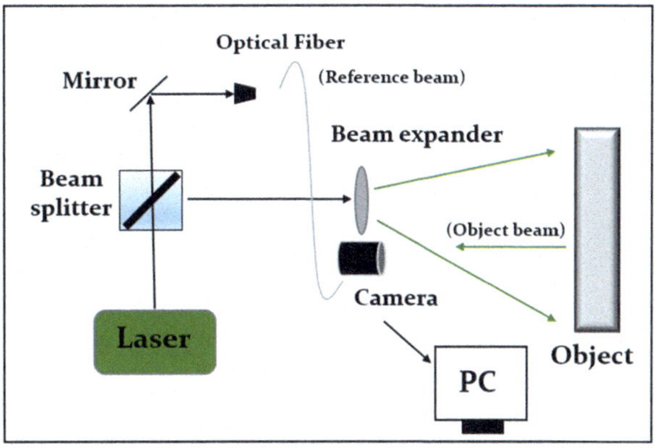

Fig. 9 Basic scheme for the ESPI system

A CCD/CMOS camera records the deformation in the form of correlation fringes, associated with a slight temperature increase in the material due to damage. A basic scheme of the system for recording speckle interferometry measurements is illustrated in Fig. 9. The laser beam is split into a reference beam and an object beam by means of a splitter, which enables control of the relationship between the reference and the object. A speckled image of the object, resulting from the interference between the wavefronts of the two beams on the plane of sensor camera, is recorded. When the specimens are deformed due to an external perturbation, thermal or mechanical, the wavefront reflected from it is slightly changed, while the wavefront from the reference beam does not change.

A speckle pattern is recorded by the camera and stored by a computer. The subtraction of the speckle patterns registered for the deformed and non-deformed conditions provides the correlation fringes. When the subtraction is computer-aided, the technique is called electronic speckle pattern interferometry. The correlation fringes are also digitally treated to remove undesired noise and enhance contrast in phase-contrast maps (Fig. 10a). From the latter, it is possible to quantitatively measure the displacement field with high accuracy. The measured displacement is "out of plane" due to the geometrical configuration of the laser illumination. Figure 10 illustrates correlation fringes acquired by a CCD camera, the unwrapped phase-contrast map, and a magnification of damaged area with the estimation of the delaminated area.

All the data collected during the tests were used to validate existing semi-empirical models for the prediction of the dynamic behaviour of composite laminates.

After the impact tests, the specimens were tested in compression, CAI, to investigate their residual compression strength after loading. The residual properties of the laminates after dynamic loading are of crucial importance to assess whether the damaged structure should be replaced or could continue to support mechanical loading. Due to the practical ramifications of this information and the complexity of existing equipment suggested by the Standard ASTM D7137, the Department of Chemical, Materials and Industrial Production Engineering of University of Naples

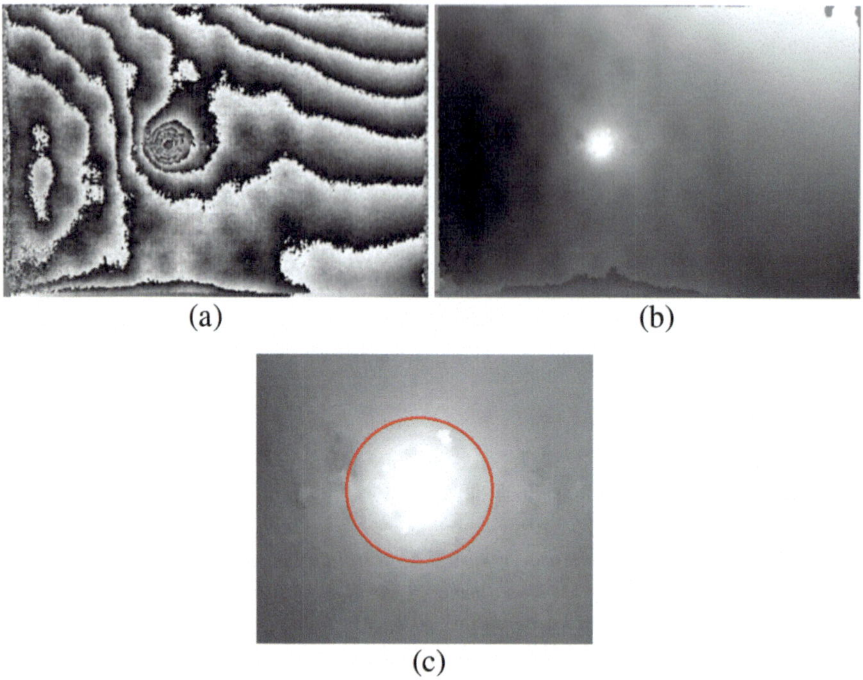

Fig. 10 (**a**) Correlation fringes (**b**) unwrapped phase-contrast map (**c**) and magnification of damaged area and the estimation of delamination

conceived a new, improved equipment that mitigates the problems of buckling and misalignment of the specimen with respect to the load. These latter phenomena often result into extended time for testing and, consequently, higher costs. The new equipment allows for the acquisition of more reliable information in a shorter time.

The existing ASTM Standard proposes the application of a compressive load through a fixture for the stabilization of the sample (Fig. 11a), which comprises several different parts, three on the front side and three on the back side, each one disconnected from the main body, (black arrows in Fig. 11b), The lateral two blocks, marked through red arrows in Fig. 4a, are disconnected from the main body, thus requiring fixing them through bolts.

In the design of the new fixture (Fig. 11b), the two lateral wings are part of the same block of the main body, and the number of the bolts to fix the sample is drastically reduced. In this way, a rigid body with a single bolt each side was realized. The sample has to be placed on a single sliding guide (marked as black arrows in Fig. 11b), fixed to the rigid main body through a single bolt on the floor. The guides ensure the alignment of the specimen and the applied load. Moreover, the three different parts are fixed to the main body through guides with a limited number of bolts. Differences and advantages of the new equipment can be readily appreciated by comparing the designs through Fig. 11.

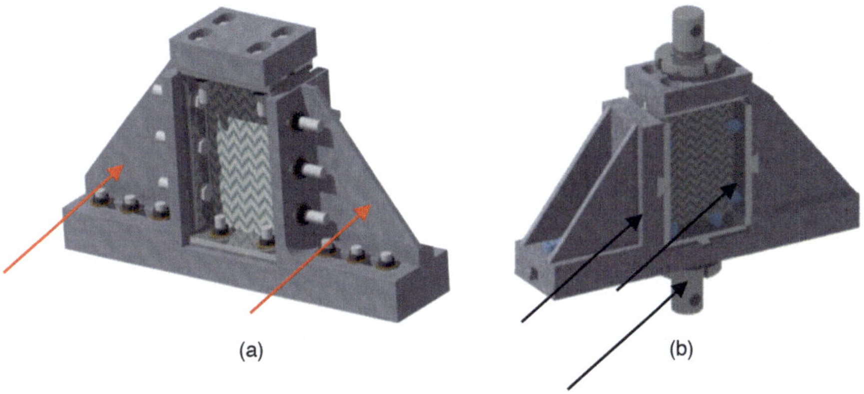

Fig. 11 (**a**) CAI tests equipment: ASTMD7137 (**b**) and new design at the University of Naples. Reproduced form [41]

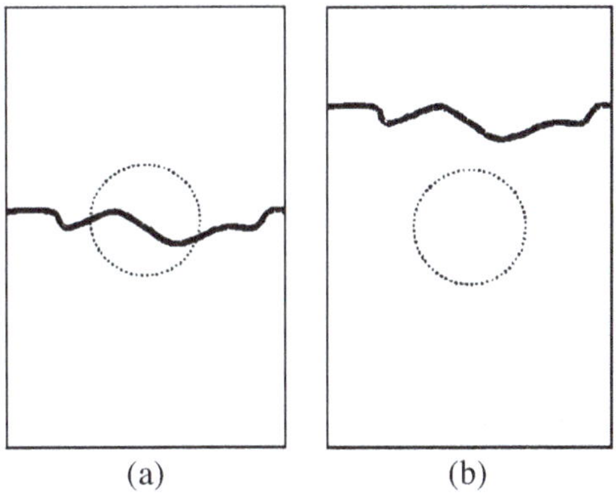

Fig. 12 Failures after CAI tests: (**a**) "correct" and (**b**) "incorrect"

Fig. 13 Fractographic analysis of the section of the specimen showing defects due to a poor compaction

However, the validity of the tests strongly depends on the ability to attain specimens with parallel lateral sides and on the quality of the laminate. Both these problems could lead to incorrect failures, which will challenge the evaluation of the residual strength. In fact, a correct failure pattern should cross the impacted damage area at the central point (Fig. 12). Nonparallel lateral sides and defects could lead to a local concentration of the load on the upper side, thereby triggering localized failure (Fig. 13).

Four strain gauges, two on the front and two on the back side, were glued on each specimen following the suggestions by the Standard.

3 Background: Residual Strength after Impact

In the literature [12–16], the residual strength after low velocity impacts was found to be influenced by the impact energy, U, following the trend in Fig. 14. Three zones are highlighted in the figure, summarizing the response of the laminate at increasing energy levels. In region I, the impact energy is below the threshold value for the damage initiation; in region II, the residual strength quickly reduces to a minimum; and in region III, we observe constant trend since the perforation has occurred.

The residual tensile strength is a function of the delaminated area and the size/shape of the impactor, whereas the residual compression strength depends only on the delamination, since it is controlled by local buckling of delaminated plies [16]. Moreover, delamination can be caused by low impact energy, so that, in correspondence of about 0.3 mm of indentation depth, large strength reduction might occur [17].

A linear elastic fracture mechanics model to predict residual strength as a function of kinetic-energy, was proposed in [12], yielding adequate correlation to experimental results:

$$\frac{\sigma_r}{\sigma_0} = \left(\frac{U_0}{U}\right)^\alpha \tag{2}$$

where U_o and α are two experimental constants and so is the strength of the non-damaged material.

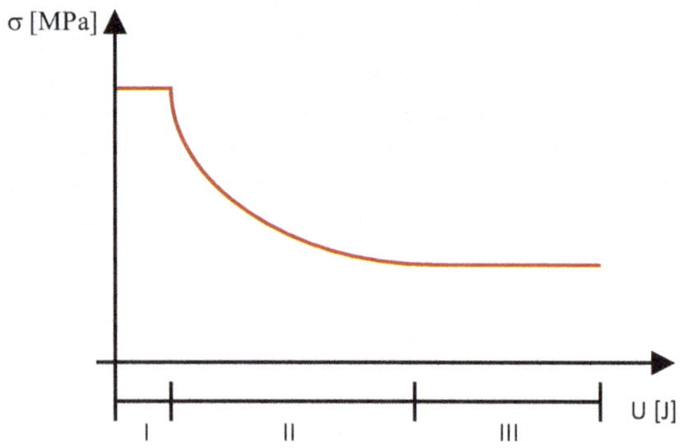

Fig. 14 Trend of the residual strength after impact, as a function of the impact energy

4 Results and Discussion

4.1 Non Destructive Damage Investigation

In Fig. 15, we compare results obtained on equivalent carbon fiber laminates using the two different non-destructive investigations introduced above. Both the US and ESPI techniques are able to capture the delamination and, more importantly, to clearly highlight the lobe shape, which is typical of the delamination between differently oriented layers. This result is particularly promising, whereby it highlights the potential of a technique, ESPI, which is not widely disseminated in the composites community, to investigate their internal damage.

In Fig. 16, we report the delaminated area against the impact energy to highlight the good agreement between the two techniques. All the data follow a linear trend suggesting that, in the studied range of impact energies, as the energy, U, increases, the internal damage, A, increases accordingly. It is worth to note that the intercept with the horizontal axis represents a threshold limit of the impact energy necessary for the delamination onset and below which no delamination is measured between the layers. This aspect will be hereafter discussed again.

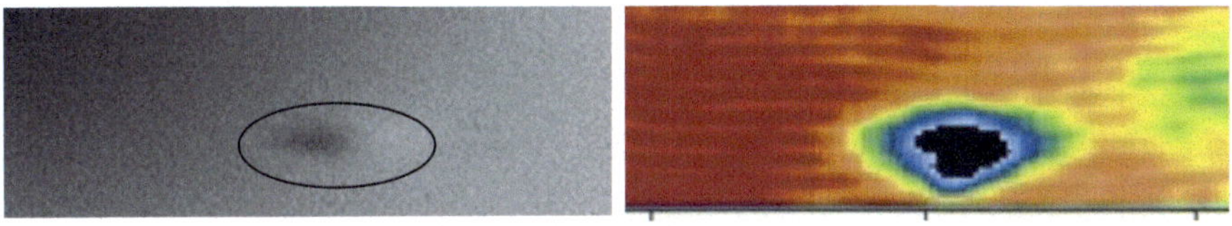

(a) (b)

Fig. 15 (**a**) Non-destructive damage quantification on the backside, for an impact of 13 J: holography (**b**) and ultra sound

Fig. 16 Delaminated area, *A*, vs impact energy, *U*, quantified using two non-destructive techniques

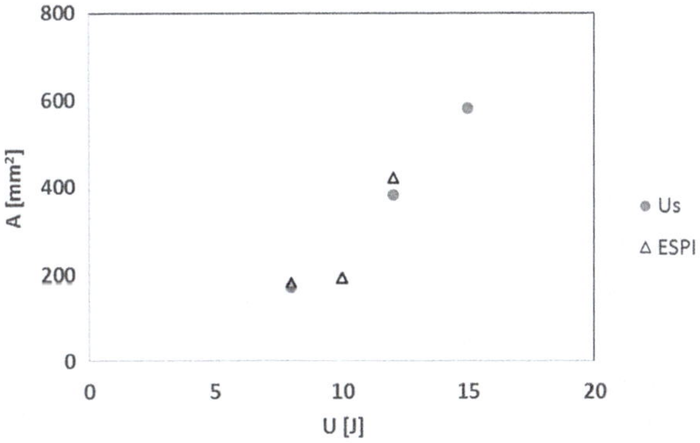

4.2 Semi-empirical Models: Influence of the Temperature

Semi-empirical models available from literature were systematically validated against the collected experimental data. The load-displacement curve at penetration, recorded during each experimental test, offers salient information about behavior of the material [30], in terms of material properties and damage mechanisms, while allowing for comparing responses from different loading conditions, such as those at room and low temperatures (Fig. 17). It displays some features that are common to all laminates, despite differences in thickness, architecture, and physical properties, affording the opportunity to identify key parameters.

From the shape of the curve it was possible to gather information about: (a) the first damage, usually delamination, which was evidenced by the first load drop or change in the slope along the rising part of the curve; (b) the first failure load, which is the load that caused the first damage; (c) the energy, measured as the area under the curve up to the first failure load; (d) other damages, such as fiber failures, from load drops near to the maximum load; (e) the absorbed energy; and (f) the penetration energy. A load drop or a change in the slope on the rising part of the curve reflects a change in the structural rigidity of the material and could be utilized as an effective parameter to reveal damage initiation and propagation.

Decreasing the temperature was found to affect the load curves, thereby supporting the influence of temperature on the dynamic response of the laminates. By comparing the curves (Fig. 17a), it was found that its effect is similar to increasing the thickness of the laminate (Fig. 17b), whereby we registered an increase in the maximum force, rigidity, and penetration energy.

The higher penetration energy and maximum load at low temperatures could be explained by the brittle mechanism of damage formation [31], which promote the formation of a large number of cracks in the matrix. At room temperature, the energy associated with the formation of these cracks is, instead, absorbed for elastic and plastic deformation. Brittle behavior at low temperature was confirmed in [32] in terms of lower measured indentation depths, where the indentation represents the

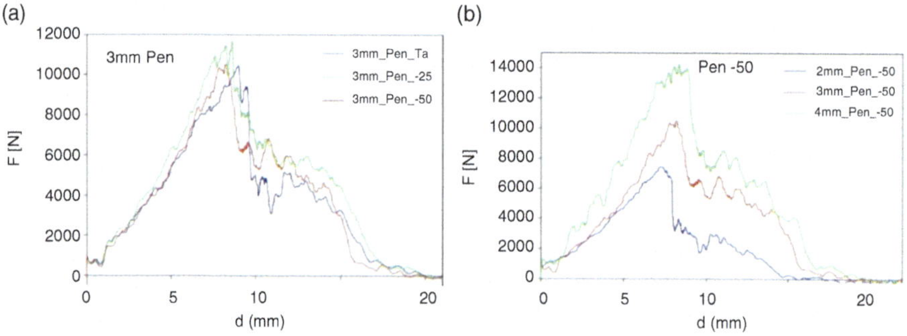

Fig. 17 (**a**) Force displacement curves at penetration for carbon fiber laminates: $n = 10$ (**b**) and $n = 7$, 10 and 13. Reproduced form [9]

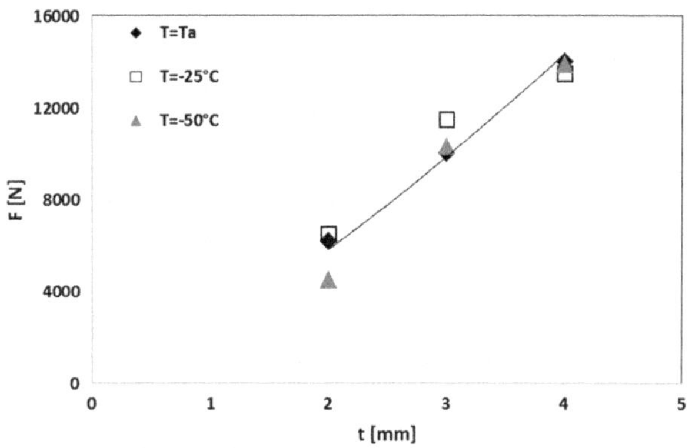

Fig. 18 Hertzian contact law validation for different temperatures

plastic deformation of the surface of the laminate due to the impactor-material contact point. Unfortunately, it was not possible to perform the same analysis on CFRP laminates of naval use studied since no indentation was measured on them. The latter was probably due to the large fiber bundles of the carbon fabrics.

The maximum load, F_{max}, inferred from the load-displacement curves, was plotted against the thickness, t, for the different testing temperatures (Fig. 18) to validate the Hertzian contact law for the prediction of F_{max}. Such a correlation might offer imporant insight into the influence of the thickness on the load which causes internal damage [30], that is,

$$F_{\max} = a \cdot t^m \tag{3}$$

where a and m are two experimental constants, which we identified as a = 2.36 N/mm^m and $m = 1.3$. From Fig. 18, we can appreciate the increase in the maximum load with decreasing temperatures; therein we use a subcript "a" to identify ambient temperature and a consistent notation is maintained throughout the chapter.

The same analysis was pursued in [33] to predict the first failure load. However, the most significant load drop was found in correspondence with the maximum force such that damage which occurs before would not significantly affect the curve.

The effect of temperature on the penetration energy, measured as the area under the complete load-displacement curve, was also studied. We found that higher energies are necessary to penetrate the same laminates for lower temperature, and, as expected, the penetration energy increases with the number of layers. We plotted the penetration energy, U_p, against the compound quantity $t{\cdot}V_f{\cdot}D_p$, (Fig. 19), where t is the thickness, V_f the fiber volume fraction, and D_p the diameter of the impactor. In agreement [24], we discovered a power law which is useful to predict U_p from known experimental quantities. Specifically, we found

$$U_p = \alpha \cdot \left(t \cdot D_p \cdot V_f\right)^{\beta} \tag{4}$$

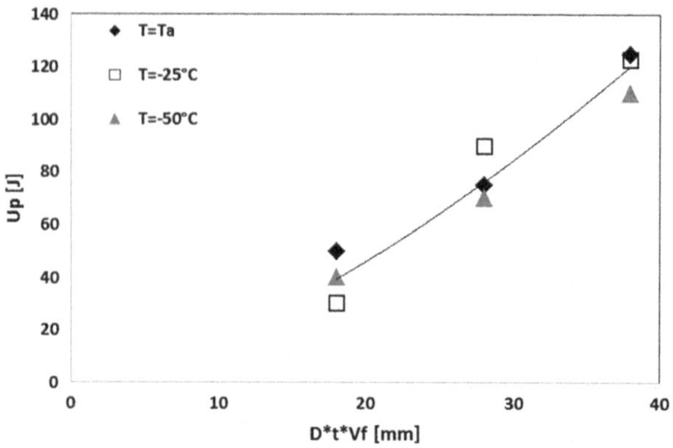

Fig. 19 Penetration energy against the compound quantity $t{\cdot}V_f{\cdot}D_p$, for different temperatures

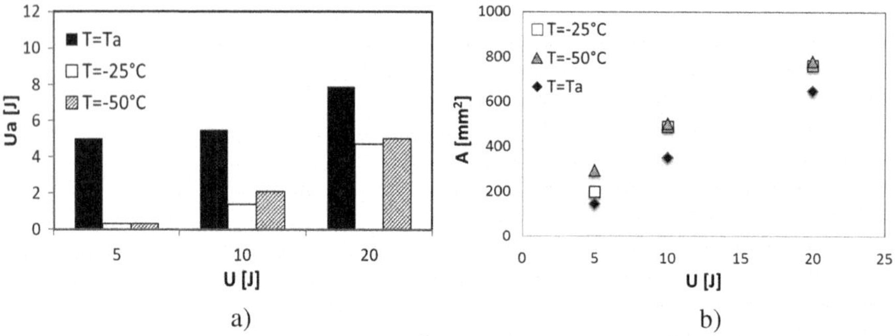

Fig. 20 Influence of the temperature on (**a**) absorbed energy and (**b**) delamination area for laminates of 3 mm in thickness

where $\alpha = 0.87$ J/mm and $\beta = 1.34$ are two constants that were experimentally determined, irrespective of the temperature. In particular, α depends on the material and β on the phenomenon being independent of the material.

This semi-empirical model highlights the importance of the fiber content in an impact phenomenon and it allows us to compare data obtained on different material systems and test conditions, as seen by its validity in extreme temperature conditions. Moreover, this model allows to predict the penetration energy as a function of the impactor diameter. Since the penetration energy is related to the residual strength through the indentation [34], one may estimate the residual strength from a simple indentation measurement.

The absorbed energy represents a crucial parameter in dynamic phenomena. It is the energy necessary to create damage and propagate it. Understanding the mechanisms of energy absorption requires elucidating damage onset and propagation, as well as damage interaction during the loading phase.

By inspecting Fig. 20a, which shows the influence of temperature on absorbed energy for a fixed impact energy, we gather that the absorbed energy decreases for

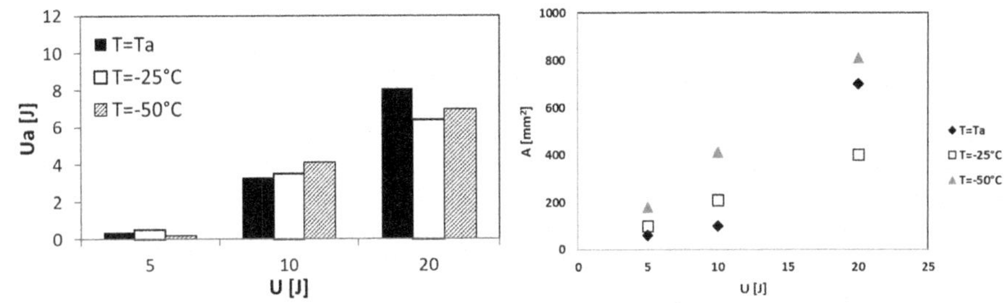

(a) (b)

Fig. 21 Influence of the temperature on (**a**) absorbed energy and (**b**) delamination area for laminates of 4 mm in thickness

lower temperature. Thus, at low temperatures, it is easier to initiate and propagate internal damage. The influence of the impact energy on the delaminated area, A, was studied at low temperature as shown in Fig. 20b. In correspondence to a decrease in the temperature, an increase in the internal damage extension was registered. This effect is more evident at larger impact energies. Overall, this finding is in line with the experiments that suggest that less energy is absorbed at low temperature [35], due to the ease of damage propagation, as confirmed in Fig. 20a.

Moreover, the delaminated area was found to increase with the impact energy, but the trend seems to become more linear at lower temperatures (Fig. 20b). While at room temperature the damage of laminates of 3 mm in thickness becomes constant after a certain impact energy value, it continues to increase at temperatures less than 0 °C, indicating the onset of more severe damage mechanisms.

A different behavior was observed for laminates of 4 mm in thickness (Fig. 21). Significant differences on the absorbed energy were noted for an impact of 20 J, thereby highlighting the role of rigidity in the dynamic response of laminates [30]. The larger delamination at low temperatures explains the lower residual strength reported and commented hereafter in Figs. 23 and 25.

The extension of the delamination with the increase in the thickness is larger also at low temperatures, confirming the difference between the bending effect, predominant in thinner laminates, and the shear one, usually affecting the response of thick laminates, as extensively discussed in the literature and by the authors in [30].

Prompted by the open debate about the relative importance of impact energy and maximum force on the delamination extension [36, 37], we identified a linear trend between the delaminated area and the maximum load, F_{max}, irrespective of the thickness. All the experimental data from Fig. 20, follow a single linear trend, irrespective of the testing temperature (Fig. 22). Therefore, we identify threshold values for the delamination initiation, below which no delamination should take place in the laminate.

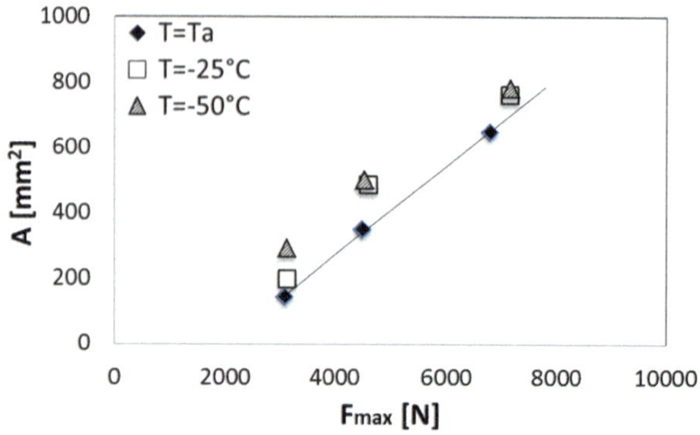

Fig. 22 Delaminated area, A, against the maximum load, F_{max}. Data from Fig. 20

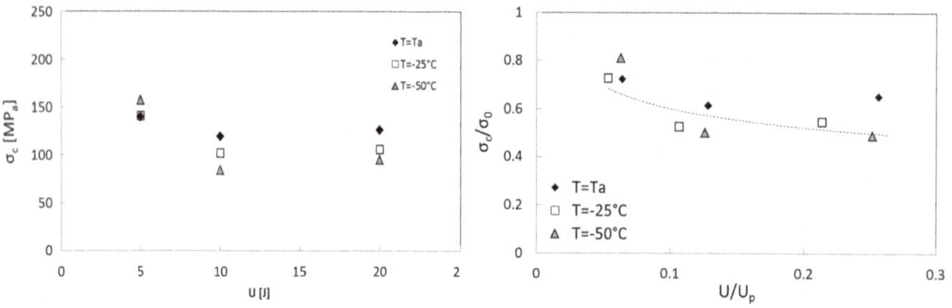

(a) (b)

Fig. 23 Residual strength of a laminate of 3 mm in thickness (**a**) influence of impact energy (**b**) and semi-empirical model in Eq. (2)

4.3 Residual Strength: Air-Backed Tests

The trend found in literature [12] regarding the dependence of the residual strength on the impact energy was confirmed at low temperatures. Specifically, Fig. 23a illustrates such a trend for carbon laminates of 3 mm in thickness. A perceptible decrease of the residual strength at decreased temperature is noted, except when the specimens were impacted at a level of 5 J in energy. In that case, we registered similar values between room temperature, −25 °C and higher values at −50 °C.

In Fig. 23b, the power law in Eq. (2) is verified using data from Fig. 23a for an exponent $\alpha = 0.25$. This important relationship between the residual strength and the impact energy should afford the prediction of the residual properties of a composite laminate that is dynamically loaded.

Differently from [3], plotting the residual strength against the delaminated area (Fig. 24), we observed the same decreasing trend reported in Fig. 23, with, perhaps, a horizontal asymptote. Such a trend is justified by the linear increase of the delaminated area for increasing impact energy, as shown above.

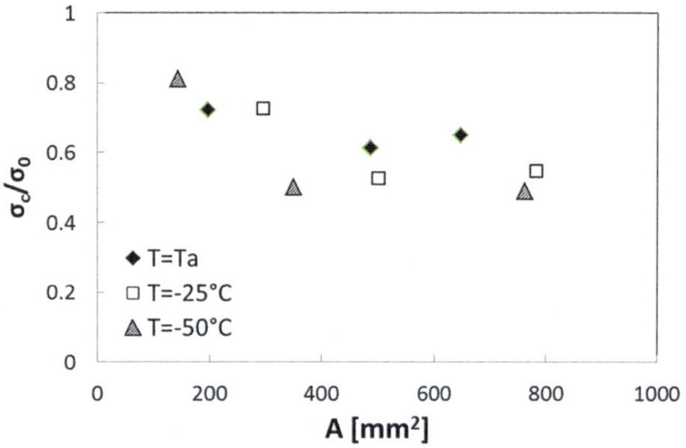

Fig. 24 Non-dimensional residual strength versus delaminated area for laminates of 3 mm in thickness

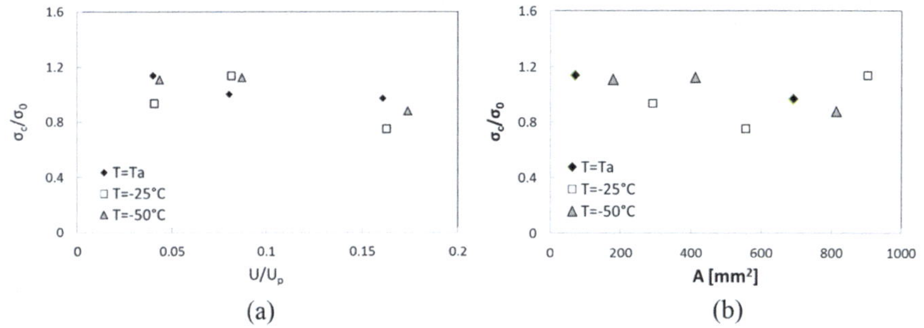

Fig. 25 Residual strength of a laminate of 4 mm in thickness (**a**) influence of impact energy (**b**) and semi-empirical model in Eq. (2)

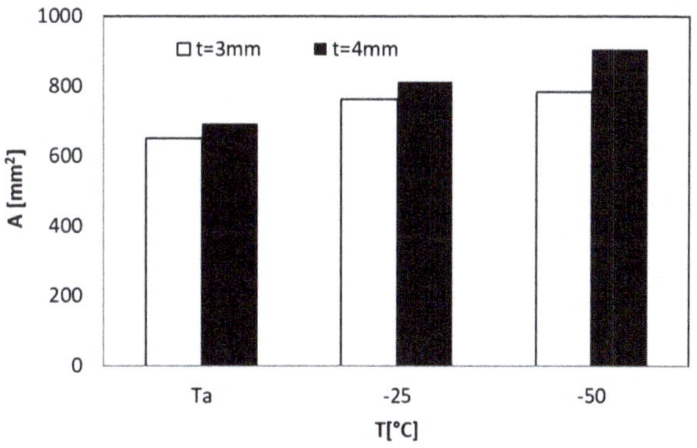

Fig. 26 Influence of the thickness on the delaminated area at different temperature

An equivalent analysis was performed on the 4 mm-thickness laminate as shown in Fig. 25. Therein, we observe a flatter trend for the residual strength as a function of impact energy, compared to the delamination analysis. This suggests a lower residual strength reduction, along with a weaker influence of the temperature for composites of higher thicknesses. Such an observation is confirmed by the lower internal damage measured in these conditions (Fig. 26).

At room temperature, increasing the thickness of the laminate results leads to an increase in the stiffness, whereby most of the absorbed energy is utilized to create internal damage that causes lather delamination. In contrast, for thinner laminates, bending allows for a larger energy absorption, thereby hampering delamination [30].

4.4 Residual Strength: Water-Backed Impact Tests

Low velocity impact tests on free water-backed samples were carried out at the same impact energies as in-air tests. In Fig. 27, the load-displacement curves obtained by impacting the specimens of 4 mm in thickness at 20 J are compared. Predictably, the shape of the curves is different as well as the maximum load and displacement experienced by the laminate.

In contrast with one's intuition, the maximum load and initial rigidity are higher in the absence of the water. This is attributed to the different boundary conditions employed in the tests, whereby water-backed panel are suspended above a water pillow that might, in principle, allow for rigid-body motion of the sample. In contrast, specimens tested in air are fixed on their sides, causing a higher resistance to the impact.

In Fig. 28, we compare the delaminated area for in-air and free water-backed impact, which confirms that the specimens are subjected to more severe loading conditions when tested in air.

Because of the lower delamination, we also register (Fig. 29) higher residual properties for water-backed specimens, except for the case of laminates of 4 mm thickness that were impacted at 10 J.

These anomalous results prompted a collaboration with Prof. Porfiri's group, who is honing a physically-based mathematical model to examine the dynamic response of water-backed panels that takes into account the elastic response of the composite at moderately-large deformations, the role of the boundary conditions, and the added mass effects from the fluid-structure interactions. The theoretical underpinnings of the model are presented later in the chapter, along with some preliminary results on homogenous, isotropic panels.

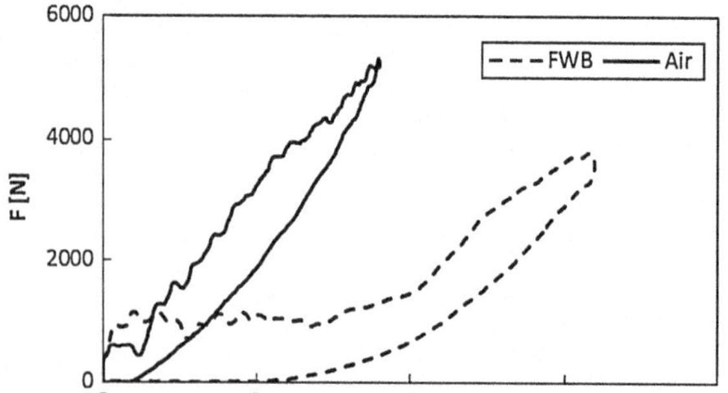

Fig. 27 Load-displacement curves comparison for laminates of 4 mm in thickness impacted at an energy of 20 J

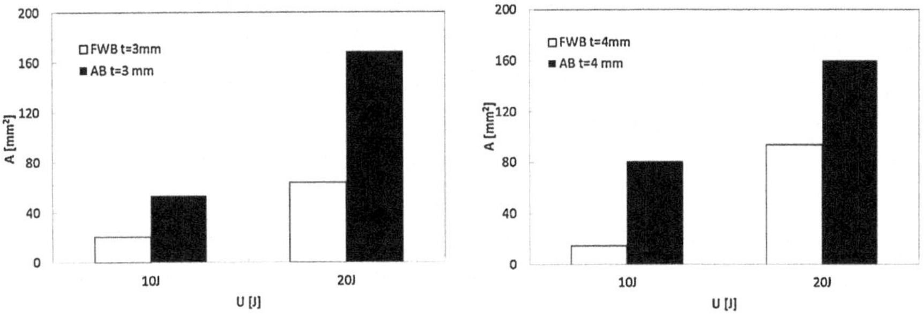

(a) (b)

Fig. 28 Delamination measured in air and free water-backed tests: (**a**) $t = 3$ mm (**b**) and $t = 4$ mm

Fig. 29 Residual strength: comparison between pure impact in air and free water-backed tests

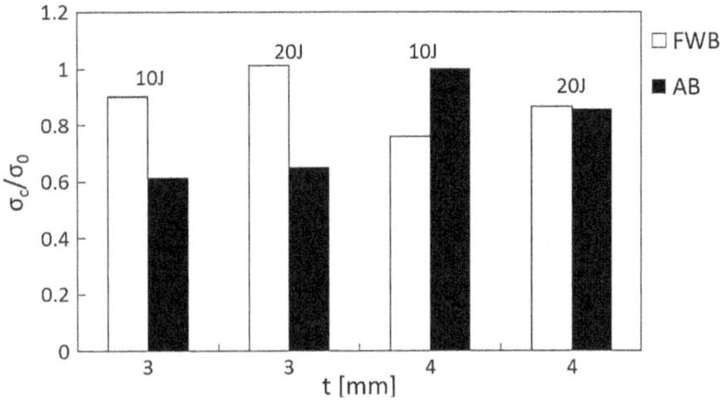

Fig. 30 Clamped water-backed and water-backed free: comparison between load curves for laminates of 3 mm in thickness, impacted at 20 J

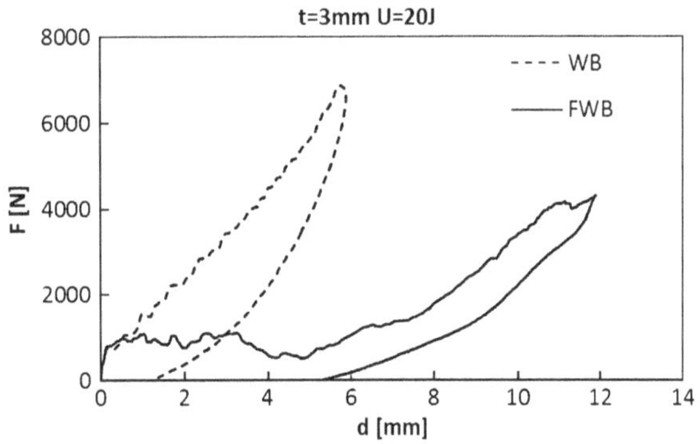

4.5 *Effect of the Clamping Device and Fluid-Material Interaction*

In Fig. 30, load curves obtained in free water-backed and clamped water-backed impact tests are compared to ascertain the critical role played by the boundary conditions. The initial stiffness and maximum load experienced by clamped specimens is considerably higher than those that were merely supported by the water column.

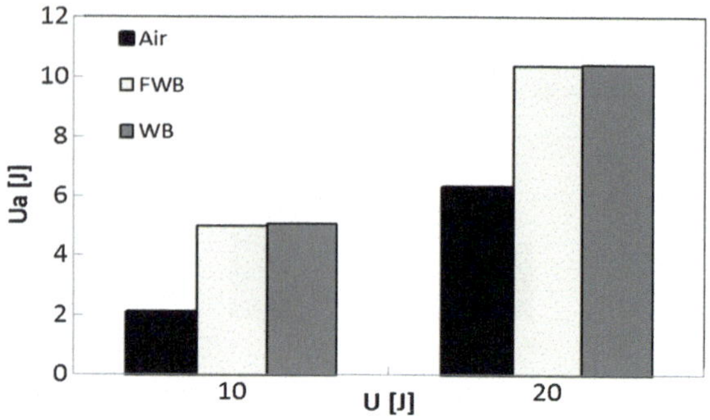

Fig. 31 Absorbed energy versus impact energy for laminates of 3 mm in thickness

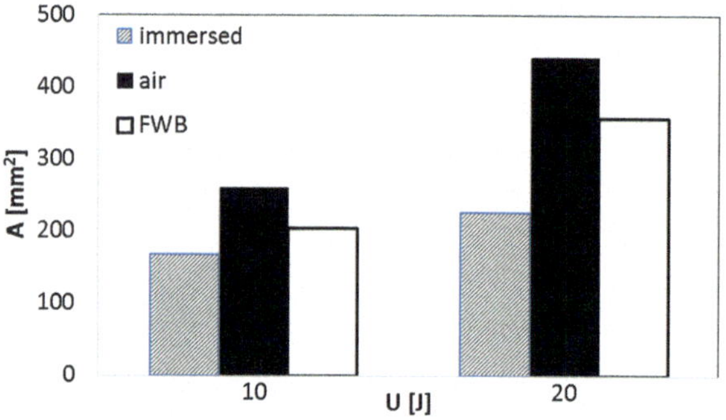

Fig. 32 Delamination versus impact energy for laminates of 3 mm in thickness

Notwithstanding these differences, the energy absorption between the two conditions is quite similar (Fig. 31), potentially pointing as a local failure process that is marginally affected by boundary conditions. Comparing both the water-backing setups with in-air impact, we confirm an increase in the absorbed energy, which might be associated with the added mass effect from the motion of the surrounding water.

Even if the absorbed energy is higher when the panel is in contact with water (Fig. 31), Fig. 32 shows that the delamination propagates more in experiments in air. This phenomenon could be again explained by the added mass of the water, which causes a redistribution of the load that is ultimately diffused throughout the extent of the panel.

4.6 *Impact on Specimen Immersed in the Water*

To simulate more closely impact conditions of marine composites, impact tests were performed on specimens fully immersed in the water. In this condition, the sample was at the bottom of a water tank, with its wet side being impacted. By comparing the results obtained in this case with in-air and water-backed samples, we noted a difference in the absorbed energy and displacement. In Fig. 33, all load curves have

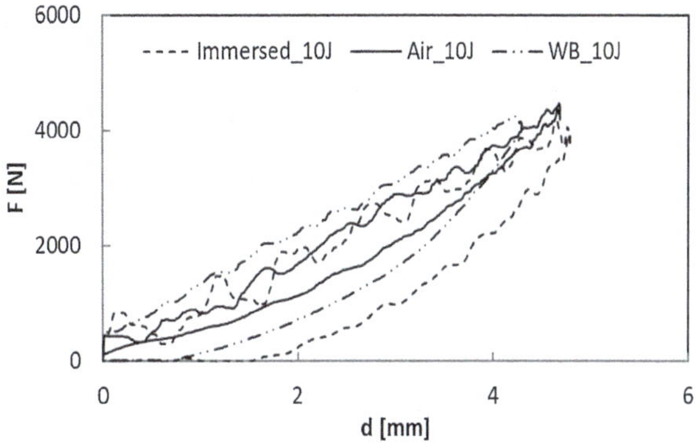

Fig. 33 Comparison at 10 J between for all testing conditions

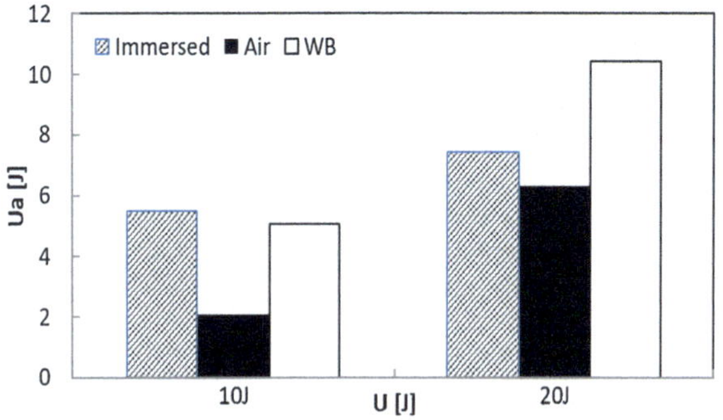

Fig. 34 Absorbed energies versus impact energy for laminates of 3 mm in thickness

similar shapes, irrespective of the support, showing similar initial rigidity and maximum loads that are very close, whereas different amounts of absorbed energy (Fig. 34) were measured.

The results suggest the presence of multiple mechanisms of damage evolution, represented by the different absorbed energy in the presence of the water. An interesting consideration can be offered by looking at Fig. 34. The lower energy of 10 J results in a higher absorbed energy when the panel is immersed in the water, even if energy absorption between water-backed and fully immersed samples are very close to each other. On the other hand, in-air panels present the lowest values of energy absorption. Increasing the impact energy to 20 J, in-air samples continue to absorb the least amount of energy, but we observe a significant difference between water-backed and immersed samples, with the former absorbing more energy. It is difficult to explain the difference between the response of WB and immersed samples. However, it is likely that the fluid-structure plays a major role. Ongoing work on mathematical modelling will hopefully be able to address this aspect.

4.7 Introduction to the Theory

A physically-principled understanding of the fluid-structure interaction occurring during the impact is currently not available. To fill this gap in knowledge, we developed a reduced-order modeling framework of water-backed impact, focusing on in-plane deformation of thin panels. We specifically analyzed the dynamic response of a beam resting on an infinitely-extended water domain, clamped at both its ends. The beam is loaded at its midspan by a concentrated force, simulating the fall of a heavy mass. The modeling framework is based on nonlinear Euler-Bernoulli beam, to account for nonlinear stiffening due to membrane stretching. Water backing is modeled using potential flow theory with a simplified linear treatment of the boundary conditions inspired by the classical work of Wagner [39]. Based on this premise, we can exactly compute the hydrodynamic pressure on the beam as a function of its acceleration. Hydrodynamic pressure acts as a nonlocal added mass effect, where the local acceleration of the beam will translate in a pressure everywhere on the structure.

The Galerkin discretization is used to transform the governing nonlinear partial integro-differential equation for the deflection field into a mathematically tractable set of nonlinear ordinary differential equations. We established two semi-analytical solutions, by using a polynomial approximation of the in-vacuum linear mode shapes of the beam (assumed modes, AM, method) and Hermitian finite element basis functions (one-dimensional finite element, 1D FE, method). A Newmark-type integration scheme is combined with the modified Newton-Raphson method to predict the dynamic response of the panel.

Results were specialized to plain vinyl ester resins panels and semi-analytical findings validated against full two-dimensional finite element simulations, performed using the commercial software COMSOL Multiphysics. The comparison with such computationally expensive simulations offers compelling evidence for the validity of the mathematically-tractable approach, which is amenable to future generalizations to complex geometries and damage models. Figure 35 illustrates the COMSOL implementation of the problem, clarifying the boundary conditions used in the simulation, along with the mesh used for the analysis.

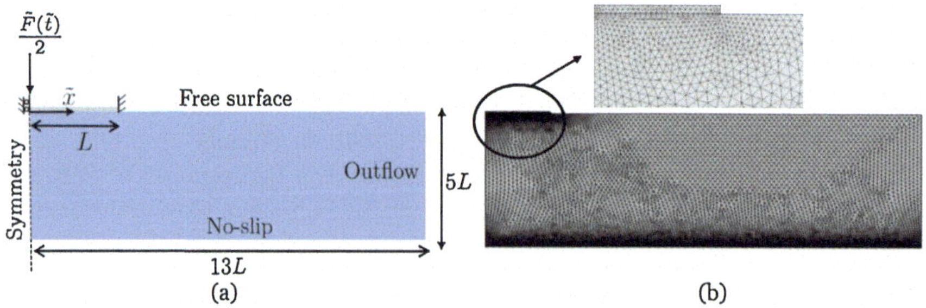

Fig. 35 (**a**) Finite element model in COMSOL. (**b**) Computational mesh used for the structure and fluid, along with a close up of the fluid-structure interface. Reproduced from [38]

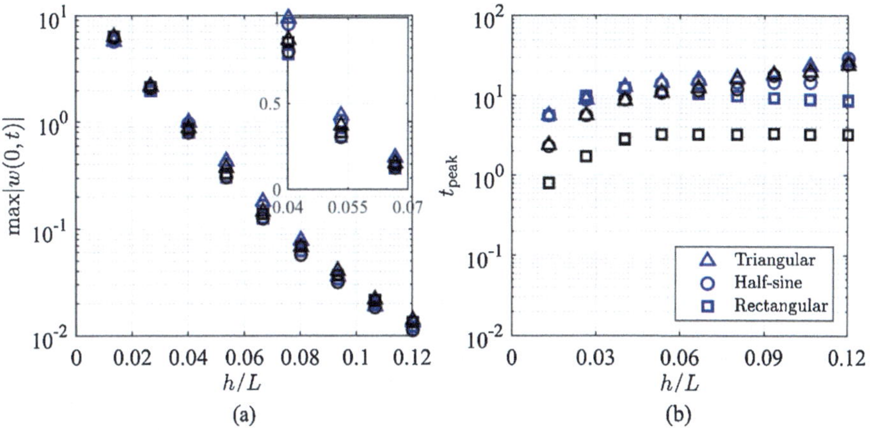

Fig. 36 (**a**) Comparison of the maximum displacement and (**b**) peak displacement time in the presence (blue) and absence (black) of water-backing for a beam with thickness ranging from 1 to 9 mm. Reproduced from [39]

We conducted a comprehensive parametric study to ascertain the role of the thickness of the beam and explore different functional forms of the impact loading, from a square to a parabolic pulse. The key findings of the study are presented in Fig. 36, which displays the peak displacement (scaled by the thickness) and the time at which it is attained (scaled by the inverse of the fundamental radian frequency for in-air vibrations) for different thickness to length ratios.

Through the mathematical model, we confirm the intuition that water-backing has a critical role on the structural dynamics, stronger for thin beams subject to rapid pulses. Such an effect is more clear in terms of the time scale of the response (Fig. 36b), rather than the severity of the vibrations (Fig. 36a). In fact, the added mass phenomenon significantly slows down the dynamic response of the panel, especially for thin structures subject to rectangular pulses. Membrane contributes a robust stiffening effect which mitigates the severity of the response and the time at which the peak vibration is attained, as seen in the nonlinear dependence of the time scale on the thickness of the panel.

Based on this preliminary evidence, we are in the process of collecting experimental data on the impact of aluminum panels at the University of Naples. These experiments require a slight modification of the existing setup for water-backing tests, such that the panel will be clamped on two opposite sides only and the impactor will have a cylindrical shape. Notably, we are conducting these experiments utilizing different liquids, beyond water, to verify the role of the added-mass phenomenon. Collecting these data will constitute a first step toward the validation of the modeling framework for two-dimensional deformation, before expanding to three-dimensional impact and including constitutive models for the composite.

Another piece of research that is being pursued is the development of a combined particle image velocimetry/digital image correlation technique to simultaneously capture the mechanical deformation of the panel and the fluid flow. Such an analysis is expected to aid in the extension of the model and its careful validation, by shedding light on the fluid-structure interaction from both the points of view of experimental solid and fluid mechanics.

5 Conclusions

Low velocity impact tests until penetration were performed on polymeric composite laminates for varying energy values. At first, we focused on the characterization of damage and on the influence of impact energy and laminate thickness on the onset and propagation of damage. Then, we examined dynamic behavior of carbon fiber reinforced plastic laminates in extreme conditions, by comparing results obtained in different temperature conditions. Results obtained from the analysis of the load displacement curve, were studied and examined within the context of existing semi-empirical models. These data are also used to support the numerical simulations presented by Porfiri in another chapter of the present book.

The internal damage was investigated through destructive and non-destructive techniques, offering insight into promising, novel experimental techniques. Results in terms of delaminated area were compared, suggesting good agreement between techniques. All the techniques evidenced the classical lobe shape of the delamination, affording simple experimental measurements. The ESPI technique, in particular, was determined to be a valid tool to investigate the impact damage on composite laminates.

Testing at low temperatures, we discovered a generally more brittle behavior with respect to experiments at room temperature. In particular, we draw the following conclusions. Higher energies were necessary to penetrate the same laminates at lower temperature, due to additional brittle damage. On the contrary, the absorbed energy was found to decrease at higher temperature, suggesting that internal damage can propagate in an easier way. Our experiments offer evidence in favor of a power law for the prediction of the penetration energy as a function of the fibre volume fraction and the impactor diameter. The brittle behaviour at low temperature was confirmed by the lower indentation depths respect to the room conditions. The indentation depth, in fact, was found to increase at higher temperature. The laminate thickness was found to be an important factor in determining changes in the damage mechanisms, by shaping the interplay between bending and shear effects. Delamination was found to be more extensive at lower temperature, although plotting the delaminated area against the maximum load, we identified a single linear trend, irrespective of the temperature.

The validation of semi-empirical models at low temperature constitutes a significant contribution of this research, toward the final aim of predicting the residual properties of a laminate after dynamic loading. Given the existing relationship between residual strength and impact energy and between indentation and nondimensional energy, it could be possible to correlate the residual strength after low velocity impacts to external damage. However, this requires knowledge o the impactor, which might represent a limitation of the approach. Through a simple damage measurement, it could be possible to gain information about the possibility to replace a mechanical part on the structure without dismantling the whole panel.

Another central contribution of this study is the systematic analysis of the impact response of composites in the presence of water, simulating a realistic loading condition in Navy-relevant applications. Specifically, we examined three complementary scenarios. First, the case in which the panel was supported by a water layer, without clamping. Second, we clamped the panel, similar to in-air testing to perform water-backing experiments. Third, we clamped the panel at the bottom of the a tank and loaded it from its wet side. By comparing energy absorption and delamination across these different scenarios, we brought forward compelling evidence substantiating the critical role played by the fluid-structure interaction. Collectively, these experiments point at the critical role played by the added mass phenomenon in redistributing the impact throughout the panel, thereby improving energy absorption and potentially reducing the severity of the damage.

Motivated by this experimental evidence, we embarked on a modeling effort to elucidate the fluid-structure interaction that underlies water-backed impact. A preliminary mathematical model was formulated in which two-dimensional deformation of a clamped panel are examined. The model takes into account nonlinear stiffening from membrane-stretching along with non-local added mass phenomena associated with the motion of the water below the panel. The reduced-order model was validated through comparison against detailed numerical simulations on a finite element model. Future work should seek to extend the model to incorporate three-dimensional effects, as well as resolving the contact between the panel and the impactor.

Acknowledgement The work has been supported by the Office of Naval Research (Grant N00014-10-1-0988, N00014-18-1-2218, N00014-14-1-0380, and N62909-16-1-2220) with Dr. Y.D.S. Rajapakse as the program manager.

References

1. Howard M, Hollaway L (1987) The characterization of the non-linear viscoelastic properties of a randomly orientated fibre/matrix composite. Composites 18:317–323
2. Shindo Y, Ueda S, Nishioka Y (1993) Mechanical behavior of woven composites at low temperatures. Fusion Eng Des 20:469–474
3. Kim MG, Kang SG, Kim CG, Kong CW (in press) Composite structures
4. Sefrani Y, Berthelot JM (2006) Temperature effect on the damping properties of unidirectional glass fibre composites. Compos Part B 37:346–355
5. Khalid A (2006) The effect of testing temperature and volume fraction on impact energy of composites. Mater Des 27:499–506
6. Ibekwe SI, Mensah PF, Li G, Pang SS, Stubblefield MA (2007) Impact and post impact response of laminated beams at low temperatures. Compos Struct 79:12–17

7. Okada T, Faudre MC, Tsuchikura N, Nishi Y (2016) Improvement of low-temperature impact value of Sandwich-structural(CFRP/ABS/CFRP) laminate plies by Homogeneous Low-EnergyElectron Beam Irradiation (HLEBI). Mater Trans 57(3):305–311
8. Kang K-w, Kim H, Chung T, Koh S (2010) Temperature effect of low velocity impact resistance of glass/epoxy laminates. Int J Mod Phys B 24(15–16):2657–2663
9. Lopresto V, Langella A, Papa I (2017) Dynamic load on composite laminates in the presence of water. Polym Eng Sci 57(6):613–620
10. Liu S, Kutlu Z, Chang FK (1993) Matrix cracking and delamination propagation in laminated composites subjected to transversely concentrated loading. J Compos Mater 27:436
11. Heimbs S, Heller S, Middendorf P, Hähnel F, Weiße J (2009) Low velocity impact on CFRP plates with compressive preload: test and modelling. Int J Impact Eng 36(10–11):1182–1193
12. Bouvet C, Rivallant S (2016) Damage tolerance of composite structures under low-velocity impact. In: Dynamic deformation, damage and fracture in composite materials and structures. Woodhead Publishing, Cambridge, pp 1–7
13. Aymerich F, Dore F, Priolo P (2008) Prediction of impact-induced delamination in cross-ply composite laminates using cohesive interface elements. Compos Sci Technol 68 (12):2383–2390
14. ABAQUS Analysis User's Manual 6.12, 2012
15. Belingardi G, Vadori R (2002) Low velocity impact tests of laminate glass–fiber–epoxy matrix composite materials plates. Int J Impact Eng 27:213–229
16. Belingardi G, Vadori R (2003) Influence of the laminate thickness in low velocity impact behaviour of composite material plate. Compos Struct 61:27–38
17. Pagliarulo V, Rocco A, Langella A, Riccio A, Ferraro P, Antonucci V, Ricciardi MR, Toscano C, Lopresto V (2015) Impact damage investigation on composite laminates: comparison among different NDT methods and numerical simulation. Meas Sci Technol 26(8):085603
18. Busse G (1979) Optoacoustic phase angle measurement for probing a metal. Appl Phys Lett 35:759–760
19. Pagliarulo V, Palummo R, Rocco A, Ferraro P, Ricciardi MR, Antonucci V (2014) Evaluation of delaminated area of polymer/carbon nanotubes fiber reinforced composites after flexural tests by ESPI, IEEE MAS Proceedings 6865922: 211–215
20. Yang SH, Sun CT (1982) In: Daniel IM (ed) Indentation law for composite materials, Composite materials: testing and design, ASTM STP 787. American society for testing and materials, Philadelphia, pp 425–449
21. Timoshenko SP (1953) Strength of materials. McGraw-Hill, New York
22. Wang H, Vu-Khanh T (1995) Fracture mechanism and mechanisms of impact-induced delamination in laminated composites. J Compos Mater 29:156–178
23. Chang FK, Choi HY, Wang HS (1990) Damage of laminated composites due to low velocity impact, 31st AIAA/ASME/ASCE/AHS/ASC structures, structures dynamics and materials conference, Long Beach, 930
24. Abrate S (2001) Modeling of impacts on composite structures. Compos Struct 51:129–138
25. Krueger (2004) Virtual crack closure technique: history, approach, and applications. Appl Mech Rev 57(2):109–143
26. Pagliarulo V, Lopresto V, Langella A, Antonucci V, Ricciardi MR, Ferraro P (2016) Non-destructive evaluation of impact damage on carbon fiber laminates: comparison between ESPI and Shearography, AIP Conference Proceedings 1740, 040002
27. Kwon YW, Owens AC (2011) Dynamic responses of composite structures with fluid–structure interaction, Advanced Composite Materials IN-TECH publisher
28. Kwon YW (2014) Dynamic responses of composite structures in contact with water while subjected to harmonic loads. Appl Compos Mater 21:227–245
29. Kwon YW, Violette MA (2012) Damage initiation and growth in laminated polymer composite plates with fluid-structure interaction under impact loading. Int J Multiphysics 6(1):29–42

30. Caprino G, Lopresto V, Scarponi C, Briotti G (1999) Influence of material thickness on the response of carbon-fabric/epoxy panels to low-velocity impact. Compos Sci Technol 59:2279–2286
31. Putić S, Stamenović M, Bajčeta B, Stajčić P, Bošnjak S (2007) The influence of high and low temperatures on the impact properties of glass–epoxy composites. J Serb Chem Soc 72(7):713–722
32. Lopresto V, Langella A (2014) Composite laminates under dynamic extreme conditions, international symposium on dynamic response and failure of composite materials, DRaF2014. Procedia Eng 88:173–179
33. Hashin Z (1980) Failure criteria for unidirectional fiber composites. J Appl Mech 47:329–334
34. Caprino G (1984) Residual strength prediction of impacted CFRP laminates. J Compos Mater 18:508–518
35. Papa I, Langella A, Lopresto V (2018) CFRP laminates impacted at low velocity: influence of the matrix and temperature, AIP Conference Proceedings, 1981
36. Jackson WC, Poe CC Jr (1993) The use of impact force as a scale parameter for the impact response of composite laminates. J Compos Tech Res 15(4):282–289
37. Lagace PA, Williamson JE, Tsang PHW, Wolf E, Thomas SA (1993) A preliminary proposition for a test method to measure (impact) damage resistance. J Reinf Plast Compos 12(5):584–601
38. Shams A, Lopresto V, Porfiri M (2017) Modeling fluid-structure interactions during impact loading of water-backed panels. Compos Struct 171:576–590
39. Wagner H (1932) Uber stoss und gleitvorgange an der oberflache von flussigkeite, ZAMM – Zeitschrift fur Ange- wandte Mathematik und. Mechanik 12(4):193–215
40. Lopresto V, Langella A, Papa I (2019) Interaction of water with carbon fiber reinforced polymer laminates under dynamic loading conditions. J Mater Eng Perform,. in press 28:3220–3227. https://doi.org/10.1007/s11665-019-03915-5
41. Lopresto V, Langella A, Papa I (2016) Residual strength evaluation after impact tests in extreme conditions on CFRP laminates. Procedia Eng 167:138–142

Inferring Impulsive Hydrodynamic Loading During Hull Slamming From Water Velocity Measurements

Maurizio Porfiri

1 Introduction

Naval and aeronautical structures are routinely exposed to impulsive loading conditions associated with slamming on the water surface. The complexity and significance of water slamming research are exemplified by the hull of a planing vessel that periodically impacts the water as the vessel rises above the water surface and then re-enters with a high velocity, or by the fuselage of an aircraft which must complete an emergency landing on the water. Understanding the physics of water impact is key to the design of high performance structures, which must withstand dramatic slamming conditions, where pressure as high as few megapascal are attained in few milliseconds, thereby eliciting severe stress levels and fatigue, potentially contributing to early failure [1–4].

The very first studies in this area of hull slamming are the seminal contributions of von Karman [5] and Wagner [6], who first shed light on the phenomenon of added mass, the onset of water pile-up, and the spatiotemporal evolution of the hydrodynamic loading. Since these early endeavors, the community has made several strides in experimental, numerical, and theoretical research, as summarized in the recent review by Abrate [7]. Our understanding of the physics of the impact is, however, far from complete and numerous questions remain open in the mechanics of slamming. Particularly elusive is the transition from stiff metal structures to lightweight composites, which opens the door for a critical reconsideration of the current practice to the design of structures that must withstand slamming loading. The compliance of the impacting structure will cause complex, unsteady fluid-structure interactions which might exacerbate the slamming conditions, but also constitute a design element for improving performance and extend service operation.

M. Porfiri (✉)
Department of Mechanical and Aerospace Engineering and Department of Biomedical Engineering, New York University Tandon School of Engineering, Brooklyn, NY, USA
e-mail: mporfiri@nyu.edu

S. W. Lee (ed.), *Advances in Thick Section Composite and Sandwich Structures*,
https://doi.org/10.1007/978-3-030-31065-3_9

I was first exposed to the topic of water entry in the annual reviews of the Office of Naval Research (ONR), Solid Mechanics Program managed by Dr. Y.D.-S. Rajapakse around 2010, where I learnt from experts in the U.S., Europe, and Oceania what were the key challenges in the design of composite materials for use in ship hulls. The more I saw about the sophisticated models by Abrate [8, 9] and Batra [10–14] and the impressive experimental setups by Battley [15–19] and Rosen [20–23], the more it became apparent to me that there was a profound need for an experimental technique that could offer spatially-resolved data about the hydrodynamic loading experienced by structures during slamming events.

Mathematical models are hungry of experimental data. Just as they need experimental data to be formulated from first physical principles, they require data for calibration of their parameters and validation against independent observations, prior to being effectively used for quantitative predictions and structural design.

In the context of slamming problems, the hydrodynamic loading experienced by the impacting solid is influenced by the motion of the solid itself (let it be a rigid body motion or a complex elastic deformation), resulting into an authentic unsteady fluid-solid interaction problem. Hence, experimental data should encompass measurements of the fluid flow, that is, the velocity and the pressure field during slamming. The limited availability of these experimental data was one of the main drivers for me and my group to embark on the experimental visualization and analysis of the flow physics associated with slamming.

Not only I realized that there was a paucity of velocity and pressure data, but also I apprehended that available measurements of hydrodynamic loading were restricted to the direct evaluation of the pressure field at specific locations on the slamming structured through mounted pressure gauges [18, 19, 24–27]. While this approach provides accurate pressure measurements, it is limited to local observations and often practically unfeasible, for the need to instrument the body and run wires throughout.

Motivated by these technical needs, some of my related work on particle image velocimetry (PIV) [28, 29], and recent advancements in the use of PIV data to indirectly reconstruct hydrodynamic pressure, I began a research program that would bring PIV into the study of slamming problems. PIV is the gold standard technique in experimental fluid mechanics. In a typical PIV experiment, the fluid is seeded with micro-particles, and is illuminated on a two-dimensional plane by a laser sheet. A high-speed camera is utilized to acquire consecutive images of the particles, which are then cross-correlated within small interrogation windows to estimate the velocity field in the fluid domain [30].

From knowledge of the velocity field, it is possible to infer the pressure field through post-processing. The recent review by van Oudheusden [31] is an excellent starting point to delve into available approaches for PIV-based pressure inference and apprehend an overview of the state-of-the art on the topic. Either starting from Navier-Stokes or Poisson equations, one can establish a linear boundary value problem for the pressure field, where the PIV velocity enters like a forcing term to the equations. By discretizing the region of interest into a grid, associated with the availability of PIV measurements, it is possible to reconstruct the pressure everywhere in the fluid domain.

Several authors have reported successful use of PIV data in the reconstruction of aerodynamic and hydrodynamic loading on immersed objects and flow channel walls [49–53] and in the study of free-surface flows [54]. However, the use of PIV for slamming problems was relatively untouched when we started our research. The only exception was the work of Nila and colleagues [55], which appeared at the very same time of our first paper on PIV-based pressure reconstruction during slamming in 2013 and is significantly different than our work in at least three aspects: the use of wedges with large deadrise angles that prevent the onset of large impulsive loadings, the constant entry velocity, and the need of additional computational data to enforce boundary conditions.

Table 1 chronologically summarizes our endeavors at New York University (NYU), under the support of the ONR Solid Mechanics Program to thoroughly demonstrate the use of PIV-based pressure reconstruction in hull slamming and tackle the analysis of a number of related fluid-solid interaction problems that were not completely clear and could benefit from PIV. Starting from the free fall symmetric water entry of a rigid wedge, we pursued several research areas, seeking to elucidate how geometry, physical properties, and impact conditions contribute to the fluid-solid interaction underpinning hull slamming.

The rest of the book chapter is organized as follows. In Sect. 2, we touch on the seminal work of von Karman [5] and Wagner [6] to offer some context regarding the physics of hull slamming, on which one can gain a better appreciation of our technical advancements which are summarized in this chapter. In Sect. 3, we describe the experimental setup which we have refined over the years to support the study of hull slamming problems through PIV. In Sect. 4, we articulate our approach to PIV-based pressure reconstruction, spanning the use of Navier-Stokes and Poisson equations, as well as extension to the study of three-dimensional phenomena. Section 5 offers validation in favor of our approach through comparison of PIV-based pressure inference with direct measurement using independent sensors and assessment against synthetic datasets. In Sect. 6, we briefly illustrate the application of our approach to two exemplary problems, for which observations were not previously available. Section 7 summarizes the main conclusions of our work, touches on ongoing work, and identifies avenues of potential future research.

2 A Jump in the Early 1900 to Gain Some Physical Insight

An excellent start to appreciate the physics underlying water entry problem is the recent review paper by Abrate [7], who masterfully laid a detailed history of early research on water entry by giants like von Karman [5] and Wagner [6] 90 years back.

With respect to the free fall of a wedge that is vertically entering the water surface, second Newton law will yield

Table 1 Synoptic presentation of ONR-supported, PIV-based studies on water impact at NYU

Ref.	Problem studied	Notes
[32]	Water entry of a rigid/compliant wedge	Electroactive polymers integrated in the wedge to measure local strain
[33]	Water entry of a rigid wedge	First NYU paper demonstrating PIV-based pressure reconstruction
[34]	Water entry of a compliant wedge	PIV-based pressure reconstruction; semi-analytical model for hydroelastic impact
[35]	Verification of PIV-based pressure reconstruction on synthetic data from computational fluid dynamics	Comparison with experimental data from [33]; parametric analysis on the effect of acquisition frequency and spatial resolution
[36]	Water entry of a series of curved bodies	Analysis of energy transfer during impact; verification of existing models to predict the pile-up coefficient
[37]	Three-dimensional physics associated with the water entry of a rigid wedge	Reconstruction of pressure field in three dimensions by integrating planar PIV data from different views; quantification of the extent of three-dimensional phenomena along the width
[38]	Three-dimensional physics associated with the water entry of a rigid body with a complex geometry	Extension of the approach in [37] to cope with asynchronous water entry of parallel image planes; analysis of energy transfer during impact
[39]	Asymmetric water entry of a rigid wedge	Inference of the pressure from Poisson's equation; comparison of multiple theories; evaluation of the effect of gravity
[40]	Three-dimensional computational fluid dynamics of the water entry of rigid body with a complex geometry	Comparison between numerical results and experimental results from [38]
[41]	Shallow water entry of a rigid wedge	PIV-based pressure reconstruction; actuation system to control the impact of the wedge; new theory to predict the flow physics associated
[42]	Water entry/exit of a compliant wedges	PIV-based pressure reconstruction; new theory to predict the hydroelastic response of a compliant wedge, based on our previous work [43]; analysis of the effect of gravity and viscosity
[44]	Water entry of a composite wedge	PIV-based pressure reconstruction; syntactic foam with different inclusion volume fraction and wall thickness
[45]	Water entry of a slender, compliant beam	New theory to predict the elastic response of the beam to impact
[46]	Water entry of a compliant cylinder	New theory to predict the elastic response of the cylinder
[47]	Asymmetric and oblique water entry of a rigid wedge	PIV-based pressure reconstruction; detailed comparison between PIV indirect measurement and direct readings from pressure sensors and accelerometers
[48]	Water entry in the presence of solid obstacles	New theory to study the interaction of a wedge with an obstacle

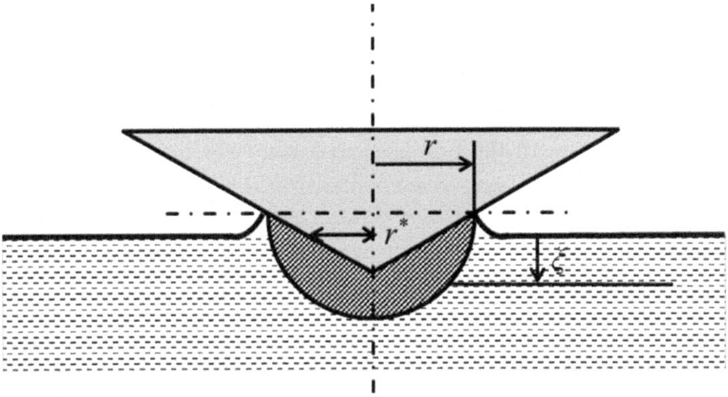

Fig. 1 Sketch of a wedge symmetrically impacting the water surface, showing the fluid mass displaced during water entry and the pile-up phenomenon. For clarity, we also display relevant symbols: entry depth ξ, wetted width r, and nominal wetted width r^*

$$\frac{d}{dt}\left[\left(M_{\text{dry}}+M_{\text{added}}\right)\dot{\xi}\right]=0 \tag{1}$$

where t is the time variable, ξ is the entry depth with respect to the unperturbed water surface, M_{dry} is the mass of the wedge, and M_{added} is the mass of water that is displaced during the impact. Eq. (1) assumes that the only force acting on the wedge during impact is given by the added mass phenomenon, neglecting gravity, drag, and surface tension, among other phenomena that might be considered during slamming.

In two-dimensions, added mass is estimated by hypothesizing that the wedge will displace half of a cylinder of water of height equal to the width of the wedge W and radius given by the wetted length of the wedge r, as shown in Fig. 1. More specifically, we can write the added mass as

$$M_{\text{added}}=\frac{\pi}{2}\rho W r^2 \tag{2}$$

where ρ is the water density. While this may seem a very crude estimation of the physics of the impact, in reality it works really well for predicting the time evolution of the entry depth of a rigid wedge, provided we can estimate properly the wetted length during impact.

Due to the piling-up of water during water entry, the wetted length will be different from the nominal wetted length r^*, which one may compute by simply taking the intersection of the wedge with the undisturbed free surface of the water. For a wedge with deadrise angle β, the nominal wetted width is $r^*=\xi/\tan\beta$ and we hypothesize that the wetted width is proportional to r^* through a constant proportionality factor γ, called pile-up coefficient.

For wedges with low deadrise angles, Wagner estimated the pile-up coefficient from classical potential flow theory to be $\pi/2$. An elegant framework for deriving the pile-up coefficient can be found in the work of Korobkin, in which a so-called

Wagner condition for the determination of the pile-up coefficient is presented in an integral form, to guarantee that the free-surface elevation at the location of the jet roots equals the vertical location of the entering wedge [56, 57]. Should one vary the shape of the impacting body or release the assumption of rigidity, one would discover different values of the pile-up coefficient, which can potentially vary in time due to hydroelastic coupling as we demonstrated in our previous work [36, 43].

For Eq. (1), initial conditions at $t = 0$ are $\xi = 0$ and $\dot{\xi} = V_0$, where V_0 is the impact velocity. Integrating once Eq. (1) with respect to time and imposing the initial condition on the velocity, we discover

$$\left(M_{\text{dry}} + M_{\text{added}}\right)\dot{\xi} = M_{\text{dry}} V_0 \tag{3}$$

Upon using Eq. (2) and replacing for the wetted length in terms of the entry depth, we determine the following first-order nonlinear differential equation with homogeneous initial condition:

$$\dot{\xi} = \frac{M_{\text{dry}} V_0}{M_{\text{dry}} + \frac{\pi}{2}\rho W\left(\frac{\gamma\xi}{\tan\beta}\right)^2} \tag{4}$$

To compute the impact force, we only need to calculate the acceleration of the wedge and multiply by the dry mass. The acceleration is computed by differentiating Eq. (4) with respect to time so that

$$\ddot{\xi} = -\frac{M_{\text{dry}} V_0}{\left[M_{\text{dry}} + \frac{\pi}{2}\rho W\left(\frac{\gamma\xi}{\tan\beta}\right)^2\right]^2}\left[\frac{\pi}{2}\rho W\left(\frac{\gamma}{\tan\beta}\right)^2\right]2\xi\dot{\xi} \equiv -\frac{\pi\rho W\gamma^2}{M_{\text{dry}} V_0 \tan^2\beta}\xi\dot{\xi}^3 \tag{5}$$

Following the line of argument by Abrate [7], one may establish a closed form solution for the peak force experienced by the wedge, as well as the time at which such a force is attained.

Upon knowledge of the entry depth as a function of time, we can embark on the computation of the hydrodynamic loading on the wetted surface of the wedge. Specifically, assuming that the wedge can be assimilated to a flat plate of semilength equal to r, one can exactly solve the potential flow problem everywhere in the fluid. In this context, one assumes that three-dimensional effects are secondary and that the flow is irrotational and incompressible. By replacing the expression of the velocity potential in the complete, nonlinear, Bernoulli's equation, one can find the following expression for the pressure:

$$\frac{p}{\rho} = \ddot{\xi}\sqrt{r^2 - x^2} + \frac{\pi}{2}\frac{\dot{\xi}^2 r}{\tan\beta\sqrt{r^2 - x^2}} - \frac{1}{2}\frac{\dot{\xi}^2 x^2}{r^2 - x^2} \tag{6}$$

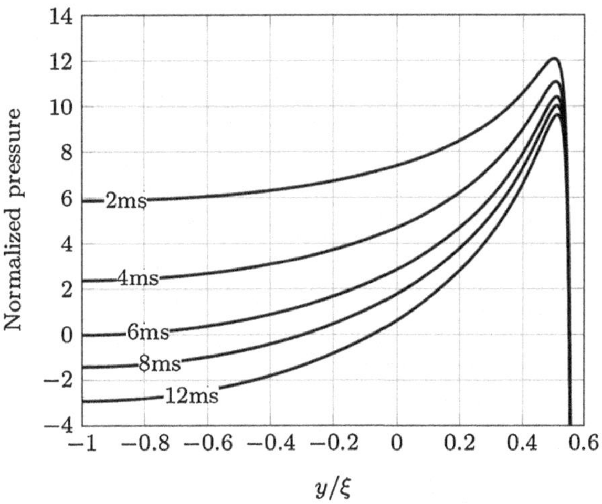

Fig. 2 Normalized pressure (scaled by ½ ρ V_0^2) as a function of the vertical coordinate y, ranging from $-\xi$ to $(r \tan\beta - \xi) \equiv \xi\,(\pi/2-1)$ for different times obtained from Wagner's solution for a wedge with 25° deadrise angle entering the water at 3 m/s. Reproduced from [33]

where x runs from the keel to the wetted length, parallel to the undisturbed water surface. In Eq. (6), the first two terms on the right hand side correspond to the time derivative of the velocity potential, while the latter term comes from the nonlinearity in Bernoulli equation.

An example of Eq. (6) is shown in Fig. 2, where we display several pressure distribution as time advances. The location of the peak pressure is nearly constant and a local minimum is always seen at the keel. We also observe a singularity as x tends to r, whereby the pressure tends to negative infinity at the intersection between the body and the free surface, so that negative pressure values are predicted therein.

3 Experimental Setup at NYU

Here, we explain our experimental approach to the study of water entry problems. We begin by illustrating the latest, most complete experimental apparatus we assembled for the study of water impact problems. Then, we detail our data acquisition system, comprising traditional sensors for the measurement of position, acceleration, and pressure, and the PIV system used for distributed velocity measurement in the fluid.

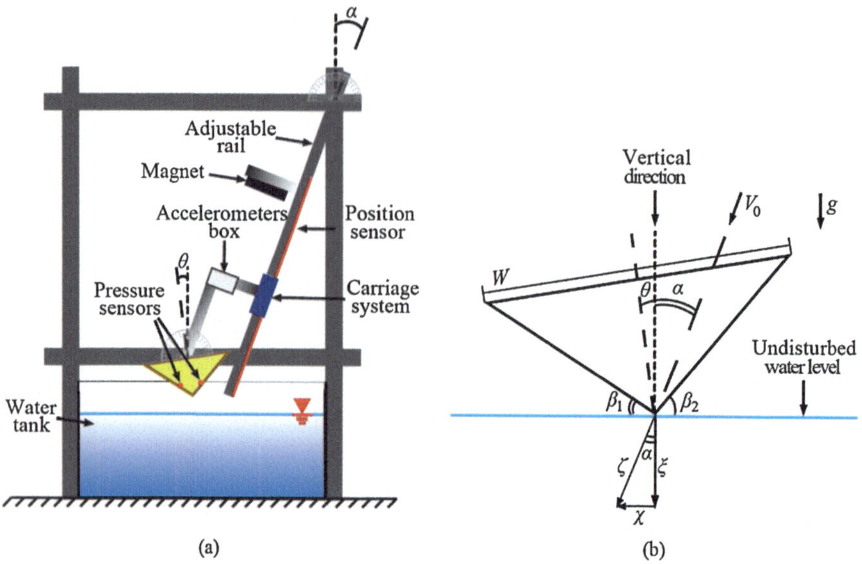

Fig. 3 Illustration of the NYU experimental apparatus: sketch of the front view (**a**), showing the automatic release system, the instrumented wedge, and the heel angle θ. Illustration of the wedge at the onset of the entry along with other relevant geometric parameters, including the inclination α, the displacement ζ and its components as χ and ξ, and the entry velocity V_0 (**b**). Reproduced from [47]

3.1 Apparatus

From the start of our research on hull slamming at NYU in 2012, we assembled several experimental apparatuses, and, as it typically happens in experimental research, it took several years to finally construct a robust, versatile, and reliable setup. The setup was put together by Drs. Jalalisendi and Russo (then NYU Ph.D. student and visiting student from Parthenope University, Italy), as described in [47]. The custom-made drop tower consisted of a transparent tempered water tank (800 mm long, 320 mm wide, and 350 mm deep) and a free fall mechanism, as shown in Fig. 3(a).

The free fall mechanism was the key element of the drop tower. It comprised of a 1.5 m aluminum rail and a carriage system, fabricated as a 3D-printed cart rigidly attached to an aluminum arm. The carriage system could be connected to a wedge or others objects with different geometries, concave or convex. Of the free fall mechanism, we could control both the inclination, α, and the heel angle, θ, defined in Fig. 3(b). The inclination was manually controlled by adjusting the inclination of the rail, and the heel angle was manually selected via an ad hoc realized junction. To control the release of the carriage system, we used an electric magnet, powered by a continuous voltage supply.

Should one be interested in water exit problems, where the impacting body shall be actuated to reverse its motion during impact, or in shallow impact problems, where there is the potential of damage to the tank, it is recommended to opt for a pneumatic control system, in place of the free fall mechanism. As detailed in [42], a viable pneumatic control system could be assembled by using an air cylinder, an electronic control valve, and an external air compressor, commanded by a DAQ card.

Throughout our work, we fabricated several specimens to simulate a variety of geometries that are representative of ship hulls, as detailed in [36]. Generally, these specimens were assembled by gluing balsa wood panels to 3D-printed models using epoxy resin. The models consisted of an array of parallel ribs to ensure that the specimen would be lightweight and to facilitate the housing of pressure sensors. The surface of the wood panels was waterproofed via a layer of epoxy resin, that minimized wood degradation over time.

3.2 Data Acquisition and Analysis

In our experiments, data acquisition was based on an array of sensors installed in the carriage system and in the drop tower, along with a time-resolved PIV system.

Three types of sensors were utilized in the experiments, namely:

- One linear position sensor was mounted on the rail to measure the displacement of the carriage system during free fall and during impact. The potentiometer was operated via a spring-loaded wiper installed on the cart, which would touch the position sensor to signal its location.
- Three accelerometers of varying ranges were mounted on the aluminum arm of the carriage system, within a sealed box. Two piezoelectric accelerometers with a dynamic range of ± 20 g and ± 200 g were utilized to quantify the acceleration during water impact, thereby affording the estimation of the total force exerted by the water. A capacitive accelerometer with a smaller ± 3 g range was used to measure the acceleration during free fall, toward the accurate inference of the entry velocity through integration in time.
- Two injection control pressure sensors were installed on the wedge to acquire a point measurement of the hydrodynamic loading during water entry. The sensors were mounted using two ABS 3D-printed supports, flushed with respect to the surface of the wedge. To mitigate thermal shock, the tip of each sensor was covered with a double layer of electrical insulating tape. The sensors were connected to a series signal conditioner, supplying the voltage bias and the current feed.

All these data were automatically acquired at a sampling frequency of 10 kHz from the release of the magnet via a DAQ board, controlled via a LabVIEW flow-sheet.

PIV requires seeding the water tank with particle tracers, whose size must be chosen according to the expected flow physics and the available hardware. In our experiments, we typically worked with hollow glass spheres or polyamide seeding particles of diameter on the range of 50 μm. Although we experimented with three-dimensional measurements as well in [37, 38], considerable physical insight into water entry can be garnered by simply using a planar system with a measurement plane at the mid-span of the sample, where three-dimensional effects are secondary. In our setup, such a plane was created by reflecting a laser sheet from the laser source via a 45°-inclined mirror which was installed below the tank.

Critical to the success of the experiments are the spatial and temporal resolutions of the high speed camera used for PIV acquisitions. From our first experiments in 2012 to the most recent ones, we improved the time resolution from 4 to 6 kHz and the spatial resolution from 980×640 pixels with 8 bits in grayscale to 2048×520 pixels and 12 bits. In our most recent experiments, we worked with a NAC MEMRECAM HX-5 high speed camera and a Dantec Dynamics Raypower 5000 laser at a wavelength of 532 nm. Earlier experiments used one or two high-speed Phantom cameras V.9.1. We typically analyzed PIV data from the instant when the keel touched the water surface for approximately 100 frames.

In our work, we extensively utilized the open source Matlab GUI "PIVlab" [58, 59] for PIV analysis. This software constitutes an excellent choice for data analysis, due to its versatility and ease of use. Typically, we have employed a fast Fourier transform multigrid scheme with a 50% interrogation window overlap, a decreasing interrogation window of 64×64, 32×32, and 16×16 pixels, and a 2×3 Gaussian scheme for subpixel interpolation. Working with a field of view of about 100 mm in length, the approximate pixel size is of the order of 0.1 mm, which will yield an interrogation region of about 1 mm^2. PIVlab will output a velocity vector for each interrogation window, partitioning the image. From the images, we also manually identify the free surface and wedge boundaries by placing an image mask on each frame via a Matlab script. The motion of the mask could, in turn, be useful to estimate the wetted width and the motion of the body.

4 PIV-Based Pressure Reconstruction

Here, we summarize our approach to estimate the pressure field from planar PIV data through incompressible Navier-Stokes equations [33] and synoptically introduce pressure reconstruction from Poisson equation, to reduce experimental uncertainties [39], as well as three-dimensional pressure inference [37, 38].

4.1 Pressure Reconstruction Using Navier-Stokes Equations

In [33], we proposed a novel experimental approach to estimate the pressure field everywhere in the fluid from planar PIV data. Neglecting gravity, viscosity, and compressibility, the incompressible Navier-Stokes equations read as follows [60]:

$$\frac{\partial p(x,y,t)}{\partial x} = -\rho\left(\frac{\partial u(x,y,t)}{\partial t} + u(x,y,t)\frac{\partial u(x,y,t)}{\partial x} + v\left(x,y,t\right)\frac{\partial u(x,y,t)}{\partial y}\right) \tag{1a}$$

$$\frac{\partial p(x,y,t)}{\partial y} = -\rho\left(\frac{\partial v(x,y,t)}{\partial t} + u(x,y,t)\frac{\partial v(x,y,t)}{\partial x} + v\left(x,y,t\right)\frac{\partial v(x,y,t)}{\partial y}\right) \tag{1b}$$

$$\frac{\partial u(x,y,t)}{\partial x} + \frac{\partial v(x,y,t)}{\partial y} = 0 \tag{1c}$$

where u and v are the velocity components along the x and y-directions, respectively; p is the pressure; ρ is fluid density; and t is time.

From PIV, one can directly estimate the velocity components in Eqs. (1a) and (1b) everywhere in the fluid domain (Fig. 4). The pressure field is then inferred from Eqs. (1a) and (1b) through the following steps [33]. First, we compute the derivatives of the velocity field in Eqs. (1a) and (1b) by using the multidimensional spline smoothing algorithm [61]. With reference to Fig. 3, we set the pressure to zero at a point A on the free surface, away from the wedge. Then, forward integration of Eq. (1a) is used to compute the pressure along $\mathcal{E}_1$ starting from A; upon reaching the end of the line, we change direction to integrate Eq. (1b) along $\mathcal{E}_2$. Once, the pressure along $\mathcal{E}_1$ and $\mathcal{E}_2$ is known, we calculate the pressure everywhere in the fluid using a spatial eroding scheme of [62]. This procedure is repeated at each time

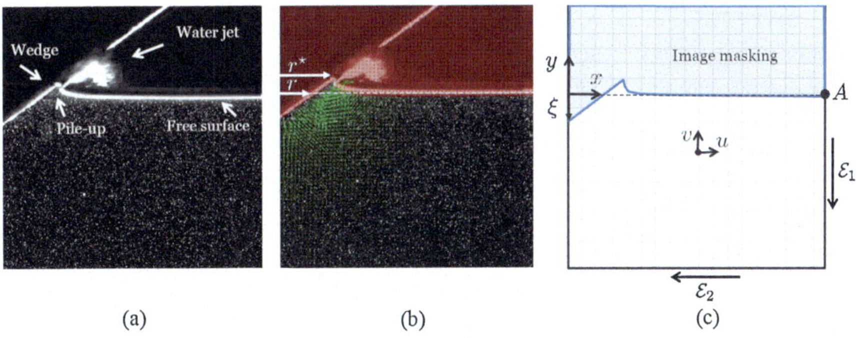

Fig. 4 Few steps for reconstructive the pressure field from PIV images. A representative PIV image during water entry of a rigid wedge with overlaid notation (**a**). PIV image with overlaid velocity vectors, and wetted, reference wetted widths (**b**). Schematic of a uniformly spaced PIV grid, illustrating the image masking, location of the Cartesian coordinate system, velocity components u and v, path for pressure integration, and entry depth ξ (**c**). Reproduced from [63]

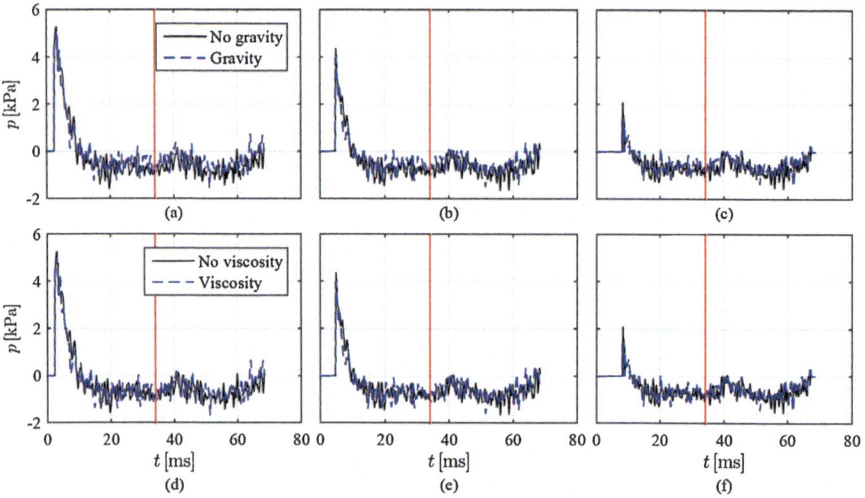

Fig. 5 Hydrodynamic loading on a compliant, aluminum wedge at ((**a**) and (**d**)) 10 mm, ((**b**) and (**e**)) 20 mm, and ((**c**) and (**f**)) 30 mm from the keel, considering the presence of ((**a**)–(**c**)) gravity and ((**d**)–(**f**)) viscosity. The vertical solid line identifies the beginning of the water exit. Reproduced from [42]

frame, as further explained in [63]. The approach is entirely based on PIV recordings, so that no extra information is required for the computation of the pressure field, beyond the physical constant ρ.

From the knowledge of the pressure field, one can determine the hydrodynamic loading by simply evaluating the pressure on the wetted surface of the wedge. The total force is computed by taking the integral of the hydrodynamic loading on the wetted surface.

The role of viscosity and gravity on the reconstructed pressure field was studied in [42] in the complex case of water entry and exit of a compliant wedge. Specifically, Fig. 5 shows the pressure reconstructed at several locations on a flexible wedge made of aluminum with a 25° deadrise angle, which is controlled to enter and exit the water surface through a pneumatic actuator. When accounting for gravity, we modify Eq. (1b) by including ρg on the right hand side. In general, the presence of gravity does not lead to significant variations in the pressure inference, except for a modest change at the onset of water exit, where the wedge attains its maximum entry depth before reverting its motion.

For the same problem of water entry and exit, Fig. 6 shows results on the total force experienced by the flexible wedge along with its rigid counterpart. Consistent with the discussion above, we do not register any appreciable effect of gravity on the total force. When accounting for viscosity, by modifying Eqs. (1a) and (1b) through the inclusion of a diffusion term, we do not find any noticeable effect.

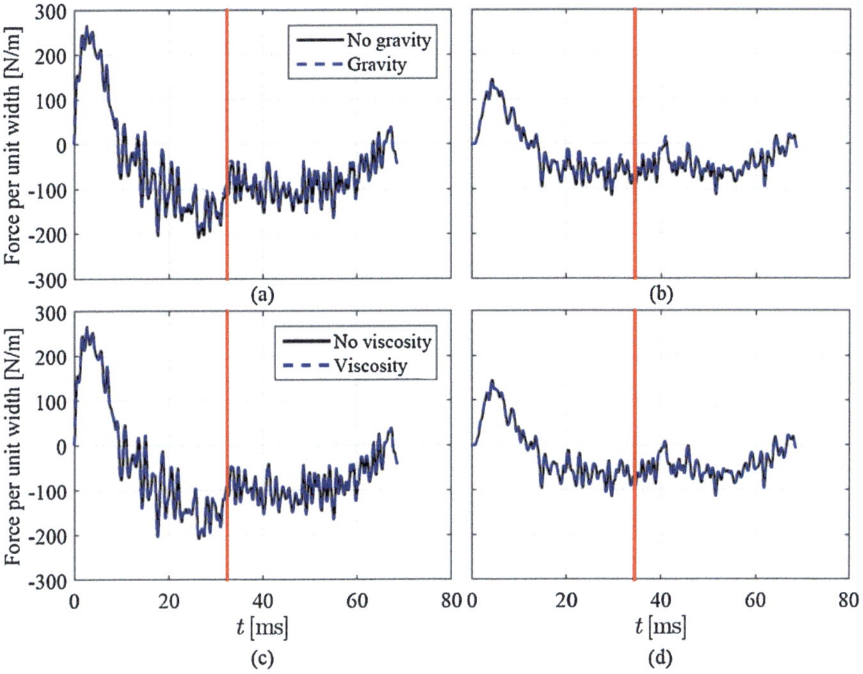

Fig. 6 Total force per unit width on ((**a**) and (**c**)) rigid and ((**b**) and (**d**)) compliant wedges, considering the presence of ((**a**) and (**b**)) gravity and ((**c**)and (**d**)) viscosity. The vertical solid line identifies the beginning of the water exit Reproduced from [42]

4.2 Advancements of the Approach

Over the years, we explored two potential avenues of improvement of the proposed pressure reconstruction scheme. In an effort to reduce uncertainty in the inference, we considered a Poisson-based numerical integration. In this vein, rather than integrating Eqs. (1a) and (1b), one would work with the classical Poisson equation

$$\frac{\partial^2 p(x,y,t)}{\partial x^2} + \frac{\partial^2 p(x,y,t)}{\partial y^2} = \\ -\rho\left(\left(\frac{\partial u(x,y,t)}{\partial x}\right)^2 + 2\frac{\partial u(x,y,t)}{\partial x}\frac{\partial v(x,y,t)}{\partial y} + \left(\frac{\partial v(x,y,t)}{\partial y}\right)^2\right) \tag{2}$$

Boundary conditions for Eq. (2) depend on which portion of the PIV image is examined. Specifically, we set the pressure to zero somewhere on the free surface (points A and B), and then we integrate Eqs. (1a) and (1b) along the rest of the free surface, $\mathcal{S}_{\mathrm{SR}}$ and $\mathcal{S}_{\mathrm{SL}}$, and then on the boundary of the recorded image in the fluid, $\mathcal{S}_{\mathrm{BR}}$, $\mathcal{S}_{\mathrm{BB}}$, and $\mathcal{S}_{\mathrm{BL}}$, as shown in Fig. 7(a). For the rest of the fluid boundary in contact

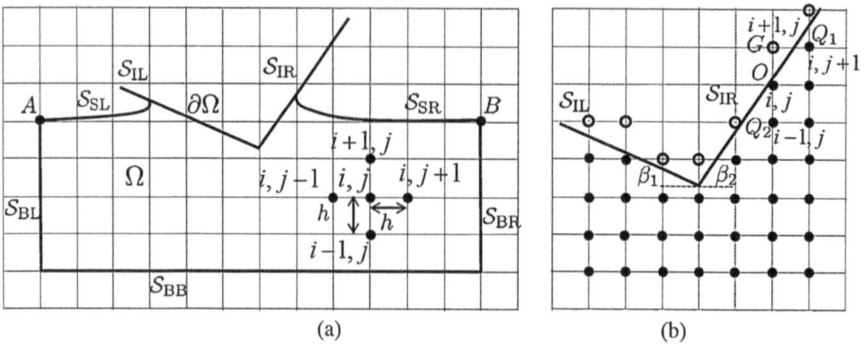

Fig. 7 Computational details for the implementation of the Poisson solver. Schematic of the stencil used to discretize the Poisson equation, along with the definition of all symbols used in the solution (**a**). Schematic of the procedure used to treat Neumann boundary conditions: the fluid points Q_1 and Q_2 (dots) are utilized to reconstruct the pressure at ghost point G (open circle) close the wedge boundary (**b**). Reproduced from [39]

with the impacting body, we set the derivative of the pressure along the normal, by using Eqs. (1a) and (1b) once more. To solve Eq. (2), we used a time marching algorithm with a grid that is consistent with the PIV data matrix, as sketched in Fig. 7 (b).

Another major research thrust has been the extension of the approach to three-dimensional problems, where no data was previously available. Specifically, we sought to clarify how much the pressure would vary along the width of a wedge to justify the classical assumption of planar flow, and, at the same time, establish a reliable approach to examine the pressure field associated with the impact of bodies of complex geometries. Using a traditional planar PIV system, we proposed to estimate the full three-dimensional velocity field in the fluid by measuring the two-dimensional velocity field on several planes along both the length and width of the body, as shown in Fig. 8. We performed two sets of PIV trials on multiple cross-sections of the wedge. In the first set, the PIV plane was orthogonal to the sample width, and its location was varied along the width to estimate the cross-sectional velocity components. In the second set, we rotated the measurement plane to record images orthogonal to the length, and its location was varied along the length to acquire the axial velocity component. From these sets of measurements, we estimated the three-dimensional velocity field everywhere in the domain via a cubic spline interpolation scheme.

Upon the knowledge of the three-dimensional velocity field, we estimate the pressure field by looking at each of the cross-sectional planes and integrating incompressible Navier-Stokes equations. Due to the three-dimensional nature of the fluid flow, Eqs. (1a) and (1b) of these equations should be changed to account for convection and, if viscosity is retained, diffusion along the width. However, all these additions are known from PIV measurements and the integration can be performed following the same procedure as described above. Notably, such a modification of Eqs. (1a) and (1b) has a minimal role on the pressure estimation

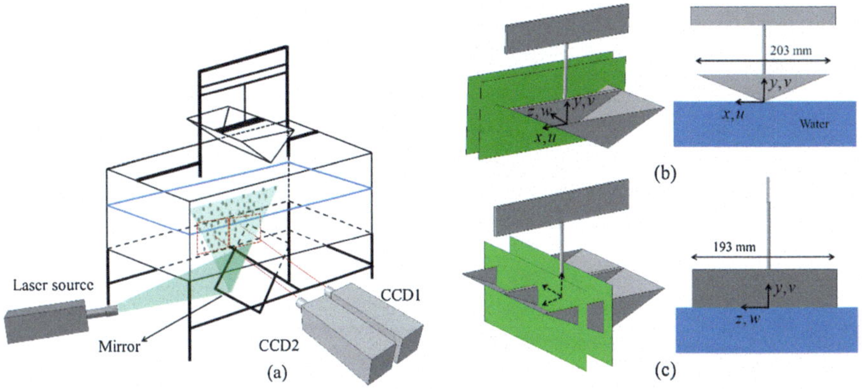

Fig. 8 Methodology to reconstruct the pressure field around a wedge in three dimensions from multiple planar PIV measurements. Schematics of the experimental apparatus with representative location of the laser source and the high-speed cameras (**a**). Illustrations of the measurement planes for PIV analysis in: cross-sectional (**b**) and axial (**c**) PIV. Reproduced from [37]

for a wedge, whereby one would obtain the same prediction for the force by simply adjusting the PIV velocity field to the one corresponding to that particular plane [41]. However, when working with solid bodies with curvature along two planes, it is important to account for the axial flow in the estimation of the pressure field, which would otherwise be overestimated [38].

5 Validation of PIV-Based Pressure Reconstruction

Our approach to validate PIV-based pressure reconstruction proceeds along two independent avenues. One avenue consists of the mere comparison between direct readings from the available sensor array. The other entails the creation of a synthetic dataset based on computational fluid dynamics (CFD), on which we could test the approach without confounds from experimental uncertainty and, potentially, unaccounted physical phenomena. Together, these two routes offer robust validation and guidance on ways to improve the process of data collection and analysis.

5.1 *Validation Through Experimental Measurements*

Experimental validation was conducted in our recent work [47] through experiments on a rigid wedge with length of 19 cm, width of 20 cm, and deadrise angle of 37°. While experiments presented therein include a very wide set of combinations of the inclination and heel angle, for validation purposes it is sufficient to focus on the case of vertical, symmetric impact (both angles equal to zero). The analysis of this specific experimental condition was, in fact, the object of a conference paper by our group [64].

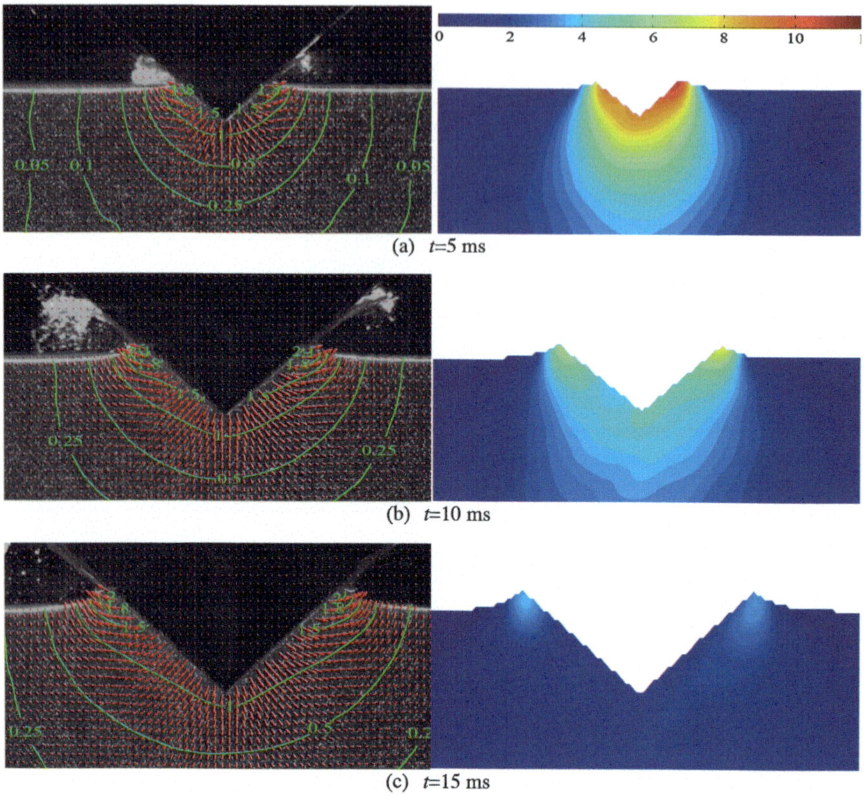

Fig. 9 Acquired high speed images overlaid with velocity vectors and contour map in m/s (left panels) and contour plots of the pressure magnitude in kPa (right panels). Reproduced from [64]

From these data, in Fig. 9, we report velocity and pressure fields at $t = 5$, 10, and 15 ms. In agreement with our intuition, we observe that both the velocity and pressure are maximized in the pile-up region during the whole entry and they decay away toward the water bulk. Consistent with the classical Wagner theory, the pressure on the wetted surface of the wedge decreases from the pile-up region to the keel, where we register the smallest pressure values. Moving from the keel to the pile-up, we discover that the vertical component of the velocity changes its sign, pointing downward toward the keel and then upward in the pile-up where the water raises to form the jets. On the other hand, the horizontal velocity component continuously increases from zero to its maximum.

To begin the assessment of PIV readings against direct sensor measurements, in Fig. 10(a), we compare the entry depth acquired by the position sensor, estimated from the integration of the accelerometer data, and estimated from masking of the PIV images. With respect to impact speed, in Fig. 10(b), we compare the wedge velocity obtained from differentiation of the position sensor data, estimated from the integration of the accelerometer data, and inferred from the PIV velocity of the fluid

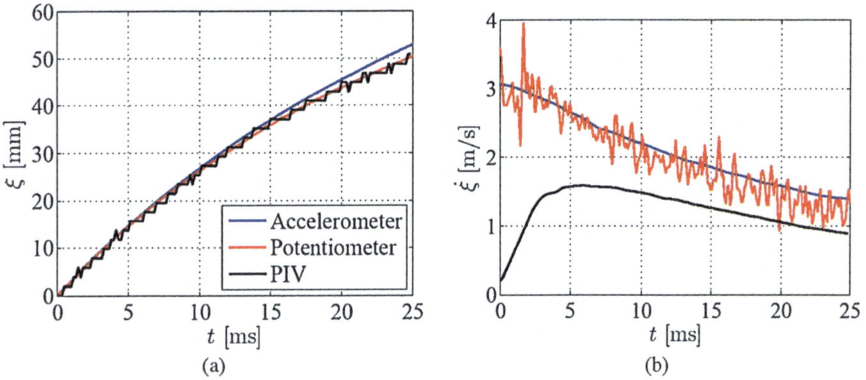

Fig. 10 Time histories of the averaged: entry depth ξ (**a**) and velocity $\dot{\xi}$ (**b**). Results are computed using three approaches, while averaging on five trials for accelerometer and position sensor. Reproduced from [64]

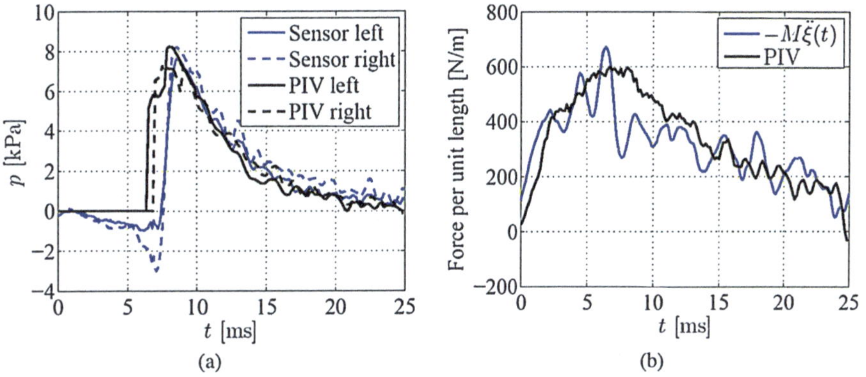

Fig. 11 Pressure signals at pressure sensors locations from PIV and pressure transducers (**a**) and time histories of the averaged force per unit length obtained from PIV and accelerometer data (**b**). Note that the M in (**b**) is the mass per unit width of the wedge. Reproduced from [64]

flow. Our results demonstrate a good agreement between PIV measurements and data obtained from the sensor array, except of the velocity estimate at the very beginning of the impact. As originally discussed in [33], such an underestimation should be ascribed to the fact that PIV provides only an average estimation of the velocity inside each of the interrogation region. Due to the large velocity gradients at the onset of the impact, such an averaging process will translate into an underestimation of the velocity throughout the fluid.

Beyond kinematic data, in Fig. 11(a), we demonstrate the accuracy of PIV-based pressure reconstruction by comparing against direct pressure measurements through pressure sensors symmetrically located 47 mm from the keel. Results indicate remarkable agreement between indirect PIV-based pressure measurements and direct observations. Interestingly, although pressure readings start at approximately

5 ms, when the velocity at the keel is still considerably underestimated by PIV, we do not register any effect on the accuracy of PIV-based pressure inference away from the keel.

As a further independent assessment, in Fig. 11(b), we compare the force per unit width obtained by integrating the PIV pressure on the wetted surface with the estimation of the total force obtain by scaling the accelerometer data by the dry mass of the carriage. In agreement with our expectations, the total force is underestimated at the onset of the impact, where we see large velocity gradients. The peak of the force is slightly smaller in PIV-based measurements, likely due to three-dimensional effects which are neglected in this analysis.

5.2 *Validation Through Synthetic Data*

While experiments are the cornerstone against which we must validate our approach to the inference of the pressure field, a number of technical limitations may suggest a complementary approach to validation. First, measurements acquired through pressure sensors can only acquire local data, which may not suffice for a comprehensive validation of the pressure field, and accelerometer data can only help to resolve the total force on the wedge. Second, pressure sensors are subject to noise, thermal shock, and potential drift, which could challenge validation. Third, without a trustworthy ground-truth it is difficult to understand the role of experimental settings, such as the spatial resolution of the camera and its acquisition frequency.

To address these limitations, we embarked on a thorough assessment of PIV-based pressure reconstruction through synthetic datasets, in which PIV parameters could be systematically varied, without experimental confounds. Toward this aim, we implemented a direct computational framework to study the two-dimensional flow physics associated with the water entry of a rigid wedge. Water and air were treated as immiscible phases and their relative motion was utilized to identify the evolution of the free surface. The entering wedge was described as a moving wall, which translated vertically following our experimental observations in [33].

Figure 12 shows velocity contours from simulations and experiments. While in the water bulk the velocity fields display patterns that are highly comparable qualitatively and quantitatively, we register some differences between simulations and experiments in the pile-up region. First PIV cannot capture the onset of the water jets, whose extent was too small to be experimentally resolved. In addition, PIV reports a lower pile-up compared to CFD and, as a consequence, the maximum velocity identified through PIV is lower than numerical predictions. Likely, this is due to the averaging process that is inherent to the finite size of the interrogation windows in PIV.

Working with synthetic data on the velocity field, one can test the feasibility of our approach to PIV-based pressure reconstruction through direct comparison with CFD predictions of the pressure field. By using the same PIV settings of [33], we

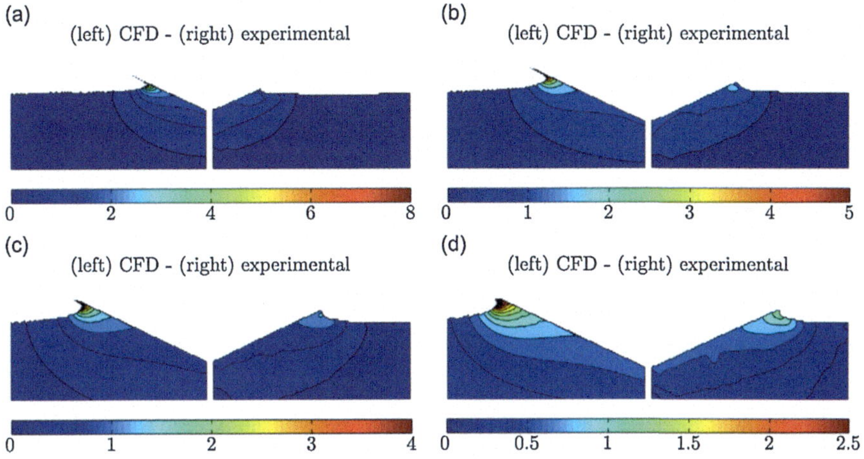

Fig. 12 Comparison between numerical and experimental velocity fields for a wedge with 25° deadrise angle. The color-bar represents the fluid velocity magnitude in m/s: $t = 5$ ms (**a**), $t = 10$ ms (**b**), $t = 15$ ms (**c**), and $t = 20$ ms (**d**). Reproduced from [35]

Fig. 13 Comparison between CFD results (solid lines) and pressure reconstruction from a synthetic dataset (dashed lines) for the normalized hydrodynamic pressure (using the entry speed and the water density) on the wetted surface of the wedge at several instants. Reproduced from [35]

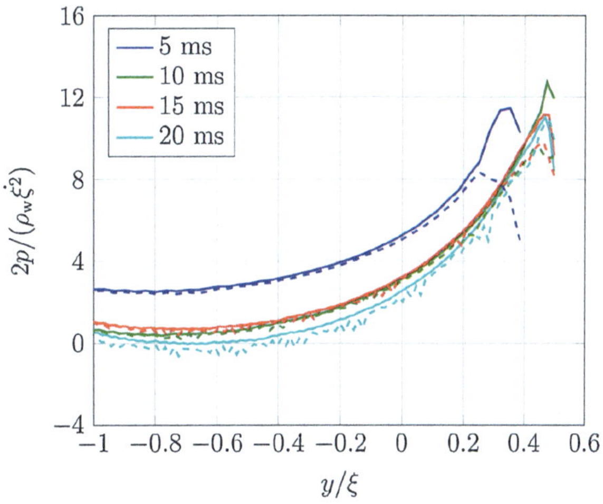

inferred the pressure distribution on the wedge and compared with CFD predictions, as shown in Fig. 13. The comparison indicates excellent agreement with CFD, although some discrepancies are registered in the early stage of the impact, when the reconstructed maximum pressure is lower than expectations. Such a behavior could be associated with the few vectors over which the pressure is reconstructed. As time progresses, the pile-up region grows, resulting in a larger number of vectors. Increasing the spatial resolution of the measurement and the acquisition frequency could help improving on the pressure inference. The latter is particularly important, whereby we documented a stronger dependence of the pressure reconstruction algorithm on the acquisition frequency than the spatial resolution in [35].

6 Exemplary Applications

Here, we demonstrate the use of the proposed PIV-based approach to tackle two distinct problems, with relevance to the design of advanced marine vessels. Specifically, we summarize our recent work on the asymmetric and oblique water entry of a rigid wedge [47] and water entry of a compliant wedge constituted by two panels of vinyl ester/glass syntactic foam [44].

6.1 Asymmetric and Oblique Impact of a Rigid Wedge

Extending the results from Sect. 5.1, we examined asymmetric and oblique impact of a rigid wedge entering the water surface in free fall. With reference to the nomenclature in Fig. 3(b), we demonstrate the velocity and pressure field in case of symmetric (Fig. 14(a)), asymmetric (Fig. 14(b)), oblique (Fig. 14(c)), and asymmetric and oblique impact (Fig. 14(d)). Results on symmetric impact are just an instance of those presented in Fig. 9, which suggest that the velocity and pressure data are maximized in the pile-up regions and decrease toward the fluid bulk.

More intriguing are the experimental observations for the other impact conditions. Inspection of the results on asymmetric water entry in Fig. 14(b) suggests that the velocity and pressure are again maximized in the pile-up regions, like the symmetric impact. However, the values of the velocity and pressure in the left pile-up are larger than those in the right pile-up. For the oblique case in Fig. 14(c), the velocity and pressure on the left side of the wedge are slightly higher than those on the right side due to the velocity angle, which causes the left side of the wedge to push more water than the right side.

Interestingly, the case of combined oblique and asymmetric impact in Fig. 14(d) is similar to an asymmetric impact, suggesting that the heel angle has a dominant role on the flow physics, more than the velocity angle. However, asymmetries in the velocity and pressure distributions are magnified by the combined effect of α and θ. Hence, the velocity at the keel reaches values close to those in the left pile-up and the pressure in the left pile-up is larger than the values shown in Fig. 10(b). The onset of ventilation that is observed in this experimental condition might be explained by the increase in the velocity at the keel.

To further detail the effect of θ and α on the impact, we examined the energy exchange during water entry. Specifically, we computed the amount of energy that the wedge imparts to the fluid during the impact, by assuming that damping was absent. Following our previous work [36], we decomposed the energy of the fluid into energy of the risen water (pile-up regions and the water jets) and the energy of the bulk, evaluated from PIV. Figure 15 shows the time evolution of the ratio between the energy of the risen water and the total energy of the system (estimated from the initial potential energy of the wedge).

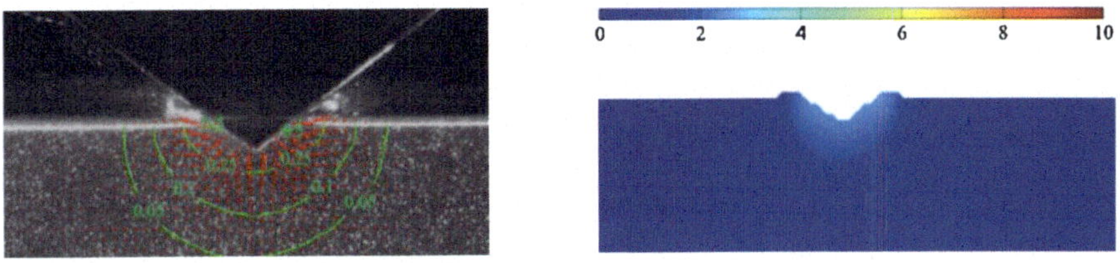

(a) $\alpha = 0°$ and $\theta = 0°$

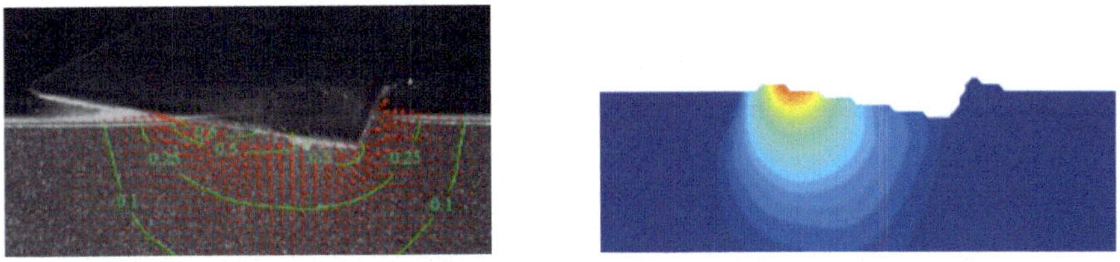

(b) $\alpha = 0°$ and $\theta = 30°$

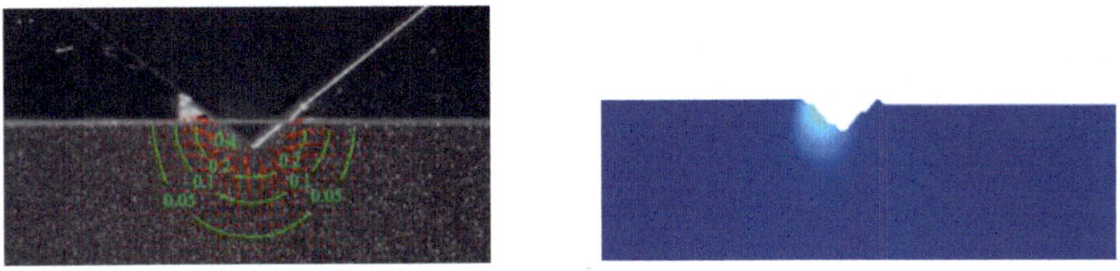

(c) $\alpha = 30°$ and $\theta = 0°$

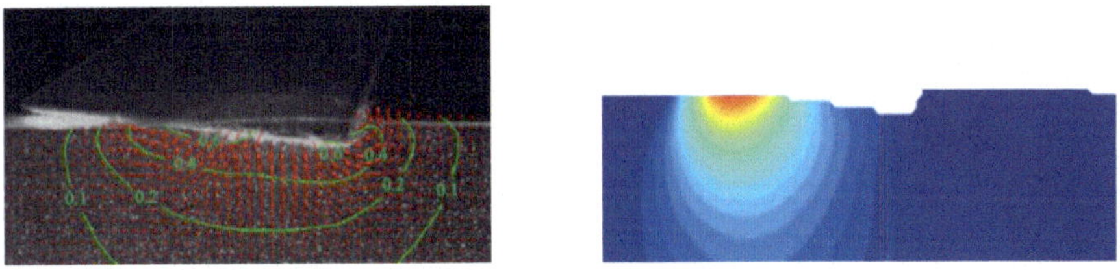

(d) $\alpha = 30°$, $\theta = 30°$

Fig. 14 Normalized velocity (scaled by the entry speed scaled by V_0; left column) and pressure (scaled by ½ ρ V_0^2; right column) fields in the fluid at $t = 3.33$ ms for different values of α and θ. The chosen time corresponds to the peak force observed for the case of combined oblique and asymmetric impact (α = 30° and θ = 30°). Reproduced from [47]

Predictably, in all the experimental conditions, the fraction of the energy released to the pile-up and water jets increases in time, reaching 40–60% of the total energy. Increasing the heel angle results in a steep increase in the energy transferred to the risen water region during the early phase of water entry.

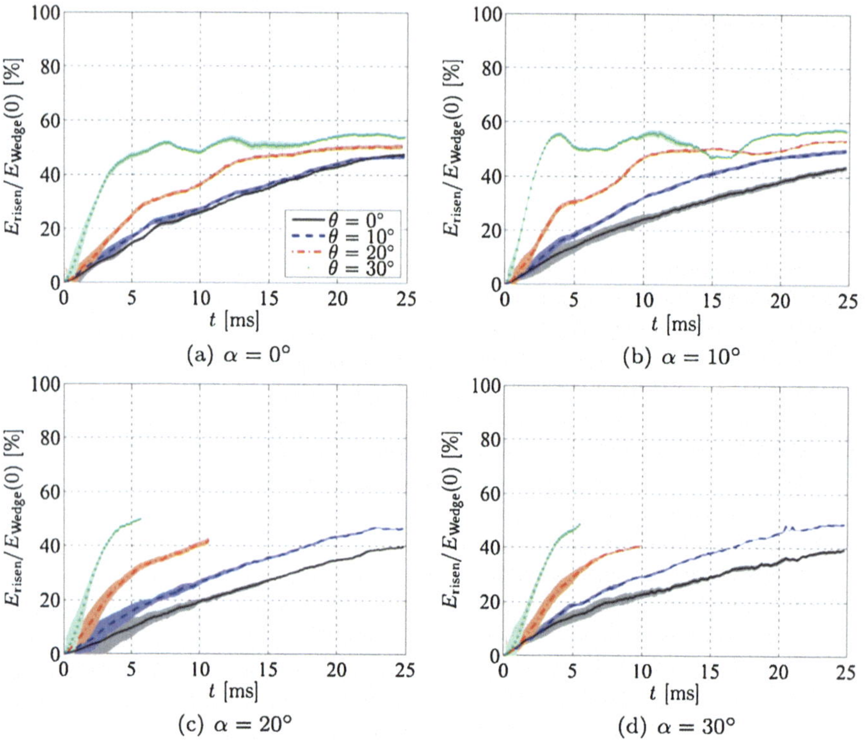

Fig. 15 Energy transferred to the risen water E_{risen} (t) calculated as the percentage of the total energy of the wedge E_{Wedge} (t) for all the considered combinations of the heel and velocity angles. Results are averaged across trials, and shaded areas indicate one standard deviation. Reproduced from [47]

6.2 *Impact of a Composite Wedge*

Recently, we have attempted at a detailed experimental analysis of the water entry of highly compliant composite wedges consisting of syntactic foam panels. Syntactic foams are lightweight composites which are fabricated by dispersing hollow microballoons in a matrix material [65]. The inclusions create a closed-cell microstructure which can be leveraged to reduce weight, improve mechanical properties, and mitigate moisture absorption. These propitious attributes have prompted the use of syntactic foams in marine vessels and aircrafts, as core materials for sandwich composites [66].

While several studies have examined the response of syntactic foams to Izod impact, Charpy Impact, and drop weight as reviewed in [67], experiments on syntactic foams subjects to slamming were not available. To fill this knowledge gap, we extended our PIV-based analysis to the study of slamming response of syntactic foams panels, with vinyl ester matrix and glass particles. The panels were

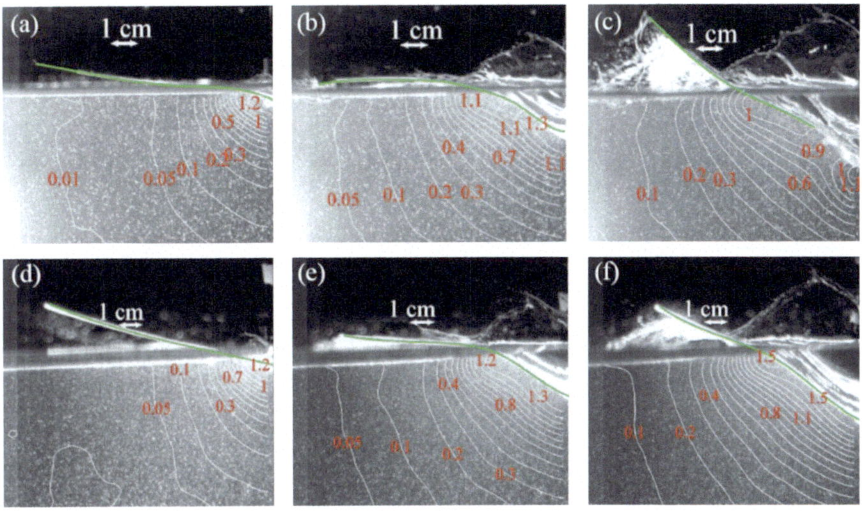

Fig. 16 High-speed images overlaid with velocity magnitude in m/s for (**a–c**) VE220–60 and (**d–f**) VE460–60, at different time instants: (**a**,**d**) $t = 1.25$ ms; (**b**,**e**) $t = 7.5$ ms; and (**c**,**f**) $t = 12.5$ ms. The green line identifies the panel shape. Reproduced from [44]

assembled in a wedge of 10° deadrise angle, which entered in free fall the water surface. Two microballoon densities were examined to elucidate the role of syntactic foam composition on slamming response. Specifically, we maintained the volume fraction of the glass particles at 60% and chose two densities (220 and 460 kg/m^3), resulting into two types of syntactic foam: VE220–60 and VE460–60.

In Fig. 16, we report PIV results for the drop of VE220–60 and VE460–60 syntactic foam wedges with an entry velocity of approximately 2.75 m/s. Interestingly, the deformation of the panel is not captured by the fundamental in-vacuum mode shape of a cantilever beam. Rather, we register the presence of higher vibration modes, which cause the curvature of the panel to change sign along its span. While the heavier panel was able to withstand the impact, the lighter one failed between 7.5 and 12.5 ms, close to the keel, prompting flow recirculation in the vicinity of the keel. The detachment of the panel from the holding frame forces the water to roll around it, creating a vortical structure that may be seen from the presence of closed velocity contours close to the keel.

For the same conditions of Fig. 16, in Fig. 17, we show PIV-based pressure reconstruction. Results confirm our intuition that pressure is maximized in the early stage of the impact close to the water jet, consistent with our previous work on aluminum wedges [34]. The pressure field is similar for the two panels at time instants $t = 1.25$ and 7.5 ms, while the pressure in the fluid for the VE220–60 panel at $t = 12.5$ ms is remarkably different from the pressure field for the VE460–60 panel, due to the failure of the panel. Consistent with the velocity recirculation noted in Fig. 16, we see that the pressure reaches negative values in the vicinity of the failed panel, while it is positive close to the keel.

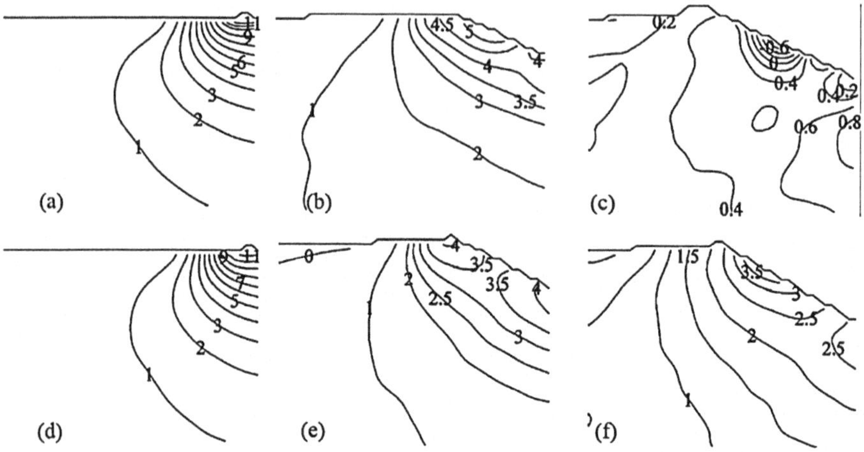

Fig. 17 Pressure distribution in the fluid in kPa for (**a–c**) VE220–60 and (**d–f**) VE460–60, at different time instants: (**a,d**) $t = 1.25$ ms; (**b,e**) $t = 7.5$ ms; and (**c,f**) $t = 12.5$ ms. Reproduced from [44]

Scanning electron micrographs of the failed panels do not show the presence of debris on the surface, indicating that failure is initiated on the tensile side of the cross-section during bending. However, different from a quasi-static three- or four-point bending test, the tensile side of the cross-section alternates during water entry between the top and bottom edge. Consistent with this claim, we also noted microballoons debonding from the matrix on both the top and bottom edges. As a result of debonding, most of the tensile stress is resisted by the matrix, whose tensile cracks contribute to the failure of the panel.

7 Conclusions

From direct measurement of the velocity field through PIV, it is possible to reconstruct the pressure field in the fluid domain by integrating Navier-Stokes or Poisson equations. In this chapter, we summarized research supported by ONR, Solid Mechanics Program, at NYU on the integration of PIV in the study of the slamming response of rigid and compliant bodies. We summarize 7 years of research on the topic, addressing the very foundations of the approach and Navy-relevant applications.

PIV-based pressure inference holds promise to advance our understanding of hull slamming by affording spatially-distributed, time-resolved measurement of the hydrodynamic loading experienced by structures that enter or exit the water surface. Several groups throughout the world working on the area of water impact have recognized the potential value of a PIV-based approach to the inference of the pressure field [68]. Some of the recent efforts in this area have directly benefited from experimental results from our line of work, which has enabled thorough experimental validation of existing theories and computer codes, while offering insight into the complex, unsteady fluid-structure interactions underpinning water entry and exit problems.

For example, in our own theoretical work [43] on mathematical modeling of hydroelastic phenomena, we utilized our experimental data to benchmark predictions of the effect of elastic compliance on the hydrodynamic loading experienced during water entry. Similarly, in [69], Izadi and colleagues used our experimental results to validate a computational model, based on the STAR CCM+ software and adopting an overset mesh.

While we have made several progresses in PIV-based pressure inference, future research is required to fill several knowledge gaps. First, further research should seek to establish tight error bounds for the estimation of the hydrodynamic loading to facilitate uncertainty analysis and parameter optimization. Some work along this directions has been recently reported in [70]. Second, future research should explore tomographic PIV for the reconstruction of a three-dimensional pressure field, without the need of multiple planar PIV experiments that might increase experimental uncertainty. Third, more research is needed to expand the characterization of the slamming response of composite materials, beyond the single instance of syntactic foams considered herein to clarify the failure response of other lightweight composites. Fourth, a potential line of inquiry entails the analysis of repeated slamming, consisting of multiple entries and exits, which characterize the typical operation of marine vessels and may bear important consequences in the context of fatigue.

Our current work is pursuing two research directions. First, we are pursuing the integration of digital image correlation (DIC) and PIV to afford the possibility of simultaneous measurement of structural deformations and flow physics in Navy-relevant problems. We have recently demonstrated the possibility of a combined PIV-DIC approach in the study of static deformations induced by a steady fluid flow [71] and we are presently seeking to extend the method to dynamic problems. Second, we are working on the fluid-structure interaction associated with the impact on a water-backed plate. We have developed a mathematical model to capture the structural response of water-backed plates in two-dimensions, which we have verified through comparison with detailed computer simulations [72]. Presently, we are working to gather experimental results that could support validation of the framework and help elucidate the physics of the impact.

Acknowledgments The work has been supported by the Office of Naval Research (Grant N00014-10-1-0988 and N00014-18-1-2218) with Dr. Y.D.S. Rajapakse as the program manager. The author would like to thank Dr. Andrea Facci, Dr. Giacomo Falcucci, Dr. Mohammad Jalalisendi, Dr. Simonluca Russo, Dr. Adel Shams, Dr. Stefano Ubertini, Dr. Peng Zhang, and Mr. Sam Zhao, who have contributed to the research summarized in this chapter.

References

1. Faltinsen OM (1990) Sea loads on ships and offshore structures. Cambridge University Press, New York
2. Faltinsen OM, Landrini M, Greco M (2004) Slamming in marine applications. J Eng Math 48 (3–4):187–217
3. Hughes K et al (2013) From aerospace to offshore: bridging the numerical simulation gaps-simulation advancements for fluid structure interaction problems. Int. J. Impact Eng 61:48–63
4. Kapsenberg G (2011) Slamming of ships: where are we now? Philos Trans R Soc A Math Phys Eng Sci 369(1947):2892–2919
5. Von Karman T (1929) The impact on seaplane floats, during landing. NACA-TN-321
6. Wagner H (1932) Uber stoss-und gleitvorgange an der ober- flache von flussigkeite. ZAMM – Zeitschrift fur Ange- wandte Mathematik und Mechanik 12(4):193–215
7. Abrate S (2011)) Hull slamming. Appl Mech Rev 64(6):060803
8. Panciroli R, Abrate S, Minak G, Zucchelli A (2012) Hydroelasticity in water-entry problems: comparison between experimental and SPH results. Compos Struct 94(2):532–539
9. Panciroli R, Abrate S, Minak G (2013) Dynamic response of flexible wedges entering the water. Compos Struct 99:163–171
10. Das K, Batra RC (2011) Local water slamming impact on sandwich composite hulls. J Fluid Struct 27(4):523–551
11. Qin Z, Batra RC (2009) Local slamming impact of sandwich composite hulls. Int J Solids Struct 46(10):2011–2035
12. Ray MC, Batra RC (2013) Transient hydroelastic analysis of sandwich beams subjected to slamming in water. Thin-Walled Struct 72:206–216
13. Xiao J, Batra RC (2012) Local water slamming of curved rigid hulls. Int J Multiphys 6 (3):305–340
14. Xiao J, Batra RC (2014) Delamination in sandwich panels due to local water slamming loads. J Fluid Struct 48:122–155
15. Allen T, Battley M (2015) Quantification of hydroelasticity in water impacts of flexible composite hull panels. Ocean Eng 100:117–125
16. Battley MA, Clark AM, Allen TD, Cameron CJ (2014) Shear strength of sandwich core materials subjected to loading rates relevant to water slamming. J Reinf Plast Compos 33 (6):506–513
17. Swidan A et al (2016) Experimental drop test investigation into wetdeck slamming loads on a generic catamaran hullform. Ocean Eng 117:143–153
18. Battley M, Allen T (2016) Characterisation of fluid-structure interaction for water impact of composite panels. Int J Multiphys 6:3
19. Battley M, Allen T (2012) Servo-hydraulic system for controlled velocity water impact of marine sandwich panels. Exp Mech 52(1):95–106
20. Razola M, Rosén A, Garme K (2014) Experimental evaluation of slamming pressure models used in structural design of high-speed craft. Int Shipbuild Prog 61(1–2):17–39
21. Stenius I, Rosén A, Battley M, Allen T (2013) Experimental hydroelastic characterization of slamming loaded marine panels. Ocean Eng 74:1–15
22. Stenius I, Rosén A, Battley M, Allen T, & Pehrson P (2011) Hydroelastic effects in slamming loaded panels. 11th. International conference on Fast Sea transportation (FAST 2011), Honolulu, Hawaii pp 644–652
23. Stenius I, Rosén A, Kuttenkeuler J (2011) Hydroelastic interaction in panel-water impacts of high-speed craft. Ocean Eng 38(2–3):371–381
24. Lewis SG, Hudson DA, Turnock SR, Taunton DJ (2010) Impact of a free-falling wedge with water: synchronized visualization pressure and acceleration measurements. Exp Fluids 42 (3):035509
25. Luo H, Wang H, Soares CG (2012) Numerical and experimental study of hydrodynamic impact and elastic response of one free-drop wedge with stiffened panels. Ocean Eng 40:1–14

26. De Backer G et al (2009) Experimental investigation of water impact on axisymmetric bodies. Appl Ocean Res 31(3):143–156
27. Charca S, Shafiq B, Just F (2009) Repeated slamming of sandwich composite panels on water. J Sandw Struct Mater 11(5):409–424
28. Aureli M, Porfiri M (2010) Low frequency and large amplitude oscillations of cantilevers in viscous fluids. Appl Phys Lett 96(16):164102
29. Jalalisendi M, Panciroli R, Cha Y, Porfiri M (2014) A particle image velocimetry study of the flow physics generated by a thin lamina oscillating in a viscous fluid. J Appl Phys 115 (5):054901
30. Raffel M, Willert C, Wereley S, Kompenhans J (2007) Particle image velocimetry: a practical guide. Springer, New York
31. Van Oudheusden BW (2013) PIV-based pressure measurement. Meas Sci Technol 24 (3):032001
32. Cha Y, Phan CN, Porfiri M (2012) Energy exchange during slamming impact of an ionic polymer metal composite. Appl Phys Lett 101(9):094103
33. Panciroli R, Porfiri M (2013) Evaluation of the pressure field on a rigid body entering a quiescent fluid through particle image velocimetry. Exp Fluids 54(12):1630
34. Panciroli R, Porfiri M (2015) Analysis of hydroelastic slamming through particle image velocimetry. J Sound Vib 347:63–78
35. Facci AL, Panciroli R, Ubertini S, Porfiri M (2015) Assessment of PIV-based analysis of water entry problems through synthetic numerical datasets. J Fluid Struct 55:484–500
36. Panciroli R, Shams A, Porfiri M (2015) Experiments on the water entry of curved wedges: high speed imaging and particle image velocimetry. Ocean Eng 94:213–222
37. Jalalisendi M, Shams A, Panciroli R, Porfiri M (2015) Experimental reconstruction of three-dimensional hydrodynamic loading in water entry problems through particle image velocimetry. Exp Fluids 56(2):1–17
38. Jalalisendi M, Osma SJ, Porfiri M (2015) Three-dimensional water entry of a solid body: a particle image velocimetry study. J Fluid Struct 59:85–102
39. Shams A, Jalalisendi M, Porfiri M (2015) Experiments on the water entry of asymmetric wedges using particle image velocimetry. Phys Fluids 27(2):027103
40. Facci AL, Porfiri M, Ubertini S (2016) Three-dimensional water entry of a solid body: a computational study. J Fluid Stuct 66:36–53
41. Jalalisendi M, Zhao S, Porfiri M (2017) Shallow water entry: modeling and experiments. J Eng Math 104(1):131–156
42. Shams A, Zhao S, Porfiri M (2017) Hydroelastic slamming of flexible wedges: modeling and experiments from water entry to exit. Phys Fluids 29(3):037107
43. Shams A, Porfiri M (2015) Treatment of hydroelastic impact of flexible wedges. J Fluid Struct 57:229–246
44. Shams A, Zhao S, Porfiri M (2017) Water impact of syntactic foams. Materials 10(3)
45. Jalalisendi M, Porfiri M (2018) Water entry of compliant slender bodies: theory and experiments. Int J Mech Sci 149:514–529
46. Jalalisendi M, Porfiri M (2018) Water entry of cylindrical shells: theory and experiments. AIAA J 56(11):4500–4514
47. Russo S, Jalalisendi M, Falcucci G, Porfiri M (2018) Experimental characterization of oblique and asymmetric water entry. Exp Thermal Fluid Sci 92:141–161
48. Jalalisendi M, Benbelkacem G, Porfiri M (2018) Solid obstacles can reduce hydrodynamic loading during water entry. Phys Rev Fluids 3(7):074801
49. Baur T & Köngeter J (1999) PIV with high temporal resolution for the determination of local pressure reductions from coherent turbulence phenomena. 3rd International Workshop on PIV'99, Santa Barbara, pp 101–106
50. Fujisawa N, Nakamura Y, Matsuura F, Sato Y (2006) Pressure field evaluation in microchannel junction flows through PIV measurement. Microfluid Nanofluid 2(5):447–453

51. Fujisawa N, Tanahashi S, Srinivas K (2005) Evaluation of pressure field and fluid forces on a circular cylinder with and without rotational oscillation using velocity data from PIV measurement. Meas Sci Technol 16(4):989–996
52. Liu X, Katz J (2006) Instantaneous pressure and material acceleration measurements using a four-exposure PIV system. Exp Fluids 41(2):227–240
53. Murai Y, Nakada T, Suzuki T, Yamamoto F (2007) Particle tracking velocimetry applied to estimate the pressure field around a savonius turbine. Meas Sci Technol 18(8):2491–2503
54. Jensen A, Sveen JK, Grue J, Richon J-B, Gray C (2001) Accelerations in water waves by extended particle image velocimetry. Exp Fluids 30(5):500–510
55. Nila A, Vanlanduit S, Vepa S, Van Paepegem W (2013) A PIV-based method for estimating slamming loads during water entry of rigid bodies. Meas Sci Technol 24(4):045303
56. Korobkin A (2004) Analytical models of water impact. Eur J Appl Math 15(06):821–838
57. Korobkin AA (1996) In: Ohkusu M (ed) Advances in marine hydrodynamics. Computational Mechanics, Boston, pp 323–371
58. Thielicke W & Stamhuis EJ (2014) PIVlab – time-resolved digital particle image velocimetry tool for MATLAB (version: 1.32)
59. Thielicke W, Stamhuis EJ (2014) PIVlab - towards user-friendly, affordable and accurate digital particle image velocimetry in MATLAB. J Open Res Softw 2(1):e30
60. Panton RL (1994) Incompressible flow. Wiley, New York
61. Garcia D (2010) Robust smoothing of gridded data in one and higher dimensions with missing values. Comput Stat Data Anal 54(4):1167–1178
62. Baur T (1999) PIV with high temporal resolution for the determination of local pressure reductions from coherent turbulence phenomena. 3rd International Workshop on PIV'99, Santa Barbara pp 101–106
63. Porfiri M, Shams A (2017) Pressure reconstruction during water impact through particle image velocimetry: methodology overview and applications to lightweight structures. In: Lopresto V, Langella A, Abrate S (eds) Dynamic response and failure of composite materials and structures. Woodhead Publishing, Duxford, pp 395–416
64. Russo S, Jalalisendi M, Falcucci G, & Porfiri M (2018) A critical assessment of PIV-based pressure reconstruction in water-entry problems. AIP Conference Proceedings, (AIP Publishing), p 420012
65. Narkis M, Kenig S, Puterman M (1984) Three-phase syntactic foams. Polym Compos 5 (2):159–165
66. Gupta N, Zeltmann SE, Shunmugasamy VC, Pinisetty D (2014) Applications of polymer matrix syntactic foams. JOM 66(2):245–254
67. Shunmugasamy VC, Anantharaman H, Pinisetty D, Gupta N (2015) Unnotched Izod impact characterization of glass hollow particle/vinyl ester syntactic foams. J Compos Mater 49 (2):185–197
68. Wang S, Soares CG (2017) Review of ship slamming loads and responses. J Mar Sci Appl 16 (4):427–445
69. Izadi M, Ghadimi P, Fadavi M, Tavakoli S (2018) Numerical modeling of the freefall of two-dimensional wedge bodies into water surface. J Braz Soc Mech Sci Eng 40(1):24
70. Pan Z, Whitehead JP, Richards G, Truscott TT, & Smith BL (2018) Error propagation dynamics of PIV-based pressure field calculation (3): what is the minimum resolvable pressure in a reconstructed field? arXiv preprint arXiv:1807.03958
71. Zhang P, Peterson SD, Porfiri M (2019) Combined particle image velocimetry/digital image correlation for load estimation. Exp Thermal Fluid Sci 100:207–221
72. Shams A, Lopresto V, Porfiri M (2017) Modeling fluid-structure interactions during impact loading of water-backed panels. Compos Struct 171:576–590

Response of Sandwich Structures to Blast Loads

Romesh C. Batra, Arka P. Chattopadhyay, and Priyal H. Shah

1 Introduction

An explosion in air or water instantaneously raises pressure in the medium adjacent to the blast that compresses the medium, results in a shock wave propagating in it, and applies an impulse load on a nearby structure. The resulting blast wave in air is called a blast wind [1]. An underwater explosion results in many subsequent phenomena that can cause catastrophic damage to a marine structure [2]. Early techniques of damaging a ship used underwater mines and torpedoes that exploded upon contacting the ship. However, by the end of the First World War, the effectiveness of a non-contact explosion in damaging a marine vessel was realized.

Rayleigh [3] studied cavitation in a gas bubble formed in an underwater explosion. Subsequently, Cole [4] characterized an underwater explosion and its interaction with marine structures. Taylor [5] studied the interaction of a blast wave from an underwater explosion with a wedge-type structure and a monolithic plate. Keil [2] discussed, in detail, the response of surface ships and submarines to underwater blasts that included the severity of damage to various components. Since then, experimental, analytical and computational research in underwater explosions has significantly advanced that has considered using composite laminates and sandwich structures to reduce the ship weight. Here, we first review some works on underwater explosion including gas bubble characteristics, shock wave propagation, cavitation phenomenon, and structural deformations, and then describe modeling and simulation of structural damage caused by shock loads.

R. C. Batra (✉) · A. P. Chattopadhyay · P. H. Shah
Department of Biomedical Engineering and Mechanics, M/C 0219, Virginia Polytechnic Institute and State University, Blacksburg, VA, USA
e-mail: rbatra@vt.edu

S. W. Lee (ed.), *Advances in Thick Section Composite and Sandwich Structures*,
https://doi.org/10.1007/978-3-030-31065-3_10

2 Loads Produced by Underwater Explosions

During the Second World War, the efficiency of a non-contact underwater explosion in disabling ships was realized thus triggering extensive research in this field.

2.1 Shock Wave

An underwater explosion results in a high-pressure superheated gas bubble centered at the explosion point. The bubble rapidly expands and creates a shock wave that initially travels at a speed much higher than the speed of sound, $c = 1.5$ km/s, in water. However, the shock wave speed rapidly drops to 1.5 km/s within a distance of 10 times the charge radius [2]. The shock wave propagates radially outwards and gets reflected from the sea free surface and the sea bottom. Rayleigh [3] characterized the velocity of bounding surfaces of an expanding bubble, the spherical cavity formed after an explosion, and the pressure in the bubble and on the surrounding fluid that was assumed to be incompressible. Deformations of the surrounding water are studied in [4] wherein it is stipulated that the exponential decay in time of the pressure, P, in the shock wave is given by

$$P = P_0 e^{-t/\theta}, \tag{1a}$$

$$P_0 = K_1 \left(\frac{W^{1/3}}{R_h} \right)^{A_1}, \tag{1b}$$

$$\theta = K_2 \left(W^{1/3} \right) \left(\frac{W^{1/3}}{R_h} \right)^{A_2} \tag{1c}$$

Here P_0 is the peak pressure at the explosion point, θ the decay constant, W the mass of the charge in kg, R_h the distance in meters of the point of interest (stand-off distance) from the explosion center, and K_1, K_2, A_1 and A_2 are constants. Values of these constants determined by Cole [4] from the test data for four explosives are listed in Table 1. Using these values, the pressure histories of the shock wave at R_h= 50 m are exhibited in Fig. 1. We note that for R_h< (10 times the charge radius) Eq. (1) considerably under-predicts P [6]. When the shock wave passes a fixed

Table 1 Values of constants in Eqs. (1b, 1c) for four explosives [4]

Explosive type	TNT	HBX – 1	PETN	Nuclear
K_1	52.12	53.31	56.21	1.06×10^4
K_2	1.18	1.144	1.194	1.13
A_1	0.0895	0.092	0.084	3.627
A_2	-0.185	-0.247	-0.257	-0.22

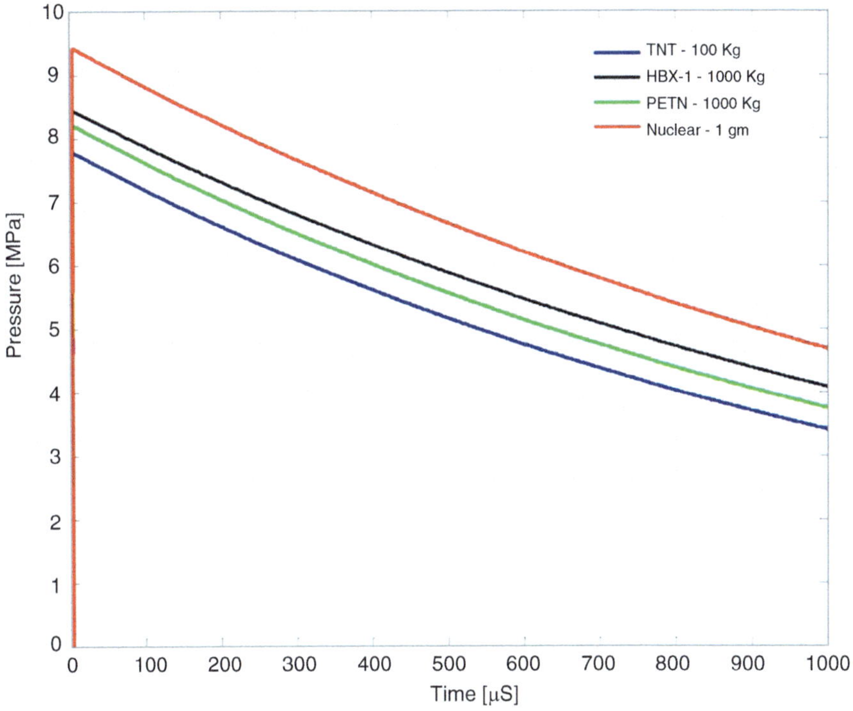

Fig. 1 For four explosives, time histories of the shock wave pressure at a stand-off distance of 50 m from the blast site

location in the liquid, particles there start flowing in the shock wave propagation direction at velocity *u(t)* given by

$$P(t) = \rho c u(t) \tag{2}$$

where ρ is the mass density of water. Keil [2] has suggested that due to the shock wave being spherical, Eq. (2) should be modified to

$$u(t) = \frac{P(t)}{\rho c} + \frac{1}{\rho R_h} \int_0^t P(t) dt \tag{3}$$

In Eq. (3), the first and the second terms, respectively, correspond to a planar and a spherical flow. In close proximity of the explosion, the second term significantly affects the particle velocity.

The total energy, because of the pressure and the particles velocity, associated with the shock wave is given by [4]

$$E_{sh} = \frac{1}{\rho c}\int_0^\infty P^2(t)dt, \tag{4a}$$

$$E_{sh} = \frac{1}{\rho c}P_0^2\theta \tag{4b}$$

where we have substituted for $P(t)$ from Eq. (1).

Besides the pressure and the energy of a shock wave, its capacity to induce damage is measured by the total impulse of the wave. The impulse of a shock wave acting on a unit area at a point until time t can be expressed as

$$I = \int_0^t P(t)dt = P_0\theta \tag{5}$$

where we have used Eq. (1).

2.2 The Gas Bubble

After emission of the initial high pressure from the gas bubble in the form of a shock wave, the pressure in the gas remains higher than the surrounding hydrostatic pressure. Thus the bubble expands. However, due to inertia effects, the expansion does not stop at the equilibrium pressure and the bubble keeps expanding until the hydrostatic pressure exceeds the pressure of the gas in the bubble [7]. This contracts the bubble until its radius becomes the minimum. The bubble continues to oscillate around the equilibrium radius for many cycles; e.g., see Fig. 2. When the bubble reaches the minimum radius, a pressure pulse in the form of an aftershock, with the peak pressure of about 10–15% of the primary shock wave, is released.

$$R_{\max} = 3.3\left(\frac{W}{Z}\right)^{1/3}, \tag{6a}$$

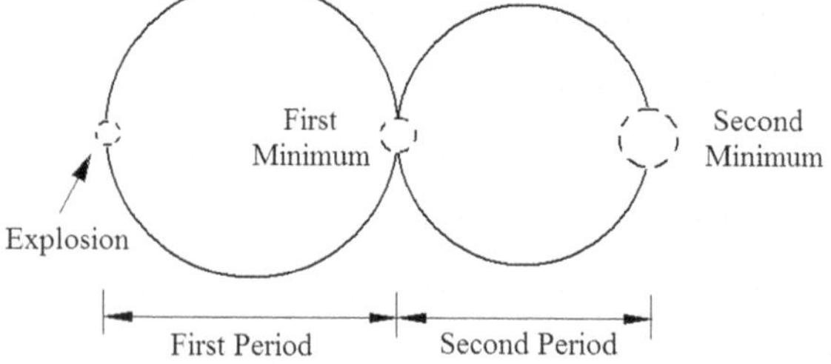

Fig. 2 Pulsation of gas bubble in an underwater explosion [4]

$$T = 2.08 \frac{W^{1/3}}{Z^{5/6}} \tag{6b}$$

In Eq. (6), $Z = D + 10$ is the total static pressure head at the explosion site, and D the depth of the explosion site in meters. Due to buoyancy force, the gas bubble migrates upwards towards the free surface. The migration rate reaches the maximum when the bubble radius reaches the minimum. For a TNT explosion, Reid [7] postulated the empirical expression (7), derived from test data, for the first migration, *mf*, from the explosion location to the place where the bubble radius reaches the first minimum.

$$mf = \frac{12.2}{Z} W^{1/3} \tag{7}$$

The migration distance from the first minimum to the second minimum is usually one-half of the distance calculated from Eq. (7). The bubble migration can cause extensive damage to ships in the migration path. The characteristic features of a gas bubble are summarized in Fig. 3.

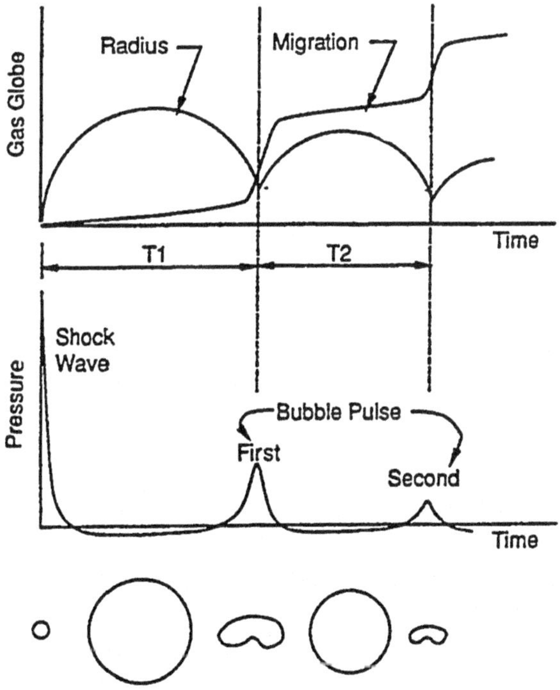

Fig. 3 Characteristics of a gas bubble at different times [4]

2.3 *Reflection of the Shock Wave from the Sea Surface and the Sea Bed*

The reflection of a shock wave from the sea surface and the sea bed is schematically represented in Fig. 4. The wave reflected from the sea surface (bed) is tensile (compressive). When the wave reflected from the sea bed (surface) interacts with the primary shock wave, it adds (subtracts) to the pressure of the primary wave. When the pressure temporarily becomes zero, the phenomenon is called surface cut-off. Keil's [2] empirical relation between the surface cut-off time, t_{zero}, and the position of the point is given by Eq. (8).

$$t_{zero} = \frac{0.122Dd}{R_h}, ms \tag{8}$$

In Eq. (8), d is the depth of the point, R_h the horizontal stand-off distance of the point from the blast location with d and R_h measured in meters.

When an explosion occurs at the sea bottom, the bubble develops as a hemisphere [4], and migrates along with the regular pulsation phenomenon [9].

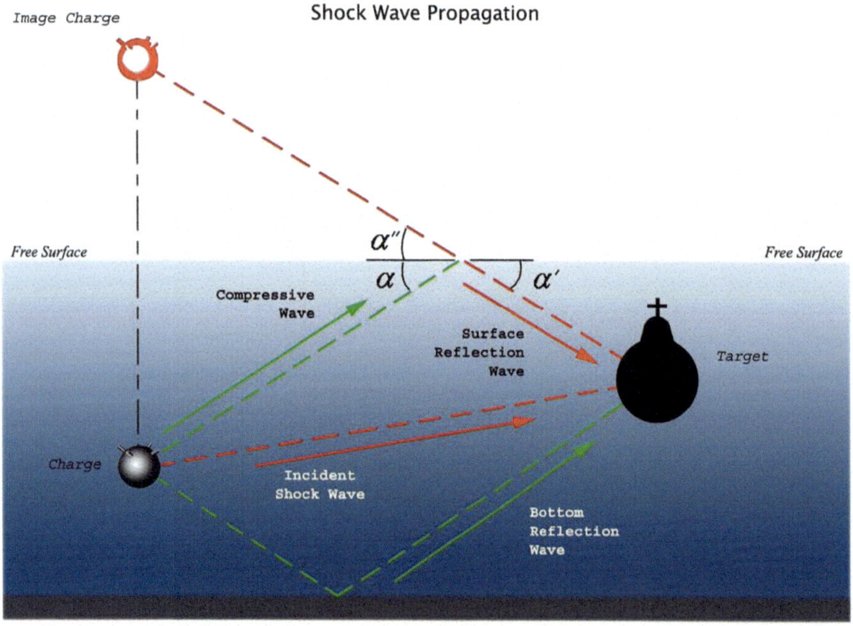

Fig. 4 Schematic representation of shock wave reflections [8]

2.4 Cavitation

Water cannot sustain tensile stresses. The bulk and the hull cavitation, respectively, ensue when stresses in a volume of water become tensile due to the interaction between the incident shock wave and that reflected from the sea surface or a structure. The bulk cavitation that is more likely to occur for an underwater explosion near the sea surface has been studied by Cole [4] and Snay [9]. Costanzo and Gordon [10] have calculated the extent and the duration of the cavitation process. The eventual closing of the cavitated region results in a water hammer and a pressure pulse called cavitation pulse [7]. The bulk cavitation occurring at the sea surface due to an underwater explosion is shown in Fig. 5.

A shock wave impinging upon a deformable ship hull accelerates the hull particles. When particles' velocities exceed the transient velocity of the adjacent volume of water, tensile tractions act on the water volume and ensue cavitation. This phenomenon is called hull cavitation, the maximum velocity of the hull is called the kick-off velocity [7], and has been studied by Cole [4] and Taylor [5].

Fig. 5 Bulk cavitation at the sea surface from an underwater explosion [4]

3 Response of Ships to an Underwater Explosion

Toward the end of the 2nd World War, it was realized that an underwater explosion can destroy ships. Keil [2] has discussed the response of ships based on experimental observations. Depending upon where an explosion occurs, they are classified as contact and non-contact type with the former (latter) having the explosion location adjacent (away) to either the ship hull or the ship bottom. In the former, either torpedoes or mines were used to tear open a 30–50 feet diameter hole on the ship exterior depending on the explosion type and size. The explosion also ruptures the ship bulkheads and the blast fragments heavily damage the machinery close to the explosion. Nurick and Martin [11, 12] have reviewed the literature on deformations of plates under loads typical of a contact explosion. Weirzbicki and Nurick [13] have experimentally and theoretically investigated deformations of plates under a localized impulsive loading. Assuming that the plate can be modeled as a membrane comprised of a perfectly plastic material and is deformed into a part of a sphere of radius ρ, we have [13]

$$(2\rho - \delta)\delta = R^2 \tag{9}$$

where δ is the maximum depth, and R the radius of the deformed plate along the undeformed surface; e.g., see Fig. 6. Cole [4] found R by equating the work, W, to the input energy, E_{in}, from the shock wave [14]. That is,

$$W = \sigma_y t_p \Delta A, \tag{10a}$$

$$R = \sqrt{\frac{2\eta E_{in}}{\pi t_p \sigma_y \varepsilon_f}}, \tag{10b}$$

$$E_{in} = WE_{TNT}Eq_{TNT}J \tag{10c}$$

where η is the fraction of the E_{in} used to deform the plate, ($E_{def} = E_{in}\eta$), t_p the plate thickness, σ_y the yield stress of the plate material, ε_f the fracture strain, E_{TNT} the energy content of the TNT, E_{TNT} the TNT equivalent of the explosive used, and J the energy conversion factor. Rajendran and Lee [15] have reviewed contact explosion problems.

Figure 7 schematically shows characteristic dimensions associated with a non-contact underwater explosion for which the damage is mainly due to a shock

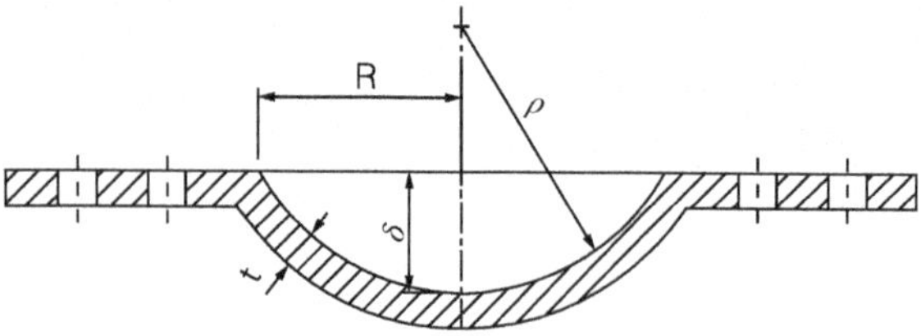

Fig. 6 Schematic representation of the deformation of a circular plate subjected to a contact explosion [14]

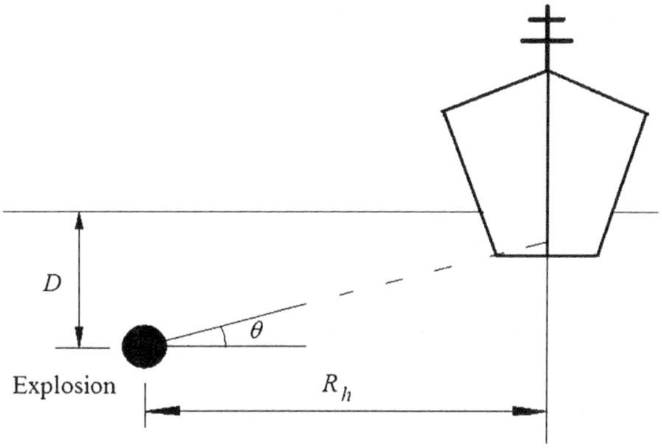

Fig. 7 Schematic representation of a non-contact underwater explosion

wave, bubble pulsation, bubble migration and cavitation. The impact of a shock wave on a ship hull requires analyzing fluid-structure interaction (FSI) between the water volume and the hull which can be simulated as an assembly of flat and curved plates.

One measure of the intensity of the shock wave at the hull is the Hull Shock Factor, HSF, [7], is given by

$$HSF = \frac{W_{eq}^{n}}{R_h} \tag{11}$$

where W_{eq} is the mass of the explosive used equivalent to the TNT and the parameter n, determined experimentally, depends on the explosive used and the explosion conditions. When the position of the explosion is measured with respect to the ship keel, the Keel Shock Factor, KSF, [7] for a shock wave making angle θ (see Fig. 7) with the horizontal line is defined as

$$KSF = HSF\left(\frac{1 + \sin\theta}{2}\right) \tag{12}$$

For explosions close to the ship hull, the damage is localized due to the spherical wave front that generally tears open a large hole similar to the damage caused by a contact explosion. For explosions at large stand-off distances, the incident shock wave is approximately planar, and deformations of different parts of the ship is generally dissimilar.

3.1 Fluid Structure Interaction of Monolithic Plates

Early works on studying the fluid-structure interaction (FSI) include those of Cole [4], Taylor [5] and Kennard [16]. When a traveling pressure pulse hits a ship hull, the pulse is reflected back with some modifications caused due to the accelerating and deforming plate of the hull [6]. It is assumed that the plate is of infinite area based on the time scale of interaction of the pulse with the plate. For the incident pressure given by Eq. (1), the reflected wave pressure can be written as [4, 5]

$$P_r = P_0\varphi(t) \tag{13}$$

The total force acting on the plate is the sum of that due to the incident and the reflected pressure pulses. Assuming no separation between the water and the solid, at the water-plate interface the normal traction and the normal velocity must be continuous. However, the tangential velocity can be discontinuous for water modeled as inviscid. The plate deformations depend on support conditions at its rear surface and edges.

In the first approximation, the plate is modeled as a point, and its motion described by,

$$m\ddot{u} + c\dot{u} + ku = P_{net} \tag{14}$$

where m is the mass of the plate per unit area, a superimposed dot denotes the time derivative, c the damping coefficient, k the plate stiffness, P_{net} the net force per unit area, and u the plate displacement in the direction of the pressure pulse. The initial conditions are

$$u(0) = \dot{u}(0) = 0 \tag{15}$$

The solution of differential Eq. (14) provides the plate displacement and the velocity histories.

For an air backed plate (ABP), shown schematically in Fig. 8, the right face of the plate is subjected to air and does not resist motion. Using continuity conditions at the water-plate interface, the total pressure acting on the plate is found to be [4, 5]

$$P_{net}^{ABP} = 2P_0 e^{-t/\theta} - \rho c\dot{u} \tag{16}$$

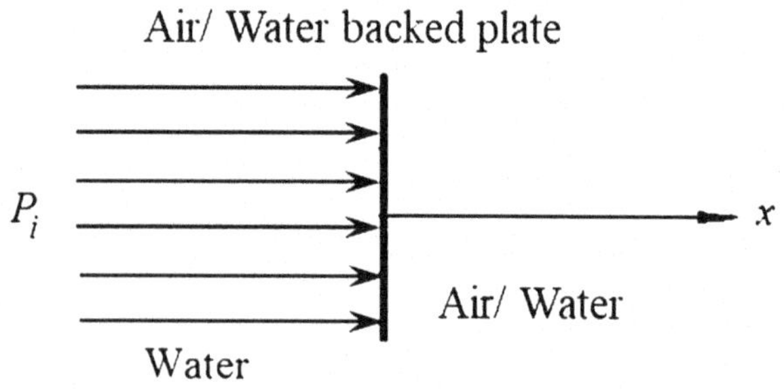

Fig. 8 Schematic representation of an air/water backed monolithic plate

where ρ is the mass density of water, and c is the sound speed in the plate. For the pressure pulse hitting a stationary rigid wall, $\dot{u}= 0$.

Substitution of the total pressure from Eq. (16) into Eq. (14) and using initial conditions described by Eq. (15) gives the following for the plate displacement and velocity [5, 17].

$$u_a = \frac{2P_0\theta}{m\psi(\psi-1)}\left[(\psi-1)+e^{-(\psi/\theta)t}-\psi e^{-t/\theta}\right], \tag{17a}$$

$$v_a = \dot{u}_a = \frac{2P_0\theta}{m(1-\psi)}\left(e^{-\psi t/\theta}-e^{-t/\theta}\right) \tag{17b}$$

The dimensionless quantity, $\psi = \rho c\theta/m$, is called the FSI parameter of an ABP. Substitution from Eq. (17) into Eq. (16) results in

$$P_{net}^{ABP} = 2P_0 e^{-t/\theta} - \frac{2P_0\psi}{(\psi-1)}\left(e^{-t/\theta}-e^{-\psi t/\theta}\right) \tag{18}$$

Recalling that the hull cavitation begins when the net pressure equals zero, the cavitation inception time, t_c^{ABP}, is [5, 17].

$$t_c^{ABP} = \theta\frac{\ln\psi}{\psi-1} \tag{19}$$

Another parameter of interest is the fraction of the maximum momentum transferred to the plate that occurs at the cavitation inception time when the plate has the maximum velocity. Thus, the plate maximum momentum, I_p, is given by

$$I_P = \zeta^{ABP} I_0, \zeta^{ABP} = \psi^{\psi/1-\psi} \tag{20}$$

where the free pulse momentum, I_0, is given by Eq. (5).

For a water backed plate (WBP), the water at the back of the plate exerts pressure on it and decreases the net force on the plate. The problem has been studied by Keil [2]. Proceeding in the same way as for the ABP, we get

$$u_w = \frac{2P_0\theta}{m\phi(\phi-1)}\left[(\phi-1)+e^{-(\phi/\theta t)}-\phi e^{-t/\theta}\right], \tag{21a}$$

$$v_w = \dot{u}_w = \frac{2P_0\theta}{m(1-\phi)}\left(e^{-\phi t/\theta}-e^{-t/\theta}\right) \tag{21b}$$

where $\phi = 2\psi = 2\rho c\theta/m$ is the FSI parameter for a WBP. Results for a WBP can be obtained from those for an ABP with ψ replaced by ϕ; e.g., see Liu and Young [18]. In Figs. 9 and 10, we have plotted time histories of the plate displacement, the plate velocity and the cavitation time for an ABP and a WBP. The characteristics of a WBP are important when designing liquid holding tanks close to a ship hull. When a

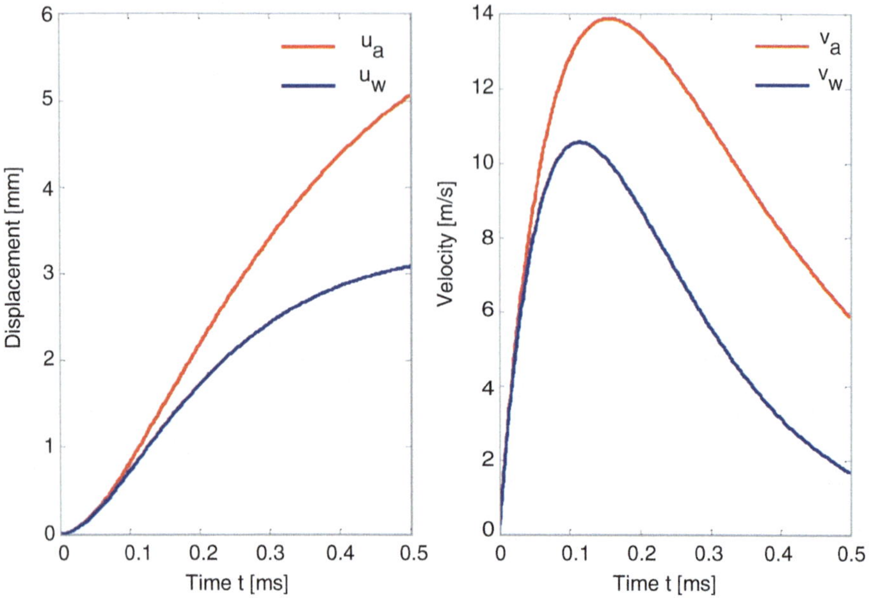

Fig. 9 The displacement and velocity histories of a plate with $\phi = 0.75$ and $\psi = 0.375$

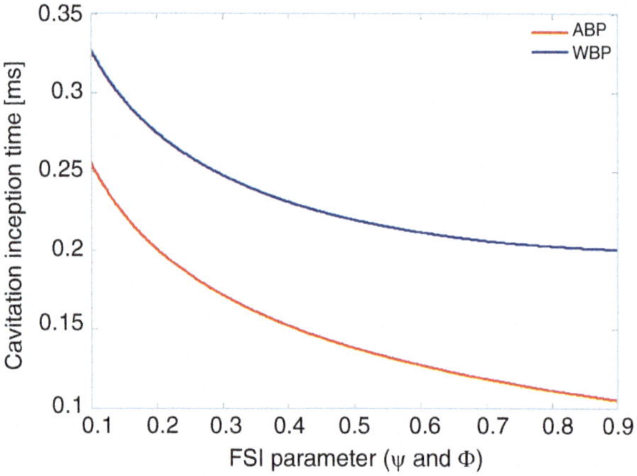

Fig. 10 Variation with the FSI parameter of the cavitation inception time for air- and water-backed plates

shock wave hits the outer wall of a tank, a part of the shock wave is transferred to the fluid inside the tank to generate another shock wave that can severely damage the internal wall. One way to avoid this is to not completely fill the tank that allows the shock wave energy to be converted to a free surface wave [7].

3.2 *Fluid Structure Interaction of Composite and Sandwich Plates*

Mouritz [19, 20] has experimentally studied the impact of glass fiber reinforced polymer (GRP) laminates including their fatigue behavior under increasing shock loads. The tests used large explosive charges at short stand-off distances for both ABP and WBP, and stitched and unstitched GRP laminates. He found that the damage caused to both stitched and unstitched laminates was similar. However, the damage and delamination in stitched laminates was less than that in unstitched specimens. At low (high) shock factors, the damage to Kevlar stitching threads was minimal (extensive).

Thick laminates respond differently to a shock wave than thin ones [21]. Morais et al. [22] experimentally studied the effect of the laminate thickness by dropping a weight on a clamped laminate. Gellert et al. [23] also experimentally studied the effect of the laminate thickness on its response to a shock wave. Several works [24–28] present experimental studies on the response of laminate plates to pressure pulses and delineate the effect on composites' deformations of fiber length, fiber/matrix interface and ply stacking sequence. Mouritz et al. [29] have reviewed composite structures used in ships and submarines.

Sandwich structures have a distinct advantage over monolithic plates in withstanding short time-duration shock wave pressure pulses [17, 30]. For the same areal density, a sandwich structure can sustain a more intense pressure pulse than a laminated plate. Fleck and Deshpande [17] used the relative time scales and the associated phenomena to divide the interaction of a metallic sandwich structure with a pressure pulse into three stages. In the first stage, the face plate hit by the shock wave is elastically compressed. In the second stage, the core is dynamically crushed and its initiation is marked by the cavitation inception time where the face plate achieves the maximum velocity which can be considered equivalent to the kick-off velocity of a monolithic plate. For an effective sandwich design, most of the incident energy is absorbed during core crushing [31]. The final stage involves global tearing and rupture of the face plate during which the remaining energy of the shock wave is absorbed. Figure 11 schematically shows phenomena occurring in the three stages [32].

Many researchers [16, 30–37] have analytically, numerically and experimentally studied deformations of a sandwich structure under a shock wave pressure pulse. In stage 1 of the interaction, the core stiffness resists motion of the face plate due to the impinging pressure pulse. For a core assumed to be rigid-perfectly plastic and having yield strength, σ_y, the motion of a unit area of the face plate is governed by [31]

$$m_f \ddot{u} = P(x, t) - \sigma_y \tag{22}$$

under null initial conditions (cf. Eq. (15)) and the resultant pressure $P(x, t)$ due to the incident and the reflected waves given by

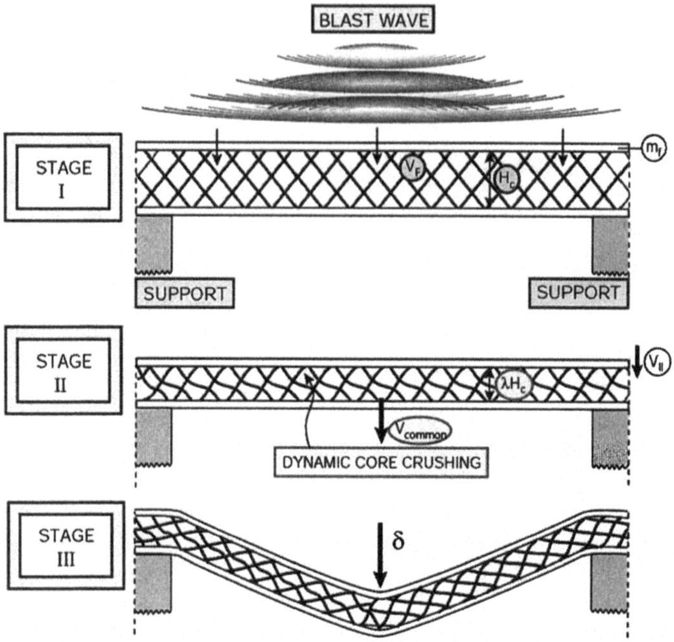

Fig. 11 Schematic representation of three stages of a blast wave-sandwich interaction [32]

$$P(x,t) = P_0\left(e^{\zeta} + e^{\xi}\right) - \rho c\dot{u}, \zeta = -t/\theta - x/c\theta, \xi = -t/\theta + x/c\theta \tag{23}$$

Fleck and Deshpande [17] have discussed the core yield strength as a function of the yield strength of the core material and the relative mass density of the core.

The solution of Eq. (22) under null initial conditions gives the following for the face plate displacement and velocity [35].

$$u = \frac{2P_0\theta}{m\psi(1-\psi)}\left[(1-\psi) - e^{-\psi t/\theta} + \psi e^{-t/\theta}\right] + \frac{\sigma_y\theta^2}{m\psi^2}\left(1 - e^{-\psi t/\theta} - \frac{\psi t}{\theta}\right) \tag{24}$$

$$\dot{u} = \frac{2P_0}{m(1-\psi)}\left(e^{-\psi t/\theta} - e^{-t/\theta}\right) + \frac{\sigma_y\theta}{m\psi}\left(e^{-\psi t/\theta} - 1\right) \tag{25}$$

For a monolithic (sandwich) plate, the spatial dependence of the pressure is ignored (considered) because the cavitation inception plane contacts (is away from) the plate surface.

The cavitation occurs when the face plate velocity is the maximum and the pressure in the fluid equals zero. By substituting from Eq. (25) into Eq. (23), we obtain the following expression for the fluid pressure as a function of space and time [35].

$$\frac{P}{P_0} = \frac{2}{1-\psi}(\cosh\widetilde{x} - \psi\sinh\widetilde{x})e^{-\widetilde{t}} + \left(\frac{2\psi}{\psi-1}\right)e^{-\psi\widetilde{t}-\psi\widetilde{x}} - \widetilde{\sigma} \tag{26}$$

Here $\widetilde{x} = x/c\theta, \widetilde{t} = t/\theta$ and $\widetilde{\sigma} = \sigma_y/P_0$. The cavitation inception occurs when $P = 0$ and $\partial P/\partial\widetilde{x} = 0$. Solving these equations, we get the following expressions for the cavitation inception time, $t_c = \widetilde{t}_c\theta$, and the cavitation inception plane, $x_c = \widetilde{x}_c c\theta$ [35]

$$\begin{aligned}\ln\left[\frac{2(1+\psi)\sinh\widetilde{x}_c}{\widetilde{\sigma}\psi}\right] &= \frac{1}{1-\psi}\left[\ln\frac{\psi\cosh\widetilde{x}_c + \sinh\widetilde{x}_c}{\psi^2 - 0.5\widetilde{\sigma}\psi(\psi-1)}\right] - \psi\widetilde{x}_c, \widetilde{t}_c \\ &= \ln\left[\frac{2(1+\psi)\sinh\widetilde{x}_c}{\widetilde{\sigma}\psi}\right]\end{aligned} \tag{27}$$

Hutchinson and Xue [31] proposed that the water layer between the cavitation inception plane and the face plate moves with the face plate as schematically shown in Fig. 12. The mass of the water layer adds to the mass of the face plate, and should be considered in the momentum and the energy calculation of the face plate. Hutchinson and Xue proposed the following empirical expression (28) for the ratio, r_w, of the added mass to the mass of the face plate for sandwich structures with identical front and back face plates and $\widetilde{\sigma} < 0.15$.

$$r_w = 0.71\psi\widetilde{\sigma} \tag{28}$$

The cavitation inception ends stage 1 of the interaction between a shock wave and a sandwich structure. With $\overline{\varepsilon}$ and C, respectively, equaling the average crushing strain and the core thickness, the energy, $\sigma_y\overline{\varepsilon}C$, consumed in crushing the core in stage 2 is found from

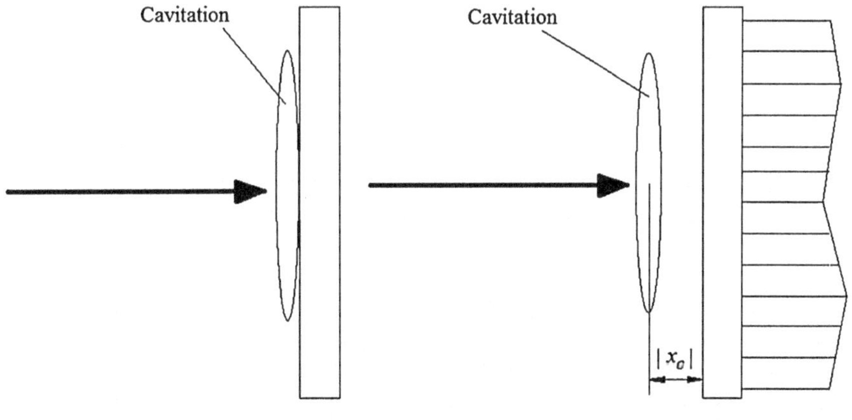

Fig. 12 Schematic representation of the cavitation phenomenon in a blast wave interacting with a (left) monolithic, and (**b**) sandwich plate

$$\sigma_y \bar{\varepsilon} C = \Delta KE = KE_F - KE_I \tag{29}$$

where KE_F (KE_I) is the kinetic energy at the end (beginning) of stage 2. KE_I equals the sum of the kinetic energy of the face plate, added water layer, the core and the back plate. That is,

$$KE_I = \frac{I_f^2}{2(m_w + m_f)} + \frac{I_b^2}{2(m_c + m_f)}, KE_F = \frac{I_t^2}{2(2m_f + m_w + m_c)} \tag{30}$$

In Eq. (30), I_f, I_b and I_t are, respectively, the face plate momentum, combined momentum of the core and the back plate, and the total momentum; m_w, m_c and m_f, respectively, equal the mass of the added water, the core, and the face plate. Furthermore, $I_f = m_f$ multiplied by the face plate velocity at the instant of cavitation. The momentum transferred to the back plate and the core, respectively, equals $\sigma_y t_c$ [31], and $I_t = I_f + I_b$. The stage 2 ends when the face plate, the deformed core and the back plate start moving at the same velocity [17].

In stage 3, the residual kinetic energy causes global bending and shearing off of the face plate. The approximate times of the beginning and the end of the three stages are ([31, 32])

$$t_I = t_c \approx t_0 \frac{\ln \beta}{\beta - 1}, t_{II} \approx \frac{I_T}{2\sigma_y^D}, t_{III} \approx L\sqrt{\rho/\sigma_y} \tag{31}$$

3.3 Bubble Pulsation, Bubble Migration and Cavitation

In addition to a primary shock wave, secondary shocks due to bubble pulsation and bubble migration can significantly damage a ship. As the bubble pulsation occurs after passing of a primary shock, secondary pulses usually affect a different region of the ship due to its motion [7]. The severity of damage due to bubble pulsation is very high when the bubble pulsation frequency is close to a natural frequency of the ship. This bends a ship structure, and the associated damage is called whipping [7]. In severe cases, the peak deflection exceeds the elastic limit causing permanent damage. This is more severe for surface ships than for submarines for which the surrounding water damps resonant vibrations. The damage caused by the bubble migration is significant for explosions occurring under a ship.

Early works in modeling of gas dynamics include those of Gudonov [38, 39] and van Leer [40] who studied one-dimensional (1-D) problems using, respectively, the Lagrangian and the Eulerian schemes with the gas modeled as compressible. An all-purpose computer code, MUSCL (monotonic upstream-centered scheme for conservation laws), was developed to numerically study the phenomenon of gas bubble formation from the explosion [41]. Colella [42] modified the MUSCL algorithm to formulate gas dynamics in a single Eulerian step instead of a remap from the Lagrangian to the Eulerian coordinates; also see [43]. In subsequent works, the boundary element method (BEM) was extensively used to study the bubble dynamics and the pulsation phenomena. For example, Blake and Gibson [44, 45] and Geers and Hunter [46, 47] studied the growth of a bubble near a free surface and the cavitation near solid boundaries using the BEM.

Bulk cavitation occurs due to the reflection of the incident shock wave off the water surface as a tensile wave. The cavitated zone lies just beneath the water free surface with the upper and the lower boundaries having their centers of curvature at the explosion point. The cavitation envelope collapses after a certain time resulting in water hammer [7] that can produce significant strains in a surface ship positioned close to the bulk cavitation zone. Cavitation also occurs at the fluid-plate interface due to the reflection of the incident pressure pulse from the plate. Due to multiple reflections of this tensile wave between the plate surface and the cavitated zone, several pressure peaks occur at the plate resulting in spray loading [14]. Also, similar to the closing of a cavitation envelope in bulk cavitation, the closing of the hull cavitated region causes water hammer loading on the plate. Although the spray loading magnitude is negligible as compared to the peak pressure, the magnitude of water hammer is significant enough to cause noticeable damage [48].

4 Computational Modeling of Ships Deformations

Early works on mathematical modeling of the FSI include that of Bleich and Sandler [49, 50] who modeled the fluid as bi-linear and studied 1-D wave propagation by using the method of characteristics. Geers [51, 52] has employed the double asymptotic approximation (DAA) based finite element method (FEM) to analyze transient interaction of a flexible structure and an infinite fluid by considering the acoustic media around the structure as a membrane that covers the wet surface of the structure [53]. A major advantage of the DAA is using parameters of the wet surface response to eliminate the need to discretize the media surrounding the structure. Two different DAA based models, characterized according to their computational cost, are presented.

Other significant works on ship-underwater explosion include those of DeRuntz et al. [54] and DeRuntz [55] who developed a numerical package, Underwater Shock Analysis (USA), based on a fluid BEM to model the fluid in a FSI problem that did not capture the cavitation phenomenon and required using an alternative fluid model to prevent the fluid pressure from becoming negative. For example, DeRuntz and Rankin [55] and Fillipa and DeRuntz [56] employed a bilinear fluid model to accomplish this. The computational code, Cavitating Fluid Analyzer (CFA), could treat the cavitation of the fluid. The CFA code is an acoustic fluid formulation based volume element processor which is based on the displacement potential. Shin and Santiago [53] coupled the USA-CFA code with NASTRAN to analyze a 2-D problem, namely deformations of a ship cross-section, and compared results with

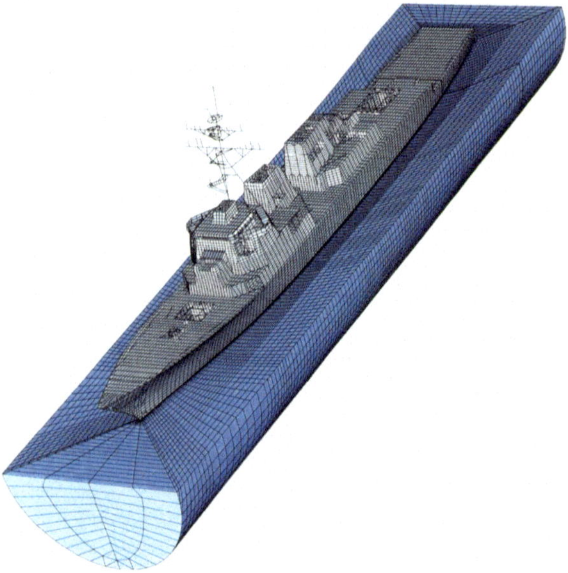

Fig. 13 Discretization of the domain of study into finite elements [58]

and without using the cavitation model, and with and without the DAA. Shin [57] used a coupled LSDYNA-USA code to simulate 3-D deformations of a ship exposed to a blast wave. Solid elements included in LS-DYNA were used to model the ship, and the fluid volume surrounding the ship was modeled using the material type MAT_90 (MAT_ACOUSTIC) that can simulate wave propagation in an irrotational, compressible and either linear or bilinear fluid but not negative pressures in the fluid [55, 57]. Shin's computed results compared well with the ship shock test data. Figure 13 taken from [57] shows the discretization of the domain studied into FEs. Kwon and Cunningham [58] employed the USA-DYNA software to study deformations of a stiffened shell subjected to dynamic impulse loading. Newton simulated the fluid cavitation by using the Cavitation Acoustic Finite Element (CAFE) method [59–62], and compared results with those of Bleich and Sandler [50]. Newton's work was later modified by Felippa and DeRuntz [56]. Rehak et al. [63] also incorporated the cavitation phenomenon in their FSI computational model. The CAFE algorithm used for modeling large scale simulations is resource intensive [64]. Sprague and Geers [65] overcame this by modifying the algorithm and named it the Cavitating Acoustic Spectral Element (CASE) [66, 64]. The CASE and the CAFÉ, respectively, use Legendre polynomials and "trilinear polynomials" as basis functions, and the CASE model uses a non-conformal FE mesh for the FSI modeling.

Klaseboer et al. [66] have compared experimental findings with numerical results for explosions close to a structure by modeling the gas bubble using the BEM and the structure using a nonlinear FE software, PEM-CRASH™. A similar work on the effect of the close proximity blasts on ship like structures was presented in [67]. Huang et al. [68] numerically modeled the FSI problem using ANSYS-AUTODYN. Mair [69] has reviewed other hydrocodes for studying the FSI problems.

4.1 Fluid Structure Interaction of Monolithic Plates

The response of monolithic plates to an underwater explosion has been experimentally analyzed in [66, 70–73], and that due to an air blast wave in [74, 75]. Deformations of a cylindrical shell interacting with a blast wave have been studied in [58, 76]. Taylor [5] proposed a theoretical model of a free standing rigid plate subjected to a pressure pulse and adopted a linear acoustic model of water. He showed that the FSI reduces the momentum transferred to the plate as compared to that for an impact loading on a free standing plate.

The analysis of transient deformations of a plate involves solving the following system of coupled ordinary differential equations generally under null initial conditions [15].

$$[\mathbf{M}]\{\ddot{x}\} + [\mathbf{C}]\{\dot{x}\} + [\mathbf{K}]\{x\} = \{P(t)\}, x(0) = \dot{x}(0) = 0 \tag{32}$$

Here **M**, **C** and **K**, respectively, represent the structural mass, the structural damping and the structural stiffness matrix, and *P*(*t*) is the time varying load that can include forces due to the bubble formation, migration, and collapse. The FSI can be accounted for by suitably modifying **M** for the added mass of water. The non-linear compressibility of water was considered in [77–79].

Fleck and Deshpande [17] studied the FSI of a plate by using an exponentially decaying forcing function as employed by Taylor [5]. Using the same forcing function, both linear and nonlinear problems were studied with the FE software ADINA in [80]. Louca et al. [81] compared the response of imperfect and stiffened monolithic plates with the two FE software DYNA3D and ABAQUS/Explicit. Ramajeyathilagam and Vendhan [82] experimentally and numerically (with DYNA3D) studied the deformation and rupture of air-backed clamped rectangular plates. Deformations and tearing of circular plates due to blast loads, studied by Gupta and Nagesh [80] using the FE software ANSYS, were found to correlate well with test observations of [70, 71].

Kwon and Cunningham [58] computationally studied the response of stiffened plates and ring stiffened cylindrical shells by using the coupled BEM and FEM code, USA-DYNA [54, 83] that models the gas bubble and the traveling pressure pulse from the explosion. Hammond and Grzebieta [84] also used USA-DYNA to delineate the structural response of an air backed clamped square plate and found that the computed plate displacements matched well with their own test results but computations under-predicted the plate velocity. They attributed this to ignoring hydrostatic effects of the water volume in the computational model and not having clamped edges in the tests. Flat monolithic plates were tested in [66] using a (4m $\times$ 4m $\times$ 4m) water tank and Detasheet explosives with DP60 detonator, and numerically analyzed with the USA-DYNA code considering the DAA [51, 52] to model the FSI.

4.2 Fluid Structure Interaction of Composite Plates and Sandwich Panels

Composite laminates have very high specific modulus and specific strength along the fiber direction than monolithic structures but their response to impact loads is not very encouraging [85] possibly due to several different failure modes. Hall [86] has experimentally studied deformations of glass-reinforced polymer and foam cored sandwich hulls. Mouritz [19, 20] experimentally studied deformations of laminated plates subjected to impulsive loads. Abrate [85] has reviewed the work on impact loading of sandwich structures having laminated face sheets and has discussed their failure mechanisms. Numerical studies [89] of the FSI of a submerged stiffened composite using USA-DYNA code showed that composite plates have better specific energy absorbing characteristics than monolithic plates. The interaction of a composite cylinder with a blast wave was studied theoretically and computationally using ABAQUS in [87]. Fu et al. [26] discussed the fracture resistance properties of impact loaded short glass/carbon fiber reinforced composites.

Hassan and Batra [88–90] developed a mathematical model and a computer code to study the damage initiation and propagation due to fiber breakage, matrix-fiber de-bonding, matrix cracking and de-lamination in AS4/PEEK composites. They approximated the blast wave load [89, 91] as the product of a function of spatial coordinates and a function of time. Values of material parameters in the damage evolution laws were found by using the test data. De Morais et al. [22] experimentally studied the effect of laminate thickness on damage induced under repeated impact loads. Espinosa et al. [27] employed contact/cohesive laws to analyze finite 3-D deformations of glass fiber reinforced polymeric laminates, and compared computed and experimental results. Nezami et al. [92] used the Rayleigh-Ritz method to study the response of a plate under different boundary conditions and subjected to an explosion blast wave. LeBlanc and Shukla [93, 94] highlighted aspects of the matrix-fiber breakage and internal de-lamination on the interaction with low and high energy blast waves of flat and curved glass-epoxy composite panels. Experimental findings were compared with those computed using LS-DYNA, and displayed differences between the two using the Russel error criteria [95].

Sandwich structures are believed to be more efficient than monolithic plates of the same areal density in their interactions with blast waves [35]. Makinen [96] has discussed dissimilarities in the response behavior of a monolithic plate and a sandwich structure by studying in them the cavitation phenomenon using a 1-D model. Deformations of sandwich structures with face and back plates made of monolithic materials have been scrutinized in [17, 30, 31, 33–35] by using the FEM. Liang et al. [32] and McMeeking et al. [97] discussed constitutive properties of prismatic metallic cores, considered effects of the added layer of water between the cavitation zone and the face plate, and found a good correlation between the numerical results computed with ABAQUS and theoretical results. Librescu and Noiser [98]. and Librescu et al. [99] studied the response of flat composite panels and sandwich plates with anisotropic composite face plates and an orthotropic core to explosive loads. Wei et al. [100] compared test findings on multi-core sandwich structures [101] with those computed using the FE code ABAQUS/Explicit. In [102], the transient response of a submerged cylindrical foam core sandwich panel subjected to shock loading was analyzed using ABAQUS and integrating fluid and structural models. They reported the circumferential strain at different points in the cylinder.

5 Summary of Batra's Team Work

We summarize below some work completed in Batra's group on structural response to extreme loads.

5.1 Homogenization of Material Properties

Gopinath and Batra [103] have recently used three micro-mechanics approaches, namely, the MoCs, the Fourier series analysis and the transformation field analysis, for homogenizing material properties of a unidirectional fiber-reinforced composite with fibers modeled as linearly elastic transversely isotropic and the matrix as a strain-rate dependent elastic-plastic material. They [105] have also characterized the sensitivity of responses from the three approaches with changes in a unit cell configuration. As shown in Fig. 14, the variation of the transverse elastic modulus versus the fiber volume fraction strongly depends upon the homogenization technique employed.

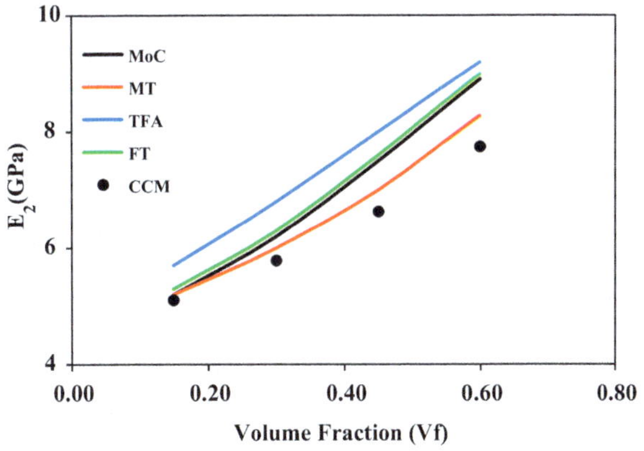

Fig. 14 Variation of the transverse Young's modulus with the fiber volume fraction using different homogenization techniques [103]

5.2 Modeling 3-D Deformations

As stated above, Batra and Hassan [91] identified damage/failure modes (fiber/matrix debonding, fiber breakage, and matrix cracking) with internal variables, used test data to find values of material parameters in their evolution laws, and considered all geometric nonlinearities incorporated in the St. Venant-Kirchhoff material. They postulated a damage surface defined in terms of the transverse normal and the transverse shear tractions at an interface and the interfacial strengths. Delamination ensued once the stress state at an interfacial point reached the damage surface. Subsequently, they introduced two nodes a tiny distance apart at the delamination point and prevented inter-penetration by connecting them with a spring very stiff in compression but weak in tension. They developed in-house a FE based software to analyze large 3-D transient deformations of 4-ply unidirectional fiber-reinforced laminates and delineated the energy dissipated in each failure mode with the blast load modeled as a spatially non-uniformly distributed pressure with the peak value exponentially decaying in time. They simulated damage progression by degrading the stiffness of an FE in which a constituent had failed, and reduced its stiffness to a very small value at complete failure. By retaining such FEs in the analysis, they considered their contributions to inertia forces. The spatial distribution of the matrix cracking and time- histories of the debonding damage at centroids of three different surfaces of the plate are illustrated in Fig. 15.

Whereas Batra and Hassan employed a mechanics of materials approach to homogenize material properties of a unidirectional fiber-reinforced composite, Batra et al. [104] used Aboudi's method of cells (MoCs) with null tangential tractions (e.g., see Robertson and Mall [105]) assumed at an interface between two adjoining cells. Furthermore, they determined stresses in the fiber and the matrix from the macro-level stresses and thus used the constituent level failure criterion but ignored fiber/matrix debonding. They ascertained failed/damaged elements and appropriately reduced their stiffness. The developed UMAT was implemented in ABAQUS. Their computed results for 4 m/s impact deformations of a clamped

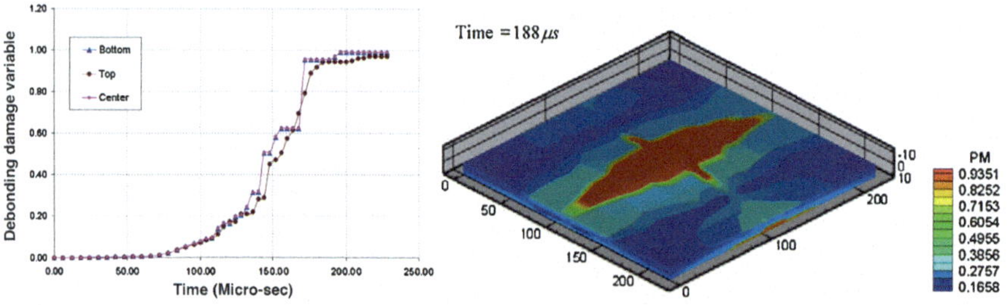

Fig. 15 (Left) Time histories of evolution of the debonding damage variable at centroids of the bottom, the middle and the top surfaces, and (right) fringe plots of the matrix cracking damage variable [91]

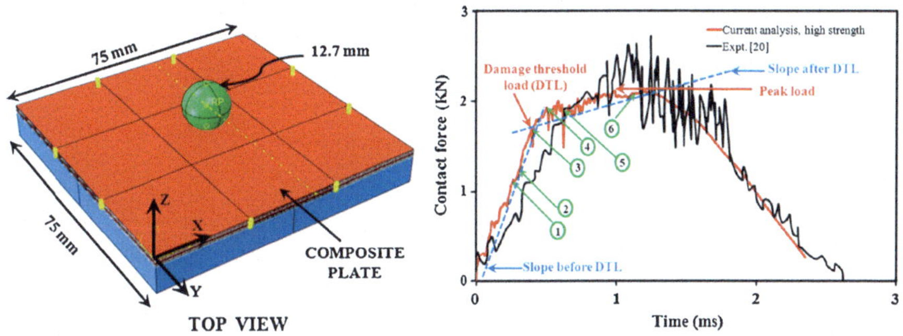

Fig. 16 (Left) Sketch of the impact problem studied with the impactor shown in green, the composite laminate in red, and the supporting steel plate in blue color; (right) contact force time history with points of significance for the damage initiation and propagation [104]

fiber-reinforced laminate qualitatively agreed with the test data. Points 1 through 4 in Fig. 16, respectively, correspond to damage initiation due to fiber compression, matrix tension, fiber tension, and accumulation of the fiber tensile damage.

5.3 Reduced-Order Models (Third-Order Shear and Normal Deformable Theory)

We have developed a third-order shear and normal deformable theory (TSNDT) for plates/ shells that accounts for all geometric nonlinearities including the von Karman nonlinearities. The plate material is assumed to be St. Venant-Kirchhoff for which the 2nd Piola-Kirchhoff stress tensor, $\boldsymbol{S}$, is a linear function of the Green-St. Venant strain tensor, $\boldsymbol{E}$. Thus the constitutive relation is objective under superimposed rigid body rotations. Whereas $\boldsymbol{S}$ is work conjugate to $\boldsymbol{E}$ there is no stress tensor that is work conjugate to the von Karman strain tensor. Below we briefly review the TSNDT, and give results for a few problems.

Figure 17 depicts a schematic sketch of a laminated doubly-curved shell composed of N layers of not necessarily equal thickness. Each ply is made of a homogeneous, orthotropic and elastic material with layers perfectly bonded to each other. We denote the total thickness and the constant two principal radii of curvature of the shell by h, R_1 and R_2, respectively. We employ orthogonal curvilinear coordinates (x, y, z) with curves x = constant and y = constant representing principal curvatures on a surface, z = constant. The shell arc lengths in the x and the y directions, respectively, equal a and b with the corresponding platform lengths l_x and l_y. The position vectors of a point p are denoted by $\mathbf{X}$ and $\mathbf{x}$ with respect to the fixed rectangular Cartesian coordinate axes in the reference and the current configuration, respectively, with the X_3- and the x_3-axes parallel to the z-axis.

The components G_{ij} of the metric tensor in the reference configuration are given by

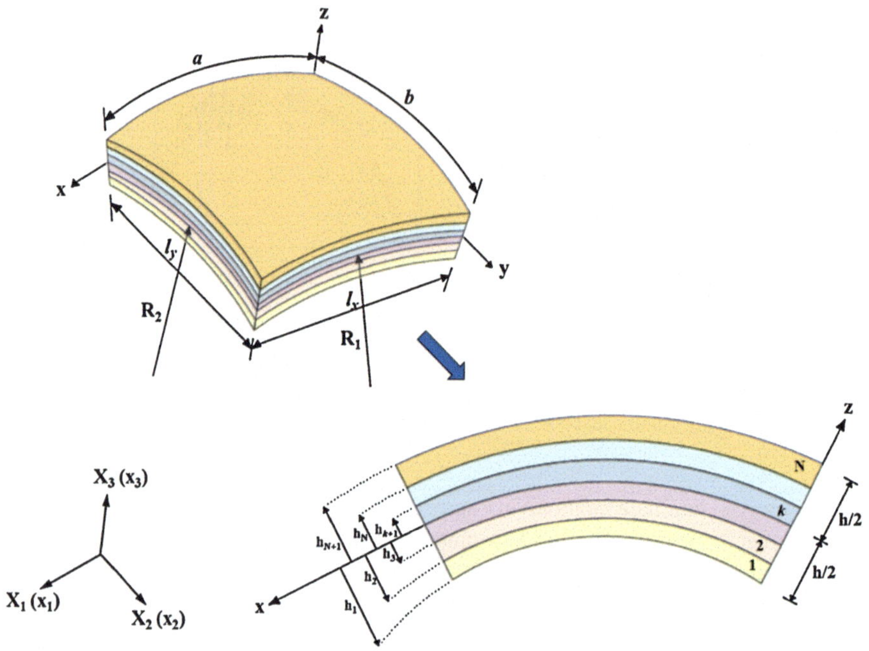

Fig. 17 Geometry and coordinate axes of a doubly-curved N-ply laminated shell

$$G_{ij} = \mathrm{A_i} \cdot \mathrm{A_j}, \mathrm{A_1} = \frac{\partial \mathrm{X}}{\partial x}, \ \mathrm{A_2} = \frac{\partial \mathrm{X}}{\partial y}, \ \mathrm{A_3} = \frac{\partial \mathrm{X}}{\partial z}, \tag{33}$$

where $\mathrm{A_i} \cdot \mathrm{A_j}$ equals the inner product between vectors $\mathrm{A_i}$ and $\mathrm{A_j}$. For orthogonal curvilinear coordinate system, G_{ij} is non-zero only when i = j. We define $\bar{e}_i = \frac{\mathbf{A}_{(i)}}{\mathrm{H}_{(i)}}$ (no sum on i), where $\mathrm{H_1} = \left(1 + \frac{\mathrm{Z}}{\mathrm{R_1}}\right)$, $\mathrm{H_2} = \left(1 + \frac{\mathrm{Z}}{\mathrm{R_2}}\right)$, $\mathrm{H_3} = 1$. For the TSNDT the displacement, $\boldsymbol{u} = \boldsymbol{x} - \boldsymbol{X}$ or (u, v, w), of a point in the shell is expressed as

$$\mathbf{d} = \mathbf{d}_t + \mathbf{Z}_\theta \mathbf{d}_\theta + \mathbf{Z}_\phi \mathbf{d}_\phi + \mathbf{Z}_\gamma \mathbf{d}_\gamma \tag{34}$$

where

$$\begin{aligned}
&\mathbf{d} = [\mathrm{u\ \ v\ \ w}]^{\mathrm{T}}, \mathrm{d_t} = [\mathrm{u_0}\quad \mathrm{v_0}\quad \mathrm{w_0}]^{\mathrm{T}}, \\
&\mathrm{d_\theta} = [\theta_{\mathrm{x}}\ \theta_{\mathrm{y}}\ \theta_{\mathrm{z}}]^{\mathrm{T}}, \ \mathrm{d_\phi} = [\phi_{\mathrm{x}}\ \phi_{\mathrm{y}}\ \phi_{\mathrm{z}}]^{\mathrm{T}}, \ \mathrm{d_\gamma} = [\gamma_{\mathrm{x}}\ \gamma_{\mathrm{y}}\ \gamma_{\mathrm{z}}]^{\mathrm{T}}, \\
&\mathrm{u_0} = \mathrm{u(x, y, 0, t)}, \quad \mathrm{v_0} = \mathrm{v(x, y, 0, t)}, \quad \mathrm{w_0} = \mathrm{w(x, y, 0, t)}, \\
&\mathrm{Z_\theta} = \mathrm{zI}, \mathrm{Z}_\phi = \mathrm{z^2 I}, \mathrm{Z}_\gamma = \mathrm{z^3 I}, \text{I is the identity matrix}
\end{aligned}$$

We first find physical components of the displacement gradient, $\boldsymbol{F}$, and then find $\boldsymbol{E} = \frac{1}{2}\left(\boldsymbol{F}^T\boldsymbol{F} - \mathbf{1}\right)$. We assume that the material of each layer is St. Venant-Kirchhoff for which the strain energy density, W, per unit reference volume, and the relation between $\boldsymbol{S}$ and $\boldsymbol{E}$ are

$$W = \frac{1}{2} E_{mn} C_{mn\alpha\beta} E_{\alpha\beta}, C_{mn\alpha\beta} = C_{\alpha\beta mn} = C_{nm\alpha\beta}, S_{mn} = \frac{\partial W}{\partial E_{mn}}. \tag{35}$$

Here $\boldsymbol{C}$ is the fourth-order elasticity tensor having 21, 9, 5 and 2 independent components for a general anisotropic, orthotropic, transversely isotropic and isotropic material, respectively. The strain energy density for the St. Venant-Kirchhoff material reduces to that of a Hookean material if the finite strain tensor $\boldsymbol{E}$ is replaced by the infinitesimal deformations strain tensor.

5.3.1 Effect of Curvature on Deformations of Shells

We use the linear theory valid for infinitesimal deformations to study transient deformations of four clamped spherical sandwich shells of equal principal radii given by $R/a = \infty$ (plate), 5, 1 and 0.5. Each shell is comprised of a soft honeycomb core modeled as an isotropic material with the top and the bottom face sheets made of an orthotropic material with fibers oriented at 45° with respect to the x-axis. The total thickness of the shell, h = 22 mm, and thicknesses of the core and each face sheet are 16 mm and 3 mm, respectively. Furthermore, the aspect ratio of the shell, $a/h = 10$ and $a = b$. Values of face sheets material parameters are

$E_1 = 251$ GPa, $E_2 = 48$ GPa, $E_3 = 7.5$ GPa, $G_{12} = 13.6$ GPa, $G_{13} = 12$ GPa, $G_{23} = 4.7$ GPa,

$\nu_{12} = 0.036$, $\nu_{13} = 0.25$, $\nu_{23} = 0.171$, $\rho = 1600$ kg/m^3

and those of the core isotropic material are, Young's modulus $E = E_1/100$ and Poisson's ratio $\nu = 0.2$.

In Table 2 we have listed the first ten lowest natural frequencies of free vibration and the maximum natural frequency of the four shells. These values suggest that each frequency increases with a decrease in R, and the fundamental mode frequency of the shell with $R/a = 1$ is nearly twice of that of the corresponding flat plate.

In Fig. 18 we have depicted time histories of the deflection of the core centroid, $w(a/2, b/2, 0)$, for the four shells having traction free bottom surface and the top surface subjected to the pressure due to an explosive load given by

$$P(r,t) = \left(-0.0005r^4 + 0.01r^3 - 0.0586r^2 - 0.001r + 1\right)P(t) \tag{36}$$

where r is the distance in cm, from the shell top surface center, and

Table 2 Natural frequencies (Hz) of four sandwich shells [106]

Mode	Plate	R/a = 5	R/a = 1	R/a = 0.5
1	3,644	3,960	7,022	8,450
2	5,826	6,093	7,383	9,889
3	6,305	6,525	8,331	10,553
4	7,946	8,281	9,475	10,630
5	8,966	9,305	10,068	12,517
6	9,214	9,632	10,450	12,986
7	10,234	10,647	11,602	13,286
8	10,981	11,533	11,660	13,736
9	11,850	11,857	12,193	14,665
10	12,252	12,860	12,560	15,524
Maximum frequency	643,872	822,912	658,275	685,535

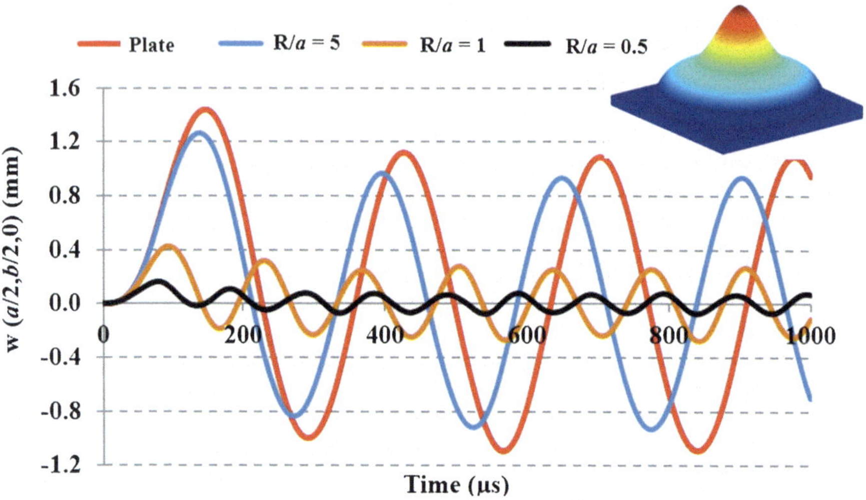

Fig. 18 Time histories of the deflection of centroid of the core for the four sandwich shells with R/$a = \infty$ (plate), R/a = 5, R/a = 1 and R/a = 0.5 [106]

$$P(t) = \begin{cases} 10t/60 \quad \text{MPa}, & t \leq 60\ \mu\text{s} \\ 10e^{-\frac{(t-60)}{30}} \quad \text{MPa}, & t > 60\ \mu\text{s} \end{cases} \tag{37}$$

The spatial distribution of the load is depicted in Fig. 18 inset. A comparison of these histories reveals that the maximum deflection decreases with an increase in the shell curvature. We note that the pressure applied at every point on the top surface linearly increases for the first 60 μs and then exponentially decreases becoming essentially zero for t > 120 μs. Thus for t > 120 μs, the shells freely vibrate.

In Fig. 19 we have displayed time histories of the axial stress, $\sigma_x(a/2, b/2, z)$, at centroids of each face sheet and the core for the shell with R/a = 5. Similar results for

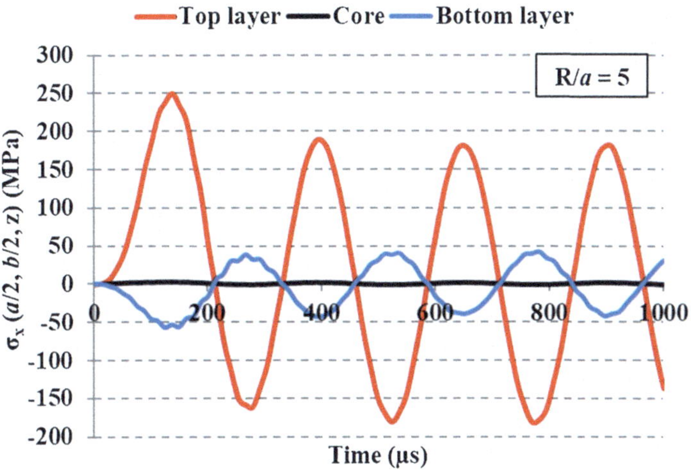

Fig. 19 Time histories of the axial stress, at centroid of each layer for the sandwich shell with R/a = 5 [106]

the other three shells suggest that, except for the flat plate, the maximum magnitude of the bending stress in the shell decreases with a decrease in the value of R or an increase in the shell curvature.

5.3.2 Effect of Geometric Nonlinearities on Orthotropic Plate's Deformations

For the transient pressure, given by Eq. (36) with the right-hand side of Eq. (37) replaced by $P_0 e^{-(t-t_r)/\theta}$, $P_O = 1\ GPa$, θ given by Eq. (1c), applied to the top surface of a clamped 22 cm × 22 cm plate made of a homogeneous and orthotropic material, we compare in Fig. 20 the centroidal deflection and the Cauchy stress σ_{11} at (a/2, b/2, h/2) time histories predicted by the linear and the nonlinear theories. Both the maximum deflection and the maximum value of σ_{11} from the nonlinear theory are considerably less than those from the linear theory. The linear theory over-predicts the peak displacement by a factor of almost 2, and the peak Cauchy stress by a factor of 2.5. However, this is not always the case as illustrated by results for the full sine wave beam studied in [107].

5.3.3 Stacking Sequence Optimization for Maximizing the Failure Initiation Load [108]

The objective is to find the fiber orientation angle in each ply of a rectangular laminate subjected to transverse loads on its top surface that maximizes the first failure load using the Tsai-Wu failure criterion, and for this stacking sequence find the ultimate failure load. We have numerically solved the problem by using the Nest-Site Selection (NeSS)

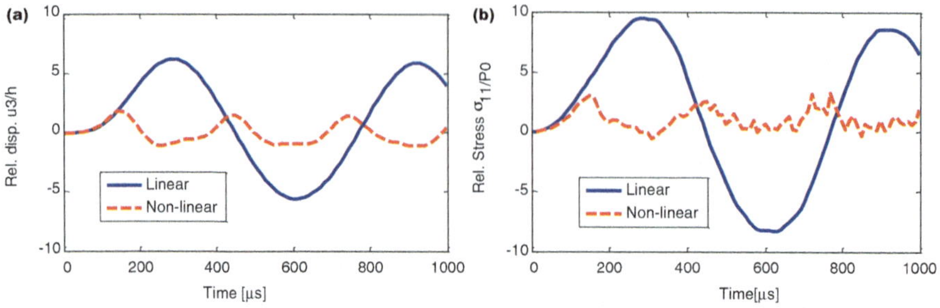

Fig. 20 Comparison of the (**a**) out of plane displacement of the plate centroid, and (**b**) σ_{11} at the centroid of the top face from the linear and the nonlinear theories [106]

optimization algorithm, the TSNDT, the FEM, and a one-step stress-recovery scheme (SRS), and shown that the computed six stress components have values close to those given by the solution of the 3-D linear elasticity equations with the FE software, ABAQUS. The number of layers and the thickness of each layer are kept fixed. The failure index, F, of the Tsai-Wu criterion is calculated at 36,000 points in the structure using stress components with respect to the material principal axes. For the first failure load to have the highest value, the fiber orientations in each ply should be such that the maximum value, F_{max}, of F for the structure has the least value. Subsequent to the initiation of the failure at a point and the degradation of material properties there, the structure becomes nonhomogeneous. The optimization problem is stated as follows.

Find $\boldsymbol{\theta} = \{\theta_1, \theta_2, \theta_3, \ldots, \theta_n\}$, $\theta_i \in [-90^{\circ}, 90^{\circ}]$ in 1° increment for $i = 1, \ldots, n$,

so that $F_{\max} = \sup\{F(X_1, X_2, X_3), (X_1, X_2, X_3) \in \Omega = [0, a] \times [0, b] \times [-\frac{h}{2}, \frac{h}{2}]\}$ has the least value. Here θ_i is the fiber orientation angle in the *ith* ply measured counter clockwise from the positive X_1-axis, and n equals the number of plies. Subsequent to ascertaining the optimal stacking sequence, we perform progressive failure analysis to find the ultimate failure load.

We analyze below the optimization problem for a 10-layer square laminate of side $a = 22$ cm, $a/h = 10$, the load distribution illustrated in Fig. 18 inset, and values of material parameters listed in Table 3.

In Table 4 we have listed the five optimal stacking sequences (corresponding to five randomly chosen values of the initial stacking sequence) for the pressure load given by Eq. (32) the first failure load L_F (or q_0) and the ultimate failure load L_U. The (L_F, L_U) for the $[0]_{10}$, $[0/90]_5$ and $[0_5/90_5]$ clamped laminates, respectively, are (12.41, 20.99), (14.67, 24.48) and (7.22, 10.28). For the three loadings studied, the $[0/90]_5$ laminate has the maximum L_U. Values in the column "% increase" are the relative difference between L_F and L_U. Laminates with non-traditional fiber orientations have higher L_F and L_U than those with commonly used fiber angles.

We note that (a) for a simply supported laminate, L_U and L_F are essentially equal to each other, (b) the L_U for a clamped plate is 32% more than that for a simply supported laminate, (c) for clamped laminates the ratio, L_U/L_F, varies from 1.4 to 1.2 and depends upon the stacking sequence and the load type, and (d) the stacking sequence for the maximum L_F need not have the maximum L_U.

Table 3 Values of material parameters for a lamina with respect to material principal axes

Elastic moduli (GPa)			Shear moduli (GPa)			Poisson's ratios		
E_1	E_2	E_3	G_{12}	G_{13}	G_{23}	ν_{12}	ν_{13}	ν_{23}
132.5	10.8	10.8	5.7	5.7	3.4	0.24	0.24	0.49
Strength parameters (MPa)								
X_T	X_C	Y_T	Y_C	Z_T	Z_C	R	S	T
1515	1697	43.8	43.8	43.8	43.8	67.6	86.9	86.9

Table 4 Optimal designs of plates for two boundary conditions with the fiber orientation angle measured counter-clockwise from the X_I-axis entries below should not be bold.

Boundary conditions	Design	Stacking sequence	First failure load (MPa)	Ultimate failure load (MPa)	% increase
	c1	[−46/43/29/79/−29/−13/−54/−79/−27/49]	20.15	25.08	24.48
	c2	[−48/27/31/−44/−32/−71/−13/3/74/−37]	20.07	23.18	15.48
	c3	[50/−24/−61/31/39/43/55/83/−49/44]	20.18	22.85	13.25
	c4	[−52/26/47/−39/−30/−22/−62/−68/−66/44]	19.54	22.63	15.81
Clamped	c5	[−50/31/68/−62/−43/−67/−12/−79/63/−46]	20.06	22.45	11.92
Simply supported	c1'	[29/−35/−37/13/−32/6/45/35/47/−32]	18.96	18.96	0
	c2'	[−63/46/40/−84/77/−81/−60/43/−44/56]	18.40	18.40	0
	c3'	[−32/52/83/76/81/21/42/−45/−45/61]	18.31	18.31	0
	c4'	[−24/29/46/16/−14/48/−17/−14/−43/33]	18.02	18.02	0
	c5'	[25/−36/30/−52/15/87/19/45/50/−26]	17.59	17.59	0

For the two boundary conditions, we have exhibited in Fig. 21 the final plate configurations with the failed material separately shown. The plate thickness has been enlarged by a factor of 12 and various layers are differently colored for ease of viewing. For a simply supported laminate, the failure initiates at the centroid of either the top or the bottom surface and rapidly propagates radially outwards and along the plate thickness creating a virtual hole (because elements in our work are not really deleted but are made very weak). The laminate ultimately fails at a load very close to the first failure load. For the clamped plate, the failure initiates at points near the edges and propagates downwards along the thickness. Some material near the plate vertical centroidal axis also has failed prior to the application of the ultimate load.

In Fig. 22, we have displayed the fraction of failed points and the maximum vertical deflection (with their scales shown on the left and the right vertical axes,

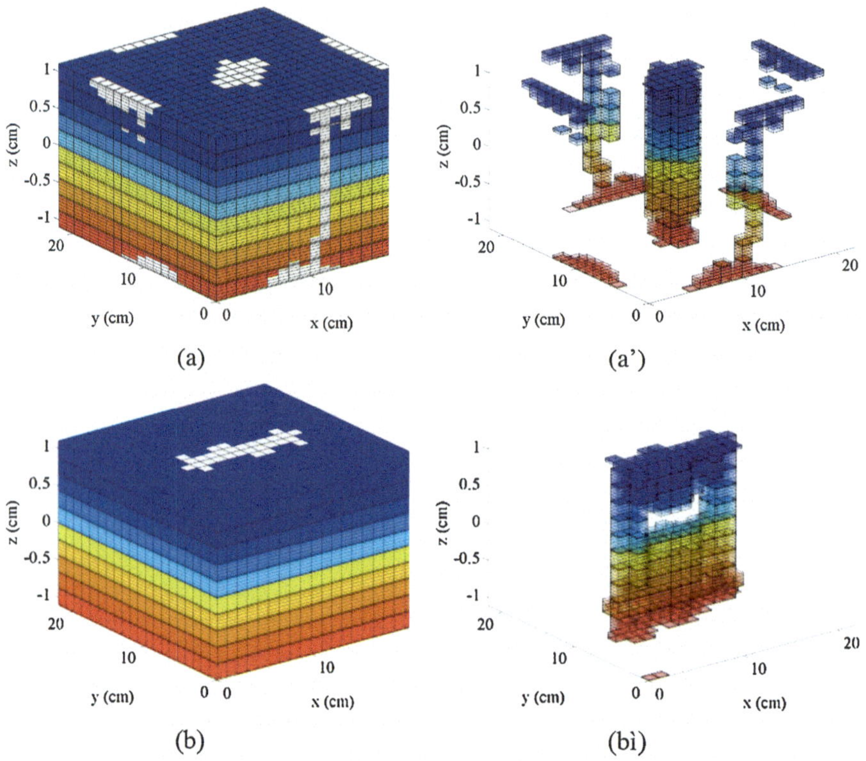

Fig. 21 The final laminate configurations for (**a**) clamped plate c1 and (**b**) simply supported plate c1' with only the failed material depicted in Figs. (**a'**) and (**b'**) [108]

respectively) versus the peak load for the clamped plates under (a) uniform, (b) sinusoidal, and (c) non-uniform distributed load given by Eq. (36) with P(t) = 1 applied on the top surface. Taking the rate of increase of the fraction of failed points as an indicator of the rate of progression of failure in the laminate, it is evident that the failure/damage propagates most (least) rapidly for the uniformly (non-uniformly) distributed load.

5.3.4 Fluid-Structure Interaction

Qin and Batra [109] developed a hydroelastic model using a {3, 2} sandwich plate theory (i.e., deformations of face sheets are simulated using the Kirchhoff plate theory and for the core in-plane and transverse displacements are, respectively, polynomials of degree 3 and 2 in the z-coordinate) and Wagner's water impact theory to study a slamming process. They found that considering wedge's deformations significantly affects the slamming pressure between the water and the wedge.

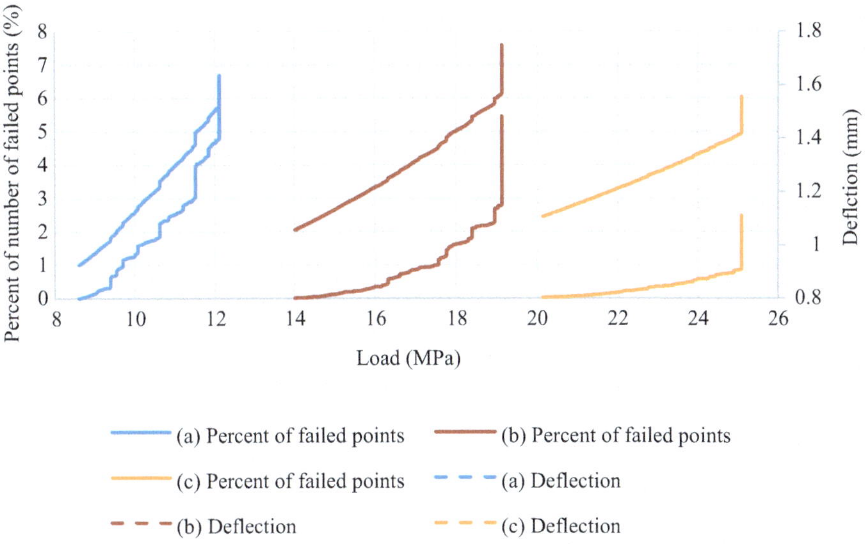

Fig. 22 The variation of the percentage of the number of failed points and the maximum vertical deflection with the peak load for the clamped plates subjected to (**a**) uniform normal traction, (**b**) sinusoidal traction, and (**c**) non-uniform traction [108]

Das and Batra [110] used the FE commercial software, LSDYNA, and the coupled Lagrangian and Eulerian (sometimes called ALE) approach to describe finite plane strain deformations of a hull panel. A linear relation between the Cauchy stress tensor and the Almansi-Hamel strain tensor was employed for the hull material and the water was modeled as an inviscid compressible fluid. The continuity of surface tractions and the inter-penetration of water into the hull was satisfied by using a penalty method. They delineated jet flows near the edges of a wetted hull, and analyzed delamination induced in a sandwich composite panel. As displayed in Fig. 23, the slamming pressure can be considered as a wave traveling along the hull with peak value gradually decreasing as the wave propagates along the hull.

Xiao and Batra [111] analyzed finite plane strain elastic deformation of a sandwich panel by using a layer wise TSNDT. They numerically solved the irrotational and isochoric flow equations by the BEM and nonlinear governing equations of the solid by the FEM. They employed a cohesive zone model to decipher mixed mode delamination at an interface between a face sheet and the core, and delineated effects of hull curvature on the hydroelastic interaction. Figure 24 depicts time-histories of a panel centroid deflection with and without considering delamination. For one of the problems studied, the energy dissipated during delamination was quantified as 8% of the total work done on the hull.

For water slamming problems studied above, the solid body need not be immersed in the fluid. Qu and Batra [112] developed a numerical algorithm for studying the interaction of inviscid and compressible flows with arbitrary shaped moving rigid solids in which the fluid flow equations are solved on a fixed

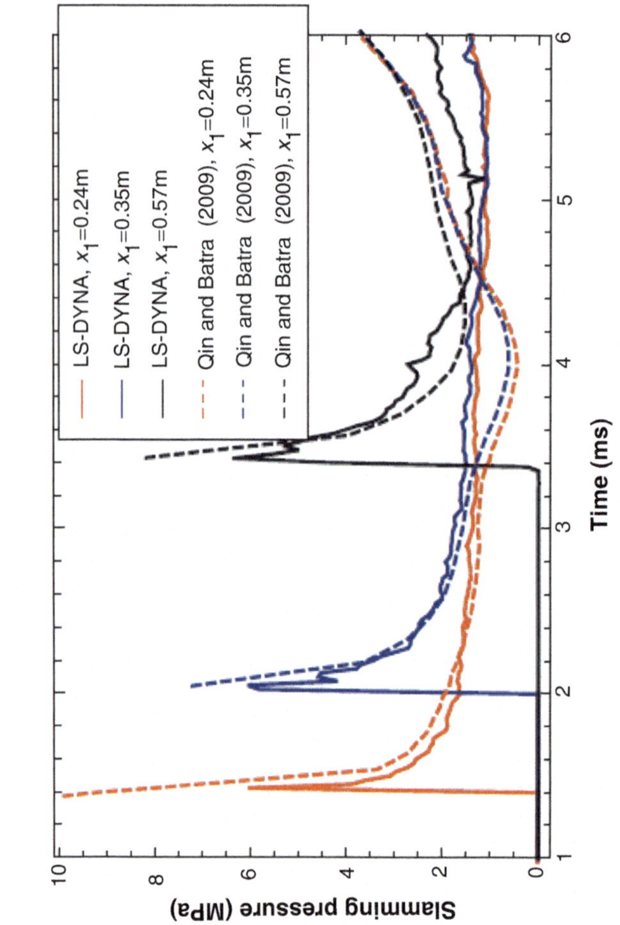

Fig. 23 Time histories of the interface pressure at 3 locations on the hull-water interface [110]

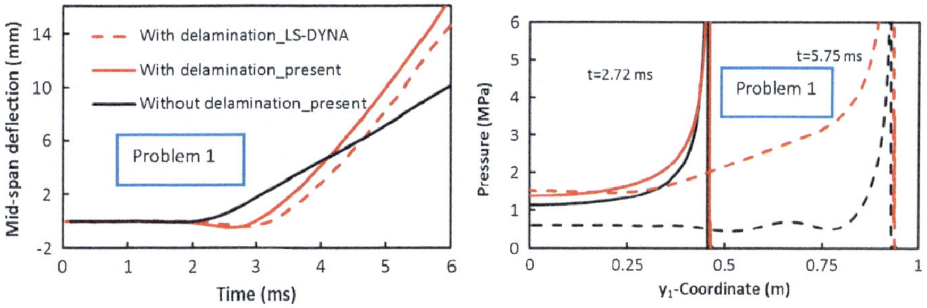

Fig. 24 Time histories of the deflection of a straight panel centroid, and the hydroelastic pressure distribution at two different times [111]

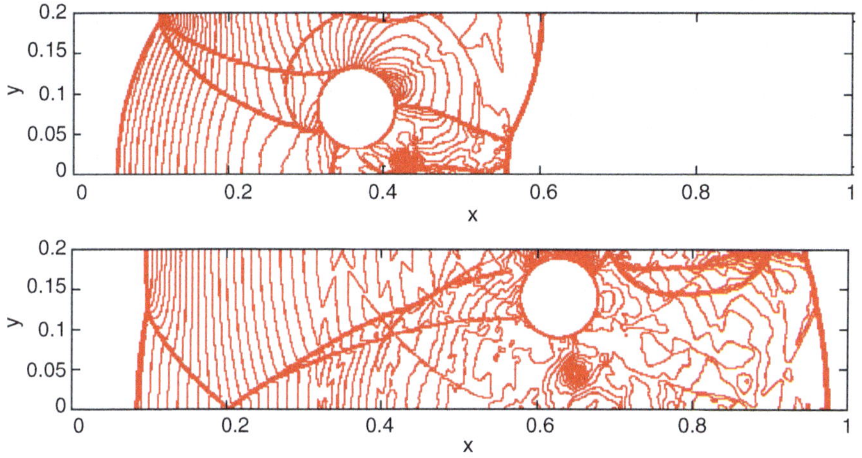

Fig. 25 Pressure contours at (top) t = 0.140 s and (bottom) t = 0.255 s (50 contours from 0 to 22) [112]

rectangular Cartesian grid using a 5th order weighted essentially non-oscillatory (WENO) scheme. The continuity of pressure at the interface was enforced with a constrained moving least-squares sharp interface method and the continuity of normal velocities by a penalty method. The solution is advanced in time by alternately using the 3rd order Runge-Kutta and the implicit Newmark integration schemes for the fluid and the solid motion equations. They studied the lift-off of a 0.1 m diameter rigid circular cylinder of mass density 7.6 kg/m^3 initially resting at (0.15 m, 0.05 m) on the lower wall of an air filled channel. The cylinder is driven by an incident shock wave with Mach number = 3 starting at 0.08 m; e.g. see [113, 114]. The instantaneous pressure contours at two times exhibited in Fig. 25 reveal the existence of a strong vortex below the cylinder that is probably associated with a Kelvin-Helmholtz instability of the contact discontinuity.

Qu et al. [115] have extended the work in [112] and [113] to high speed compressible viscous flows modeled by the Navier-Stokes equations that are discretized using a low-diffusion flux splitting method for the inviscid fluxes and

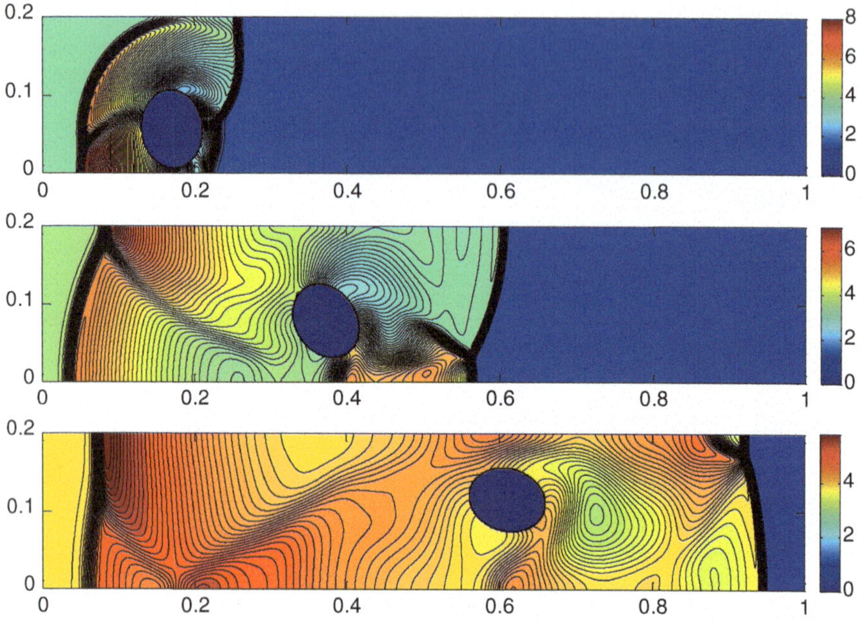

Fig. 26 Contours of mass density from top to bottom: t = 0.05 (100 contours from 0 to 8.11), t = 0.15 (100 contours from 0 to 7.12), and t = 0.25 (100 contours from 0 to 5.83) [115]

conservative higher-order central differences for the viscous components. Referring the reader to [115] for details, we have exhibited in Fig. 26 the instantaneous pressure contours for the problem of the lifting off of an elliptic cylinder.

6 Conclusions

We have shown that a third-order shear and normal deformable plate/shell theory coupled with a one-step stress recovery scheme provides reasonably accurate values of the three displacement components and the six stress components. The blast/explosion load has been approximated as the product of a function of time and a function of the spatial coordinates. The optimal stacking sequence for a laminate that maximizes the first failure load has plies with fiber angles quite different from those commonly studied, and do not conform to a functionally graded laminate. The analyses of water slamming problems suggest that the pressure on the hull surface, modeled as a wedge, can be regarded as a traveling pressure wave with the peak value decaying as it travels away from the keel. The hull curvature and the hydroelastic interaction strongly influence the pressure wave. The developed sharp-interface immersed boundary method for numerically studying the interaction between high speed viscous flows and rigid solids can be extended to deformable solids.

Acknowledgments The work was supported by the Office of Naval Research grant, N000141812548, to Virginia Polytechnic Institute and State University. Views expressed in this paper are those of the authors and neither of their organizations nor of the ONR.

References

1. Cullis I (2001) Blast waves and how they interact with structures. J R Army Med Corps 147 (1):16–26
2. Keil A (1961) The response of ships to underwater explosions. David Taylor Model Basin, Washington
3. Rayleigh L (1917) VIII. On the pressure developed in a liquid during the collapse of a spherical cavity. Lond Edinb Dubl Phil Mag & J Sci 34(200):94–98
4. Cole RH, Weller R (1948) Underwater explosions. Phys Today 1:35
5. Taylor G (1963) The pressure and impulse of submarine explosion waves on plates, The scientific papers of GI Taylor 3:287–303
6. Rajendran R, Narasimhan K (2006) Deformation and fracture behaviour of plate specimens subjected to underwater explosion—a review. Int J Impact Eng 32(12):1945–1963
7. Reid W (1996) The response of surface ships to underwater explosions, department of defence. Defence Science and Technology Organisation, Melbourne
8. Shin Y (1996) Naval ship shock and design analysis. Course notes for underwater shock analysis, Naval Postgraduate School, Monterey 87
9. Snay HG (1957) Hydrodynamics of underwater explosions. In symposium on naval hydrodynamics
10. Costanzo FA, Gordon JD (1980) An analysis of bulk cavitation in deep water, DTNSRDC, UERD Report
11. Nurick G, Martin J (1989) Deformation of thin plates subjected to impulsive loading—a review Part 1: Theoretical considerations. Int J Impact Eng 8(2):159–170
12. Nurick G, Martin J (1989) Deformation of thin plates subjected to impulsive loading—a review part II: experimental studies. Int J Impact Eng 8(2):171–186
13. Wierzbicki T, Nurick G (1996) Large deformation of thin plates under localised impulsive loading. Int J Impact Eng 18(7–8):899–918
14. Rajendran R, Narasimhan K (2001) Damage prediction of clamped circular plates subjected to contact underwater explosion. Int J Impact Eng 25(4):373–386
15. Rajendran R, Lee J (2009) Blast loaded plates. Mar Struct 22(2):99–127
16. Kennard E (1994) The effect of a pressure wave on a plate or a diaphragm. David Taylor Model Basin, Washington
17. Fleck N, Deshpande V (2004) The resistance of clamped sandwich beams to shock loading. J Appl Mech 71(3):386–401
18. Liu Z, Young YL (2008) Transient response of submerged plates subject to underwater shock loading: an analytical perspective. J Appl Mech 75(4):044504
19. Mouritz A (1995) The damage to stitched GRP laminates by underwater explosion shock loading. Compos Sci Technol 55(4):365–374
20. Mouritz A (1996) The effect of underwater explosion shock loading on the flexural properties of GRP laminates. Int J Impact Eng 18(2):129–139
21. Wen H (2001) Penetration and perforation of thick FRP laminates. Compos Sci Technol 61 (8):1163–1172
22. De Morais W, Monteiro S, d'Almeida J (2005) Effect of the laminate thickness on the composite strength to repeated low energy impacts. Compos Struct 70(2):223–228

23. Gellert E, Cimpoeru S, Woodward R (2000) A study of the effect of target thickness on the ballistic perforation of glass-fibre-reinforced plastic composites. Int J Impact Eng 24 (5):445–456
24. Sierakowski RL, Chaturvedi SK (1997) Dynamic loading and characterization of fiber-reinforced composites, dynamic loading and characterization of fiber-reinforced composites, by Robert L, Sierakowski, Shive K, Chaturvedi, pp. 252. ISBN 0-471-13824-X. Wiley-VCH: pp 252
25. Sjögren B (2001) Static strength of CFRP laminates with embedded fiber-optic edge connectors. Compos Part A Appl Sci Manuf 32(2):189–196
26. Fu S, Lauke B, Mäder E, Hu X, Yue C (1999) Fracture resistance of short-glass-fiber-reinforced and short-carbon-fiber-reinforced polypropylene under charpy impact load and its dependence on processing. J Mater Process Tech 89:501–507
27. Espinosa H, Dwivedi S, Lu HC (2000) Modeling impact induced delamination of woven fiber reinforced composites with contact/cohesive laws. Comput Methods Appl Mech Eng 183 (3–4):259–290
28. Roy R, Sarkar B, Bose N (2001) Impact fatigue of glass fibre–vinylester resin composites. Compos Part A Appl Sci Manuf 32(6):871–876
29. Mouritz AP, Gellert E, Burchill P, Challis K (2001) Review of advanced composite structures for naval ships and submarines. Compos Struct 53(1):21–42
30. Xue Z, Hutchinson JW (2004) A comparative study of impulse-resistant metal sandwich plates. Int J Impact Eng 30(10):1283–1305
31. Hutchinson JW, Xue Z (2005) Metal sandwich plates optimized for pressure impulses. Int J Mech Sci 47(4–5):545–569
32. Liang Y, Spuskanyuk AV, Flores SE, Hayhurst DR, Hutchinson JW, McMeeking RM, Evans AG (2007) The response of metallic sandwich panels to water blast. J Appl Mech 74(1):81–99
33. Xue Z, Hutchinson JW (2003) Preliminary assessment of sandwich plates subject to blast loads. Int J Mech Sci 45(4):687–705
34. Xue Z, Hutchinson JW (2006) Crush dynamics of square honeycomb sandwich cores. Int J Numer Methods Eng 65(13):2221–2245
35. Deshpande V, Fleck N (2005) One-dimensional response of sandwich plates to underwater shock loading. J Mech Phys Solids 53(11):2347–2383
36. Qiu X, Deshpande V, Fleck N (2003) Finite element analysis of the dynamic response of clamped sandwich beams subject to shock loading. Eur J Mech A-Solid 22(6):801–814
37. Rabczuk T, Samaniego E, Belytschko T (2007) Simplified model for predicting impulsive loads on submerged structures to account for fluid-structure interaction. Int J Impact Eng 34 (2):163–177
38. Godunov SK (1959) A difference method for numerical calculation of discontinuous solutions of the equations of hydrodynamics. Mat Sb 89(3):271–306
39. Godunov SK, Zabrodin AV, Prokopov GP (1962) A computational scheme for two-dimensional non stationary problems of gas dynamics and calculation of the flow from a shock wave approaching a stationary state. USSR Comput Math & Math Phys 1 (4):1187–1219
40. Van Leer B (1979) Towards the ultimate conservative difference scheme. V. A second-order sequel to Godunov's method. J Comput Phys 32(1):101–136
41. Van Leer B, Woodward P (1979) The MUSCL code for compressible flow: philosophy and results. TICOM conference
42. Colella P (1985) A direct Eulerian MUSCL scheme for gas dynamics. SIAM J Sci Comput 6 (1):104–117
43. Wardlaw AB, Mair HU (1998) Spherical solutions of an underwater explosion bubble. Shock Vib 5(2):89–102
44. Blake JR, Gibson D (1981) Growth and collapse of a vapour cavity near a free surface. J Fluid Mech 111:123–140

45. Blake JR, Gibson D (1987) Cavitation bubbles near boundaries. Annu Rev Fluid Mech 19 (1):99–123
46. Geers TL, Hunter KS (2002) An integrated wave-effects model for an underwater explosion bubble. J Acoust Soc Am 111(4):1584–1601
47. Hunter KS, Geers TL (2004) Pressure and velocity fields produced by an underwater explosion. J Acoust Soc Am 115(4):1483–1496
48. Rajendran R, Satyanarayana K (1997) Interaction of finite amplitude acoustic waves with a plane plate. JASI 25:V5
49. Bleich H, Sandler IS (1968) Dynamic interaction between structures and bilinear fluids. Columbia University, New York
50. Bleich H, Sandler I (1970) Interaction between structures and bilinear fluids. Int J Solid Struct 6(5):617–639
51. Geers TL (1971) Residual potential and approximate methods for three-dimensional fluid-structure interaction problems. J Acoust Soc Am 49(5B):1505–1510
52. Geers TL (1978) Doubly asymptotic approximations for transient motions of submerged structures. J Acoust Soc Am 64(5):1500–1508
53. Shin YS, Santiago LD (1998) Surface ship shock modeling and simulation: two-dimensional analysis. Shock Vib 5(2):129–137
54. DeRuntz J, Geers T, Felippa C (1980) The underwater shock analysis code (USA-version 3): a reference manual. Lockheed Missiles and Space Co Inc, Palo Alto
55. DeRuntz Jr, JA, Rankin C (1989) Applications of the USA-STAGS-CFA code to nonlinear fluid-structure interaction problems in underwater shock of submerged structures, In Proceedings of the 60th shock and vibration symposium
56. Felippa C, DeRuntz J (1984) Finite element analysis of shock-induced hull cavitation. Comput Methods Appl Mech Eng 44(3):297–337
57. Shin YS (2004) Ship shock modeling and simulation for far-field underwater explosion. Comput Struct 82(23–26):2211–2219
58. Kwon Y, Cunningham R (1998) Comparison of USA-DYNA finite element models for a stiffened shell subject to underwater shock. Comput Struct 66(1):127–144
59. Newton R (1978) Effects of cavitation on underwater shock loading—axisymmetric geometry, technical ReportNPS-69-78-017PR. Naval Postgraduate School, Monterey
60. Newton RE (1978) Effects of cavitation on underwater shock loading. Part 1. Naval Postgraduate School, Monterey
61. Newton R (1980) Finite element analysis of shock-induced cavitation. ASCE Spring Convention, Portland
62. Newton RE (1981) Effects of cavitation on underwater shock loading-plane problem. Naval Postgraduate School, Monterey
63. Rehak ML, DiMaggio FL, Sandler IS (1985) Interactive approximations for a cavitating fluid around a floating structure. Comput Struct 21(6):1159–1175
64. Shin YS, Schneider NA (2003) Ship shock trial simulation of USS Winston S. Churchill (DDG 81): modeling and simulation strategy and surrounding fluid volume effects, Shock and vibration symposium
65. Sprague M, Geers T (2004) A spectral-element method for modelling cavitation in transient fluid–structure interaction. Int J Numer Methods Eng 60(15):2467–2499
66. Klaseboer E, Hung K, Wang C, Wang C, Khoo B, Boyce P, Debono S, Charlier H (2005) Experimental and numerical investigation of the dynamics of an underwater explosion bubble near a resilient/rigid structure. J Fluid Mech 537:387–413
67. Webster KG (2007) Investigation of close proximity underwater explosion effects on a ship-like structure using the multi-material arbitrary lagrangian eulerian finite element method, Virginia Tech
68. Huang H, Jiao QJ, Nie JX, Qin JF (2011) Numerical modeling of underwater explosion by one-dimensional ANSYS-AUTODYN. J Energ Mater 29(4):292–325

69. Mair HU (1999) Hydrocodes for structural response to underwater explosions. Shock Vib 6 (2):81–96
70. Teeling-Smith R, Nurick G (1991) The deformation and tearing of thin circular plates subjected to impulsive loads. Int J Impact Eng 11(1):77–91
71. Nurick G, Gelman M, Marshall N (1996) Tearing of blast loaded plates with clamped boundary conditions. Int J Impact Eng 18(7–8):803–827
72. Espinosa HD, Lee S, Moldovan N (2006) A novel fluid structure interaction experiment to investigate deformation of structural elements subjected to impulsive loading. Exp Mech 46 (6):805–824
73. Rajendran R, Paik J, Lee J (2007) Of underwater explosion experiments on plane plates. Exp Tech 31(1):18–24
74. Houlston R, Slater J (1991) Global and local modelling of naval panels subjected to shock loads. Comput Struct 40(2):353–364
75. Houlston R, Slater J (1993) Damage analysis with ADINA of naval panels subjected to a confined air-blast wave. Comput Struct 47(4–5):629–639
76. Jiang J, Olson M (1991) Nonlinear dynamic analysis of blast loaded cylindrical shell structures. Comput Struct 41(1):41–52
77. Kambouchev N, Noels L, Radovitzky R (2006) Nonlinear compressibility effects in fluid-structure interaction and their implications on the air-blast loading of structures. J Appl Phys 100(6):063519
78. Kambouchev N, Noels L, Radovitzky R (2007) Numerical simulation of the fluid–structure interaction between air blast waves and free-standing plates. Comput Struct 85 (11–14):923–931
79. Kambouchev N, Radovitzky R, Noels L (2007) Fluid–structure interaction effects in the dynamic response of free-standing plates to uniform shock loading. J Appl Mech 74 (5):1042–1045
80. Gupta N (2007) Deformation and tearing of circular plates with varying support conditions under uniform impulsive loads. Int J Impact Eng 34(1):42–59
81. Louca L, Pan Y, Harding J (1998) Response of stiffened and unstiffened plates subjected to blast loading. Eng Struct 20(12):1079–1086
82. Ramajeyathilagam K, Vendhan C (2004) Deformation and rupture of thin rectangular plates subjected to underwater shock. Int J Impact Eng 30(6):699–719
83. DeRuntz J (1989) The underwater shock analysis code and its applications, In 60th shock and vibration symposium proceedings
84. Hammond L, Grzebieta R (2000) Structural response of submerged air-backed plates by experimental and numerical analyses. Shock Vib 7(6):333–341
85. Abrate S (1997) Localized impact on sandwich structures with laminated facings. Appl Mech Rev 50(2):69–82
86. Hall D (1989) Examination of the effects of underwater blasts on sandwich composite structures. Compos Struct 11(2):101–120
87. McCoy R, Sun C (1997) Fluid-structure interaction analysis of a thick-section composite cylinder subjected to underwater blast loading. Compos Struct 37(1):45–55
88. Hassan N, Batra R (2008) Modeling damage in polymeric composites. Compos Part B-Eng 39 (1):66–82
89. Batra R, Hassan N (2008) Blast resistance of unidirectional fiber reinforced composites. Compos Part B-Eng 39(3):513–536
90. Batra RC, Hassan NM (2009) Modeling of progressive damage in high strain—rate deformations of fiber-reinforced composites. In: Major accomplishments in composite materials and sandwich structures. Springer, Dordrecht, pp 89–111
91. Batra R, Hassan N (2007) Response of fiber reinforced composites to underwater explosive loads. Compos Part B-Eng 38(4):448–468

92. Nezami M, Jam J, Nia N (2009) Dynamic response of free–free laminated curved panels subjected to explosive blast. Polym Compos 30(9):1199–1203
93. LeBlanc J, Shukla A (2010) Dynamic response and damage evolution in composite materials subjected to underwater explosive loading: an experimental and computational study. Compos Struct 92(10):2421–2430
94. LeBlanc J, Shukla A (2011) Dynamic response of curved composite panels to underwater explosive loading: experimental and computational comparisons. Compos Struct 93 (11):3072–3081
95. Russell DM (1997) Error measures for comparing transient data: Part I, development of a comprehensive error measure. In: Proceedings of the 68th shock and vibration symposium. Shock and Vibration Exchange, Hunt Valley
96. Mäkinen K (1999) The transverse response of sandwich panels to an underwater shock wave. J Fluid Struct 13(5):631–646
97. McMeeking RM, Spuskanyuk A, He M, Deshpande V, Fleck N, Evans A (2008) An analytic model for the response to water blast of unsupported metallic sandwich panels. Int J Solid Struct 45(2):478–496
98. Librescu L, Nosier A (1990) Response of laminated composite flat panels to sonic boom and explosive blast loadings. AIAA J 28(2):345–352
99. Librescu L, Oh SY, Hohe J (2004) Linear and non-linear dynamic response of sandwich panels to blast loading. Compos Part B-Eng 35(6–8):673–683
100. Wei Z, Dharmasena K, Wadley H, Evans A (2007) Analysis and interpretation of a test for characterizing the response of sandwich panels to water blast. Int J Impact Eng 34 (10):1602–1618
101. Wadley H, Dharmasena K, Chen Y, Dudt P, Knight D, Charette R, Kiddy K (2008) Compressive response of multilayered pyramidal lattices during underwater shock loading. Int J Impact Eng 35(9):1102–1114
102. Panahi B, Ghavanloo E, Daneshmand F (2011) Transient response of a submerged cylindrical foam core sandwich panel subjected to shock loading. Mater Des 32(5):2611–2620
103. Gopinath G, Batra RC (2018) A common framework for three micromechanics approaches to analyze elasto-plastic deformations of fiber-reinforced composites. Int J Mech Sc 148 (2018):540–553
104. Batra RC, Gopinath G, Zheng JQ (2012) Damage and failure in low energy impact of Fiber-reinforced polymeric composite laminates. Compos Struct 94:540–547
105. Robertson DD, Mall S (1994) Micromechanical analysis for thermo-viscoplastic behavior of unidirectional fibrous composites. Compos Sci Technol 50:483–496
106. Batra RC et al. (2018) Unpublished results
107. Batra RC, Xiao J (2013) Finite deformations of curved laminated St. Venant-Kirchhoff beam using layer-wise third order shear and Normal deformable beam theory (TSNDT). Compos Struct 97:147–1614
108. Taetragool U, Shah PH, Halls VA, Zheng JQ, Batra RC (2017) Stacking sequence optimization for maximizing the first failure initiation load followed by progressive failure analysis until the ultimate load. Compos Struct 180:1007–1021
109. Qin Z, Batra RC (2009) Local slamming impact of sandwich composite hulls. Int J Solid Struct 46(10):2011–2035
110. Das K, Batra RC (2011) Local water slamming impact on sandwich composite hulls. J Fluid Struct 27(4):523–551
111. Xiao J, Batra RC (2014) Delamination in sandwich panels due to local water slamming loads. J Fluid Struct 48:122–155
112. Qu Y, Batra RC (2017) Constrained moving least-squares immersed boundary method for fluid-structure interaction analysis. Int J Numer Methods Fluids 85(12):675–692

113. Monasse L, Daru V, Mariotti C, Piperno S, Tenaud C (2012) A conservative coupling algorithm between a compressible flow and a rigid body using an embedded boundary method. J Comput Phys 231:2977–2994
114. Pasquariello V, Hammerl G, Örley F et al (2016) A cut-cell finite volume–finite element coupling approach for fluid–structure interaction in compressible flow. J Comput Phys 307:670–695
115. Qu Y, Shi R, Batra RC (2018) An immersed boundary formulation for simulating high-speed compressible viscous flows with moving solids. J Comput Phys 354:672–691

The Extended High Order Sandwich Panel Theory for the Static and Dynamic Analysis of Sandwich Structures

George A. Kardomateas

1 Introduction

Sandwich construction is a structural concept that results in a very stiff but lightweight structure. In addition, these structures normally possess a high-energy absorption capability. These attributes are achieved due to the existence of a relatively soft and lightweight core, typically made out of polymeric or metallic foam or honeycomb. The core is between two stiff metallic or composite thin face sheets, which provide the stiffness. As a result, sandwich structures have found applications in aerospace vehicles, including satellites, helicopter and fixed-wing aircraft components, as well as naval vehicles, wind turbines and civil infrastructure. The initial studies on sandwich panels were done by neglecting the transverse deformation of the core (see the textbooks [1–3]). The core of a sandwich structure was considered as infinitely rigid in the thickness direction and only its shear stresses could be taken into account while the in-plane stresses were neglected as a result of its low rigidity in this direction relative to that of the face sheets. Plantema [1] and Allen [2] well summarized the work done in the 1960s. The earliest models are called the classical (no transverse shear of the core) or first order shear deformation (FOSD) theory (transverse shear of the core included). These models are based on the Euler-Bernoulli (classical) or Timoshenko beam (first order shear) theories. These assumptions could potentially be quite inaccurate esp. for dynamic loading; in particular, sudden loading experiments conducted by Wang et al. [4] showed large amounts of core compression. Consideration of the core compressibility implies that the displacements of the upper and lower face sheets may not be identical.

The first theory to consider the core compressibility is the High Order Sandwich Panel Theory (HSAPT) formulated by Frostig et al. [5] in the 1990s; in this theory, the resulting shear strain in the core is constant and the resulting transverse normal

G. A. Kardomateas
School of Aerospace Engineering, Georgia Institute of Technology, Atlanta, GA, USA
e-mail: george.kardomateas@aerospace.gatech.edu

S. W. Lee (ed.), *Advances in Thick Section Composite and Sandwich Structures*,
https://doi.org/10.1007/978-3-030-31065-3_11

strain in the core is linear in the transverse coordinate, as a result of the assumption that the in-plane rigidity of the core is negligible. In the 2000s Hohe et al. [6] proposed a model for sandwich plates by assuming the transverse normal strain to be constant along the transverse coordinate, while the shearing strains were assumed to be linear in the transverse coordinate.

The accuracy of any of these models can be readily assessed if corresponding elasticity solutions exists. To this extent, for a three dimensional sandwich plate consisting of orthotropic material, static elasticity solutions were developed by Vlasov [7] for isotropic plates and by Pagano [8] for a restricted case of material sandwich combination. And these solutions were extended to cover all possible orthotropic face sheet and core combinations of a sandwich beam/wide plate by Kardomateas and Phan [9] and for a plate of arbitrary aspect ratio by Kardomateas [10]. Regarding the blast loading case, a sandwich beam/wide plate elasto-dynamic solution was developed by Kardomateas et al. [11]. The latter work was extended to the three-dimensional elasticity sandwich plate of arbitrary aspect ratio case by Kardomateas et al. [12]. Besides flat panels, an elasticity solution was developed by Kardomateas et al. [13] for the geometry of curved sandwich beams/panels. Regarding buckling, an elasticity solution for the global buckling of a sandwich beam/wide plate was presented by Kardomateas [14] and for the case of wrinkling of a sandwich beam/wide, a corresponding elasticity solution was presented by Kardomateas in [15].

The Extended High Order Sandwich Panel theory (EHSAPT) was introduced in 2012 by Phan et al. [16], and is a theory that allows for the transverse shear distribution in the core to acquire the proper distribution as the core stiffness increases as a result of non-negligible in-plane stresses in the core; thus it is valid for weak or stiff cores. This theory is an extension of the high-order sandwich panel theory [5]; its novelty is that it allows for three generalized coordinates in the core (the axial and transverse displacements at the centroid of the core, and the rotation at the centroid of the core) instead of just one (mid-point transverse displacement) commonly adopted in other available theories. The theory was formulated for a sandwich panel with a general layout. The major assumptions of the theory are as follows: (1) the face sheets satisfy the Euler-Bernoulli assumptions, and their thicknesses are small compared with the overall thickness of the sandwich section; they can be made of different materials and can have different thicknesses; they undergo large displacements with moderate rotations; (2) the core is compressible in the transverse and axial directions (transverse displacement is 2nd order in the transverse coordinate (z) and axial displacement is 3rd order in z); it has in-plane, transverse and shear rigidities; and it undergoes large displacements but with kinematic relations of small deformations due to its low in-plane rigidity as compared with that of the face sheets; and (3) the face sheets and core are perfectly bonded at their interfaces. Subsequently, the dynamic version of the Extended High-Order Sandwich Panel Theory was formulated in its full nonlinear version [17]. A simply supported sandwich beam subjected to a sinusoidal distributed blast load on the top face was studied and the results were compared to the dynamic elasticity solution in [11]; it was shown that the EHSAPT is very accurate and can capture the complex dynamic phenomena observed during the initial, transient phase of blast loading.

The EHSAPT was applied to the problem of global buckling of a sandwich wide plate/beam in [18]. Three different solution approaches were presented to investigate the effect of simplifying the loading case: (a) axial load applied exclusively to the face sheets and the geometric nonlinearities in the core are neglected (linear core); (b) uniform axial strain applied through the entire thickness and, again, linear core; and (c) uniform axial strain applied through the entire thickness but now the geometric nonlinearities in the core are included (non-linear core). The results were also compared with these from a benchmark elasticity solution [14] and, furthermore, from the simple sandwich buckling formula of Allen (thick faces version) [2] and the High Order Sandwich Panel Theory [5]. It was found that all three theories are close to the elasticity solution for "soft" cores with core over face modulus ratio less than 0.001. However, for the more "moderate" cores, i.e. with core over face modulus above 0.001, the theories diverge from each other, with the EHSAPT being the most accurate, i.e. the closest to elasticity.

A similar study was conducted on the wrinkling problem [19] and the results were, again, compared to the ones from a benchmark elasticity solution [15]. In addition, edgewise compression experiments were conducted on Glass Face/Nomex Honeycomb Core specimens and the ensuing wrinkling point was compared with the theoretical predictions. A comparison was also made with earlier edgewise compression experiments on Aluminum face/Granulated-cork core reported in literature. Other wrinkling formulas that were included in the comparison are: the Hoff-Mautner [20], and the High-Order Sandwich Panel Theory (HSAPT) [5]. In all cases the EHSAPT was the closest to both the Elasticity predictions and the experimental data. The HSAPT was in significant error for the relatively thinner faces. The large discrepancy between HSPAT and EHSAPT for very low ratios of face over total thickness (when the beam is most susceptible to wrinkling), and the associated smaller discrepancy for higher such ratios (when the beam tends to buckle globally), indicates that including the axial rigidity of the core (as in EHSAPT) is very important for wrinkling.

Recently, a linear finite element was formulated based on the EHSAPT [21]. It was proven that the finite element version of the EHSAPT constitutes a very powerful analytical tool for sandwich panels. Furthermore, the effects of geometric non-linearities were studied in detail in [22]. A critical assessment of including the various nonlinear terms in the faces and the core was conducted. The nonlinear buckling response of sandwich panels was subsequently studied [23]. It was shown that the axial rigidity of the core has a pronounced effect on both the critical load and the buckling mode. The corresponding non-linear post-buckling response was studied in [24]. It was found that due to the interaction between faces and core, localized effects may be easily initiated by imperfections after the sandwich structure has buckled globally. Furthermore, this could destabilize the post-buckling response. It was also found that the axial rigidity of the core, although it is very small compared to that of the faces, has a significant effect on the post-buckling response. This underscores the need to include it, as is done in the formulation of the EHSAPT.

Finally, the EHSAPT was also extended recently to the geometry of a curved panel [25]. Two distinct core displacement fields were proposed and investigated. One is a logarithmic (it includes terms that are linear, inverse, and logarithmic functions of the radial coordinate). The other is a polynomial (it consists of second and third order polynomials of the radial coordinate) and it is an extension of the corresponding field for the flat panel. The relative merits of these two approaches were assessed by comparing the results to an elasticity solution [13]. It was shown that the logarithmic formulation is more accurate than the polynomial especially for the stiffer cores and for curved panels of smaller radius.

In this review article, we present the basic premises, the formulation, and a series of accuracy studies for the Extended High Order Sandwich Panel Theory (EHSAPT).

2 Formulation of the Extended High Order Sandwich Panel Theory

Let us consider a sandwich panel of length a with a core of thickness $2c$ and top and bottom face sheet thicknesses f_t and f_b, respectively (Fig. 1). A Cartesian coordinate system (x,y,z) is defined at one end of the beam and its origin is placed at the middle of the core. Only loading in the x-z plane is considered to act on the beam which solely causes displacements in the x and z directions designated by u and w,

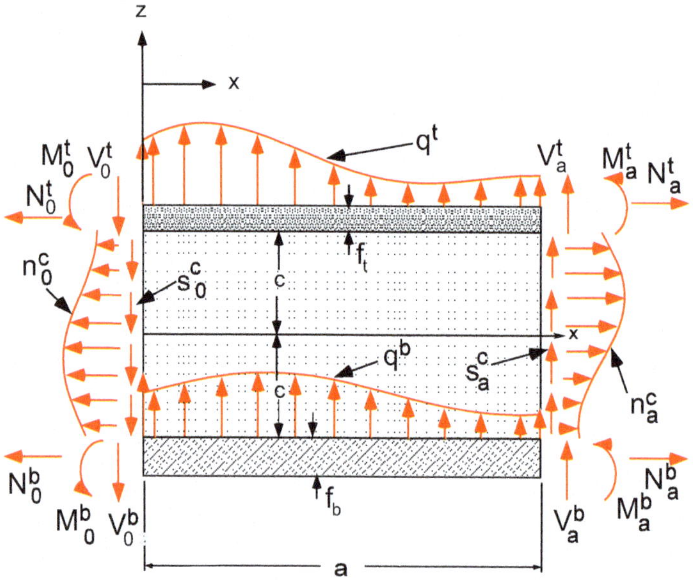

Fig. 1 Definition of the geometry and the loading

respectively. The superscripts t, b, and c shall refer to the top face sheet, bottom face sheet, and core, respectively. We should also note that in our formulation the rigidities and all applied loadings are per unit width.

The displacement field of the top and bottom face sheets are assumed to satisfy the Euler-Bernoulli assumptions: Therefore, the displacement field for the top face, $c \le z \le c + f_t$, is:

$$w^t(x,z) = w_0^t(x); \quad u^t(x,z) = u_0^t(x) - \left(z - c - \frac{f_t}{2}\right) w_{0,x}^t(x), \tag{1a}$$

and for the bottom face, $-(c + f_b) \le z \le -c$:

$$w^b(x,z) = w_0^b(x); \quad u^b(x,z) = u_0^b(x) - \left(z + c + \frac{f_b}{2}\right) w_{0,x}^b(x), \tag{1b}$$

In these relations, the sub script 0 refers to the centroid (middle surface).

The only nonzero strain in the faces is the axial strain, which in the general nonlinear case (necessary, for example, for buckling) is:

$$\varepsilon_{xx}^{t,b}(x,z) = u_{,x}^{t,b}(x,z) + \frac{1}{2}\left[w_{0,x}^{t,b}(x)\right]^2, \tag{1c}$$

If a linear analysis is pursued, the second (squared) term in (1c) is neglected.

While the face sheets can change their length only longitudinally, the core can change its height and length. The displacement fields considered for the core are of the form:

$$w^c(x,z) = w_0^c(x) + z w_1^c(x) + z^2 w_2^c(x), \tag{2a}$$

$$u^c(x,z) = u_0^c(x) + z\phi_0^c(x) + z^2 u_2^c(x) + z^3 u_3^c(x), \tag{2b}$$

where $w_0{}^c$ and $u_0{}^c$ are the transverse and in-plane displacements, respectively, $\phi_0{}^c$ is the slope at the centroid of the core, while $w_1{}^c$, $w_2{}^c$, and $u_2{}^c$, $u_3{}^c$, are transverse and in-plane unknown functions to be determined by the transverse and the in-plane compatibility conditions applied at the upper, $z = c$, and lower, $z = -c$, face-core interfaces:

$$w^c(x,c) = w_0^t(x); \quad u^c(x,c) = u_0^t(x) - \frac{f_t}{2} w_{0,x}^t(x), \tag{2c}$$

$$w^c(x,-c) = w_0^b(x); \quad u^b(x,-c) = u_0^b(x) + \frac{f_b}{2} w_{0,x}^b(x), \tag{2d}$$

Hence, using the first Eqs. in (2c, d) and Eqs. (1a) and (2a) yields the following distribution of the transverse displacement in the core:

$$w^c(x,z) = \left(-\frac{z}{2c} + \frac{z^2}{2c^2}\right) w_0^b(x) + \left(1 - \frac{z^2}{c^2}\right) w_0^c(x) + \left(\frac{z}{2c} + \frac{z^2}{2c^2}\right) w_0^t(x). \quad (3a)$$

The axial displacement of the core, $u^c(x,z)$, is determined through the fulfillment of the continuity conditions in the in-plane direction [see second Eqs. in (2c,d)]. Hence, after some algebraic manipulation:

$$\begin{aligned} u^c(x,z) = & z\left(1 - \frac{z^2}{c^2}\right)\phi_0^c(x) + \frac{z^2}{2c^2}\left(1 - \frac{z}{c}\right)u_0^b(x) + \left(1 - \frac{z^2}{c^2}\right)u_0^c(x) \\ & + \frac{z^2}{2c^2}\left(1 + \frac{z}{c}\right)u_0^t(x) + \frac{f_b z^2}{4c^2}\left(-1 + \frac{z}{c}\right)w_{0,x}^b + \frac{f_t z^2}{4c^2}\left(1 + \frac{z}{c}\right)w_{0,x}^t \,. \end{aligned} \quad (3b)$$

Therefore, this theory is in terms of seven generalized coordinates (unknown functions of x): two for the top face sheet, w_o^t, u_o^t, two for the bottom face sheet, w_o^b, u_o^b, and three for the core, w_o^c, u_o^c, and ϕ_o^c.

The strains can be obtained from the displacements using the linear strain-displacement relations. For the core, the transverse normal strain is:

$$\varepsilon_{zz}^c(x,z) = \frac{\partial w^c}{\partial z} = \left(\frac{z}{c^2} - \frac{1}{2c}\right)w_0^b(x) - \frac{2z}{c^2}w_0^c(x) + \left(\frac{z}{c^2} + \frac{1}{2c}\right)w_0^t(x), \quad (4a)$$

and the shear strain

$$\begin{aligned} \gamma_{xz}^c(x,z) = & \frac{\partial u^c}{\partial z} + \frac{\partial w^c}{\partial x} = \left(1 - \frac{3z^2}{c^2}\right)\phi_0^c(x) + \left(\frac{z}{c^2} - \frac{3z^2}{2c^3}\right)u_0^b(x) - \frac{2z}{c^2}u_0^c(x) \\ & + \left(\frac{z}{c^2} + \frac{3z^2}{2c^3}\right)u_0^t(x) + \left[-\frac{(c+f_b)}{2c^2}z + \frac{(2c+3f_b)}{4c^3}z^2\right]w_{0,x}^b \\ & + \left(1 - \frac{z^2}{c^2}\right)w_{0,x}^c + \left[\frac{(c+f_t)}{2c^2}z + \frac{(2c+3f_t)}{4c^3}z^2\right]w_{0,x}^t \,. \end{aligned} \quad (4b)$$

There is also a nonzero linear axial strain in the core $\varepsilon_{xx}{}^c = \partial u^c/\partial x$, which has the same structure as Eq. (3b), but with the generalized function coordinates replaced by one order higher derivative with respect to x.

In the following we use the notation 1 ≡ x, 3 ≡ z, and 55 ≡ xz. We assume orthotropic face sheets, thus the non-zero stresses for the faces are:

$$\sigma_{xx}^{t,b} = C_{11}^{t,b}\varepsilon_{xx}^{t,b}; \quad \sigma_{zz}^{t,b} = C_{13}^{t,b}\varepsilon_{xx}^{t,b}, \quad (5a)$$

where, in terms of the extensional (Young's) modulus, $E_1{}^{t,b}$, and the Poisson's ratio $\nu_{31}{}^{t,b}$, the stiffness constants for a beam are: $C_{11}{}^{t,b} = E_1{}^{t,b}$ and $C_{13}{}^{t,b} = \nu_{31}{}^{t,b}\, E_1{}^{t,b}$. Notice that the $\sigma_{zz}{}^{t,b}$ does not ultimately enter into the variational equation because the corresponding strain $\varepsilon_{zz}{}^{t,b}$ is assumed to be zero.

We also assume an orthotropic core with stress-strain relations:

$$\begin{bmatrix} \sigma^c_{xx} \\ \sigma^c_{zz} \\ \tau^c_{xz} \end{bmatrix} = \begin{bmatrix} C^c_{11} & C^c_{13} & 0 \\ C^c_{13} & C^c_{33} & 0 \\ 0 & 0 & C^c_{55} \end{bmatrix} \begin{bmatrix} \varepsilon^c_{xx} \\ \varepsilon^c_{zz} \\ \gamma^c_{xz} \end{bmatrix}, \tag{5b}$$

where $C_{ij}^{\ c}$ are the stiffness constants for the core. These constants are determined from the inverse of the compliance matrix. In particular, they are [26]

$$C^c_{11} = E^c_1 \frac{\left(1 - \nu^c_{23}\nu^c_{32}\right)}{C_0};\; C^c_{13} = E^c_3 \frac{\left(\nu^c_{13} - \nu^c_{12}\nu^c_{23}\right)}{C_0};\; C^c_{33} = E^c_3 \frac{\left(1 - \nu^c_{12}\nu^c_{21}\right)}{C_0}, \tag{5c}$$

$$C_0 = 1 - \left(\nu^c_{12}\nu^c_{21} + \nu^c_{23}\nu^c_{32} + \nu^c_{13}\nu^c_{31}\right) - \left(\nu^c_{12}\nu^c_{23}\nu^c_{31} + \nu^c_{21}\nu^c_{13}\nu^c_{32}\right), \tag{5d}$$

$$C^c_{55} = G^c_{31}. \tag{5e}$$

In the following we'll formulate the problem for the general case of dynamic loading. If the problem is static, the time terms in the equations just need be neglected. The governing equations and boundary conditions are derived from Hamilton's principle:

$$\int_{t_1}^{t_2} \delta(U + V - T)dt = 0, \tag{6a}$$

where U is the strain energy of the sandwich beam, V is the potential due to the applied loading, and T is the kinetic energy. The first variation of the strain energy per unit width of the sandwich beam is:

$$\delta U = \int_0^a \left[\int_{-c-f_b}^{-c} \sigma^b_{xx}\delta\varepsilon^b_{xx}dz + \int_{-c}^{c} \left(\sigma^c_{xx}\delta\varepsilon^c_{xx} + \sigma^c_{zz}\delta\varepsilon^c_{zz} + \tau^c_{xz}\delta\gamma^c_{xz}\right)dz + \int_{c}^{c+f_t} \sigma^t_{xx}\delta\varepsilon^t_{xx}dz \right] dx, \tag{6b}$$

and the first variation of the external potential per unit width is:

$$\begin{aligned} \delta V = &-\int_0^a \left(q^t\delta w^t_0 + q^b\delta w^b_0 + p^t\delta u^t_0 + p^b\delta u^b_0 + m^t\delta w^t_{0,x} + m^b\delta w^b_{0,x}\right) \\ &- \left[\int_{-c}^{c} \left(n^c\delta u^c + s^c\delta w^c\right)dz\right]_{x=0}^{a} - \left[N^t\delta u^t_0\right]_{x=0}^{a} - \left[N^b\delta u^b_0\right]_{x=0}^{a} - \left[V^t\delta w^t_0\right]_{x=0}^{a} \\ &- \left[V^b\delta w^b_0\right]_{x=0}^{a} - \left[M^t\delta w^t_{0,x}\right]_{x=0}^{a} - \left[M^b\delta w^b_{0,x}\right]_{x=0}^{a}, \end{aligned} \tag{6c}$$

where, by reference to Fig. 1, $q^{t,b}$ is the distributed transverse (along z) force per unit width, $p^{t,b}$ is the distributed in-plane (along x) force per unit width and $m^{t,b}$ is the distributed moment per unit width on the top and bottom faces. Moreover $N^{t,b}$ is the end axial force per unit width, $V^{t,b}$ is the end shear force per unit width and $M^{t,\,b}$ is the end moment per unit width at the top and bottom face sheets, at the ends $x = 0,\, a$. In addition, n^c is the end axial force per unit width and s^c is the end shear force per unit width at the core, at the ends $x = 0,\, a$.

In the following, we assume that n^c and s^c are constant. In this case,

$$\int_{-c}^{c} n^c \delta u^c dz = n^c c\left[\frac{1}{3}\left(\delta u_0^b + \delta u_0^t\right) + \frac{4}{3}\delta u_0^c - \frac{f_b}{6}\delta w_{0,x}^b + \frac{f_t}{6}\delta w_{0,x}^t\right], \tag{6d}$$

$$\int_{-c}^{c} s^c \delta w^c dz = s^c c\left[\frac{1}{3}\left(\delta w_0^b + \delta w_0^t\right) + \frac{4}{3}\delta w_0^c\right]. \tag{6e}$$

Of course, the theory can admit any variation of n^c and s^c along z; for example, a bending moment on the core would correspond to a linear variation of n^c with respect to z. However, for most practical purposes, loads are applied to the skins and not the core.

The kinetic energy term is:

$$\delta T = \int_0^a \left[\int_{(c+f_b)}^{-c} \rho^b\left(\dot{u}^b \delta\dot{u}^b + \dot{w}^b \delta\dot{w}^b\right)dz + \int_{-c}^{c} \rho^c\left(\dot{u}^c \delta\dot{u}^c + \dot{w}^c \delta\dot{w}^c\right)dz \right.$$
$$\left. + \int_{c}^{c+f_t} \rho^t\left(\dot{u}^t \delta\dot{u}^t + \dot{w}^t \delta\dot{w}^t\right)dz\right] dx. \tag{6f}$$

For the sandwich plates made out of orthotropic materials, we can substitute the stresses in terms of the strains from the constitutive relations, Eqs. (5a), and then the strains in terms of the displacements and the displacement profiles, Eqs. (1–4), and finally apply the variational principle, Eqs. (6a); thus we can write a set of non-linear governing differential Eqs. (D.Es) in terms of the seven unknown generalized coordinates as follows:

Top face sheet D.Es (two nonlinear):

$$\delta u_0^t: \quad -\left(\frac{4}{5}C_{55}^c + \frac{2c^2}{35}C_{11}^c\frac{\partial^2}{\partial x^2}\right)\phi_0^c - \left(\frac{7}{30c}C_{55}^c + \frac{c}{35}C_{11}^c\frac{\partial^2}{\partial x^2}\right)u_0^b$$
$$-\left(\frac{4}{3c}C_{55}^c + \frac{2c}{15}C_{11}^c\frac{\partial^2}{\partial x^2}\right)u_0^c + \left[\frac{47}{30c}C_{55}^c - \alpha_1^t\frac{\partial^2}{\partial x^2}\right]u_0^t$$
$$-\left(\alpha_2^b\frac{\partial}{\partial x} - \frac{cf_b}{70}C_{11}^c\frac{\partial^3}{\partial x^3}\right)w_0^b + \left(\beta_1\frac{\partial}{\partial x}\right)w_0^c + \left(\alpha_3^t\frac{\partial}{\partial x} - \frac{3cf_t}{35}C_{11}^c\frac{\partial^3}{\partial x^3}\right)w_0^t$$
$$+D_u^t = p^t + F_u^t, \tag{7a}$$

where $F_u^{\,t}$ is the nonlinear term:

$$F_u^t = C_{11}^t f_t \, w_{0,x}^t \, w_{0,xx}^t, \tag{7b}$$

and $D_u^{\,t}$ is the dynamic term:

$$\begin{aligned} D_u^t = & \rho^c \frac{2c^2}{35}\frac{\partial^2 \phi_0^c}{\partial t^2} + \rho^c \frac{c}{35}\frac{\partial^2 u_0^b}{\partial t^2} + \rho^c \frac{2c}{15}\frac{\partial^2 u_0^c}{\partial t^2} + \left(\rho^c \frac{6c}{35} + \rho^t f_t\right)\frac{\partial^2 u_0^t}{\partial t^2} \\ & - \rho^c \frac{c f_b}{70}\frac{\partial^3 w_0^b}{\partial x \partial t^2} + \rho^c \frac{3 c f_t}{35}\frac{\partial^3 w_0^t}{\partial x \partial t^2}. \end{aligned} \tag{7c}$$

and

$$\begin{aligned} \delta w_0^t: \quad & \left(\alpha_4^t \frac{\partial}{\partial x} + \frac{c^2 f_t}{35} C_{11}^c \frac{\partial^3}{\partial x^3}\right)\phi_0^c + \left(\alpha_5^t \frac{\partial}{\partial x} + \frac{c f_t}{70} C_{11}^c \frac{\partial^3}{\partial x^3}\right) u_0^b \\ & + \left(\alpha_6^t \frac{\partial}{\partial x} + \frac{c f_t}{15} C_{11}^c \frac{\partial^3}{\partial x^3}\right) u_0^c + \left(-\alpha_3^t \frac{\partial}{\partial x} + \frac{3 c f_t}{35} C_{11}^c \frac{\partial^3}{\partial x^3}\right) u_0^t \\ & + \left(\frac{1}{6c} C_{33}^c + \beta_2 \frac{\partial^2}{\partial x^2} - \frac{c f_b f_t}{140} C_{11}^c \frac{\partial^4}{\partial x^4}\right) w_0^b + \left(-\frac{4}{3c} C_{33}^c + \alpha_7^t \frac{\partial^2}{\partial x^2}\right) w_0^c \\ & + \left(\frac{7}{6c} C_{33}^c + \alpha_8^t \frac{\partial^2}{\partial x^2} + \alpha_9^t \frac{\partial^4}{\partial x^4}\right) w_0^t + D_w^t = q^t - m_{,x}^t + F_w^t, \end{aligned} \tag{8a}$$

where $F_w^{\,t}$ is the nonlinear term:

$$F_w^t = C_{11}^t f_t \left[w_{0,x}^t u_{0,xx}^t + u_{0,x}^t w_{0,xx}^t + \frac{3}{2}\left(w_{0,x}^t\right)^2 w_{0,xx}^t \right], \tag{8b}$$

and $D_w^{\,t}$ is the dynamic term:

$$\begin{aligned} D_w^t = & -\rho^c \frac{f_t c^2}{35}\frac{\partial^3 \phi_0^c}{\partial x \partial t^2} - \rho^c \frac{f_t c}{70}\frac{\partial^3 u_0^b}{\partial x \partial t^2} - \rho^c \frac{f_t c}{15}\frac{\partial^3 u_0^c}{\partial x \partial t^2} - \rho^c \frac{3 f_t c}{35}\frac{\partial^3 u_0^t}{\partial x \partial t^2} \\ & + \left(-\rho^c \frac{c}{15}\frac{\partial^2}{\partial t^2} + \rho^c \frac{c f_b f_t}{140}\frac{\partial^4}{\partial x^2 \partial t^2}\right) w_0^b + \rho^c \frac{2c}{15}\frac{\partial^2 w_0^c}{\partial t^2} \\ & + \left[\left(\rho^c \frac{4c}{15} + \rho^t f_t\right)\frac{\partial^2}{\partial t^2} - \left(\rho^c \frac{3 c f_t^2}{70} + \rho^t \frac{f_t^3}{12}\right)\frac{\partial^4}{\partial x^2 \partial t^2}\right] w_0^t. \end{aligned} \tag{8c}$$

Core D.Es (three linear):

$$\delta u_0^c: \quad -\left(\frac{4}{3c}C_{55}^c + \frac{2c}{15}C_{11}^c\frac{\partial^2}{\partial x^2}\right)u_0^b + \left(\frac{8}{3c}C_{55}^c - \frac{16c}{15}C_{11}^c\frac{\partial^2}{\partial x^2}\right)u_0^c$$
$$-\left(\frac{4}{3c}C_{55}^c + \frac{2c}{15}C_{11}^c\frac{\partial^2}{\partial x^2}\right)u_0^t + \left(\alpha_6^b\frac{\partial}{\partial x} + \frac{cf_b}{15}C_{11}^c\frac{\partial^3}{\partial x^3}\right)w_0^b \tag{9a}$$
$$-\left(\alpha_6^t\frac{\partial}{\partial x} + \frac{cf_t}{15}C_{11}^c\frac{\partial^3}{\partial x^3}\right)w_0^t + D_u^c = 0,$$

where $D_u^{\ c}$ is the dynamic term:

$$D_u^c = \rho^c\left[\frac{2c}{15}\frac{\partial^2 u_0^b}{\partial t^2} + \frac{16c}{15}\frac{\partial^2 u_0^c}{\partial t^2} + \frac{2c}{15}\frac{\partial^2 u_0^t}{\partial t^2} - \frac{cf_b}{15}\frac{\partial^3 w_0^b}{\partial x\partial t^2} + \frac{cf_t}{15}\frac{\partial^3 w_0^t}{\partial x\partial t^2}\right]. \tag{9b}$$

$$\delta\phi_0^c: \quad \left(\frac{8c}{5}C_{55}^c - \frac{16c^3}{105}C_{11}^c\frac{\partial^2}{\partial x^2}\right)\phi_0^c + \left(\frac{4}{5}C_{55}^c + \frac{2c^2}{35}C_{11}^c\frac{\partial^2}{\partial x^2}\right)u_0^b$$
$$-\left(\frac{4}{5}C_{55}^c + \frac{2c^2}{35}C_{11}^c\frac{\partial^2}{\partial x^2}\right)u_0^t - \left(\alpha_4^b\frac{\partial}{\partial x} + \frac{c^2 f_b}{35}C_{11}^c\frac{\partial^3}{\partial x^3}\right)w_0^b + \left(\beta_3\frac{\partial}{\partial x}\right)w_0^c$$
$$-\left(\alpha_4^t\frac{\partial}{\partial x} + \frac{c^2 f_t}{35}C_{11}^c\frac{\partial^3}{\partial x^3}\right)w_0^t + D_\phi^c = 0, \tag{10a}$$

where $D_\phi^{\ c}$ is the dynamic term:

$$D_\phi^c = \rho^c\left[\frac{16c^3}{105}\frac{\partial^2\phi_0^c}{\partial t^2} - \frac{2c^2}{35}\frac{\partial^2 u_0^b}{\partial t^2} + \frac{2c^2}{35}\frac{\partial^2 u_0^t}{\partial t^2} + \frac{f_b c^2}{35}\frac{\partial^3 w_0^b}{\partial x\partial t^2} + \frac{f_t c^2}{35}\frac{\partial^3 w_0^t}{\partial x\partial t^2}\right], \tag{10b}$$

and

$$\delta w_0^c: \quad -\left(\beta_3\frac{\partial}{\partial x}\right)\phi_0^c + \left(\beta_1\frac{\partial}{\partial x}\right)u_0^b - \left(\beta_1\frac{\partial}{\partial x}\right)u_0^t + \left(-\frac{4}{3c}C_{33}^c + \alpha_7^b\frac{\partial^2}{\partial x^2}\right)w_0^b$$
$$+\left(\frac{8}{3c}C_{33}^c - \frac{16c}{15}C_{55}^c\frac{\partial^2}{\partial x^2}\right)w_0^c + \left(-\frac{4}{3c}C_{33}^c + \alpha_7^t\frac{\partial^2}{\partial x^2}\right)w_0^t + D_w^c = 0, \tag{11a}$$

where $D_w{}^c$ is the dynamic term:

$$D_w^c = \rho^c \left[\frac{2c}{15} \frac{\partial^2 w_0^b}{\partial t^2} + \frac{16c}{15} \frac{\partial^2 w_0^c}{\partial t^2} + \frac{2c}{15} \frac{\partial^2 w_0^t}{\partial t^2} \right]. \tag{11b}$$

Bottom face sheet D.Es (two nonlinear):

$$\begin{aligned} \delta u_0^b: &\left(\frac{4}{5} C_{55}^c + \frac{2c^2}{35} C_{11}^c \frac{\partial^2}{\partial x^2} \right) \phi_0^c + \left[\frac{47}{30c} C_{55}^c - \alpha_1^b \frac{\partial^2}{\partial x^2} \right] u_0^b \\ &- \left(\frac{4}{3c} C_{55}^c + \frac{2c}{15} C_{11}^c \frac{\partial^2}{\partial x^2} \right) u_0^c - \left(\frac{7}{30c} C_{55}^c + \frac{c}{35} C_{11}^c \frac{\partial^2}{\partial x^2} \right) u_0^t \\ &+ \left(-\alpha_3^b \frac{\partial}{\partial x} + \frac{3cf_b}{35} C_{11}^c \frac{\partial^3}{\partial x^3} \right) w_0^b - \left(\beta_1 \frac{\partial}{\partial x} \right) w_0^c + \left(\alpha_2^t \frac{\partial}{\partial x} - \frac{cf_t}{70} C_{11}^c \frac{\partial^3}{\partial x^3} \right) w_0^t \\ &+ D_u^b = p^b + F_u^b, \end{aligned} \tag{12a}$$

where $F_u{}^b$ is the nonlinear term:

$$F_u^b = C_{11}^b f_b w_{0,x}^b w_{0,xx}^b, \tag{12b}$$

and $D_u{}^b$ is the dynamic term:

$$\begin{aligned} D_u^b = &-\rho^c \frac{2c^2}{35} \frac{\partial^2 \phi_0^c}{\partial t^2} + \left(\rho^c \frac{6c}{35} + \rho^b f_b \right) \frac{\partial^2 u_0^b}{\partial t^2} + \rho^c \frac{2c}{15} \frac{\partial^2 u_0^c}{\partial t^2} + \rho^c \frac{c}{35} \frac{\partial^2 u_0^t}{\partial t^2} \\ &-\rho^c \frac{3f_b c}{35} \frac{\partial^3 w_0^b}{\partial x \partial t^2} + \rho^c \frac{f_t c}{70} \frac{\partial^3 w_0^t}{\partial x \partial t^2}, \end{aligned} \tag{12c}$$

and

$$\begin{aligned} \delta w_0^b: &\left(\alpha_4^b \frac{\partial}{\partial x} + \frac{c^2 f_b}{35} C_{11}^c \frac{\partial^3}{\partial x^3} \right) \phi_0^c + \left(\alpha_3^b \frac{\partial}{\partial x} - \frac{3cf_b}{35} C_{11}^c \frac{\partial^3}{\partial x^3} \right) u_0^b \\ &- \left(\alpha_6^b \frac{\partial}{\partial x} + \frac{cf_b}{15} C_{11}^c \frac{\partial^3}{\partial x^3} \right) u_0^c - \left(\alpha_5^b \frac{\partial}{\partial x} + \frac{cf_b}{70} C_{11}^c \frac{\partial^3}{\partial x^3} \right) u_0^t \\ &+ \left(\frac{7}{6c} C_{33}^c + \alpha_8^b \frac{\partial^2}{\partial x^2} + \alpha_9^b \frac{\partial^4}{\partial x^4} \right) w_0^b + \left(-\frac{4}{3c} C_{33}^c + \alpha_7^b \frac{\partial^2}{\partial x^2} \right) w_0^c \\ &+ \left(\frac{1}{6c} C_{33}^c + \beta_2 \frac{\partial^2}{\partial x^2} - \frac{cf_b f_t}{140} C_{11}^c \frac{\partial^4}{\partial x^4} \right) w_0^t + D_w^b = q^b - m_{,x}^b + F_w^b, \end{aligned} \tag{13a}$$

where ${F_w}^b$ is the nonlinear term:

$$F_w^b = C_{11}^b f_b \left[w_{0,x}^b u_{0,xx}^b + u_{0,x}^b w_{0,xx}^b + \frac{3}{2} \left(w_{0,x}^b \right)^2 w_{0,xx}^b \right], \tag{13b}$$

and ${D_w}^b$ is the dynamic term:

$$\begin{aligned} D_w^b = & -\rho^c \frac{f_b c^2}{35} \frac{\partial^3 \phi_0^c}{\partial x \partial t^2} + \rho^c \frac{3 f_b c^2}{35} \frac{\partial^3 u_0^b}{\partial x \partial t^2} + \rho^c \frac{f_b c}{15} \frac{\partial^3 u_0^c}{\partial x \partial t^2} + \rho^c \frac{f_b}{70} \frac{\partial^3 u_0^t}{\partial x \partial t^2} \\ & + \left[\left(\rho^c \frac{4c}{15} + \rho^b f_b \right) \frac{\partial^2}{\partial t^2} - \left(\rho^c \frac{3 f_b^2 c}{70} + \rho^b \frac{f_b^3}{12} \right) \frac{\partial^4}{\partial x^2 \partial t^2} \right] w_0^b \\ & + \rho^c \frac{2c}{15} \frac{\partial^2 w_0^c}{\partial t^2} + \rho^c \left(- \frac{c}{15} \frac{\partial^2}{\partial t^2} + \frac{c f_b f_t}{140} \frac{\partial^4}{\partial x^2 \partial t^2} \right) w_0^t. \end{aligned} \tag{13c}$$

In the above expressions, the following constants are defined:

$$\alpha_1^i = \frac{6c}{35} C_{11}^c + f_i C_{11}^i; \quad \alpha_2^i = \frac{1}{30} C_{13}^c + \left(\frac{1}{30} - \frac{7 f_i}{60c} \right) C_{55}^c, \tag{14a}$$

$$\alpha_3^i = -\frac{11}{30} C_{13}^c + \left(\frac{19}{30} + \frac{47 f_i}{60c} \right) C_{55}^c; \quad \alpha_4^i = \frac{4c}{15} C_{13}^c + \left(\frac{4c}{15} + \frac{2 f_i}{5} \right) C_{55}^c, \tag{14b}$$

$$\alpha_5^i = -\alpha_2^i; \quad \alpha_6^i = \frac{2}{3} C_{13}^c + \left(\frac{2}{3} + \frac{2 f_i}{3c} \right) C_{55}^c; \ \alpha_7^i = -\frac{f_i}{5} C_{13}^c - \left(\frac{2c}{15} + \frac{f_i}{5} \right) C_{55}^c; \tag{14c}$$

$$\alpha_8^i = \frac{11 f_i}{30} C_{13}^c - \left(\frac{4c}{15} + \frac{19 f_i}{30} + \frac{47 f_i^2}{120c} \right) C_{55}^c; \quad \alpha_9^i = \frac{f_i^3}{12} C_{11}^i + \frac{3 c f_i^2}{70} C_{11}^c, \tag{14d}$$

and

$$\beta_1 = \frac{2}{5} \left(C_{13}^c + C_{55}^c \right); \quad \beta_3 = \frac{8c}{15} \left(C_{13}^c + C_{55}^c \right), \tag{14e}$$

$$\beta_2 = \frac{f_b + f_t}{60} C_{13}^c + \left(\frac{c}{15} + \frac{f_b + f_t}{60} - \frac{7 f_b f_t}{120c} \right) C_{55}^c. \tag{14f}$$

The corresponding boundary conditions (B.Cs) at $x = 0, a$, read as follows (at each end there are nine boundary conditions, three for each face sheet and three for the core):

Top face sheet B.Cs (three):

(i) *Either* $\delta u_0{}^t = 0$ *or,*

$$\left(\frac{2c^2}{35}C_{11}^c\frac{\partial}{\partial x}\right)\phi_0^c + \left(\frac{c}{35}C_{11}^c\frac{\partial}{\partial x}\right)u_0^b + \left(\frac{2c}{15}C_{11}^c\frac{\partial}{\partial x}\right)u_0^c + \left(\alpha_1^t\frac{\partial}{\partial x}\right)u_0^t - \left(\frac{2}{5}C_{13}^c\right)w_0^c$$
$$+\left(\frac{1}{30}C_{13}^c - \frac{cf_b}{70}C_{11}^c\frac{\partial^2}{\partial x^2}\right)w_0^b + \left(\frac{11}{30}C_{13}^c + \frac{3cf_t}{35}C_{11}^c\frac{\partial^2}{\partial x^2}\right)w_0^t = N^t + \frac{n^c c}{3} + B_u^t, \tag{15a}$$

where N^t is the end axial force per unit width at the top face and n^c is the (uniformly distributed) end axial force per unit width at the core (at the end $x = 0$ or $x = a$) and $B_u{}^t$ is the nonlinear term

$$B_u^t = -\frac{f_t}{2}C_{11}^t\left(w_{0,x}^t\right)^2. \tag{15b}$$

(ii) *Either* $\delta w_0{}^t = 0$ *or,*

$$-\left[\frac{2(2c+3f_t)}{15}C_{55}^c + \frac{c^2 f_t}{35}C_{11}^c\frac{\partial^2}{\partial x^2}\right]\phi_0^c + \left[\frac{(2c-7f_t)}{60c}C_{55}^c - \frac{cf_t}{70}C_{11}^c\frac{\partial^2}{\partial x^2}\right]u_0^b$$
$$-\left[\frac{2(c+f_t)}{3c}C_{55}^c + \frac{cf_t}{15}C_{11}^c\frac{\partial^2}{\partial x^2}\right]u_0^c + \left[\frac{(38c+47f_t)}{60c}C_{55}^c - \frac{3cf_t}{35}C_{11}^c\frac{\partial^2}{\partial x^2}\right]u_0^t$$
$$+\left[\left(\frac{f_b}{60}C_{13}^c - \beta_2\right)\frac{\partial}{\partial x} + \frac{cf_b f_t}{140}C_{11}^c\frac{\partial^3}{\partial x^4}\right]w_0^b - \left(\alpha_7^t\frac{\partial}{\partial x}\right)w_0^c$$
$$+\left[\left(\frac{11f_t}{60}C_{13}^c - \alpha_8^t\right)\frac{\partial}{\partial x} - \alpha_9^t\frac{\partial^3}{\partial x^3}\right]w_0^t = V^t + m^t + \frac{s^c c}{3} + B_w^t + L_w^t, \tag{16a}$$

where V^t is the end shear force per unit width at the top face and s^c is the (assumed constant) end shear force per unit width at the core (at the end $x = 0$ or $x = a$) and $B_w{}^t$ is the nonlinear term:

$$B_w^t = -\frac{f_t}{2}C_{11}^t w_{0,x}^t\left[2u_{0,x}^t + \left(w_{0,x}^t\right)^2\right], \tag{16b}$$

and $L_w{}^t$ is the dynamic term:

$$L_w^t = \frac{f_t}{420}\left[\begin{array}{c}\rho^c 12c^2\dfrac{\partial^2\phi_0^c}{\partial t^2} + \rho^c 6c\dfrac{\partial^2 u_0^b}{\partial t^2} + \rho^c 28c\dfrac{\partial^2 u_0^c}{\partial t^2} + \rho^c 36c\dfrac{\partial^2 u_0^t}{\partial t^2} \\ -\rho^c 3cf_b\dfrac{\partial^3 w_0^b}{\partial x\partial t^2} + \left(\rho^c 18cf_t + \rho^t 35f_t^2\right)\dfrac{\partial^3 w_0^t}{\partial x\partial t^2}\end{array}\right]. \tag{16c}$$

(iii) *Either* $\delta w_{0,x}{}^{t} = 0$ *or,*

$$\left(\frac{c^2 f_t}{35} C_{11}^c \frac{\partial}{\partial x}\right)\phi_0^c + \left(\frac{c f_t}{70} C_{11}^c \frac{\partial}{\partial x}\right)u_0^b + \left(\frac{c f_t}{15} C_{11}^c \frac{\partial}{\partial x}\right)u_0^c + \left(\frac{3 c f_t}{35} C_{11}^c \frac{\partial}{\partial x}\right)u_0^t$$
$$+\left(\frac{f_t}{60} C_{13}^c - \frac{c f_b f_t}{140} C_{11}^c \frac{\partial^2}{\partial x^2}\right) w_0^b - \left(\frac{f_t}{5} C_{13}^c\right) w_0^c + \left(\frac{11 f_t}{60} C_{13}^c + \alpha_9^t \frac{\partial^2}{\partial x^2}\right) w_0^t$$
$$= M^t + \frac{n^c c f_t}{6}, \tag{17}$$

where M^t is the end moment per unit width at the top face (at the end $x = 0$ or $x = a$).

Core B.Cs (three):

(i) *Either* $\delta u_0{}^{c} = 0$ *or,*

$$\left(\frac{2c}{15} C_{11}^c \frac{\partial}{\partial x}\right)u_0^b + \left(\frac{16c}{15} C_{11}^c \frac{\partial}{\partial x}\right)u_0^c + \left(\frac{2c}{15} C_{11}^c \frac{\partial}{\partial x}\right)u_0^t$$
$$-\left(\frac{2}{3} C_{13}^c + \frac{c f_b}{15} C_{11}^c \frac{\partial^2}{\partial x^2}\right) w_0^b + \left(\frac{2}{3} C_{13}^c + \frac{c f_t}{15} C_{11}^c \frac{\partial^2}{\partial x^2}\right) w_0^t = \frac{4 n^c c}{3}. \tag{18}$$

(ii) *Either* $\delta \phi_0{}^{c} = 0$ *or,*

$$\left(\frac{16c^3}{105} C_{11}^c \frac{\partial}{\partial x}\right)\phi_0^c - \left(\frac{2c^2}{35} C_{11}^c \frac{\partial}{\partial x}\right)u_0^b + \left(\frac{2c^2}{35} C_{11}^c \frac{\partial}{\partial x}\right)u_0^t$$
$$+\left(\frac{4c}{15} C_{13}^c + \frac{c^2 f_b}{35} C_{11}^c \frac{\partial^2}{\partial x^2}\right) w_0^b - \left(\frac{8c}{15} C_{13}^c\right) w_0^c + \left(\frac{4c}{15} C_{13}^c + \frac{c^2 f_t}{35} C_{11}^c \frac{\partial^2}{\partial x^2}\right) w_0^t = 0. \tag{19}$$

(iii) *Either* $\delta w_0{}^{c} = 0$ *or,*

$$C_{55}^c \left[\frac{8c}{15}\phi_0^c - \frac{2}{5} u_0^b + \frac{2}{5} u_0^t + \frac{(2c + 3f_b)}{15} w_{0,x}^b + \frac{16c}{15} w_{0,x}^c + \frac{(2c + 3f_t)}{15} w_{0,x}^t\right] = \frac{4}{3} s^c c. \tag{20}$$

Bottom face sheet B.Cs (three):

(i) *Either* $\delta {u_0}^b = 0$ *or,*

$$-\left(\frac{2c^2}{35}C_{11}^c\frac{\partial}{\partial x}\right)\phi_0^c+\left(\alpha_1^b\frac{\partial}{\partial x}\right)u_0^b+\left(\frac{2c}{15}C_{11}^c\frac{\partial}{\partial x}\right)u_0^c+\left(\frac{c}{35}C_{11}^c\frac{\partial}{\partial x}\right)u_0^t$$
$$+\left(\frac{2}{3}C_{13}^c\right)w_0^c-\left(\frac{11}{30}C_{13}^c+\frac{3cf_b}{35}C_{11}^c\frac{\partial^2}{\partial x^2}\right)w_0^b \tag{21a}$$
$$+\left(-\frac{1}{30}C_{13}^c+\frac{cf_t}{70}C_{11}^c\frac{\partial^2}{\partial x^2}\right)w_0^t=N^b+\frac{n^c c}{3}+B_u^b,$$

where N^b is the end axial force per unit width at the bottom face and ${B_u}^b$ is the nonlinear term,

$$B_u^b=-\frac{f_b}{2}C_{11}^b\left(w_{0,x}^b\right)^2. \tag{21b}$$

(ii) *Either* $\delta {w_0}^b = 0$ *or,*

$$-\left[\frac{2(2c+3f_b)}{15}C_{55}^c+\frac{c^2f_b}{35}C_{11}^c\frac{\partial^2}{\partial x^2}\right]\phi_0^c+\left[-\frac{(38c-47f_b)}{60c}C_{55}^c+\frac{3cf_b}{35}C_{11}^c\frac{\partial^2}{\partial x^2}\right]u_0^b$$
$$+\left[\frac{2(c+f_b)}{3c}C_{55}^c+\frac{cf_b}{15}C_{11}^c\frac{\partial^2}{\partial x^2}\right]u_0^c+\left[\frac{(-2c+7f_b)}{60c}C_{55}^c+\frac{cf_b}{70}C_{11}^c\frac{\partial^2}{\partial x^2}\right]u_0^t$$
$$-\left(\alpha_7^b\frac{\partial}{\partial x}\right)w_0^c+\left[\left(\frac{11f_b}{60}C_{13}^c-\alpha_8^b\right)\frac{\partial}{\partial x}-\alpha_9^b\frac{\partial^3}{\partial x^3}\right]w_0^b$$
$$+\left[\left(\frac{f_t}{60}C_{13}^c-\beta_2\right)\frac{\partial}{\partial x}+\frac{cf_bf_t}{140}C_{11}^c\frac{\partial^3}{\partial x^3}\right]w_0^t+L_w^b$$
$$=V_0^b+m^b+\frac{s^c c}{3}+B_w^b, \tag{22a}$$

where V^b is the end shear force per unit width at bottom face and ${B_w}^b$ is the nonlinear term:

$$B_w^b=-\frac{f_b}{2}C_{11}^b w_{0,x}^b\left[2u_{0,x}^b+\left(w_{0,x}^b\right)^2\right], \tag{22b}$$

and $L_w^{\,b}$ is the dynamic term:

$$L_w^b = \frac{f_b}{420}\left[\begin{array}{c}\rho^c 12c^2\dfrac{\partial^2\phi_0^c}{\partial t^2} - \rho^c 36c\dfrac{\partial^2 u_0^b}{\partial t^2} - \rho^c 28c\dfrac{\partial^2 u_0^c}{\partial t^2} - \rho^c 6c\dfrac{\partial^2 u_0^t}{\partial t^2} \\ +\left(\rho^c 18cf_b + \rho^b 35f_b^2\right)\dfrac{\partial^3 w_0^b}{\partial x\partial t^2} - \rho^c 3cf_t\dfrac{\partial^3 w_0^t}{\partial x\partial t^2}\end{array}\right]. \tag{22c}$$

(iii) *Either* $\delta w_{0,x}{}^{\mathrm{b}} = 0$ *or,*

$$\begin{aligned}&\left(\frac{c^2 f_b}{35}C_{11}^c\frac{\partial}{\partial x}\right)\phi_0^c - \left(\frac{3cf_b}{70}C_{11}^c\frac{\partial}{\partial x}\right)u_0^b - \left(\frac{cf_b}{15}C_{11}^c\frac{\partial}{\partial x}\right)u_0^c - \left(\frac{cf_b}{70}C_{11}^c\frac{\partial}{\partial x}\right)u_0^t \\ &-\left(\frac{f_b}{5}C_{13}^c\right)w_0^c + \left(\frac{11f_b}{60}C_{13}^c + \alpha_9^b\frac{\partial^2}{\partial x^2}\right)w_0^b + \left(\frac{f_b}{60}C_{13}^c - \frac{cf_b f_t}{140}C_{11}^c\frac{\partial^2}{\partial x^2}\right)w_0^t \\ &= M^b - \frac{n^c cf_b}{6},\end{aligned} \tag{23}$$

where M^b is the end moment per unit width at the bottom face.

Hamilton's principle results in 7 coupled partial differential equations, Eqs. (7a) to (13a), four of which are nonlinear due to the consideration of nonlinear axial strains in the face sheets. The order of the equations of motion is 18. Therefore, there are 18 boundary conditions, 9 at each end at $x = 0$ and $x = a$, given by Eqs. (15a) to (23). Notice that since the rotations of the face sheets are assumed to be the derivative of the transverse displacement with respect to x, there exist inertial terms $L_w^{\,t}$ and $L_w^{\,b}$ in the boundary conditions in equations, (16a) and (22a). The 7 unknowns of EHSAPT are: $u_0^t(x,t)$, $u_0^c(x,t)$, $u_0^b(x,t)$, $\phi_0^c(x,t)$, $w_0^t(x,t)$, $w_0^c(x,t)$ and $w_0^b(x,t)$.

3 Accuracy Study I: A Statically Loaded Simply Supported Sandwich Panel

In this section we shall study the linear response of a simply supported sandwich panel under transversely applied loading of the form:

$$q^t(x) = q_0 \sin\frac{\pi x}{a}. \tag{24a}$$

In this case, the boundary conditions for $x = 0$, a (Fig. 1) are the three kinematic conditions

$$w_0^t = w_0^b = w_0^c = 0. \tag{24b}$$

and the 6 natural boundary conditions in (15), (17), (18), (19), (21) and (23).

All these are satisfied by displacements in the form:

$$u_0^{t,b,c} = U_0^{t,b,c} \cos \frac{\pi x}{a}; \qquad w_0^{t,b,c} = W_0^{t,b,c} \sin \frac{\pi x}{a}; \qquad \phi_0^c = \Phi_0^c \cos \frac{\pi x}{a}. \tag{24c}$$

We consider the linear problem, which means that the nonlinear terms $F_{u,w}{}^{t,b}$ in the governing differential Eqs and the nonlinear terms $B_w{}^{t,b}$ in the boundary conditions are neglected.

Substituting Eq. (25a) into Eq. (10–14) results in a system of seven linear equations for the seven unknown constants $U_0{}^t$, $U_0{}^c$, $U_0{}^b$, $\Phi_0{}^c$, $W_0{}^t$, $W_0{}^c$, and $W_0{}^b$.

We shall consider sandwich configurations where the two face sheets are assumed identical with thickness $f_t = f_b = f = 2$ mm. The core thickness is $2c = 16$ mm. The total thickness of the beam/plate is defined as $h_{tot} = 2f + 2c$ and the length of the beam is $a = 20h_{tot}$.

Regarding materials, the faces are made out of graphite-epoxy with moduli (GPa): $E_1{}^f = 181.0$; $E_2{}^f = E_3{}^f = 10.3$; $G_{23}{}^f = 5.96$; $G_{31}{}^f = G_{12}{}^f = 7.17$ and Poisson's ratios: $\nu_{32}{}^f = 0.40$; $\nu_{31}{}^f = 0.016$; $\nu_{12}{}^f = 0.277$.

The core is made out of glass-phenolic honeycomb with moduli (GPa): $E_1{}^c = E_2{}^c = 0.032$; $E_3{}^f = 0.300$; $G_{23}{}^c = G_{31}{}^c = 0.048$; $G_{12}{}^c = 0.013$ and Poisson's ratios: $\nu_{32}{}^c = \nu_{31}{}^c = \nu_{12}{}^c = 0.25$.

In the following results, the displacements are normalized with $w_{norm} = 3q_0a^4/(2\pi^4 f^3 E_f)$ and the stresses with q_0.

Plotted in Fig. 2 is the normalized displacement at the top face sheet as a function of x. In this figure, we also show the predictions of the simple Classical beam theory, which does not include transverse shear, as well as the First Order Shear theories; for the latter, there are two versions: one that is based only on the core shear stiffness and one that includes the face sheet stifnesses. In addition, we show the predictions of the High Order sandwich panel theory [5]. This theory, which is based on an assumption that the in-plane rigidity of the core is neglected and yields a constant shear stress and zero axial stress in the core. Finally, we also show the predictions from a finite element method (FEM) study [21].

From Fig. 2, we can see that both the Classical and First Order Shear (both versions) seem to be inadequate. The Classical theory is too non-conservative and the First Order Shear theory with face sheets included can hardly make a difference. On the other hand, the First Order Shear theory where shear is assumed to be carried exclusively by the core is too conservative; this clearly demonstrates the need for higher order theories in dealing with sandwich structures. In this regard, both the Frostig et al. [5] and the Extended High Order theories give a displacement profile which is essentially identical to the Elasticity solution. In Fig. 2 we can also readily observe the large effect of transverse shear, which is an important feature of sandwich structures.

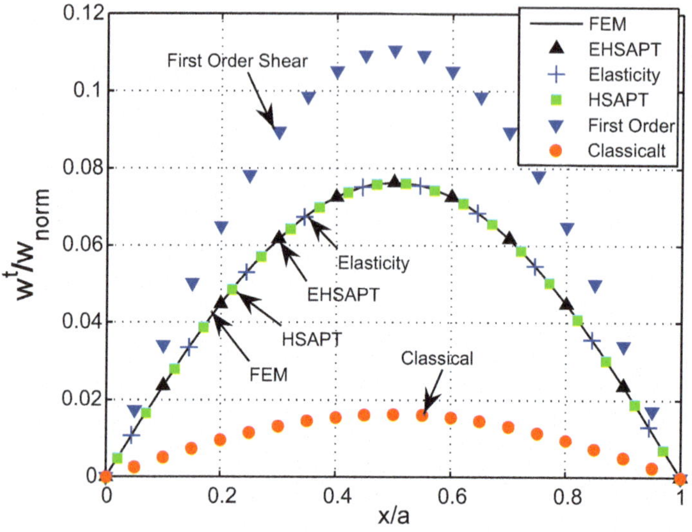

Fig. 2 Transverse displacement, w, at the top

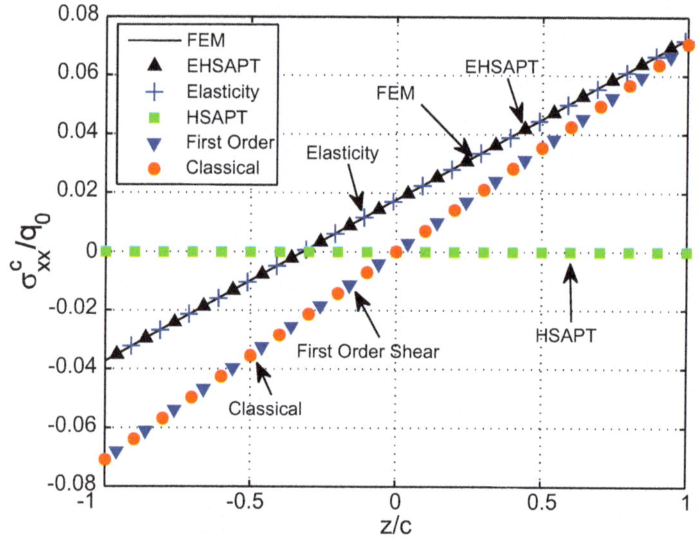

Fig. 3 Through-thickness distribution in the core of the axial stress, σ_{xx}, at midspan

The distribution of the axial stress in the core, σ_{xx}, as a function of z at the midspan location, $x = a/2$ (where the bending moment is maximum), is plotted in Fig. 3. The Extended High Order theory predicts a stress profile practically identical to the Elasticity. Note that the HSAPT [5] neglects the in-plane rigidity of the core,

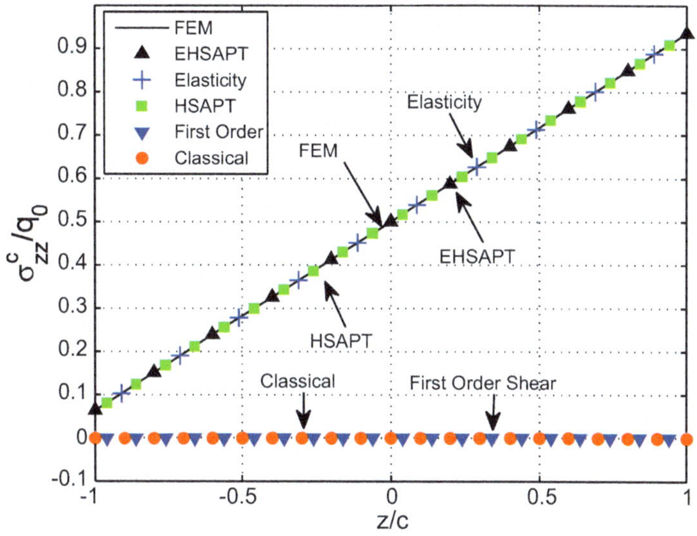

Fig. 4 Through-thickness distribution in the core of the transverse normal stress, σ_{zz}, at midspan

yielding a zero axial stress. The Classical and First Order Shear theories give practically identical predictions but they are in appreciable error by comparison to the Elasticity, the error increasing towards the lower end of the core ($z = -c$). All curves are linear. Notice also that for the Elasticity and the Extended High Order theory there is not a symmetry with regard to the mid line ($z = 0$) unlike the Classical and First Order Shear theories.

The through-thickness distribution of the transverse normal stress in the core, σ_{zz}, at the midspan location, $x = a/2$, is shown in Fig. 4. Note that the First Order Shear theory and the Classical theory consider the core incompressible. Both high order theories are practically coinciding with the Elasticity curve and all are nearly linear.

The transverse shear profile, τ_{xz}, is investigated in detail by considering a sandwich construction in which both the face sheets and the core are isotropic. By varying the moduli ratio, we can accordingly increase the shear stress range in the core. Thus, we assume that the face sheets are made out of isotropic Aluminum Alloy with $E_f = 100$ GPa and the core is made out of isotropic material having a modulus E_c such that the ratio E_f/E_c assumes the values of: 50, 5 and 2. The Poisson's ratios are assumed $\nu_f = \nu_c = 0.30$. Fig. 5 shows the shear stress distribution through the thickness near the support wher the shear force is large, $x = a/10$. For the moduli ratio of 2 the range is very large, with the maximum over minimum shear stress ratio being about 2. On the contrary, for the moduli ratio of 50, the shear stress range is very small, with the corresponding maximum over minimum shear stress ratio being only about 1.04. The Extended High Order theory is capable of capturing the shear stress profile in all cases, even the most demanding case of $E_f/E_c = 2$, and in all cases is very close to the Elasticity. On the contrary, a constant shear stress assumption as is [5] would be applicable only for the large ratios of E_f/E_c.

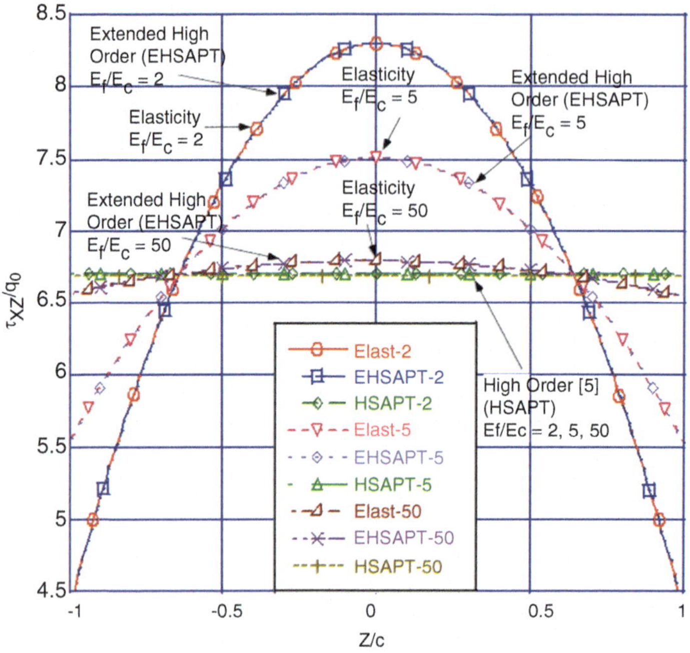

Fig. 5 Through-thickness distribution in the core of the transverse shear stress, τ_{xz}, at $x = a/10$ for the case of isotropic aluminum alloy faces and a wide range of isotropic cores

4 Accuracy Study II: Wrinkling of a Sandwich Panel

The EHSAPT formulation for predicting the critical wrinkling load for a simply supported sandwich was done in [19]. The simplest way to apply the compressive loading is for concentrated compressive loads to be applied on the top and bottom faces such that they sum up to the total applied compressive load with the pre-buckling axial strains being equal on the top and bottom faces. The core is considered to have linear strains. Results in the following will be presented for this simpler case. Another case considered was applying uniform strain loading throughout the thickness of the panel and a nonlinear core assumption. The results from this more complicated case of load application are not much different than the simpler loading case. A perturbation approach was used resulting in an eigenvalue problem formulation [19].

Tables 1 and 2 give the critical loads (normalized with the Euler load) for symmetric sandwich beams with length ratio $a/h_{tot} = 5$ and varying face thickness ratios f/h_{tot}, where $f = f_t = f_b$ is the face thickness and $h_{tot} = 2(f + c)$ is the total beam thickness.

Table 1 Critical Loads for $E_f/E_c = 500$; normalized with the Euler load (no shear) A and S in the wave numbers stand for anti-symmetric and symmetric, respectively

f/h_{tot}	Elasticity (n)	Hoff (n) (Error%)	HSAPT (n) (Error%)	EHSAPT (n) (Error%)
0.01	0.1222 (A30)	0.0631 (34) (−48.4%)	0.0370 (S24) (−69.8%)	0.1370 (A26) (+12.1%)
0.02	0.1210 (A15)	0.0654 (17) (−45.9%)	0.0548 (S14) (−54.7%)	0.1162 (A15) (−4.0%)
0.03	0.1211 (A10)	0.0672 (11) (−44.5%)	0.0698 (S11) (−42.3%)	0.1143 (A10) (−5.6%)
0.04	0.1188 (A6)	0.0687 (9) (−42.1%)	0.0836 (S9) (−29.6%)	0.1128 (A7) (−5.0%)
0.05	0.1027 (A1)	0.0703 (7) (−31.6%)	0.0962 (S7) (−6.3%)	0.1003 (A1) (−2.3%)

Table 2 Critical Loads for $E_f/E_c = 1000$; normalized with the Euler load (no shear) A and S in the wave numbers stand for anti-symmetric and symmetric, respectively

f/h_{tot}	Elasticity (n)	Hoff (n) (Error%)	HSAPT (n) (Error%)	EHSAPT (n) (Error%)
0.01	0.07381 (A24)	0.04038 (27) (−45.3%)	0.02654 (S20) (−64.0%)	0.07909 (A22) (+7.2%)
0.02	0.7393 (A12)	0.04154 (13) (−43.8%)	0.03902 (S12) (−47.2%)	0.07080 (A12) (−4.2%)
0.03	0.07288 (A7)	0.04251 (9) (−41.7%)	0.04945 (S9) (−32.2%)	0.06967 (A8) (−4.4%)
0.04	0.06489 (A1)	0.04345 (7) (−33.0%)	0.05900 (S7) (−9.1%)	0.06389 (A1) (−1.5%)
0.05	0.05411 (A1)	0.04439 (5) (−18.0%)	0.05336 (A1) (−1.4%)	0.05336 (A1) (−1.4%)

The sandwich beam is made of isotropic face and core with $E_f/E_c = 500$ and 1000 and Poisson's ratios $\nu_f = 0.35$ and $\nu_c = 0$. These tables compare the Elasticity results to the wrinkling predictions from EHSAPT, the HSAPT [5], and the Hoff-Mautner (semi-empirical constant = 0.5) [20]. The tables also show the mode and percent Error with respect to Elasticity.

It can be concluded that the EHSAPT is the most accurate theory and esp. by comparison to the HSAPT [5]. Indeed, the HSAPT is inaccurate in predicting wrinkling loads for sandwiches with very thin faces, under-predicting the critical load by as much as about 70% for the more moderately stiffer core configuration with $E_f/E_c = 500$ and $f/h_{tot} = 0.01$. In addition, the HSAPT predicts symmetric wrinkling modes, while the EHSAPT predicts anti-symmetric wrinkling modes, similar to Elasticity.

5 Accuracy Study III: Blast Loading of a Simply Supported Sandwich Panel

In this section, the dynamic response of a simply supported sandwich beam, initially at rest, then subjected to a temporal blast load that exponentially decays in time and has a half-sine spatial profile along the beam is studied. The applied load in kN/m (with time t in milli-sec) is:

$$q^t(x,t) = q_0 \sin\frac{\pi x}{a} e^{-\beta t}, \tag{25a}$$

where $q_0 = 510$ KN/m and $\beta = 1.25$ milli-sec^{-1} which decays to less than 0.1% of its original magnitude after 5.5 milli-sec. The above blast load parameters, as well as the material and geometry data were taken from the experimental investigations of Gardner et al. [27]. The faces are E-glass vinyl-ester composite: Young's modulus $E_1^c = 13{,}600$ MPa, density $\rho^f = 1800$ kg/m^3, and the isotropic core is Corecell™ A300 styrene acrylonitrile (SAN) foam: Young's modulus $E^c = 32$ MPa, $\rho^c = 58.5$ kg/m^3, Poisson's ratio $\nu^c = 0.3$, and shear modulus $G^c = E^c/[2(1 + \nu^c)]$. The geometry of the sandwich configuration is: face thickness $f_t = f_b = 5$ mm, core thickness $2c = 38$ mm, width $b = 102$ mm, and span of beam $a = 152.4$ mm.

In this case, the displacement functions that satisfy the boundary conditions are [17]:

$$u_0^{t,b,c} = U_0^{t,b,c}(t)\cos\frac{\pi x}{a}; \quad w_0^{t,b,c} = W_0^{t,b,c}(t)\sin\frac{\pi x}{a}; \quad \phi_0^c = \Phi_0^c(t)\cos\frac{\pi x}{a}. \tag{25b}$$

Substituting (25) into (7a–13c) (neglecting the nonlinear terms but including the dynamic terms), turns the partial differential equations of motion into linear ordinary differential equations in time:

$$[M]\{\ddot{U}_0(t)\} + [K]\{U_0(t)\} = \{F(t)\}, \tag{25c}$$

where the 7 × 7 matrix [M] and [K] are the mass matrix containing the inertial terms and the stiffness matrix, respectively. The vector of the unknown generalized coordinates are

$$\{U_0(t)\} = \left\{U_0^b(t), U_0^c(t), \Phi_0^c(t), U_0^t(t), W_0^b(t), W_0^c(t), W_0^t(t)\right\}^T, \tag{25d}$$

and the load vector

$$\{F(t)\} = e^{-\beta t}\{0, 0, 0, 0, 0, 0, q_0\}^T. \tag{25e}$$

The ordinary differential equations can be solved using standard numerical integration methods.

The transverse displacements w_0^t, w_0^c and w_0^b, at the mid-span location $x = a/2$ versus time are shown in Fig. 6. In this figure we show the results from Elasticity, EHSAPT, and HSAPT. The two high-order sandwich panel theories are practically on top of each other and display the same trend in behavior of the top, core, and bottom displacements as Elasticity, i.e. that the top face travels down first, followed by the core, then the bottom face sheet. EHSAPT and HSAPT match the mid-core transverse displacement of Elasticity. The high order theories over estimate the maximum displacement of the top face by a modest amount, no more than 5%. The bottom face transverse displacements from EHSAPT and HSAPT do not exactly follow Elasticity, but give values within less than 6% error over the time range in Fig. 2.

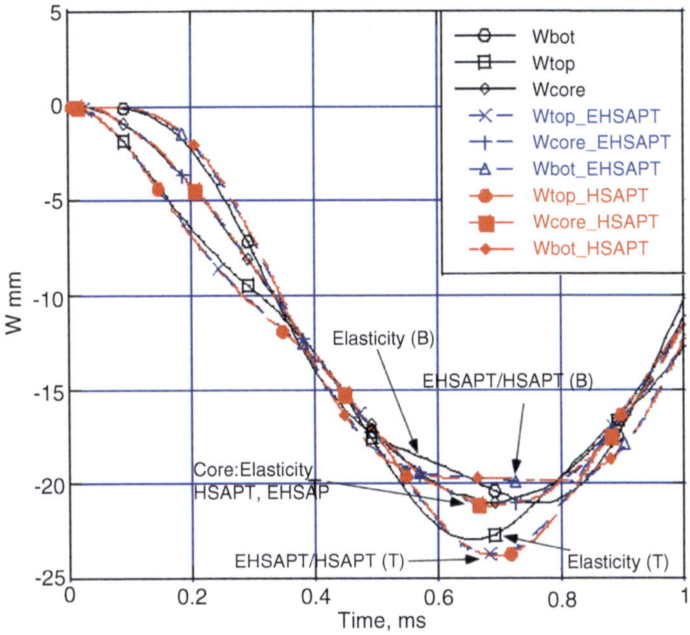

Fig. 6 Transverse displacement at the top face, middle of core and bottom face at the mid-span location for Elasticity, EHSAPT and HSAPT during the initial phase of blast

It should also be noted that the First Order Shear Deformation Theory (FOSDT) can be very inaccurate in its prediction of transverse displacement, as shown in Sect. 3, and, of course, cannot capture the differences in the displacements of the face sheets and the core.

Figure 7 shows the axial displacements u_0^t, u_0^c and u_0^b, at the edge x = 0 versus time. EHSAPT and HSAPT capture the high cyclic behavior of u_0^c that Elasticity displays, with EHSAPT being closer in value to Elasticity than HSAPT. The first peak in the core axial displacement, u_0^c of EHSAPT, is 10% under Elasticity, while the first peak in u_0^c of HSAPT is 32% under Elasticity. Both high-order theories and Elasticity predict very similar behavior with time of the top and bottom face sheet axial displacements, u_0^t and u_0^b.

The shear stress at the top and bottom face/core interfaces at $x = 0$ is shown in Fig. 8. EHSAPT is the only theory that can show the differences in the shear stresses at the top and bottom face/core interfaces like Elasticity, while HSAPT predicts that the shear stress is constant throughout the thickness and seems to be about the average value of EHSAPT and Elasticity. EHSAPT gives a minimum shear stress (most negative shear stress) at the top and bottom face/core interface under the minimum Elasticity values by just 0.5%.

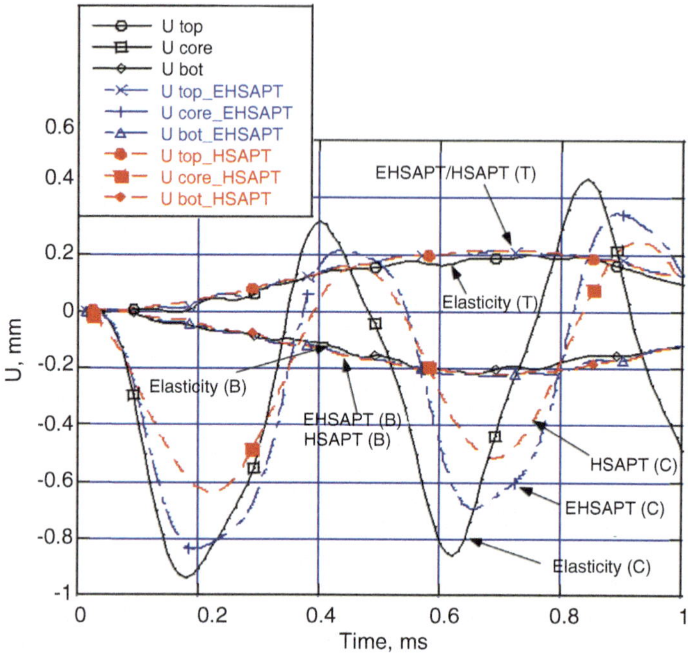

Fig. 7 Axial displacement at the top face, middle of core and bottom face at the support location ($x = 0$) for Elasticity, EHSAPT and HSAPT during the initial phase of blast

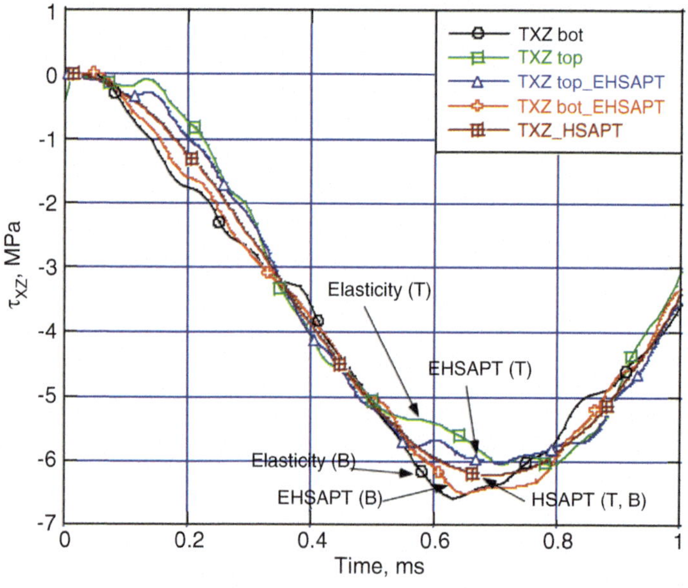

Fig. 8 The shear stress, τ_{xz}, at the top (T) and bottom (B) face/core interfaces during the initial phase of blast

6 Conclusion

The extended high order sandwich beam theory is capable of including the unique features of sandwich construction, i.e. large transverse shear and core compressibility. In this paper, the basic premises and the formulation of this theory, in both its static and dynamic versions, is described in detail. In this theory, which is derived for a general asymmetric construction, all displacement continuity conditions at the interface of the core with the top and bottom face sheets are enforced. A comparison to the elasticity solution shows that this extended high-order theory can be used with any combinations of core and face sheets and not only the very "soft" cores that the other high order sandwich theories demand.

Results have been presented for the case of static transverse loading of a simply supported sandwich beam by comparison to the Elasticity, the Classical sandwich beam theory, the First Order Shear theory and the HSAPT model [5]. The results show that the extended high order theory is very close to the Elasticity solution in terms of both the displacements and the transverse stress or strain, as well as axial stress through the core, and, in addition, the shear stress distributions in the core for core materials ranging from very soft to almost half the stiffness of the faces. In particular, it captures the very large range of core shear stress and the nearly parabolic profile in the cases of cores that are not "soft".

A case study involving an exponentially decaying blast load with a spatial half-sine profile across the top of the beam is used to compare the dynamic EHSAPT to a Dynamic Elasticity benchmark. The case study showed that the EHSAPT captures very well the complex behavior of the transverse and axial displacements in the faces and the core, as well as the stresses, and in particular, the lag in the bottom versus top displacements versus time.

In addition, the wrinkling predictions of the EHSAPT are compared with predictions from Elasticity and in all cases the EHSAPT was very close to the Elasticity predictions. On the contrary, the HSAPT [5] was in significant error for the relatively thinner faces. The large discrepancy between HSAPT and EHSAPT for very low face over total thickness ratios (when the beam is most susceptible to wrinkling) indicates that including the axial rigidity of the core (as is done in the EHSAPT) is very important during wrinkling.

Acknowledgments The financial support of the Office of Naval Research, Grants N00014-11-1-0597 and N00014-16-1-2831, and the interest and encouragement of the Grant Monitor, Dr. Y.D.S. Rajapakse, are both gratefully acknowledged.

References

1. Plantema FJ (1966) Sandwich construction. Wiley, New York
2. Allen HG (1969) Analysis and design of structural sandwich panels. Oxford, Pergamon
3. Vinson JR (1999) The behavior of sandwich structures of isotropic and composite materials. Technomic Publishing Company, Lancester

4. Wang E, Gardner N, Shukla A (2009) The blast resistance of sandwich composites with stepwise graded cores. Int J Solids Struct 46(18–19):3492–3502
5. Frostig Y, Baruch M, Vilnay O, Sheinman I (1992) High-order theory for Sandwich-beam behavior with transversely flexible Core. J Eng Mech 118(5):1026–1043
6. Hohe J, Librescu L, Oh SY (2006) Dynamic buckling of flat and curved sandwich panels with transversely compressible core. Compos Struct 74:10–24
7. Vlasov BF (1957), On one case of bending of rectangular thick plates, Vestnik Moskovskogo Universiteta Serie¨ii'a Matematiki, mekhaniki, astronomii, fiziki, khimii, 2: 25–34
8. Pagano NJ (1969) Exact solutions for composite laminates in cylindrical bending. J Compos Mater 3:398–411
9. Kardomateas GA, Phan CN (2011) Three dimensional elasticity solution for Sandwich beams/ wide plates with orthotropic phases: the negative discriminant case. J Sandw Struct Mater 13 (6):641–661
10. Kardomateas GA (2009) Three dimensional elasticity solution for Sandwich plates with orthotropic phases: the positive discriminant case. J Appl Mech 76:014505–0141–4
11. Kardomateas GA, Frostig Y, Phan CN (2013) Dynamic elasticity solution for the transient blast response of Sandwich beams/wide plates. AIAA J 51(2):485–491
12. Kardomateas GA, Rodcheuy N, Frostig Y (2015) Transient blast response of Sandwich plates by dynamic elasticity. AIAA J 53(6):1424–1432
13. Kardomateas GA, Rodcheuy N, Frostig Y (2017) Elasticity solution for curved Sandwich beams/panels and comparison with structural theories. AIAA J 55(9):3153–3160. 2017
14. Kardomateas GA (2010) An elasticity solution for the global buckling of Sandwich beams/wide panels with orthotropic phases. J Appl Mech 77(2):021015–0211–7
15. Kardomateas GA (2005) Wrinkling of wide Sandwich panels/beams with orthotropic phases by an elasticity approach. J Appl Mech 72:818–825
16. Phan CN, Frostig Y, Karomateas GA (2012) Analysis of Sandwich panels with a compliant Core and with in-plane rigidity-extended high-order Sandwich panel theory versus elasticity. J Appl Mech 79:041001–041–11
17. Phan CN, Kardomateas GA, Frostig Y (2013) Blast response of a Sandwich beam/wide plate based on the extended high-order Sandwich panel theory (EHSAPT) and comparison with elasticity. J Appl Mech 80:061005–061–11
18. Phan CN, Karomateas GA, Frostig Y (2012) Global buckling of a Sandwich wide panel/beam based on the extended high order theory. AIAA J 50(8):1707–1716
19. Phan CN, Bailey NW, Kardomateas GA, Battley MA (2012) Wrinkling of Sandwich wide panels/beams based on the extended high order Sandwich panel theory: formulation, comparison with elasticity and experiments. Arch Appl Mech. (special issue in honor of prof. Anthony Kounadis) 82:1585–1599
20. Hoff NJ, Mautner SF (1945) The buckling of Sandwich-type panels. J Aeronaut Sci 12 (3):285–297
21. Yuan Z, Kardomateas GA, Frostig Y (2015) Finite element formulation based on the extended high order Sandwich panel theory. AIAA J 53(10):3006–3015
22. Yuan Z, Kardomateas GA, Frostig Y (2016) Geometric nonlinearity effects in the response of Sandwich wide panels. J Appl Mech 83(9):091008–091–10
23. Yuan Z, Kardomateas GA (2018) Nonlinear stability analysis of sandwich wide panels – part I: buckling behavior, J Appl Mech 85. 081006-1-11–18
24. Yuan Z, Kardomateas GA (2018) Non linear stability analysis of sandwich wide panels – part II: post-buckling response, J Appl Mech 85. 081007-1-9–18
25. Rodcheuy N, Frostig Y, Kardomateas GA (2017) Extended high order theory for curved Sandwich panels and comparison with elasticity. J Appl Mech 84, 84(8). 081002-1–16). https://doi.org/10.1115/1.4036612
26. Carlsson L, Kardomateas GA (2011) Structural and failure mechanics of Sandwich composites. Springer
27. Gardner N, Wang E, Kumar P, Shukla A (2012) Blast mitigation in a Sandwich composite using graded Core and Polyurea interlayer. Exp Mech 52(2):119–133. https://doi.org/10.1007/s11340-011-9517-9

Mechanics Based Modeling of Composite and Sandwich Structures in the Naval Environment: Elastic Behavior, Fracture and Damage Evolution

Roberta Massabò

1 Introduction

Navy ships operate in extreme environments and under extreme loadings. The use of composite materials for load bearing components has a number of advantages over traditional materials, for instance in terms of strength/stiffness-weight ratios and maintenance costs. Composites can be designed to optimize certain performances or to mitigate the effects of loadings or environmental conditions. However, their inhomogeneous structure and inherently brittle nature strongly affect the response to mechanical and environmental loadings, and this substantially complicates the formulation of physically based predictive models and the optimal design. The layered architecture, with discontinuities in the elastic/thermal properties at the layer interfaces, and the presence of imperfections generate complex local fields, affect the propagation of stress waves and favor the initiation and propagation of cracks at the interfaces between layers. The post-elastic response is dominated by multiple damage interaction and evolution. In addition, the material properties, e.g. strengths, stiffnesses and fracture properties, are influenced by high-strain loadings and by extreme temperatures and sea water, which also modify the local fields due to thermal and moisture induced strains and stresses caused by the inhomogeneous material structure.

The above mentioned phenomena and behaviors need to be understood, through carefully chosen experiments and mechanical models, in order to be then accounted for in the formulation of predictive models of the response of the systems under various loadings and environmental conditions. The chapter reviews selected work conducted during the last decade by the author and collaborators within the Solid Mechanics Program of the U.S. Office of Naval Research.

R. Massabò (✉)
Department of Civil, Chemical and Environmental Engineering,
University of Genova, Genova, Italy
e-mail: roberta.massabo@unige.it

S. W. Lee (ed.), *Advances in Thick Section Composite and Sandwich Structures*,
https://doi.org/10.1007/978-3-030-31065-3_12

Section 2 deals with the propagation of stress waves in layered elastic plates. Stress waves are generated by sudden disturbances, such as blasts or impacts and their propagation is controlled by geometrical constraints, material properties and architecture and the presence of perfect or imperfect interfaces and flaws. In [1–3], frequency equations and dispersion curves of plane-strain harmonic waves propagating in layered plates with imperfect interfaces have been obtained in closed-form using the multiscale structural model formulated in [4]. The model captures the effects of the inhomogeneous material structure and the presence of interfacial imperfections through local enrichments and homogenized equilibrium equations which depend on the same number of kinematic variables of an equivalent single layer theory. The solutions are more accurate than those of classical structural theories and capture the effects of the presence of soft inter-layers, adhesives or damaged interfaces. They accurately define the cut-off frequencies of the lowest cut-off frequency modes, which are useful matching frequencies for the definition of correction factors of low order equivalent single layer theories.

Section 3 presents a matrix technique formulated to derive explicit 2D and 3D elasticity solutions for stresses and displacements in simply supported layered plates subjected to thermo-mechanical loading [5–7]; thermal and mechanical contact between the layers may be perfect or imperfect. The formulation uses classical elasticity solutions in the layers, introduces interfaces which relate the interfacial tractions/thermal-flux to the interfacial relative displacement/temperature-jump, and a matrix technique which allows to treat plates with many layers as if they were single layered. The explicit solutions are applicable for parametric analyses and benchmarking.

Section 4 presents semi-analytic solutions for the fracture parameters, energy release rate and mode mixity phase angle, of composite sandwich beams with face/core debonds [8]. The solutions extend the original models in [9–13] in order to account for the three-layer structure and the effects of shear. The fracture parameters are defined in terms of crack tip force and moment resultants which can be defined in actual beams through equilibrium or structural mechanics models. The solutions depend on a few numerical coefficients which are presented in tabular form for a wide range of core/face material combinations. They are applicable to fracture specimens used for the characterization of the mixed-mode fracture properties. Using these solutions complex, case specific numerical calculations can be avoided, while maintaining accuracy. An application is presented to a DCB sandwich specimen in order to highlight the important effects of shear.

Section 5 presents a theoretical investigation on the effects of the interaction of the damage mechanisms for face sheet delamination and core crushing in a sandwich beam subjected to time-dependent loading [14, 15]. The beam is continuously supported by a rigid plane and the problem is described using a 1D approximation which treats the impacted face as a set of Timoshenko beams joined by cohesive interfaces and resting on a nonlinear Winkler foundation. The face/core interactions generates energy barriers to the propagation of cracks in the face sheet and their location and magnitude depend on the core nonlinear behavior and on duration and magnitude of the applied load. The model captures the experimentally observed transitions in damage evolution on varying the impact energy in low velocity impact tests [16].

Section 6 presents a novel modeling approach to study mode II dominant delamination fracture in layered structures [7, 17, 18]. The model uses a multiscale structural strategy formulated in [4, 19, 20] for structures with arbitrary numbers of layers and imperfect interfaces. The layered structure and the delaminations are accounted for by enriching the displacement field of a classical first-order plate theory and using homogenization to define the local enrichments as functions of the global variables. The equilibrium equations, which are derived using a variational technique, then depends on the same displacement variables of the global theory and this facilitates the closed form solution of problems that would otherwise require numerical solution. Applications are presented which show the capability of the model to analyze single and multiple crack propagation in unidirectional composite and layered beams using only in-plane discretization. The model captures the effects of the layered structure on energy release rate and macrostructural response, the unstable behavior and local instabilities which characterize the propagation of cracks in mode II fracture specimens and the local and global effects of the interaction of multiple cracks, including multiple snap-back and snap-through instabilities, hyper-strength and crack pull-along phenomena.

2 Propagation of Plane Strain Harmonic Waves in Layered Plates

When a layered structure is hit by a sudden disturbance, stress waves are generated and propagate through the body. These stress waves may cause damage, either directly or indirectly through the impulse they deliver. Predicting how stress waves propagate in the phase that precedes damage is a prerequisite to the solution of dynamic problems, the design of blast/impact mitigation strategies and the design and application of damage detection techniques. The layered material structure with layers having different elastic properties and discontinuities in the elastic constants at the layer interfaces, influences the modes of propagation and dispersion characteristics. The propagation of Rayleigh-Lamb waves in a model plate with two isotropic layers with different properties was first studied using exact two-dimensional elasticity in [21, 22]. As for the solution of the homogeneous plate (Lamb 1917), the characteristic equations of the problem were derived in closed-form and their solution required a numerical method. The exact solution of the problem becomes daunting in plates with more layers, especially if propagation is not in one of the principal directions. The problem is then typically solved using a matrix formalism, Thomson [23] (see Sect. 3 for an application of the method to another elasticity problem). It is difficult using these approaches to infer qualitative information about the dispersion relations without solving the characteristic equation, something which remains complex. Some limiting explicit solutions, in the regimes of small and large wave-numbers, may be derived using asymptotic methods and perturbation techniques [24]. The problem of guided wave propagation in layered composite structures is instead quite easily solved using equivalent single-layer theories, which also allow the efficient solution of more complex problems of forced-vibrations in finite structures with boundaries [22, 25]. These approximate structural solutions, however, are limited in frequency range, do not typically account for interfacial imperfections, and at the lower order require the case specific calculation of shear and normal correction factors.

In [1] the propagation of plane-strain harmonic waves in multilayered plates has been studied through an enhanced structural theory, which uses a homogenization technique to derive governing equations which are still tractable, as for classical plate theories, but contains some of the features of the exact theories to improve the solution for higher frequencies and account for various material inhomogeneities. The approach is similar to that proposed in [25] for plates with fully bonded layers.

The multiscale model has been formulated in [19], for the solution of general plate problems, and particularized in [4] to plane-strain problems; it is based on the original zig-zag theory developed by Di Sciuva [26] for structures with fully bonded layers. The theory has been extended to account for the presence of interfacial imperfections, delaminations and thin interlayers. The ability of the proposed model to predict, with explicit closed-form solutions, the complex local fields and displacement discontinuities in multilayered plates with continuous imperfect interfaces and subjected to steady-state thermo-mechanical loading has been verified in [20]. The implementation of the model to study fracture problems will be discussed in Sect. 6.

Sections 2.1 and 2.2 review the wave propagation model and show selected results.

2.1 Multiscale Structural Model for Layered Structures with Interfacial Imperfections

The multiscale model in [1] refers to the multilayered plate of infinite in plane extent and thickness h shown in Fig. 1. The plate consists of n layers joined by $n-1$ zero-thickness interfaces. The thickness of the layer k, is ${}^{(k)}h$ and the coordinates $x_3{}^{k-1}$ and

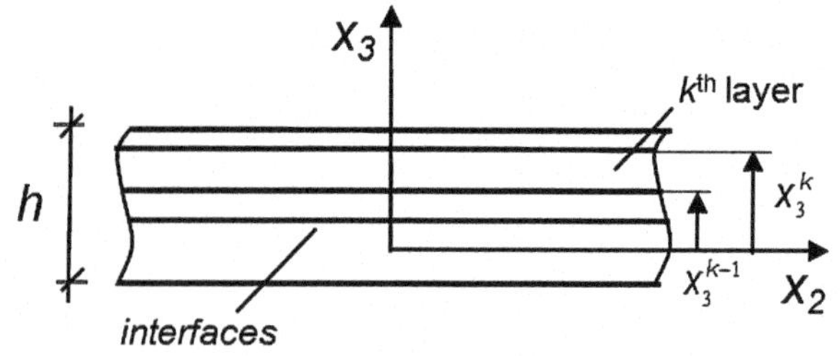

Fig. 1 Layered plate with imperfect interfaces used to study the propagation of plane-strain waves

${x_3}^k$ define lower and upper interfaces. The interfaces are mathematical surfaces where the interfacial tractions are continuous while materials properties and displacements may change discontinuously. The term *perfect interface* is used here and in the rest of the chapter to indicate full bonding of the adjacent layers while *imperfect interface* indicates an interface which allows controlled relative displacements of the adjacent layers; a *fully debonded interface* defines traction free surfaces and allows free relative displacements.

The model relies on the formulation in [4] and on two fundamental equations, which define the small-scale and macro-scale displacement fields in the layers. For plates deforming in cylindrical bending, where the transverse displacements are assumed to be constant through the thickness (no transverse compressibility or interfacial openings) and the transverse normal stresses are negligible, the small-scale displacements are assumed as:

$$\begin{aligned} {}^{(k)}v_2(x_2,x_3) &= v_{02}(x_2) + x_3\varphi_2(x_2) + \sum_{i=1}^{k-1}\Omega_2^i(x_2)\left(x_3 - {x_3}^i\right) + \sum_{i=1}^{k-1}\hat{v}_2^i(x_2) \\ {}^{(k)}v_3(x_2,x_3) &= w_0(x_2). \end{aligned} \tag{1}$$

where $v_{02}(x_2)$, $\varphi_2(x_2)$, $w_0(x_2)$ are global variables, which directly define the in-plane displacement, bending rotation and transverse displacement of the reference surface, at $x_3 = 0$, when this falls into the lowest layer. The local terms which are introduced to enrich the global field are the zigzag functions, $\Omega_2^i(x_2)(x_3 - {x_3}^i)$, and the relative sliding displacements at the layer interfaces, $\hat{v}_2^i(x_2) = {}^{(k+1)}v_2\left(x_2, x_3 = x_3^k\right) - {}^{(k)}v_2\left(x_2, x_3 = x_3^k\right)$. They are used to describe the inhomogeneous material structure, which gives rise to "zig-zag" effects in the through-thickness direction, and the discontinuities generated at the layer interfaces by the presence of damage, delaminations or thin inter-layers, e.g. adhesives.

In [1] the interfacial sliding displacements are related to the interfacial tangential tractions through a linear function:

$$\hat{\sigma}_S^k\left(x_2, x_3 = x_3^k\right) = K_S^k \hat{v}_2^k(x_2) \tag{2}$$

where K_S^k is the interfacial stiffness. This simple linear model allows the description of imperfect interfaces with partial bonding of the layers, and of the two asymptotic limits of perfectly bonded interfaces, which are defined by $1/K_S^i \to 0$ and $\hat{v}_2^i \to 0$, and fully debonded interfaces, which are defined by $K_S^i \to 0$ and $\hat{\sigma}_S^i \to 0$ (different laws are assumed to analyze delamination fracture in Sects. 5 and 6). The small-scale displacement field in the plate is then defined by a total of three global variables and $2(n - 1)$ local variables.

The macro-scale displacement field depends on the global variables only and is obtained from the small-scale field using a homogenization technique which imposes continuity on the tractions at the interfaces (the tractions are related to the displacements in Eq. (1) through constitutive and compatibility equations). The procedure is detailed in [1, 4]. The macro-scale displacements in the layer k take the form:

$$
\begin{aligned}
{}^{(k)}v_2(x_2,x_3) &= v_{02}(x_2) + x_3\varphi_2(x_2) + [w_{0},_2(x_2) + \varphi_2(x_2)]R^k_{S22}(x_3) \\
{}^{(k)}v_3(x_2) &= w_0(x_2).
\end{aligned}
\tag{3}
$$

with

$$
\begin{aligned}
&R^k_{S22}(x_3) = \sum_{i=1}^{k-1} [\Lambda^i_{22}(x_3 - x_3{}^i) + \Psi^i_{22}] \\
&\Lambda^k_{22} = {}^{(1)}G_{23}\left(\frac{1}{{}^{(k+1)}G_{23}} - \frac{1}{{}^{(k)}G_{23}}\right), \quad \Psi^k_{22} = \frac{{}^{(k+1)}G_{23}}{K^k_S}\left(1 + \sum_{j=1}^{k} \Lambda^{(j)}_{22}\right)
\end{aligned}
\tag{4}
$$

with ${}^{(k)}G_{23}$ the shear modulus of layer k. Strains and stresses in the layers are then defined from the macro-scale displacements and a variational technique defines the equilibrium equations. The equations for wave propagation are:

$$
\begin{aligned}
L_{22}v_{02} + L_{23}w_0 + L_{24}\varphi_2 &= 0 \\
L_{32}v_{02} + L_{33}w_0 + L_{34}\varphi_2 &= 0 \\
L_{42}v_{02} + L_{43}w_0 + L_{44}\varphi_2 &= 0
\end{aligned}
\tag{5}
$$

with

$$
\begin{aligned}
&L_{22} = A_{22}(\,),_{22} - R_0(\,),_{tt};\ L_{23} = L_{32} = C^{0S}_{22}(\,),_{222} - R^{0S}(\,),_{2tt} \\
&L_{24} = L_{42} = B_{22}(\,),_{22} - R_1(\,),_{tt};\ L_{33} = -A_{44}(\,),_{22} + C^{S2}(\,),_{2222} + R_0(\,),_{tt} - R^{S2}(\,),_{22tt} \\
&L_{34} = L_{43} = -A_{44}(\,),_2 + (C^{1S}_{22} + C^{S2})(\,),_{222} - (R^{S2} + R^{1S})(\,),_{2tt}; \\
&L_{44} = D_{22}(\,),_{22} - A_{44}(\,) - R_2(\,),_{tt}
\end{aligned}
\tag{6}
$$

where A_{22}, B_{22}, D_{22}, A_{44}, R_0, R_1, R_2 are stiffness and inertia terms similar to those of the first order shear deformation theory but modified to account for the local enrichments (given the low order of the global model, A_{44} depends of a shear correction factor $k_{44} = \pi^2/12$); $R^{rS}, R^{S2}, C^{rS}_{22}, C^{S2}, C^S$ are additional coefficients depending on layup and status of the interfaces [1]; t is the time variable.

The propagation of plane-strain harmonic waves is studied in [1] by assuming solutions of the form $v_{02}=V_{02}expi(kx_2-\omega t)$ and similar equations for the other variables, with k the wave number and ω the radian frequency. Frequency equations and dispersion curves are obtained through the non-trivial solution of the homogeneous system of equations obtained by imposing the vanishing of the determinant:

$$\begin{vmatrix} \left(\frac{\omega^2}{k^2}-\frac{A_{22}}{R_0}\right)k^2R_0 & \left(\frac{\omega^2}{k^2}-\frac{B_{22}}{R_1}\right)k^2R_1 & \left(\frac{\omega^2}{k^2}-\frac{C_{22}^{0S}}{R^{0S}}\right)k^3R^{0S} \\ \left(\frac{\omega^2}{k^2}-\frac{B_{22}}{R_1}\right)k^2R_1 & \left(\frac{\omega^2}{k^2}-\frac{D_{22}}{R_2}\right)k^2R_2-A_{44} & \left(\frac{\omega^2}{k^2}-\frac{C_{22}^{1S}+C^{S2}}{R^{1S}+R^{S2}}\right)k^3\left(R^{1S}+R^{S2}\right)-A_{44}k \\ \left(\frac{\omega^2}{k^2}-\frac{C_{22}^{0S}}{R^{0S}}\right)k^3R^{0S} & \left(\frac{\omega^2}{k^2}-\frac{C_{22}^{1S}+C^{S2}}{R^{1S}+R^{S2}}\right)k^3\left(R^{1S}+R^{S2}\right)-A_{44}k & \left(\frac{\omega^2}{k^2}-\frac{C^{S2}}{R^{S2}}\right)k^4R^{S2}+\left(\frac{\omega^2}{k^2}-\frac{A_{44}}{R_0}\right)k^2R_0 \end{vmatrix} \quad (7)$$

The determinant is particularized to describe different problems. The determinant of a classical first order beam theory is obtained by imposing $\Lambda_{22}^{(1;k)} = 0$ and $1/K_S^k \to 0$, for $k = 1,.., n - 1$. For an unidirectionally reinforced laminate with imperfect interfaces, $\overline{E}_2 = \overline{E}_L$ the effective Young's moduli, G_{LT} the shear modulus and ρ the density, the determinant simplifies and the frequency equation is:

$$\left[\left(\frac{\omega}{k}\right)^2-\frac{\overline{E}_L}{\rho}\right]\left\{\left(\frac{\omega}{k}\right)^4-\left[\frac{\overline{E}_L}{\rho}\Theta_1+\Theta_2\right]\left(\frac{\omega}{k}\right)^2+\left(\frac{\overline{E}_L}{\rho}\right)^2(\Theta_1-1)+\frac{\overline{E}_L}{\rho}\Theta_3\right\}=0 \quad (8)$$

The form of the equation is similar to that of first order shear deformation theory; the effects of the interfacial imperfections are described by the coefficients Θ_1, Θ_2, Θ_3, which depend on material/geometrical properties and are given in [1].

2.2 The Influence of Interfacial Imperfections on Wave Propagation and Dispersion

The diagrams in Fig. 2 show dispersion curves obtained using Eq. (8) for the lowest and the second lowest flexural modes in a layered plate with six unidirectionally reinforced layers and interfaces with stiffness K_S. The limiting curves define the response of plates with fully bonded layers, $1/K_S \to 0$, and fully debonded layers, $K_S \to 0$. Closed form solutions for these limits have been obtained in [1] through a perturbation analysis. The second lowest spectrum, which is dominated by thickness-shear deformations, has cut-off frequency $\omega_{co} = \sqrt{k_{44}G_{LT}h/\left(\rho\overline{E}_LI\right)}$ in the fully bonded plate and $\omega_{co} = 0$ in the fully debonded plate, with $k_{44} = \pi^2/12$ and $I = h^3/12$. For the imperfect interfaces the curves vary between the limiting solutions with important effects on the second lowest mode and its cut-off frequency.

In [2, 3] the thickness-shear and -stretch modes of propagations have been studied using the homogenized model (and its extended version in [19]) and the cut-off frequencies of the lowest cut-off frequency modes have been defined in closed form. The cut-off frequency of the shear-thickness mode is $\omega_{co} = [A_{44}/(R_2 - R_1{}^2/R_0)]^{1/2}$, with A_{44}, R_0, R_1, R_2 stiffness and inertia terms defined explicity in [1]. The accuracy of the expression has been verified with comparison with exact elasticity solutions

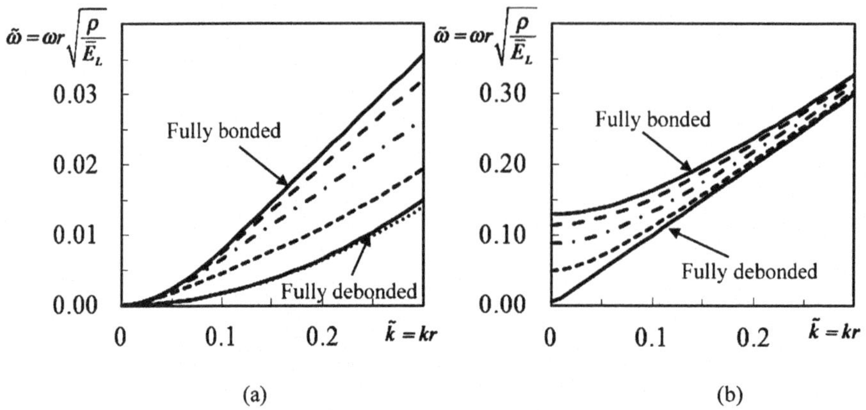

Fig. 2 Dispersion curves, relating wave frequency and wave number, of the two lowest flexural modes in a plate with six unidirectionally reinforced layers and imperfect interfaces with stiffness K_S. Results are presented for $\overline{E}_L/k_{44}G_{LT} = 60$ and $K_Sh/\mathrm{G}_{LT} = 0$, 1, 5, 20, with $k_{44} = \pi^2/12$ and $r = h/\sqrt{12}$ (lower solid, small dashed, dash-dot, dashed lines); upper curve corresponds to fully bonded case with $\mathrm{G}_{LT}/K_Sh = 0$

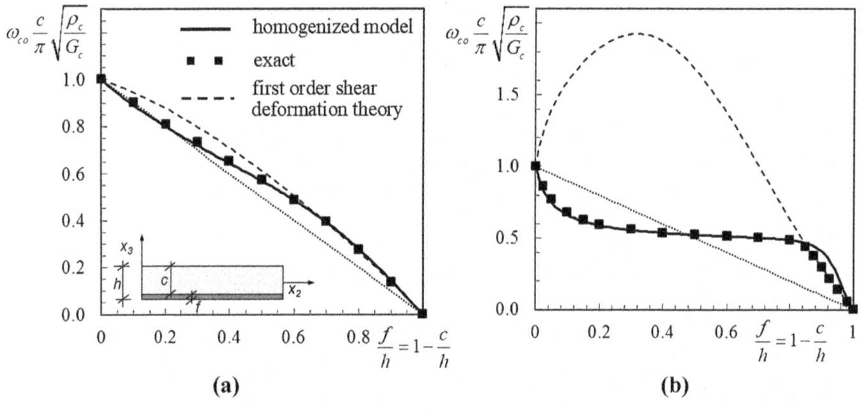

Fig. 3 Dimensionless cut-off frequency of the thickness-shear mode in a bilayer with f and c the thicknesses of the layers, $h = f + c$, and G_f and G_c the corresponding shear moduli. Results of FOSDT obtained assuming a shear correction factor $k_{44} = \pi^2/12$. (**a**) $G_c = G_f/2$ and $\rho_c = \rho_f$ (e.g. cross-ply laminate); (**b**) $G_c = 0.02G_f$ and $\rho_c = 0.15\rho_f$ (e.g. composite sandwich with foam core) (after [3])

for a two-layer problem, see Fig. 3. The cut-off frequencies can be used as matching frequencies for the definition of correction factors of low order single layer theories [2, 3].

2.3 Conclusions on Wave Propagation Analysis

A homogenized structural model has been formulated to study plane-strain wave propagation in layered plates. The closed form solutions have been used to investigate the effects of interfacial imperfections on dispersion curves and cut-off frequencies. Accurate cut-off frequencies corresponding to the lowest cut-off frequency modes have been derived which can be used as matching frequencies for the derivation of correction factors of low order equivalent single layer theories.

3 Elasticity Solutions for Plates with Imperfect Thermal/ Mechanical Contact of the Layers

Multilayered structures are frequently used as load-bearing components, also for applications which necessitate withstanding severe thermo-mechanical loadings. These structures typically have complex geometries, boundary and loading conditions, and numerical models are applied to determine their response in the elastic and post-elastic regimes. 2D and 3D thermo-elasticity solutions are limited to simple geometries and boundary conditions; however, these solutions are valuable because they exactly predict the field variables and can be used to assess the accuracy of numerical and approximate models and perform parametric analyses. The exact solutions are not always easily obtainable or usable. An example is the problem of layered plates with mechanical and thermal interfacial imperfections. In the 2D elasticity formulation in [20] for simply supported plates with n layers, the algebraic system resulting from the imposition of the continuity and boundary conditions depend on the number of layers ($4 \times n$ coupled equations) and typically require a numerical solution but for plates with few layers.

In [5–7] a matrix technique has been formulated to obtain exact solutions for stationary thermo-mechanical problems in simply supported multilayered plates with an arbitrary number of imperfectly bonded layers in imperfect thermal contact. Solutions for stresses and displacements are presented in explicit expanded forms for direct application, with a clear advantage over previous derivations [20, 27–30]. The solutions refer to plates subjected to sinusoidally distributed transverse surface tractions and thermal gradients and are applicable to load distributions other than sinusoidal using Fourier series. The technique uses the theoretical formulation in [20], with a classical displacement approach, and the transfer matrix method [23]. The formulation of the model for plates under plane strain condition is summarized in Sect. 3.1 and explicit expressions for displacement and stresses are presented in Sect. 3.2. Formulation and solution for general 3D plates are in [7].

Fig. 4 Simply supported layered plate subject to thermo-mechanical loading $p = \pi/L$ $(m = 1)$

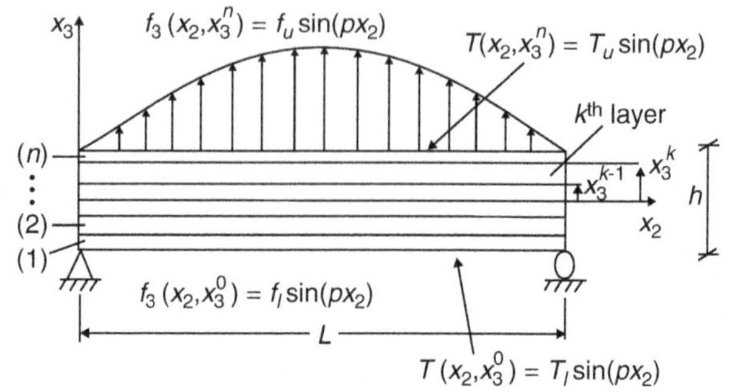

3.1 Problem and Model

The schematic in Fig. 4, describes a multilayered plate under plane-strain conditions parallel to the plane $x_2 - x_3$. The plate is simply supported, has global thickness h and is composed of n layers of thickness ${}^{(k)}h$. The coordinates of the lower and upper surfaces of the layer k are x_3^{k-1} and x_3^k. The plate is subjected to thermo-mechanical loadings: $f_3(x_2, x_3 = x_3^n) = f_u \sin(px_2)$ and $f_3(x_2, x_3 = x_3^0) = f_l \sin(px_2)$ are normal surface tractions acting on the upper and lower surfaces, and $T(x_2, x_3 = x_3^n) = T_u \sin(px_2)$ and $T(x_2, x_3 = x_3^0) = T_l \sin(px_2)$ are applied temperatures, with $p = m\pi/L$ and m a natural constant. The layers are linearly elastic, homogenous and orthotropic with principal material axes parallel to the geometrical axes (this assumption is relaxed in the 3D solution in [7]). Coupling between elastic deformations and heat transfer is neglected and the thermal conditions are assumed to be stationary. The stiffness coefficients in the layer k are ${}^{(k)}C_{ij}$ and the coefficients of thermal expansion along the x_i direction are ${}^{(k)}\alpha_i$; ${}^{(k)}T = {}^{(k)}T(x_2, x_3)$ is the temperature increment in the layer k.

Imperfect thermal and mechanical contact of the layers is introduced using interfaces, which are zero-thickness mathematical surfaces where interfacial tractions and heat flux are continuous while displacements and temperature may be discontinuous. The constitutive behavior of the interface is described by linear uncoupled interfacial traction laws [31], $\hat{\sigma}_2^k(x_2, x_3 = x_3^k) = K_S^k \hat{v}_2^k(x_2, x_3 = x_3^k)$ and $\hat{\sigma}_3^k(x_2, x_3 = x_3^k) = K_N^k \hat{v}_3^k$, where $\hat{\sigma}_2^k$ and $\hat{\sigma}_3^k$ are the interfacial tangential and normal tractions acting on the upper surface of the layer k with unit positive normal vector, K_S^k and K_N^k are tangential and normal stiffnesses and $\hat{v}_2^k$ and $\hat{v}_3^k$ are the relative displacements between the layers k and $k + 1$ at the interface, $\hat{v}_i^k(x_2, x_3 = x_3^k) = {}^{(k+1)}v_i(x_2, x_3 = x_3^k) - {}^{(k)}v_i(x_2, x_3 = x_3^k)$. A perfectly bonded interface is described by $1/K_S^k = 1/K_N^k = 0$, which implies $\hat{v}_2^k = \hat{v}_3^k = 0$; an interface which allows free sliding in constrained contact has $1/K_N^k = K_S^k = 0$, which yields $\hat{\sigma}_2^k = 0$ and $\hat{v}_3^k = 0$.

The thermal behavior of the interface is described by a linear relationship between the heat flux through the interface and the jump in the temperatures of the layers:

$$^{(k)}q_3\left(x_2, x_3 = x_3^k\right) = -\frac{1}{R^k}\left[^{(k+1)}T\left(x_2, x_3 = x_3^k\right) - {}^{(k)}T\left(x_2, x_3 = x_3^k\right)\right] \tag{9}$$

where R^k is the interfacial thermal resistance, which is assumed to be independent of the interfacial displacements; $^{(k)}q_3(x_2, x_3) = -{}^{(k)}K_3{}^{(k)}T(x_2, x_3)_{,3}$ is the heat flux in the layer k with $^{(k)}K_i$ the thermal conductivity in the x_i direction. The interfacial thermal resistance accounts for the various modes of heat transfer through the interface, namely conduction at contact spots and conduction/radiation through the air gaps; when the interface is used to describe a thin interlayer, the resistance depends on its conductivity/thickness. Perfect thermal contact, where the temperature is continuous, is described by $R^k = 0$, and an impermeable interface, where the heat flux through the interface vanishes, by $1/R^k = 0$. The equations governing the response of the layer k are listed below.

Heat conduction with no heat generation:

$$\frac{^{(k)}K_2}{^{(k)}K_3}\frac{\partial^2\,{}^{(k)}T(x_2, x_3)}{\partial x_2^2} + \frac{\partial^2\,{}^{(k)}T(x_2, x_3)}{\partial x_3^2} = 0 \tag{10}$$

Continuity of heat flux and interfacial thermal response:

$$\begin{aligned} &\frac{^{(k)}K_3}{^{(k+1)}K_3}\,{}^{(k)}T\left(x_2, x_3^k\right)_{,3} = {}^{(k+1)}T\left(x_2, x_3^k\right)_{,3} \\ &^{(k+1)}T\left(x_2, x_3^k\right) - {}^{(k)}T\left(x_2, x_3^k\right) = -{}^{(k)}q_3\left(x_2, x_3^k\right)R^k \end{aligned} \tag{11}$$

Thermal boundary conditions:

$$\begin{aligned} &^{(n)}T\left(x_2, x_3 = x_3^n\right) = T_u \sin\left(px_2\right), \quad {}^{(1)}T\left(x_2, x_3 = x_3^0\right) = T_l \sin\left(px_2\right) \\ &^{(k)}T(x_2 = 0 \text{ and } L, x_3) = 0, \quad \text{for } k = 1, \ldots, n \end{aligned} \tag{12}$$

Equilibrium, Navier equations:

$$\begin{aligned} &^{(k)}C_{22}{}^{(k)}v_{2,22} + {}^{(k)}(C_{23} + C_{55})^{(k)}v_{3,23} + {}^{(k)}C_{55}{}^{(k)}v_{2,33} = {}^{(k)}(C_{12}\alpha_1 + C_{22}\alpha_2 + C_{23}\alpha_3)^{(k)}T_{,2} \\ &^{(k)}C_{33}{}^{(k)}v_{3,33} + {}^{(k)}(C_{23} + C_{55})^{(k)}v_{2,23} + {}^{(k)}C_{55}{}^{(k)}v_{3,22} = {}^{(k)}(C_{13}\alpha_1 + C_{23}\alpha_2 + C_{33}\alpha_3)^{(k)}T_{,3} \end{aligned} \tag{13}$$

Continuity conditions and interfacial constitutive response:

$$\begin{aligned} &^{(k)}\sigma_{23}\left(x_2, x_3^k\right) = {}^{(k+1)}\sigma_{23}\left(x_2, x_3^k\right), \quad {}^{(k)}\sigma_{33}\left(x_2, x_3^k\right) = {}^{(k+1)}\sigma_{33}\left(x_2, x_3^k\right) \\ &^{(k+1)}v_2\left(x_2, x_3^k\right) - {}^{(k)}v_2\left(x_2, x_3^k\right) = \frac{1}{K_S^k}\,{}^{(k)}\sigma_{23}\left(x_2, x_3^k\right), \quad {}^{(k+1)}v_3 - {}^{(k)}v_3 = \frac{1}{K_N^k}\,{}^{(k)}\sigma_{33} \end{aligned} \tag{14}$$

Boundary conditions:

$$\begin{aligned}&{}^{(n)}\sigma_{33}\left(x_2, x_3 = x_3^n\right) = f_u \sin\left(px_2\right), \quad {}^{(n)}\sigma_{23}\left(x_2, x_3 = x_3^n\right) = 0\\&{}^{(1)}\sigma_{33}\left(x_2, x_3 = x_3^0\right) = -f_l \sin\left(px_2\right), \quad {}^{(1)}\sigma_{23}\left(x_2, x_3 = x_3^0\right) = 0\\&{}^{(k)}\sigma_{22}(x_2 = 0 \text{ and } L, x_3) = 0, \quad {}^{(k)}v_3(x_2 = 0 \text{ and } L, x_3) = 0, \text{for } k = 1, \ldots, n\end{aligned} \tag{15}$$

The general solution of the heat transfer problem, which satisfies the boundary conditions (12) at the plate edges is:

$$\begin{aligned}&{}^{(k)}T(x_2, x_3) = {}^{(k)}F(x_3) \sin\left(px_2\right)\\&{}^{(k)}F(x_3) = \left({}^{(k)}c_1 e^{{}^{(k)}sx_3} + {}^{(k)}c_2 e^{-{}^{(k)}sx_3}\right); \quad {}^{(k)}s = p\sqrt{\frac{{}^{(k)}K_2}{{}^{(k)}K_3}}, \quad p = \frac{m\pi}{L}\end{aligned} \tag{16}$$

where ${}^{(k)}c_1$ and ${}^{(k)}c_2$ are integration constants. Eq. (16) and the conditions in Eqs. (11), (12) yield an algebraic system of $2 \times n$ coupled equations for the $2 \times n$ unknown constants ${}^{(k)}c_1$ and ${}^{(k)}c_2$, for $k = 1, \ldots, n$. The solution of the thermo-elastic problem which satisfies the boundary conditions at the plate edges, Eq. (15), and with temperature in Eq. (16) are:

$$ {}^{(k)}v_i(x_2, x_3) = {}^{(k)}v_{ip}(x_2, x_3) + {}^{(k)}v_{ic}(x_2, x_3) \tag{17}$$

where:

$$\begin{aligned}{}^{(k)}v_{2p}(x_2, x_3) &= {}^{(k)}\left(B_1 e^{sx_3} + B_2 e^{-sx_3}\right) \cos\left(px_2\right)\\{}^{(k)}v_{3p}(x_2, x_3) &= {}^{(k)}\left(D_1 e^{sx_3} + D_2 e^{-sx_3}\right) \sin\left(px_2\right)\\{}^{(k)}v_{2c}(x_2, x_3) &= {}^{(k)}V(x_3) \cos\left(px_2\right)\\{}^{(k)}v_{3c}(x_2, x_3) &= {}^{(k)}W(x_3) \sin\left(px_2\right)\end{aligned} \tag{18}$$

with ${}^{(k)}B_1$, ${}^{(k)}B_2$, ${}^{(k)}D_1$ and ${}^{(k)}D_2$ constants which are derived substituting Eq. (18) into Eq. (13) [5]. The functions ${}^{(k)}V(x_3)$, ${}^{(k)}W(x_3)$ depend on 4 constants, ${}^{(k)}a_{11}$, ${}^{(k)}a_{21}$, ${}^{(k)}a_{12}$ and ${}^{(k)}a_{22}$, which leads to a total of $4 \times n$ unknowns for the plate (see [5, 20]). In the classical approach, the unknowns are obtained by imposing continuity/boundary conditions, which lead to a system of $4 \times n$ coupled algebraic equations.

The transfer matrix method allows an efficient closed-form solution of the thermal and mechanical problem for any numbers of layers.

3.2 *Explicit Expressions for Stresses and Displacements in Layered Plates*

The transfer matrix method is the first attempt in the literature to methodically extend the solution derived for a layer to a multilayered medium [23]. The method has been applied to obtain solutions for the heat conduction problem and for the thermo-elastic problem described in Sect. 3.1 in [5]. The 3D problem has been solved in [7]. The basic steps of the method are outlined below.

First a local transfer matrix is defined which relates the temperature (displacements and stresses) at the bottom and top surfaces of a generic layer using Eqs. (11), (14). The local transfer matrix is then related to that of the adjacent layer by imposing continuity conditions at the interfaces, (11), (14). Starting from the uppermost layer and using the local matrices and continuity conditions at the interfaces, a global transfer matrix is then derived which relates the field variables at the bottom surface of the medium to those at the top. The solution for a medium with any arbitrary number of layers is then obtained through the imposition of the boundary conditions only, Eqs. (12), (15). Using the approach detailed above the following explicit expressions have been obtained for temperature, stresses and displacements in the layers of the plate in Fig. 1:

$$
\begin{aligned}
&{}^{(k)}T(x_2,x_3)=\left[\left({}^{(k)}F_{11}{}^{(1)}c_1+{}^{(k)}F_{12}{}^{(1)}c_2\right)e^{{}^{(k)}sx_3}+\left({}^{(k)}F_{21}{}^{(1)}c_1+{}^{(k)}F_{22}{}^{(1)}c_2\right)e^{-{}^{(k)}sx_3}\right]\sin(px_2)\\
&{}^{(1)}c_1=\frac{T_u-T_l e^{{}^{(1)}sx_3^0}\left({}^{(n)}F_{12}e^{sx_3^n}+{}^{(n)}F_{22}e^{-sx_3^n}\right)}{\left({}^{(n)}F_{11}e^{sx_3^n}+{}^{(n)}F_{21}e^{-sx_3^n}\right)-e^{2{}^{(1)}sx_3^0}\left({}^{(n)}F_{12}e^{sx_3^n}+{}^{(n)}F_{22}e^{-sx_3^n}\right)};\quad {}^{(1)}c_2=\frac{T_l-{}^{(1)}c_1e^{{}^{(1)}sx_3^0}}{e^{-{}^{(1)}sx_3^0}}
\end{aligned}
\tag{19}
$$

$$
\begin{aligned}
{}^{(k)}v_2(x_2,x_3)&=\left\{{}^{(k)}Q_1(x_3)+\sum_{l=1}^{4}\left[\sum_{r=1}^{4}{}^{(k)}P_{1r}(x_3){}^{(k)}\Omega_{rl}\right]\left({}^{(1)}M_l(x_3^0)-{}^{(1)}Q_l(x_3^0)\right)+\sum_{t=1}^{4}{}^{(k)}P_{1t}(x_3){}^{(k)}S_t\right\}\cos(px_2)\\
{}^{(k)}v_3(x_2,x_3)&=\left\{{}^{(k)}Q_2(x_3)+\sum_{l=1}^{4}\left[\sum_{r=1}^{4}{}^{(k)}P_{2r}(x_3){}^{(k)}\Omega_{rl}\right]\left({}^{(1)}M_l(x_3^0)-{}^{(1)}Q_l(x_3^0)\right)+\sum_{t=1}^{4}{}^{(k)}P_{2t}(x_3){}^{(k)}S_t\right\}\sin(px_2)\\
{}^{(k)}\sigma_{33}(x_2,x_3)&=\left\{{}^{(k)}Q_3(x_3)+\sum_{l=1}^{4}\left[\sum_{r=1}^{4}{}^{(k)}P_{3r}(x_3){}^{(k)}\Omega_{rl}\right]\left({}^{(1)}M_l(x_3^0)-{}^{(1)}Q_l(x_3^0)\right)+\sum_{t=1}^{4}{}^{(k)}P_{3t}(x_3){}^{(k)}S_t\right\}\sin(px_2)
\end{aligned}
\tag{20}
$$

$$
\begin{aligned}
&{}^{(k)}\sigma_{23}(x_2,x_3)=\left\{{}^{(k)}Q_4(x_3)+\sum_{l=1}^{4}\left[\sum_{r=1}^{4}{}^{(k)}P_{4r}(x_3){}^{(k)}\Omega_{rl}\right]\left({}^{(1)}M_l(x_3^0)-{}^{(1)}Q_l(x_3^0)\right)+\sum_{t=1}^{4}{}^{(k)}P_{4t}(x_3){}^{(k)}S_t\right\}\cos(px_2)\\
&{}^{(k)}\sigma_{22}(x_2,x_3)=\left\{-p^{(k)}C_{22}\left[{}^{(k)}Q_1(x_3)+\sum_{l=1}^{4}\left[\sum_{r=1}^{4}{}^{(k)}P_{1r}(x_3){}^{(k)}\Omega_{rl}\right]\left({}^{(1)}M_l(x_3^0)-{}^{(1)}Q_l(x_3^0)\right)\right.\right.\\
&\left.+\sum_{t=1}^{4}{}^{(k)}P_{1t}(x_3){}^{(k)}S_t\right]+{}^{(k)}C_{23}\left[{}^{(k)}Q_2(x_3)_{,3}+\sum_{l=1}^{4}\left[\sum_{r=1}^{4}{}^{(k)}P_{2r}(x_3)_{,3}{}^{(k)}\Omega_{rl}\right]\left({}^{(1)}M_l(x_3^0)-{}^{(1)}Q_l(x_3^0)\right)\right.\\
&\left.\left.+\sum_{t=1}^{4}{}^{(k)}P_{2t}(x_3)_{,3}{}^{(k)}S_t\right]\right\}\sin(px_2)\quad {}^{(k)}(C_{12}\alpha_1+C_{22}\alpha_2+C_{23}\alpha_3){}^{(k)}T(x_2,x_3)
\end{aligned}
$$

where ${}^{(k)}F_{ij}$, ${}^{(1)}M_l\left(x_3^0\right)$, ${}^{(k)}\Omega_{rl}$, ${}^{(k)}S_t$, ${}^{(k)}Q_l(x_3)$, ${}^{(k)}P_{ir}(x_3)$, B^k, ${}^{(k)}E_{ij}(x_3)$ depend on the geometrical and material properties of layers and interfaces through the formulas in [5]. Applications to sandwich and laminated plates can be found in [6, 7].

3.3 *Conclusions on Elasticity Solutions through the Transverse Matrix Method*

Explicit 2D and 3D solutions have been derived for stresses and displacements in simply supported layered plates with interfacial imperfections subject to thermo-mechanical loading. The solutions are exact and can be applied for accurate parametric analyses and as benchmark solutions for the verification of approximate models.

4 Linear Elastic Fracture Mechanics Solutions for Sandwich Beams with Face-Core Delaminations

Sandwich structures are highly susceptible to manufacturing flaws and in-service damage. Debonds may arise at the face/core interfaces which will degrade the load carrying capacity and integrity of the sandwich structure, and may even grow catastrophically during both quasi-static and fatigue loading. In order to assess the criticality of debond flaws, the fracture properties of the face/core interface have to be measured and used as input properties in fracture mechanical analysis models. Several mixed mode interface fracture tests have been proposed in the literature for the characterization of sandwich face/core interfaces: e.g. Double Cantilever Beam test (DCB), Cracked Sandwich Beam test (CSB), Mixed-Mode Bending test (MMB), DCB Uneven Bending Moment test (DCB-UBM). The analysis of most of the proposed test specimens relies on approximate structural theories and/or numerical finite element analyses to define energy release rate and mode mixity phase angle and consequently the fracture toughness for different mode mixity conditions.

Semi-analytic linear elastic fracture mechanics solutions have been derived in [8, 65] for symmetric sandwich beams with face/core debonds subjected to arbitrary end loading, Fig. 5. They define energy release rate and mode mixity angle in terms of crack tip force and moment resultants and are applicable to experimental fracture tests. Solutions for sandwich systems where the debonded arms are subjected to bending moments and axial forces are available in [9, 13]. However, most of the fracture specimens are subjected to shear, e.g. the CSB and the DCB tests, which strongly affects the fracture parameters [11, 12, 34]. The expressions in [8] account for all effects of shear on the fracture parameters and in particular for those associated with the near tip deformations produced by the crack tip moments,

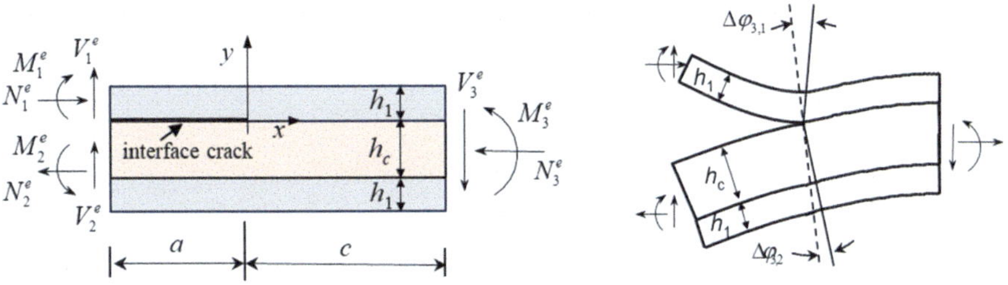

Fig. 5 Sandwich beam with a face/core debond subjected to generalized end forces

which are important and cannot be neglected also for very long crack lengths. The expressions derived in [8] are also applicable to complement finite element or structural mechanics calculations in beams and wide-plates subjected to various loading conditions, [17, 20, 32, 33] avoiding the case specific calculation of the fracture parameters. Recently the procedure in [8] has been extended to model the decohesion of thin layers from infinitely thick substrates [66]. The derivation in [8] is reviewed in Sect.4.1 and an application to a DCB sandwich beam is presented in Sect. 4.2.

4.1 Semi-Analytic Solutions for Energy Release Rate and Mode Mixity Phase Angle

A sandwich beam with an interface edge-crack of length a, subjected to force and moment resultants applied per unit width at the end cross sections is shown in Fig. 5. The layers are assumed to be isotropic and linearly elastic, with h_1 the thickness of the upper and lower face-sheets and h_c the thickness of the core. The external force and moment resultants are applied at distances $a,c >> h_1, h_c$ (minimum distances for the application of the solutions to laboratory specimens are given in [8] and will not be discussed here).

Crack tip force and moment resultants are derived from Fig. 5 and shown in Fig. 6(a); they are separated into (b) normal forces and bending moments and (c) shear forces, Fig. 6. Following the methodology in [10], the load system in Fig. 6b is conveniently decomposed into the elementary systems: (b1) pure bending moments, M, and normal forces, P, acting on the two delaminated arms plus the required compensating moment, M_*; (b2) pure bending moments and normal forces, N_3 and M_3, acting on the intact element. The sub-system (b2) is not involved in the calculation of the fracture parameters. Following [12], the loads in Fig. 6.c are decomposed into the elementary systems: (c1) double-shear, V_D, acting on the two arms and (c2) single-shear, V_S, acting on the upper arm plus the compensating shear force acting on the base section. The elementary loads are:

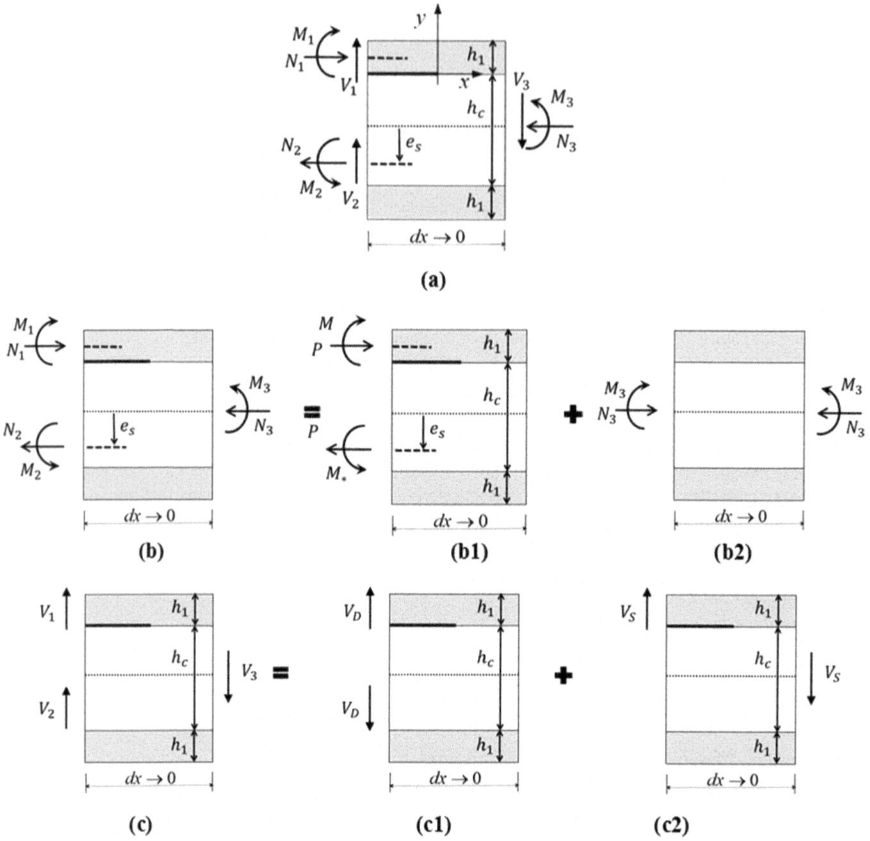

Fig. 6 (**a**) Crack tip forces. (**b**) Load decomposition: bending moments and axial forces. (**c**) Load decomposition: shear forces

$$
\begin{aligned}
&P = N_1 - N_3 C_1 - \frac{M_3}{h_1} C_2, && M = M_1 - M_3 C_3 \\
&M_* = M + P h_1 \left(\frac{1}{2} + \frac{1}{2\eta} + \tilde{e}_S \right), && V_D = -V_2, \quad V_S = V_3
\end{aligned}
\tag{21}
$$

with:

$$
C_1 = \frac{\Sigma\eta}{1 + 2\Sigma\eta}, \quad C_2 = \frac{1}{\tilde{D}_b} \left(\frac{1}{2} + \frac{1}{2\eta} \right), \quad C_3 = \frac{1}{12\tilde{D}_b}
\tag{22}
$$

where $\eta = h_1/h_c$ is the relative thickness of face sheets and core and $\Sigma = \overline{E}_1/\overline{E}_c$ is the ratio of the effective Young's moduli of the layers, which is related to the first Dundurs' parameter $\alpha = (\Sigma - 1)/(\Sigma + 1)$, with $\overline{E}_i = E_i$ for plane-stress and

$\overline{E}_i = E_i/(1-\nu_i^2)$, for plane-strain; $\tilde{e}_s$ defines the position of the neutral axis of the substrate arm and $\tilde{D}_b$ is the dimensionless bending stiffness of the base cross section (see [8]).

The energy release rates associated to the elementary load systems in Fig. 6. are [8, 10]:

$$\begin{aligned} \mathcal{G}_P &= (f_P P)^2/(\overline{E}_1 h_1), \quad \mathcal{G}_M = (f_P M)^2/(\overline{E}_1 {h_1}^3), \\ \mathcal{G}_{VD} &= (f_{VD} V_D)^2/(\overline{E}_1 h_1), \quad \mathcal{G}_{VS} = (f_{VS} V_S)^2/(\overline{E}_1 h_1) \end{aligned} \tag{23}$$

where

$$\begin{aligned} f_M(\Sigma,\eta) &= \left[\frac{1}{2}\left(12+\frac{1}{\tilde{D}_s}\right)\right]^{1/2} \\ f_P(\Sigma,\eta) &= \left[\frac{1}{2}\left(1+\frac{\Sigma\eta}{1+\Sigma\eta}+\frac{1}{\tilde{D}_s}\left(\frac{1}{2}+\frac{1}{2\eta}+\tilde{e}_s\right)^2\right)\right]^{1/2} \end{aligned} \tag{24}$$

The functions f_{VD}, f_{VS}, depend also on the second Dundurs' parameter, which accounts for the compressibility of the layers, $\beta = \frac{G_1(\kappa_c-1)-G_c(\kappa_1-1)}{G_1(\kappa_c+1)+G_c(\kappa_1+1)}$, with G_i the shear modulus and $\kappa_i = 3-4\nu_i$ for plane strain and $\kappa_i = (3-\nu_i)/(1+\nu_i)$ for plane stress. The functions f_{VD}, f_{VS} cannot be defined in closed form under general conditions and require a numerical derivation (an analytical solution for the problem of a thin layer on a semi-infinite substrate has been recently obtained in [35]). They have been defined in [8] using a finite element model implementing the Crack Surface Displacement Extrapolation CSDE method formulated in [36]. Tabulated results for different geometrical/material properties of the layers are in [8]. Table 1 presents exemplary results.

The debond crack is typically under mixed-mode conditions and the interfacial toughness is strongly affected by the mode mixity. A mixed-mode phase angle is associated to the elementary loads in Fig. 6. The angle measures the ratio of normal and shear stresses at a fixed distance ahead of the crack tip, $r = h_1$, and is given by:

$$\psi = \psi_{h1} = \tan^{-1}\left(\mathrm{Im}\left[K{h_1}^{i\varepsilon}\right]/\,\mathrm{Re}\left[K{h_1}^{i\varepsilon}\right]\right) \tag{25}$$

where $K = K_1 + iK_2$ is the complex stress intensity factor and $\varepsilon = 1/2\pi \ln[(1-\beta)/(1+\beta)]$ the oscillatory index. The definition of the reference distance, $r = h_1$, is irrelevant since the transformation rule $\psi_{\hat{r}} = \psi_{h1} + \varepsilon \ln(\hat{r}/h_1)$ holds [37]. For incompressible materials with $\beta = 0$ and $\varepsilon = 0$, the phase angle takes the usual form, $\psi = \tan^{-1}(K_{II}/K_I)$ and is uniquely defined.

The absolute values of the complex stress intensity factors associated with the elementary load systems, $P + M_*$ and M, are $|K_P| = \sqrt{(1-\alpha)(1-\beta^2)^{-1}} f_P P {h_1}^{-1/2}$

Table 1 Dimensionless functions f_{VD}, f_{VS} and phase angles ψ_{VD}, ω and ψ_{VS} for an interface crack in a symmetric sandwich beam with $\eta = h_1/h_c = 0.25$ and $\alpha \le 0.8$ (more data in [8])

$\eta = h_1 / h_c = 0.25$										
α \ β	0		0.2		0.4		0.6		0.8	
	f_{VD}	f_{VS}	f_{VD}	f_{VS}	f_{VD}	f_{VS}	f_{VD}	f_{VS}	f_{VD}	f_{VS}
0	1.937	1.788	2.078	1.907	2.280	2.089	2.608	2.397	3.294	3.071
0.1	1.967	1.806	2.107	1.922	2.310	2.100	2.639	2.402	3.328	3.060
0.2	1.992	1.821	2.133	1.934	2.336	2.107	2.665	2.401	3.354	3.043
0.3					2.357	2.109	2.685	2.395	3.373	3.018
0.4									3.385	2.987

	ψ_{VD}	ω	ψ_{VS}	ψ_{VD}	ω	ψ_{VS}	ψ_{VD}	ω	ψ_{VS}	ψ_{VD}	ω	ψ_{VS}	ψ_{VD}	ω	ψ_{VS}
0	–0.7°	52.0°	–3.5°	–2.2°	54.0°[a]	–5.8°	–3.9°	56.1°[a]	–8.3°	–5.8°	59.4°[a]	–11.3	–8.3°	63.5°[a]	–15.0°
0.1	–3.1°	47.6°	–5.9°	–4.4°	49.7°[a]	–8.0°	–5.8°	52.0°[a]	–10.2	–7.3°	55.6°[a]	–12.8	–9.2°	60.0°[a]	–16.1°
0.2	–5.6°	43.1°	–8.4°	–6.7°	45.3°[a]	–10.2°	–7.8°	47.8°[a]	–12.2	–9.0°	51.7°[a]	–14.6	–10.2	56.5°[a]	–17.3°
0.3							–10.0	43.4°[a]	–14.4	–10.8	47.6°[a]	–16.3	–11.4	52.7°[a]	–18.6°
0.4													–12.9	48.4°[a]	–20.0°

[a] Interpolation/extrapolation on α, β of the ω values in [9]. Uncertainties on the tabulated values: ±0.2° on ψ_{VD}, ψ_{VS}; ±0.007 on f_{VD}, f_{VS}

and $|K_M| = \sqrt{(1-\alpha)(1-\beta^2)^{-1}} f_M M h_1^{-3/2}$. The associated mode mixity angles are ψ_M and $\psi_P = \omega$, which are related through:

$$\psi_M = \omega + \gamma_M - \pi/2,\ \gamma_M(\Sigma, \eta) = \sin^{-1}\left[\frac{1}{2\tilde{D}_s f_P f_M}\left(\frac{1}{2} + \frac{1}{2\eta} + \tilde{e}_s\right)\right] \tag{26}$$

The equation above shows that only one of the phase angles associated to the elementary loads in Fig. 6b remain undefined and must be derived through a single numerical analysis of the problem. The values of $\psi_P = \omega$ have been tabulated in [8] on varying geometrical and material properties and exemplary results are presented in Table 1.

The absolute values of the complex stress intensity factors and phase angles associated to the elementary loads of single and double shear are $|K_{VD}| = \sqrt{(1-\alpha)(1-\beta^2)^{-1}} f_{VD} V_D h_1^{-1/2}$, $|K_{VS}| = \sqrt{(1-\alpha)(1-\beta^2)^{-1}} f_{VS} V_S h_1^{-1/2}$, $\psi_{VD} = \tan^{-1}(\mathrm{Im}[K_{VD} h_1^{i\varepsilon}]/\ \mathrm{Re}\ [K_{VD} h_1^{i\varepsilon}])$ and $\psi_{VS} = \tan^{-1}(\mathrm{Im}[K_{VS} h_1^{i\varepsilon}]/\ \mathrm{Re}\ [K_{VS} h_1^{i\varepsilon}])$. The phase angles have been calculated numerically in [8] and exemplary results are in Table 1.

The stress intensity factors are combined in the complex plane of the stress intensity factors to define general expressions for the energy release rate and the mode mixity phase angle under general loading conditions:

$$\begin{aligned}
\mathcal{G} = {} & \frac{f_M{}^2 M^2}{\overline{E}_1 h_1^3} + \frac{f_P{}^2 P^2}{\overline{E}_1 h_1} + 2\frac{f_M f_P}{\overline{E}_1 h_1{}^2} \sin\gamma_M MP \\
& + \frac{f_{VD}{}^2 V_D{}^2}{\overline{E}_1 h_1} + \frac{f_{VS}{}^2 V_S{}^2}{\overline{E}_1 h_1} + 2\frac{f_{VD} f_{VS}}{\overline{E}_1 h_1} \cos(\psi_{VD} - \psi_{VS}) V_D V_S \\
& + 2\frac{f_M f_{VD}}{\overline{E}_1 h_1{}^2} \cos(\psi_M - \psi_{VD}) M V_D + 2\frac{f_M f_{VS}}{\overline{E}_1 h_1{}^2} \cos(\psi_M - \psi_{VS}) M V_S \\
& + 2\frac{f_P f_{VD}}{\overline{E}_1 h_1} \cos(\omega - \psi_{VD}) P V_D + 2\frac{f_P f_{VS}}{\overline{E}_1 h_1} \cos(\omega - \psi_{VS}) P V_S
\end{aligned} \tag{27}$$

$$\psi = \psi_{h_1} = \tan^{-1}\left(\frac{f_M M \sin\psi_M + f_P P h_1 \sin\omega + f_{VD} V_D h_1 \sin\psi_{VD} + f_{VS} V_S h_1 \sin\psi_{VS}}{f_M M \cos\psi_M + f_P P h_1 \cos\omega + f_{VD} V_D h_1 \cos\psi_{VD} + f_{VS} V_S h_1 \cos\psi_{VS}}\right) \tag{28}$$

A second method has been used in [8] to define the energy release rate in the cracked sandwich beam in Fig. 5. The beam has been described using a structural mechanics approximation and introducing crack tip root-rotations to represent the near tip deformations which arise due to the different crack tip loads, Fig. 5. These deformations affect the energy release rate and mode mixity angle in the presence of shear loading [11, 12]. Referring to [8] for details, the energy release rate is given by:

$$
\begin{aligned}
\mathcal{G} = {} & \frac{{f_M}^2 M^2}{\overline{E}_1 h_1^3} + \frac{{f_P}^2 P^2}{\overline{E}_1 h_1} + \frac{MP}{\overline{E}_1 h_1^2} 2 f_M f_P \cdot \sin(\gamma_M) \\
& + \frac{{V_S}^2}{\overline{E}_1 h_1}\left(a_1^{VS} + \frac{1}{2}\left(\frac{1}{\tilde{A}_{Vd}} - \frac{1}{\tilde{A}_{Vb}}\right)\right) + \frac{{V_D}^2}{\overline{E}_1 h_1}\left(a_1^{VD} - a_2^{VD} + \frac{1}{2}\left(\frac{1}{\tilde{A}_{Vd}} + \frac{1}{\tilde{A}_{Vs}}\right)\right) \\
& + \frac{V_S P}{\overline{E}_1 h_1} a_1^P + \frac{V_S M}{\overline{E}_1 h_1^2} a_1^M + \frac{V_D P}{\overline{E}_1 h_1}\left(a_1^P - a_2^P\right) + \frac{V_D M}{\overline{E}_1 h_1^2}\left(a_1^M - a_2^M\right) \\
& + \frac{V_D V_S}{\overline{E}_1 h_1}\left(a_1^{VD} + a_1^{VS} - a_2^{VS} + \frac{1}{\tilde{A}_{Vd}}\right)
\end{aligned}
\tag{29}
$$

where $\tilde{A}_{Vd} = A_{Vd}/\overline{E}_1 h_1$ and $\tilde{A}_{Vs} = A_{Vs}/\overline{E}_1 h_1$ are the dimensionless shear stiffnesses of the beam arms; $a_i^M, a_i^P, a_i^{VS}, a_i^{VD}$ for i = 1,2 are the root-rotation compliance coefficients, which define the relative crack tip rotations between the delaminated/substrate arms and the base arm generated by unit bending moments, axial forces and shear forces: $\Delta\varphi_{3,1} = \varphi_3 - \varphi_1$ and $\Delta\varphi_{3,2} = \varphi_3 - \varphi_2$ with φ_1, φ_2, φ_3 the bending rotations of the cross sections of the debonded, base and substrate arms, around the z axis (positive if counterclockwise), Fig. 5. The relative rotations depend on the compliance of the material at and ahead of the crack tip and, if a and c are sufficiently long, are described as linear combinations of crack tip force and moment resultants through the compliance coefficients:

$$
\Delta\varphi_{3,i} = \frac{1}{\overline{E}_1 h_1}\left(\frac{a_i^M}{h_1} M + a_i^P P + a_i^{VS} V_S + a_i^{VD} V_D\right) \tag{30}
$$

The two expressions of the energy release rate in Eqs. (27) and (29) can be compared to define the relationships between the root rotation coefficients, $a_i^M, a_i^P, a_i^{VS}, a_i^{VD}$, for $i = 1{,}2$, and the dimensionless coefficients of Eq. (27). The terms which multiply $V_S M$ and $V_S P$ in the two expressions of $\mathcal{G}$,

$$
2\cos(\psi_M - \psi_{VS}) f_M f_{VS} = a_1^M, \; 2 f_P f_{VS} \sin\gamma_{VS} = a_1^P, \tag{31}
$$

directly define the root rotations of the delaminated arm, $\Delta\varphi_{3,1} = a_1^M$ and $\Delta\varphi_{3,1} = a_1^P$, due to unit bending moments $M/\left(\overline{E}_1 h_1^2\right) = 1$ and unit normal forces $P/\left(\overline{E}_1 h_1\right) = 1$, respectively. Similarly, the terms which multiply $V_D M$ and $V_D P$ define the differences between the root rotations of the delaminated and substrate arms, $\left(a_1^M - a_2^M\right)$ and $\left(a_1^P - a_2^P\right)$, generated respectively by unit bending moments and unit normal forces:

$$
2\cos(\psi_M - \psi_{VD}) f_M f_{VD} = a_1^M - a_2^M, \; 2 f_P f_{VD} \sin\gamma_{VD} = a_1^P - a_2^P \tag{32}
$$

Similar relations are derived for the other compliance coefficients:

$$f_{VD}{}^2 = a_1^{VD} - a_2^{VD} + \frac{1}{2}\left(\frac{1}{\tilde{A}_{Vd}} + \frac{1}{\tilde{A}_{Vs}}\right) f_{VS}{}^2 = a_1^{VS} + \frac{1}{2}\left(\frac{1}{\tilde{A}_{Vd}} - \frac{1}{\tilde{A}_{Vb}}\right)$$
$$2f_{VD}f_{VS}\cos(\psi_{VD} - \psi_{VS}) = a_1^{VD} + a_1^{VS} - a_2^{VS} + \frac{1}{\tilde{A}_{Vd}} \tag{33}$$

The seven equations above can be used to define the root rotation coefficients, or their combinations which appear in the expression of the energy release rate in Eq. (29), a_1^{VS}, $\left(a_1^{VD} - a_2^{VD}\right)$, a_1^P, a_1^M, a_2^M, a_2^P, $\left(a_1^{VD} - a_2^{VS}\right)$, from the dimensionless coefficients in Table 1 (or the tables in [8] for other material combinations). Or, if the root rotation coefficients are known, for instance through finite element calculations or structural mechanics approximations [11], the equations can be used to define the dimensionless coefficients in the expression of the energy release rate in Eq. (27).

Equation (33) shows that the functions $f_{VD}{}^2$ and $f_{VS}{}^2$ account for different effects: (i) the root rotations due to unit double, $\left(a_1^{VD} - a_2^{VD}\right)$, and single, a_1^{VS}, shear and (ii) the shear strains which arise along the beam arms and depend on their stiffnesses, $\gamma_d = (V_D + V_S)/A_{Vd}$, $\gamma_S = V_D/A_{Vs}$ and $\gamma_b = V_S/A_{Vb}$. The shear stiffnesses can then be defined through accurate numerical analyses or using the Jourawsky approximation [8]. If the shear stiffnesses of the arms are assumed to be infinite, as in Euler Bernoulli beam theory (or classical plate theory) then all terms with $\tilde{A}_{Vi}$ would vanish in the equations above and the root rotations can be assumed to account for all contribution of shear; this assumption has been recently used in [38].

4.2 *Application to a DCB Sandwich Specimen: Influence of Shear*

In a DCB sandwich specimen subjected to transverse forces F (inset of Fig. 7a), the elementary crack tip loads are $M = Fa$ and $V_D = F$, while P, $V_S = 0$. The energy release rate and mode mixity angle, Eqs. (27), (28), (29) simplify as:

$$\mathcal{G}_{DCB} = \frac{F^2}{\bar{E}_1 h_1}\left[\left(\frac{a}{h_1}\right)^2 f_M^2 + 2\left(\frac{a}{h_1}\right) f_{VD} f_M \cos(\psi_M - \psi_{VD}) + f_{VD}^2\right] \tag{34}$$

$$\mathcal{G}_{DCB} = \frac{F^2}{\bar{E}_1 h_1}\left[\left(\frac{a}{h_1}\right)^2 f_M^2 + \left(\frac{a}{h_1}\right)\left(a_1^M - a_2^M\right) + a_1^{VD} - a_2^{VD} + \frac{1}{2}\left(\frac{1}{\tilde{D}_{Vd}} + \frac{1}{\tilde{D}_{Vs}}\right)\right] \tag{35}$$

$$\psi_{DCB} = \tan^{-1}\left(\frac{a/h_1 f_M \sin\psi_M + f_{VD}\sin\psi_{VD}}{a/h_1 f_M \cos\psi_M + f_{VD}\cos\psi_{VD}}\right) \tag{36}$$

The energy release rate has terms which depend on the crack length at different powers. The dominant term, with the crack length squared, is the contribution of the crack tip bending moments. The second term, which depend linearly on the crack length describes the effects of the shear forces on the root rotations generated by the crack tip bending moments, which are given by $\left(a_1^M - a_2^M\right)Fa/\left(E_1 h_1{}^2\right)$. The remaining terms, which are independent of the crack length, and become less important on increasing a/h_1, describe the effects of the root rotations due to shear forces, $\left(a_1^{VD} - a_2^{VD}\right)F/\left(E_1 h_1\right)$ and shear strains.

An application of the results is presented in Fig. 7a-d. Assuming $\eta = h_1/h_c = 0.25$, $\Sigma = \overline{E}_1/\overline{E}_c = 9$ (or $\alpha = 0.8$) and $\beta = 0$ yields $f_M = 2.49$ and $\gamma_M = 0.159$, Eqs. (24), (26); the numerical coefficients are obtained from Table 1: $\omega = 63.5^\circ, \psi_M = -17.4^\circ, f_{VD} = 3.294$ and $\psi_{VD} = -8.3^\circ$. The dimensionless energy release rate and mode mixity angle are shown in the diagrams in Fig. 7a-d on varying the normalized length of the crack. The results are compared with solutions which neglect root rotations and shear deformations. The solid curves in (b)(d) define the absolute relative percent error on energy release rate and mode mixity angle if the effects of shear are fully neglected, but for the bending moment generated at the crack tip, and $\mathcal{G} = \mathcal{G}_M = F^2/\overline{E}_1 h_1 (a/h_1)^2 f_M^2$ and $\psi = \psi_M$. The error decays slowly and on $\mathcal{G}$ it goes from 37% for $a/h_1 = 5$ to 8% for $a/h_1 = 30$.

4.3 Conclusions on Interfacial Fracture Mechanics Solutions for Sandwich Beams With Face/Core Delaminations

Semi-analytic solutions have been derived using 2D elasticity and interfacial fracture mechanics for energy release rate and mode mixity phase angle in sandwich beams with face/core debonds. The solutions account for the important effects of shear. They are applicable to support experimental tests and avoid complex, case specific, finite element calculations for the characterization of the fracture properties.

5 Interaction of Multiple Damage Mechanisms in Sandwich Beams Subjected to Time-Dependent Loading

The high strength and stiffness and light weight of polymer composite sandwiches, explains their expanding use for structural applications of the naval, aerospace, civil and transportation industries. The major drawback of these materials is their sensitivity to different forms of damage: crushing and fracture of the core; face-sheet delamination, face-sheet/core interface fracture, matrix cracking and fibre failure and face-sheet instabilities. Studies have been conducted to analyze the fundamental failure modes and define failure maps in order to optimize design and understand the effects of pre-existing damage on the response (e.g [15, 39–43]). Experimental results show that damage in sandwich structures, especially under dynamic loads, is usually the result of a combination of several mechanisms [44, 45]. Understanding

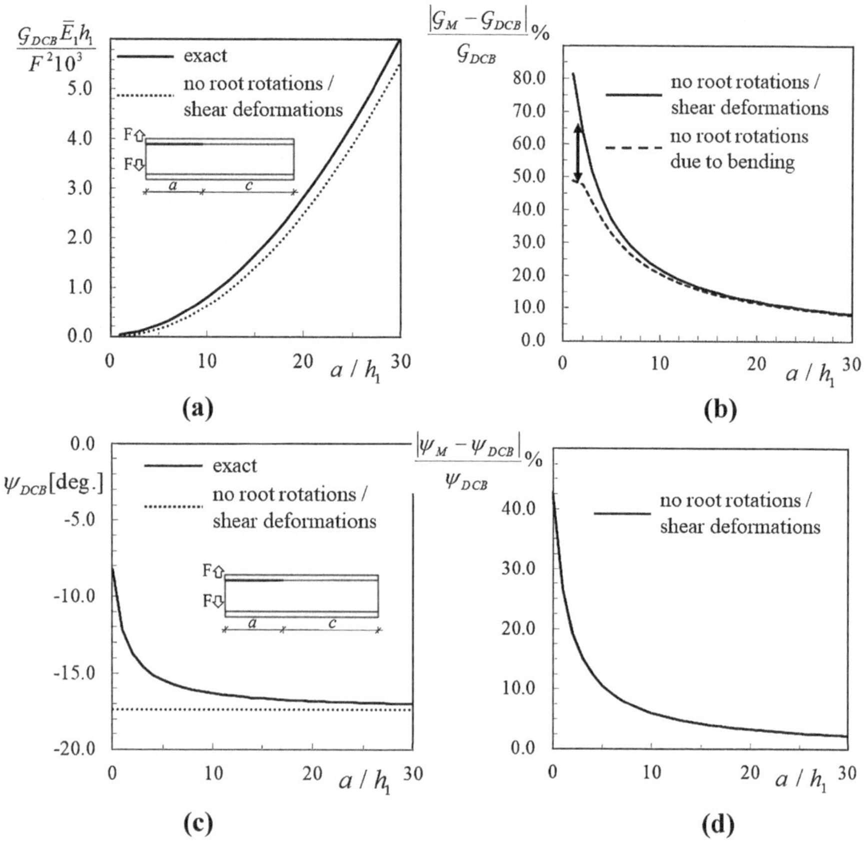

Fig. 7 (**a**, **c**) Energy release rate and mode mixity phase angle in a DCB specimen; (**b**, **d**) relative error on energy release rate and mode mixity phase angle using approximate theory which neglects the effects of shear (after [8])

damage interaction and evolution may give useful indications for the optimal design of these systems.

In [14] theoretical models have been formulated to explain evolution and interaction of the damage mechanisms for delamination of the face-sheets and core crushing in a composite sandwich beam subjected to dynamically applied out-of-plane loading and continuously supported by a rigid plane. The characteristic features of the problem and transitions in damage progression have been explored. In Sect. 5.1 the formulation of the models is reviewed and in Sect. 5.2 some results are presented and discussed.

5.1 Models

Consider a sandwich beam with a rectangular cross section of unit width, laminated face sheets of thickness h and a homogeneous core of thickness c, subjected to an out-of-plane time dependent load per unit width $p(t)$, with t the time variable, Fig. 8. The laminated face sheets are assumed to be linearly elastic and perfectly brittle with the layers between the laminae the potential fracture surfaces; the core is assumed to be elastic (in tension) and elastic-perfectly plastic (in compression) to describe materials that plastically indent under the applied load, namely the polymeric foams used for marine applications [42]. The beam is continuously supported by a rigid plane so that overall bending and shear are absent.

The upper face of the sandwich is modelled as a layered composite beam resting on a nonlinear elastic-plastic Winkler foundation, which approximates the core. The skin, which may have arbitrarily distributed delaminations, is discretized and decomposed into sub-beams using the two decomposition schemes in [33, 46]. In the first scheme, Fig. 8b, the decomposition is defined by longitudinal sections along the pre-existing delamination planes and vertical sections at the crack tip positions and at all coordinates where there are changes in the state of contact/cohesion. In the second scheme, Fig. 8c, the subdivision is defined by longitudinal sections at all pre-existing and potential delamination planes; the longitudinal sections extend also in the intact portion of the beam. All pre-existing and potential delaminations are treated as cohesive interfaces.

The sub-beams are described by first order shear deformation theory and normal and tangential tractions T^N and T^S are assumed to act at the lower and upper surfaces of each sub-beam (Fig. 9a). The tractions describe externally applied tractions on the skin upper surface, the action developed by the core on the skin lower surface and interfacial tractions due to contact or cohesion between the sub-beams.

The cohesive tractions acting between the sub-beams in Fig. 8c are assumed to be rate independent and are introduced to describe the unfailed interfaces and reproduce perfectly brittle fracture. The cohesive traction laws are:

$$\begin{aligned} T^N_{i,i+1} &= -k^N_{i,i+1} w^N_{i,i+1} \quad \text{for } 0 \le w^N_{i,i+1} \le w^N_0 \\ T^S_{i,i+1} &= -k^S_{i,i+1} w^S_{i,i+1} \quad \text{for } 0 \le w^S_{i,i+1} \le w^S_0 \end{aligned} \tag{37}$$

where $w^N_{i,i+1}$ and $w^S_{i,i+1}$ are the interfacial opening/sliding displacements between beam i and i + 1. The energy release rate is calculated as

$$\mathcal{G} = \mathcal{G}_I + \mathcal{G}_{II} = -\int_0^{w^N_{i,i+1}} T^N_{i,i+1} d\left(w^N_{i,i+1}\right) - \int_0^{w^S_{i,i+1}} T^S_{i,i+1} d\left(w^S_{i,i+1}\right)$$

and the interface fracture energies are $\mathcal{G}_{Icr} = 1/2k^N_{i,i+1}\left(w^N_0\right)^2$ and $\mathcal{G}_{IIcr} = 1/2k^S_{i,i+1}\left(w^S_0\right)^2$.

Contact tractions are introduced in both models in Fig. 8 to describe normal frictionless contact between the sub-beams:

$$T^N_{i,i+1} = -k^{contact}_{i,i+1} w^N_{i,i+1},\; T^S_{i,i+1} = 0,\; k^{contact}_{i,i+1} = H\left(-w^N_{i,i+1}\right)\frac{2E_T}{h_i + h_{i+1}} \tag{38}$$

where $k^{contact}_{i,i+1}$ is the through-thickness stiffness of the contacting beam segments and $H\left(-w^N_{i,i+1}\right)$ is the Heaviside step function.

Interfacial tractions are also introduced between the lower sub-beams and the rigid foundation to describe the elasto-plastic response of the core:

$$\begin{aligned} T^N_{i,i+1} &= T^f = kw(z) && \text{for}\, w(z) \le w_{cr} \\ T^N_{i,i+1} &= T^f_{cr} && \text{for}\, w(z) \ge w_{cr} \end{aligned} \tag{39}$$

and $T^S_{i,i+1} = 0$, for all $w(z)$, with k the foundation elastic modulus per unit width, $w(z) = w_i(z)$ the transverse displacement of the lowest sub-beam, T^f_{cr} and w_{cr} the core yielding strength and critical displacement. In thin sandwiches, the core is uniformly compressed and the modulus of the foundation is $k = E_c/c$, with E_c the transverse Young's modulus.

The solution methods for the two models in Fig. 8. are detailed in [14]. They use the formulations in [33, 46] for layered beams and apply an iterative procedure to

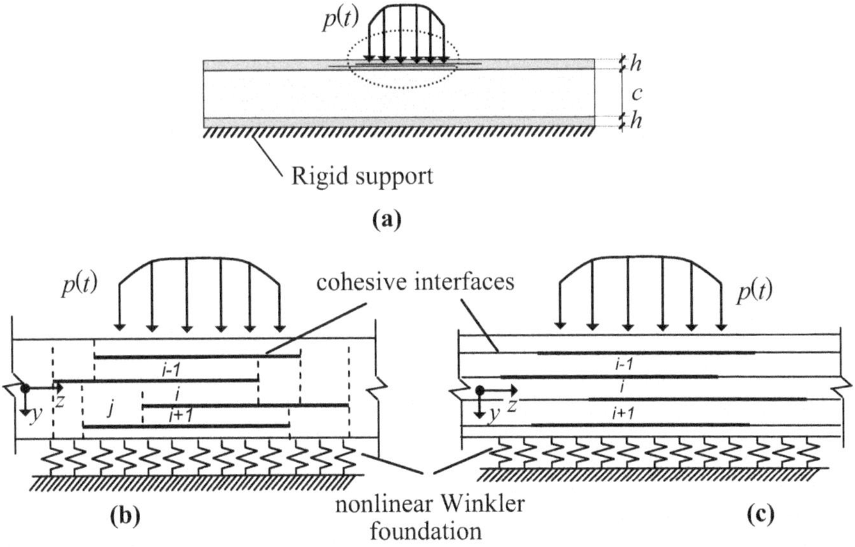

Fig. 8 (**a**) Composite sandwich beam continuously supported by a rigid plane and subjected to a time dependent load. Theoretical model for (**b**) quasi-static and (**c**) dynamic loading

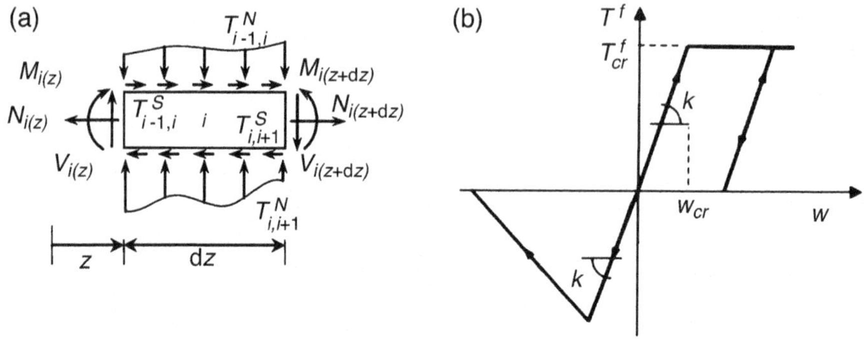

Fig. 9 (**a**) Infinitesimal element of a typical sub-beam *i* in the skin. (**b**) Interfacial tractions used to describe elasto-plastic response of of the core

follow the evolution of crack propagation in the skin and plasticity of the core. A discussion on the limitations of the methods is given in [14].

5.2 Results for Quasi-Static and Dynamic Loadings

The applications in [14] refer to a homogeneous, orthotropic/isotropic and perfectly brittle beam with central, equally spaced delaminations of half lengths a_i loaded by a concentrated point force per unit width P at the mid-span. The beam, of length 2 L, is assumed to be sufficiently long so that the solution is unaffected by the boundary conditions at the beam ends, which are assumed to be perfectly clamped (Fig. 10).

5.2.1 Elastic-Brittle System under Quasi-static Loading – Discussion and Conclusions

Figure 11 shows the critical load for the propagation of a single delamination in the skin, Fig. 10, which is described by the following relation:

$$\frac{P_{cr}}{\sqrt{\mathcal{G}_{cr}Eh}}\left(\frac{E}{hk}\right)^{1/4} = f_1\left(\frac{a}{\lambda_r}\right) \tag{40}$$

obtained assuming that the skin behaves as an elementary Euler-Bernoulli beam on a Winkler elastic foundation, and that the crack propagates according to a global energy release rate criterion $\mathcal{G} = \mathcal{G}_I + \mathcal{G}_{II} = \mathcal{G}_{cr}$.

The wavelength of the oscillations of the beam on the elastic foundation is the characteristic length scale of the problem, $\lambda_r = 2\sqrt{2}\pi(k_{br}/k)^{1/4}$ with k_{br} the bending stiffness of the delaminated beam. The dashed curve in the diagram defines the response of a clamped-clamped beam in the absence of the foundation, given by $P_{cr}/\sqrt{\mathcal{G}_{cr}Eh} = 2\sqrt{2}/3(h/a)$. The response of the beam tends to this limit for short cracks or large values of λ_r, which occur in the case of beams with soft cores and $k \to 0$. Crack propagation in this limit is unstable and unaffected by the presence of the core.

For larger values of a/λ_r, the foundation has an important influence on the response of the system and the critical load for crack propagation becomes a monotonically increasing function of the crack length up to a maximum, which corresponds approximately to $a_b \approx \lambda_r/2$; this behavior is due to the shielding produced by the reactions of the elastic foundation on the crack tip stress field. Crack propagation occurs in this range in pure mode II conditions, as detailed in the mode ratio diagram of Fig. 11b. After the peak, for $a \approx a_b$, the response becomes unstable and dominated by delamination openings.

The energy barrier acts as a crack arrester and its presence is confirmed by experimental observations showing that multiple delamination damage in the face-sheet produced by low velocity impact typically remains localized near the applied load [45].

The general conclusions, applicable also to faces with multiple delaminations are the following. Face/core interactions in elastic-brittle systems, shield the crack tips from the applied load and crack propagation is rendered stable (the critical applied load becomes an increasing function of the crack length). In addition, the core/face-sheet interactions generate energy barriers to the propagation of cracks in the face-sheet. The position of the barriers, which act as crack arresters, depends on the wavelength of the oscillations of the face-sheet on the elastic foundation. In beams with soft cores or when the delaminations are short, shielding of the fracture parameters is negligible and the unstable propagation can be controlled only by increasing bending stiffness or fracture toughness of the face-sheet only.

5.2.2 Influence of Core-plasticity on the Response under Quasi-static Loading – Discussion and Conclusions

Core plasticity diminishes the positive effects of the energy barriers on damage control. The critical load for the propagation of a single delamination in the skin now becomes:

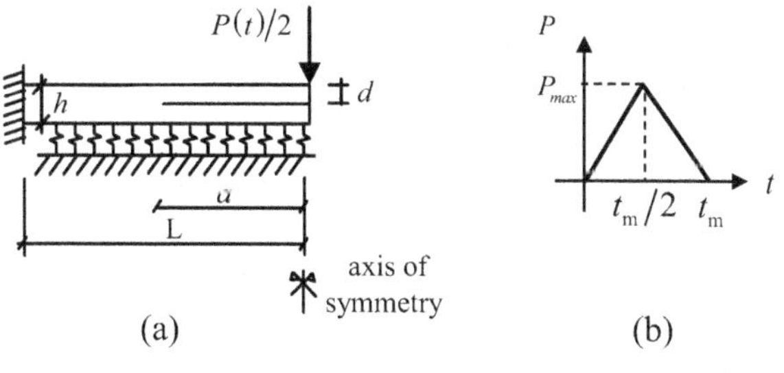

Fig. 10 (**a**) Schematics of the beam in the examples. (**b**) Time dependent load with shape and duration

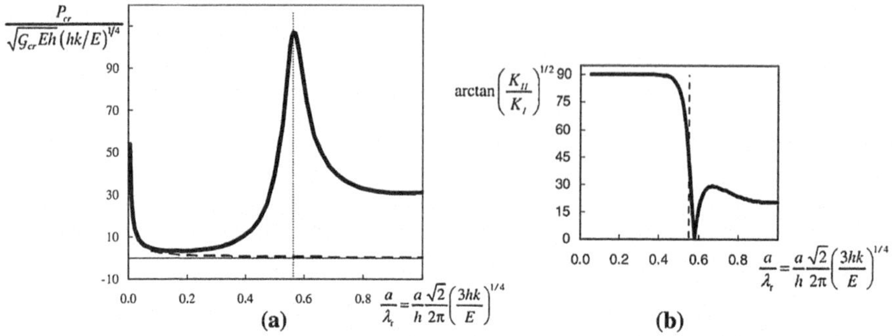

Fig. 11 (**a**) Dimensionless critical load for the propagation of a central, mid-plane delamination in the skin as a function of the delamination length in a sandwich beam with an elastic core (**b**) Relative amount of mode II to mode I stress intensity factors (after [14])

$$\frac{P_{cr}}{\sqrt{\mathcal{G}_{cr}Eh}}\left(\frac{E}{hk}\right)^{1/4} = f_3\left(\frac{a}{\lambda_r}, \frac{T^f_{cr}}{\sqrt{\mathcal{G}_{cr}k}}\right) \tag{41}$$

and is depicted in Fig. 12 for different values of the dimensionless yielding strength $T^f_{cr}/\sqrt{\mathcal{G}_{cr}k}$ (thick lines). The dashed curve in the diagram refers to the elastic foundation in Fig. 11. Compressive yielding of the core reduces the shielding of the fracture parameters with respect to the elastic-brittle system, the maximum load decreases and the energy barrier moves away from the applied load when the dimensionless yielding strength is reduced.

Magnitude and position of the energy barriers can be optimized in order to improve mechanical performance. To reduce damage extension, higher barriers closer to the applied load are sought by increasing the density of the core, or by reducing the bending stiffness of the face-sheet; on the other hand, softer cores, with reduced stiffness and strength, favour energy dissipation through plastic deformation and the formation of fracture surfaces in the skin leading to improved energy absorption.

The models in [14] also predict transitions in damage evolution in beams subjected to low velocity impact on varying geometrical/material parameters and energy input. If the geometry is kept unchanged, beams with soft cores damage by core plasticity, for low input energies, and by skin delamination coupled with core plasticity, for larger input energies; beams with dense cores damage by skin delamination with no core plasticity, for low input energies (though much higher than those producing core plasticity in the beams with softer cores), and by skin delamination coupled with core plasticity, for larger input energies. The transitions in damage evolution confirm experimental observations in [16].

5.2.3 Dynamic Loading – Discussion and Conclusions

In elastic-brittle systems, where the dimensionless yielding strength $T^f_{cr}/\sqrt{\mathcal{G}_{cr}k}$ is sufficiently high and the foundation remain elastic during and after loading, dynamic effects induce amplification of the fracture parameters with respect to the quasi static solution for all load durations $t_m < 2t_1$, with t_1 the fundamental period of vibration of an infinitely long intact beam, Fig. 10b. Dynamic amplification effects are very high in the free vibration phase, after the load has been removed, due to delamination openings.

The diagrams (a,b) in Fig. 13 are time histories of the displacements of the delaminated arms at the load point and of the crack length. The duration of the load pulse is $t_m = t_l$ and the applied dimensionless peak load is $P_{max}/\sqrt{\mathcal{G}_{cr}Eh} \times \sqrt[4]{E/(hk)} = 3.56$; the dashed lines refer to an elastic-brittle system and the thick solid lines to an elastic-plastic-brittle system with $T^f_{cr}/\sqrt{\mathcal{G}_{cr}k}$=1.89.

Consider first the elastic-brittle system. In the loading phase, the displacements of the sub-beams at the load point are larger than those of a stationary system due to a progressive lengthening of the delamination that grows under pure mode II conditions. In the free vibration phase, after the load has been removed, important mode I openings are generated and the crack continues to propagate under mixed-mode conditions up to a length $a_{max}/\lambda_r \approx 0.7$. Dynamic crack propagation occurs at a similar speed in both phases. Under quasi-static conditions and for the same value of the maximum load, the crack would propagate up to a length $a_{st}/\lambda_r \approx 0.22$ and then arrest while approaching the energy barrier.

The following conclusions can be drawn for elastic-brittle systems. The positive effects of the energy barriers predicted in the quasi-static solution are much reduced

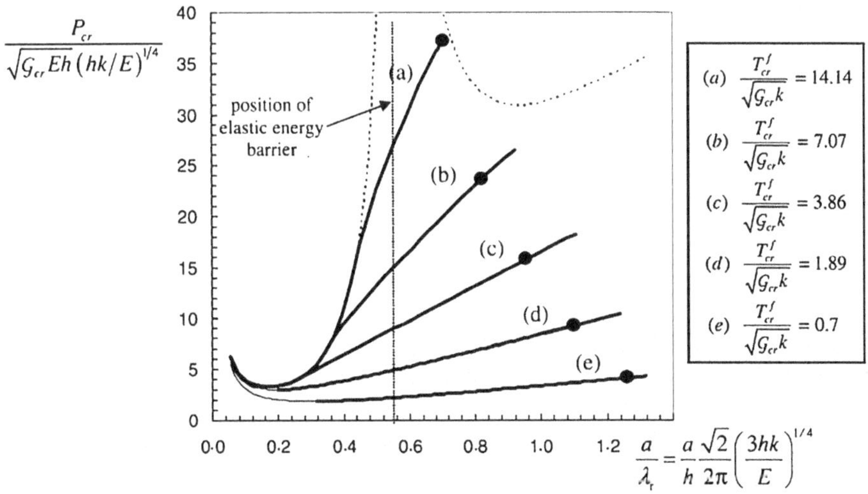

Fig. 12 Influence of the plasticity of the core on the critical load for the propagation of a mid-plane delamination in the skin (after [14])

during the loading phase and the barriers disappear in the phase that follows the removal of the load. Crack extension in the face-sheets of elastic-brittle systems is always amplified with respect to the quasi-static solution ($t_m \geq 2t_1$) but for the shortest time durations, and maximum amplification is for load durations in the range $t_m \approx 0.5 - 1.0t_1$

In the elastic-plastic-brittle system, thick lines in (13), plastic deformations take place in the foundation during loading and the consequent stiffness reduction leads to larger displacements and extended mode II crack growth with respect to the elastic-brittle system. Mixed-mode growth, in the free vibration phase, does occur at reduced speed and over a length similar to that of the elastic system. The figure highlights the irreversible plastic part of the displacement that will remains as a residual dent in the beam. The use of techniques that reduce delamination openings after the removal of the load, e.g. a through thickness reinforcement or 3D weaving [46–49] is expected to restore the important action of the energy barriers and improve performance in all cases.

In elastic-plastic-brittle systems where the core yields during the dynamic application of the load, the fracture parameters in the face-sheet are always amplified with respect to those of corresponding elastic systems and reducing the foundation yielding strength causes extended mode II crack propagation. However, core plasticity may reduce delamination openings and mixed-mode crack growth after the removal of the load. This behavior is controlled by duration and maximum value of the applied load: higher applied loads or shorter load durations reduce the extension of skin damage in the elastic-plastic system with respect to that of elastic systems.

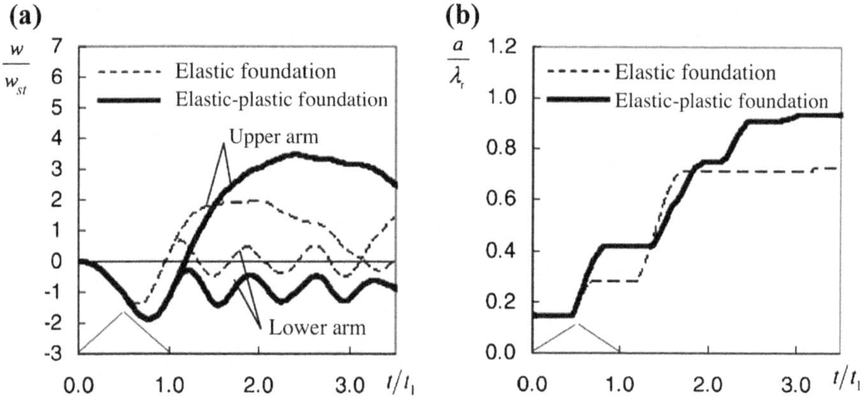

Fig. 13 (**a**, **b**) Time histories of the displacements of the delaminated arms at the load point and crack length in a beam with an central mid-plane delamination of initial length $a_0/\lambda = 0.1$ on an elastic (dashed lines) and an elastic-plastic (thick lines, $T^f_{cr}/\sqrt{\mathcal{G}_{cr}k}$=1.89) core, subjected to dynamic loading (after [14])

6 Multiscale Structural Modeling of Delamination Fracture in Layered Beams

Layered composite materials are extensively being used for construction of structural components. These components may be subjected to impacts and dynamic or repeated loadings, and suffer damage at several scales. Knowing the extent of damage, through inspection methods and theoretical modeling, is critical to determine the residual life and for damage tolerance assessments.

Delamination of the layers is one of the dominant failure mechanisms and modeling its evolution in layered structures requires proper representation of various aspects. Delamination fracture is a discrete and localized damage event, which typically occurs simultaneously at the interface between different layers and may often be catastrophic. Delaminations may grow from manufacturing flaws or defects or may form, in service, due to impacts and the presence of through-thickness stresses. Delamination may be brittle or controlled by cohesive mechanisms. The most common numerical technique to analyze the evolution of delaminations in layered structures uses the finite element method and cohesive interface elements [50–53]. Damage within the layers may be accounted for using continuum damage approaches. Discrete-layer cohesive-crack models and layer-wise theories are also used, often coupled with analytical solution methods or numerical techniques others than the finite element method [31–33, 46, 54, 55]. Another common approach to study delamination fracture within finite element models uses the Virtual Crack Closure Technique [56] and requires remeshing techniques to follow delamination evolution.

One of the drawbacks of the classical approaches is the fine discretization which is necessary in the through-thickness direction to predict and model onset and growth of delaminations at the different interfaces. This complicates the finite element discretization and increases computational cost, especially since the problem is nonlinear. In addition, the requirement of a fine discretization limits the analytical/semi-analytical solutions to systems with a small number of layers/delaminations and simple loading and boundary conditions.

In [17, 18, 57] we have investigated the use of a homogenized structural approach, already recalled in Sect. 2 for the analysis of wave propagation, to model mode II dominant delamination fracture in layered structures. The model does not require a through-thickness discretization and uses concepts developed in the original zigzag theory in [26], to homogenize the problem, and the multiscale strategy formulated in [4] to account for the presence of delaminations. The unknowns of the problem are substantially reduced, are independent of the numbers of layers and coincide with those of classical first-order shear deformation theory.

In [58–63] other strategies have been applied to model the presence and evolution of mode II delaminations within a zigzag description of the problem. They are represented as thin and very compliant layers which are added to the regular layers of the system (compliant layer concept) whose properties may progressively degrade to simulate fracture. The work presented in [17, 18] differs considerably from the previously published work. The delaminations are modelled using cohesive interfaces, which are then included into the homogenization, and fracture mechanics methods are applied to rigorously analyze and propagate cracks. In addition, the solutions presented in [17, 18] are analytical or semi-analytical to highlight the effects of the homogenization on the fracture parameters.

In Sect. 6.1 the formulation of the model is briefly recalled and in Sect. 6.2 exemplary results are presented.

6.1 Multiscale Model

The models refer to the schematic shown in Fig. 14 which describes a beam or wide-plate (deforming in cylindrical bending) with n layers and multiple delaminations subjected to transverse load. The plate is studied using the multiscale strategy developed in [4] which assume the small-scale displacement field in Eq. (1) (Sect. 2) and uses a homogenization technique to derive the macro-scale displacement field of Eq. (3). Equilibrium equations are then derived using classical variational techniques:

$$\begin{aligned}
&A_{22}v_{02,22}+\left(B_{22}+C^{0S}\right)\varphi_{2,22}+C^{0S}w_{0,222}=0\\
&\left(B_{22}+C^{0S}\right)v_{02,22}+\left(D_{22}+2C^{1S}+C^{S2}\right)\varphi_{2,22}+\left(C^{1S}+C^{S2}\right)w_{0,222}-A_{44}\left(w_{0,2}+\varphi_2\right)=0\\
&C^{0S}v_{02,222}+\left(C^{1S}+C^{S2}\right)\varphi_{2,222}+C^{S2}w_{0,2222}-A_{44}\left(w_{0,22}+\varphi_{2,2}\right)-f_3=0
\end{aligned}\qquad(42)$$

They depend on only three global displacement variables, which coincide with those of first order shear deformation theory, $v_{02}(x_2)$, $\varphi_2(x_2)$,$w_0(x_2)$, and define the in-plane displacement, bending rotation and transverse displacement of the reference surface, at $x_3 = 0$, when this falls into the lowest layer. The coefficients depend on layup and status of the interface and are given in [17]. The formulation, as already explained in Sect. 2, assumes the layers to be homogeneous and orthotropic, principal material directions parallel to the geometrical axes, neglects transverse compressibility of the layers and normal stresses.

The formulation is limited to mode II dominant problems and the cracks are assumed to be in constrained contact (no openings). The interfacial traction law, which relates the interfacial tractions to the relative sliding displacements of the layers at their interfaces, is shown in Fig. 14. The law is piecewise linear in order to describe the intact portions of the plate, where $1/K_S \rightarrow 0$, and the delaminated portions, where $K_S \rightarrow 0$.

The energy release rate for the propagation of the delaminations in the beam is derived using different techniques: the J-integral calculated along different paths and the compliance method. For a homogeneous edge-cracked layers and bi-material beams with a single delamination, the J-integral can be expressed in closed form in terms of the sub-resultants acting along the cross sections of the different beam arms, calculated a posteriori through the solution of the homogenized model, and the solution coincide with predictions made using classical discrete layer models. More details on the model and applications to a number of relevant problems are in [17].

6.2 *Single and Multiple, Mode II Dominant Delamination Fracture of Layered Beams*

Here some exemplary applications of the multiscale model in [17] are presented to highlight the capability of the model to analyze single and multiple mode II dominant delamination fracture in layered systems, Fig. 15. The importance of these results is in the capability of the model, which uses only three displacement variables as a classical single layer theory, to describe a discrete event, the brittle delamination, to reproduce the interaction effects of multiple delaminations and to follow the evolution of multiple cracks.

Figure 16 shows the macro-structural response of a unidirectional End-Notched Flexural specimen, Fig. 15a, through the critical load for crack propagation versus load-point deflection. The theoretical results have been obtained using the compliance method, to calculate the energy release rate from the results of the homogenized structural model, and the fracture criterion $\mathcal{G}_{II} = \mathcal{G}_{IIC}$. The dimensionless critical load is $P_{cr}/\sqrt{\mathcal{G}_{IIC} h \overline{E}_2} = 4/3(h/a)$ and coincides with the load obtained using classical structural mechanics predictions. The exact solution of the problem has two additional terms, which account for the effects of crack tip shear on the near tip deformations and that are important when the crack is short (see [17] for a discussion on these results).

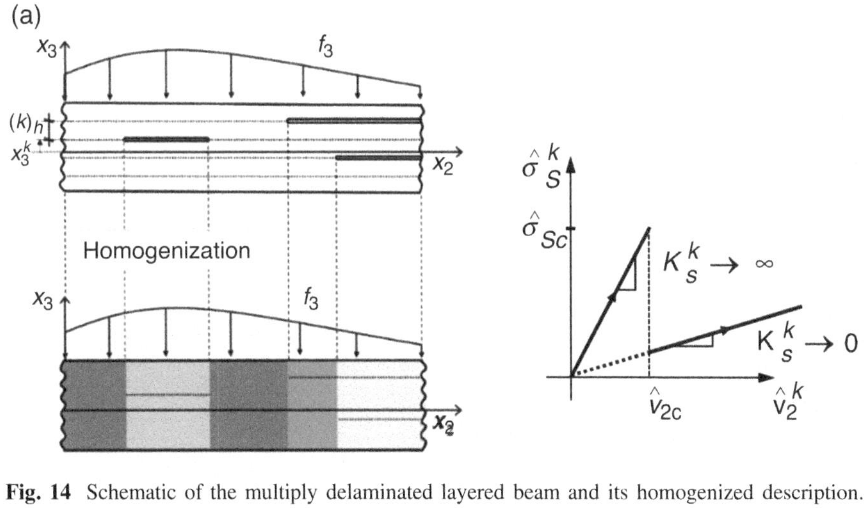

Fig. 14 Schematic of the multiply delaminated layered beam and its homogenized description. Interfacial traction law used to reproduce perfectly brittle fracture

The response diagrams are compared with the experimental results on two Graphite/epoxy $[0]_{24}$ laminates tested in [64]. The geometry is defined by: L = 50 mm, 2 h = 3.4 mm, a_0 = 25mm, b = 25 mm (width). The material of the beam in the first diagram is graphite/epoxy AS-4/828 with $E_2 = 139 \pm 16.7$ GPa and G_{23} = 6GPa and Mode II fracture energy $\mathcal{G}_{IIC} = 1.04 \pm 0.17$ N/mm, calculated in using the compliance method. The material of the beam in the second diagram is a graphite/epoxy AS-4C/828 with $E_2 = 158 \pm 5.06$ GPa and G_{23} = 6GPa and $\mathcal{G}_{IIC} = 1.15 \pm 0.13$ N/mm.

Two theoretical curves are shown in each diagrams. The curves in Fig. 16a have been obtained using the average values of the elastic constants and energy release rate, red dashed lines Model (a), and the maximum values, black solid lines Model (b). In Fig. 16b the red curve corresponds to the average values, Model (a), and the black curve has been obtained using the average value of energy release rate and the maximum value of the Young modulus, Model (b). The experimental results, under displacement control show a load drop in the critical load at the onset of propagation; this is due to an unstable propagation of the crack which grows catastrophically and arrests near the mid-span. The homogenized model, which is under crack-length control, is able to capture the snap-back instability and follow the virtual branch where crack growth is associated to a reduction of the load-point displacement. Crack propagation is modelled also in the region beyond the mid-span to show that the curve stably approaches the limiting solution (dotted line) corresponding to two fully delaminated layers.

Figure 17 highlights the capability of the homogenized model to analyze multiple delamination fracture and reproduce the effects of the interaction between delaminations. The dimensionless diagrams depict the critical load for the propagation of the cracks in the Cantilever Beam in Fig. 15b versus load-point deflection. The diagram (a) refers to a homogeneous isotropic beam where ${}^{(2)}h/{}^{(1)}h = 2$, ${}^{(3)}h/{}^{(1)}h = 1$ and initial crack lengths are $a_{U0}/h = 5.5$ and $a_{L0}/h = 6$. The diagram (b) is for a homogeneous isotropic beam where ${}^{(2)}h/{}^{(1)}h = {}^{(3)}h/{}^{(1)}h = 1/3$ and the initial crack lengths are $a_{U0}/h = 5$ and $a_{L0}/h = 4$. The delaminations are assumed to propagate

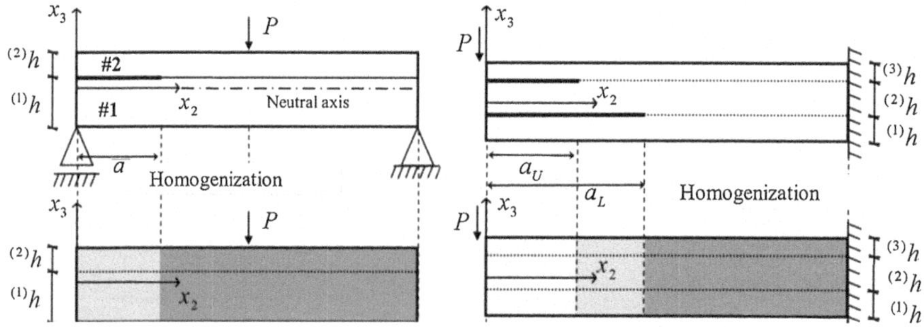

Fig. 15 (**a**) Three-point bend-beam and homogenized model, used for the comparison with the experimental results in Fig. 16. (**b**) Cantilever beam with two delaminations and homogenized model, used to study the interaction of multiple delaminations in Fig. 17

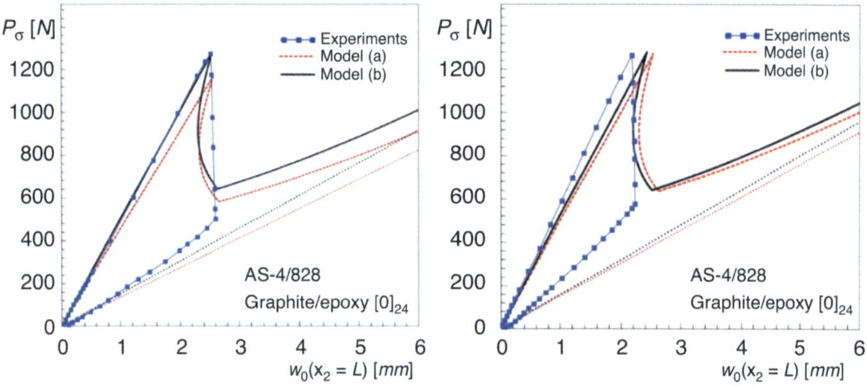

Fig. 16 Dimensionless diagrams of the critical load for crack propagation as function of the load point displacement in unidirectional ENF fracture specimens, Fig. 15a, tested in [64]

when the energy release rate, calculated using the compliance method, approaches the critical value, $\mathcal{G}_{II} = \mathcal{G}_{IIC}$. The results of the homogenized model are compared with the results of the discrete-layer cohesive-crack model with spring-contact in [33]. A local snap-through instability is observed in the diagram (a) when the upper crack starts to propagate in A and approaches the lower crack tip, in B. Then the load to propagate the crack must be increased, due to a shielding phenomenon, up to point C where the two cracks propagate together unstably. In the diagram (b) the lower crack, which is shorter, starts to propagate at the maximum load, point A; crack propagation is unstable and characterized by a snap-back instability up to point B. Then there is a sudden drop in the load, to point C, caused by a sudden amplification discontinuity. After point C the lower crack continues to propagate alone.

6.3 Conclusions on Multiscale Homogenized Modeling of Mode II Dominant Fracture

A multiscale homogenized model has been formulated to analyze mode II dominant multiple delamination fracture in layered composite and sandwich beams. The model enriches the displacement field of a first order shear deformation theory to account for zig-zag effects and the presence of delaminations and homogenized equations are derived which depend on the global variables only. The through thickness discretization is therefore unnecessary in the solution of fracture problems. Unstable crack propagation and the local effects of delamination interaction on the macrostructural response of the beams are successfully reproduced.

Acknowledgments The support by the U.S. Navy, Office of Naval Research, ONR, grant N00014-17-1-2914 and the useful discussions and suggestions of the program manager Dr. Y.D.S. Rajapakse are gratefully acknowledged.

References

1. Massabò R (2017) Propagation of Rayleigh-lamb waves in multilayered plates through a multiscale structural model. Int J Solids Struct 124:108–124. https://doi.org/10.1016/j.ijsolstr.2017.06.020

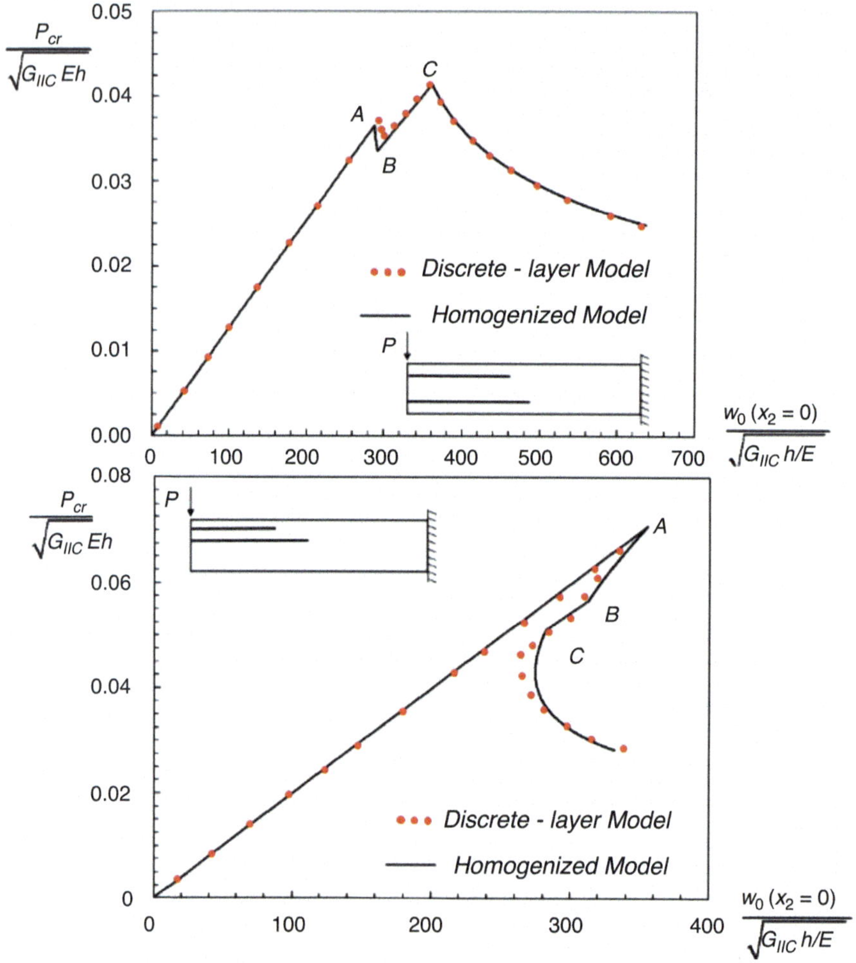

Fig. 17 Dimensionless diagrams of the critical load for crack propagation versus load point displacement in the cantilever beam with two delaminations in Fig. 15b. Comparison between the results of the homogenized model and those of a discrete-layer model in [32]

2. Massabò R (2017) Cut-off frequencies and correction factors of equivalent single layer theories. Procedia Eng 199:1466–1471. https://doi.org/10.1016/j.proeng.2017.09.406
3. Massabó R (2018) Wave propagation and dynamic correction factors for composite structures In: Gopalkrishnan, S., Rajapakse Y, editor. Blast mitigation strategies in marine composite sandwich structures Springer, ISBN 9789811071706; p. 191–208
4. Massabò R, Campi F (2014) An efficient approach for multilayered beams and wide plates with imperfect interfaces and delaminations. Compos Struct 116:311–324. https://doi.org/10.1016/j.compstruct.2014.04.009
5. Darban H, Massabò R (2018) Thermo-elastic solutions for multilayered wide plates and beams with interfacial imperfections through the transfer matrix method. Meccanica 53:553–571. https://doi.org/10.1007/s11012-017-0657-6
6. Darban H, Massabò R (2017) 2D thermo-elastic solutions for laminates and sandwiches with interlayer delaminations and imperfect thermal contact. In: Lo Presto V, Langella A, Abrate S (eds) Dynamic response failure composite materials structures. Elsevier Inc., pp 3–46. https://doi.org/10.1016/B978-0-08-100887-4.00001-9
7. Darban H (2018) Multiscale modeling of delamination fracture in multilayered structures. Ph.D. Thesis, University of Genova
8. Barbieri L, Massabò R, Berggreen C (2018) The effects of shear and near tip deformations on interface fracture of symmetric sandwich beams. Eng Fract Mech 201:1–24. https://doi.org/10.1016/j.engfracmech.2018.06.039
9. Østergaard RC, Sørensen BF (2007) Interface crack in sandwich specimen. Int J Fract 143:301–316. https://doi.org/10.1007/s10704-007-9059-4
10. Suo Z, Hutchinson JW (1990) Interface crack between two elastic layers. Int J Fract 43:1–18. https://doi.org/10.1007/BF00018123
11. Andrews MG, Massabò R (2007) The effects of shear and near tip deformations on energy release rate and mode mixity of edge-cracked orthotropic layers. Eng Fract Mech 74:2700–2720. https://doi.org/10.1016/j.engfracmech.2007.01.013
12. Li S, Wang J, Thouless MD (2004) The effects of shear on delamination in layered materials. J Mech Phys Solids 52:193–214. https://doi.org/10.1016/S0022-5096(03)00070-X
13. Kardomateas GA, Berggreen C, Carlsson LA Energy-release rate and mode Mixity of face/Core Debonds in Sandwich beams. AIAA J 51(2913):885–892. https://doi.org/10.2514/1.J051765
14. Massabò R, Cavicchi A (2012) Interaction effects of multiple damage mechanisms in composite sandwich beams subject to time dependent loading. Int J Solids Struct 49:720–738. https://doi.org/10.1016/j.ijsolstr.2011.11.012
15. Campi F, Massabò R (2011) An analytical assessment of the influence of skin imperfections on the indentation collapse mechanism in composite sandwich beams. Compos Struct 94:299–311. https://doi.org/10.1016/j.compstruct.2011.05.006
16. Daniel IM (2010) Impact response and damage tolerance of composite sandwich structures. Proceedings of the ICSS-9, Pasadena. 1–10
17. Massabò R, Darban H (2019) Mode II dominant fracture of layered composite beams and wide-plates: a homogenized structural approach. Eng Fract Mech 213:280–301. https://doi.org/10.1016/j.engfracmech.2019.03.002
18. Massabò R (2014) Influence of boundary conditions on the response of multilayered plates with cohesive interfaces and delaminations using a homogenized approach. Frattura ed Integrità Strutturale 8:230–240. https://doi.org/10.3221/IGF-ESIS.29.20
19. Massabò R, Campi F (2015) Assessment and correction of theories for multilayered plates with imperfect interfaces. Meccanica 50:1045–1071. https://doi.org/10.1007/s11012-014-9994-x
20. Pelassa M, Massabò R (2015) Explicit solutions for multi-layered wide plates and beams with perfect and imperfect bonding and delaminations under thermo-mechanical loading. Meccanica 50:2497–2524. https://doi.org/10.1007/s11012-015-0147-7
21. Jones JP (1964) Wave propagation in a two-layered medium. J Appl Mech Trans 31:213–222. https://doi.org/10.1115/1.3629589

22. Yang PC, Norris CH, Stavsky Y (1966) Elastic wave propagation in heterogeneous plates. Int J Solids Struct 2:665–684. https://doi.org/10.1016/0020-7683(66)90045-X
23. Thomson WT (1950) Transmission of elastic waves through a stratified solid medium. J Appl Phys 21:89–93. https://doi.org/10.1063/1.1699629
24. Liu L, Bhattacharya K (2009) Wave propagation in a sandwich structure. Int J Solids Struct 46:3290–3300. https://doi.org/10.1016/j.ijsolstr.2009.04.023
25. Whitney JM, Sun CT (1973) A higher order theory for extensional motion of laminated composites. J Sound Vib 30:85–97. https://doi.org/10.1016/S0022-460X(73)80052-5
26. di Sciuva M (1986) Bending, vibration and buckling of simply supported thick multilayered orthotropic plates: an evaluation of a new displacement model. J Sound Vib 105:425–442. https://doi.org/10.1016/0022-460X(86)90169-0
27. Chen WQ, Cai JB, Ye GR (2003) Exact solutions of cross-ply laminates with bonding imperfections. AIAA J 41:2244–2250. https://doi.org/10.2514/2.6817
28. Fan J, Ye J (1990) An exact solution for the statics and dynamics of laminated thick plates with orthotropic layers. Int J Solids Struct 26:655–662. https://doi.org/10.1016/0020-7683(90)90036-U
29. Qian H, Zhou D, Liu W, Fang H (2014) 3-D elasticity solutions of simply supported laminated rectangular plates in uniform temperature field. J Therm Stresses 37:661–677. https://doi.org/10.1080/01495739.2014.885329
30. Qian H, Zhou D, Liu WQ, Fang H, Lu WD (2015) 3-D elasticity solutions of layered rectangular plates subjected to thermo-loads. J Therm Stresses 38:377–398. https://doi.org/10.1080/01495739.2014.985570
31. Williams TO, Addessio FL (1997) A general theory for laminated plates with delaminations. Int J Solids Struct 34:2003–2024. https://doi.org/10.1016/S0020-7683(96)00131-X
32. Andrews MG, Massabò R (2008) Delamination in flat sheet geometries with material imperfections and thickness variations. Compos Part B Eng 39:139–150. https://doi.org/10.1016/j.compositesb.2007.02.017
33. Andrews MG, Massabò R, Cox BN (2006) Elastic interaction of multiple delaminations in plates subject to cylindrical bending. Int J Solids Struct 43:855–886. https://doi.org/10.1016/j.ijsolstr.2005.04.025
34. Brandinelli L, Massabò R (2006) Mode II weight functions for isotropic and orthotropic double cantilever beams. Int J Fract 139:1–25
35. Ustinov K (2015) On separation of a layer from the half-plane: elastic fixation conditions for a plate equivalent to the layer. Mech Solids 50:62–80. https://doi.org/10.3103/S0025654415010070
36. Berggreen C, Simonsen BC, Borum KK (2007) Experimental and numerical study of interface crack propagation in foam-cored sandwich beams. J Compos Mater 41:493–520. https://doi.org/10.1177/0021998306065285
37. Rice JR (1988) Elastic fracture mechanics concepts for interfacial cracks. J Appl Mech 55:98–103. https://doi.org/10.1115/1.3173668
38. Thouless MD (2018) Shear forces, root rotations, phase angles and delamination of layered materials. Eng Fract Mech 191:153–167. https://doi.org/10.1016/j.engfracmech.2018.01.033
39. Abrate S (1997) Localized impact on Sandwich structures with laminated facings. Appl Mech Rev 50:69. https://doi.org/10.1115/1.3101689
40. Shenhar Y, Frostig Y, Altus E (1996) Stresses and failure patterns in the bending of sandwich beams with transversely flexible cores and laminated composite skins. Compos Struct 35:143–152. https://doi.org/10.1016/0263-8223(96)00016-5
41. Shipsha A, Zenkert D (2005) Compression-after-impact strength of sandwich panels with core crushing damage. Appl Compos Mater 12:149–164. https://doi.org/10.1007/s10443-005-1119-1
42. Steeves CA, Fleck NA (2004) Collapse mechanisms of sandwich beams with composite faces and a foam core, loaded in three-point bending. Part I: analytical models and minimum weight design. Int J Mech Sci 46:561–583. https://doi.org/10.1016/j.ijmecsci.2004.04.003
43. Wu CL, Sun CT (1996) Low velocity impact damage in composite sandwich beams. Compos Struct 34:21–27. https://doi.org/10.1016/0263-8223(95)00127-1

44. Jackson M, Shukla A (2011) Performance of sandwich composites subjected to sequential impact and air blast loading. Compos Part B Eng 42:155–166. https://doi.org/10.1016/j.compositesb.2010.09.005
45. Schubel PM, Luo JJ, Daniel IM (2004) Low velocity impact behavior of composite sandwich panels. Compos Part A Appl Sci Manuf 2005. https://doi.org/10.1016/j.compositesa.11.014
46. Andrews MG, Massabò R, Cavicchi A, Cox BN (2009) Dynamic interaction effects of multiple delaminations in plates subject to cylindrical bending. Int J Solids Struct 46:1815–1833. https://doi.org/10.1016/j.ijsolstr.2008.11.027
47. Brandinelli L, Massabò R (2003) Free vibrations of delaminated beam-type structures with crack bridging. Compos Struct 61:129–142. https://doi.org/10.1016/S0263-8223(03)00035-7
48. Lundsgaard-Larsen C, Massabò R, Cox BN (2009) The design of dynamic tests to infer rate dependence in large-scale crack bridging. Soc Exp Mech – SEM Annu Conf Expo Exp Appl Mech 1:2009
49. Sridhar N, Massabò R, Cox BN, Beyerlein IJ (2002) Delamination dynamics in through-thickness reinforced laminates with application to DCB specimen. Int J Fract 118:119–144. https://doi.org/10.1023/A:1022884410968
50. Allix O, Ladevèze P (1992) Interlaminar interface modelling for the prediction of delamination. Compos Struct 22:235–242. https://doi.org/10.1016/0263-8223(92)90060-P
51. Alfano G, Crisfield MA (2001) Finite element interface models for the delamination analysis of laminated composites: mechanical and computational issues. Int J Numer Methods Eng 50:1701–1736. https://doi.org/10.1002/nme.93
52. Turon A, Camanho PP, Costa J, Renart J (2010) Accurate simulation of delamination growth under mixed-mode loading using cohesive elements: definition of interlaminar strengths and elastic stiffness. Compos Struct 92:1857–1864. https://doi.org/10.1016/j.compstruct.2010.01.012
53. Lundsgaard-Larsen C, Massabò R, Cox BN (2012) On acquiring data for large-scale crack bridging at high strain rates. J Compos Mater 46:949–971. https://doi.org/10.1177/0021998311413622
54. Bruno D, Greco F (2001) Mixed mode delamination in plates: a refined approach. Int J Solids Struct 38:9149–9177. https://doi.org/10.1016/S0020-7683(01)00179-2
55. Barbero EJ, Reddy JN (1991) Modeling of delamination in composite laminates using a layer-wise plate theory. Int J Solids Struct 28:373–388. https://doi.org/10.1016/0020-7683(91)90200-Y
56. Rybicki EF, Kanninen MF (1997) A finite element calculation of stress intensity factors by a modified crack closure integral. Eng Fract Mech 9:931–938. https://doi.org/10.1016/0013-7944(77)90013-3
57. Darban H, Massabò R (2017) A multiscale structural model for cohesive delamination of multilayered beams, AIMETA 2017 – Proceedings of the 23rd conference of the Italian association of theoretical and applied mechanics, 2, pp. 1785–1792
58. Icardi U, Zardo G (2005) C0plate element for delamination damage analysis, based on a zig-Zag model and strain energy updating. Int J Impact Eng 31:579–606. https://doi.org/10.1016/j.ijimpeng.2004.02.002
59. Groh RMJ, Tessler A (2017) Computationally efficient beam elements for accurate stresses in sandwich laminates and laminated composites with delaminations. Comput Methods Appl Mech Eng 320:369–395. https://doi.org/10.1016/j.cma.2017.03.035
60. Eijo A, Oñate E, Oller S (2014) Delamination in laminated plates using the 4-noded quadrilateral QLRZ plate element based on the refined zigzag theory. Compos Struct 108:456–471. https://doi.org/10.1016/j.compstruct.2013.09.052
61. Eijo A, Oñate E, Oller S (2013) A numerical model of delamination in composite laminated beams using the LRZ beam element based on the refined zigzag theory. Compos Struct 104:270–280. https://doi.org/10.1016/j.compstruct.2013.04.035
62. Groh RMJ, Weaver PM, Tessler A (2015) Application of the refined zigzag theory to the modeling of delaminations in laminated composites. https://doi.org/10.13140/RG.2.1.3147.0804

63. Averill RC (1994) Static and dynamic response of moderately thick laminated beams with damage. Compos Eng 4:381–395. https://doi.org/10.1016/S0961-9526(09)80013-0
64. Madhukar MS, Drzal LT (1992) Fiber-Matrix adhesion and its effect on composite mechanical properties: IV. Mode I and mode II fracture toughness of graphite/epoxy composites. J Compos Mater 26:936–968. https://doi.org/10.1177/002199839202600701
65. Massabò R, Barbieri L, Berggreen C (2017) Energy release rate and mode mixity of face/core debonds in sandwich beams, AIMETA 2017 – Proceedings of the 23rd conference of the Italian association of theoretical and applied mechanics, 2, pp. 1913–1919
66. Massabò R, Ustinov K, Barbieri L, Berggreen C (2019) Fracture mechanics solutions for interfacial cracks between compressible thin layers and substrates, Coatings, 9(3), art. no. 152., https://doi.org/10.3390/coatings9030152

On Characterizing Multiaxial Polymer Foam Properties in Sandwich Structures

Michelle S. Hoo Fatt, Chong Zhong, and Xiaolong Tong

1 Introduction

Research over the last several years has shown that the PVC foam cores in composite sandwich panels have blast mitigation effects when these panels are subjected to nearby explosions in air and water [1]. Sandwich panels with softer more ductile foam cores can offer better blast resistance than panels with stiffer and stronger foams because of the associated shock attenuation and energy dissipation due to core crushing. This is especially true in water blast cases where fluid-structure interaction causes triaxial plasticity of the foam, specifically, transverse shear, and transverse and in-plane compression [2].

Triaxial stresses in the core also occur in transversely loaded curved sandwich panels because of membrane stresses that are induced by shell curvature [3]. Figures 1 (a) and (b) show the distribution of core stresses during plastic deformation of the foam in flat and curved sandwich panels subjected to air and water blasts, respectively. In the air blast, transverse or out-of-plane shear stresses (S13 and S23) are the most dominant in the flat panel while in the curved panel, transverse compressive stresses (S33) are comparable to transverse shear stresses. Since there are also membrane compressive stresses (S11 and S22) in the air blast curved panel, the foam yielding occurs with triaxial stresses. In the water blast case, foam plasticity is dominated by triaxial compressive stresses rather than transverse shear in the curved sandwich panel. A water blast flat sandwich panel experiences transverse shear and compression stresses during core plasticity. These are some examples where the crushing of foam is multiaxial. However, many analytical and computational models of sandwich panels do not consider multiaxial properties of the foam. In most cases,

M. S. Hoo Fatt (✉) · C. Zhong · X. Tong
Department of Mechanical Engineering, The University of Akron, Akron, OH, USA
e-mail: hoofatt@uakron.edu; cz29@zips.uakron.edu; xt7@zips.uakron.edu

S. W. Lee (ed.), *Advances in Thick Section Composite and Sandwich Structures*,
https://doi.org/10.1007/978-3-030-31065-3_13

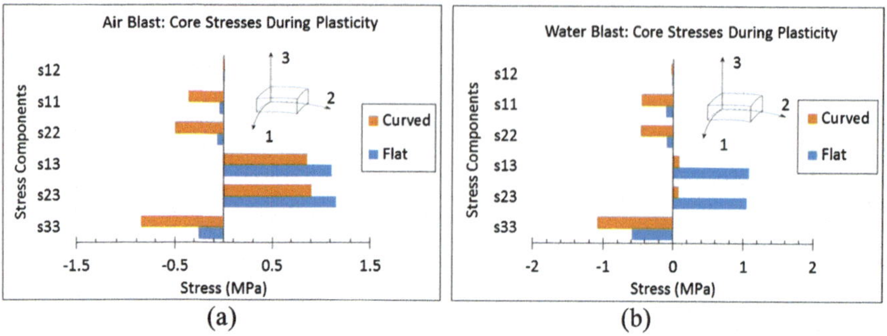

Fig. 1 Distribution of stresses during Divinycell PVC H100 foam core plasticity in sandwich panels during (**a**) air blast and (**b**) water blasts

the foam is assumed to be isotropic with elastic and inelastic properties that are assigned to be in the out-of-plane or transverse direction.

There is a lack of triaxial material tests and material models that address plastic deformation, hysteresis and damage of the foam. Current foam inelastic or crushing models are based on experiments involving monotonic loading [4–7], while the subsequent unloading and hysteresis behavior of the foam are necessary to address plasticity, viscoelasticity and damage. Furthermore, many of the inelastic foam models do not address anisotropic behavior, which is an issue when there are triaxial stress states in the foam. In situations where the foam core experiences triaxial stress states, anisotropic behavior of the foam must be taken into consideration because there is risk of over-predicting the load carrying capacity of a sandwich structure. Polymer foams are transversely isotropic because cells become more elongated in the foam's rise or out-of-plane direction during processing, and the out-of-plane stiffness and strength of the foam can be much higher than its in-plane values. If during analysis foams are treated as being isotropic with out-of-plane properties, one inadvertently assumes the in-plane stiffness and strength of the foam to be higher than it actually is and this results in design loads that are higher than what they should be.

The objective of this work is to develop more accurate material models to describe multiaxial behavior of polymeric foams in the core of sandwich panels. Our approach is two-fold: (1) to design pressure vessel experiments to obtain and characterize triaxial hysteresis properties of Divinycell PVC H100 foam and (2) to analyze the data and develop material model that would be suitable for FEA implementation. The following sections describe the pressure vessel experiments and experimental results that were obtained from them. Based on experimental observations, a material model is developed and comparisons are made between the predicted material response and the experimental results. Finally, concluding remarks are given in the last section.

2 Pressure Vessel Experiments

An aluminium pressure vessel apparatus was made to be used in conjunction with an MTS servohydraulic machine to determine triaxial material properties, including plastic deformations, damage and hysteresis of Divinycell PVC H100 foam. As shown in Fig. 2(a), nozzles in the pressure vessel allowed the piston rods to move freely through the body of the pressure vessel and also provided alignment for Arcan holders and butterfly specimens shown in Fig. 2(b). The nozzle/piston interface was sealed with a custom-designed, low friction lip seals. The Arcan butterfly specimen was used to obtain biaxial (shear-compression) material properties when the vessel was not pressurized. By pressurizing the vessel, triaxial material properties in shear, tension and/or compression could be obtained. The pressure transient was controlled by two solenoid valves and a pressure controller programmed to comply with the MTS machine controller.

One inch-thick glass windows were mounted at opposite ends of the pressure vessel so that video photography of the specimen deformation could be taken. Two high-speed cameras and light source enabled video photography of the specimen while it deformed. Digital Image Correlation (DIC) was used to determine specimen strains, while the load cell of the MTS machine and a pressure transducer were used to determine specimen stresses.

Inside the pressure vessel, uniaxial tensile dumbbell specimens and Arcan butterfly specimens were exposed to pressurized air. The 0 deg Arcan butterfly

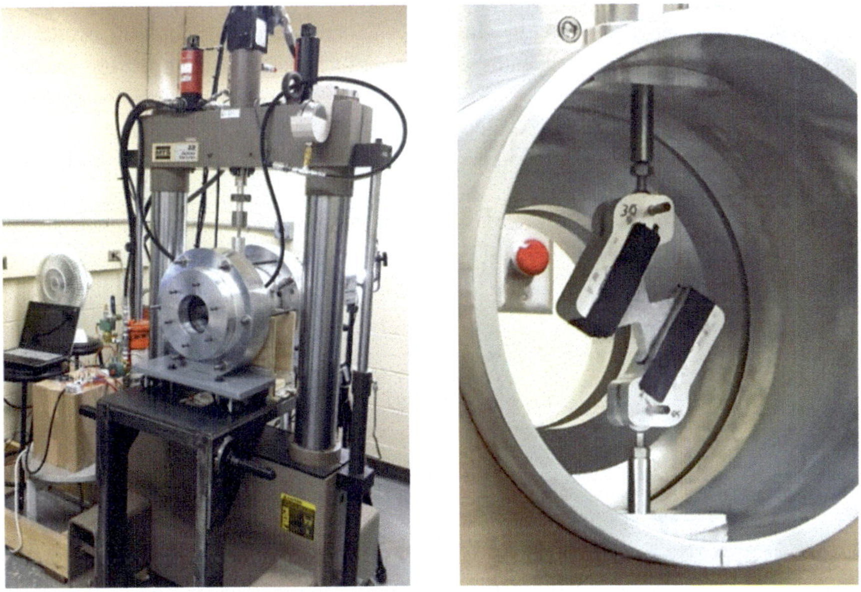

Fig. 2 Experimental setup: (**a**) pressure vessel mounted in MTS machine and (**b**) 30 deg Arcan specimen in vessel

specimen was used for shear, while the 15, 30, 45, 60 and 75 deg Arcan butterfly specimens were used for combined normal and shear modes (biaxial tests). A 90 deg Arcan butterfly specimen was used for compression, while a 90 deg dogbone specimen was used for tension. Pressurizing the cylindrical vessel allowed triaxial stress states to be superimposed on the 0 deg shear and 90 deg compression Arcan butterfly specimens as well as the dumbbell tension specimen.

Each specimen was subjected to cyclic displacement- and pressure-control loading. Hysteresis loops were acquired after plastic deformation had occurred at pre-determined maximum displacements and pressures. This was done for each mode of loading and for several maximum displacement amplitudes. The following section describes experimental data that were acquired from these tests.

3 Transversely Isotropic Properties

The hysteresis behavior of the PVC H100 foam in uniaxial compression, tension and shear are shown in Figs. 3, 4 and 5. Regardless of the mode of loading, the foam is roughly double the stiffness and strength in the out-of-plane direction when compared to the in-plane directions. Thus PVC H100 foam is a transversely isotropic material. Furthermore, the inelastic response of the material can be characterized by plastic deformation, viscoelastic hysteresis and damage. Plastic strain hardening is also anisotropic, due the different micro-mechanisms during yielding. In compression, there is negligible or very low strain hardening because macroscopic plasticity is caused by the buckling of cells. In contrast to this, strain hardening occurs in tension and shear because cell walls are forced to bend and rotate during macroscopic yielding.

The biaxial hysteresis tests revealed very similar response to these tests, and are not given here for brevity. The reader may find details of them in Ref. [8].

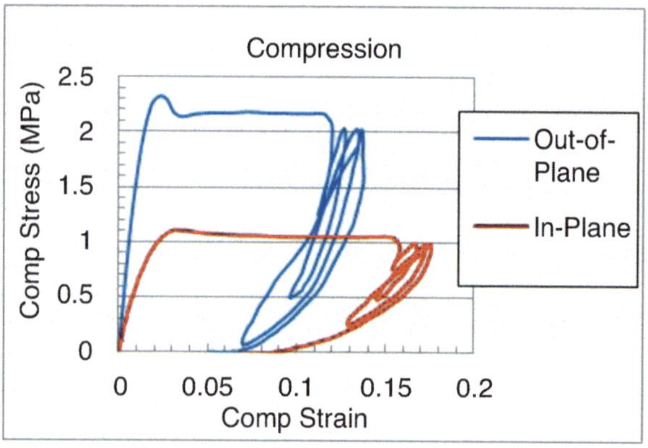

Fig. 3 Compressive hysteresis behavior of PVC H100

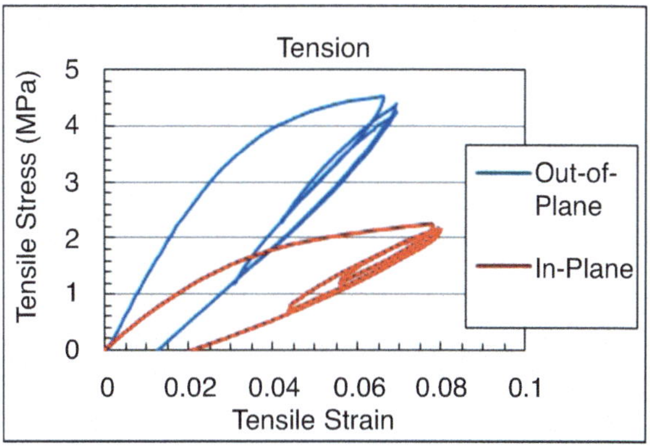

Fig. 4 Tensile hysteresis behavior of PVC H100

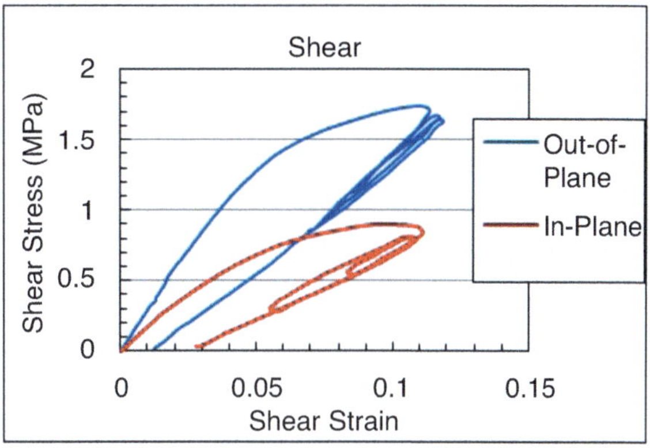

Fig. 5 Shear hysteresis behavior of PVC H100

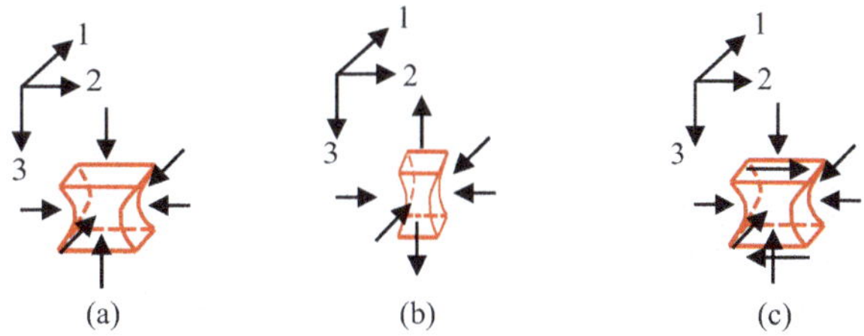

Fig. 6 Out-of-plane triaxial loading configurations: (**a**) triaxial compression, (**b**) tension-biaxial compression, (**c**) triaxial compression- shear

4 Triaxial Response

Triaxial stress-strain hysteresis curves were obtained in the out-of-plane (3-direction) and in-plane (1–2 plane) of the PVC H100 foam. Tests were done in the six loading configurations listed below and shown in Figs. 6 and 7:

1. Triaxial compression: Out-of-plane displacement-control (Fig. 6(a))
2. Triaxial compression: In-plane displacement-control (Fig. 7(a))

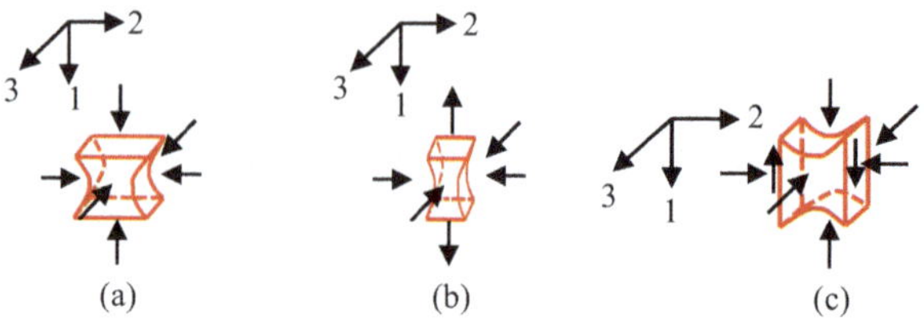

Fig. 7 In-plane triaxial loading configurations: (**a**) triaxial compression, (**b**) tension-biaxial compression, (**c**) triaxial compression- shear

3. Triaxial tension/compression: Out-of-plane tension; in-plane compression (Fig. 7(b))
4. Triaxial tension/compression: In-plane tension and compression; out-of-plane compression (Fig. 6(b))
5. Triaxial compression and out-of-plane shear (Fig. 6(c))
6. Triaxial compression and in-plane shear (Fig. 7(c))

In these experiments, cyclic displacement and pressure were simultaneously applied to each specimen. In each loading state, (a) the pressure was varied while the displacement amplitude was kept constant and (b) the displacement was varied while the pressure loading rate was held constant. By varying the pressure amplitude, a multitude of tri-axial stress states in shear, compression and tension could be realized. Varying the displacement amplitude was necessary to obtain plasticity and damage evolution data for development of a material constitutive model. For brevity, only the foam's in-plane response is given here. The out-of-plane response is similar except that it occurs at roughly double stress amplitude.

4.1 Triaxial Compression

Two types of triaxial compression tests were done: one in which displacements were controlled in the out-of-plane direction and the other in which displacements were controlled in the in-plane direction. Only the tests for which in-plane displacements were controlled are discussed here. In these tests, the primary crushing was done in the material 1-direction, while pressure controlled stresses in the 2- and 3-directions. Stress-strain hysteresis curves with varying pressure amplitudes and a fixed displacement amplitude of 2.03 mm are given in Figs. (8a-c). The material crushing behavior with no pressure is shown as a black line in Fig. 8(a). With triaxial compression, the plastic flow stress decreases or softens with increasing strain rather than taking on the plateau often seen in uniaxial compression tests. Softening becomes more dominant as the pressure increases. There is also a reduction in modulus as the pressure increases, which signifies damage is occurring in the foam.

Figures 9(a)-(c) show how hysteresis curves change with displacement amplitude, while the pressure rate was kept constant at 110 kPa/s. Negligible out-of-plane compressive modulus reduction after yielding was noticed with increasing displacement amplitude.

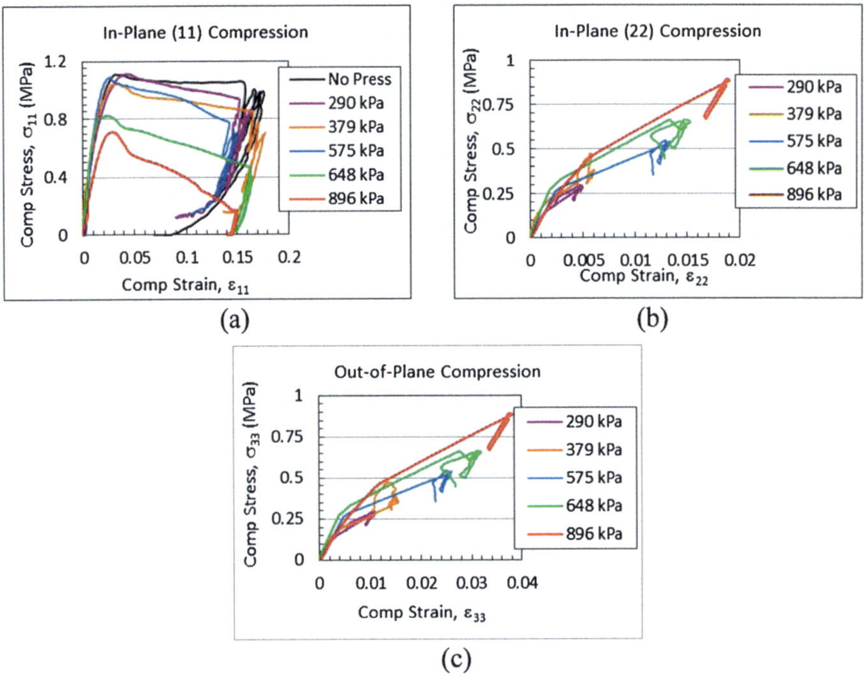

Fig. 8 Effect of pressure amplitude on compression response: (**a**) compressive σ_{11} (controlled), (**b**) compressive σ_{22} and (**c**) compressive σ_{33}

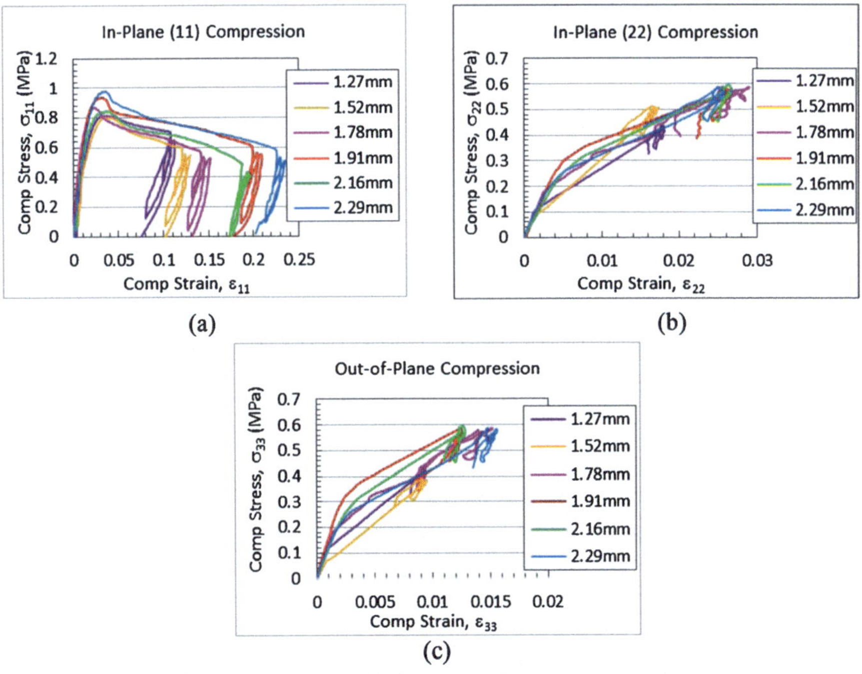

Fig. 9 Effect of displacement amplitude on compression response: (**a**) compressive σ_{11} (controlled), (**b**) compressive σ_{22} and (**c**) compressive σ_{33}

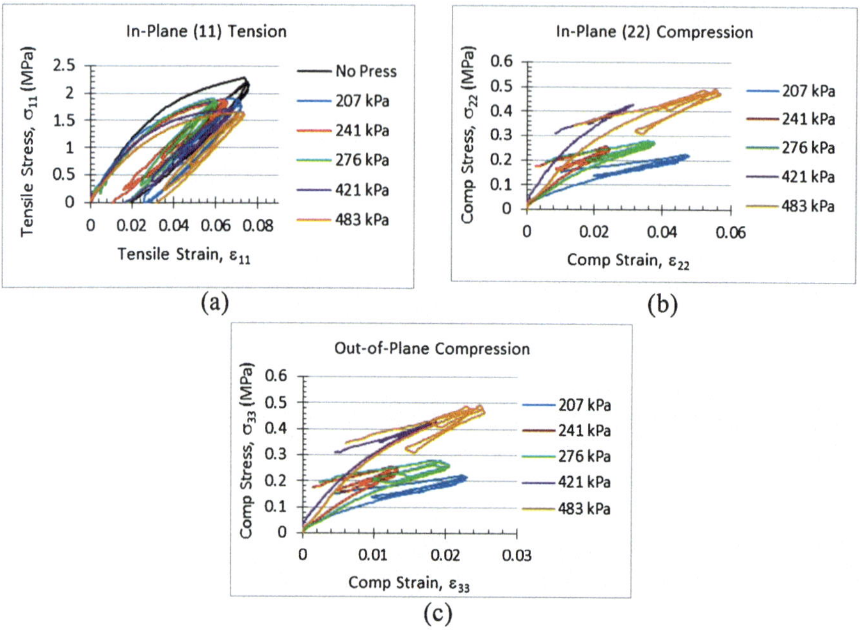

Fig. 10 Effect of pressure amplitude on tension/compression responses: (**a**) tensile σ_{11} (controlled), (**b**) compressiveσ_{22} and (**c**) compressive σ_{33}

4.2 Triaxial Tension/Compression

The dumbbell tension specimen was used with foam material 1-axis parallel loading axis (see Fig. 7(b)). Hysteresis stress-strain curves with various pressure amplitudes and a constant displacement amplitude of 1.52 mm are givem in Figs. 10(a-c). The tension response without pressure is indicated in black in Fig. 10(a) for reference. As in the case of triaxial compression, increasing pressure causes the tensile stress-strain curves to fall because of monotonically increasing compressive stresses induced in the in-plane directions.

In (Figs. 11a-c) the pressure loading rate is kept constant at 108 kPa/s while the displacement amplitude is varied. Unlike the out-of-plane tension tests, very little modulus reduction after yielding and during unloading was noticed in these experiments.

4.3 Triaxial Compression and In-Plane Shear

The 0 deg Arcan butterfly was used with foam material 1–2 axes parallel and perpendicular to the butterfly specimen (see Fig. 7(c)). Full stress-strain hysteresis curves for this specimen for a displacement amplitude of 2.03 mm and various pressure amplitudes are shown in (Figs. 12a-d). Figure 12(a) shows that increasing

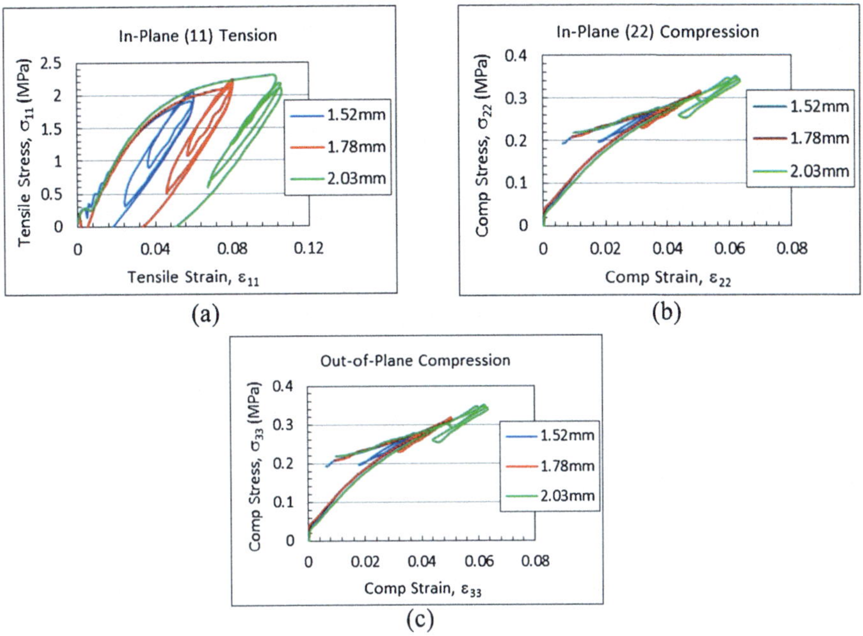

Fig. 11 Effect of displacement amplitude on tension/compression responses: (**a**) tensile σ_{11} (controlled), (**b**) compressive σ_{22} and (**c**) compressive σ_{33}

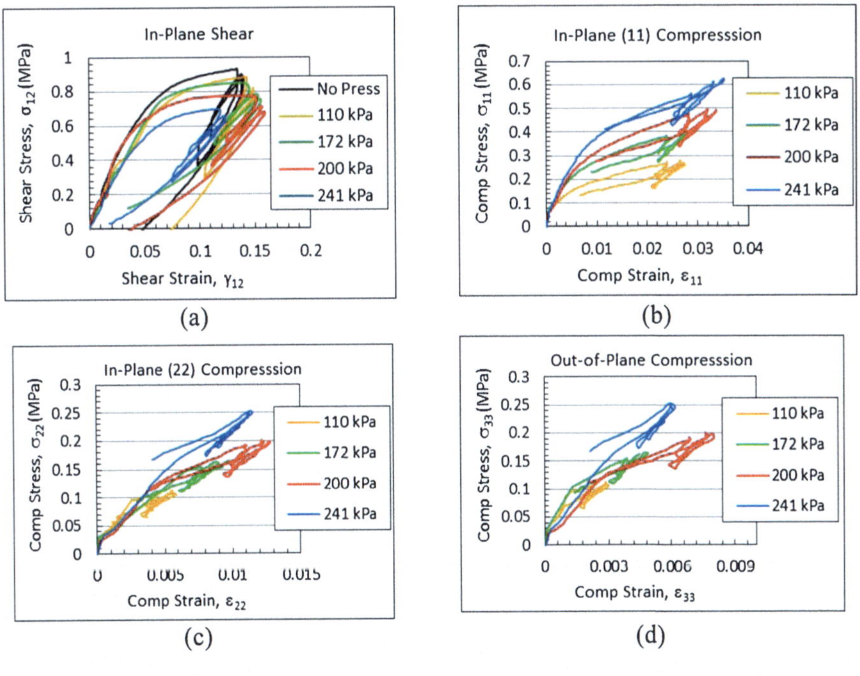

Fig. 12 Effect of pressure amplitude on in-plane shear and compression: (**a**) shear τ_{12}, (**b**) compressive σ_{22}, (**c**) compressive σ_{22} and (**d**) compressive σ_{33}

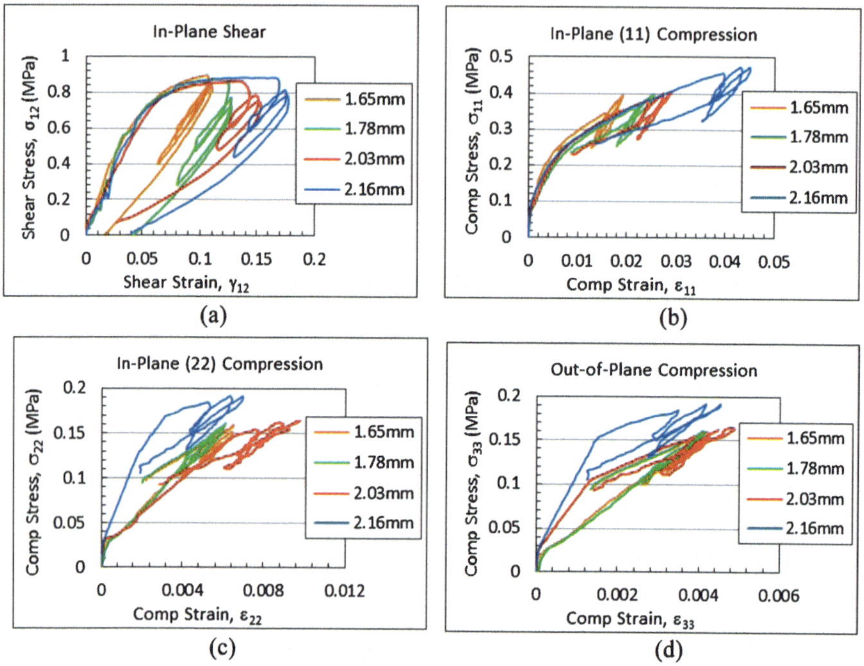

Fig. 13 Effect of displacement amplitude on in-plane shear and compression: (**a**) shear τ_{12}, (**b**) compressive σ_{11}, (**c**) compressive σ_{22} and (**d**) compressive σ_{33}

the pressure amplitude reduces the in-plane shear plastic flow stress. The in-plane shear response with no pressure is indicated in black for reference. The hysteresis loops themselves appear to be relatively unchanged in terms of average modulus after unloading. This indicates that damage is the same for a fixed displacement amplitude. The corresponding three normal compressive stress responses are shown in (Figs. 12b-d). Here the stresses are controlled by pressure, which was cyclic to match the displacement history. A definitive hysteresis loop was not observed in any of these compressive stress-strain curves. In Figs. 13(a-d) the displacement amplitude was varied, while the pressure rate was kept constant at 60 kPa/s. Damage or shear modulus reduction after yielding with increasing displacement amplitude was even more evident in the in-plane material direction.

5 Tsai-Wu Yield Criterion

As a first step in deriving an inelastic material model, one needs to determine the appropriate yield criterion. The triaxial tests were used to find an initial yield criteria for PVC H100 foam. Here yielding is defined at the onset of inelastic behavior, i.e., the point at which the foam does not experience hysteresis and permanent

Table 1 Yield strengths of Divinycell PVC H100 foam

S_{12} (MPa)	$X_c = Y_c$ (MPa)	$X_t = Y_t$ (MPa)	$S_{13} = S_{23}$ (MPa)	Z_c (MPa)	Z_t (MPa)
0.65	0.85	0.95	1.0	1.9	2.35

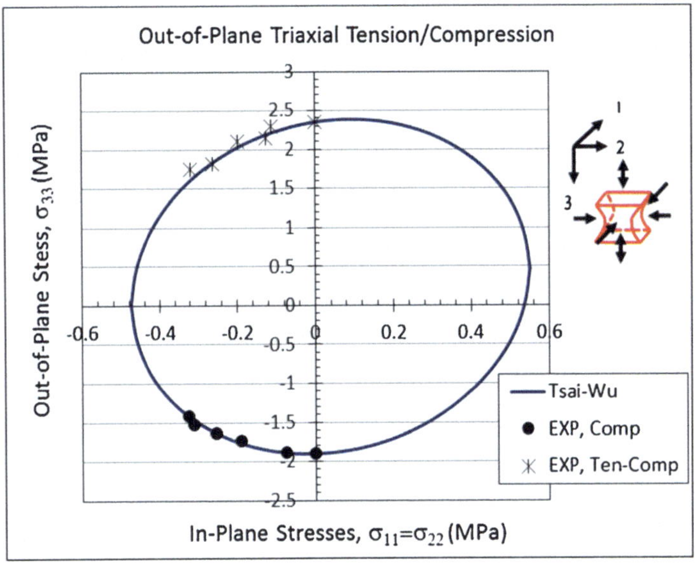

Fig. 14 Yielding under triaxial compression and triaxial tension/compression (out-of-plane control with $\sigma_{11} = \sigma_{22}$)

deformation. Yield strengths under shear and uniaxial tension and compression are specified in Table 1. Three of the most commonly used orthotropic, quadratic yield criteria were considered: Hill [9], Hoffman [10] and Tsai-Wu [11]. The Hill criteria is a pressure-independent orthotropic yield criterion, while the Hoffman and Tsai-Wu criteria are both pressure-dependent by virtue of the fact that they distinguish between the tensile and compressive yield strengths. As shown in Table 1, PVC H100 has different yield strengths in tension and compression and the Hill criterion did not give a good fit to predict initial yielding of this material under triaxial stress. In fact, both the Hill and Hoffman yield criteria did not provide the closed or elliptical yield surfaces, which are shown in Figs. 14, 15, 16 and 17. Only the Tsai-Wu quadratic criterion was able to provide an ellipsoidal yield surface that is consistent with the experimental results, and it is discussed in more details below. Gdoutos et al. [12] were the first to show that the Tsai-Wu criterion could be successfully used to fit biaxial test data on PVC foams, but this is the first time that the Tsai-Wu criterion has been fitted to triaxial test results.

The Tsai – Wu failure criterion is given in terms of 12 coefficients by

$$X_1\sigma_{11} + X_2\sigma_{22} + X_3\sigma_{33} + X_{11}\sigma_{11}^2 + X_{22}\sigma_{22}^2 + X_{33}\sigma_{33}^2 + X_{44}\tau_{23}^2 + X_{55}\tau_{13}^2 + X_{66}\tau_{12}^2$$
$$+2X_{12}\sigma_{11}\sigma_{22} + 2X_{13}\sigma_{11}\sigma_{33} + 2X_{23}\sigma_{22}\sigma_{33} \leq 1 \tag{1}$$

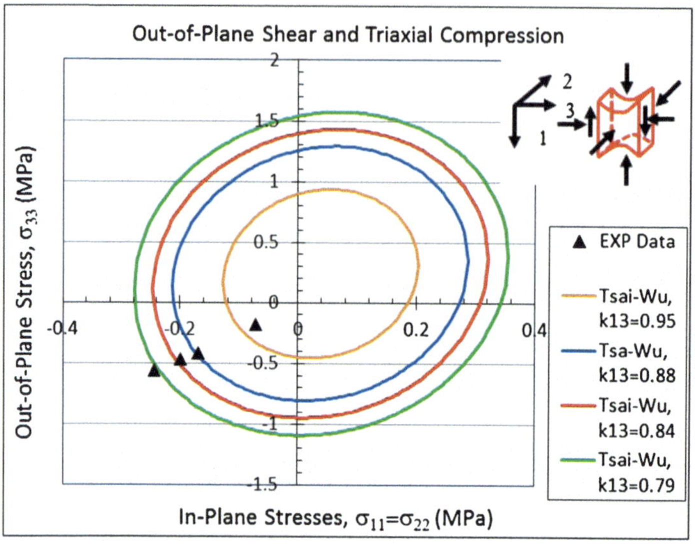

Fig. 15 Yielding under triaxial compression and out-of-plane shear ($k13 = \sigma_{13}/S_{13}$)

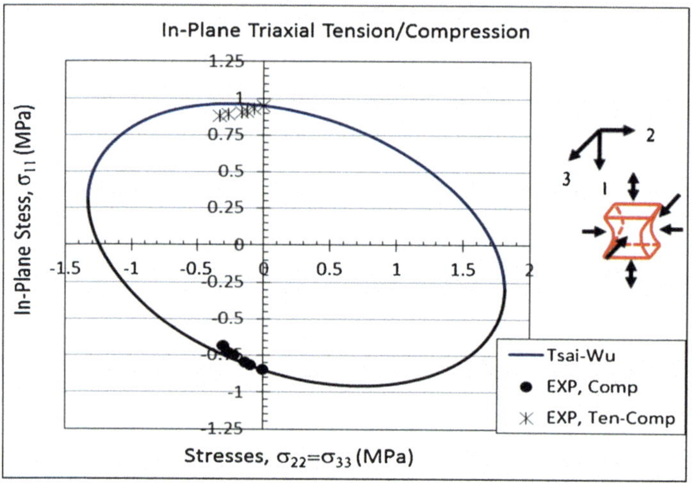

Fig. 16 Yielding under triaxial compression and triaxial tension/compression (in-plane control with $\sigma_{22} = \sigma_{33}$)

where

$$X_1 = \frac{1}{X_t} - \frac{1}{X_c}, X_2 = \frac{1}{Y_t} - \frac{1}{Y_c}, X_3 = \frac{1}{Z_t} - \frac{1}{Z_c}, X_{11} = \frac{1}{X_t X_c}, X_{22} = \frac{1}{Y_t Y_c},$$

$$X_{33} = \frac{1}{Z_t Z_c}, X_{44} = \left(\frac{1}{S_{23}}\right)^2, X_{55} = \left(\frac{1}{S_{13}}\right)^2, X_{66} = \left(\frac{1}{S_{12}}\right)^2$$

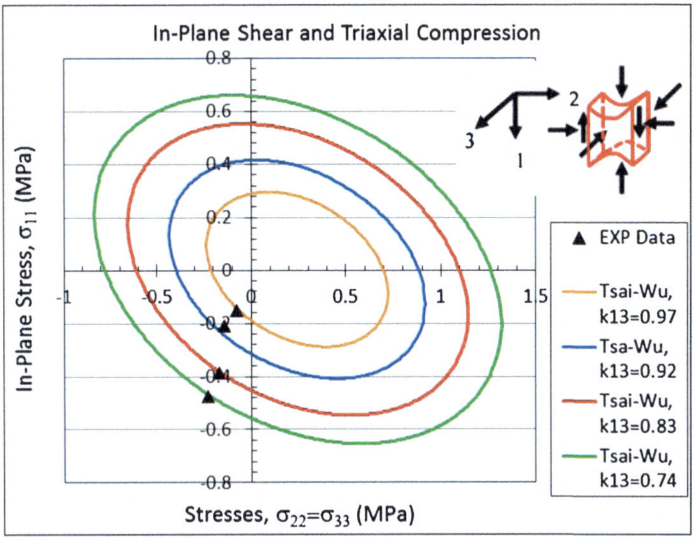

Fig. 17 Yielding under triaxial compression and in-plane shear ($k12 = \sigma_{12}/S_{12}$)

and X_{12}, X_{13} and X_{23} are interaction coefficient terms which can generally be obtained from biaxial tests. For a transversely isotropic material, $X_2 = X_1$, $X_{22} = X_{11}$, $X_{55} = X_{44}$ and the interaction coefficients reduce to $X_{12} = X_{11} - X_{66}/2$, and $X_{23} = X_{13}$. One may use terms in Table 1 to calculate the first interaction term as $X_{12} = X_{11} - X_{66}/2 = 0.055$. Rather than biaxial tests, the triaxial results were used to obtain the other interaction coefficients, $X_{23} = X_{13} = -0.02$. Yielding of the PVC foam under triaxial stresses as predicted by the Tsai-Wu criterion with these interaction terms were shown to compare very well with experimental results. These included results from the triaxial compression and triaxial tension-compression with out-of-plane control and $\sigma_{11} = \sigma_{22}$ tests (Fig. 14); triaxial compression and out-of-plane shear tests (Fig. 15); triaxial compression and triaxial tension-compression tests with in-plane control and $\sigma_{22} = \sigma_{33}$ tests (Fig. 16); and triaxial compression and in-plane shear tests (Fig. 17). Shear factor k13 and k12 are defined in Figs. 15 and 17 so that the shear yield stress could be added to the triaxial stresses in these curves. In all cases, the Tsai-Wu yield function was able to fit experimental results remarkably well.

6 Material Model

The behavior of the PVC foam is described by elasticity, plasticity, viscoelasticity and damage, as shown in Fig. 18. The phenomenological model in Fig. 19 was developed to predict this behavior. Before yielding, the foam was described as a

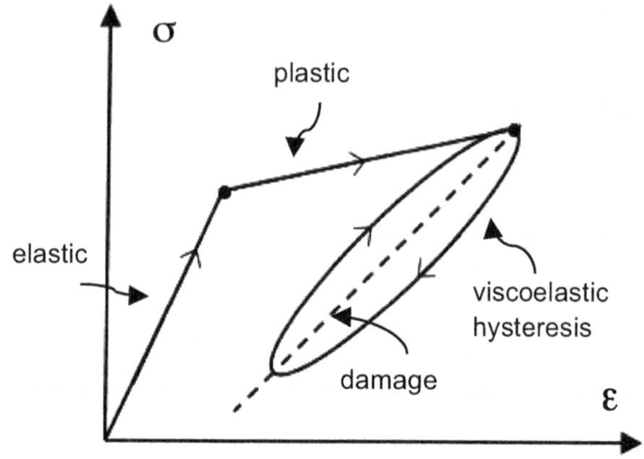

Fig. 18 Elastic and inelastic response of PVC foam

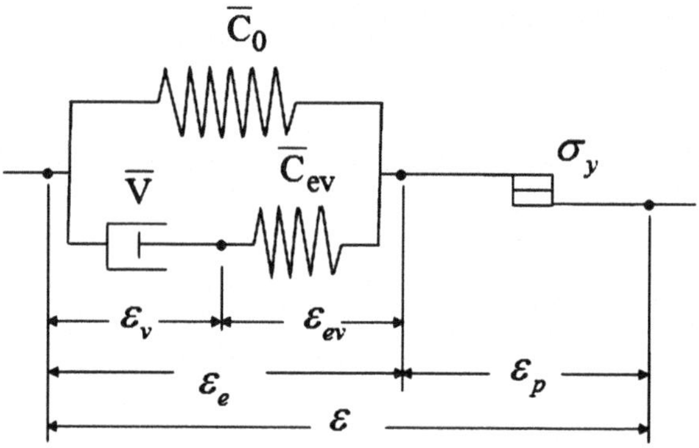

Fig. 19 Model for viscoelastic damage and plastic flow after initial yield

linear elastic with transversely isotropic properties. After yielding, plastic flow evolved with viscoelasticity and damage.

For computational purposes, the Tsai-Wu yield criterion is conveniently expressed as

$$\frac{1}{2}\boldsymbol{\sigma}\mathbf{P}\boldsymbol{\sigma}^{\mathrm{T}} + \mathbf{q}^{\mathrm{T}}\boldsymbol{\sigma} = 1 \tag{2}$$

where

$$\mathbf{P} = 2\begin{bmatrix} X_{11} & X_{12} & X_{13} & 0 & 0 & 0 \\ X_{12} & X_{22} & X_{23} & 0 & 0 & 0 \\ X_{13} & X_{23} & X_{33} & 0 & 0 & 0 \\ 0 & 0 & 0 & X_{44} & 0 & 0 \\ 0 & 0 & 0 & 0 & X_{55} & 0 \\ 0 & 0 & 0 & 0 & 0 & X_{66} \end{bmatrix}$$

and

$$\mathrm{q} = [X_1\ \ X_2\ \ X_3\ \ 0\ \ 0\ \ 0]^T$$

The Tsai-Wu plastic potential function, including both isotropic and kinematic hardening, is then given by

$$\Phi = \frac{1}{2}\boldsymbol{\sigma}'\mathbf{P}\boldsymbol{\sigma}'^{\mathrm{T}} + \mathbf{q}^{\mathrm{T}}\boldsymbol{\sigma}' - \overline{\sigma}^2 = 0 \tag{3}$$

where $\boldsymbol{\sigma}' = \boldsymbol{\sigma} - \boldsymbol{\alpha}, \boldsymbol{\alpha}$ is the back stress, $\overline{\sigma} = \overline{\sigma}(\overline{\varepsilon}_{\mathrm{p}})$ and $\overline{\varepsilon}_{\mathrm{p}}$ is the equivalent plastic strain. The equivalent stress σ_e is defined from the Tsai-Wu potential as

$$\sigma_e^2 = \frac{1}{2}\boldsymbol{\sigma}'\mathbf{P}\boldsymbol{\sigma}'^{\mathrm{T}} + \mathbf{q}^{\mathrm{T}}\boldsymbol{\sigma}' \tag{4}$$

The plastic strain rate $\dot{\boldsymbol{\varepsilon}}_{\mathbf{p}}$ is given by an associate flow rule

$$\dot{\boldsymbol{\varepsilon}}_{\mathbf{p}} = \dot{\lambda}\frac{\partial\Phi}{\partial\boldsymbol{\sigma}} \tag{5}$$

where $\dot{\lambda}$ is the plastic multiplier. In terms of the **P** and **q**, Eq. (5) reduces to

$$\dot{\boldsymbol{\varepsilon}}_{\mathbf{p}} = \dot{\lambda}\frac{(\boldsymbol{\sigma}^{\mathrm{T}}\mathbf{P}\boldsymbol{\sigma} + \mathbf{q}^{\mathrm{T}}\boldsymbol{\sigma})}{\sqrt{\frac{1}{2}\boldsymbol{\sigma}'^{\mathrm{T}}\mathbf{P}\boldsymbol{\sigma}' + \mathbf{q}^{\mathrm{T}}\boldsymbol{\sigma}'}} \tag{6}$$

The back stress is given in terms of a kinematic hardening law

$$\dot{\boldsymbol{\alpha}} = \mathbf{H}\dot{\boldsymbol{\varepsilon}}_{\mathbf{p}} \tag{7}$$

where **H** is the kinematic hardening coefficient matrix. Anisotropic hardening functions were derived based on experimental results. Plastic evolution curves with these anisotropic hardening values are shown in Fig. 20.

During unloading the total stress is given as the sum of equilibrium stress and overstress:

$$\boldsymbol{\sigma} = \boldsymbol{\sigma}_{\mathbf{eq}} + \boldsymbol{\sigma}_{\mathbf{ov}} \tag{8}$$

The equilibrium stress $\boldsymbol{\sigma}_{\mathbf{eq}}$ is

$$\boldsymbol{\sigma}_{\mathbf{eq}} = \overline{\mathbf{C}}_0(\boldsymbol{\varepsilon}\text{-}\boldsymbol{\varepsilon}_{\mathbf{p}}) \tag{9}$$

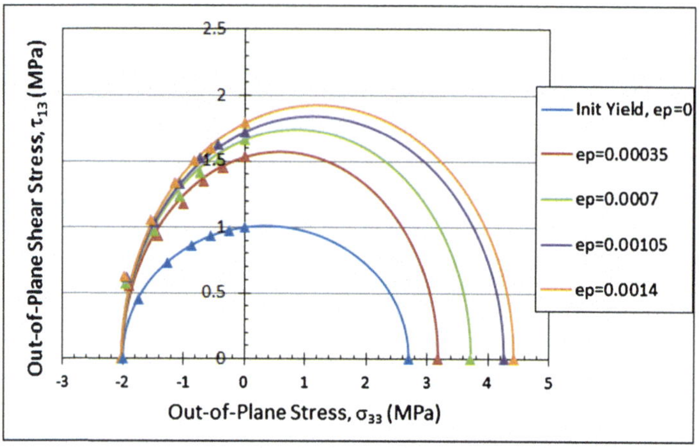

Fig. 20 Plastic evolution curves for out-of-plane biaxial shear-compression

where $\overline{\mathbf{C}}_{\mathbf{0}}$ is the damage stiffness matrix of the equilibrium spring given by

$$\overline{\mathbf{C}}_{\mathbf{0}} = \begin{bmatrix} \overline{C}_{11} & \overline{C}_{12} & \overline{C}_{13} & 0 & 0 & 0 \\ \overline{C}_{12} & \overline{C}_{22} & \overline{C}_{23} & 0 & 0 & 0 \\ \overline{C}_{13} & \overline{C}_{23} & \overline{C}_{33} & 0 & 0 & 0 \\ 0 & 0 & 0 & \overline{C}_{44} & 0 & 0 \\ 0 & 0 & 0 & 0 & \overline{C}_{55} & 0 \\ 0 & 0 & 0 & 0 & 0 & \overline{C}_{66} \end{bmatrix}$$

and

$$\overline{C}_{11} = \frac{\left(\overline{E}_{22} - \nu_{23}^2 \overline{E}_{33}\right)\overline{E}_{11}^2}{\Omega} \qquad \overline{C}_{12} = \frac{\left(\nu_{12}\overline{E}_{22} + \nu_{13}\nu_{23}\overline{E}_{33}\right)\overline{E}_{11}\overline{E}_{22}}{\Omega}$$

$$\overline{C}_{13} = \frac{\left(\nu_{12}\nu_{23} + \nu_{13}\right)\overline{E}_{11}\overline{E}_{22}\overline{E}_{33}}{\Omega} \qquad \overline{C}_{22} = \frac{\left(\overline{E}_{11} - \nu_{13}^2 \overline{E}_{33}\right)\overline{E}_{22}^2}{\Omega}$$

$$\overline{C}_{23} = \frac{\left(\nu_{23}\overline{E}_{11} + \nu_{12}\nu_{13}\overline{E}_{22}\right)\overline{E}_{22}\overline{E}_{33}}{\Omega} \qquad \overline{C}_{33} = \frac{\left(\overline{E}_{11} - \nu_{12}^2 \overline{E}_{22}\right)\overline{E}_{22}\overline{E}_{33}}{\Omega}$$

$$\overline{C}_{44} = \overline{G}_{23} \qquad \overline{C}_{55} = \overline{G}_{13} \qquad \overline{C}_{66} = \overline{G}_{12}$$

and

$$\Omega = \overline{E}_{11}\overline{E}_{22} - \nu_{12}^2\overline{E}_{22}^2 - \nu_{13}^2\overline{E}_{22}\overline{E}_{33} - \nu_{23}^2\overline{E}_{11}\overline{E}_{33} - 2\nu_{12}\nu_{13}\nu_{23}\overline{E}_{22}\overline{E}_{33}$$

The overstress σ_{ov} is

$$\sigma_{ov} = \overline{C}_{ev}\varepsilon_{ev} \tag{10}$$

where $\overline{\mathbf{C}}_{\mathbf{ev}}$ is the stiffness matrix of the intermediate spring given by

$$\overline{\mathbf{C}}_{\mathbf{ev}} = \begin{bmatrix} C_{ev11} & C_{ev12} & C_{ev13} & 0 & 0 & 0 \\ C_{ev12} & C_{ev22} & C_{ev23} & 0 & 0 & 0 \\ C_{ev13} & C_{ev23} & C_{ev33} & 0 & 0 & 0 \\ 0 & 0 & 0 & C_{ev44} & 0 & 0 \\ 0 & 0 & 0 & 0 & C_{ev55} & 0 \\ 0 & 0 & 0 & 0 & 0 & C_{ev66} \end{bmatrix}$$

and

$$C_{ev11} = \frac{\left(E_{ev22} - \nu_{23}^2 E_{ev33}\right)E_{ev11}^2}{\Delta} \qquad C_{ev12} = \frac{(\nu_{12}E_{ev22} + \nu_{13}\nu_{23}E_{ev33})E_{ev11}E_{ev22}}{\Delta}$$

$$C_{ev13} = \frac{(\nu_{12}\nu_{23} + \nu_{13})E_{ev11}E_{ev22}E_{ev33}}{\Delta} \qquad C_{ev22} = \frac{\left(E_{ev11} - \nu_{13}^2 E_{ev33}\right)E_{ev22}^2}{\Delta}$$

$$C_{ev23} = \frac{(\nu_{23}E_{ev11} + \nu_{12}\nu_{13}E_{ev22})E_{ev22}E_{ev33}}{\Delta} \qquad C_{ev33} = \frac{\left(E_{ev11} - \nu_{12}^2 E_{ev22}\right)E_{ev22}E_{ev33}}{\Delta}$$

$$C_{ev44} = G_{ev23} \qquad C_{ev55} = G_{ev13} \qquad C_{ev66} = G_{ev12}$$

and

$$\Delta = E_{ev11}E_{ev22} - \nu_{12}^2 E_{ev22}^2 - \nu_{13}^2 E_{ev22}E_{ev33} - \nu_{23}^2 E_{ev11}E_{ev33} - 2\nu_{12}\nu_{13}\nu_{23}E_{ev22}E_{ev33}$$

Compatibility of strain requires that

$$\boldsymbol{\varepsilon} = \boldsymbol{\varepsilon}_{\mathbf{p}} + \boldsymbol{\varepsilon}_{\mathbf{ev}} + \boldsymbol{\varepsilon}_{\mathbf{v}} \tag{11}$$

Substituting Eq. (11) into Eq. (9) gives

$$\boldsymbol{\sigma}_{\mathbf{ov}} = \overline{\mathbf{C}}_{\mathbf{ev}}\left(\boldsymbol{\varepsilon} - \boldsymbol{\varepsilon}_{\mathbf{p}} - \boldsymbol{\varepsilon}_{\mathbf{v}}\right) \tag{12}$$

The overstress is also governed by a linear viscosity law:

$$\sigma_{\mathbf{ov}} = \overline{\mathbf{V}}\dot{\boldsymbol{\varepsilon}}_{\mathbf{v}} \tag{13}$$

where $\mathbf{V}$ is the viscosity matrix given by

$$\mathbf{V} = \begin{bmatrix} V_{11} & V_{12} & V_{13} & 0 & 0 & 0 \\ V_{12} & V_{22} & V_{23} & 0 & 0 & 0 \\ V_{13} & V_{23} & V_{33} & 0 & 0 & 0 \\ 0 & 0 & 0 & V_{44} & 0 & 0 \\ 0 & 0 & 0 & 0 & V_{55} & 0 \\ 0 & 0 & 0 & 0 & 0 & V_{66} \end{bmatrix}$$

and

$$V_{11} = \frac{(\eta_{22} - \nu_{23}^2\eta_{33})\eta_{11}^2}{\Psi} \qquad V_{12} = \frac{(\nu_{12}\eta_{22} + \nu_{13}\nu_{23}\eta_{33})\eta_{11}\eta_{22}}{\Psi}$$

$$V_{13} = \frac{(\nu_{12}\nu_{23} + \nu_{13})\eta_{11}\eta_{22}\eta_{33}}{\Psi} \qquad V_{22} = \frac{(\eta_{11} - \nu_{13}^2\eta_{33})\eta_{22}^2}{\Psi}$$

$$V_{23} = \frac{(\nu_{23}\eta_{11} + \nu_{12}\nu_{13}\eta_{22})\eta_{22}\eta_{33}}{\Psi} \qquad V_{33} = \frac{(\eta_{11} - \nu_{12}^2\eta_{22})\eta_{22}\eta_{33}}{\Psi}$$

$$V_{44} = \eta_{23} \qquad V_{55} = \eta_{13} \qquad V_{66} = \eta_{12}$$

and

$$\Psi = \eta_{11}\eta_{22} - \nu_{12}^2\eta_{22}^2 - \nu_{13}^2\eta_{22}\eta_{33} - \nu_{23}^2\eta_{11}\eta_{33} - 2\nu_{12}\nu_{13}\nu_{23}\eta_{22}\eta_{33}.$$

Combining Eqs. (12) and (13) gives an evolution equation for $\boldsymbol{\varepsilon}_v$:

$$\overline{\mathbf{V}}\dot{\boldsymbol{\varepsilon}}_v = \overline{\mathbf{C}}_{ev}(\boldsymbol{\varepsilon} - \boldsymbol{\varepsilon}_p - \boldsymbol{\varepsilon}_v) \tag{14}$$

The total stress is represented as

$$\boldsymbol{\sigma} = \overline{\mathbf{C}}_0\boldsymbol{\varepsilon} + \overline{\mathbf{C}}_{ev}(\boldsymbol{\varepsilon} - \boldsymbol{\varepsilon}_p - \boldsymbol{\varepsilon}_v) \tag{15}$$

The above equations was used to simulate some of the test results. Good agreement can be seen (Fig. 21) between the predicted response based on the above constitutive model and the test data of the 45 deg Arcan specimen in Figs. 3(a) and (b).

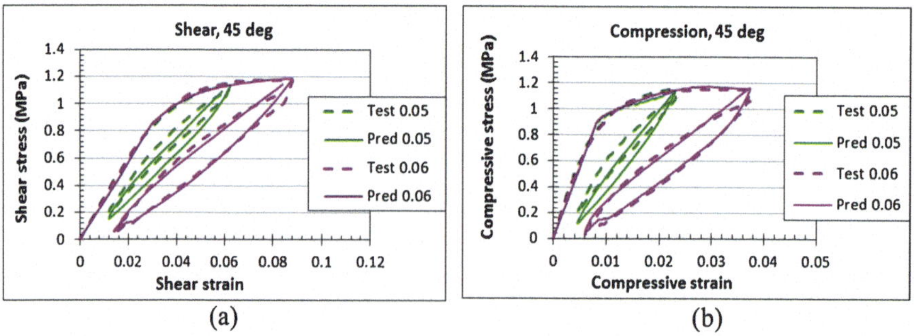

Fig. 21 Comparison between predicted and experimental results for the out-of-plane 45 deg. Arcan specimens: (**a**) shear and (**b**) compression

7 Concluding Remarks

Pressure vessel experiments were done to determine the multiaxial, elastic-plastic and hysteresis behavior of Divinycell PVC H100 foam. The foam, which was transversely isotropic, exhibited elastic-plastic response followed by damage and viscoelastic hysteresis. Tsai-Wu plasticity, including combined kinematic and isotropic hardening, was combined with a linear viscoelastic, damage mechanism to describe multiaxial yielding of the foam. Good agreement was found between the proposed elastic-plastic-viscoelastic-damage constitutive model and experimental results. The constitutive equation will be used to simulate the underwater blast response of sandwich structures in future work.

Acknowledgments This research was supported under ONR Grant N00014-16-1-2840. The authors thank Dr. Yapa D. S. Rajapakse, Solid Mechanics Program Manager at the Office of Naval Research, for making this work possible, and DIAB for supplying PVC H100 foam panels used in the tests. We also thank Dale Ertely and Bill Wenzel, Engineering Machine Shop at The University of Akron, for machining the pressure vessel, specialty fixtures and specimens for the experiments.

References

1. Gopalakrishnan S, Rajapakse YDS (2019) Blast mitigation strategies for marine composite and Sandwich structures. Springer Nature Singapore Pte Ltd, Singapore
2. Hoo Fatt MS, Sirivolu D (2017) Marine composite sandwich plates under air and water blasts. Mar Struct 56:63–185
3. Hoo Fatt MS, Sirivolu D (2015) Blast response of double curvature, composite sandwich shallow shells. Eng Struct 100:696–706
4. Deshpande VS, Fleck NA (2001) Multi-axial yield behavior of polymer foams. Acta Mater 49:1859–1186
5. Tagarielli VL, Desphande VS, Fleck NA, Chen C (2005) A constitutive model for transversely isotropic foams, and its application to the indentation of balsa wood. Int J Mech Sci 47 (4–5):666–686
6. Gielen AWJ (2008) A PVC-foam material model based on a thermodynamically elasto-plastic-damage framework exhibiting failure and crushing. Int J Solids Struct 45:1896–1917
7. Li P, Guo YB, Shim VPW (2018) A constitutive model for transversely isotropic material with anisotropic hardening. Int J Solids Struct 138:40–49
8. Zhong C (2019) Pressure chamber experiments to determine triaxial material properties of polymer foams, PhD dissertation, The University of Akron, May 2019
9. Hill R (1948) A theory of the yielding and plastic flow of anisotropic metals. Proc Roy Soc London 193:281–297
10. Hoffman O (1967) The brittle strength of orthotropic materials. J Compos Mater 1(2):200–206
11. Tsai SW, Wu EM (1971) A general theory of strength for anisotropic materials. J Compos Mater 5:58–80
12. Gdoutos EE, Daniels IM, Wang KA (2002) Failure of cellular foams under multiaxial loading Composites A Appl Sci Manuf 33:163–176

3D Printing of Syntactic Foams for Marine Applications

Nikhil Gupta and Mrityunjay Doddamani

1 Introduction

Additive manufacturing (AM), also commonly known as 3D printing, is revolutionizing many fields including medical, aerospace and automotive [1–3]. The capability of manufacturing complex shaped parts in net shape, creating internal structural details in a part and optimizing the weight of a part are among the advantages provided by AM methods. Weight sensitive aircraft and spacecraft structures are especially benefitting from the advancements in the AM methods. Marine structural applications also represent the category of weight sensitive structures, where a combination of weight and mechanical properties needs to be optimized for efficient design and operation of the structure. AM methods can help in making significant progress in developing next generation marine structures that are optimized for the complex loading conditions encountered in such applications.

AM represents a suite of methods that range from depositing a thermoplastic filament to jetting of thermosetting resins. Some of the commonly known AM methods include Fused Filament Fabrication (FFF), Stereolithography (SLA), Selective Laser Sintering/Melting (SLS/SLM), Direct Metal Laser Sintering (DMLS), Electron-beam Melting (EBM), Polymer Jetting (PolyJet), and Laminated Object Manufacturing (LOM) [1, 4]. These techniques can be categorized into seven groups based on the process types [4]:

N. Gupta (✉)
Composite Materials and Mechanics Laboratory, Mechanical and Aerospace Engineering Department, New York University, Tandon School of Engineering, Brooklyn, NY, USA
e-mail: ngupta@nyu.edu

M. Doddamani
Advanced Manufacturing Laboratory, Department of Mechanical Engineering, National Institute of Technology Karnataka, Surathkal, Karnataka, India
e-mail: mrdoddamani@nitk.edu.in

S. W. Lee (ed.), *Advances in Thick Section Composite and Sandwich Structures*,
https://doi.org/10.1007/978-3-030-31065-3_14

(a) Binder Jetting
(b) Directed Energy Deposition
(c) Material Extrusion
(d) Material Jetting
(e) Powder Bed Fusion
(f) Sheet Lamination
(g) Vat Photopolymerization

Many of these AM methods are now commercially viable and 3D printers are available based on these technologies. Material extrusion and powder bed fusion based methods are being widely used now for industrial production of parts from metals and thermoplastic polymers [5]. Printers based on the material extrusion method have the lowest cost, which has helped in making them widely available. Printers based on powder bed fusion and polymerization are relatively more expensive but can provide application grade parts right out of the build plate. In the present case, FFF method, which is also called Fused Deposition Modeling (FDM), is used for syntactic foam printing.

1.1 AM Process Chain

The computer aided design (CAD) files are not the direct input for 3D printers [6]. These files go through a series of transitions before printing the parts using a 3D printer. Major steps of a typical additive manufacturing process chain are shown in Fig. 1. The steps include:

(a) Preliminary CAD model development
(b) Finite element analysis, computational fluid dynamics and other analyses to optimize the design
(c) Conversion of CAD file to STL (StereoLithography) format
(d) Slicing of the model into 2D slices
(e) Generation of toolpath in the form of G-code
(f) Printing of 3D object on a printer using G-code
(g) Testing, imaging and non-destructive evaluation of the printed objects for quality and reliability

The additive manufacturing process chain starts with the CAD model development of the part using software programs such as SolidWorks or AutoCAD Inventor. It is a standard practice to refine the design by finite element analysis, computational fluid dynamics or multi-physics analysis to obtain optimized design. The final CAD model is exported in the STL format, which contains a triangulated representation of the surface geometry of the part. These triangulated surfaces, or facets, are described by the unit normal vector and three vertices ordered by the right-hand rule using the Cartesian coordinate system. The STL file is then imported in a slicing software that converts the geometry into a stack of 2D slices. Next a toolpath is generated to build

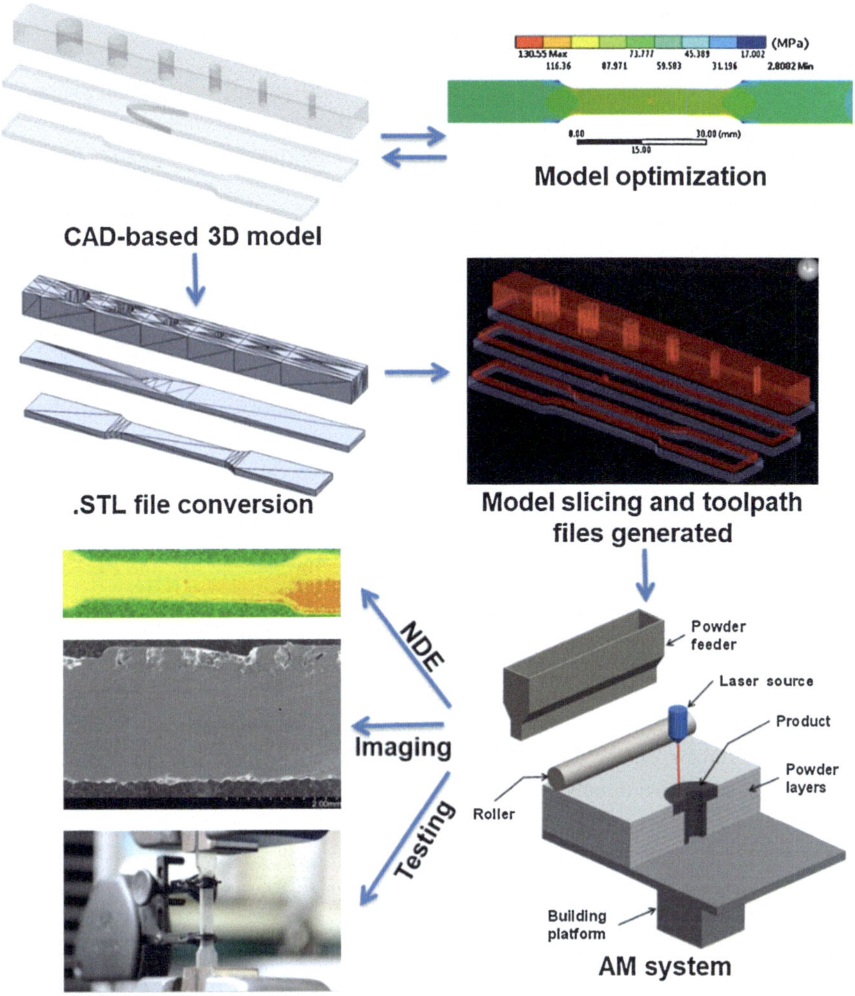

Fig. 1 Typical process chain of additive manufacturing methods

these slices on the 2D plane of the 3D printer's build plate. The toolpath file, called G-code, also contains the machine parameters such as print head speed, print head movement coordinates and temperature. Finally, G-code is sent to the 3D printer to print the part. The printed part goes through a series of inspections and tests to ensure the quality. The tests can include both destructive and non-destructive tests to ensure that the part confirms to the specifications required for the application.

The completely digital process chain of AM from the CAD model development to final printing utilizes computers and cloud-based resources. Protection of this process chain from cybersecurity threats is an important consideration in the field [7, 8], however, this topic is extensively discussed elsewhere in the published literature and the focus of this chapter is on the technical aspects of 3D printing of syntactic foam components.

1.2 Syntactic Foams

Hollow particle filled composites are called syntactic foams [9]. A typical syntactic foam microstructure is shown in Fig. 2a, where thin walled glass hollow particles are embedded in a polymeric matrix to create lightweight syntactic foams. A CAD model representation of syntactic foam microstructure is presented in Fig. 2b, where the matrix is made transparent to observe the dispersion of hollow particles throughout the composite. Thermosetting syntactic foams with epoxy and vinyl ester matrices have been studied for mechanical, thermal and electrical properties in a wide range of published literature [10–16]. An overview of some of the thermoplastic matrix syntactic foams can be found in chapters of a book [17, 18] and relevant publications [19–22].

Syntactic foams are used in many structural applications in aerospace industry [23]. The close-cell microstructure of these foams provides low density and high compressive strength as well as low moisture absorption coefficient for marine structural applications. These foams are used for well-known underwater explorer vehicles such as Alvin, Jason and Deepsea Challenger (Fig. 3). These vehicles have enabled developing ocean surface topography, studying marine life and exploration of the deepest part of the ocean – Mariana Trench. Apart from the marine vessel structures, syntactic foams are also widely used for thermal insulation purposes in underwater oil exploration pipes.

An example of a small remotely operated vehicle for underwater exploration is shown in Fig. 4. Such vehicles may have a variety of different sensors such as temperature sensor, camera, and water quality sensors mounted in them. In addition, they also have communication equipment to enable their remote operation. These

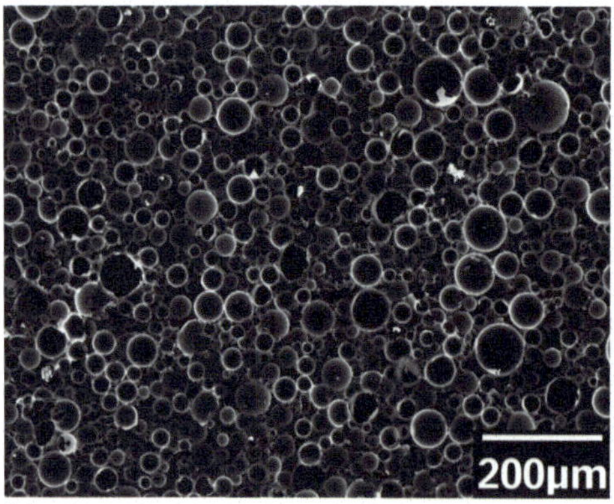

(a)

(b)

Fig. 2 (**a**) A typical syntactic foam microstructure where hollow glass particles are embedded in a polymer matrix to realize a lightweight composite and (**b**) a CAD model representation of syntactic foam microstructure. The matrix material is made transparent in the model to observe the particle distribution inside the material structure

Fig. 3 Deepsea Challenger during construction phase, where syntactic foam parts are assembled to create the body of the underwater vehicle. Photo credit: James Cameron and Dianna Bisset

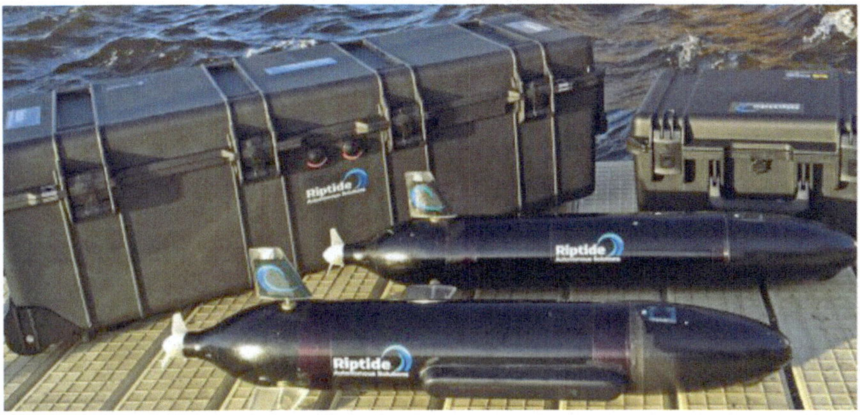

Fig. 4 Underwater remotely operated vehicles. Image courtesy: Riptide autonomous solutions (www.riptideas.com)

vehicles have been used for exploration of sunken ships and airplanes, oil fields, and marine life in ocean. Vehicles are also being developed to remotely inspect chemical storage tanks and water tanks.

Hollow particles of glass and fly ash have been used widely for manufacturing syntactic foams. Methods are available to manufacture hollow particles of almost any ceramic material. Most widely used particles are in the range 10–100 μm diameter and wall thickness of 0.3–1.5 μm. Selection of hollow particles of desired material, diameter and wall thickness allows tailoring the properties of syntactic foams over a wide range, which provides flexibility in designing their applications. Novel variants of syntactic foams are also used for electromagnetic interference (EMI) shielding applications. Use of syntactic foams in USS Zumwalt construction allows reducing the radar signature of the ship to make it difficult to detect.

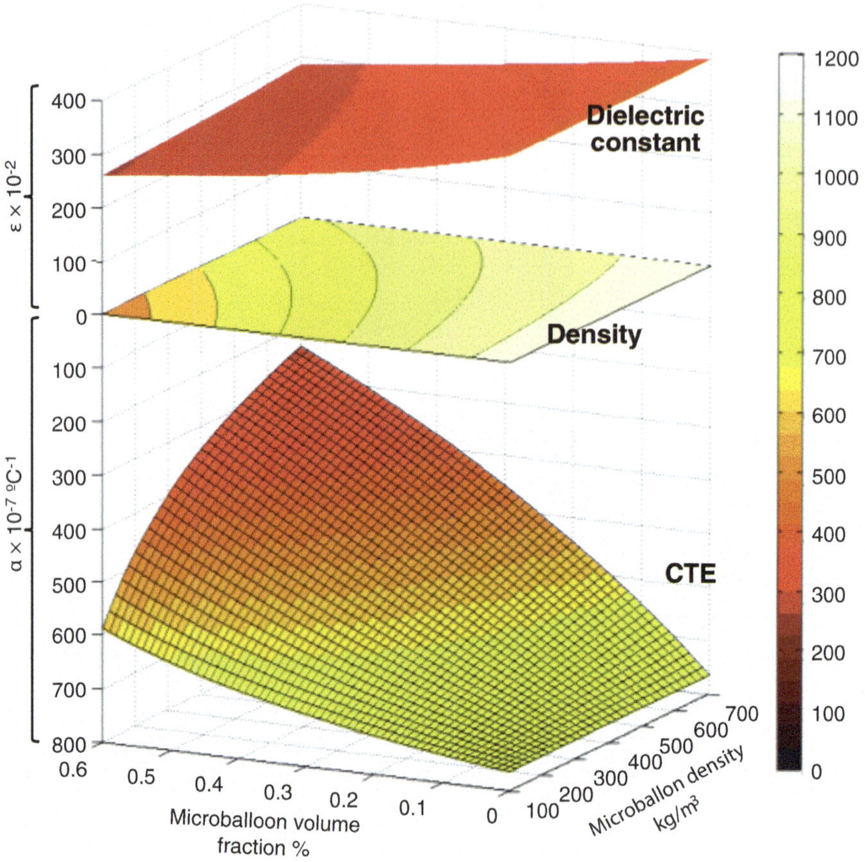

Fig. 5 Design of syntactic foams using muti-criteria optimization scheme where composition is determined based on the desired dielectric constact, density and coefficient of thermal expansion (CTE). Similar optimization scheme can be used for other properties of interest

1.3 Need for 3D Printing of Syntactic Foams for Marine Applications

One of the major advantages of syntactic foams is their ability to be tailored for a wide range of properties. Figure 5 shows that use of hollow particle wall thickness and volume fraction as two independent parameters allows tailoring the syntactic foam properties over a wide range. In comparison, only volume fraction can be varied in solid particle filled composites for tailoring their properties. Selecting a particular volume fraction freezes all the properties according to that proportion. In the case of syntactic foams, simultaneous optimization of two parameters allows tailoring several properties at the same time within a desired range based on the design criteria [9, 23]. For example, Fig. 5 shows that for a particular value of density, dielectric constant and coefficient of thermal expansion (CTE) can be tailored over a range of values and the best possible composition can be selected.

A number of methods are already available for manufacturing syntactic foams. Two commonly used methods are

- Stir mixing and casting: This method is widely used for thermosetting resin syntactic foams, for example, epoxy and vinyl ester matrix syntactic foams. Particles are mixed with the resin and then hardener is added to the mix, followed by curing. Thermoplastic resin matrix syntactic foams have also been developed by this method.
- Injection molding: This method is used for thermoplastic syntactic foams. Initially the particles are compounded with the resin and then the mixture is molded using an injection molding machine. Injection temperature, pressure and die design are important parameters in this process. Excess pressure and high temperature can lead to mold overfill and bleed out of the mixture from the joints, while low pressure and temperature can lead to under fill.

Apart from these methods, infiltration of particle bed, compression molding and several other methods have also been used for manufacturing syntactic foams but their use has been limited [20, 24]. Each of these methods has some advantages and limitations, which decide their application domain. For example, pressure or vacuum infiltration method requires a packed bed of particles and cannot be used to fabricate syntactic foams with less than 50 vol.% hollow particles.

Viscosity of the resin plays an important role in obtaining high quality composites in both these methods. While low viscosity of the resin is useful in uniformly mixing the particles, it also leads to floatation of particles due to their significantly lower density than the resin if the particle volume fraction is lower than 30% in the mixture. If the resin viscosity is increased to reduce the particle floatation, then the particle fracture increases due to increased shear. In many applications, syntactic foams with less than 30 vol.% hollow particles are desired, which is difficult to fabricate with both casting and injection molding methods as the uniform quality of the product cannot be obtained due to particle floating during solidification. In such case, AM methods are expected to provide advantage of uniform hollow particle distribution in the entire part.

There is also a second reason for developing AM methods for syntactic foams. In creating underwater vehicle structures, it is a common practice to adhesively join several blocks of syntactic foams and then machine the desired shape [25]. These foams work well in deep sea environment. However, cycling of pressurization and depressurization can lead to fracture of foams either at the adhesive joints or next to the joints due to stress concentration effects as shown in Fig. 6. In order to manufacture complex shapes and contours and eliminate the need for adhesive bonding, AM methods are considered promising for this application.

Simultaneous optimization of density and mechanical properties has become even more challenging as the next generation underwater vehicles are being developed to explore 4000–6000 m depths. In addition, it is desired to add functionality to

Fig. 6 A syntactic foam block cracked near the adhesive joint during pressure testing due to stress concentration effects

syntactic foam components such as damping, EMI shielding, and electrical conductivity [26, 27]. AM methods can present options to develop syntactic foam components in such a challenging landscape.

In the present case, methods are developed to 3D print syntactic foams using FFF technology [28]. FFF printers are low cost and versatile. It is envisioned that if a syntactic foam filament is developed that can be used in commercial FFF printers without any hardware modification, then the adoption of such filaments by industry may have low barriers. Both filament manufacture and 3D printing processes are essentially extrusion processes, where minimization of hollow particle fracture is desired by processing parameter optimization.

2 Materials and Methods

2.1 *Filament Material*

High density polyethylene (HDPE) of HD50MA180 grade is procured from Reliance Polymers, Mumbai, India, in the form of 3 mm diameter granules of mean molecular weight 97,500 g/mol. The resin has 20 g/10 min melt flow index. Cenospheres (CIL-150 grade) are supplied by Cenospheres India Pvt. Ltd., Kolkata, India. Chemical analysis details of cenospheres are presented elsewhere [20]. The particle size analysis of cenospheres is presented in Fig. 7. The figure shows results for both as received and silane coated cenospheres. Silane coating is expected to provide improved bonding with the polymeric resin. 3-amino propyl triethoxy silane is used as the coupling agent to treat the particle surface. This figure shows that the curve for coated particles shifts to larger particle sizes. The upper end of the particle size increased considerably from 170 μm to 475 μm. The HDPE matrix is

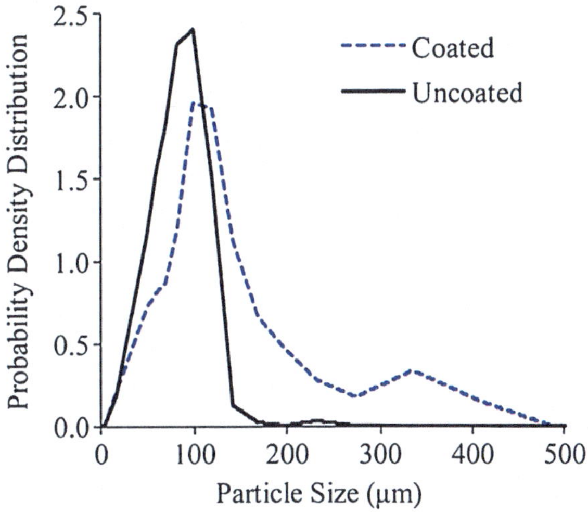

Fig. 7 Particle size analysis of coated and silane treated cenospheres

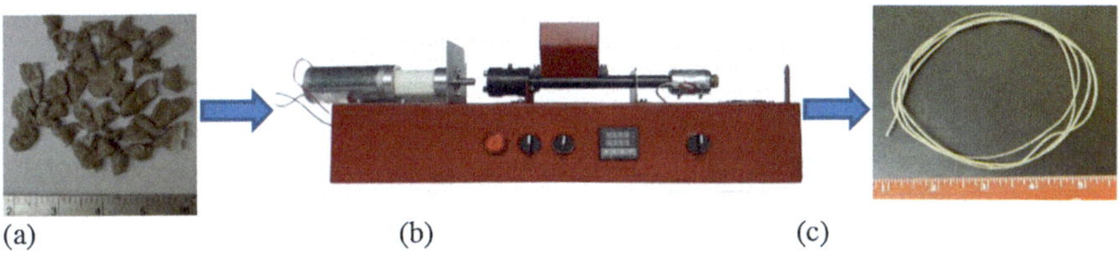

Fig. 8 (**a**) Pellets created using brabender by mixing HDPE resin with cenospheres, (**b**) laboratory scale extruder and (**c**) filament extruded for 3D printing

compatibilized using dicumyl peroxide to allow coupling with the silane coating on the cenospheres and functionalized with dibutyl maleate to improve blending. A small peak around 330 μm may indicate agglomeration of particles due to coating, providing a concentration of larger size particles. Some of these clusters are expected to break during further process of the material because shear will be applied during compounding and extrusion stages.

Syntactic foams with 40 wt.% cenospheres are blended and are designated by HDPE40. In previous studies, HDPE40 composition is found to be the highest filling level of cenospheres without substantial fracture [20, 24].

2.2 *Pellet Manufacturing and Filament Extrusion*

HDPE resin-cenosphere mixture is produced using Brabender (CMEI, MODEL-16 CME SPL, Western Company Keltron). The mixture is melted to form approximately $4 \times 4 \times 3\ \text{mm}^3$ pellets. These pellets are fed into a laboratory scale filament extruder having two heating zones. The extruder design is developed based on the

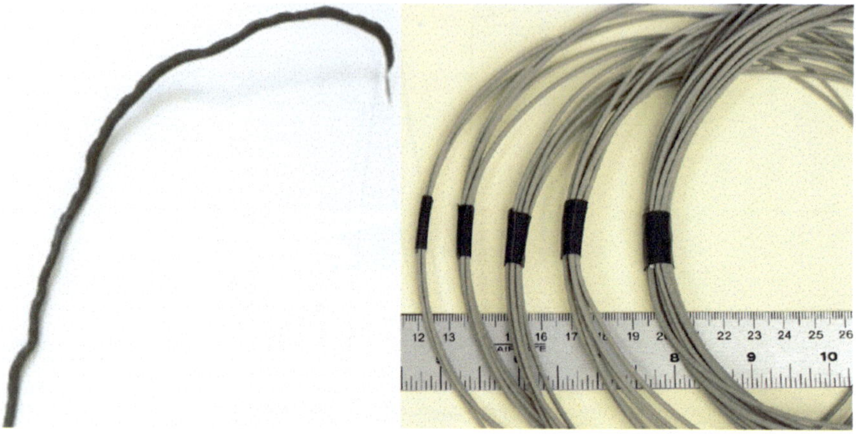

(a) (b)

Fig. 9 (**a**) Filament made in initial round of extrusion before optimization shows poor diameter control and (**b**) filaments produced with optimized parameters has a uniform appearance

initial pilot experiments. The process is depicted in Fig. 8, where Fig. 8a shows the pellets that are created from Brabender to feed into the extruder that is shown in Fig. 8b. Two heaters mounted on the barrel of this extruder are controlled separately using solid-state relays that are switched on/off by an Arduino using the input from two K-type thermocouples attached to each heater. The primary and secondary heaters are set at 127 and 120 °C, respectively. After pre-heating of HDPE40 pellets for 10 min, the material is extruded through a die with a 2.2 mm hole to obtain the filament. An example of the extruded filament is shown in Fig. 8c. The filament is pulled using an automatic winding system that can generate enough tension in the extruded filament to keep the diameter constant. Cooling shrinkage is significant in thermoplastics. Therefore, the extruded diameter of the filament is larger than the desired 1.75 mm.

A comparison of a filament made in the initial rounds before optimization is shown in Fig. 9a, which can be compared with filaments manufactured with optimized parameters shown in Fig. 9b. The diameter was not uniform in the filament that was initially manufactured. Once the parameters are optimized, any length of filament can be extruded from the system. The quality of the filament is assessed for uniformity of diameter and then the parameters such as temperature of extrusion and tension in the extruded filament are adjusted to improve the quality in the next iteration. Success in the extrusion parameter optimization has resulted in filament that appears to be much more uniform in diameter.

2.3 Filament Recycling

The first extruded filament is termed as 1X filament. For the purpose of assessing the recyclability of syntactic foam filaments, the 1X filament is cut into pieces of about 5 mm length and then fed into the hopper to again extrude the filament that is termed as 2X. Finally, this filament is cut and extruded again the create 3X filament. These recycled filaments are also tested for mechanical properties and used in the 3D printer for printing tensile test coupons. Neat HDPE resin filaments are also recycled under the same conditions to develop a baseline for mechanical properties of the neat resin. At least five specimens are cut from random sections of each HDPE and HDPE40 filaments and the mass and volume are measured to calculate the density.

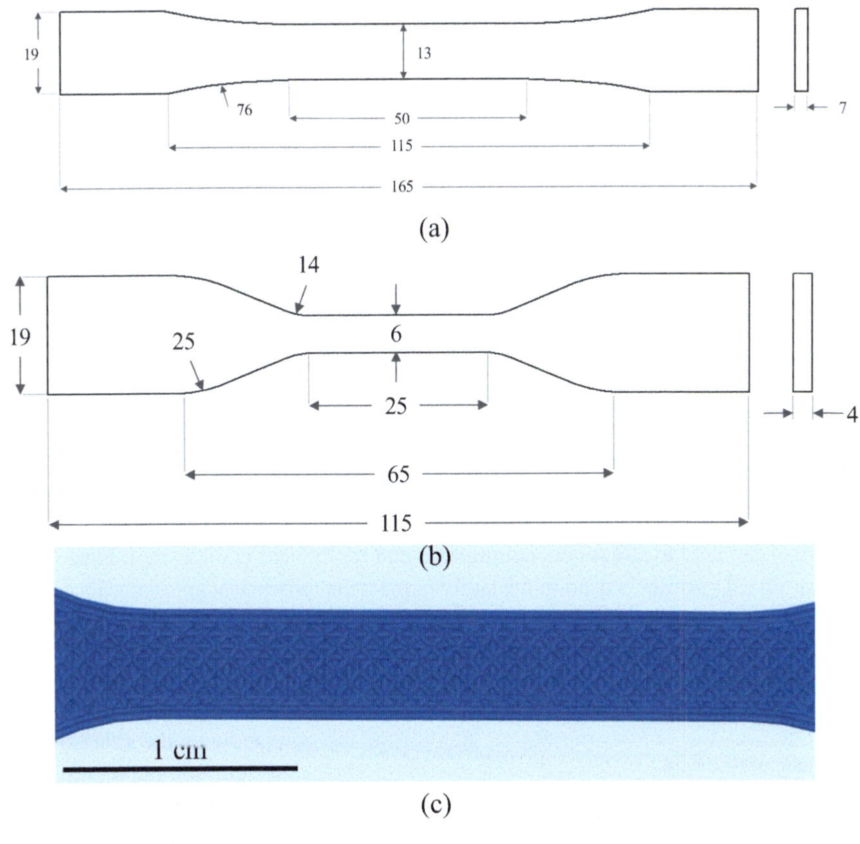

Fig. 10 ASTM D638 (**a**) Type I and (**b**) Type IV tensile specimen dimensions. (**c**) Gauge section of tensile bar showing infill pattern used in printing HDPE40 specimens in this work

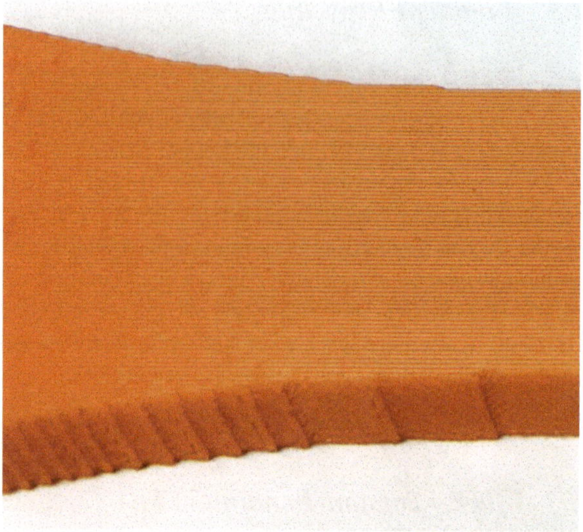

Fig. 11 An example of coarse STL file of a tensile test specimen printed without the shell around the specimen. The contour lines can be observed in this magnified view around the shoulder of the specimen. These lines can raise stress concentration in this region

2.4 CAD Modeling and 3D Printing

ASTM D638 tensile specimens (Types I and IV) are modeled using Solidworks. ASTM D638 Type I tensile specimens are printed using HDPE resin and Type IV specimens are printed for HDPE40 syntactic foams. The specimen geometries are shown in Fig. 10a and 10b, respectively. The solid model files are converted to STL format with fine resolution setting. The STL resolution is an important factor in printing the specimens with curve sections. Fine resolution results in larger file size but better curvatures in the printed part. The STL models are sliced to generate the tool path using ReplicatorG software. The specimen is oriented horizontally along the x-axis of the print bed. One shell and 60% infill are chosen through optimization for the print. The infill pattern for the specimens is shown in Fig. 10c, which shows a shell around the specimen boundary and an infill pattern.

If the shell is not used around the specimen and coarse STL resolution is used, then the 3D printer may create contours along the curvature similar to the example shown in Fig. 11. Presence of these lines along the shoulder of the tensile test specimen can lead to stress concentration in that region and is undesired. Flashforge Creator Pro 3D printer is used in the study to print the tensile test coupons. To ensure adhesion of the bottom layer, an HDPE plate was used as the build plate. Matching of CTE of the modeling material and build plate material helps in reducing the effect of warpage of the layers when the material cools down after deposition.

2.5 *Imaging*

A Hitachi S-3400 N scanning electron microscope (SEM), equipped with backscattered and secondary electron detectors, is used to study the filaments microstructure. Specimens are coated with a conducting layer of gold using Cressington 108 Auto Sputter Coater prior to imaging.

The internal structure of filaments is studied by using a Bruker Skyscan 1172 micro-CT scanner equipped with a digital X-ray camera having 4000 × 2664-pixel array (corresponding to 4 K resolution). The specimens are placed 36.55 and 308.36 mm from the source and camera, respectively, resulting in a pixel size of 0.95 μm. A rotation step of 0.2° is used in scanning. Samples are exposed for 1420 ms at a source voltage of 50 kV. The micro-CT scan results are used to analyze the cenosphere distribution and voids in the filaments and printed parts.

2.6 *Tensile Testing*

An Instron 4467 Universal Testing System with a 30 kN load cell is used to perform tensile tests at $10^{-4}\ s^{-1}$, $10^{-3}\ s^{-1}$ and $10^{-2}\ s^{-1}$ strain rates. Thermoplastic resins are viscoelastic in nature and show strain rate sensitivity even at quasi-static strain rate range. Therefore, testing at various strain rates is required to obtain sufficient information about their behavior. An Instron extensometer of 1-inch gauge length is used to measure strain. At least five specimens of each filament type are tested at each strain rate.

3 Syntactic Foam Filament Manufacturing

In the first step, a filament of syntactic foams is developed and tested for quality and mechanical properties. In the second step, the filament is used for 3D printing using a Flashforge commercial 3D printer. As per the goal of the study, no hardware modification is conducted to the printer. Only the printing parameters are adjusted and HDPE plate is used as the build plate.

3.1 *Filament Quality*

HDPE and HDPE40 filaments manufactured in this study are tested for quality and consistency. It is observed that all 1X, 2X and 3X filaments have 1.72 ± 0.01 mm diameters, suggesting consistency in their cross section dimensions. One of the major concerns in processing syntactic foams is the possibility of excessive particle fracture as a result of applied shear forces, which needs to be studied by means of microstructural observations and density measurements.

The blended HDPE40 pellets and 1X, 2X and 3X filaments have densities of 0.807 ± 0.016, 0.905 ± 0.076, 0.966 ± 0.018 and 0.975 ± 0.022 g/cm^3, respectively. The theoretical density of HDPE40 pellets, estimated using rule of mixtures, is 0.943 g/cm^3. Compared to the pellets, syntactic foam filaments have higher densities. Since the theoretical density of the pellets is higher, it is expected that entrapment of air voids in the matrix during pellet manufacture is leading to the lower density. 1X filaments have lower density than the theoretical value, while 2X and 3X filaments have higher than the theoretical density. It is expected that with each pass, the matrix air voids have reduced, while some particles have fractured. The exact quantitative assessment of both these factures is not possible at this stage due to the limitation of the imaging equipment. The balance between entrapped matrix air voids and broken particles provides the final density for syntactic foams. This uncertainty in density measurement is similar in nature to the cast syntactic foams, which has been discussed in the published literature [10, 29]. Cenospheres have wall thickness below the resolution limit of micro-CT scan and resolving the particle walls to correctly count each particle is not possible with this method.

It is observed that the density of the foam increases after each extrusion pass, although the values are in a small range. The standard deviation is higher for HDPE40-1X filament, which suggests more prominent effect of entrapped air voids. The entrapped air void size is larger than the particle diameter. Hence, small sections taken for density measurements are affected by the presence of air voids in the localized regions. The entrapped air porosity is expected to reduce with each pass, making the density more consistent. Due to the larger standard deviation hinting at quality inconsistencies in HDPE-1X filament, they are subjected to tensile testing but not used for 3D printing in this work.

Density of syntactic foams does not show any significant change between the second and the third extrusion pass. Since the density of the cenosphere wall material is around 3.5 g/cm^3, stabilization of density below 1 g/cm^3 shows cenosphere survival in syntactic foams. Use of lighter particles in further studies can help in developing syntactic foams with even lower densities to obtain greater buoyancy.

3.2 *Filament Microstructure*

Figure 12a shows a low magnification micrograph of the HDPE40-1X filament. The filament is freeze-fractured and the fracture surface is imaged. It is observed that cenospheres are uniformly distributed in the entire cross-section. Similar uniform distribution of cenospheres is observed in other cross-sections of HDPE40-1X filaments. A close observation of the filament microstructure at higher magnification in Fig. 12b shows some broken fragments of cenospheres embedded in the resin among dispersed cenospheres, which implies that particle fracture has taken place as expected from the density calculations. However, significant amount of intact cenospheres observed in the syntactic foam filament help in obtaining the low density.

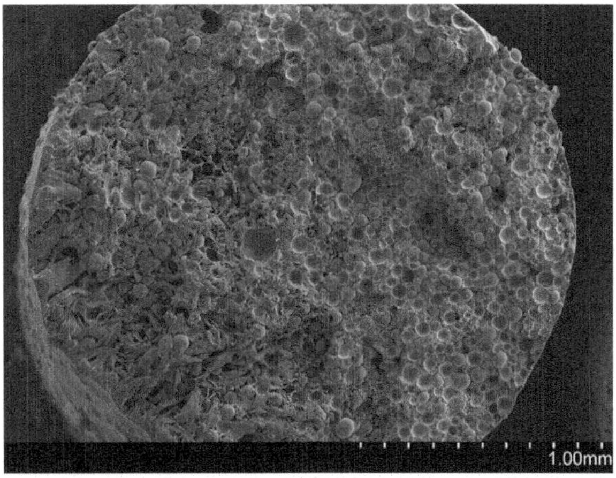

(a)

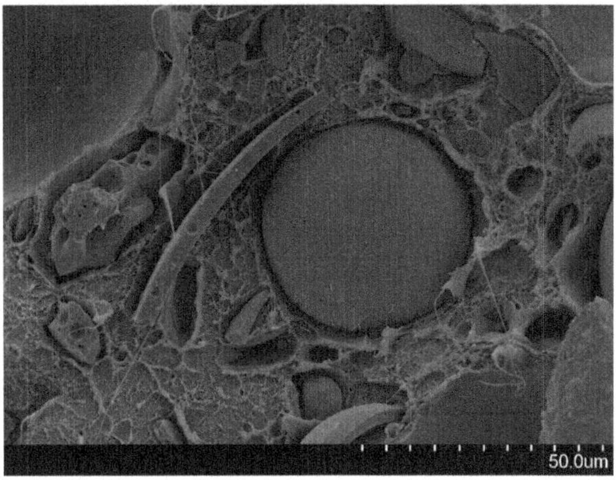

(b)

Fig. 12 (**a**) Cross section of a freeze fractured HDPE40-1X filament showing uniform distribution of cenospheres. (**b**) Debris of broken cenospheres can be observed embedded in the resin

A close observation of cenosphere fragments also reveals that the walls of cenospheres have significant porosity, which makes it difficult to determine their true particle density. Air voids are also dispersed in the HDPE matrix of the filament as observed in Fig. 13a. Usually, the size of these voids is much larger than the cenosphere diameter. Some of these voids may not find their way into the 3D printed part because the material flow during extrusion process of 3D printing may fill some of them. Figure 13b shows a fractured cenosphere embedded in the matrix resin. Some cenospheres seem to be damaged but are holing their place instead of getting dispersed as fragments. These damaged cenospheres may support compressive loads but may be dispersed as fragments under shear loading. The extrusion process of 3D printing may result in shear loads; hence, these broken cenospehres are not expected to be useful in the final printed part.

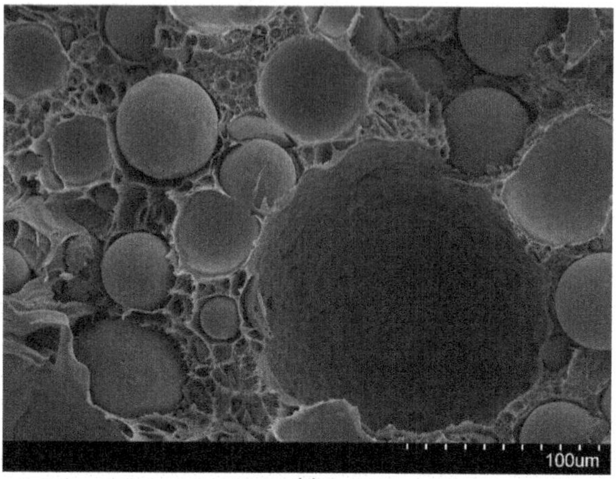

(a)

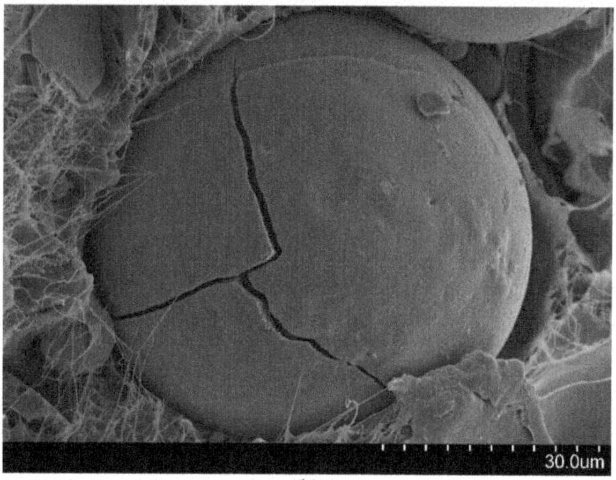

(b)

Fig. 13 (**a**) An air void observed in one of the filament cross sections. The size of air voids is generally larger than the cenosphere diameter. Some of these voids may be filled by material flow during 3D printing process. (**b**) A cracked cenospehre embedded in the matrix resin. The cenospehre is still holding the shape despite cracking across its entire diameter

Figures 14a and 14b show the microstructure of HDPE40-2X and 3X specimens, respectively. Since the density of HDPE40-2X and 3X filaments is the same, their microstructures also reflect that fact and appear to be very similar. In both cases, particles are uniformly distributed in the cross-section. It appears that the second and third extrusion passes do not lead to significant additional breakage of particles. This observation is also important because it shows that the recycling of the syntactic foam filament is possible. The filaments are cut into small pieces and extruded again for the purpose of recycling, which appears not to have damaged significantly more cenospheres.

Scanning electron micrographs show the features on fracture plane of the specimens. They require destructive surface preparation and do not show the internal structure of the specimens. Therefore, micro-CT scan observations are taken on the specimens to observe the distribution of cenospehres and air voids inside the filaments.

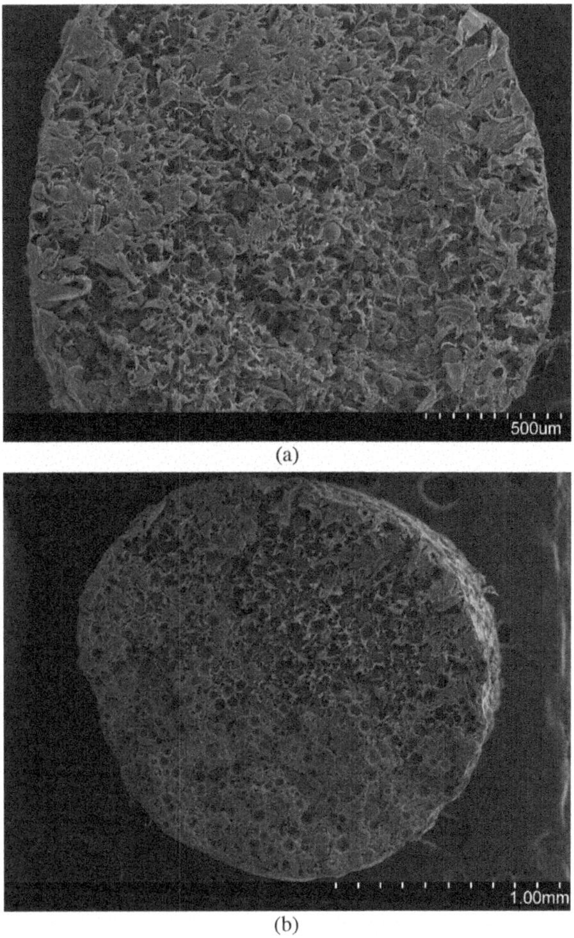

Fig. 14 Microstructure of (**a**) HDPE40-2X and (**b**) HDPE40-3X filaments showing uniform distribution of cenospheres in the cross-section

Figure 15 presents micro-CT image slices of one representative HDPE40-1X specimen [30]. The filament specimens are mounted vertically with the length oriented along the z-axis. The top right figure shows the three planes along which the images i, ii, and iii are obtained. Cenospheres are observed to be uniformly distributed across all sections in the filament. These images also show air voids that are distributed in the filament and are of larger size than the cenospheres. Such observations are typical of any other cross-section taken in the filaments.

Figures 16a and 16b show that cenosphere walls have porosity embedded in them. It is also observed that the cenosphere wall thickness is not constant and varies throughout the particle. In addition, the cenosphere wall thickness is less than 5 μm. Due to the wall porosity, some sections in the wall have less than 1 μm thickness, which is well below the measurement resolution of micro-CT scan. These features impose some limitations on the micro-CT scan analysis.

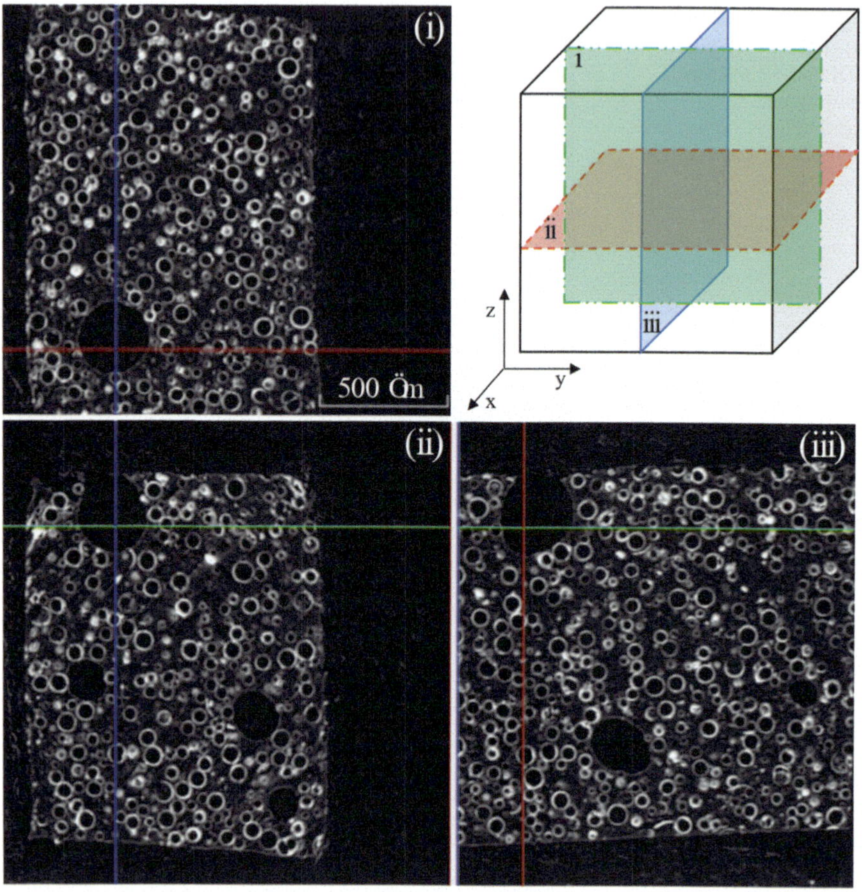

Fig. 15 Micro-CT scan image showing reconstructed (**i**) sagittal (*y*-*z*), (**ii**) trans-axial (*x*-*y*) and (**iii**) coronal (*z*-*x*) views of HDPE40-1X specimen. The three planes of observation are shown in the cube represented in the top right fig [30]

Figure 17c presents successive trans-axial images (i, ii, iii and iv) that are 2-pixel distance apart. The circles highlight a portion of the same cenosphere that appears to be broken in image (iii), and solid in image (iv). It is likely that this discontinuity is caused by one of the defects similar to those shown in Figs. 16a and 16b. Some of these defective regions may make the walls appear broken in an intact cenosphere in micro-CT images, which makes it difficult to obtain reliable results from image analysis. Such deviations lead to erroneous data and thereby experimental density estimations to quantify hollow particle breakage is the relatively better approach in case of cenospheres.

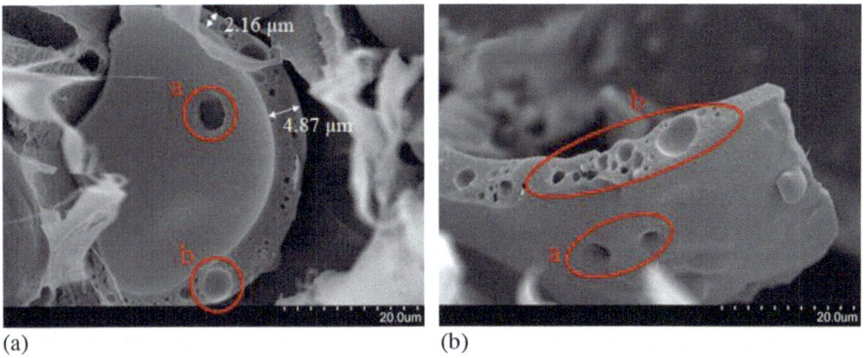

Fig. 16 (**a**) and (**b**) micrographs showing cenosphere walls with porosities and variable thickness

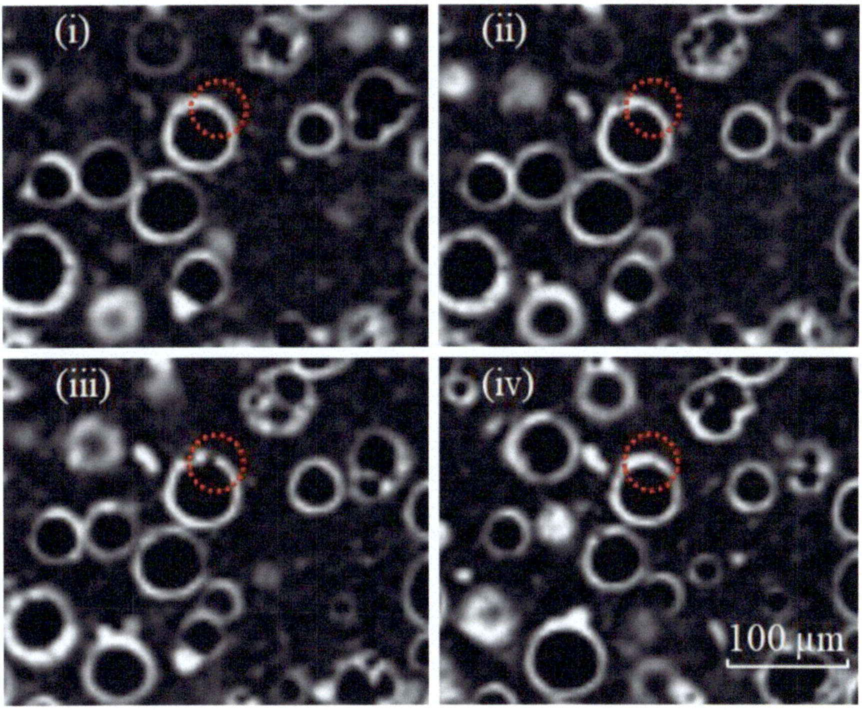

Fig. 17 Voids present in cenosphere walls can appear to be discontinuities in some slices in the micro-CT scan images of HDPE40. Such discontinuities make it difficult to resolve porosity from matrix because polymer molecules have low atomic number

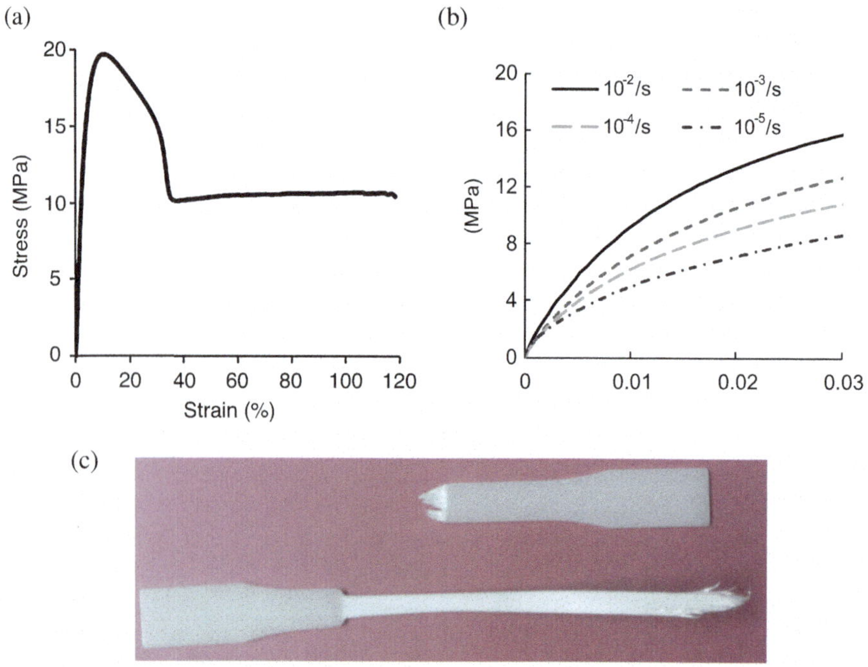

Fig. 18 (**a**) Tensile stress-strain curves of an injection molded neat HDPE resin specimen showing very large strain. (**b**) Initial portion of the tensile stress-strain curves of HDPE resin showing strong strain rate sensitivity within the quasi-static strain rate range. (**c**) A failed specimen showing large plastic deformation

3.3 *Tensile Behavior*

Tensile testing is conducted on the extruded filaments of HDPE resin and HDPE40 syntactic foams. Previously, it has been observed for injection molded HDPE tensile test specimens that the resin shows a highly ductile and strain rate sensitive behavior as observed in Fig. 18 [19]. The failure strain of over 120% was obtained in the injection molded neat resin specimens (Fig. 18a). In addition, the specimens showed significant strain rate sensitivity in tensile properties even in the quasi-static strain rate range of 10^{-2}–10^{-5} s^{-1} as observed in Fig. 18b.

Strong dependence of mechanical properties on strain rate and temperature makes it very labor intensive to conduct enough tensile tests to measure properties over a wide range of condition. Therefore, new methods are being developed to determine their properties from dynamic mechanical analysis using a single specimen subjected to frequency and temperature sweep and such results on HDPE are available in the literature [31–34]. However, the present work uses traditional tensile test method for all fabricated materials. Figure 18c shows a specimen that failed at 10^{-2} s^{-1} strain rate after significant plastic deformation. The features of this injection molded specimen can be compared with those of the extruded filaments and 3D printed specimens.

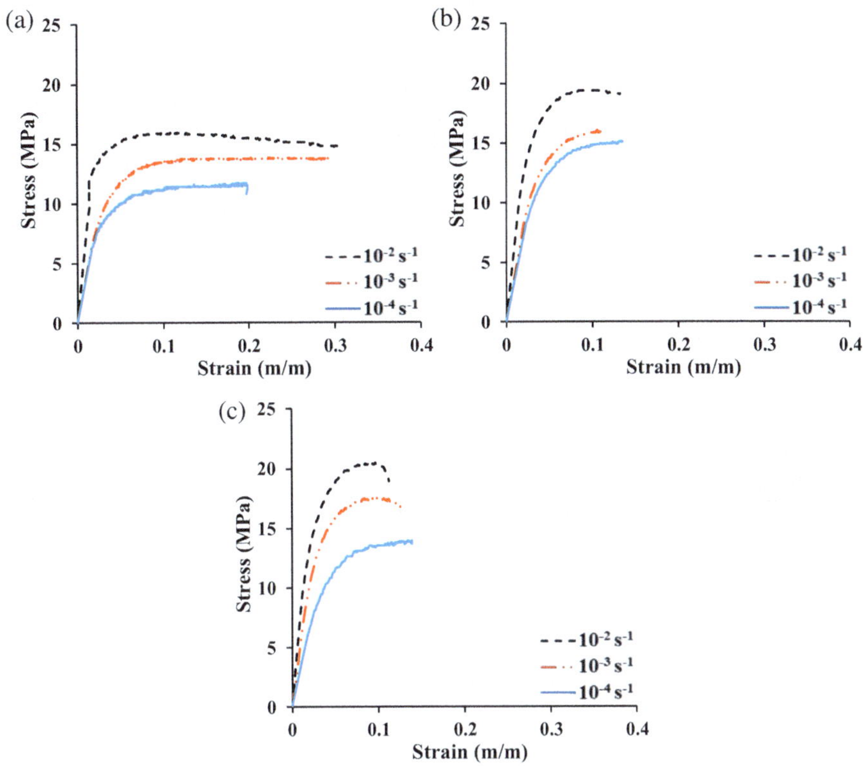

Fig. 19 A representative set of tensile stress-strain graphs for filaments of (**a**) HDPE-1X, (**b**) HDPE-2X and (**c**) HDPE-3X

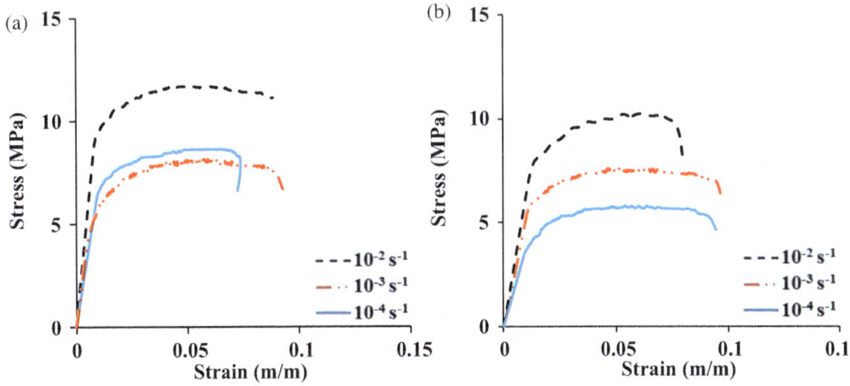

Fig. 20 A representative set of tensile stress-strain graphs for filaments of (**a**) HDPE40-2X and (**b**) HDPE40-3X

Table 1 Tensile properties of extruded HDPE and HDPE40 filaments [30]

Filament	Strain rate (s^{-1})	Modulus (MPa)	Yield strength[a] (MPa)	Ultimate tensile strength (MPa)
HDPE-1X	10^{-4}	353 ± 28	6.7 ± 1.5	11.5 ± 0.6
	10^{-3}	536 ± 26	8.0 ± 0.2	12.7 ± 0.9
	10^{-2}	687 ± 33	13.3 ± 0.7	16.1 ± 0.3
HDPE-2X	10^{-4}	380 ± 39	8.8 ± 0.8	14.6 ± 0.8
	10^{-3}	521 ± 10	10.0 ± 0.4	17.3 ± 0.6
	10^{-2}	719 ± 13	12.3 ± 0.3	20.3 ± 1.0
HDPE-3X	10^{-4}	397 ± 55	6.1 ± 1.7	13.9 ± 1.3
	10^{-3}	545 ± 70	10.3 ± 1.6	17.8 ± 0.9
	10^{-2}	732 ± 59	13.1 ± 1.2	21.0 ± 1.0
HDPE40-2X	10^{-4}	744 ± 71	7.2 ± 0.2	8.6 ± 0.3
	10^{-3}	897 ± 9	5.1 ± 0.2	8.0 ± 0.1
	10^{-2}	1020 ± 11	9.8 ± 0.4	11.7 ± 0.0
HDPE40-3X	10^{-4}	403 ± 83	5.6 ± 1.6	7.0 ± 0.8
	10^{-3}	620 ± 84	6.4 ± 0.6	8.8 ± 0.7
	10^{-2}	722 ± 11	6.9 ± 1.5	9.0 ± 1.3

[a]0.2% strain offset method

A representative set of stress-strain diagrams for HDPE filaments is shown in Fig. 19. HDPE-1X filament shows over 20% strain at all strain rates. However, the failure strain decreases when the filament is recycled. In all cases, the material shows strain rate sensitivity and the strength and modulus increase with strain rate. Reduction in strain and increase in strength as the filament is subjected to greater number of extrusion cycles is the general trend for HDPE as per these results. Successive extrusion cycles through small diameter nozzle lead to alignment of polymer chains in the filament, which results in increased strength and reduced strain. Similar effects may also be observed in the syntactic foam matrix.

The stress-strain graphs for HDPE40-2X and 3X syntactic foams are presented in Fig. 20. HDPE40-1X syntactic foam prints are not tested due to their larger standard deviation in density values, indicating uneven distribution of matrix voids in the filament. Figure 20 shows that syntactic foams have lower failure strain than the matrix resin. Inclusion of stiffer cenospheres in relatively soft and compliant HDPE matrix leads to such a behavior. The strength of these filaments is also observed to be lower than that of the neat resin.

The effect of strain rate is also prominet on the syntactic foam filaments. The measured tensile properties of the neat resin and syntactic foams are compiled in Table 1. The modulus is 11, 2 and 6% higher for HDPE-3X filament compared to HDPE-1X at strain rates of 10^{-4}, 10^{-3} and 10^{-2}, respectively. Yield strengths of neat resin and syntactic foams are comparable but the ultimate strength of syntactic foams is lower than that of the neat resin.

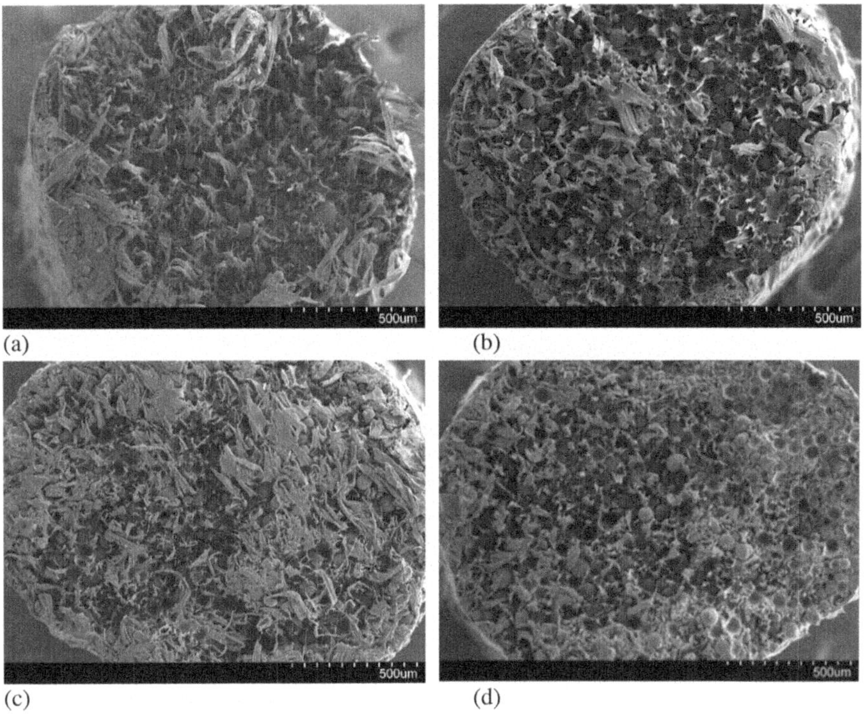

Fig. 21 HDPE40-2X at (**a**) 10^{-4} s^{-1} and (**b**) 10^{-2} s^{-1} strain rates. HDPE40-3X at (**a**) 10^{-4} s^{-1} and (**b**) 10^{-2} s^{-1} strain rates. In both materials slower strain rate shows greater plastic deformation of matrix on the failed cross-section

Although this change is very small, the yield strength values are higher for all strain rates for filaments compared to the injection molded specimens. Syntactic foams show reduced tensile properties when the material is recycled, which is likely due to the increased particle damage. Specifically, modulus has decreased by about 45% at 10^{-4} s^{-1} strain rate from second to third pass of the material. HDPE extrusion cycles leads to alignment of polymer chains [35], resulting in increase in yield strength and modulus. Also, thermal processing of HDPE leads to crosslinking in the individual polymer chains, increasing stiffness as reported previously by other groups [36]. These effects can be further investigated by means of microstructure of the failed specimens.

Tensile failure features of syntactic foam filaments are compared in Fig. 21 for HDPE40-2X and 3X specimens tested at 10^{-4} and 10^{-2} s^{-1} strain rates. In both cases, failure of cenospheres is not prominent. At slower strain rates, plastic deformation of matrix resin can be observed to be higher than the faster strain rates for both material types. These micrographs show that matrix deformation and failure are the main contributors to the failure of syntactic foams.

The fabricated filaments are used in a 3D printer to print tensile test specimens of HDPE resin and HDPE40 syntactic foams. Mechanical properties of 3D printed tensile test specimens will be compared with those of the filament and of injection molded syntactic foams.

4 3D Printing of Syntactic Foam

CTE of syntactic foams is lower than that of the neat resin due to the presence of ceramic particles in the polymeric resin [37, 38]. 3D printing by material extrusion requires printing at high temperature, where the polymer is in the viscous state. The temperature of the deposited material decreases by the time the print head deposits adjacent or the top layer. Change in the temperature causing material to shrink can result in peeling of the deposited layer from the build plate and is a major challenge. Matching of CTE of the deposition material and build plate material is important for successfully printing the part. In addition, increasing the temperature of the printing chamber may help in avoiding large shrinkage.

Pilot studies are conducted to obtain the optimum settings for printing with HDPE40 filament. Similar studies are also conducted for HDPE specimens but the results are not shown here. Specimens of size 25 × 12 × 6 mm^3 are printed using parameters listed in Table 2 for each trial setting. At each setting several specimens are printed to determine quality parameters such as dimensions and density. Some of the printed specimens are shown in Fig. 22. Shrinkage in the outer surfaces of the part during cooling causes the envelope of the part to shrink at a faster rate than the

Table 2 Pilot study print settings

Parameter	Trial 1	Trial 2	Trial 3	Trial 4
Infill (%)	100	100	60	60
Layer thickness (mm)	0.27	0.35	0.35	0.35
Feed rate (mm/s)	35	35	30	30
Print head travel rate (mm/s)	35	35	30	30
Print temperature (°C)	250	250	250	250
Surface temperature (°C)	110	110	125	115
Air cooling	on	on	on	off

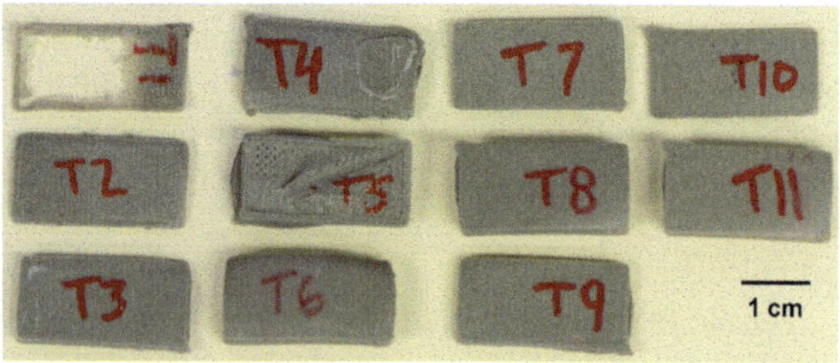

Fig. 22 Specimens printed in the pilot study to determine the best possible printing conditions

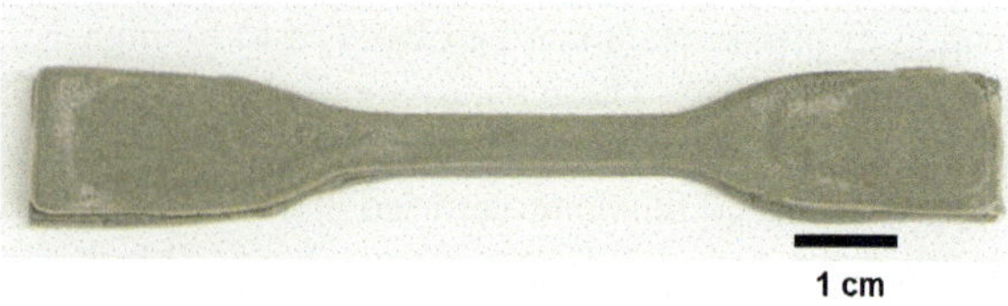

Fig. 23 3D printed HDPE40-2X tensile test bar

material at the center of the part, which causes a bulge and poor dimensional control. Parameters set by Trial 4 listed in Table 2 with higher layer thickness and very slow cooling of the part worked the best for this material system.

In addition, the infill percentage was lowered from 100 to 60% to further reduce the amount of extruded material within the solidified shell. These trials resulted in understanding of the effect of each parameter on the build quality and allowed developing a parameter set for printing syntactic foams using commercial 3D printers without making any changes to the printer hardware. A tensile test bar printed based on the optimized parameters is presented in Fig. 23. The gauge section of the bar has close dimension control to obtain reliable results from the testing. Hence these specimens are used for tensile testing in this work.

4.1 Density

Density of 3D printed HDPE specimens is 0.855 ± 0.012 g/cm^3, which is 19% lower than the injection molded HDPE (1.056 g/cm^3) reported in earlier studies. Lower density is also observed for 3D printed syntactic foams as well compared to their injection molded counterparts. Density of HDPE40-2X is 0.95 g/cm^3, which is 6 and 2% lower than injection molded HDPE40 and HDPE40-2X filaments [30], respectively. The decrease in density of 3D printed specimens is attributed to the presence of porosity between layers. It is noted that 60% infill is used because of coefficient of thermal expansion problem that results in swelling of the part if fully dense part is printed. However, lower infill also has potential of embedding porosity in the printed specimen. These effects can be observed in the specimen microstructure.

4.2 Microstructure of 3D Printed Syntactic Foams

One of the major concerns in 3D printed specimens was fracture of cenospheres during extrusion through the nozzle of 3D printer. Such possibility could have provided high density in the printed part. It was also possible that ceramic particles could clog the nozzle and reduce the size of the orifice to let only polymer extrude out and block cenospheres. Therefore, observing uniformity of cenosphere dispersion in the 3D printed part is desired. Extensive scanning electron microscopy is conducted on the 3D printed specimens to observe the microstructure.

Microstructural studies on HDPE40 specimens show a uniform distribution of cenospheres throughout the specimen cross-section (Fig. 24a). Intact cenospheres are visible in high magnification micrographs indicating high survival rate through two and three rounds of extrusion during filament manufacturing and the 3D printing process. Cenospheres of mean diameter 76 ± 32 μm are used with the printing nozzle diameter of 200 μm without any clogging in the present case [22].

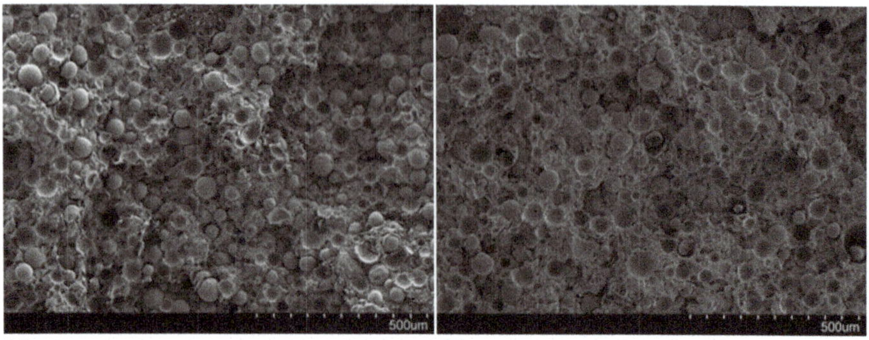

(a) (b)

Fig. 24 Micrographs showing uniform distribution of cenospheres in 3D printed (**a**) HDPE40-2X and (**b**) HDPE40-3X syntactic foam specimen cross-sections

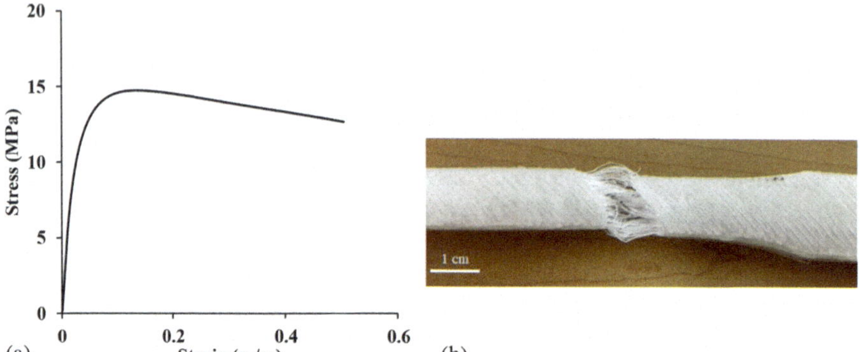

(a) (b)

Fig. 25 (**a**) Representative stress strain curve for HDPE tensile samples at $10^{-3}\ s^{-1}$ strain rate, and (**b**) failure mode for HDPE tensile test specimen at 50% strain

4.3 *Tensile Behavior*

A representative tensile stress-strain graph for 3D printed HDPE specimen is shown in Fig. 25a. Tests are stopped at 50% strain to observe the failure mode even though the specimen has not fractured. The curve shows a peak at 13% strain followed by a gradual drop in stress. The elastic modulus, yield strength and ultimate tensile strength for the 3D printed HDPE specimens are 632 $\pm$ 76, 6.6 $\pm$ 0.6 and 14.3 $\pm$ 0.3 MPa, respectively. 3D printed HDPE shows 18% higher modulus and 17.5% lower yield strength compared to the extruded filament at $10^{-3}\ s^{-1}$ strain rate. Compared to injection molded HDPE, modulus of 3D printed HDPE is 16% higher [20]. These results show that the mechanical properties of 3D printed HDPE part are competitive for applications.

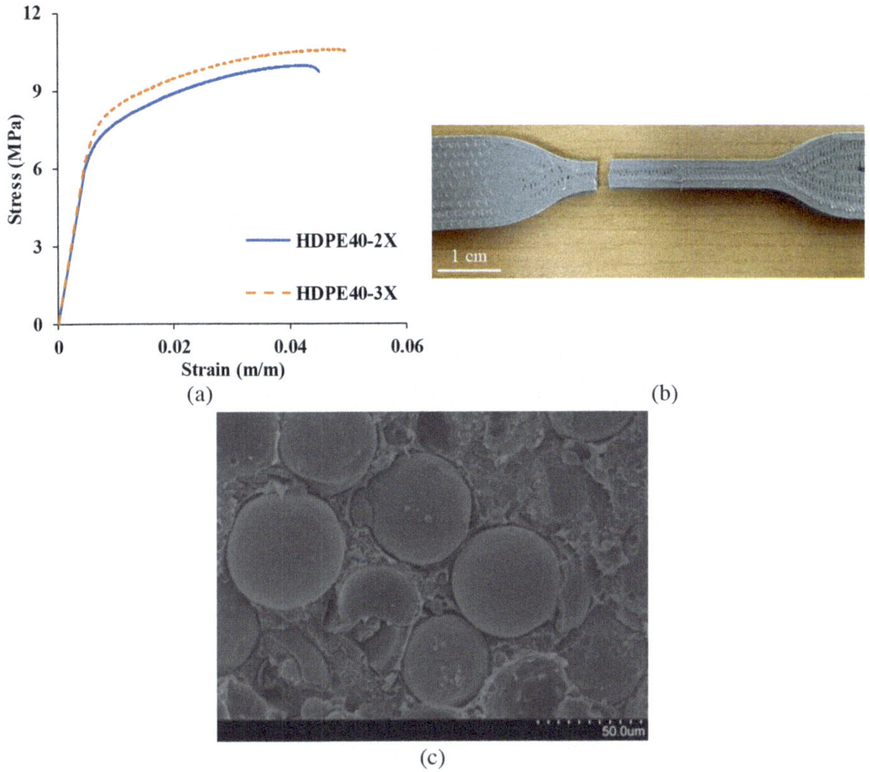

Fig. 26 (**a**) Representative stress-strain plots for 3D printed HDPE40-2X and HDPE40-3X tensile specimens, (**b**) failure mode in HDPE40-2X specimen and (**c**) intact cenospheres on the fracture surface of a tensile tested HDPE40-3X specimen

Failure mode of a 3D printed HDPE specimen at 50% strain is shown in Fig. 25b. Debonding between various deposited layers leads to a fibrous failure in this specimen. Long chains of HDPE are not bonded well with each other and tend to separate under shear because the infill direction is different for each fiber and is not aligned with the direction of applied tensile force. These individual filaments are strong because of chain alignment but bonding between them is less than perfect. The specimen could have sustained more tensile strain before complete fracture. Although 3D printed specimens perform comparably to the molded specimens, further process improvements can be conducted to obtain monolithic specimens that do not show deboning between deposited layers.

The tensile behavior of a representative set of 3D printed HDPE40-2X and 3X specimens is compared in Fig. 26a. Unlike HDPE samples, stress-strain curves do not show a peak followed by a decreasing trend in the stress value. In this case, the stress rises with strain until failure due to higher brittleness of the material caused by the presence of ceramic particles. Failure strain is less than 6% for all 3D printed HDPE40 specimens, which is 30–40% lower than HDPE40 filaments [30],

Table 3 Density, elastic modulus and yield strength of 3D printed HDPE40-2X and 3X specimens

Material	Density (g/cm^3)	Elastic modulus (MPa)	Yield strength (MPa)
HDPE40-2X	0.950	1337 ± 109	7.0 ± 0.4
HDPE40-3X	0.959	1569 ± 143	7.4 ± 0.5
Molded HDPE40 [20]	1.0078	723 ± 27	–

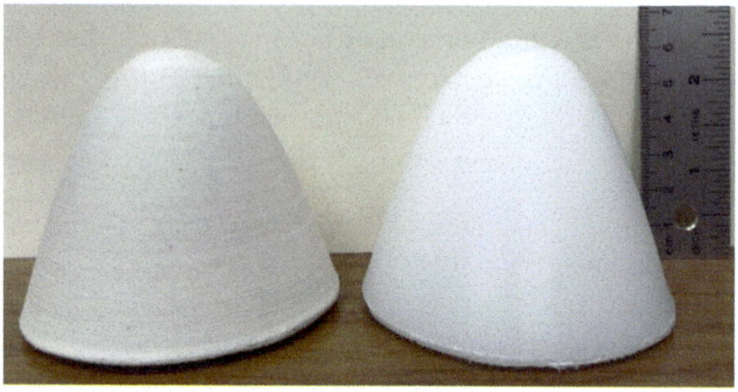

Fig. 27 A nose cone printed for an underwater vehicle using syntactic foam (left) and HDPE (right) filament using a commercial FFF printer

indicating that the material becomes brittle by the 3D printing process. Elongation of individual 3D printed layers under the applied tensile force is hindered by the presence of high volume fraction cenospheres. Failure of HDPE40-2X is brittle in nature as shown in Fig. 26b, which is under brittle mode. Cenosphere failure is not observed on the fracture surface of the specimens in Fig. 26c because of low failure strain of the specimens.

Tensile stress-strain curves are used to calculate the mechanical properties. Tensile moduli of 3D printed HDPE40-2X and 3X, presented in Table 3, are 1.6 and 2.6 times higher than that of 2X and 3X filaments, respectively [30]. This enhancement is likely due to additional crosslinking and realignment of polymer chains during high temperature extrusion process of 3D printing [35, 39]. In addition, crystallization of HDPE resin due to thermal gradients can also contribute to enhancement in mechanical properties [40]. These results show that 3D printing is a viable method for processing syntactic foam components. Further improvements in materials and printing technology for this specific purpose may help in moving forward with developing high quality parts for deployment in load bearing applications.

4.4 3D Printing of a Component

Design of a nose cone of an underwater remotely operated vehicle provided by Riptide Autonomous Solutions was printed for observing the quality of the component. Both HDPE and syntactic foams were used for printing the part geometry. Complex contour on the top poses challenge in printing the geometry, especially with the neat HDPE resin. Figure 27 shows successful prints using syntactic foam and neat resin. For this particular part, a glass hollow particle filled syntactic foam filament was used and the 3D printed part geometry was found to be in confirmation with the design. The syntactic foam part was about 30% lighter than the HDPE part. Development of specific 3D printer hardware that can print long tubular designs with internal features can be very useful for printing the entire body of the vehicle.

5 Conclusions

The present work is focused on developing HDPE and HDPE-fly ash cenosphere syntactic foam filaments for use in commercial FFF 3D printers and optimizing the 3D printing conditions. Filament recyclability potential is also addressed. Mechanical properties of the filaments and 3D printed tensile test specimens are characterized at multiple strain rates within the quasi-static strain rate regime. The measured mechanical properties are compared with those of injection molded specimens of the same composition. Results show that:

- Syntactic foam filaments show increasing trend in modulus and strength with strain rate.
- Syntactic foam filaments show higher mechanical properties than the HDPE specimens under comparable processing conditions.
- Specific modulus values of extruded filaments are higher than those of injection molded samples at the same strain rates.

These results show that the developed filaments have potential to be used in 3D printing as the properties are comparable to the injection molded specimens. Results also showed that the filaments can be recycled without significant cenosphere fracture. Use of 60 vol.% less polymer and possibility of recycling make these filaments economically viable despite the added processing steps of extrusion. Further results on mechanical properties of 3D printed specimens show that

- Warping of the specimen was reduced and quality was improved by optimizing printer speed, layer thickness, print temperature, and cooling conditions.
- 3D printed HDPE has 18% higher modulus and 17.5% lower yield strength than the HDPE filament.
- Microstructure of 3D printed and injection molded syntactic foams is similar. Cenospheres are uniformly distributed and survived the 3D printing process.
- Comparison of specific mechanical properties shows promise in 3D printed syntactic foam components in replacing some of the molded specimens for weight-sensitive applications.

Overall, the success of 3D printing of syntactic foam specimens shows that the technology can be used in the future to develop jointless structures for underwater marine applications.

6 Future Perspectives

AM technologies continue to develop rapidly with the aim of improving the printing resolution, developing better control over part dimension tolerances, and improving the surface finish of the printed part. Technologies are also being developed to print components using multiple materials, complex shapes, and larger part sizes. Many of these advancements are directly relevant to 3D printing of marine components. Technologies of printing thermosetting resin parts are still very expensive compared to the FFF based technologies. However, it is expected that with increased appreciation of the capabilities of AM methods in providing functionalities and structures that cannot be realized by other means, wider adoption of these methods will take place for marine applications.

Increased awareness on the total life cycle of materials and post-use recycling are also factors that will become more mainstream, keeping pace with automotive and aircraft industries. The possibility of recycling studied in the present work shows favorable results that the thermoplastic syntactic foam components can be recycled without a significant loss in their properties. Further work is required in this area that are specifically tailored to study the use conditions and end of use recycling possibilities.

Marine structures are exposed to complex loading conditions, including temperatures ranging from arctic conditions to high values, repeated loading of waves, fluid-structure interactions, and combined effects of moisture and strain rates, among others. The 3D printed structures and components need to be characterized for these conditions to develop a confidence in their performance. Modeling and simulation studies also need to keep up with the developments in this area so that predictive capabilities can be developed for such materials. It is observed that alignment of polymer chains in the direction of extrusion can result in directional properties in 3D printed parts. Such effects need to be understood better in conjunction with the processing parameters such as extrusion temperature and speed to design part with desired performance under marine relevant loading conditions.

Acknowledgments Nikhil Gupta acknowledges the Office of Naval Research grant N00014-10-1-0988 grant with Dr. Yapa D. Rajapakse as the program manager. Mrityunjay Doddamani acknowledges Department of Science and Technology, India, grant DST/TSG/AMT/2015/394/G and Visiting Scientist Fellowship grant VSP 17-7-001 by the U.S. Office of Naval research – Global (Program manager: Dr. Ramesh Kolar) to visit NYU for part of this work. The views expressed in this work are of authors, not of funding agency. Riptide Autonomous Solutions (www.riptideas.com) is thanked for providing designs and useful discussions.

References

1. Gibson I, Rosen DW, Stucker B (2015) Additive manufacturing technologies: 3D printing, rapid prototyping, and direct digital manufacturing, 2nd edn. Springer, New York
2. Wohlers TT (2016) Wohlers report 2016: 3D printing and additive manufacturing state of the industry annual worldwide progress report. Wohlers Associates, Fort Collins
3. Attaran M (2017) The rise of 3-D printing: the advantages of additive manufacturing over traditional manufacturing. Bus Horiz 60:677–688
4. ASTM International (2015) ASTM ISO/ASTM52900–15 standard terminology for additive manufacturing – general principles – terminology, West Conshohocken
5. Sing SL, An J, Yeong WY, Wiria FE (2016) Laser and electron-beam powder-bed additive manufacturing of metallic implants: a review on processes, materials and designs. J Orthop Res 34:369–385
6. Zeltmann SE, Gupta N, Tsoutsos NG, Maniatakos M, Rajendran J, Karri R (2016) Manufacturing and security challenges in 3D printing. JOM 68:1872–1881
7. Chen F, Mac G, Gupta N (2017) Security features embedded in computer aided design (CAD) solid models for additive manufacturing. Mater Des 128:182–194
8. Chen F, Luo Y, Tsoutsos NG, Maniatakos M, Shahin K, Gupta N (2019) Embedding tracking codes in additive manufactured parts for product authentication. Adv Eng Mater. Paper# 1800495
9. Gupta N, Pinisetty D, Shunmugasamy VC (2013) Reinforced polymer matrix syntactic foams: effect of Nano and Micro-scale reinforcement. SpringerBriefs in Materials, Springer
10. Gupta N, Woldesenbet E, Mensah P (2.004) Compression properties of syntactic foams: effect of cenosphere radius ratio and specimen aspect ratio. Compos A: Appl Sci Manuf 35:103–111
11. Gupta N, Nagorny R (2006) Tensile properties of glass microballoon-epoxy resin syntactic foams. J Appl Polym Sci 102:1254–1261
12. Porfiri M, Gupta N (2009) Effect of volume fraction and wall thickness on the elastic properties of hollow particle filled composites. Compos Part B 40:166–173
13. Gupta N, Ye R, Porfiri M (2010) Comparison of tensile and compressive characteristics of vinyl ester/glass microballoon syntactic foams. Compos Part B 41:236–245
14. Kishore SR, Sankaran S (2005) Short beam three point bend tests in syntactic foams. Part I: microscopic characterization of the failure zones. J Appl Polym Sci 98:673–679
15. Kishore SR, Sankaran S (2005) Short-beam three-point bend tests in syntactic foams. Part II: effect of microballoons content on shear strength. J Appl Polym Sci 98:680–686
16. Kishore SR, Sankaran S (2005) Short-beam three-point bend test study in syntactic foam. Part III: effects of interface modification on strength and fractographic features. J Appl Polym Sci 98:687–693
17. Yalcin B (2015) Chapter 7 - hollow glass microspheres in polyurethanes. In: Amos SE, Yalcin B (eds) Hollow glass microspheres for plastics, elastomers, and adhesives compounds. William Andrew Publishing, Oxford, pp 175–200
18. Yalcin B, Amos SE (2015) Chapter 3 – hollow glass microspheres in thermoplastics. In: Amos SE, Yalcin B (eds) Hollow glass microspheres for plastics, elastomers, and adhesives compounds. William Andrew Publishing, Oxford, pp 35–105
19. Bharath Kumar BR, Doddamani M, Zeltmann SE, Gupta N, Ramesh MR, Ramakrishna S (2016) Data characterizing tensile behavior of censophere/HDPE syntactic foam. Data Brief 6:933–941
20. Bharath Kumar BR, Doddamani M, Zeltmann SE, Gupta N, Ramesh MR, Ramakrishna S (2016) Processing of cenosphere/HDPE syntactic foams using an industrial scale polymer injection molding machine. Mater Des 2016:414–423
21. Bharath Kumar BR, Doddamani M, Zeltmann SE, Gupta N, Uzma GS, Sailaja RRN (2016) Effect of particle surface treatment and blending method on flexural properties of injection-molded cenosphere/HDPE syntactic foams. J Mater Sci 51:3793–3805

22. Bharath Kumar BR, Singh AK, Doddamani M, Luong DD, Gupta N (2016) Quasi-static and high strain rate compressive response of injection-molded cenosphere/HDPE syntactic foam. JOM 68:1861–1871
23. Gupta N, Zeltmann SE, Shunmugasamy VC, Pinisetty D (2014) Applications of polymer matrix syntactic foams. JOM 66:245–254
24. Jayavardhan ML, Bharath Kumar BR, Doddamani M, Singh AK, Zeltmann SE, Gupta N (2017) Development of glass microballoon/HDPE syntactic foams by compression molding. Compos Part B 130:119–131
25. Bardella L, Perini G, Panteghini A, Tessier N, Gupta N, Porfiri M (2018) Failure of glass-microballoons/thermoset-matrix syntactic foams subject to hydrostatic loading. Eur J Mech A-Solid 70:58–74
26. Sankaran S, Dasgupta S, Sekhar KR, Kumar MJ (2006) Thermosetting polymer composites for EMI shielding applications. 2006 9th International Conference on Electromagnetic Interference and Compatibility (INCEMIC 2006), 1–6
27. Singh AK, Shishkin A, Koppel T, Gupta N (2018) A review of porous lightweight composite materials for electromagnetic interference shielding. Compos Part B 149:188–197
28. Mohamed OA, Masood SH, Bhowmik JL (2015) Optimization of fused deposition modeling process parameters: a review of current research and future prospects. Adv Manuf 3:42–53
29. Gupta N, Woldesenbet E (2004) Microballoon wall thickness effects on properties of syntactic foams. J Cell Plast 40:461–480
30. Singh AK, Patil B, Hoffmann N, Saltonstall B, Doddamani M, Gupta N (2018) Additive manufacturing of syntactic foams: part 1: development, properties, and recycling potential of filaments. JOM 70:303–309
31. Xu X, Gupta N (2018) Determining elastic modulus from dynamic mechanical analysis: a general model based on loss modulus data. Materialia 4:221–226
32. Xu X, Koomson C, Doddamani M, Behera RK, Gupta N (2019) Extracting elastic modulus at different strain rates and temperatures from dynamic mechanical analysis data: a study on nanocomposites. Compos Part B 159:346–354
33. Xu X, Gupta N (2018) Determining elastic modulus from dynamic mechanical analysis data: reduction in experiments using adaptive surrogate modeling based transform. Polymer 157:166–171
34. Xu X, Gupta N (2019) Artificial neural network approach to predict the elastic modulus from dynamic mechanical analysis results. Adv Theor Simul 2(4):1800131
35. Picot JJC (1984) Molecular orientation in film extrusion of high-density polyethylene. Polym Eng Sci 24:415–420
36. Yin S, Tuladhar R, Shi F, Shanks R, Combe M, Collister T (2015) Mechanical reprocessing of polyolefin waste: a review. Polym Eng Sci 55:2899–2909
37. Shunmugasamy VC, Pinisetty D, Gupta N (2012) Thermal expansion behavior of hollow glass particle/vinyl ester composites. J Mater Sci 47:5596–5604
38. Labella M, Shunmugasamy VC, Strbik OM III, Gupta N (2014) Compressive and thermal characterization of syntactic foams containing hollow silicon carbide particles with porous shell. J Appl Polym Sci 131. Paper #40689
39. Yamaguchi M, Suzuki K-I (2002) Enhanced strain hardening in elongational viscosity for HDPE/crosslinked HDPE blend. II. processability of thermoforming. J Appl Polym Sci 86:79–83
40. Bin Y, Yan-Li D, Gui-Jing L, Ji-Bin M, Ru X, Jia-Sheng Q, Peng C, Jing-Wang L (2015) Solidification behavior of high-density polyethylene (HDPE) during injection molding: Correlation between crystallization kinetics and thermal gradient field. IOP Conference Series: Materials Science and Engineering 87, Paper #012020

Damage Tolerance Assessment of Naval Sandwich Structures with Face-Core Debonds

Christian Berggreen and Brian Hayman

1 Introduction

Sandwich structures offer superior stiffness/weight and strength/weight ratios compared with monolithic composites and traditional metallic structural concepts, and are thus of great interest for lightweight structural applications in naval ships. For marine applications, the sandwich face sheets have traditionally been made of glass fibre reinforced polymer (GFRP), the matrix being polyester, vinylester or epoxy, while the core has been either balsa wood or a polymer foam such as polyvinyl chloride (PVC). More recently carbon fibre reinforcement has also been introduced, while for naval ships the use of balsa cores has been limited to some superstructure components and internal structures.

Such sandwich structures contain imperfections resulting from either the production processes or subsequent handling and use. Many fibre reinforced polymer (FRP) structures have traditionally been designed without explicitly addressing such defects and damage; production defects have been allowed for by applying high factors of safety, and a certain degree of robustness in relation to accidental events has been built-in by such means as the minimum thickness or minimum reinforcement requirements in classification society rules.

In contrast, the aircraft industry has focused more explicitly on defects and damage by requiring fail-safe or damage tolerant design. As more optimised, lightweight composite designs have become established in marine applications, particularly for naval and high-speed vessels, the need for adopting damage tolerance principles in these applications has become evident. However, because the operating conditions, in-service inspection regimes and safety considerations differ

C. Berggreen (✉) · B. Hayman
Department of Mechanical Engineering, Technical University of Denmark,
Kongens Lyngby, Denmark
e-mail: cbe@mek.dtu.dk; bhayman@mek.dtu.dk

S. W. Lee (ed.), *Advances in Thick Section Composite and Sandwich Structures*,
https://doi.org/10.1007/978-3-030-31065-3_15

from those for aircraft, there must be differences in the way a damage tolerance approach is implemented.

A start towards establishing a consistent damage tolerance approach for naval sandwich structures was made several years ago in the project THALES JP3.23: Inspection and Repair of Sandwich Structures in Naval Ships (SaNDI), a collaboration between the Norwegian, Swedish, Danish, Finnish and British navies, research institutions, shipyards and other suppliers of materials and services [1]. Attention was focused on the types of defects and damage that need to be taken into account when naval ships built in FRP sandwich are designed, built, maintained and operated, on the methods available to detect such defects and damage, on the ways in which their influence on structural performance may be assessed, and on the methods available for their repair. The project developed an integrated methodology [2–4] for the inspection, assessment and repair of defects and damage, with attention to the important questions of whether, when, where and how defects and damage should be repaired and, in the case of in-service damage, whether measures should be taken to impose operational restrictions in the interim period while a repair is pending. An overview of some of the main results was presented by Hayman [5], together with discussion of the implications for introducing more damage tolerant design in FRP structures in ships and other applications.

The terms "defect" and "damage" are not uniquely defined for all contexts, and are sometimes used synonymously. To make a major distinction that is relevant in practice, it is convenient to use the terms "production defect" and "in-service damage". In all FRP composite structures it is important to be able to detect and rectify both production defects and in-service damage. For single-skin (monolithic) laminates and face laminates of sandwich structures, production defects include dry zones, voids, delaminations, wrinkles, misalignment of fibres, and poor curing (giving reduced physical properties). For sandwich structures, the following types of production defects must also be considered: face-core debonds, voids and inclusions in the core, and lack of bond (edge-to-edge and face-to-face) between core sheets. In aircraft structures the main type of in-service damage is impact damage, typically from bird strikes, hail and objects thrown up from the runway during landing and takeoff. Ship structures, however, may encounter a wider range of contact damage (quasi-static contact, e.g. during berthing, as well as impact), incidences of heat damage, and numerous types of damage resulting from overloading. This last category includes core fracture or crushing, face-core debonds, laminate rupture, delamination either within a laminate or at a joint made by secondary lamination, and failure at equipment fastenings. Impact damage may be confined to the impacted face laminate (with or without penetration of the laminate), or may involve crushing and/or cracking of the core. In extreme cases, penetration of the entire sandwich may occur. More substantial damage cases include the removal of whole panels or assemblies by fire or collision. For naval vessels some types of damage may be caused by weapon effects such as air blast, underwater explosions, and fragment or missile hits.

A common type of defect or damage that can seriously affect the integrity of a sandwich structure is a lack of adhesion between the face sheets and the core commonly known as a "debond" (or typically as a "disbond" in the aerospace industries). In connection with the SaNDI project, Berggreen [6], in a doctoral thesis at the Technical University of Denmark (DTU), studied the damage tolerance of debonded sandwich structures and initiated a long series of major contributions to this field of research. Several years previously Zenkert had also considered the problem of face-core debonds [7] in connection with his own doctoral thesis [8] at the Royal Institute of Technology (KTH) in Sweden that concerned damage tolerance of foam-cored sandwich structures. At about the same time, with support of the US Office of Naval Research (ONR), Carlsson and co-workers were also studying the fracture characterisation of face-core interfaces and the effect of debonds on the in-plane compressive strength of sandwich columns and panels. In subsequent years, immense progress has been made in this area in both Europe and the USA, much of the work being supported by the ONR, but with some close links also being established in recent years with the aircraft industry and also the wind turbine industry, where sandwich components with material configurations similar to those in marine vessels are applied in large quantities. In particular, while the SaNDI project and many other earlier studies were limited to the effects of defects and damage on static strength, recent studies of sandwich debonds have increasingly focused on debond crack propagation under fatigue loadings.

To provide a coherent, damage tolerance-based approach to sandwich structures each of the following aspects must be addressed:

- Understanding of the fracture mechanics of sandwich debonds, including especially the effects of mode mixity, i.e. combinations of the fracture modes I and II (and possibly III).
- Establishment of experimental techniques to determine the interface properties for the respective fracture modes and degrees of mode mixity.
- Development of methods of modelling crack propagation under increasing load and under repeated (fatigue) loading
- Experimental studies, and the use of experiments to validate modelling techniques, both at the level of simple sandwich beam specimens used for material characterisation and at the level of structural components such as sandwich panels.
- Development of practical ways of implementing results in damage assessment schemes.

These aspects are addressed in the following sections. While the main focus is on recent research at DTU and collaborating organisations that has been supported by the ONR, some earlier research that provided a basis for this more recent work is also outlined. Note that some of these topics are addressed in Chap. 12 of the textbook by Carlsson and Kardomateas [9].

2 Fracture Mechanics of Sandwich Face/Core Interfaces

2.1 Griffith Criterion and Use of LEFM

Analysis of the development and propagation of face-core debonds in sandwich structures requires the use of fracture mechanics principles taking account of the difference in properties between the face sheet and core materials.

In sandwich beams and panels with thin face sheets the main stress conditions of relevance in structural design are the following:

- In-plane tension, compression and shear stresses in the face sheets. Compressive and shear stresses may lead to local buckling of the face sheets, which in turn may encourage debonding.
- Transverse (out-of-plane) shear stresses, predominantly in the core. The complementary shear stresses at the face-core interface lead to mode II deformation conditions for potential debonds (Fig. 1).
- Local out-of-plane stresses generated especially at joints and equipment fixings; these can involve significant peel stresses at the face-core interface, giving predominantly mode I deformation conditions (Fig. 1).

In virtually all these cases there is some degree of mode mixity combining mode I and mode II deformations. Furthermore, in a three-dimensional sandwich panel mode III may also be present (Fig. 1).

Much research effort has been focused on

- establishing the extent to which simple tests intended to provide pure mode I or mode II data in fact are influenced by the presence of mode mixity, and how the mode mixity varies as the crack length increases during a test;
- clarifying how an interface crack actually propagates, i.e. within an adhesive layer, on the face sheet side or on the core side, and whether a crack kinks away from the interface, into the face sheet or the core;
- devising and optimising test methods that provide as closely as possible desired pure or mixed mode data;

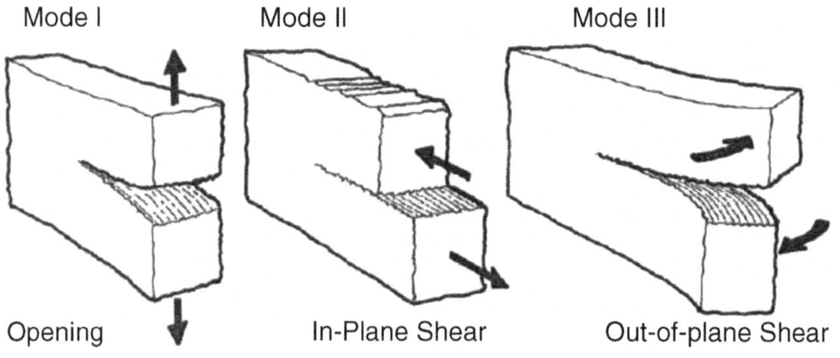

Fig. 1 The basic fracture modes

- adapting test methods originally designed for static conditions to follow crack growth under fatigue loading;
- adapting test methods to accommodate high and low temperature testing.

In the analysis of the stress state in the neighbourhood of a face-core interface, special challenges arise from the greatly differing stiffnesses between the face sheet and core materials.

For many problems in debond crack propagation inelastic deformation is confined to a very small fracture process zone, and linear elastic fracture mechanics (LEFM) can be used. The starting point for LEFM is the Griffith criterion [10], which states that a crack will propagate if the available energy release rate, G (i.e. the energy dissipated during fracture per unit of newly created fracture surface area), is greater than or equal to the fracture toughness, G_c, which is a material property that must be obtained experimentally. The fracture toughness has different values for the three main fracture modes (Fig. 1). For mixed mode fracture, at any stage in the crack propagation the strain energy release rate G and the mode mixity must first be established from analysis of the stress state in the neighbourhood of the crack, and the value of G_c for that mode mixity used in determining whether the crack propagates. The problem is complicated by the fact that in many cases the mode mixity changes as the crack propagates.

LEFM can be used to characterise the crack growth not only in a homogeneous material, but also in the interface between dissimilar materials as shown in Fig. 2.

Suo [11, 12] expressed the stress field and the displacement field at the distance x behind the crack tip as a function of the stress intensity factor:

$$\sqrt{\frac{H_{11}}{H_{22}}}\delta_y + i\delta_x = \frac{2H_{11}K|x|^{\frac{1}{2}+i\varepsilon}}{\sqrt{2\pi}(1+2i\varepsilon)\cosh(\pi\varepsilon)} \tag{1}$$

$$\sqrt{\frac{H_{22}}{H_{11}}}\sigma_{yy} + i\sigma_{xy} = \frac{K|x|^{i\varepsilon}}{\sqrt{2\pi x}} \tag{2}$$

$$K = K_1 + iK_2 \ ; \ i = \sqrt{-1} \tag{3}$$

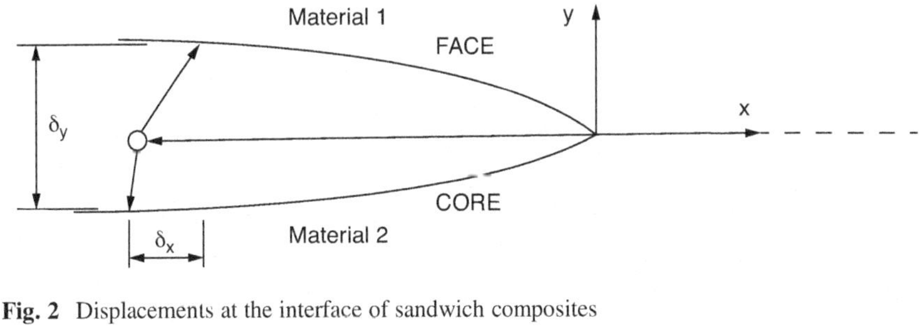

Fig. 2 Displacements at the interface of sandwich composites

Here the equations are complex, δ_y and δ_x are the relative opening and sliding displacements (see Fig. 2) of the crack flanks, σ_{yy} and σ_{xy} are the normal and shear stresses and H_{11}, H_{22} and ε are bimaterial constants given by:

$$H_{11} = \left[2n\lambda^{1/4}\sqrt{S_{11}S_{22}}\right]_1 + \left[2n\lambda^{1/4}\sqrt{S_{11}S_{22}}\right]_2 \tag{4}$$

$$H_{22} = \left[2n\lambda^{-1/4}\sqrt{S_{11}S_{22}}\right]_1 + \left[2n\lambda^{-1/4}\sqrt{S_{11}S_{22}}\right]_2 \tag{5}$$

$$\varepsilon = \frac{1}{2\pi} \ln\left(\frac{1-\beta}{1+\beta}\right) \tag{6}$$

$$\beta = \frac{\left[S_{12} + \sqrt{S_{11}S_{22}}\right]_2 - \left[S_{12} + \sqrt{S_{11}S_{22}}\right]_1}{\sqrt{H_{11}H_{22}}} \tag{7}$$

where n and λ are non-dimensional orthotropic constants, S_{ij} are components of the compliance matrix and β is Dundur's bi-material parameter [13]. The mode mixity phase angle and the strain energy release rate are described as [14, 15]:

$$\psi_K = \arctan\left[\frac{\mathrm{Im}\left(Kh^{i\varepsilon}\right)}{\mathrm{Re}\left(Kh^{i\varepsilon}\right)}\right] \tag{8}$$

$$G = \frac{H_{11}|K|^2}{4\cosh^2(\pi\varepsilon)} \tag{9}$$

Here $\mathrm{Im}(Kh^{i\varepsilon})$ and $\mathrm{Re}(Kh^{i\varepsilon})$ are the imaginary and real parts of the complex expression and h is an arbitrarily chosen characteristic length. In face/core interface crack problems, the characteristic length is often taken equal to the thickness of the debonded face sheets. The strain energy release rate and the mode mixity phase angle can also be expressed in terms of the relative opening and sliding displacements of the crack flanks:

$$G = \frac{\pi(1+4\varepsilon^2)}{8H_{11}|x|}\left(\frac{H_{11}}{H_{22}}\delta_y^2 + \delta_x^2\right) \tag{10}$$

$$\psi_F = \tan^{-1}\left(\sqrt{\frac{H_{22}}{H_{11}}}\frac{\delta_x}{\delta_y}\right) - \varepsilon \ln\left(\frac{|x|}{h}\right) + \tan^{-1}(2\varepsilon) \tag{11}$$

$$\psi = \psi_R = \tan^{-1}\left(\frac{\delta_x}{\delta_y}\right) \tag{12}$$

Here ψ_F and ψ_R are mode mixity phase angles respectively in the full and the reduced formulations. The reduced formulation of phase angle enables the correlation of phase angle values and the fracture modes; $\psi_R = 0°$ corresponds to the pure mode I and $\psi_R = \pm 90^\circ$ corresponds to pure mode II. The interface crack tends to propagate at the interface when $\psi_R \leq 0^\circ$. In contrast, positive ψ_R promotes the deviation of the crack path, also known as kinking, into the core [16, 17].

2.2 *Compliance and J-Integral Methods*

For establishing the strain energy release rate from an interface material characterisation test with simple geometry it is often possible to use the compliance method. The compliance, *C* (the rate of increase of displacement with increase of applied load), is measured for a series of crack lengths, *L*, and the rate at which *C* increases with *L*, *dC*/*dL*, is estimated. The strain energy release rate is proportional to this, the relationship depending on the loading arrangement and the geometry of the specimen. For simple test arrangements the relationship is easily established analytically, but more complex geometries and arrangements require either a more complex analysis or a numerical model, such as a finite element model (FEM).

In the latter case, when FEM is used, it is often more convenient to use instead a method based on the path-independent J-integral, first defined by Rice [18]. An example of its application is described in Sect. 3.4.

For practical purposes, the compliance method is generally confined to problems where, except for a very local process zone at the crack tip, the specimen can be treated as linear-elastic. An advantage of the J-integral method is that, while it gives identical results to the compliance method for linear-elastic problems, it can also be used for non-linear and inelastic crack propagation problems, such as those involving fibre bridging.

2.3 *Crack Surface Displacement Extrapolation (CSDE) Method*

The Crack Surface Displacement Extrapolation (CSDE), introduced by Berggreen et al. [19, 20], is a relatively new finite element (FE) based mode mixity method to determine the energy release rate and mode mixity at a crack tip. In the analysis of cracked interfaces with large stiffness difference, the CSDE method has been shown by Berggreen [6], and also by the recent study by Barbieri et al. [55] described in Sect. 3.9, to be more robust and stable than other FE based mode mixity methods such as the Virtual Crack Closure Technique (VCCT) [21, 22]. The principle of the CSDE method is to calculate the energy release rate and mode mixity phase angle values based on the relative opening and sliding displacements of the crack flanks, Eqs. (10, 11 and 12), over a region behind the crack tip. The method subsequently finds a linear sub-region and extrapolates the values linearly over the numerical error

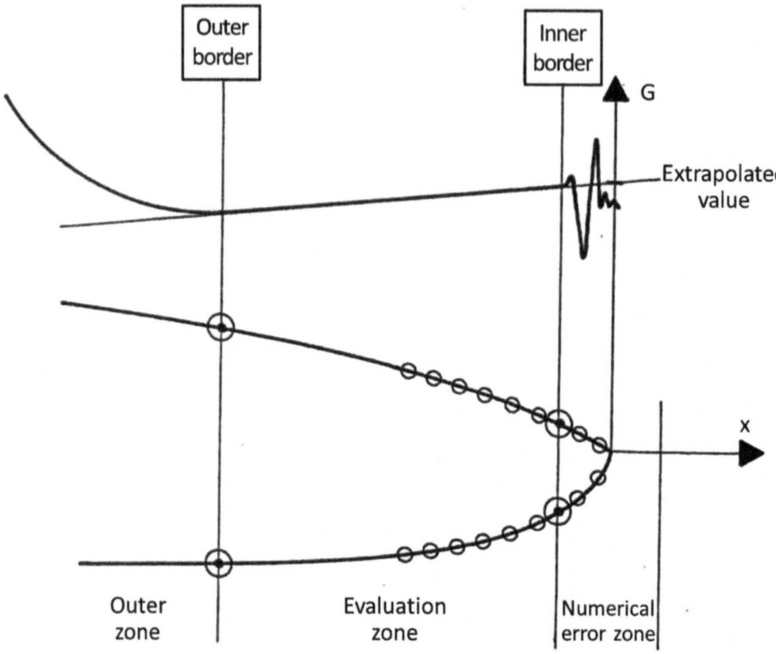

Fig. 3 The CSDE method presented schematically

zone to determine them at the crack tip, see Fig. 3. The energy release rate and mode mixity have been reported to express a physically impossible oscillatory behaviour close to the crack tip [23, 24]. The CSDE mode mixity method circumvents this behaviour, by extrapolating over the oscillatory region, which is often associated with large numerical errors in a finite element analysis.

2.4 Fatigue Crack Growth

The Paris law [25] can be expressed as a relationship between the rate of growth of a crack, da/dN, where a is the crack length and N the number of loading cycles, and the strain energy released per cycle, ΔG:

$$\frac{da}{dN} = C(\Delta G)^m \tag{13}$$

where C and m are constants that can be found from experiments.

2.5 The Cycle Jump Technique for Fatigue Crack Growth Calculation

In principle, once the fracture and crack growth properties have been obtained experimentally (Sect. 3) it is possible to model a real structure, with or without defects and damage, and estimate its fatigue life. However, in practice this can be extremely demanding and time-consuming. To overcome this problem, Moslemian et al. [26] modified the "cycle jump" technique for numerical analysis of fatigue crack propagation in a sandwich face-core interface that significantly reduces the computational effort. This is illustrated in Fig. 4, and consists essentially of the following steps:

1. Perform FE analyses of the damaged structure for a few cycles (using available mixed mode fracture data from fracture characterisation tests).
2. Fit spline curves to the development of deformations and key damage parameters with increasing number of cycles.
3. Extrapolate over a "jump" range of cycles to give a new, more advanced, assumed state of damage.
4. Repeat the process with this state as initial state.

The method was tested on 3D analysis of square sandwich panels with centrally located circular and elliptical debonds, and a concentrated normal force pulling the centre of the debonded face sheet away from the core [27], and validated against physical tests for a case with an initially circular debond [28]. It was also used by Martakos et al. [72, 74] in their studies of crack arresting devices reported in Sect. 4.6.

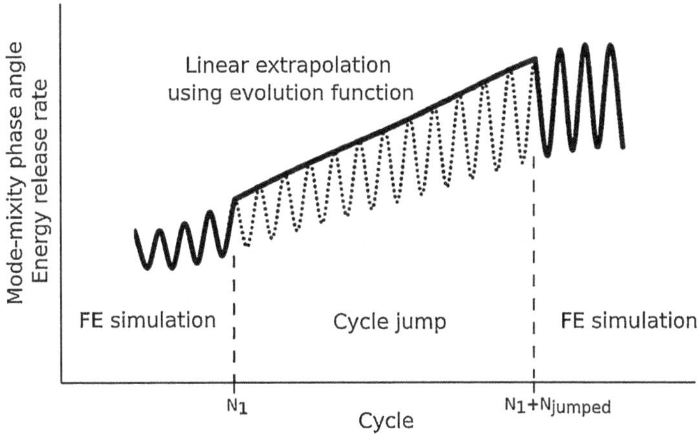

Fig. 4 Schematic of cycle jump technique

3 Experimental Fracture Characterisation Methods for Face/Core Sandwich Interfaces

3.1 Preliminary Remarks

Experimental methods for determining the fracture properties of face-core interfaces have been the subject of extensive research and development over the past 30 years; a significant part of this development has been supported by the ONR. The following subsections attempt to give an overview of the methods and the ways in which they have developed.

3.2 Cracked Sandwich Beam (CSB) Test

In sandwich panels under lateral pressure loading caused, for example, by slamming loads on the hulls of marine craft, the primary loading on the core is transverse shear. The ASTM C393 test [29] had long been used to determine the shear strength of sandwich core materials, but these tests do not consider the strength of the bond between the face sheet and core under the same loading. To address this, the cracked sandwich beam (CSB) specimen was developed and studied by Carlsson and co-workers [30, 31]. The test may be regarded as a development of the end-notch flexure (ENF) test used for determining the mode II interlaminar fracture properties of laminates. However, whereas the ENF specimen uses a crack at the mid-plane of a laminate, the CSB specimen requires the crack to be at the face-core interface and thus requires a different set of equations to extract the fracture properties.

Carlsson et al. [30] derived an expression for the compliance C of a sandwich beam under three-point bending with central load P, length $2L$ and width b having an end crack of length a. The mode II strain energy release rate G_{II} was then obtained using the relation

$$G_{II} = \frac{P^2}{2b}\frac{dC}{da} \tag{14}$$

3.3 Double Cantilever Beam (DCB) Test

Fracture at the face-core interface under peel loading (mode I) was addressed by Prasad and Carlsson [16, 17]. In [16], sandwich beam specimens were analysed using the finite element method. Peel fracture was studied using a modified double cantilever beam (DCB) sandwich specimen (Fig. 5) with a pre-crack between the face sheet and core, while shear fracture was studied with a modification of the

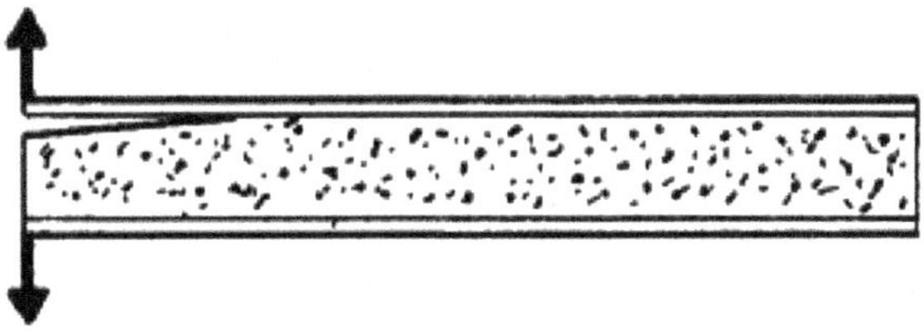

Fig. 5 Basic DCB specimen

ASTM C273 block shear test [32] to include a face/core pre-crack. Complex and conventional stress intensity factors were calculated for bimaterial cracks located between the face sheet and the bond layer and between the bond layer and the core over a large range of core moduli. Overall, much larger stress intensity factors were observed for an interfacial crack between the face sheet and bond layer than for a crack between the bond layer and core for both types of specimens. Crack kinking analysis of the DCB specimen revealed that the debond tends to remain interfacial for stiff core materials, but may deflect into the core for compliant core materials.

The same authors [17] experimentally examined bonding and crack kinking in sandwich beams, and also performed analysis using the finite element method. DCB and shear fracture specimens employing aluminium faces bonded to a wide range of PVC and PMI foam cores using two types of adhesives were considered. It was found that the Young's modulus of the core had a profound effect on the tendency of the face core interfacial crack to deflect (kink) into the core in DCB testing, while in shear testing crack kinking occurred for all core materials considered. Measured kink angles compared favourably with kink angles calculated based on the interfacial stress intensity factors prior to kinking.

Much later, Carlsson et al. [33] further investigated the crack path in foam cored DCB sandwich fracture specimens. A simple laminated beam analysis was presented for analysis of the propagation path of a core crack in both symmetric and unsymmetrical DCB sandwich fracture specimens with a polymer foam core. The analysis determined the sign of the shear stress ahead of the crack tip from the discontinuity of bending strain in the legs of the DCB specimen, and hence the kinking direction. The beam analysis, furthermore, assessed the stability of the crack path and provided the steady-state location of the crack after kinking (in the event of kinking). The analysis was in overall agreement with predictions of kink angle obtained from detailed finite element modelling and experimental crack paths in DCB specimens with an H100 foam core.

Aviles and Carlsson [34] presented an analysis of the compliance and energy release rate of the sandwich DCB specimen. They assumed that there was a starter crack at the upper face-core interface and that the crack remained at or near this interface during crack propagation. Beam, elastic foundation, and finite element analyses were presented and compared to experimentally measured compliance data, and compliance calibrated energy release rate (adjusted to eliminate the effect of the compliance of the test rig) over a range of crack lengths for foam cored specimens. It was found that the beam analysis provided a conservative estimate of the compliance and energy release rate. The elastic foundation model was in agreement with finite element analysis and experimental compliance data. Recommendations for specimen design and an expression for an upper limiting crack length were provided.

3.4 Double Cantilever Beam Loaded with Uneven/Unequal Bending Moments (DCB-UBM)

3.4.1 Description

The double cantilever beam specimen with uneven, (or, unequal) applied bending moments (DCB-UBM) was first introduced by Sørensen et al. [35] for multilayer materials and later modified for sandwich composites by Lundsgaard-Larsen et al. [36], see Fig. 6. This specimen enables the investigation of fracture toughness at different mode mixity phase angles. Fracture analysis under linear elastic assumptions is carried out to calculate the energy release rate and mode mixity at the crack tip for isotropic and orthotropic materials. Pure moments are applied at the crack flanks while the other end of the specimen is held between the rollers. The design allows the application of loads with no transverse forces, so that the bending moment distribution is uniform throughout each of the three separate regions of the specimen: in the debonded part of the upper face sheet, in the combined core and lower face sheet over the same part of the length, and in the intact part of the sandwich beam. Thus stable crack growth is obtained. The DCB-UBM specimen is therefore G-controlled by nature, in the sense that the strain energy release rate is independent of the crack length.

Lundsgaard-Larsen et al. [36] applied the DCB-UBM test to the measurement of mixed mode cohesive laws for interfaces in sandwich structures. They increased the bending stiffness of the sandwich faces by bonding steel bars onto the sandwich faces. This stiffening reduces rotations and ensures that the method is applicable for thin face sandwich specimens. The J-integral was employed, and the opening of the pre-crack tip was measured using a commercial optical measurement system, from which mixed mode cohesive laws were extracted.

Closed-form expressions for the energy release rate, G, and mode mixity, ψ, for the DCB-UBM specimen have been derived by Kardomateas et al. [37] by means of a tri-material system similar to a sandwich construction. For the case of a

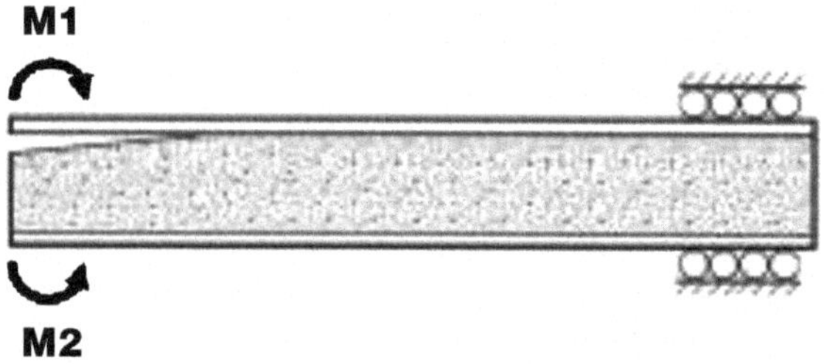

Fig. 6 Schematic illustration of DCB-UBM specimen

symmetrical layup, with identical, stiffened top and bottom face sheets, this was further developed more recently by Saseendran et al. [38], using the J-integral approach combined with laminate beam theory, to a five-layer model that includes the stiffener bars. The mode mixity, ψ, is derived in terms of a single scalar quantity ω which depends on the stiffnesses and thicknesses of the respective stiffener, face and core layers. The scalar parameter ω is obtained using, for example, finite element analysis (FEA). Therefore, for a specific face/core (or stiffener/face/core) interface system, FEA needs to be performed only once to map out ω for one loading configuration and the same ω can then be used for all other loadings as well. Effectively ω needs to be obtained for a range of typical component stiffness and thickness ratios and interpolation can then be used to obtain ω for a specific stiffness and thickness configuration.

The energy release rate and mode mixity for the DCB-UBM specimen can also be obtained using the CSDE method [20]. Geometrically linear analysis is performed on the specimen modelled with four-noded linear and eight-noded parabolic elements. The crack-tip mesh is highly densified and consists of numerically robust linear elements surrounded by a region of parabolic elements. Pure moments are applied at the ends of the crack flanks via master nodes in conjunction with multi-point constraint (MPC) elements, see Fig. 7.

The CSDE method calculates the strain energy release rate (ERR) and mode mixity phase angle ψ from the relative normal and shear crack flank displacements along the crack flanks within the parabolic elements close to the crack tip, and then extrapolates these values into the crack tip. The loading configuration for the experimental evaluation is selected according to the mode mixity at the crack tip obtained using the CSDE method.

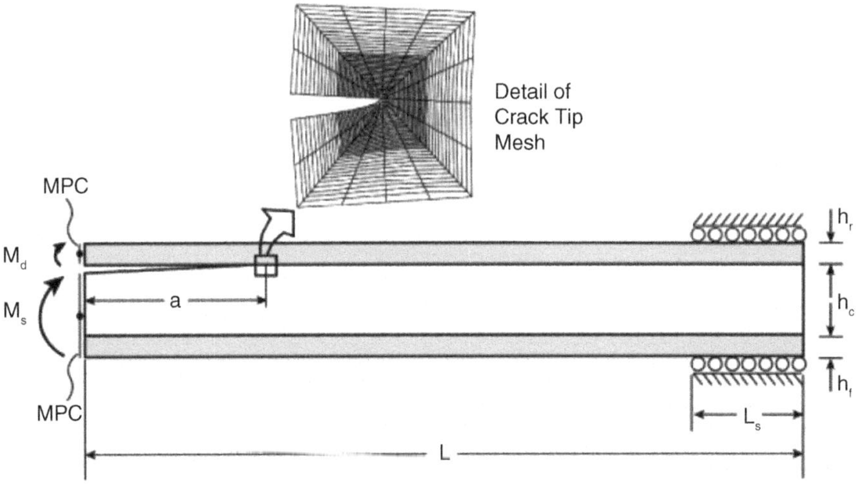

Fig. 7 FE model of DCB-UBM specimen

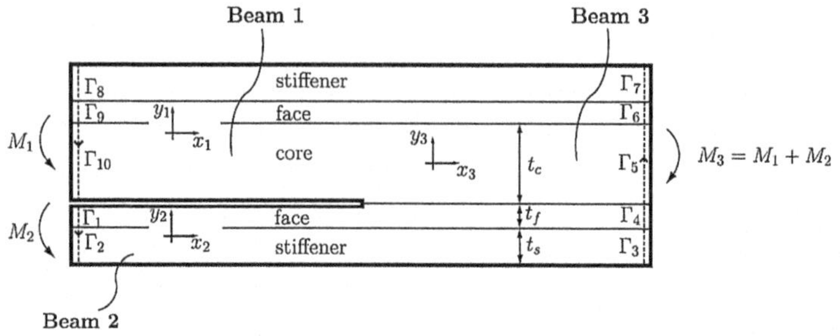

$$J = \sum_{p=1}^{10} \frac{\overline{E}_p M_b^2}{6(A_b D_b - B_b^2)^2}\left[A_b^2(y_{p-1}^3 - y_p^3) - 3A_b B_b(y_{p-1}^2 - y_p^2) + 3B_b^2(y_{p-1} - y_p)\right]$$

Fig. 8 Closed J-contour integral path for DCB-UBM multilayer specimen

3.4.2 DCB-UBM Specimen Design and Analysis

The thicknesses of the doubler layers are selected such that plastic deformation is prevented under the expected loads. In their study of cohesive laws (see Sect. 4.6) for interfaces with large-scale fibre bridging which violates the LEFM assumptions, Lundsgaard-Larsen et al. [36] selected the design based on the J-integral equation for the DCB-UBM specimen with doubler layers for a fixed moment ratio. However, in the LEFM regime the J-integral is equal to the strain energy release-rate, G. The J-integral was formulated as a function of extension, bending and coupling terms A, B and D, respectively, see Fig. 8 (which uses a slightly different configuration and notation from Saseendran et al. [38]).

To measure a J-integral value, lower strength steel can be used with thick doubler layers. Using a thick doubler would, however, require the rig to apply high bending moments. Taking this into consideration, a trade-off is made between the strength and the thickness of the doubler layers. A thickness of 6 mm is chosen with a steel type with a yield strength of 750 MPa to characterise an interface with a fracture toughness of up to $J = 1500$ J/m^2.

3.4.3 Novel DCB-UBM Test Rig

The original DCB-UBM rig was loaded by wires and required a tall test frame. It was unable to perform fracture testing with cyclic loading conditions at a reasonable and practical frequency, and allowed only a limited range of phase angles. Berggreen et al. [39] presented a novel test set up that overcomes these shortcomings. A schematic of this test rig is shown in Fig. 9 and an assembly diagram in Fig. 10. The rig is capable of applying pure moments such that the asymptotic stress-field developed at the crack tip reflects the corresponding mode mixity. A conditional control system is implemented such that the ratio of the moments applied across the

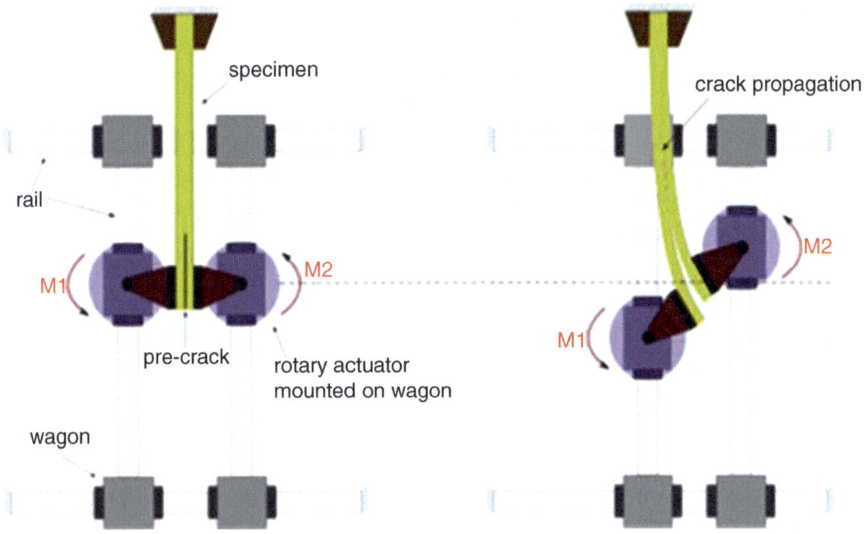

Fig. 9 Schematic illustration of DCB-UBM test rig

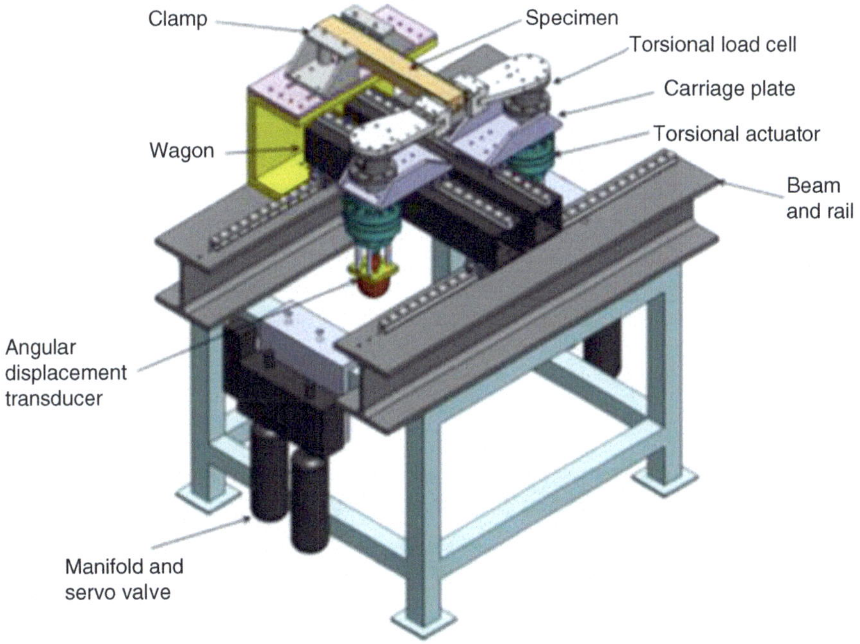

Fig. 10 Assembly drawing of the DCB-UBM test rig (isometric view)

arms remains constant, when controlling the test by increasing the angular rotations at the moment application points.

To apply the desired pure moments on each crack flank, two hydraulic rotational actuators are employed on either side of the specimen. These are equipped with torsional load cells and sensors to control the angular position of the actuators. A bi-axial servo-hydraulic controller is employed. To apply pure moments, the crack flanks must be able to deform without any constraints in the plane of deformation. This is achieved by mounting the actuators on pairs of rails which allow them to slide in the specimen plane (length-wise and width-wise). The specimen is held between roller supports, allowing it to slide in the length direction. This sliding ensures that there are no vertical forces on the specimen. The fixture does not have any restriction with regard to the type of specimens that can be tested. The steel doubler layers, however, require higher moments to bend the specimen to enable crack propagation for a particular moment ratio (MR). The extended steel arms engage in the fixture made on top of the torsional load cells (see Fig. 10). The moments are applied on each arm based on the defined moment ratios. The controller manipulates one rotational actuator such that the desired moment ratio is achieved. A conditional control ensures that, for a constant applied rotation rate in that actuator, the other actuator rotates keeping the moment ratio constant.

Note that, depending on the direction of opening of the arms, moment ratios can be positive or negative. A clear definition of the applied moment and sign convention is provided in Fig. 11, in which M_d is the moment applied on the debonded part and M_s is the moment applied at the substrate part. The FE model in Fig. 7 also shows the boundary condition due to the roller supports. The nomenclature for moments used here is the same as that used by Kardomateas et al. [37] in deriving the analytical expression for G for the DCB-UBM specimen. Thus, based on the direction of M_d and M_s, the sign of MR = M_d/M_s changes. It should be noted that MR is positive if both M_d and M_s are clockwise or if they are both counter-clockwise, see Fig. 11.

Further details of the test rig are provided by Berggreen et al. [39]. They performed fracture testing on a typical marine grade sandwich configuration consisting of H45 PVC foam core and glass fibre face sheets to demonstrate the

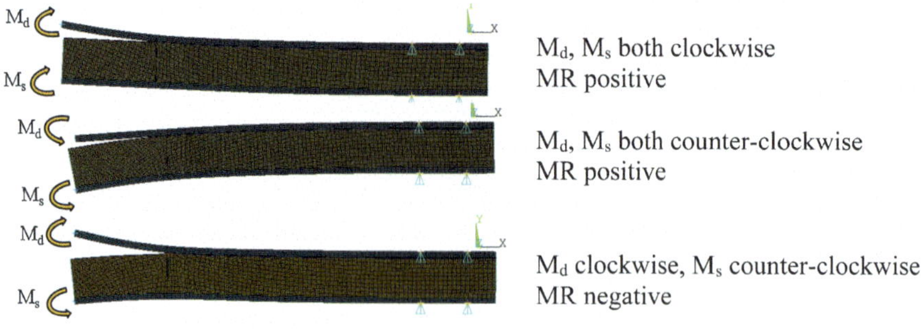

Fig. 11 Sign convention for moment ratio MR depending on applied moments M_d and M_s

applicability of the test method. The obtained fracture toughness agreed with interface toughness values in the literature measured using other test methods with the same material system. The effective kinematics of the test rig were measured using digital image correlation (DIC) by studying the rotations and moments of a specially designed calibration specimen and comparing the results against finite element results. The DCB-UBM specimen and test method were shown to be a promising candidate for obtaining face sheet/core interface fracture toughnesses in sandwich composites.

3.5 Tilted Sandwich Debond (TSD) and Modified TSD Specimens

For some face-core combinations and loading configurations a debond crack starting at the face-core interface deflects into the core, known as "kinking". To avoid this, a new test specimen for characterisation of debond failure of foam core sandwich structures, the tilted sandwich debond specimen (TSD), was proposed by Li and Carlsson [40–42] in a project carried out in collaboration between Florida Atlantic University (FAU) and KTH, and supported by the ONR Solid Mechanics Program. In the TSD test the bottom face of a sandwich is attached to an incline, and a vertical load is applied to the partially debonded top face (Fig. 12). In [40], Li and Carlsson examined the debond progression experimentally. For a sandwich consisting of glass/vinylester face sheets and PVC foam core, a critical angle of the incline, $\theta = \theta_c$, was identified. When $\theta < \theta_c$, it was found that an existing debond kinked into the core. The desired face/core debond failure occurred when $\theta \geq \theta_c$. It was observed that θ_c decreases with increasing crack length. The fracture toughness for debonding, expressed as the critical strain energy release rate, G_c, increased with increasing magnitude of the shear component.

In [41] the same authors analysed the TSD specimen using elastic foundation theory. The top face sheet was considered as a cantilever beam in the debonded region of the specimen and as a beam of finite length supported by an elastic foundation in the bonded region. An analytical model for the beam deflection was developed for design of the specimen, fracture analysis, and data reduction of face/

Fig. 12 The TSD specimen

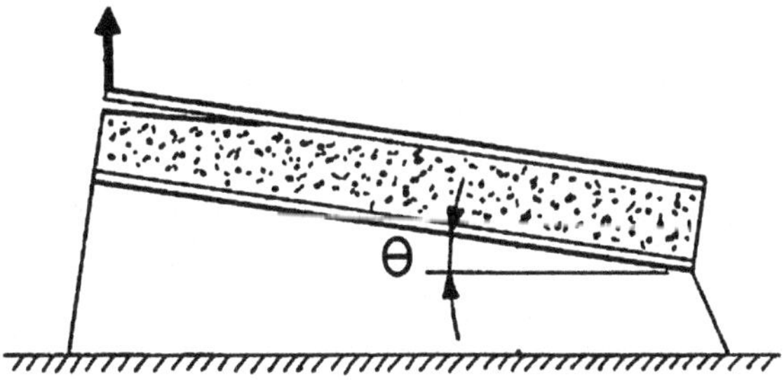

core debond toughness. A parametric investigation was conducted to examine how material and geometrical parameters of the constituents of the sandwich and test configuration influence the compliance of the TSD specimen. Testing was performed on a sandwich consisting of glass/vinylester face sheets over an H200 PVC foam core. The compliance predictions for this and a previously examined sandwich agreed favourably with experimental data over a large range of crack lengths and tilt angles. The debond toughnesses reduced from experimentally calibrated compliance and from the foundation model were consistent.

In [42] Li and Carlsson presented finite element analysis of several configurations of the TSD specimen, viz. a homogeneous and isotropic specimen, an interface crack specimen, and a sub-interface crack specimen. Stress intensity factors were determined using various mesh refinements over a range of tilt angles, and the mode mixity was related to the potential for kinking into the core. Irrespective of the tilt angle, it was found that cracks located at the face/core interface displayed a mode mixity favourable for kinking into the core. Experiments on a sandwich with an H100 PVC foam core revealed kinking at an angle in reasonable agreement with the prediction. After kinking, analysis of sub-interface crack configurations showed that the crack loading becomes highly mode I dominated, which indicates crack propagation parallel to the face/core interface, again in agreement with experimental observations. It was noted, however, that fracture mechanics analysis of foam materials, based on the assumption of a homogeneous material, becomes questionable when the singular domain becomes comparable to the cell size of the foam.

Viana and Carlsson [43], subsequently studied mode mixity and crack tip yield zones in TSD specimens with PVC foam core. Fracture mechanics analysis was conducted on core and debond fracture specimens. Five different PVC foam cores were examined. Stress intensity factors and the size of the plastic zone around the crack tip in foam core single-edge notch bend (SENB) and TSD specimens were estimated from the elastic displacements and stress fields near the crack tip. Analysis of the influence of core thickness and crack depth on the plastic zone size was performed on TSD specimens with H100 and R400 cores. It was found that the crack loading in the TSD specimen was essentially mode I, and that shear loading could not explain the higher toughness in the debond than the core. At the onset of fracture, the plastic zone height in the TSD specimen was much larger than that in a corresponding SENB specimen. It was believed that the plastic zone enlargement was a major factor explaining the elevation of debond toughness over the core toughness.

Viana and Carlsson [44] determined face/core debond toughness, expressed as the critical strain energy release rate (G_c), for sandwich specimens with composite and aluminium face sheets over various cross-linked PVC foam cores. Crack propagation occurred in the core, near the face/core interface for specimens with adequate face/core adhesion. Only for the highest density foam was it possible to propagate the crack in the immediate vicinity of the actual core/resin interface. The face/core debond toughness increased with increased foam density and reduced core thickness. For thick cores, the face/core toughness was independent of the core thickness.

A proposal for a modified TSD specimen was made several years later by Berggreen and Carlsson [45]. Analysis had shown that the range of phase angles that can be realised by altering the tilt angle and other parameters of the test is quite limited. To extend the range of mode mixities of the TSD specimen, a larger amount of transverse shear was introduced by reinforcing the loaded upper face with a stiff metal plate. Analysis showed that this method would extend the range of phase angles to a practically useful range. Guidelines on selection of thicknesses of the reinforcement, and design considerations for further modifications were provided.

Berggreen et al. [46] used the TSD test method, modified as above by reinforcing the loaded face sheet with a steel bar, to examine face/core debond fracture toughness of sandwich specimens with glass/polyester face sheets and H45 and H100 PVC foam cores over a large range of mode mixities. Fracture testing of the test specimens was conducted over a range of tilt angles. The fracture toughness exhibited mode mixity phase angle dependence, especially for mode II dominated loadings, while the fracture toughness remained quite constant for mode I dominated crack loadings. The fracture process was inspected visually during and after testing. For specimens with H45 cores the crack propagated in the core. For specimens with H100 cores, the crack propagated between the resin-rich layer and the face sheet.

3.6 *Mixed Mode Bending (MMB) Specimen and Test*

The mixed mode bending (MMB) specimen, originally developed and widely used for mixed mode delamination fracture characterisation of unidirectional monolithic composites, was extended by Quispitupa et al. [47] to the study of debond propagation in foam cored sandwich specimens (Fig. 13). The compliance and strain energy release rate expressions for the MMB sandwich specimen were derived based

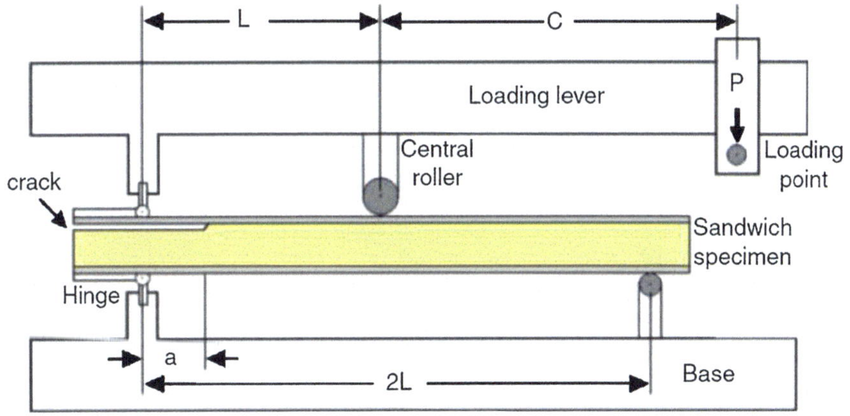

Fig. 13 MMB sandwich specimen and test arrangement

on a superposition analysis of solutions for the double cantilever beam (DCB) and cracked sandwich beam (CSB) specimens by combining a kinematic relationship for the specimen deformation with that of the loading provided by the test rig. This analysis provided also expressions for the global mode mixities.

The MMB compliance and energy release rate depend on the crack length, a, face and core thicknesses, h_f and h_c respectively, mechanical properties of the sandwich constituents (i.e. elastic modulus of the face sheet, E_f, and elastic modulus of the core, E_c), geometry of the specimen and loading conditions controlled by the lever arm distance, c. The analytical expressions for the MMB compliance C and energy release rate G can be written:

$$C = \left(\frac{c}{L} C_{DCB\_upper} + \frac{c-L}{2L} C_{DCB\_lower}\right)\left(\frac{c}{L} - \alpha\frac{c+L}{2L}\right) + \left(\frac{c+L}{L}\right)^2 C_{CSB} \tag{15}$$

$$G = \frac{P^2}{2b^2}\left(\begin{array}{l} \frac{c}{L}\left(\frac{c}{L} - \alpha\frac{c+L}{2L}\right)\frac{12}{E_f h_f^3}\left(a^2 + 2a\eta^{1/4} + \eta^{1/2}\right) + \\ \frac{c-L}{2L}\left(\frac{c}{L} - \alpha\frac{c+L}{2L}\right)\left[\frac{1}{h_c G_{xz}} + \frac{a^2}{(D - B^2/A)}\right] + \left(\frac{c+L}{L}\right)^2 \frac{a^2}{8}\left(\frac{1}{D_{debonded}} - \frac{1}{D_{intact}}\right) \end{array}\right) \tag{16}$$

where b is the specimen width, L is the span between supports, and the decomposed compliance components are:

$$C_{DCB\_lower} = \frac{a}{b}\left[\frac{1}{h_c G_{xz}} + \frac{a^2}{3(D - B^2/A)}\right] \tag{17}$$

$$C_{DCB\_upper} = \frac{4}{E_f h_f^3 b}\left(a^3 + 3a^2\eta^{1/4} + 3a\eta^{1/2} + \frac{3}{2}\eta^{3/4}\right) \tag{18}$$

$$C_{CSB} = \frac{L^3}{6bD_{intact}} + \frac{L}{2h_c b G_{xz}} + \frac{a^3}{12b}\left(\frac{1}{D_{debonded}} - \frac{1}{D_{intact}}\right) \tag{19}$$

The load partitioning factor and elastic foundation modulus parameter can be expressed as:

$$\alpha = \left[\frac{\frac{a^3}{3}\frac{1}{(D-B^2/A)} + \frac{a}{k}\frac{1}{(G_f h_f + G_{xz} h_c)}}{\frac{a^3}{3}\frac{1}{(D-B^2/A)} + \frac{a}{k}\frac{1}{(G_f h_f + G_{xz} h_c)} + \frac{a^3}{3}\frac{1}{(E_f h_f^3/12)} + \frac{a}{k}\frac{1}{G_f h_f}}\right] \tag{20}$$

$$\eta = \frac{h_f^3 h_c E_f}{6E_c} \tag{21}$$

where k is the shear correction factor ($k = 1.2$). The stiffnesses A, B and D can be written as:

$$A = E_f h_f + E_c h_c \tag{22}$$

$$B = h_f h_c \left(\frac{E_c - E_f}{2} \right) \tag{23}$$

$$D = \frac{1}{12} \left[E_f \left(h_f^3 + 3 h_f h_c^2 \right) + E_c \left(h_c^3 + 3 h_c h_f^2 \right) \right] \tag{24}$$

$$D_{intact} = \frac{E_f h_f}{2} (h_c + h_f)^2 + \frac{E_f h_f^3}{6} + \frac{E_c h_c^3}{12} \tag{25}$$

$$D_{debonded} = (1 - \alpha)\left(D - B^2/A\right) \tag{26}$$

A, *B* and *D* are the extensional, coupling and bending stiffnesses respectively, and $D_{debonded}$ and D_{intact} are the flexural stiffnesses of the debonded region and intact region of the cracked beam respectively. Further details are provided in [47].

An extensive parametric analysis to improve the understanding of the influence of loading conditions, specimen geometry and mechanical properties of the face and core materials was performed using the derived expressions and finite element analysis. The mixed mode bending compliance and energy release rate predictions were in good agreement with finite element results. The numerical crack surface displacement extrapolation (CSDE) method implemented in a finite element analysis was applied to determine the local mode mixity at the tip of the debond.

The same authors [48] presented a design analysis of the MMB sandwich specimen for face-core interface fracture characterisation. An analysis of the competing failure modes in the foam cored sandwich specimens was performed in order to ensure face-core debond fracture prior to other failure modes. This facilitates selection of the appropriate geometry for the MMB sandwich specimen to promote debond failure. An experimental study was performed using MMB sandwich specimens with an H100 PVC foam core and E-glass-polyester faces. Debond propagation was successfully achieved for the chosen geometries and mixed mode loading conditions.

In a third paper [49] the same authors presented an experimental study on face/core debond fracture of foam core sandwich specimens under a wide range of mixed mode loading conditions, using the MMB test arrangement. Sandwich beams with E-glass fibre-reinforced face sheets and PVC H45, H100 and H250 foam core materials were evaluated. A method was outlined to perform pre-cracking on fracture specimens in order to achieve a sharp and representative crack front. The mixed mode loading was controlled in the MMB test rig by changing the loading application point (and lever arm distance). Finite element analysis was performed to determine the mode mixity at the crack tip. The results showed that the face/core interface fracture toughness increased with increased mode II loading. Post-failure analysis of the fractured specimens revealed that the crack path depends on the mode mixity at the crack tip, the face sheet properties and the core density.

Manca et al. [50] examined face/core fatigue crack growth in foam-cored sandwich composites using the MMB test method. The mixed mode loading at the debond crack tip was controlled by changing the load application point in the MMB test fixture. Sandwich specimens were manufactured using H45 and H100 PVC foam cores and E-glass/polyester face sheets. All specimens were pre-cracked in order to define a sharp crack front. The static debond fracture toughness for each material configuration was measured at different mode mixity phase angles. Fatigue tests were performed at 80% of the static critical load, at load ratios of $R = 0.1$ and 0.2. The crack length was determined during fatigue testing using the analytical compliance expression and verified by visual measurements. Fatigue crack growth results revealed higher crack growth rates for mode I dominated loading. For specimens with H45 core, the crack grew just below the face/core interface on the core side for all mode mixities, whereas for specimens with H100 core, the crack propagated in the core or in the face laminate depending on the mode mixity at the debond crack tip.

3.7 G-Control

A problem that arises with fatigue testing using the test methods described so far is that the cyclic energy release rate and crack propagation rate vary during the test. Manca et al. [51] presented a computer controlled testing methodology called "The G-control Method" which allows cyclic crack growth testing using real-time control of the cyclic energy release rate. The advantages of using this approach were described and compared with traditional fatigue testing methods (displacement or load control). The compliance based analytical formulation for G-control was explained for the DCB specimen and then applied to experimental testing of the sandwich MMB specimen. Experimental results were presented for sandwich MMB specimens with glass-epoxy face sheets and PVC foam core, showing that the G-control method allows fatigue testing at a constant range of energy release rates leading to a constant crack propagation rate.

Manca et al. [52] proceeded further to present experimental results from cyclic crack propagation tests performed on sandwich specimens with glass-epoxy face sheets and PVC foam cores using the G-controlled cyclic energy release rate (ΔG) test procedure. The face material was tested in tension, compression and shear to determine in-plane and out-of-plane mechanical properties, such as Young's modulus, Poisson's ratio and shear modulus. These properties were then used in an analytical model of the mixed mode bending sandwich specimen to calculate the compliance and energy release rate. Finite element analysis was used to determine the mode mixity of the crack loading. Experimental crack growth cyclic tests were carried out on pre-cracked MMB sandwich specimens with H45, H100 and H160 PVC foam cores under two mode mixities (mode I and mode II dominant). Post-mortem analysis was performed on tested specimens, highlighting the influence of mode mixity and foam density on the crack path. Crack propagation diagrams showing da/dN versus ΔG curves were obtained to establish the Paris-Erdogan relation for each material combination tested at the two mode mixities (Fig. 14).

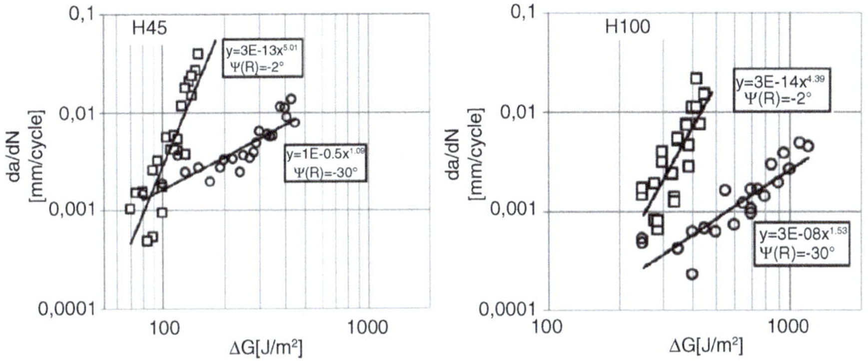

Fig. 14 *da*/*dN* versus ΔG for sandwich MMB specimens with GFRP faces and H45 and H100 cores tested at $\psi = 2°$ (mode I dominant) and $\psi = 30°$ (mode II dominant), from [52]

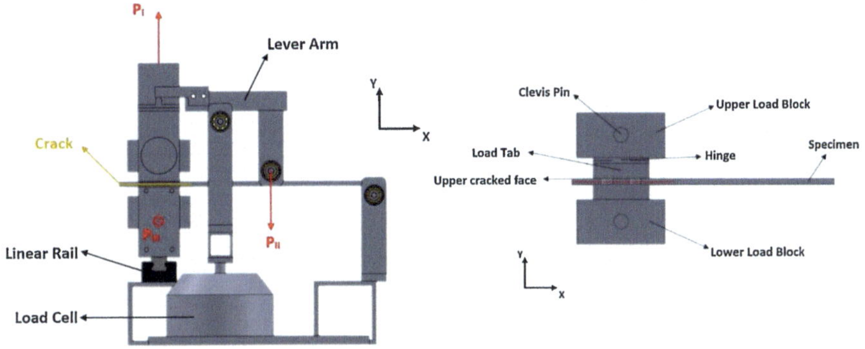

Fig. 15 Test rig for mode I-II-III testing of monolithic and sandwich specimens (schematic)

The results showed constant crack growth rates for all the materials tested and revealed the influence on mode mixity on crack propagation speed and foam density (higher foam density, slower crack propagation).

3.8 Shear-Torsion-Bending (STB) Test for Mixed Mode I-II-III Testing

Sabbadin et al. [53] report on a test machine, with associated test procedures, that has been under development to enable mixed mode I-II-III testing of cracked monolithic and sandwich specimens for fracture characterisation purposes. Thus it will cover the most general loading scenario where all three types of loadings at the crack tip are present. The test rig geometry (Fig. 15) is inspired by the STB test rig presented by Davidson and Sediles [54], whose experimental results agreed well with FEM analyses regarding pure Mode III and multiaxial (Mode I + III, II and II + III) fracture characterisation of delaminated composite laminates.

3.9 *Effects of Shear and Near Tip Deformations on Interface Fracture*

In an ONR-supported collaboration between the University of Genoa, Italy, and DTU, Barbieri et al. [55] performed a theoretical study of the effects of shear and near tip deformations on the interface fracture properties of symmetric sandwich beams. The effects of shear on the energy release rate and mode mixity in a beam with isotropic layers and a debond crack at the face-sheet/core interface were investigated through a semi-analytical approach based on two-dimensional elasticity and LEFM. Expressions for the shear components of the energy release rate and mode mixity phase angle were obtained for sandwich beams with the necessary numerical coefficients derived through finite element analyses. The expressions were combined with earlier results for sandwich beams subjected to bending moments and axial forces to obtain solutions for general loading conditions and for an extensive range of geometrical and material properties tabulated in terms of non-dimensional, numerically obtained coefficients. The physical and mechanical significance of the terms of the energy release rate which depend on the shear forces were explained using structural mechanics concepts and introducing crack tip root rotations to account for the main effects of the near tip deformations. The results are applicable to laboratory specimens used for the characterisation of the fracture properties of sandwich composites, provided the lengths of the crack and the ligament ahead of the crack tip are above minimum lengths which are defined in the paper.

3.10 *Low (Arctic) Temperatures*

Farshidi [56] experimentally investigated the effects of low temperature on the face/core interface fracture toughness and the fatigue debond growth rate in foam core sandwich composites. Mixed mode bending (MMB) specimens were statically and cyclically tested using a compact testing fixture inside a climatic chamber at a low temperature (−20 °C) and at room temperature (23 °C). Testing was conducted at a mode mixity phase angle $\psi = 4.4°$, very close to pure mode I (opening), and a mixed mode I/II (opening-sliding) mixed mode with $\psi = 23.9°$. The fatigue test results were presented according to the modified Paris-Erdogan relation. The results showed substantial fracture toughness reduction at the low temperature (Fig. 16), on average 32% for the mode I case and 12% for the mixed mode case. The low temperature also increased the cyclic crack growth rate (Fig. 17) very significantly for the mixed mode case, but somewhat less for the mode I case.

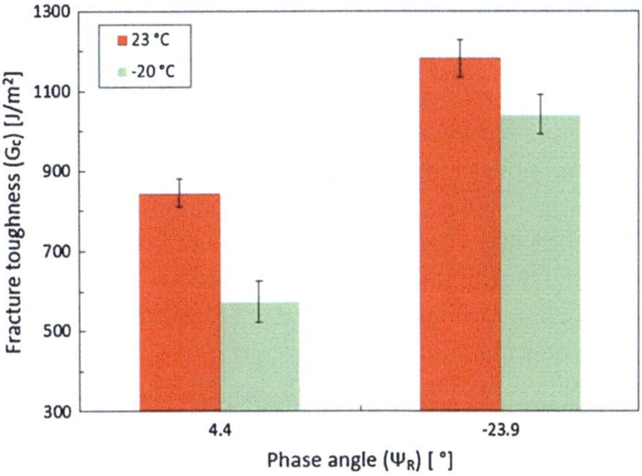

Fig. 16 Fracture toughness for two mode mixity phase angles and two temperatures

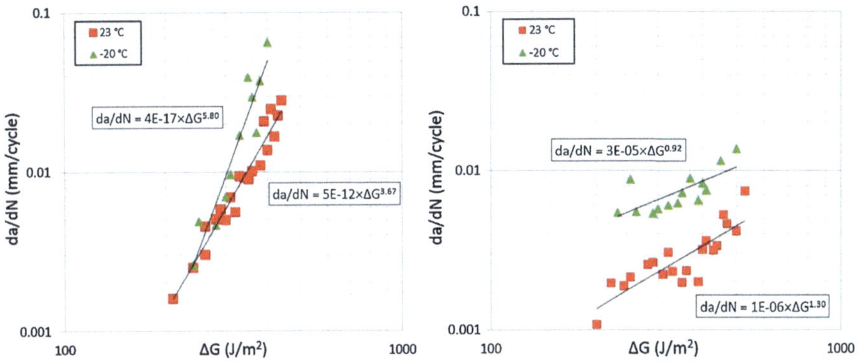

Fig. 17 *da*/*dN* versus ΔG results at $\psi_R = 4.4°$ (left) and $\psi_R = -23.9°$ (right)

4 Modelling and Testing of Sandwich Structural Components with Debonds

4.1 Curved Beams with Debonds

Layne and Carlsson [57] considered the problem of a curved sandwich beam under loading that tends to straighten the beam. Each beam was made with glass-polyester faces and a PVC foam core, and consisted of a 90° circular arc with straight legs to either side. The beams were tested in four-point bending in the direction tending to straighten the beam. The mode of failure was debonding of the face sheet on the convex (outer) side, suggesting that weaknesses in the bond were critical. The same authors [58] then tested similar curved beams with artificial debonds under similar loading. These beams suffered marginal reductions in stiffness, but substantial reductions in strength due to the presence of a debond.

4.2 Debonded Sandwich Columns in Axial Compression

In an early ONR-supported study, Aviles and Carlsson [59] developed an elastic foundation model for analysis of the local buckling behaviour of foam core sandwich columns containing a full-width face/core debond. They performed parametric studies to evaluate the effect of debond length, core thickness, core density, face sheet thickness, face sheet modulus and boundary conditions, on the local buckling load. They compared model predictions with experimental results obtained for sandwich specimens made from glass/vinylester face sheets and various PVC foam cores containing face/core debonds. The predictions were in reasonable agreement with measured buckling loads. Note that this study was confined to buckling of the face sheet, and did not consider crack propagation.

As part of the work performed under the Network of Excellence on Marine Structures (MARSTRUCT), partially funded by the European Union, Moslemian et al. [60] examined the failure of compression loaded sandwich columns with a full-width face/core debond. Tests were conducted on sandwich columns with glass/epoxy face sheets and H45, H100, and H200 PVC foam cores in a specially designed test rig. The strains and out-of-plane displacements of the debonded region were monitored using the digital image correlation technique. Finite element analysis and LEFM were employed to predict the critical instability load and compression strength of the columns. The energy release rate and mode mixity were determined and compared to fracture toughness data obtained from TSD tests, predicting propagation loads. Instability loads of the columns were determined from the out-of-plane displacements using the Southwell method. The finite element estimates of debond propagation and instability loads were in overall agreement with experimental results. The proximity of the debond propagation loads and the instability loads showed the importance of instability in connection with the debond propagation of sandwich columns.

Most of the columns with H200 core and some columns with H100 failed by debond propagation at the face/core interface towards the column ends. Bifurcation buckling instability of the debonded face sheet was not observed before the debond propagation initiated. It is believed that the initial imperfections are mostly responsible for this behaviour, which is similar to compression loading of a curved beam. Slight kinking of the debond into the core was another failure mechanism which occurred in columns with a low density H45 core. Compression failure of the face sheet occurred in all specimens with H200 cores and a 25.4 mm debond, which can be explained by the proximity between the debond propagation and the compression failure load of the face sheet. Instability and crack propagation loads of the columns were predicted based on geometrically nonlinear finite element analysis and linear elastic fracture mechanics. Testing of modified TSD specimens was conducted to measure the fracture toughness of the interface at the calculated phase angles for the column specimens associated with the debond propagation. Comparison of the measured out-of-plane deflection, instability, and debond propagation loads from experiments and finite element analyses showed fair agreement. For most of the investigated column specimens, it was shown that the instability and debond propagation loads are very reasonable estimates of the ultimate failure load, unless other failure mechanisms occur prior to buckling instability.

4.3 Debonded Sandwich Panels in Axial Compression

As part of the SaNDI Project referred to elsewhere in this chapter, Nøkkentved et al. [61] studied face/core debond-damaged sandwich panels exposed to non-uniform, in-plane compression loads. The study primarily concerned experimental methods, but simple finite element calculations were also applied. The specimens were rectangular, 580 mm $\times$ 800 mm, with 3.2 mm thick faces and 45 mm thick cores. Eight specimens had Divinycell H80 (80 kg/m^3) structural cross-linked PVC foam cores and two had H200 (200 kg/m^3) cores. The faces each had two 300 g/m^2 layers of chopped strand mat and two of 850 g/m^2 non-crimp quadri-axial fabrics, placed symmetrically about the midplane of each face. Seven of the specimens were fabricated with circular debonds between the core and one face sheet with diameters 100, 200 and 300 mm. Three additional specimens were intact prior to testing. The debonded panel area was manufactured by substituting the adhesive polyester with uncured polyester, covered by a thin paper sheet before applying the face layer. All the panels were reinforced at the edges. The complexity of applying a controlled non-uniform compressive load to the test panels required a strong focus on the development of a suitable test rig. The experimental results showed a considerable strength reduction with increasing debond diameter, with failure mechanisms varying between fast debond propagation and wrinkling-induced face compression failure for large and small debonds, respectively. A comparison between a simple numerical model and the experimental results showed fair agreement.

Berggreen and Simonsen [19] described the development, validation and application of a FEM based numerical model for prediction of residual strength of damaged sandwich panels. At the heart of the theoretical approach was the CSDE method (Sect. 2.3). The method was able to predict the maximum load carrying capacity of real-life panels with debond damage, where the failure is governed by face-sheet buckling followed by debond growth. The method was first developed in 2D and then extended to 3D by use of a number of realistic assumptions for the considered configurations. Comparison of the theoretical predictions with the series of panel experiments described in [61] showed that the model was able to predict the failure modes and the residual strength of damaged panels with accuracy sufficient for practical applications.

In an extension of the SaNDI Project aimed at generating strength reduction data for a specific series of naval vessels, with support of the Royal Norwegian Navy, Moslemian et al. [62] studied the failure of compression-loaded sandwich panels with implanted, circular face/core debonds. Uniform compression tests were conducted on intact sandwich panels with three different core materials (H130, H250 and PMI) and on similar panels with circular face/core debonds having three different diameters.

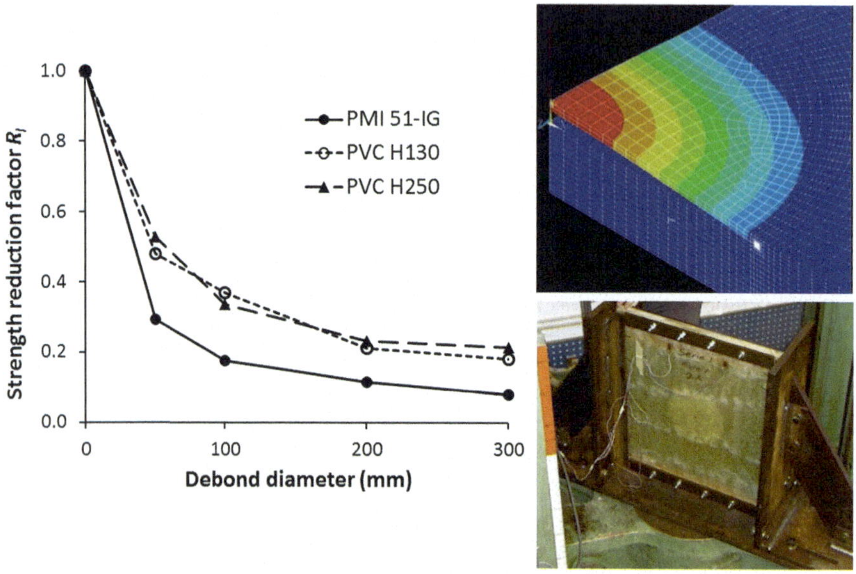

Fig. 18 Local strength reduction factors for face-core debond damage on sandwich layups with GFRP face sheets and three different foam core materials, with in-plane compressive loading [62]

The strains and out-of-plane displacements of the panel surface were monitored using the DIC technique. Mixed mode bending (MMB) tests were conducted to determine the fracture toughness of the face/core interface of the panels. Finite element analysis and LEFM were employed to determine the critical buckling load and compression strength of the panels. Numerically determined crack propagation loads in most of the cases showed a fair agreement with experimental results, but in a few cases up to 45% deviation was seen between numerical and experimental results. This can be ascribed to several factors such as the large scatter in the measured interface fracture toughness, and differing crack tip details and crack growth mechanisms between the panels and the MMB specimens. Tentative strength reduction curves were presented (Fig. 18), but uncertainty concerning the intact strengths of the materials used needs to be removed before these can be utilised with confidence.

At approximately the same time as the above studies, Aviles and Carlsson [63–66], with ONR support, also conducted studies of the compression strength of sandwich panels containing debonds. In [63] they conducted an experimental study of local buckling and failure behaviour of compression-loaded sandwich panels containing one-sided, centrally located circular and square face/core debonds. Sandwich panels with 2 mm thick glass/epoxy face sheets and a range of PVC foam cores were prepared (and also some with balsa cores). The core thickness was 25 mm in most cases, but some had 12.5 mm and 50 mm thick cores. The panels were 150 mm wide and 200 mm long. Note that these panels were appreciably smaller in scale than those considered by Moslemian et al. [62]. The face modulus and Poisson's ratio were: $E_f = 20.6$ GPa and $\nu_f = 0.42$. The foams were assumed to be isotropic with a Poisson's ratio $\nu_c = 0.32$, and Young's moduli of 40, 100 and 200 MPa for the H45, H100 and H200 foams, respectively. A face/core debond was introduced in each panel by implanting a thin (30 μm thick) Teflon film between one face sheet and the core, at the centre of the panel. The circular debonds were 50, 63, 75, and 100 mm in diameter.

A test fixture similar to that used by Moslemian et al. [62] (see Fig. 18) was developed for uniaxial compression loading of the test panels, of nominal size 200 × 150 × 30 mm (length × width × thickness) with some flexibility to accommodate other dimensions. Strain gauges were bonded on both face sheets at the panel centre to enable monitoring of the strains during loading. Three replicate panel specimens were tested.

The local buckling load was determined from the bifurcation point in the load-strain response. At the point of local (face) instability, a strain reversal was observed for the debonded face sheets. A rapid decrease in load after face buckling was indicative of a lack of post-buckling strength. Hence, this type of damage can be very serious in compression-loaded panels in a sandwich structure.

The same authors presented first a 2D elastic foundation [64] and then a 3D finite element [65] analysis of the buckling behaviour. In the latter, experimentally measured local buckling loads for the panels with H45, H100, and H200 cores and a range of debond sizes from [63] were compared to predictions from the 3D finite element analysis. For the panels with H45 core, the agreement between FEA and experiments was quite favourable. For higher density cores there was agreement for smaller debonds, but for larger debonds the FEA was conservative. The under-prediction of the critical load for large debonds was believed to be due to the way the debonds were artificially generated in the experimental study by insertion of a Teflon film. Such films may adhere to the resin and core after resin infusion resulting in tractions preventing the opening of the debond. The tendency for such problems was more pronounced for large debonds.

The experimental observations revealed that failure was initiated by local buckling of the debonded face sheet. After buckling, the debond rapidly propagated transversely to the loading direction followed by the collapse of the panel.

Aviles and Carlsson [66] also analysed the face/core debond propagation using a fracture mechanics approach. A three-dimensional, geometrically nonlinear finite element model of the debonded panel was combined with LEFM to determine the stress intensity factors and energy release rate at the debond (crack) front parallel and perpendicular to the applied load to predict initiation of debond propagation. The solution for the near tip crack flank displacements derived by Suo [11] was utilised to extract the stress intensity factors K_{I} and K_{II} from the relative opening and sliding displacements of the crack flanks. A range of core densities and debond sizes were analysed. Mode I was found to dominate the fracture process. The critical load for crack propagation predicted using fracture mechanics concepts was found to agree with measured collapse loads for smaller debonds, but fell below measured debond propagation loads for larger debonds. In all cases the predicted direction of crack propagation was perpendicular to the loading direction, in agreement with the experimental observations.

4.4 Debonded Sandwich Panels under Lateral Pressure Loading

As part of the SaNDI Project, Jolma et al. [67] developed and applied a tool for assessing residual strength of debond damaged laterally loaded sandwich panels. The analysis tool consisted of a parametric finite element model and a fracture mechanics calculation procedure to determine the residual strength. The parametric approach allowed variation of all geometric and material entities. The fracture mechanics calculation used crack flank displacements obtained from the finite element analysis solution and experimentally measured mixed mode fracture toughness values to determine the ultimate failure load. The analysis tool was validated with a number of different ship type panels by comparing the results from the tool with those from panel experiments. The analysis tool predicted both failure load and failure mode well. It was noted that the tool could be used to determine the residual strength of different damage cases and had a considerable potential for further development.

4.5 X-Joints under Fatigue Loading: STT Test Feature

In a Danish project concerned with improving damage tolerance of sandwich structures, Moslemian and Berggreen [68, 69] investigated face/core fatigue crack growth in sandwich X-joints numerically and experimentally. In the first part of the study [68] they presented an experimental investigation of fatigue crack growth and characterisation of the face/core interface of the joints. Sandwich tear test (STT) specimens with a face/core debond representing a debonded sandwich X-joint were tested under cyclic loading. Fatigue tests were conducted on the sandwich tear test specimens with H45, H100 and H250 PVC cores and glass/polyester face sheets. The digital image correlation (DIC) technique was used to locate the crack tip and monitor the crack growth. For the specimens with H45 core, unstable crack growth took place initially. Following the unstable propagation, the crack propagated in the core underneath the resin-rich cell layer approaching the interface. However, the crack did not kink into the interface. For the specimens with H100 core, the crack propagated initially in the core and then returned into the interface and continued to propagate in the interface. For the specimens with H250 core, the crack initially propagated in the core and then kinked into the interface. The interface crack eventually kinked into the face sheet, resulting in large-scale fibre bridging. Finally, mixed mode bending (MMB) tests were conducted to measure crack growth rates of the face/core interface at mode mixity phase angles similar to those calculated for the STT specimens.

In the second part of the study [69] the same authors attempted to simulate the fatigue crack growth in the same tested STT specimens using the finite element method. To accelerate the simulation, the cycle jump method (Moslemian et al. [26])

described in Sect. 2.5 was utilised and implemented in the finite element routine. This method is based on conducting finite element analysis for a set of cycles to establish a trend line, extrapolating the trend line spanning many cycles, and using the extrapolated state as initial state for additional finite element simulations. The measured *da/dN* relations of the face/core interface in Part I [68] were used as input to the fatigue crack growth simulation routine. A fair accuracy with 99% saving in computation time was achieved in the simulation of the STT specimens with H100 core. However, for the specimens with H45 core a large deviation between the numerical and experimental results was seen due to a partially unstable fatigue crack growth which was not taken into account in the simulations.

4.6 Improving Damage Tolerance

As noted in earlier sections, a face/core debond in a sandwich structure may propagate in the interface or kink into either the face or core. In a Danish project, partly funded by the European MARSTRUCT Network of Excellence, Lundsgaard-Larsen et al. [70, 71] examined various modifications of the face/core interface in foam core sandwich specimens and found that certain modifications of the face/core interface region influenced the kinking behaviour. They used finite element analysis and testing of sandwich DCB-UBM specimens with a range of mode mixities (between modes I and II) to develop a cohesive zone model of the fracture process including large-scale fibre bridging. Use of the DCB-UBM specimen enabled accurate measurement of the J-integral as the crack propagated. By altering the mode mixity of the loading, the crack path could be changed so that it deflected from the interface into the adjacent face or core. The transition points for crack kinking were identified and the influence of four interface design modifications on the propagation path and fracture resistance were investigated.

The authors investigated the use of a chopped strand mat (CSM) and of a continuous filament mat (CFM) placed between face and core, a layer of woven fabric between the face and CSM, and stitching of the layers of the face laminate. In addition, the effects of core material (comparing H200 with H130 density PVC foam) and specimen width were examined. The highest overall toughness was achieved with the CFM layer, although kinking into the face occurred at relatively low mode mixity. The mode mixity range where the crack propagated in the face/core interface region could be extended by the woven fabric layer placed adjacent to the face. Stitching prevented kinking of the crack into the face layers and provided overall high toughness. The specimens with a lower density core (H130) failed by crack kinking into the core at low mode mixities, and overall the debond toughness was much less than for the H200 core. The fracture toughness increased with specimen width as a result of more bridging fibres participating in the fracture process.

In a collaboration between DTU, Aalborg University, the University of Southampton, Siemens Wind Power A/S, and LM Wind Power Blades A/S, Martakos et al. [72] embedded a novel, pre-moulded polyurethane resin crack arresting (peel stopping) device in sandwich panels. They tested the panels under out-of-plane loading applied to the sandwich panel face-sheets. Fatigue crack propagation was induced in the face-core interface of the sandwich panels which met the crack arrester. The effect of the embedded crack arresters was evaluated in terms of the achieved enhancement of the damage tolerance of the tested sandwich panels. A finite element model of the experimental setup was used for predicting propagation rates and direction of the crack growth. The FE simulation was based on LEFM and a Paris law for crack propagation to predict the residual fatigue life-time and behaviour of the test specimens. Finally, a comparison between the experimental results and the numerical simulations was made to validate the numerical predictions as well as the overall performance of the crack arresters. In [73, 74] they investigated and optimised, using both numerical modelling and experiments, an alternative, simpler device using the STT tearing specimen.

5 Damage Tolerance and Assessment Procedures for Naval Sandwich Vessels

5.1 Introduction

Fundamental research concerning face-core debonds in sandwich structures mainly considers the development of damage in simple beam specimens of various types, and derivation of material and interface fracture properties from these tests, as described in Sect. 3. These simple beams only partially resemble real sandwich structures in ships, which generally consist of an assembly of sandwich panels (flat and/or curved), in some cases stiffened by beam-like structural components. For static loading, and to some extent impact and shock loads, fairly extensive physical testing and numerical modelling have been performed on sandwich panels under appropriate in-plane and out-of-plane loading, some of which has been described in Sect. 4. However, under cyclic fatigue loading, only very limited testing and modelling have so far been performed on real, three-dimensional structures such as panels or more complex configurations.

In principle, once the fracture properties have been obtained it is possible (e.g. for the vessel owner or operator) to model a real structure, with or without defects and damage, and estimate its residual strength and/or fatigue life. However, in practice this can be an extremely demanding and time-consuming task, especially with regard to fatigue life. Furthermore, the owner/operator needs to be able to assess the consequences for the operation of the ship when subjected to both local and global loads. Thus there is a need for an integrated defect/damage assessment system that can be used by others than research specialists.

Based on consideration of residual static strength immediately following damage, procedures for evaluating damage tolerance of sandwich hull structures have been established previously in the SaNDI Project referred to in Sect. 1 and subsequent extensions of that work. These procedures have varying degrees of simplification, and some can be readily used by the crew on board a ship for decision-making.

In Sects. 5.2, 5.3 and 5.4 an overview is provided of the damage assessment scheme developed in the SaNDI Project (2001–4) for static strength, and of simplifications and formulations made subsequently for on-board use in a specific naval application. Then, in Sects. 5.5 and 5.6, possible procedures to evaluate damage growth under cyclic loading following accidental damage are discussed.

5.2 *The SaNDI Project: Background and Aims*

The SaNDI Project, JP3.23 Inspection and Repair of Sandwich Structures in Naval Ships, has been briefly described in Sect. 1. The objectives were to develop methodologies for both production control and damage inspection for naval sandwich structures, and in particular to give improved knowledge of how defects and damage affect structural performance and to establish acceptance criteria for defects and damage. In addition, detailed studies were made of relevant NDI (non-destructive inspection) and monitoring techniques as well as repair methods. The main deliverables were two manuals that could be used as a basis for developing inspection and repair handbooks for specific vessels, focusing respectively on production defects and in-service damage [2–5].

Production defects and in-service damage types that are relevant for naval sandwich structures were listed in Sect. 1; most of these were addressed in the SaNDI project. Face-core debonds may occur during production as a result of contamination of the face and core surfaces or poor control of ambient conditions. They may also occur (or become evident) during service as a result of unfavourable events or environmental conditions. In particular, debonds may result from impact damage (if the core is damaged), from core shear cracking or from pull-off effects at out-of-plane joints between panels (such as sometimes occur where a deckhouse is attached to a deck). This last case is essentially the same as the situation represented by the STT specimen in the study of X-joints in Sect. 4.5.

If a naval vessel experiences damage while in operation, it may be necessary to make urgent decisions regarding corrective actions. Analysis of damage and its structural consequences can be a time-consuming and demanding task that takes many hours or days to perform. Thus it is desirable to have as much data already available that can be quickly accessed and processed. With this in mind, the participants in the SaNDI Project developed procedures that could use almost exclusively pre-calculated information, stored in a systematic format and readily available in decision support tools and/or manuals on board a ship and at its onshore support facilities. The data and procedures have since been further refined and developed in the context of a specific series of all-sandwich composite vessels. Similar data and procedures can also be applied to defects arising during production.

5.3 Details of the SaNDI Approach to Damage Assessment Based on Residual Strength

In proposing ways of assessing the consequences of damage it was found convenient to define four levels of damage in a structure made up principally of sandwich panels connected at their edges:

- *Level 1 damage*: Small local damage. Covers a small part of an individual panel. The influence of the damage on the panel stiffness and the stresses at remote points on the panel can be neglected.
- *Level 2 damage:* Medium local damage. Confined to one panel but does not meet the requirements for level 1 damage. The stiffness of the panel is not significantly influenced by damage. Redistribution of stresses in the global structure can be neglected in assessing the reduction of global ship strength.
- *Level 3 damage*: Large local damage. Confined to one panel but does not meet requirements for level 1 or level 2 damage. The stiffness of the panel is significantly influenced by the damage. Redistribution of stresses in the global structure cannot be neglected.
- *Level 4 damage*: Extensive damage. Affects two or more panels and/or supporting structure. Generally leads to extensive redistribution of stresses in the remaining structure and possibly severe overloading of adjacent parts. Must be analysed at the global ship level.

When considering the consequences of damage it is relevant to consider three different scales in the structure: the local scale (only relevant for level 1 damage), the panel scale and the global ship scale. At each of these scales it is possible to define a strength reduction factor defined as in Table 1.

For several level 1 and some level 2 damage cases, plots were developed showing the way the residual strength decreases with the size of the damage. These plots were derived for specific sandwich layups using analytical/numerical modelling and supported by laboratory tests. Such plots are conveniently constructed with the dimensionless local strength reduction factor R_l or panel strength reduction factor R_p plotted against damage size, though in some cases it can be difficult to decide on

Table 1 Three scales to be considered when assessing damage in a naval sandwich structure

Schematic	Scale	Strength reduction factor
	Local	$R_l = \frac{\text{Nominal (far field) stress or strain to cause failure with damage}}{\text{Nominal (far field) stress or strain to cause failure without damage}}$
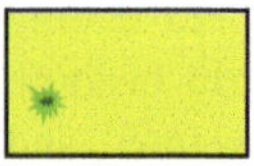	Panel	$R_p = \frac{\text{Maximum allowable load on damaged panel}}{\text{Maximum allowable load on intact panel}}$
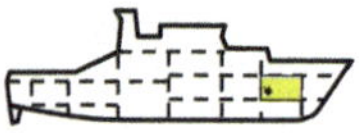	Ship	$R_s = \frac{\text{Maximum allowable load on damaged ship}}{\text{Maximum allowable load on intact ship}}$

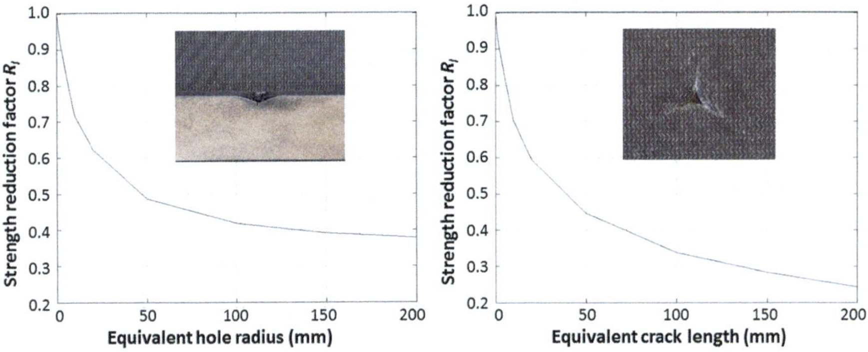

Fig. 19 Local strength reduction factors for face sheet impact damage on a sandwich layup with CFRP face sheet and PVC foam core, with in-plane compressive loading. Blunt impact (left) and sharp impact (right). The curves were obtained with the aid of equivalent hole and crack models

an intact strength value to use in the denominator. Figure 19 shows such curves plotted for two cases with local CFRP face sheet impact damage [5], while Fig. 18 in Sect. 4.3 shows corresponding curves for debond damage on sandwich layups with GFRP faces and three types of foam core, from a later, more comprehensive study [62]. Both figures, and a study at DTU of the effects of face sheet wrinkle defects [75] relate to in-plane compressive loading, which was the main focus in the SaNDI Project. However, some studies on tensile loading have been investigated more recently and the results for face sheet impact damage extended to GFRP face sheets [76]. In the studies of face sheet impact damage, blunt and sharp impacts were studied with the aid of equivalent hole and crack models, respectively, combined with laboratory testing.

Note that use of the local strength reduction factor R_l assumes that the damage size is much smaller than the in-plane panel dimensions. When the local factor R_l is used, it can be combined with a local location and load type sensitivity factor S_p, to give the panel strength reduction factor R_p:

$$R_p = R_l S_p \text{ with a maximum value of } 1.0. \tag{27}$$

The factor S_p is defined as the ratio of the value of the load on the panel that would cause the critical stress or strain component at the damage location to reach its maximum allowable value, *ignoring the damage*, to the maximum allowable value of load on the intact panel (i.e. the load that causes the critical stress or strain component to reach its maximum allowable value at the most highly stressed or strained location). This reflects the fact that, even if parts of the panel are at their design strength limit under the maximum design loading, some other parts may be less severely stressed and thus able to sustain damage without consequences for the panel as a whole. Note that the variation of S_p over the panel depends on the type of loading to which the panel is subjected and also the panel boundary conditions. By definition,

$$S_p \geq 1 \tag{28}$$

Thus it will always be on the conservative side to assume $S_p = 1.0$.

In the case of debonds, it may be appropriate to treat the damage as level 1 and use the local factor R_l, but in some cases the debond size may be too large for this and have to be treated as level 2, or even level 3 damage.

Any panel that experiences damage should be checked by comparing R_p with a minimum allowable value R_{pa}:

$$R_p \geq R_{pa} \text{ where } R_p = \min\left(R_l S_p, 1\right) \tag{29}$$

R_{pa} represents the minimum residual load-carrying capacity that can be allowed for the panel concerned, based on consideration of maintaining the panel's functionality. Such a limit can be defined for either local or global loading cases, or both.

If the damaged panel contributes to the global strength of the ship, as, for example, with a hull bottom panel amidships, the overall consequences for the ship must also be assessed. For this purpose the ship strength reduction factor R_s is estimated for one or more global loading cases. For level 1 and 2 damage types, this can be done using a panel location and load type factor S_s that is analogous to the local location and load type sensitivity factor S_p so that

$$R_s = R_p S_s \text{ with a maximum value of } 1.0. \tag{30}$$

The factor S_s represents the reserve of strength at the panel in which the damage occurs, calculated for the intact condition. Overall analysis of the intact ship can be used to establish a map showing reserves of strength in its respective panels, and an S_s value attached to each panel.

For level 3 damage, an approximate method has been proposed [5] for estimating R_s based on assumptions about the reduction of panel stiffness induced by the damage. If this is not appropriate, and for level 4 damage, a direct assessment of the damaged ship must be performed.

Finally the ship strength reduction R_s has to be compared with an allowable value R_{sa} based on evaluation of the required margin of safety for the design of the ship as a whole.

$$R_s \geq R_{sa} \text{ where } R_s = \min\left(R_p S_s, 1\right) \tag{31}$$

The total assessment process is illustrated in Fig. 20.

The global strength criterion can be checked directly, but experience has shown that it may be more convenient to transform it to either the panel or the local level, so that R_{sa} is used to establish an allowable panel strength reduction factor R_{paG} and, where appropriate, an allowable local strength reduction factor R_{laG} for global loads. This will ensure that the values of R_p and R_l will not lead to violation of the global

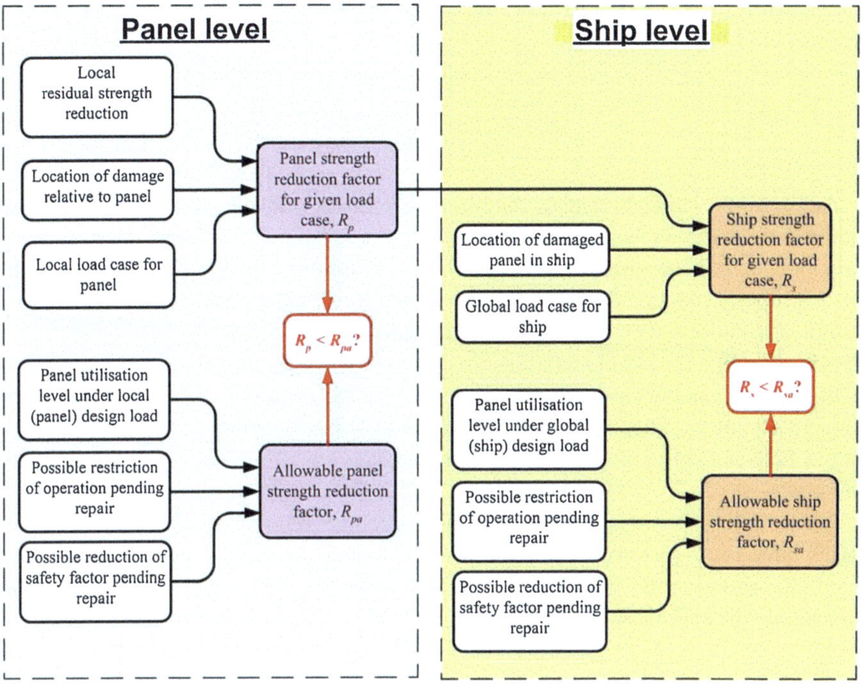

Fig. 20 Schematic diagram illustrating the strength assessment process in the presence of damage

ship strength reduction criterion. From Eqs. 27, 28, 29, 30 and 31 it can be shown that

$$R_{paG} = \min\left(\frac{R_{sa}}{S_s}, \frac{1}{S_s}\right) \quad \text{and} \quad R_{laG} = \min\left(\frac{R_{sa}}{S_pS_s}, \frac{1}{S_pS_s}\right) \tag{32}$$

Then the condition $R_s \geq R_{sa}$can be applied at either the panel or the local level:

$$R_p \geq R_{paG} = \min\left(\frac{R_{sa}}{S_s}, \frac{1}{S_s}\right) \quad \text{or} \quad R_l \geq R_{laG} = \min\left(\frac{R_{sa}}{S_pS_s}, \frac{1}{S_pS_s}\right) \tag{33}$$

Since normally $R_{sa} \leq 1$ these criteria normally reduce to

$$R_p \geq R_{paG} = \frac{1}{S_s} \quad \text{or} \quad R_l \geq R_{laG} = \frac{1}{S_pS_s} \tag{34}$$

Application at the panel level is often most convenient. However, the formulation at the local level is relevant for the extension to damage growth under cyclic loading (see Sect. 5.5). The allowable strength reduction factors for local load cases are designated R_{paL} and R_{laL}, and those for global load cases R_{paG} and R_{laG}. By similar arguments to the above,

$$R_{laL} = \min\left(\frac{R_{paL}}{S_p}, \frac{1}{S_p}\right) \tag{35}$$

Apart from establishment of the strength reduction at the lowest level, all the factors required can be obtained from prior analysis of the intact ship and its components, considering the relevant loading conditions on the ship in service. If this is properly planned during the design process, it can be performed as part of the design calculations with only modest extra effort.

Note also that the simple relationships between the local, panel and ship scales for level 1 and 2 damage cases are made possible by the assumption that small cases of damage do not significantly influence the panel stiffness, so that the loading distribution between different parts of the structure are not significantly affected by the damage.

Once the checks against the acceptable strength reductions at the appropriate scales have been performed, a decision can be made on further actions bearing in mind the situation in which the vessel is operating. The options may include (in roughly ascending order of severity:

- No repair, no action
- No immediate repair:
  - Maintain operational profile and
    - o Monitor damage, decide on repair later
    - o Repair at next scheduled docking (with or without monitoring of damage)
    - o Proceed to safe harbour and repair there (or make new assessment)
    - o Proceed to dry dock and repair there
  - Change operational profile and repair later (same options as above)
- Immediate repair at sea
  - Permanent repair, retain operational profile
  - Emergency/temporary repair. Options then as for "No immediate repair"
- Abandon ship

5.4 Simplified Procedure Developed for On-Board Use

In connection with the development of on-board inspection and repair manuals for a specific series of naval vessels, a simplified procedure was developed for level 1 type damage that used a classification system to represent the damage severity and the load and location factors. Level 1 damage is divided into five severity classes

1A to 1E according to their R_l values. For use in the panel assessment for local load cases, a series of SPL categories is similarly defined according to values of S_{pL}. A panel strength reduction category RPL0 to RPL5 can then be found from a simple table for any given combination of damage level (1A-1E) and SPL category.

The RPL categories are defined in accordance with minimum values of the factor R_{pL}. In a similar way, a set of categories for minimum allowable panel strength reduction factor is defined.

For each type of damage (e.g. sharp or blunt impact, debond) a basic data sheet is provided that includes how to identify and measure the damage and place it in the correct level.

The damage assessment is applied at the panel scale for both local and global load cases. Each panel on the ship (or a group of similar panels) is provided with a data sheet that indicates its main functions and type of loading, and, together with the damage data sheet, provides all the necessary data and categories for use in the damage assessment. Numerical values of the various factors are provided in addition to category codes.

5.5 *Application to Fatigue Loading*

Hayman and Berggreen [77] recently proposed a way in which the SaNDI approach might be extended to cover fatigue loading. With cyclic loading on level 1 damage, the local strength reduction curve should in principle be replaced by a residual life reduction curve. However, this would have to be drawn for a given load level (as amplitude of load cycles) and given minimum/maximum load or stress ratio. As the damage shape is likely to change during crack growth, it is most appropriate to use the minimum/maximum ratio for the far-field stress. Figure 21 shows schematically a set of such curves. Unlike the residual strength curves for static loading, these curves intersect with the horizontal axis at the points representing the condition

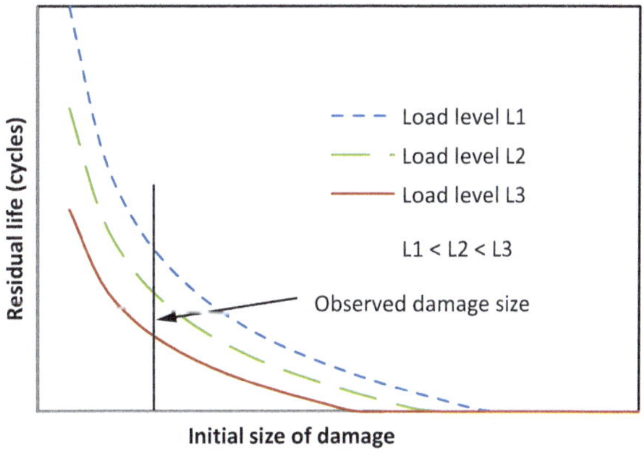

Fig. 21 Residual life reduction curves (schematic)

where the applied maximum load is equal to the residual strength of the component in its initial damaged state.

How to define the residual life can only be decided when some experience has been acquired regarding the way in which the damage growth occurs for the particular type of damage. It seems likely that the rate of growth will accelerate as the residual strength is reduced toward the level of the applied loading cycles.

For a given observed damage size and load amplitude, it is possible to read off the expected residual life. This can be compared with the minimum acceptable residual life and a decision made on further actions bearing in mind the situation in which the vessel is operating. The options will be similar to those listed in Sect. 5.3, but now the decision can be based on an estimate of the time it will take before the damage becomes critical.

A weakness of the above description is that it assumes that the load level is known. This information may be available for the scenario in question, but if it is not it is still possible to use R_{laG} given by Eq. (32), or the value of R_{laL} given by Eq. (35), to give a maximum allowable load value. This maximum allowable load is given by multiplying the intact static strength by R_{laG} or R_{laL}. As load values above this would violate the static strength acceptance criterion, this must give a conservative estimate for the residual fatigue life.

A further weakness lies in the fact that, although the initial damage may be small enough to class it as level 1, growth under cyclic loading might possibly increase the size beyond level 1. However, if the same type of damage has been considered under static loading and found to give failure while the size is within level 1, it is unlikely this will be a problem, though change of shape of the damage during growth might of course invalidate this assumption. Such situations will, however, become clear during the analysis or testing at the local scale.

Another approach is to use the same data as presented in Fig. 21 but plot the load level (amplitude) against the initial size of damage, for a series of values of residual life. In such a plot the load level can be made dimensionless by dividing it by the intact strength, as shown schematically in Fig. 22. This is the inverse of what is normally calculated, because the residual life would be calculated based on a given

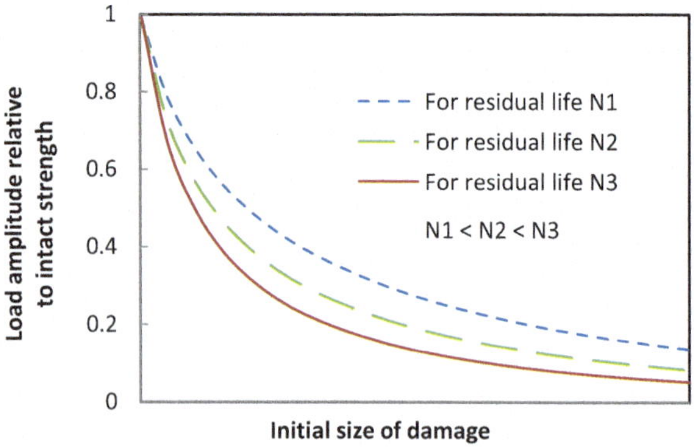

Fig. 22 Strength reduction curves for different numbers of fatigue cycles (schematic)

load amplitude. The required inversion process would necessarily involve interpolation between actual calculated cases. A major advantage of this approach is that the SaNDI damage assessment procedure can be applied directly, with use of the load type and location factors as for static loading. In effect we now have a series of local strength reduction curves for given desired values of residual life. For level 2 damage types, similar curves can be drawn at the panel level.

5.6 *Direct Estimation of Residual Fatigue Life: Integrated Fatigue Prediction System*

The generation of residual fatigue life curves of the type described in Sect. 5.5 is an extremely demanding task. An alternative to this approach is to provide a tool that will allow direct estimation of the residual life for a specific debond damage case when detected. Such a tool has long been sought after by the aircraft industry. A significant step towards the establishment of such a tool was recently reported by Farshidi et al. [78]. The system consists of a series of scripts for use with the ABAQUS software system, and is largely based on the principles for predicting fatigue life of debonded sandwich panels described earlier in this chapter. However, it has so far only been applied to a typical aircraft structural component and loading, namely a sandwich panel with honeycomb core, subjected to ground-air-ground (GAG) load cycles giving defined cyclic pressure changes inside the core. This can in principle be applied to naval sandwich structures by replacing the honeycomb with an appropriate foam core and implementing an appropriate loading. However, an appropriate, representative fatigue loading for a naval vessel will have to be defined: for a sandwich hull or deck panel it might result from local loads giving lateral pressure on the panel or global hull loads giving mainly in-plane tension or compression, or a combination of these. Note that such a tool could also be used for parametric studies to pre-calculate residual life curves as described in Sect. 5.5.

6 Conclusion

Modelling and physical testing of sandwich structures with face-core debonds at both the small, characterisation test specimen scale and the larger, ship structural component scale, have progressed considerably during the past few decades. Knowledge gained from such research is essential for the systematic implementation of damage tolerance principles in the design and operation of naval ships employing FRP sandwich construction. Much of the underlying research in recent years has been carried out at the Technical University of Denmark and at collaborating universities and research institutions. A significant amount of this research has been supported by the ONR's Solid Mechanics Research Program. This chapter has attempted to provide an overview of this research, and especially the progress made during the past 10–15 years. Previously developed schemes for assessing damage experienced by sandwich structures in naval ships and their influence on residual structural strength have been described and discussed. Possibilities for devising similar procedures for assessing residual fatigue life following a damage event have been explored.

Acknowledgements Compilation of this chapter was supported by Office of Naval Research Grant N00014-16-1-2977. The support, interest and encouragement of the ONR Solid Mechanics Program Manager, Dr. Y.D.S. Rajapakse, in this and the other ONR-funded research activities described in the chapter, are gratefully acknowledged.

References

1. Hayman B (2003) Inspection and repair of sandwich structures based on damage tolerance principles. In: Vinson JR, Rajapakse YDS, Carlsson LA (eds) 6th International Conference on Sandwich Structures. CRC Press, Boca Raton, pp 955–963
2. Hayman B, Zenkert D (2004) The influence of defects and damage on the strength of FRP sandwich panels for naval ships. In: Keil H, Lehmann E (eds) 9th International Symposium on Practical Design of Ships and Other Floating Structures (PRADS 2004). Seehafen Verlag GmbH, Hamburg, pp 719–726
3. Hayman B (2004) Defect and damage assessment for ships built in FRP sandwich. In: RINA Conference on High Speed Craft, Royal Institution of Naval Architects, London, November 2004
4. Zenkert D, Shipsha A, Bull P et al (2005) Damage tolerance assessment of composite sandwich panels with localised damage. Compos Sci Technol 65(15–16):2597–2611
5. Hayman B (2007) Approaches to damage assessment and damage tolerance for FRP sandwich structures. J Sandw Struct Mater 9(6):571–596
6. Berggreen C (2004) Damage tolerance of debonded sandwich structures, Ph.D. Thesis, Department of Mechanial Engineering, Technical University of Denmark
7. Zenkert D (1991) Strength of sandwich beams with interface debondings. Compos Struct 17:331–350
8. Zenkert D (1990) Damage tolerance of foam core sandwich constructions. Doctoral thesis published as Report No. 90–8, Royal Institute of Technology (KTH), Stockholm, Sweden 1990
9. Carlsson LA, Kardomateas GA (2011) Structural and failure mechanics of sandwich composites. Springer, Netherlands
10. Griffith AA (1921) The phenomena of rupture and flow in solids. Philos Trans R Soc Lond 221:163–198
11. Suo Z (1990) Singularities, interfaces and cracks in dissimilar media. Proc R Soc Lond A 427:331–358
12. Suo Z, Hutchinson JW (1990) Interface crack between two elastic layers. Int J Fract 43(1):1–18
13. Dundurs J (1969) Discussion on "Edge-bonded dissimilar orthogonal elastic wedges under normal and shear loading". J Appl Mech 36:650–652
14. Hutchinson JW, Suo Z (1991) Mixed mode cracking in layered materials. Adv Appl Mech 29:63–191
15. Suo Z (1989) Singularities interacting with interfaces and cracks. Int J Solids Struct 25 (10):1133–1142
16. Prasad S, Carlsson LA (1994) Debonding and crack kinking in foam core sandwich beams – I: Analysis of fracture specimens. Eng Fract Mech 47:813–824

17. Prasad S, Carlsson LA (1994) Debonding and crack kinking in foam core sandwich beam – II: Experimental investigation. Eng Fract Mech 47:825–841
18. Rice JR (1968) A path-independent integral and the approximate analysis of strain concentration by notches and cracks. J Appl Mech 35:376–386
19. Berggreen C, Simonsen BC (2005) Non-uniform compressive strength of debonded sandwich panels - II. Fracture mechanics investigation. J Sandw Struct Mater 7(6):483–517
20. Berggreen C, Simonsen BC, Borum KK (2007) Experimental and numerical study of interface crack propagation in foam cored sandwich beams. J Compos Mater 41(4):493–520
21. Rybicki EF, Kanninen MF (1977) A finite element calculation of stress intensity factors by a modified crack closure integral. Eng Fract Mech 9(4):931–938
22. Raju IS (1987) Calculation of strain-energy release rates with higher order and singular finite elements. Eng Fract Mech 28(3):251–274
23. Williams M (1959) The stresses around a fault or crack in dissimilar media. Bull Seismol Soc Am 49(2):199–204
24. Erdogan F (1963) Stress distribution in a nonhomogeneous elastic plane with cracks. J Appl Mech 30(2):232–236
25. Paris P, Erdogan F (1963) A critical analysis of crack propagation laws. J Basic Eng 85 (4):528–534
26. Moslemian R, Karlsson AM, Berggreen C (2011) Accelerated fatigue crack growth simulation in a bimaterial interface. Int J Fatigue 33(12):1526–1532
27. Moslemian R, Berggreen C, Karlsson AM (2012) Face/core debond propagation in sandwich panels under cyclic loading - Part I: Numerical modeling. In: 10th International Conference on Sandwich Structures. Nantes, France, pp 41–42
28. Moslemian R, Berggreen C, Karlsson AM (2012) Face/core debond propagation in sandwich panels under cyclic loading - Part II: Experimental validation. In: 10th International Conference on Sandwich Structures. Nantes, France, pp 43–44
29. ASTM standard C393-62 Standard method of flexure test of flat sandwich constructions. American Society for Testing and Materials, Philadelphia
30. Carlsson LA, Sendlein LS, Merry SL (1991) Characterization of face sheet/core shear fracture of composite sandwich beams. J Compos Mater 25:101–116
31. Carlsson LA (1991) On the design of the cracked sandwich beam (CSB) specimen. J Reinf Plast Compos 10:434–444
32. ASTM C273-61 (1991) Shear test in flatwise plane of flat sandwich construction and sandwich cores. American Society for Testing and Materials, Philadelphia
33. Carlsson LA, Matteson RC, Aviles F, Loup DC (2005) Crack path in foam cored DCB sandwich fracture specimens. Compos Sci Technol 65:2612–2621
34. Aviles F, Carlsson LA (2007) Analysis of the sandwich DCB specimen for debond characterization. Eng Fract Mech 75:153–168
35. Sørensen BF, Jørgensen K, Jacobsen TK, Østergaard RC (2006) DCB-specimen loaded with uneven bending moments. Int J Fract 141:163–176
36. Lundsgaard-Larsen C, Sørensen BF, Berggreen C, Østergaard RC (2008) A modified DCB sandwich specimen for measuring mixed mode cohesive laws. Eng Fract Mech 75 (8):2514–2530
37. Kardomateas GA, Berggreen C, Carlsson LA (2013) Energy release rate and mode mixity of a face/core debond in a sandwich beam. AIAA J 51(4):885–892. https://doi.org/10.2514/1.J051765
38. Saseendran V, Berggreen C, Carlsson LA (2018) Fracture mechanics analysis of reinforced DCB sandwich debond specimen loaded by moments. AIAA J 56(1):413–422. https://doi.org/10.2514/1.J056039
39. Berggreen C, Saseendran V, Carlsson LA (2018) A modified DCB-UBM test method for interfacial fracture toughness characterization of sandwich composites. Eng Fract Mech 203:208–223. https://doi.org/10.1016/j.engfracmech.2018.06.036

40. Li X, Carlsson LA (1999) The tilted sandwich debond (TSD) specimen for face/core interface fracture characterization. J Sandw Struct Mater 1(1):60–75
41. Li X, Carlsson LA (2000) Elastic foundation analysis of tilted sandwich debond (TSD) specimen. J Sandw Struct Mater 2:3–32
42. Li X, Carlsson LA (2001) Fracture mechanics analysis of tilted sandwich debond (TSD) specimen. J Compos Mater 35:2145–2168
43. Viana GM, Carlsson LA (2002) Mode mixity and crack tip yield zones in TSD sandwich specimens with PVC foam core. J Sandw Struct Mater 4:141–155
44. Viana GM, Carlsson LA (2003) Influences of foam density and core thickness on debond toughness of sandwich specimens with PVC foam core. J Sandw Struct Mater 5:103–118
45. Berggreen C, Carlsson LA (2010) A modified TSD specimen for fracture toughness characterization – fracture mechanics analysis and design. J Compos Mater 44(15):1893–1912
46. Berggreen C, Quispitupa A, Costache A, Carlsson LA (2014) Face/core mixed mode debond fracture toughness characterization using the modified TSD test method. J Compos Mater 48 (16):1939–1945. https://doi.org/10.1177/0021998313492358
47. Quispitupa A, Berggreen C, Carlsson LA (2009) On the analysis of a mixed mode bending sandwich specimen for debond fracture characterization. Eng Fract Mech 76(4):594–613
48. Quispitupa A, Berggreen C, Carlsson LA (2010) Design analysis of the mixed mode bending sandwich specimen. J Sandw Struct Mater 12(2):253–272
49. Quispitupa A, Berggreen C, Carlsson LA (2011) Face/core interface fracture characterization of mixed mode bending sandwich specimens. Fatigue Fract Eng Mater Struct 34(11):839–853
50. Manca M, Quispitupa A, Berggreen C, Carlsson LA (2012) Face/core debond fatigue crack growth characterization using the sandwich mixed mode bending specimen. Compos Part A 43:2120–2127
51. Manca M, Berggreen C, Carlsson LA (2015) G-control fatigue testing for cyclic crack propagation in composite structures. Eng Fract Mech 149:375–386. https://doi.org/10.1016/j.engfracmech.2015.06.059
52. Manca M, Berggreen C, Carlsson LA, Bortolotti P (2016) Fatigue characterization of poly vinyl chloride (PVC) foam core sandwich composite using the G-control method. J Sandw Struct Mater 18(3):374–394. https://doi.org/10.1177/1099636215603049
53. Sabbadin P, Berggreen C, Nygard B (2018) Development of a mode I/II/III test fixture for composite laminates and sandwich face/core fracture characterization. In: 12th International Conference on Sandwich Structures, Lausanne, Switzerland, pp 35–37
54. Davidson BD, Sediles FO (2011) Mixed-mode I-II-III delamination toughness determination via a shear-torsion-bending test. Compos Part A 42:589–603
55. Barbieri L, Massabo R, Berggreen C (2018) The effects of shear and near tip deformations on interface fracture of symmetric sandwich beams. Eng Fract Mech 201:298–321. https://doi.org/10.1016/j.engfracmech.2018.06.039
56. Farshidi A, Berggreen C, Carlsson LA (2019) Low temperature mixed-mode debond fracture and fatigue characterisation of foam core sandwich. J Sandwich Struct Mater in press:109963621877942. https://doi.org/10.1177/1099636218779420
57. Layne AM, Carlsson LA (2002) Test method for measuring strength of a curved sandwich beam. Exp Mech 42:194–199
58. Layne AM, Carlsson LA (2002) Flexural strength of curved sandwich beams with face/core debond. J Sandw Struct Mater 4:203–217
59. Aviles F, Carlsson LA (2005) Elastic foundation analysis of local face buckling in debonded sandwich columns. Mech Mater 37:1026–1034
60. Moslemian R, Berggreen C, Carlsson LA, Avilés F (2009) Failure investigation of debonded sandwich columns: an experimental and numerical study. J Mech Mater Struct 4 (7–8):1469–1487
61. Nøkkentved A, Lundsgaard-Larsen C, Berggreen C (2005) Non-uniform compressive strength of debonded sandwich panels – I. Experimental investigation. J Sandw Struct Mater 7 (6):461–448

62. Moslemian R, Quispitupa A, Berggreen C, Hayman B (2012) Failure of uniformly compression loaded debond damagedsandwich panels - an experimental and numerical study. J Sandw Struct Mater 14(3):297–324
63. Aviles F, Carlsson LA (2006) Experimental study of debonded sandwich panels loaded in compression. J Sandw Struct Mater 8:7–30
64. Aviles F, Carlsson LA (2005) Face sheet buckling of debonded sandwich panels using a 2D elastic foundation approach. Mech Adv Mater Struct 12(5):349–361
65. Aviles F, Carlsson LA (2006) Three-dimensional finite element buckling analysis of debonded sandwich panels. J Compos Mater 40(11):993–1008
66. Aviles F, Carlsson LA (2007) Post-buckling and debond propagation in sandwich panels subject to in-plane loading. Eng Fract Mech 74:794–806
67. Jolma P, Segercrantz S, Berggreen C (2007) Ultimate failure of debond damaged sandwich panels loaded with lateral pressure – an experimental and fracture mechanical study. J Sandw Struct Mater 9(2):167–196
68. Moslemian R, Berggreen C (2013) Interface fatigue crack propagation in sandwich X-joints – Part I: Experiments. J Sandw Struct Mater 15(4):1–22
69. Moslemian R, Berggreen C (2013) Interface fatigue crack propagation in sandwich X-joints – Part II: Finite element modeling. J Sandw Struct Mater 15(4):23–36
70. Lundsgaard-Larsen C, Berggreen C, Carlsson LA (2010) Tailoring sandwich face/core interfaces for improved damage tolerance - Part I: Finite element analysis. Appl Compos Mater 17 (6):609–619
71. Lundsgaard-Larsen C, Berggreen C, Carlsson LA (2010) Tailoring sandwich face/core interfaces for improved damage tolerance - Part II: Experiments. Appl Compos Mater 17 (6):621–637
72. Martakos G, Andreasen JH, Berggreen C, Thomsen OT (2017) Interfacial crack arrest in sandwich panels with embedded crack stoppers subjected to fatigue loading. Appl Compos Mater 24:55–76. https://doi.org/10.1007/s10443-016-9514-3
73. Martakos G, Andreasen JH, Berggreen C, Thomsen OT (2019) Experimental investigation of interfacial crack arrest in sandwich beams subjected to fatigue loading using a novel crack arresting device. J Sandw Struct Mater 21(2):401–421. https://doi.org/10.1177/1099636217695057
74. Martakos G, Andreasen JH, Berggreen C, Thomsen OT (2019) Interfacial crack arrest in sandwich beams subjected to fatigue loading using a novel crack arresting device - numerical modelling. J Sandw Struct Mater 21(2):422–438. https://doi.org/10.1177/1099636217695058
75. Hayman B, Berggreen C, Petterson R (2007) Tests and analyses of FRP sandwich structures with face sheet wrinkles. J Sandw Struct Mater 9(4):377–404
76. Hayman B, Echtermeyer AT (2019) Reduction of strength of GFRP sandwich panels in naval ships by face sheet holes, cracks and impact damage. J Sandw Struct Mater 21:1621–1653. https://doi.org/10.1177/1099636219836357
77. Hayman B, Berggreen C (2018) Damage assessment schemes for naval sandwich structures with face-core debonds considering residual strength and fatigue life. In: 12th International Conference on Sandwich Structures, Lausanne, Switzerland, pp 32–34
78. Farshidi A, Berggreen C (2019) Analysis of disbonded aircraft sandwich panels with unvented honeycomb cores. Compos Part B, submitted 2019

Modeling Nonlinear and Time-Dependent Behaviors of Polymeric Sandwich Composites at Various Environmental Conditions

Bentolhoda Davoodi, Antonio Gomez, Brian Pinto, Anastasia Muliana, and Valeria La Saponara

1 Introduction

Sandwich structures consist of two thin facesheets (skins) with high stiffness and strength that are bonded to a relatively thick core. This sandwich system forms a lightweight structure with high strength and stiffness, and predominantly loaded under bending and/or twisting. The facesheets are usually made of metals or fiber-reinforced composite laminates, and the core is typically made of balsa wood, foams or honeycombs of polymeric or metallic material. The skins carry the tensile and compressive loads, and the core sustains the shear loads and holds the skins in positions away from the neutral axis of the structure, which maximizes the flexural stiffness of the structure [1]. Foams are among the most efficient core materials for weight-saving applications. The core and skin materials can be altered during manufacturing, which give sandwich constructions flexibility in design for various deflection requirements, bending stiffness and impact resistance requirements (see for example [2–4]).

Sandwich structures are widely used in many engineering applications, such as aircraft, naval structures and wind turbine blades. In wind turbine blades, sandwich composites with polymeric foam or honeycomb core and fiber-reinforced polymer (FRP) skins are a promising solution to obtain sufficiently lightweight blades with high bending stiffness and strength. In naval structures, sandwich composites with

B. Davoodi · A. Muliana (✉)
Department of Mechanical Engineering, Texas A&M University, College town, TX, USA
e-mail: amuliana@tamu.edu

A. Gomez · B. Pinto · V. La Saponara
Department of Mechanical and Aerospace Engineering, University of California, Davis, CA, USA
e-mail: vlasaponara@ucdavis.edu

S. W. Lee (ed.), *Advances in Thick Section Composite and Sandwich Structures*,
https://doi.org/10.1007/978-3-030-31065-3_16

foam core and fiber-reinforced composite skins are used to create a light, corrosion-resistant and stiff structure. In these applications, sandwich constructions are often subjected to various environmental conditions, which eventually alter their mechanical properties. Siriruk et al. [3, 4] studied the degradation in the mechanical behavior due to exposure to sea water in polymeric sandwich composites. They also showed that fracture toughness decreases after sea water exposure. Kolat et al. [5] studied the influence of core material selection and environmental conditions on the fracture toughness of sandwich structures with sea water conditioning, reporting that fracture toughness of systems with polyurethane and coremat core increases, while it decreases for wood and plywood core systems. Joshi et al. [6] analyzed the influence of moisture diffusion on viscoelastic sandwich composites deformation, and assumed the elastic and time-dependent properties of the foam core to depend on the moisture concentration. They conducted coupled analyses of moisture diffusion and deformation to predict the viscoelastic sandwich systems performance.

Due to the presence of polymers in sandwich composites, these structures show quite pronounced viscoelastic response, which can influence their overall life performance. Du et al. [7] determined the creep properties of sandwich composites, and showed that higher relative humidity produced a significant acceleration in creep strain. Shenoi et al. [8] studied creep for a typical sandwich structure using a Burger model and a power law model, and compared the model with experimental results. Garrido et al. [9] conducted experimental and analytical studies on the creep behavior of sandwich composite to propose a creep model to simulate the sandwich panel's creep deformations in long-term.

Scudamore and Cantwell [10] showed that long term sea water exposure caused degradation of the bond between the epoxy matrix and the aluminum core, and led to skin-core interface cracks. Li and Weitsman [11] also characterized the fracture toughness of dry and wet foams (after exposure to sea water), and showed that absorption of sea water increases the toughness of the foam materials due to the softening of the wet polymeric foam, as the glass transition temperature was reduced. Ishai et al. [12] and Ishiaku et al. [13] showed the significant strength reduction of foams and polymeric sandwich composite due to moisture absorption. Belingardi et al. [14] characterized polymeric sandwich composites properties using a series of static/quasi-static tests and showed the dependency of the structural response of the sandwich to the foam core strength properties. Kim et al. [15] and Jeon et al. [16] investigated a multi-scale experimental and modeling approaches on the time dependent response of different constituents of sandwich composite. They characterized how their interaction impacts the overall creep performance of smart polymer sandwich constructions.

In the current study, we conduct experiments to investigate the mechanical behaviors of polymeric sandwich composites and their constituents, with the goal of understanding the effect of the constituents' response on the overall performance of sandwich composites. The sandwich composites and their constituents are also subjected to different environmental conditions (dry and wet due to immersion in fluid, at the temperature of 50 °C). See discussion in Fan et al. [17–19]. We also model the mechanical response of sandwich composites and their constituents. A previously developed nonlinear viscoelastic material model is used to simulate bending test for foam beams. The constitutive model is implemented in a user subroutine (UMAT) in ABAQUS FE analysis. FE analysis is then used to describe mechanical responses of the sandwich composites and their constituents.

2 Experiments

Tests were conducted on materials that are known to be time-dependent: polymeric foams and sandwich composites made with those foams. In particular, there were two sandwich composite systems, which were manufactured using out-of-autoclave Vacuum Assisted Resin Transfer Molding. The layups were selected as $[90]_4$, to ensure a higher viscoelastic response of the material. The first system is glass fiber-reinforced polymer (GFRP) having E-glass fibers (quasi-unidirectional Vectorply E-LR 0908) and epoxy (Pro-set LAM 125/LAM237, with an infusion ratio of 100:28 resin:hardener by weight ratio for the skins), and polyurethane (PU) foam, (General Plastics FR-3704, nominal thickness 19.1 mm) for the core. We refer to this system as "GFRP/PU". The resin cure cycle consisted of 14 h at room temperature, followed by 8 h at 82 °C in a convection oven. This system was studied as an example of a sandwich for wind turbine applications. The second system is a sandwich with skins made of carbon fibers (quasi unidirectional Torayca T700S) and vinylester (Hetron FR 992, with additives cobalt naphthenate and methyl ethyl ketone peroxide, with an infusion ratio of 100:0.15:1.25 by volume), and a polyvinylchloride (PVC) foam core (DIAB Divinycell H100, nominal thickness 25.5 mm). We refer to this system herein as "CFRP/PVC". The vinylester had a cure cycle of 45 min at room temperature and 4 h at 82 °C. This system was considered for naval application.

The two sandwich systems were subjected to two different conditions: the GFRP/PU sandwich and its constituents (GFRP skins and PU) were immersed in deionized water at 50 °C (herein named as "Condition 1"), while the CFRP/PVC sandwich and its constituents (CFRP skins and PVC foam) were immersed in artificial sea water at 50 °C (herein named as "Condition 2"). The expression "artificial sea water" stands for a solution of deionized water and 3.5% wt. content of NaCl. Table 1 provides the immersion testing conditions for the samples, with different durations dictated by experimental constraints. "Condition 1" and "Condition 2" are meant to simplify the discussion in this paper, with the understanding that each condition includes a time range for skins, foam and sandwich samples.

Table 1 Immersion testing conditions

Temperature	Immersion fluid	Sample	Duration (days)
50 °C	Deionized water	GFRP skins, PU foam	150
50 °C	Deionized water	GFRP/PU sandwich	188
50 °C	Artificial Sea water	CFRP skins, PVC foam	107
50 °C	Artificial Sea water	CFRP/PVC sandwich	53

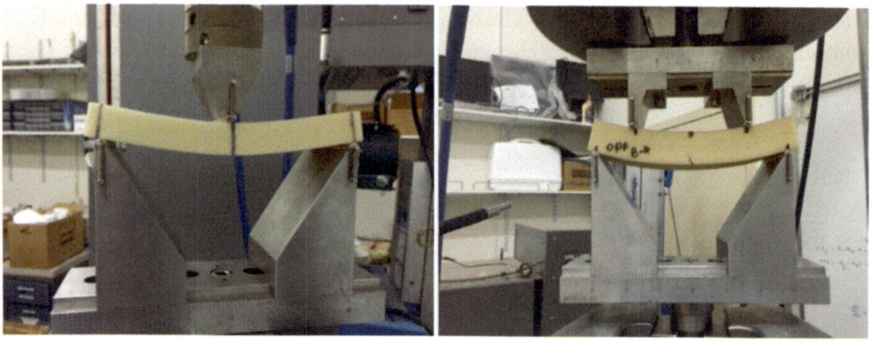

Fig. 1 (Left) three-point bending test on PU foam; (right) four-point bending test on PVC foam

Baseline and conditioned sandwich composites and their constituents were subjected to several types of mechanical tests on a hydraulic axial machine (MTS 810). Typically, bending tests (quasi-static, creep and stress relaxation) were conducted on foam and sandwich samples, while quasi-static axial tests were run for monolithic FRP samples, following the requirements of the appropriate ASTM standard. The selection between creep and stress relaxation was based on the load capacity of the sample under exam. The bending test setups are shown in Fig. 1. The bending tests consist of three- and four-point bending tests, depending on the top part of the fixture (wedge with a roller, or fixture with two rollers). In some cases, the wedge became unavailable, leading to a number of four-point bending tests on conditioned samples, while baseline samples had been tested under three-point bending. However, the model presented in this paper is capable of capturing the different bending conditions, and the softening of the samples as a result of conditioning is successfully computed.

3 Constitutive Material Models for the Constituents

3.1 Nonlinear Viscoelastic Model for Foam

The nonlinear viscoelastic model for the polymeric foams is based on the quasi-linear viscoelastic (QLV) model of Muliana et al. [20]. The QLV constitutive model was originally proposed by Fung [21]. The polymers are assumed to be isotropic and homogeneous. From the experimental observations, the magnitude of strain is relatively small, thus engineering stress and strain measures are considered in the model.

The three-dimensional (3D) QLV model for isotropic materials is shown in Eq. (1):

$$\sigma_{\mathrm{ij}}(\mathrm{t}) = \int_{0^-}^{\mathrm{t}} 2G(t-s)\frac{\mathrm{d}F_{\mathrm{ij}}}{\mathrm{d}s}\mathrm{d}s + \delta_{\mathrm{ij}}\int_{0^-}^{\mathrm{t}} \lambda(t-s)\frac{\mathrm{d}F_{\mathrm{kk}}}{\mathrm{d}s}\mathrm{d}s \tag{1}$$

where δ_{ij} is the Kronecker delta and λ is a material parameter, one of the two Lame's constants, defined as:

$$\lambda = K_B - (2/3)G \tag{2}$$

where G and K_B are the shear modulus and bulk modulus, respectively. These material parameters depend on time. The corresponding Poisson's ratio, ν_0, is assumed to be constant, and this leads to:

$$G(t) = \frac{E(t)}{2(1 + \nu_0)} \tag{3}$$

$$K_B(t) = \frac{E(t)}{3(1 - 2\nu_0)} \tag{4}$$

where *E*(*t*) is the extensional relaxation modulus of the polymer, expressed as:

$$E(t) = E_\infty + \sum_{n=1}^{N} E_n e^{\frac{-t}{\tau_{Rn}}} \tag{5}$$

τ_{Rn} is the characteristic of relaxation time, E_n is the coefficient in the time-dependent part, N is the number of terms in the Prony series, and E_∞ is the relaxed modulus. The nonlinear strain measure in a generalized 3D QLV model is defined as:

$$F(\bar{\varepsilon}(t)) = A\left[e^{B\bar{\varepsilon}(t)} - 1\right] \tag{6}$$

$$\bar{\varepsilon}(t) = \sqrt{\varepsilon_{ij}\varepsilon_{ij}} \tag{7}$$

$$\frac{dF_{ij}}{ds} = ABe^{B\bar{\varepsilon}(t)}\frac{d\varepsilon_{ij}}{ds} \tag{8}$$

$$\frac{dF_{kk}}{ds} = ABe^{B\bar{\varepsilon}(t)}\frac{d\varepsilon_{kk}}{ds} \tag{9}$$

With the above nonlinear strain measures and $E_o = AB$, the constitutive relation becomes

$$\sigma_{ij}(t) = 2C_1\int_{0^-}^{t} K(t-s)E_0 e^{B\bar{\varepsilon}}\frac{d\varepsilon_{ij}}{ds}ds + \delta_{ij}C_2\int_{0^-}^{t} K(t-s)E_0 e^{B\bar{\varepsilon}}\frac{d\varepsilon_{kk}}{ds}ds \tag{10}$$

where

$$C_1 = \frac{1}{2(1+\nu_0)} \tag{11}$$

$$C_2 = \frac{1}{3(1+2\nu_0)} - \frac{1}{3(1+\nu_0)} \tag{12}$$

The normalized time dependent function is

$$K(t) = K_\infty + \sum_{n=1}^{N} K_n e^{\frac{-1}{\tau_{Rn}}} \tag{13}$$

$$K(0) = K_\infty + \sum\nolimits_{n=1}^{N} K_n = 1.0 \tag{14}$$

The above model is implemented in a user material subroutine (UMAT) of ABAQUS FE analyses. The numerical algorithm is discussed in Muliana et al. [20].

3.2 Elastic-Plastic Model for Skins

The elastic-plastic material model [22, 23] is used to simulate the behavior of FRP skins in the sandwich composite. A rate-independent plasticity model is considered in order to capture the elastic-plastic response of the FRP skins. From the experimental observation for mechanical deformation of the FRP skins, the strains are relatively small, which allow for additive decompositions of the elastic and plastic deformations. Thus, the total strain rate is given as:

$$\dot{\boldsymbol{\varepsilon}} = \dot{\boldsymbol{\varepsilon}}^{el} + \dot{\boldsymbol{\varepsilon}}^{pl} \tag{15}$$

where $\dot{\boldsymbol{\varepsilon}}$ is the total (mechanical) strain rate, $\dot{\boldsymbol{\varepsilon}}^{el}$ is the elastic strain rate, and $\dot{\boldsymbol{\varepsilon}}^{pl}$ is the plastic strain rate. Equation (15) is an approximation when the elastic strains are infinitesimal. The rate of deformation tensor is work-conjugate to the Cauchy stress tensor and is used to define the strain rate:

$$\dot{\boldsymbol{\varepsilon}} = \frac{1}{2}\left(\frac{\partial v}{\partial x} + \left(\frac{\partial v}{\partial x}\right)^T\right) \tag{16}$$

The linear elastic isotropic constitutive equation can be written as

$$\boldsymbol{\sigma} = \boldsymbol{C} : \boldsymbol{\varepsilon}^{el} \tag{17}$$

where C denotes the fourth-order elastic stiffness tensor.

The yield function is often expressed in terms of an equivalent stress, i.e. a scalar measure of the magnitude of the Cauchy stress tensor. Von Mises equivalent stress is:

$$\overline{\sigma} = \overline{\sigma}[\boldsymbol{\sigma}] = \sqrt{\frac{3}{2}\boldsymbol{S} : \boldsymbol{S}} \tag{18}$$

where the deviatoric tensor is

$$\boldsymbol{S} = dev[\boldsymbol{\sigma}] = \boldsymbol{\sigma} - \frac{tr[\boldsymbol{\sigma}]}{3}\boldsymbol{I} \tag{19}$$

Using Von Mises equivalent stress definition, the yield function is written as:

$$f\left[\boldsymbol{\sigma}, \overline{\varepsilon}^{pl}\right] = \overline{\sigma}[\boldsymbol{\sigma}] - k\left[\overline{\varepsilon}^{pl}\right] \tag{20}$$

and the yield surface is

$$f\left[\boldsymbol{\sigma}, \overline{\varepsilon}^{pl}\right] = 0 \tag{21}$$

The $k\left[\overline{\varepsilon}_{pl}\right]$ term shows the isotropic hardening. The hardening parameters are state variables that are introduced to allow the model to describe some of the complexity of the inelastic response of real materials. In perfect plasticity, which is the simplest plasticity model, the yield surface acts as a limit surface and there are no hardening parameters. In this study, from observed experimental tests of FRP skins, an elastic-perfectly plastic deformation model is adopted.

4 Results and Discussion

As mentioned earlier, two systems of sandwich composites (GFRP/PU and CFRP/PVC) and their constituents have been tested. The sandwich composites and foams were tested under bending, while the fiber reinforced polymeric skins were tested under uniaxial tension. The constitutive models discussed above are used to describe the mechanical response of the skins and foams. These models are implemented in finite element (FE); the response of baseline and conditioned sandwich composites under bending is simulated and compared to the experimental results. Quasi-static ramp tests were performed on skins (under tension), foam and sandwich composites (under bending). In addition, creep/relaxation tests under bending were performed for the foam and sandwich composites. The overall goals are: (1) to understand the different response of the constituents in sandwich composites and their effects on the overall mechanical response of sandwich composites; (2) to investigate the effect of environmental conditions on the mechanical response of composites and their constituents.

4.1 Uniaxial Tension Response of Skins

Uniaxial static tests are performed on baseline (dry) GFRP and CFRP skins, and conditioned ("wet", removed from immersion tanks and not dried) specimens (GFRP skins under Condition 1, CFRP skins under Condition 2). Samples had a 25.7 mm nominal width, consistent with ASTM standard for uniaxial testing, and a nominal thickness of 1.5 mm. The experimental results and simulations are shown in Figs. 2, 3, 4 and 5. The calibrated material parameters are given in Table 2. From Table 2, it is seen that, due to degradation caused by the immersion in fluid at 50 °C,

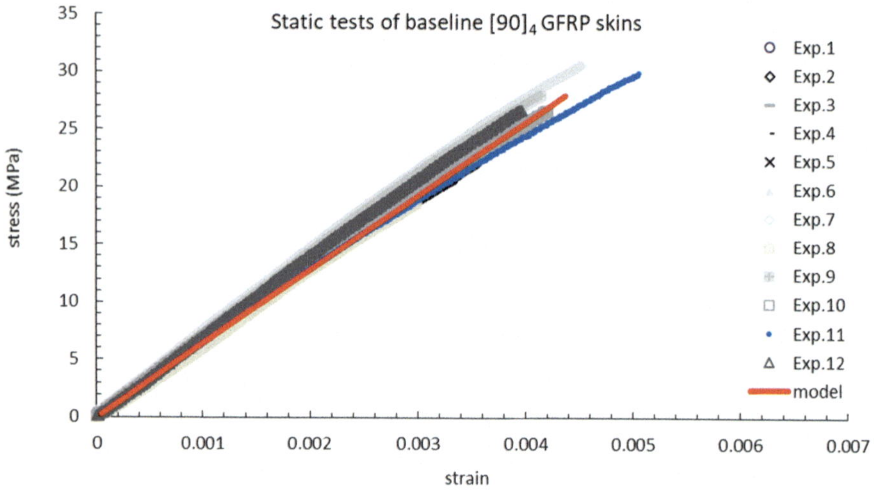

Fig. 2 Static uniaxial tests and model of baseline GFRP skins with $[90]_4$ layup

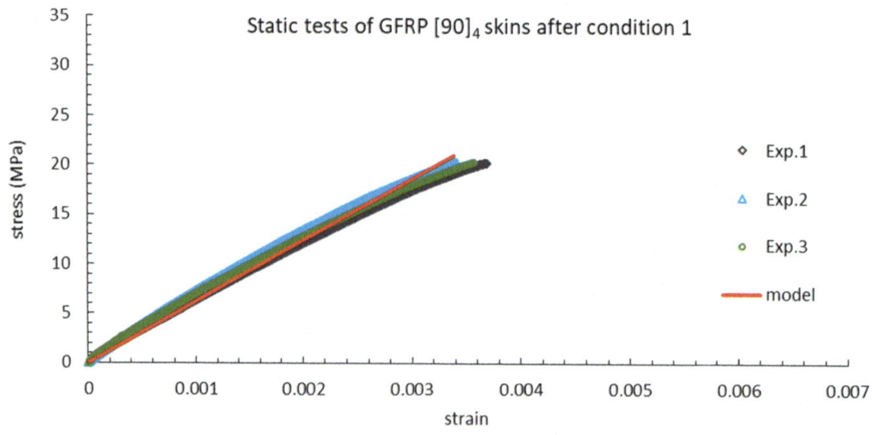

Fig. 3 Static uniaxial tests and model of conditioned GFRP skins with $[90]_4$ layup

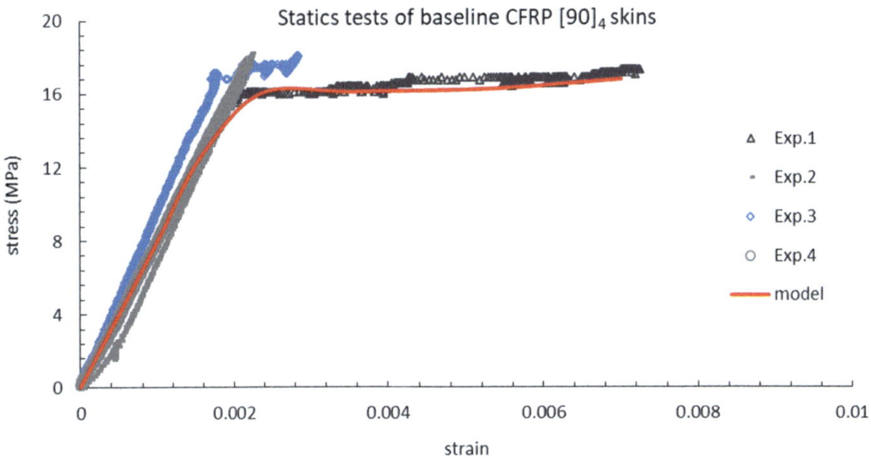

Fig. 4 Static uniaxial tests and model of baseline CFRP skins with $[90]_4$ layup

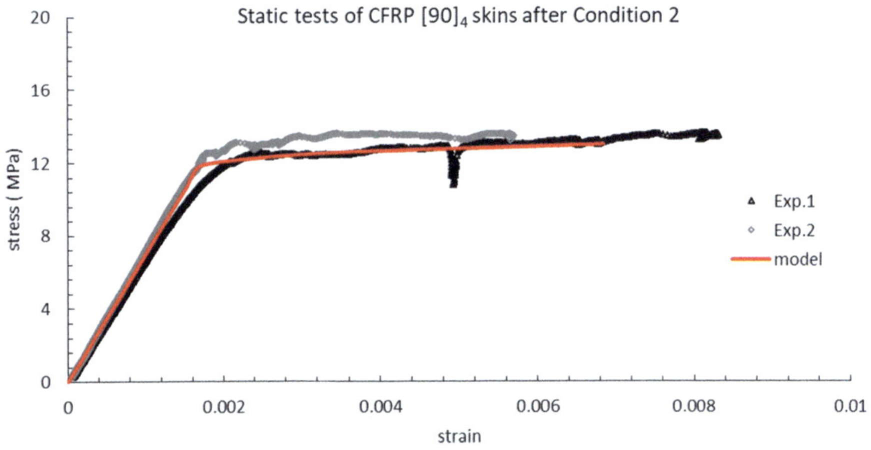

Fig. 5 Static uniaxial tests and model of conditioned CFRP skins with with $[90]_4$ layup

Table 2 Mechanical properties of GFRP and CFRP skins

Material		E (MPa)	σ_y (MPa)
Glass/epoxy	Baseline	6800	–
	After Condition 1	6200	–
Carbon/vinylester	Baseline	8000	16
	After Condition 2	7000	12

the monolithic composites become softer, which is indicated by a reduction in the elastic modulus. The CFRP composite shows inelastic deformations (elastic-plastic behavior) before failure, while the GFRP composite shows a nearly linear elastic response. Fluid immersion also decreases the yield stress in GFRP.

4.2 Bending Tests on Foams

The FE analysis is used to simulate bending in polymeric foams. The nominal dimensions of the foam test specimens are 150 mm length × 57.3 mm width × 19 mm thickness for PU and 150 mm length × 57.2 mm width × 25.5 mm thickness for PVC. Loading was applied consistently with the geometry of the fixture (either at the middle of the beam, for three-point bending tests, or at two loading locations, at 30 mm from beam supports, for the four-point bending tests, Fig. (6). The simply supported mechanical boundary conditions are assumed for both three- and four-point bending models, based on the presence of rollers in the actual fixture. In order to model a simply supported beam, one side of the beam is constrained to prevent displacement in vertical/lateral (y) direction and on the other side of the beam the displacements in axial (x) and lateral (y) directions are restricted. The FE mesh of the beam is generated using the nonlinear three dimensional continuum elements (C3D20) and a convergence study has been performed in order to determine the number of elements required in the analyses.

The load-displacement plots for the foam samples subjected bending tests, and their model for the baseline state and the conditioned state (after immersion) are shown in Figs. 7, 8, 9 and 10. The overall responses of foam under bending are nonlinear. Using the experiments (Figs. 7, 8, 9 and 10), material parameters are calibrated. The material properties are shown in Table 3. The nonlinear constitutive model based on QLV discussed above is used to describe the nonlinear quasi-static ramp response, without incorporating the time-dependent parameters. From Table 3, we can see that by immersion in fluid at elevated temperature, the stiffness of foam samples is reduced.

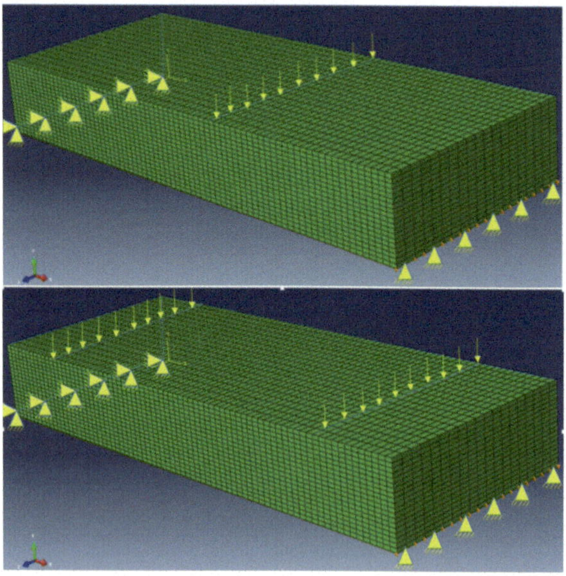

Fig. 6 Boundary conditions of three- and four-point bending tests

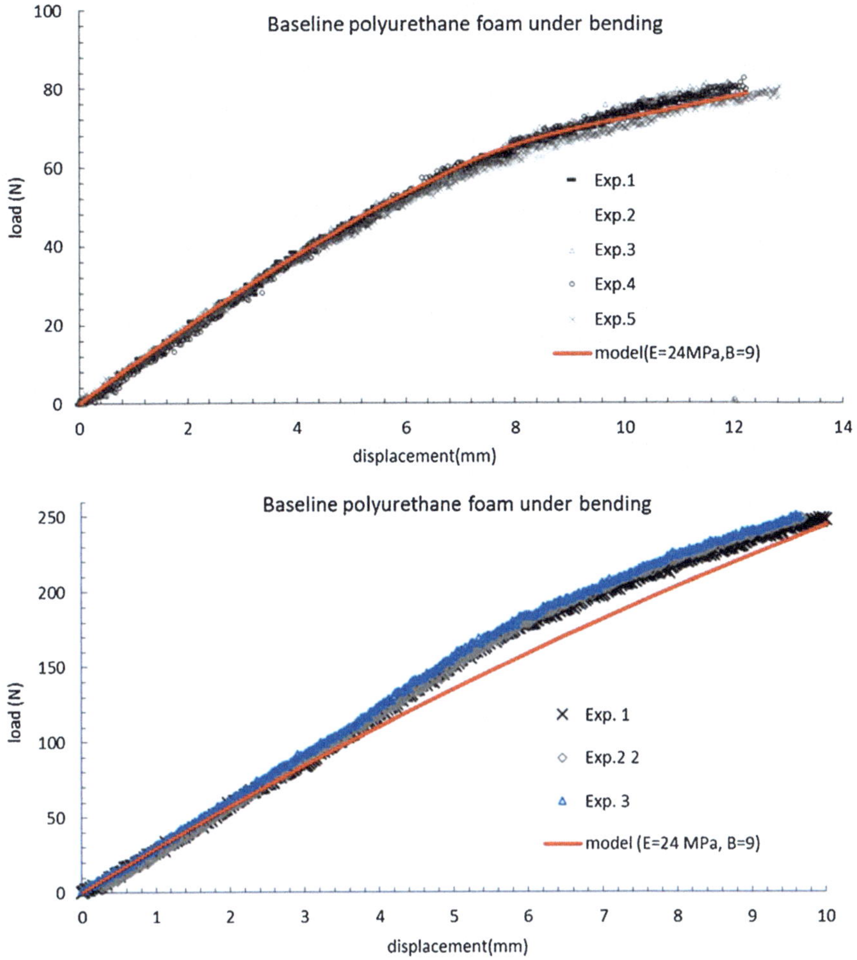

Fig. 7 Static bending tests and model of baseline polyurethane foam under three-point bending (top) and four-point bending (bottom)

4.3 Bending Response of Sandwich Composites

In the previous sections, FE analyses were performed for skins and foam cores. In this part, the response of the sandwich composite under static bending will be modeled. The specimen is a beam consisting of foam core and fiber-reinforced polymer skin, with loading perpendicular to fiber direction. The first system is a GFRP/PU sandwich composite with nominal dimensions $150 \times 26 \times 22$ mm^3 for baseline conditions, and $150 \times 57 \times 22$ mm^3 for immersion conditions, with a skin thickness of 2 mm. The second system is a CFRP/PVC sandwich composite with nominal dimensions $150 \times 26 \times 27.8$ mm^3 for baseline condition, and

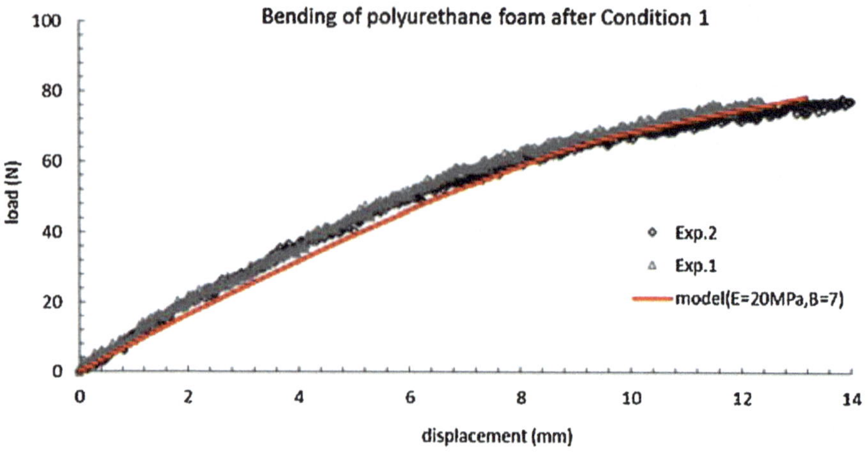

Fig. 8 Static bending tests (3 point bending) and model of conditioned polyurethane foam

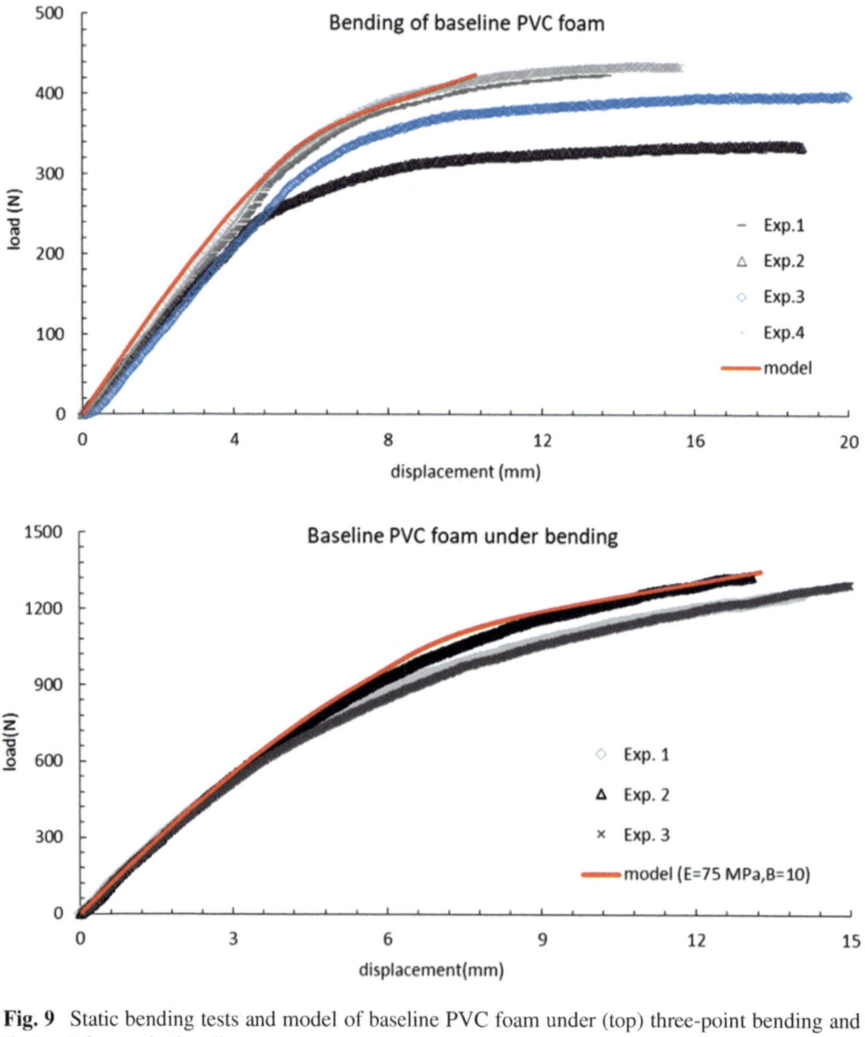

Fig. 9 Static bending tests and model of baseline PVC foam under (top) three-point bending and (bottom) four-point bending

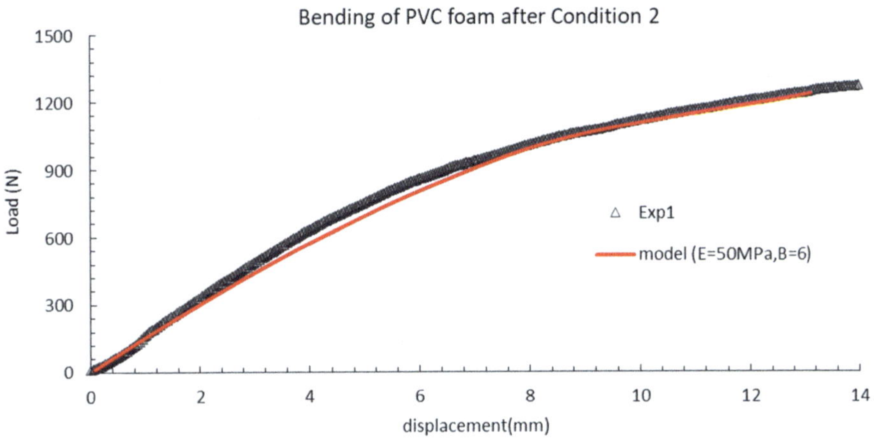

Fig. 10 Static bending tests (four-point bending) and model of conditioned PVC foam

Table 3 Computed material properties for foams

Material		*E* (MPa)	B
PU foam	Baseline	24	9
	After Condition 1	20	7
PVC foam	Baseline	75	7.2
	After Condition 2	50	6

$150 \times 57.327.8$ mm^3 for immersion conditions, with a skin thickness of 1 mm. To simulate the sandwich composite in FE, the foam and skin are separated using partitioning. The nonlinear three-dimensional continuum elements (C3D20) are used, and a convergence study is carried out for the proper number of elements. Note that the width of the condition specimens is about twice the width of the baseline specimens. This difference in width will affect the load magnitude applied to the sandwich composites.

The nonlinear responses of the sandwich composites are shown in Figs. 11, 12, 13 and 14 for the two different systems. In the quasi-static tests for baseline GFRP/PU sandwich composites, there is good agreement between model and experimental results before failure at a deflection of about 4 mm. In the conditioned GFRP/PU sandwich composites, there is also a good agreement between model and experiments before a crack occurs in the sandwich, with a resulting drop in load around a 4 mm displacement. The model of the quasi-static test on baseline CFRP/PVC sandwich composites is in acceptable agreement with the experiments. The result of experimental tests on conditioned CFRP/PVC sandwiches shows that a delamination occurs after loading. By taking into account the delamination around 350 N loading, the model captures the experimental behavior. In the GFRP/PU sandwich composites, failure is observed mostly due to foam cracking, while in the CFRP/PVC sandwich composites failure is governed by skin-core delamination. The different responses in these two sandwich composites are associated with the responses of the constituents. The PVC foam core shows a much stronger and stiffer

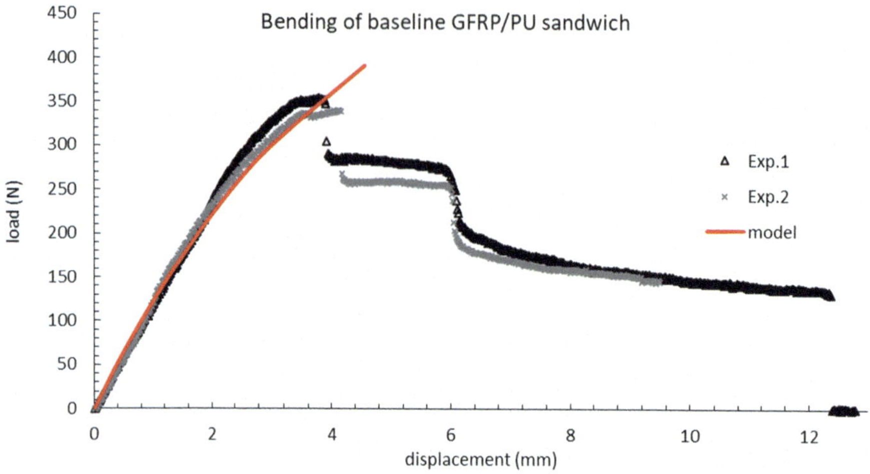

Fig. 11 Static bending tests (three-point bending) and model of baseline GFRP/PU sandwich composites

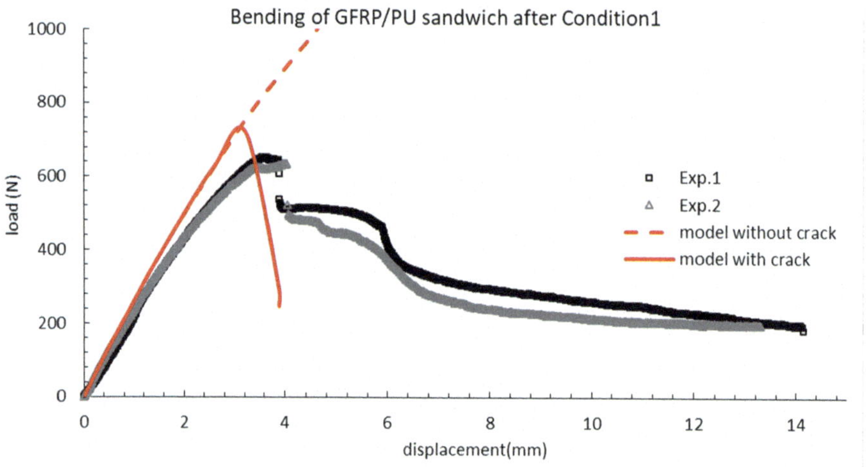

Fig. 12 Static bending tests (three-point bending) and models of conditioned GFRP/PU sandwich composites

response than the PU core, which explains that cracking is unlikely to occur in PVC foam core. The CFRP skins show a ductile behavior; thus, higher deformation in the skins combined with foam core of high stiffness and strength leads to skin-core delamination. The GFRP skins indicate a brittle-like behavior, which explains that failure in GFRP/PU sandwich is due to breaking of the skin on the tension face,

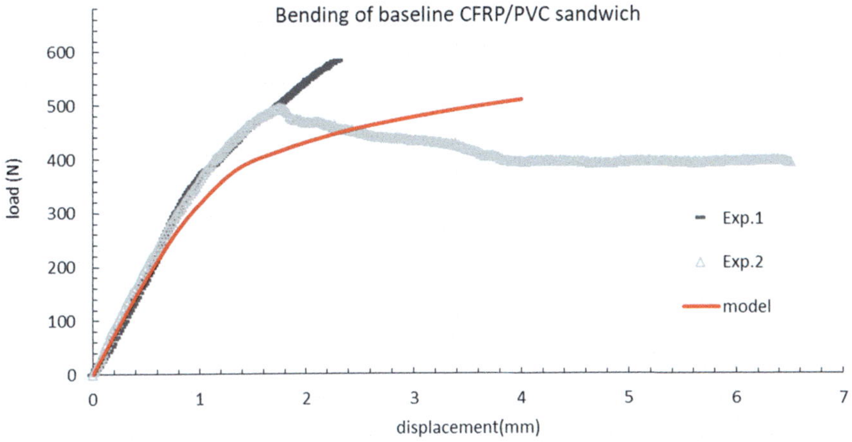

Fig. 13 Static bending tests (three-point bending) and model of baseline CFRP/PVC sandwich composites

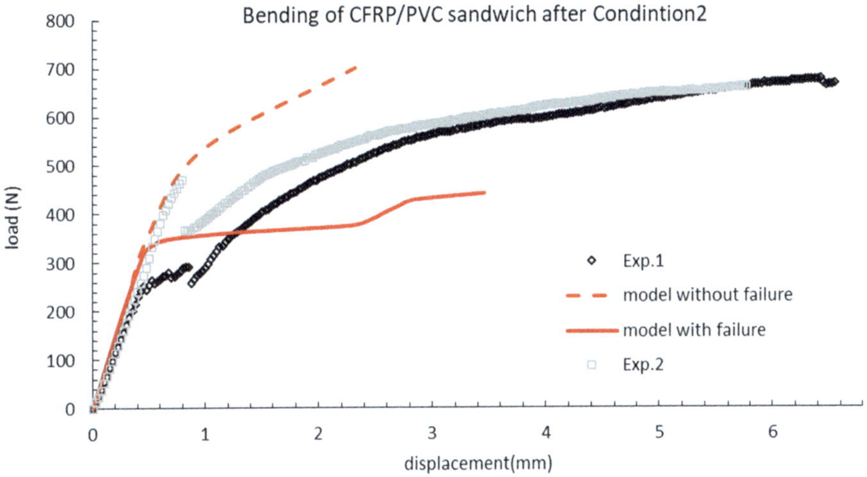

Fig. 14 Static bending tests (three-point bending) and model of baseline CFRP/PVC sandwich composites

followed by cracking of the foam. Figures 15 and 16 illustrate the axial stress and strain contours in the GFRP/PU sandwich composite and CFRP/PVC sandwich composite, respectively, from FE simulation of conditioned sandwich specimens. Due to cracks, the tensile stress in the GFRP skin drops, leading to increase in the axial strain in the foam. Delamination in the CFRP/PVC sandwich composites leads to much higher strain in the foam core.

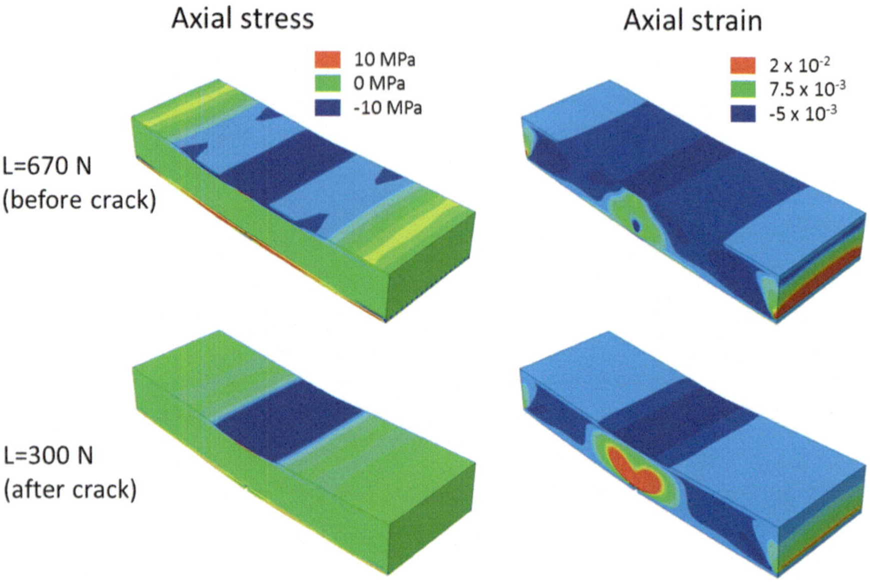

Fig. 15 Axial stress and strain contours in GFRP/PU sandwich composite before and after crack (from simulation)

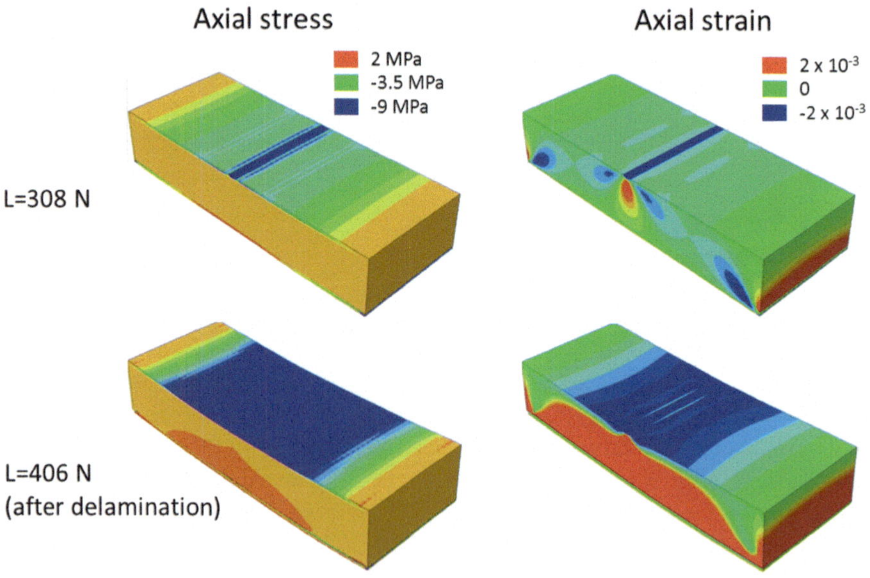

Fig. 16 Axial stress and strain contours in CFRP/PVC sandwich composite before and after delamination (from simulation)

4.4 *Time-Dependent Response of Foams and Sandwich Composites*

To determine the time-dependent properties of the foam core, creep and stress relaxation tests under bending were conducted (Figs. 17, 18, 19 and 20). The calibrated time-dependent parameters for the PU and PVC cores in terms of the normalized time-dependent parameters in Eq. (13) are given in Table 4. It is seen that

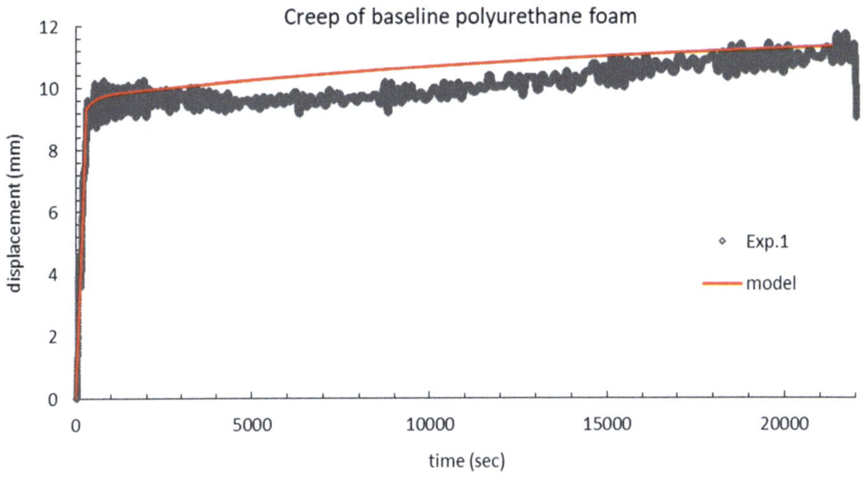

Fig. 17 Creep response under bending (three-point bending) of baseline polyurethane foam

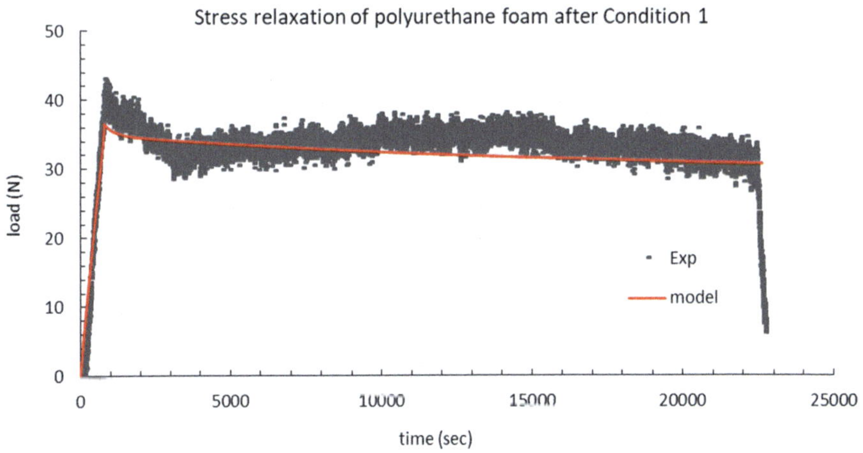

Fig. 18 Stress relaxation response under bending (three-point bending) and model of conditioned polyurethane foam

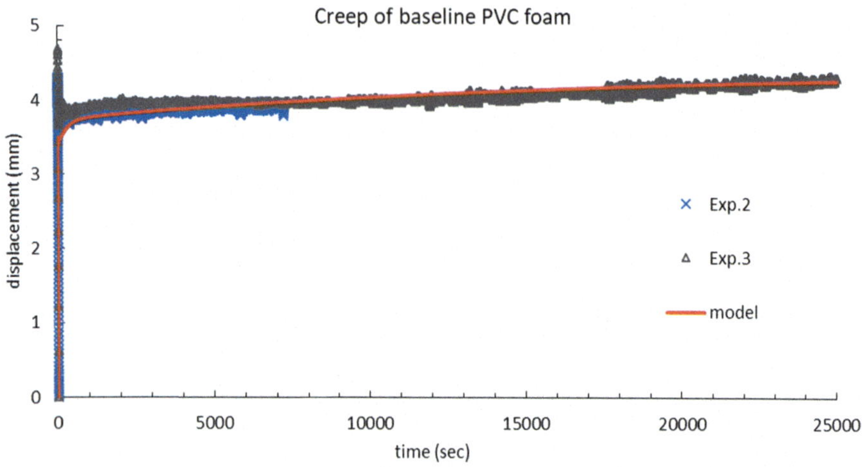

Fig. 19 Creep response under bending (three-point bending) and model of baseline PVC foam

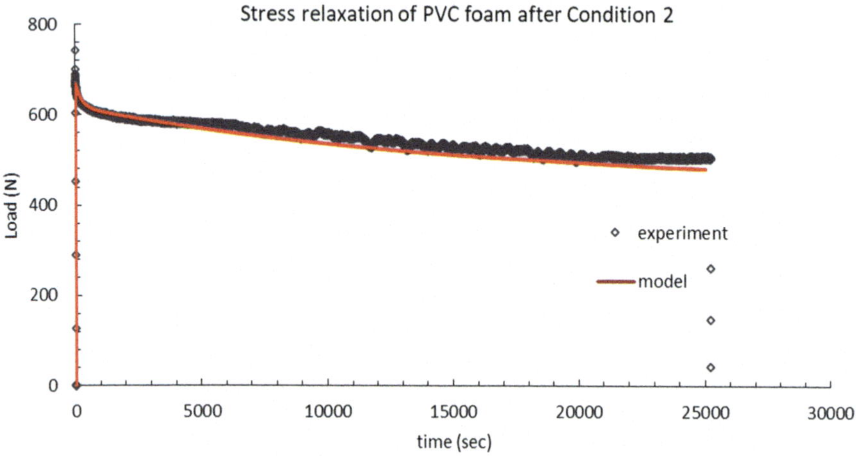

Fig. 20 Stress relaxation under bending (four-point bending) and model of conditioned PVC foam

Table 4 Normalized time-dependent foam properties

Material/Condition	K_∞	K_1	K_2	τ_1(sec)	τ_2(sec)
Baseline PU foam	0.83	0.05	0.12	250	15,000
Conditioned PU foam	0.74	0.1	0.16	350	15,000
Baseline PVC foam	0.75	0.1	0.15	250	15,000
Conditioned PVC foam	0.5	0.12	0.38	250	15,000

the long-term relaxation modulus decreases in both PU and PVC foams upon conditioning in their respective fluids at 50 °C.

FE analyses are used to simulate the creep/relaxation response in sandwich composites, and are compared to experimental results. In this part, we used the material properties in Table 4 for capturing the time-dependent response of sandwich composites. The stress relaxation test and simulation results are shown in Figs. 21 and 22 for the GFRP/PU sandwich composites, and in Figs. 23 and 24 for CFRP/

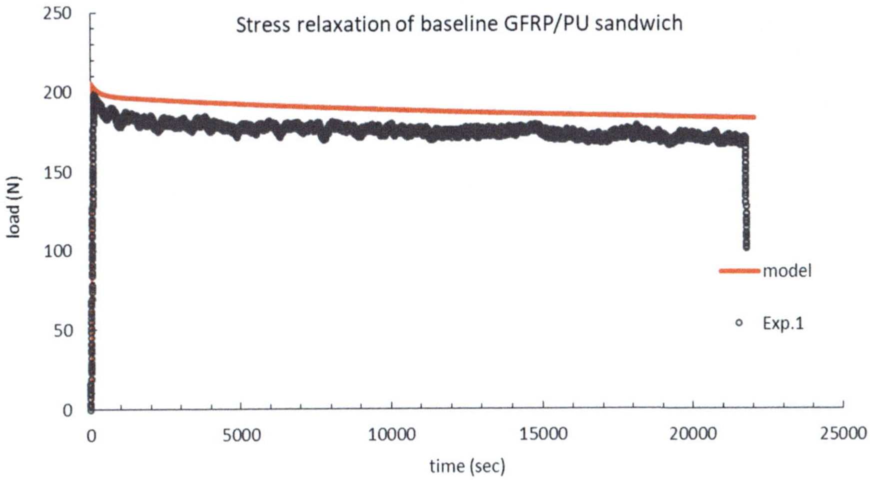

Fig. 21 Stress relaxation under bending tests (three-point bending) and model of baseline GFRP/PU sandwich composites

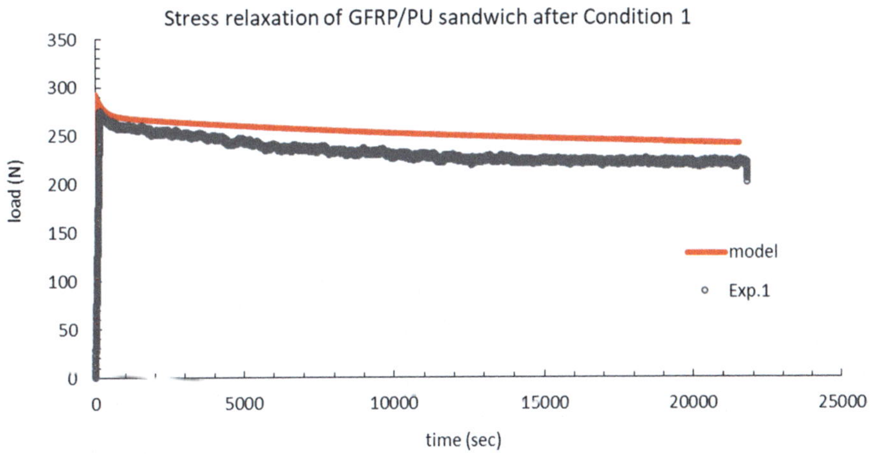

Fig. 22 Stress relaxation under bending tests (three-point bending) and model of conditioned GFRP/PU sandwich composites

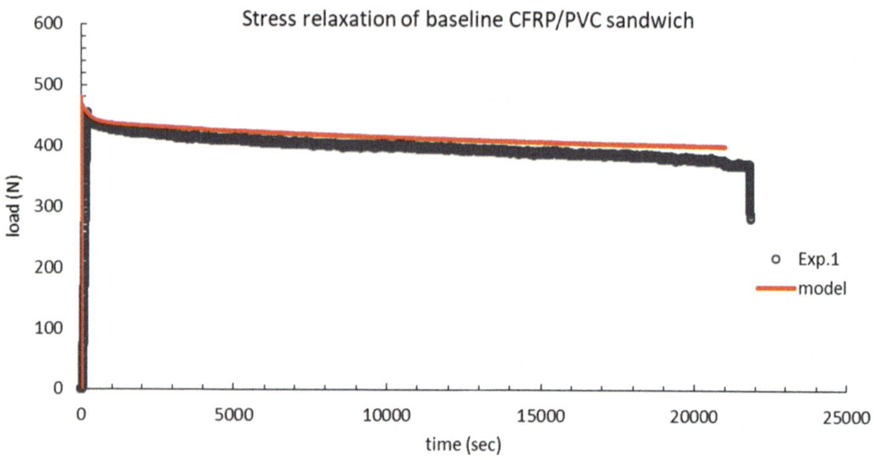

Fig. 23 Stress relaxation under bending tests (three-point bending) and model of baseline CFRP/PVC sandwich composites

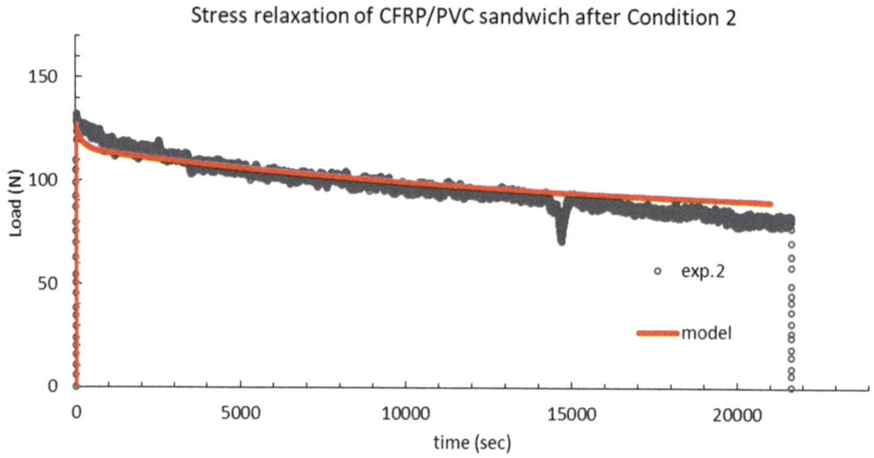

Fig. 24 Stress relaxation under bending tests (three-point bending) and model of conditioned CFRP/PVC sandwich composites

PVC sandwich composites. Stress relaxation tests and simulations for both baseline and conditioned GFRP/PU and CFRP/PVC sandwich composites are in good agreement with each other. The experimental and FE models of sandwich composites show that we can use the material properties of foam and skin to model the behavior of sandwich composite.

5 Conclusions

We have presented a viscoelastic model for predicting the response of sandwich composites subjected to mechanical loading in different environmental conditions, by incorporating different responses of the constituents (skins and foam cores). Experimental tests have been conducted on GFRP and CFRP skins and on PU and PVC foam cores. The purpose of the testing is to calibrate material parameters used in the model. Experimental tests done on sandwich composites are used for validating the model. The model is capable of predicting the behavior of the sandwich composite under mechanical loadings in different environmental conditions, both under quasi-static ramp and creep/relaxation loading conditions.

By investigating the result of the tests and model, we can conclude that the time-dependent response in sandwich composites is mainly due to the viscoelastic response of the foam, while the viscoelastic response of skin is less significant. Also immersion in fluid can alter the mechanical properties of the constituents in sandwich composites, consistent with results in the literature. The mechanical response of baseline sandwich composites can be quite different with respect to that experienced after immersion. Delamination happens in conditioned CFRP/PVC sandwich, while delamination is not seen for the baseline specimen. Furthermore, deformations in the sandwich composites are also governed by the different responses of the constituents (skin and foam core). The GFRP/PU sandwich composites experience foam cracking prior to complete failure, both for the baseline and conditioned samples. As mentioned above CFRP/PVC conditioned sample shows skin-core delamination prior to complete failure. It is shown in this study that the proposed multi-scale model together with a nonlinear viscoelastic constitutive model is capable of describing the overall mechanical response of sandwich composites reasonably well.

Acknowledgments We acknowledge the Texas A&M Supercomputing Facility (http://sc.tamu.edu/) for providing computing resources useful in conducting the research reported in this paper. The authors would like to thank Prof. Patrick Homen for access to selected testing facilities of Sacramento State University, graduate student researchers Steven Kern, Marie Marinkovich, José de Jesús Kú-Herrera and research engineer Serena Ferraro for assistance with the mechanical properties testing of the FRP skins and neat resin, undergraduate student researchers Alyssa "Nikki" Yambao, Robert Pires, Erik Quiroz, Destiny Garcia, Stephanie Zhu, Nicolas Marinkovich, Angelo Magliola, Miguel Duran, Daniel Reid, Chris O′ Keefe, who assisted with the tank manufacturing and measurements. This study is supported by the National Science Foundation (NSF) under grant CMMI-1266037 and Office of Naval Research (ONR) under grant N00014-13-1-0604 (managed by Dr. Y. Rajapakse).

References

1. Sharma N, Gibson RF, Ayorinde EO (2006) Fatigue of foam and honeycomb Core composite Sandwich structures: a tutorial. J Sandw Struct Mater 8:263–319
2. Mouritz AP, Gardiner CP (2002) Compression properties of fire-damage polymer sandwich composites. Compos 33:609–620

3. Siriruk A, Weitsman YJ, Penumadu D (2009) Polymeric foams and sandwich composites: material properties, environmental effects, and shear-lag modeling. Compos Sci Technol 69:814–820
4. Siriruk A, Penumadu D, Weitsman YJ (2009) Effect of sea environment on interfacial delamination behavior of polymeric sandwich structures. Compos Sci Technol 69:821–828
5. Kolat K, Neser IG, Özes C (2007) The effect of sea water exposure on the interfacial fracture of some sandwich systems in marine use. Compos Struct 78:11–17
6. Joshi N, Muliana A (2010) Deformation in viscoelastic sandwich composites subject to moisture diffusion. Compos Struct 92:254–264
7. Du Y, Yan N, Kortschot MT (2013) An experimental study of creep behavior of lightweight natural fiber-reinforced polymer composite/honeycomb core sandwich panels. Compos Struct 106:160–166
8. Shenoi RA, Allen HG, Clark SD (1997) Cyclic creep and creep–fatigue interaction in sandwich beams. J Strain Anal Eng Des 32:1–18
9. Garrido M, Correia JR, Branco FA, Keller T (2014) Creep behaviour of sandwich panels with rigid polyurethane foam core and glass-fibre reinforced polymer faces: experimental tests and analytical modelling. J Compos Mater 48(18):2237–2249
10. Scudamore RJ, Cantwell WJ (2002) The effect of moisture and loading rate on the interfacial fracture properties of Sandwich structures. Polym Compos 23(3):406–417
11. Li X, Weitsman YJ (2004) Sea-water effects on foam-cored composite sandwich lay-ups. Compos Part B 35:451–459
12. Ishai O, Hiel C, Luft M (1995) Long-term hygrothermal effects on damage tolerance of hybrid composite sandwich panels. Compos 26:47–55
13. Degrieck J, Paepegem WV (2001) Fatigue damage modeling of fibre-reinforced composite materials: review. Appl Mech Rev 54(4):279–300
14. Belingardi G, Cavatorta MP, Duella R (2003) Material characterization of a composite–foam sandwich for the front structure of a high speed train. Compos Struct 61:13–25
15. Kim JS, Arronche L, Farrugia A, Muliana A, La Saponara V (2011) Multi-scale modeling of time-dependent response of smart Sandwich constructions. Compos Struct 93:2196–2207
16. Jeon J, Muliana A, La Saponara V (2014) Thermal stress and deformation analyses in Fiber reinforced polymer composites undergoing heat conduction and mechanical loading. Compos Struct 111:31–44
17. Fan Y, Gomez A, Ferraro S, Pinto B, Muliana A, La Saponara V (2017) The effects of temperatures and volumetric expansion on the diffusion of fluids through solid polymers. J Appl Polym Sci 134(31)
18. Fan Y, Gomez A, Ferraro S, Pinto B, Muliana A, La Saponara V (2018) Diffusion of water in glass Fiber reinforced polymer composites at different temperatures. J Compos Mater, in press
19. Fan Y, Gomez A, Muliana A, La Saponara V (2019) Multi-scale analysis of diffusion of fluid in Sandwich composites. J Polym Compos,. in press 40:3520–3532
20. Muliana A, Rajagopal KR (2012) Modeling the response of nonlinear viscoelastic biodegradable polymeric stents. Int J Solids Struct 49:989–1000
21. Fung YC (1981) Biomechanics. In: Mechanical properties of living tissues. Springer, New York
22. Mohr D (2015) "Three dimensional rate-independent plasticity". [Online]. Available: https://www.ethz.ch/
23. "Abaqus Theory anual 4.2.1, Plasticity models: general discussion," [Online]. Available: https://www.sharcnet.ca/Software/Abaqus610/Documentation/docs/v6.10/books/stm/default.htm

Towards More Representative Accelerated Aging of Marine Composites

Peter Davies

1 Introduction

The use of composite materials for critical structural applications at sea is increasing, particularly for energy recovery applications. The required lifetimes are typically from 5 to 25 years, but experience has shown that these may be extended further when end of life approaches. Civil engineering structures such as bridges and coastal infrastructure are designed for much longer lifetimes. The nature of many marine installations, in deep water for offshore oil & gas production, or in highly energetic ocean conditions for renewable marine energy, makes intervention and maintenance costly and dangerous. The end user requires guarantees that either the material properties will be maintained over these periods or the possible degradation is sufficiently well understood to allow quantitative estimation of property changes.

A major challenge for materials laboratories today is therefore to predict with confidence the long term durability of composite materials immersed in seawater and exposed to a range of mechanical loads. In many projects the time available to evaluate a material is short, a few months at best, so accelerated testing is unavoidable. This is not a new subject. Accelerated aging tests have been used in many previous composite studies e.g [1], and particularly for marine structures [2, 3], but it is attracting considerable interest today for a wide range of other applications. Recent examples include the wind turbine [4], civil engineering [5], aerospace [6] and automobile [7] industries. For all these applications there is a need to rapidly quantify the influence of degradation on long term performance. It is therefore essential to examine how this can be achieved and the limitations of the accelerating parameters.

P. Davies (✉)
Marine Structures Laboratory, IFREMER Centre Bretagne, Plouzané, France
e-mail: peter.davies@ifremer.fr

S. W. Lee (ed.), *Advances in Thick Section Composite and Sandwich Structures*,
https://doi.org/10.1007/978-3-030-31065-3_17

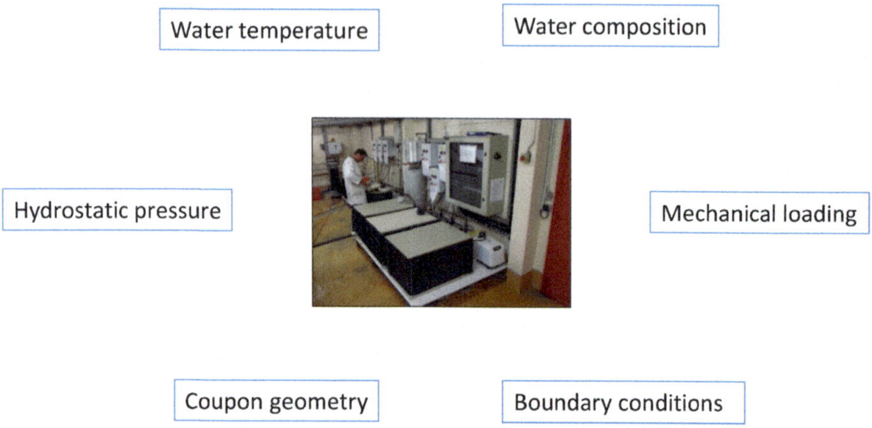

Fig. 1 Seawater aging test facility, showing heated renewed seawater tanks, with parameters affecting aging time

This chapter will first examine the various parameters that can be used to speed up material aging mechanisms, temperature, pressure, aging medium and sample thickness. There are few relevant standard test methods, so published data are based on a wide range of test conditions. Figure 1 shows the main parameters which can be used to reduce aging time. These will be considered in detail in the next section.

Most published work on long term aging is based on data generated over a few months, at most a year. This provides no opportunity to verify the extrapolations made to predict behavior after longer times. However, some samples have been aged in the laboratories at IFREMER for up to 8 years, at different temperatures, and data for these can help to validate (or not) the extrapolation of data from accelerated tests. Two examples will be discussed below, involving glass and carbon reinforced composites.

2 Acceleration by Increasing Temperature

The most widely applied approach to evaluate the influence of long-term exposure of a material to seawater is to increase the water temperature. This approach generally assumes an Arrhenius dependency of the properties of interest, namely the diffusion coefficients, so that by immersion at different temperatures the accelerating factors for water ingress can be determined experimentally [8]. Short immersion diffusion data at low temperature can thus be extended to longer times. This approach, potentially very powerful, poses a number of questions.

First, it must be verified that the Arrhenius approach actually applies to the material under the conditions of interest. Figure 2 shows an example of a set of epoxy resin weight gain plots for an immersion period of 400 days.

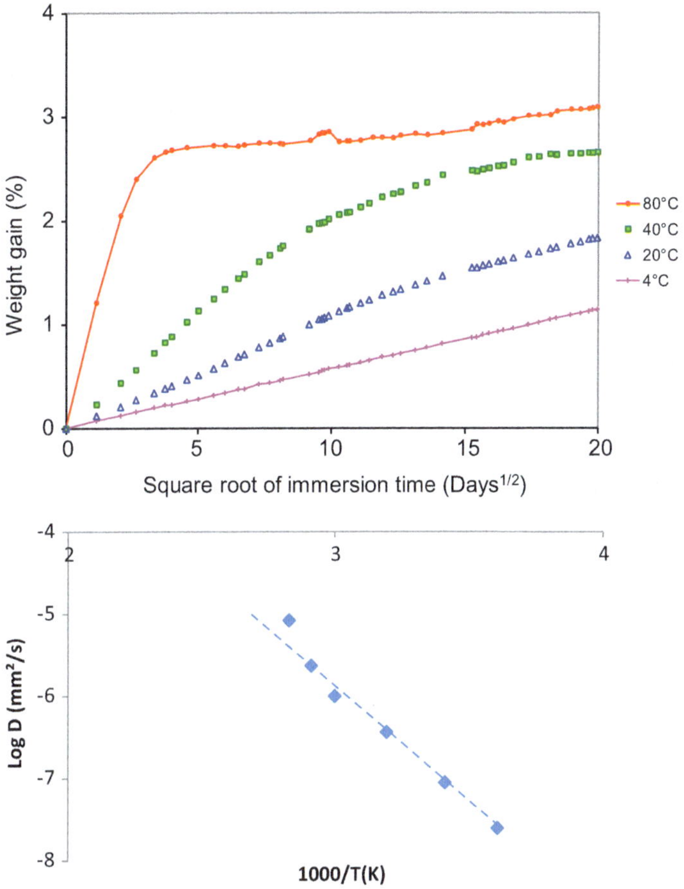

Fig. 2 Immersion of behavior of epoxy resin in sea water at different temperatures Upper: Weight gain versus square root of immersion time, Lower: Arrhenius plot

Figure 2a shows that for 5 mm thick coupons of this amine cured epoxy resin, a grade widely used for wind turbine blade composites, a reasonably stable resin weight gain can be achieved in around 2 weeks at 80 °C, but it requires over a year at 40 °C to reach the same weight gain level. If it is assumed that the weight at saturation is the same at all temperatures (2.75% here) then diffusion coefficients at different temperatures (here 4, 20, 40, 60, 70 and 80 °C) can be estimated and plotted on an Arrhenius plot. A reasonably linear fit can be obtained up to 70 °C, then for the highest aging temperature (80 °C) the diffusion coefficient is higher than expected from extrapolation of aging at lower temperatures. Longer exposure at 80 °C also results in further weight increase above the plateau value. The dry Tg of this resin is around 80 °C, so this increase occurs at a temperature above Tg.

This example already shows that in order to affirm with any confidence that an Arrhenius approach can be applied it is preferable to perform tests at a large number of different temperatures. Celina and colleagues have discussed this point for rubbers in [9]. There is no a priori reason that increasing temperature will not introduce additional diffusion mechanisms so this must be checked.

A second important question is by how much can the temperature be increased? The basic aim is to accelerate diffusion while not changing any other aspects of the material behavior. A prerequisite is therefore that the immersion temperature will not change the physical state of the material, by post-curing for example, so this must also be checked. A preliminary step in any accelerated aging study therefore involves physico-chemical analysis, DSC (differential scanning calorimetry) for example, before selecting the temperature to be used. This provides information on polymer transitions and in particular the glass transition temperature of the dry polymer, Tg. Once this has been identified, a common rule-of-thumb is to work at a maximum temperature of Tg-20 °C. However, this margin is based on the hypothesis that the Tg of most epoxy matrix resins will only drop by a few degrees after prolonged immersion and this may not be justified. Table 1 shows some examples of Tg values for epoxy matrix resins. A high temperature cured epoxy shows a drop in Tg after saturation of 36 °C, while a low temperature cured marine resin shows a drop of around 15 °C.

A more exhaustive study by Wright suggested an empirical relationship between weight gain and drop in Tg with a reduction in Tg of around 20 °C for each percent increase in moisture content [13]. For other materials, the sensitivity of the Tg to water may also be significant. Figure 3 shows the dramatic drop in Tg of polyamide

Table 1 Influence of moisture on Tg of matrix resins

Resin/hardener	Tg dry	Tg at saturation	Weight gain at saturation, %	Reference
Epoxy/Amine 1(HT cure)	170	134	2.5	[10]
Epoxy/Amine 2(LT cure)	75	60	2.8	[11]
Polyamide 6	66	−9	10	[12]

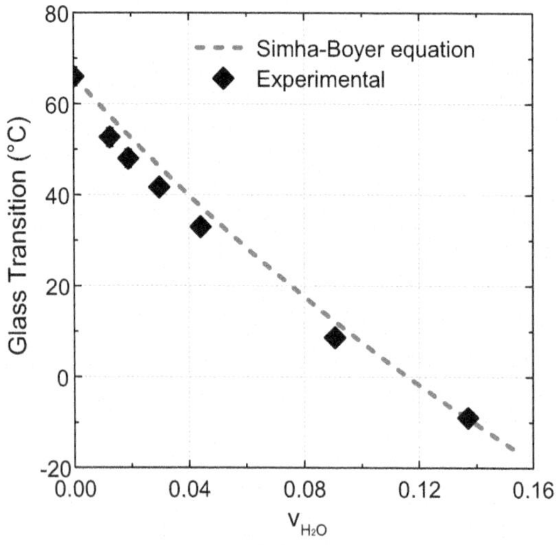

Fig. 3 Influence of Tg of polyamide 6 on water content [12]

6 with water content, and it is clear that a test on this polymer which starts with the test temperature below Tg will become a test above Tg as the material state changes with time. This has been addressed in [12], where different diffusion models were needed to simulate diffusion behavior above and below Tg.

If the Arrhenius approach is valid and allows saturation of the material in a reduced time this can enable subsequent mechanical tests to be performed on specimens with a homogeneous water profile; this is the test protocol which many test laboratories use today to evaluate the durability of materials for marine applications. While this is certainly a useful first step, and allows a first material selection to be made, it gives no information on the lifetime of the material. The definition of end-of-life will depend on the application, but if resin molecular chain breakage (hydrolysis) is possible, which will cause a significant loss in mechanical performance, then tests must be sufficiently long either to provoke this mechanism or to show that it will not occur in the period of interest. This is discussed in more detail in Sect. 8 below.

3 Acceleration by Applied Mechanical Loads (Stress-Diffusion Coupling)

While temperature is widely used to accelerate diffusion, the previous section indicates that there are a number of potential difficulties. It is therefore tempting to examine whether other factors can be used, and in particular mechanical loading. The simplest approach is to apply a tensile stress to a specimen in water, but this requires a dedicated test bench, such as the one shown in Fig. 4. Early studies on epoxy resins by Fahmy and Hunt [14] in four point flexure indicated that tension increases the diffusion coefficient while compression reduces it. They explained this in terms of free volume changes but even at the highest stress level, the difference was only a few percent.

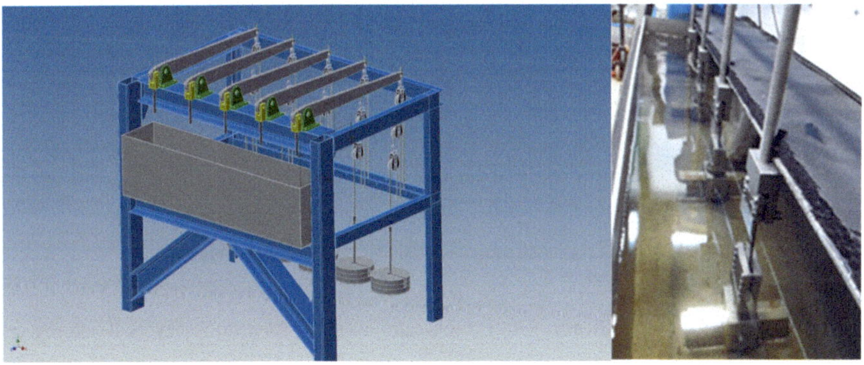

Fig. 4 Test set up to examine how applied tensile stress affects water ingress, and specimens in place

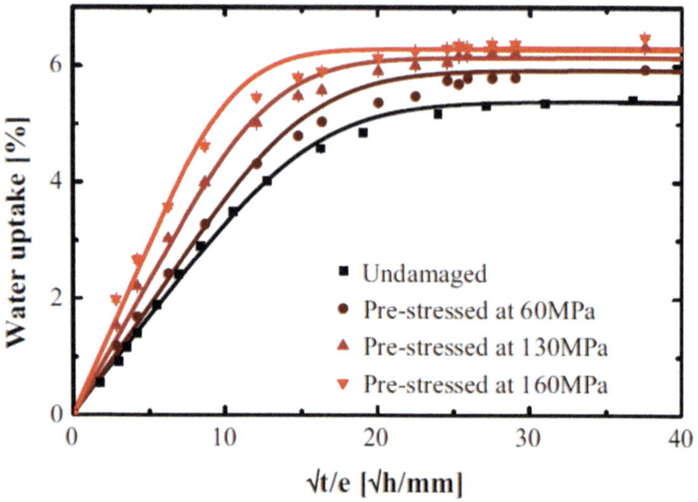

Fig. 5 Influence of different tensile preload stresses on weight gain in water at 60 °C of ±45° specimen

To examine this effect for composites it is important to define load levels with respect to damage initiation. Low applied loads, which remain in the elastic response region of the polymer and below the damage threshold, will generally have a small influence on diffusion. Higher loads applied to composites will introduce damage, in particular resin cracking in off-axis plies, which can strongly affect water entry. Several authors have examined these effects [15–17].

A recent study has used the set-up shown in Fig. 4 to examine the influence of both pre-loading (applying a load before immersion but no load applied during aging in water), and loading during aging in water, on weight gain. Figure 5 shows that there is a small effect of preloading on weight increase. The matrix resin of a ± 45° specimen immersed in water will pick up about 5% by weight at saturation, and this may increase to around 6% if the specimen has been preloaded to a stress close to its break load [18], Fig. 5.

A more significant effect is noted when similar specimens are loaded in water at 60 °C, Fig. 6, In this case weight gain may reach 9% by weight and some specimens fail prematurely.

4 Acceleration by Increasing Hydrostatic Pressure

An alternative approach to accelerate the water ingress mechanically is to increase the hydrostatic pressure during immersion, in order to try to "push" the water into the material more quickly. Pressure increases with immersion depth, so data on pressure effects are also of direct interest for underwater applications of composites. Various authors have examined their influence [19–21].

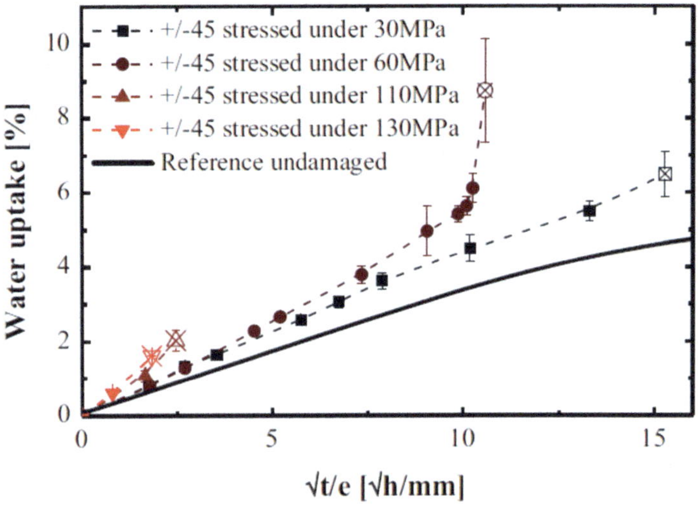

Fig. 6 Example of the influence of applied load on diffusion kinetics for ±45° specimen in water at 60 °C

Fig. 7 Pressure vessel to examine influence of pressure on weight gain. 100 MPa capacity vessels placed in temperature-controlled ovens, together with immersed reference coupons without pressure

Two effects on water diffusion might be envisaged; the first, pressure to accelerate water entry, acts against the second, a polymer volume reduction due to pressure which will reduce available volume for water to fill. However, very high pressures are needed to produce significant volume reductions. Figure 7 shows test equipment developed to perform such tests.

This facility has been used to study a number of composites, some results have been published [22, 23]. While in one particular case, in the presence of high void content, a strong influence of pressure was noted; in all other cases the effect was quite small. Figure 8 shows two examples of results [23], for weight gain at pressures up to 50 MPa.

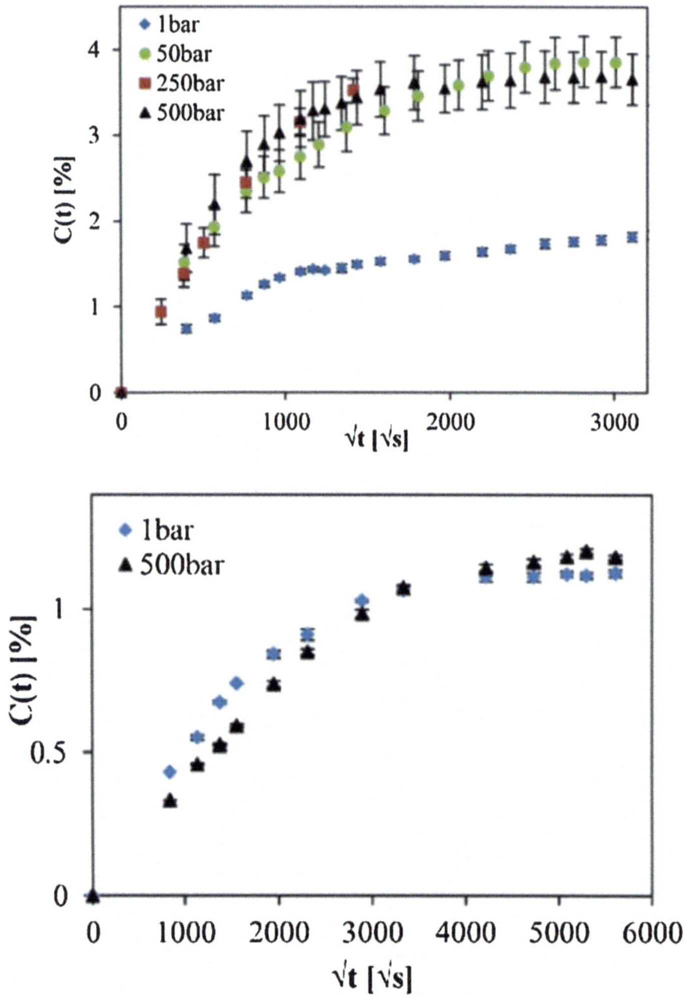

Fig. 8 Influence of hydrostatic pressure on weight gain for two glass/epoxy composites. Upper: Hand lay-up, Lower: Infused

These results indicate that it is unlikely that hydrostatic pressure can be very helpful in accelerating water entry into well-made composites.

5 Acceleration by Modifying Coupon Geometry

Another approach is to optimize the coupon geometry, and three modifications may help to saturate composites more quickly. The first, and most effective, is to reduce the material thickness. The formulation of Fick's law indicates that weight gain

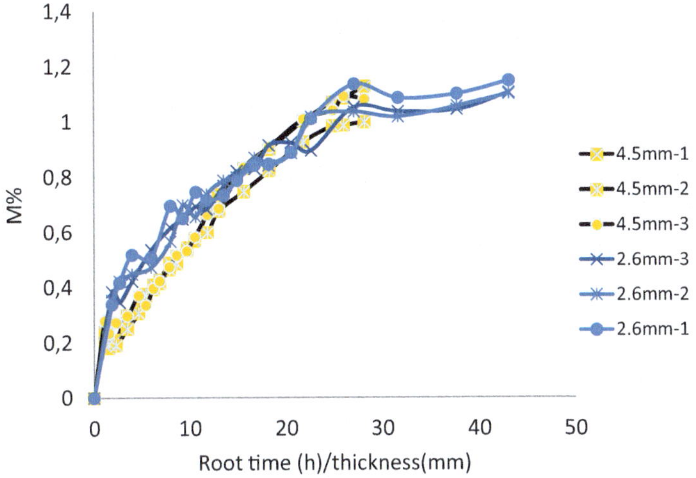

Fig. 9 Influence of coupon thickness on weight gain for carbon/epoxy composites, immersion 18 months in sea water at 40 °C. Three coupons for each thickness

depends in a linear manner on the inverse square root of sample thickness. Thus if the material is indeed Fickian it should be possible to reach saturation in a quarter of the time by halving sample thickness. Wang et al. suggested a similar approach recently for epoxy resin [24]. This is an attractive way to reduce aging time, although for composites there is a practical limit to the extent to which thickness can be reduced without changing the material microstructure. Thin composites tend to have higher fibre volume fractions than thicker materials, and thermal conditions during cure may vary with thickness. The strong relationships between manufacturing, structure and properties in composite materials must also be considered.

Figure 9 shows weight gain plots for quasi-unidirectional carbon/epoxy coupons of 2.6 and 4.5 mm thicknesses, produced by infusion, immersed in seawater at 40 °C. When the data are normalized with respect to coupon thickness the curves superpose quite closely.

A second possibility is to reduce the size of the coupon, encouraging water to enter through the specimen edges. Early work by Springer and colleagues discussed the influence of specimen geometry on diffusion measurements, and a simple expression was proposed for edge effects for isotropic materials [23]:

$$D = D_x\left(1 + \frac{h}{l} + \frac{h}{w}\right)2 \tag{1}$$

where D is the measured value from the initial slope of the weight gain versus square root of time plot, D_x is the through-thickness diffusion coefficient, h is the coupon thickness and l and w are the length and width. It should be noted that these corrections are not negligible; for a 50 mm × 50 mm coupon they indicate an over-estimation of D_x by 8% for a 1 mm thick resin sample and by 25% for 3 mm.

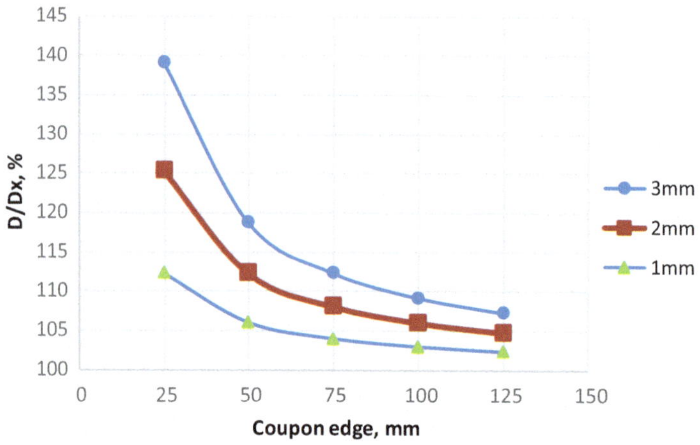

Fig. 10 Effect of square coupon size on water ingress rate for different coupon thicknesses, taking $D_y = 4\ D_x$ and $D_z = D_x$ from [26]

Reducing coupon size may still be helpful to reach saturation of a thick resin sample more quickly, but for fiber reinforced composites the situation is more complex, as diffusion is anisotropic, with a preferred orientation along the fiber direction. The expression proposed for diffusion into composites in [25] is:

$$D = D_x\left(1 + \frac{h}{l}\sqrt{\frac{D_y}{D_x}} + \frac{h}{w}\sqrt{\frac{D_z}{D_x}}\right)2 \tag{2}$$

Ratios between in-plane (y-z) and through thickness (x) diffusion coefficients can vary considerably, but in order to illustrate how coupon size can affect the D/D_x ratio and lead to overestimations of through-thickness diffusion coefficients the values given for a glass/epoxy composite in [26] will be used here. Figure 10 shows some examples of the influence of coupon dimensions.

A third possibility to accelerate water ingress is to adjust the fiber orientation. This has been discussed above, and various authors have used the fact that diffusion in the fiber direction is faster than in the other directions to characterize the different coefficients [27, 28]. Based on this difference in certain cases, for example when unidirectional composites are being studied, it may be possible to choose a coupon geometry which increases water access along the fiber direction in order to saturate them more quickly. A rectangular coupon cut with its longer dimension perpendicular to the fiber direction would be expected to saturate more quickly than one with the same surface area but a standard square geometry. However, the possibilities for using such an approach to accelerate aging are quite limited, particularly if the aim is to measure residual properties after aging rather than diffusion kinetics; geometries are then defined by standard test methods.

6 Influence of Aging Medium on Acceleration

For marine applications, the environment of interest is natural seawater. However, many aging tests are performed in either tap water or distilled water, more readily available in most laboratories than natural seawater. The former have a higher chemical activity than seawater, so it might be possible to use this difference to accelerate water ingress. Figure 11 shows an example of results from a 2 year immersion test performed at 40 °C on the same pre-preg glass/epoxy composite in deionized water and natural seawater. The initial parts of the plots are quite similar, but there is clearly a faster weight gain in deionized water later. After 2 years, the samples in seawater have reached a weight gain similar to that which those in deionized water had reached after only 1 year. However, as neither have achieved a stable weight it is not possible to conclude on whether they would reach the same final weight gain. Results for an infused glass/epoxy composite indicated slightly lower saturation values in seawater.

Various authors have discussed the influence of specific interactions between polymers and water on diffusion, and these can explain many of the anomalies noted for diffusion plots. Colin has discussed these recently [29], and showed that it is interactions between water molecules and polar groups in the polymer matrix, rather than free volume, which determine the amount of water which enters a polymer. There is a good correlation between solubility parameters and mass at saturation for a range of polymers including different epoxy/amine networks. However, the relationship between polymer structure and diffusivity is less clear. Merdas et al. [30] have found that diffusion rate is slowed down by molecular interactions in amine cured epoxies.

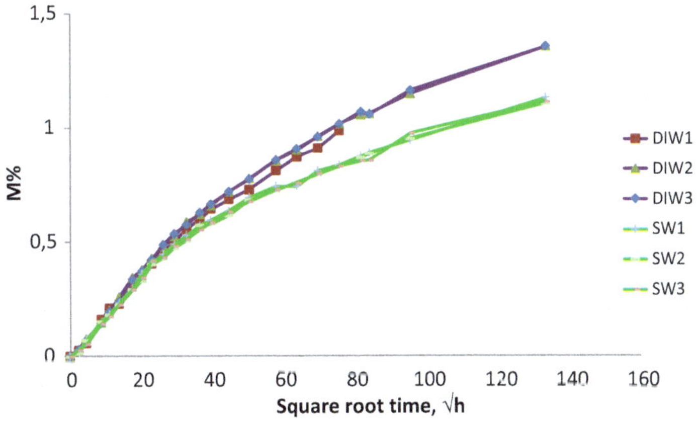

Fig. 11 Influence of immersion medium on weight gain for two composites, 3 coupons for each medium

Table 2 Examples of accelerating factors for weight gain, with respect to an immersion in seawater at 25 °C without mechanical loading, thermoset matrix composites

Temperature: 40 °C	Temperature: 60 °C	Applied strain: <0.2%	Hydrostatic pressure: 50 MPa	Medium: Deionized water	Thickness: reduced by 50%
×3.5	×10	×1.5	×1	×2	×4

7 Summary of Acceleration Factors

Table 2 shows a very simplified summary of the preceding sections, providing orders of magnitude of the acceleration in weight gain that can be achieved. Order of magnitude values are shown, to give an estimate of how the time to reach a given weight gain can be reduced. This clearly shows that the two parameters likely to be most effective in shortening test time are raising temperature and reducing coupon thickness. The other parameters may be of interest for specific applications such as deep sea structures, but their benefits in terms of aging acceleration are likely to be minor.

This table also shows the limits of accelerated aging strategies. Even under the most favorable conditions, and taking into account all the risks described above, it is unlikely that an accelerated aging campaign lasting less than a year will be able to provide useful information on the long term behavior of composites.

8 Examples of Long Duration (> 5 Year) Immersion Studies

The duration of the majority of published studies on wet aging is less than 1 year. A few longer studies have been reported. For example, Kootsookos & Mouritz describe a study of polyester and vinylester composites immersed in seawater at 30 °C for over 2 years [31], but this is unusual.

In this section, two examples of much longer immersions will be presented, in order to discuss the validity of predictions of weight gain after accelerated aging, by comparing with long duration ambient temperature measurements.

8.1 Carbon Fiber Reinforced Composite

In this first example, carbon fiber reinforced composite specimens were immersed in natural seawater for 6 years. The interesting point here is that not only three aging temperatures were used (25, 40 and 60 °C) but also that different coupon thicknesses were investigated. This has allowed both of the most promising accelerating strategies discussed above to be studied. Table 3 describes the materials studied. Fiber volume fractions were estimated to be in the range 59–61%.

Table 3 Carbon/epoxy samples for long term aging study

Fiber	Matrix	Cure cycle	Tg dry °C	Density	Thickness, mm	Orientation
T700	SR1710 epoxy, amine hardener	10 h	79	1.556	2.5	4x quasi-UD
		60 °C	80	1.545	4.5	7x quasi-UD
			82	1.537	11	18x Quadriaxial

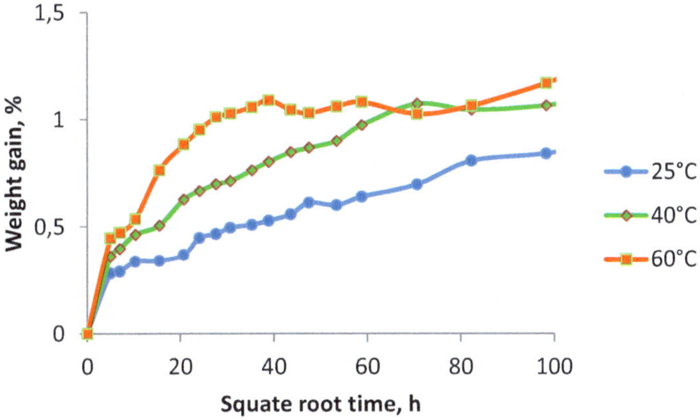

Fig. 12 Weight gain plots, carbon/epoxy, 2.5 mm thick coupons, 1 year immersion natural seawater. Each point corresponds to the mean of measurements from three samples

Figure 12 shows examples of the basic initial weight gain plots, for 2.5 mm thick coupons at each temperature.

These clearly show the accelerating effect of temperature. The 60 °C reached a reasonably stable weight gain around 1% after 2 months, the 40 °C sample takes around 7 months to get to a similar value, and after 1 year in seawater at 25 °C the weight gain is lower, around 0.85%. These are quite "well-behaved" materials, despite an initially high weight gain just after immersion. The samples were dried before immersion.

Longer immersion times, up to 6 years, revealed additional information. Figure 13 shows weight gain measurements for specimens of three thicknesses immersed at 60 °C. A plot for a coupon at 40 °C is also shown for comparison.

First, these plots show the influence of specimen thickness, with the time to saturation of the thinnest coupons being significantly shorter than for the thicker specimens. As an example, the approximate times to reach 1% weight gain are shown in Table 4 below. By reducing the thickness to about 25% (from10.5 to 2.5 mm) the time to reach 1% weight has been reduced from two and a half years to 1 month.

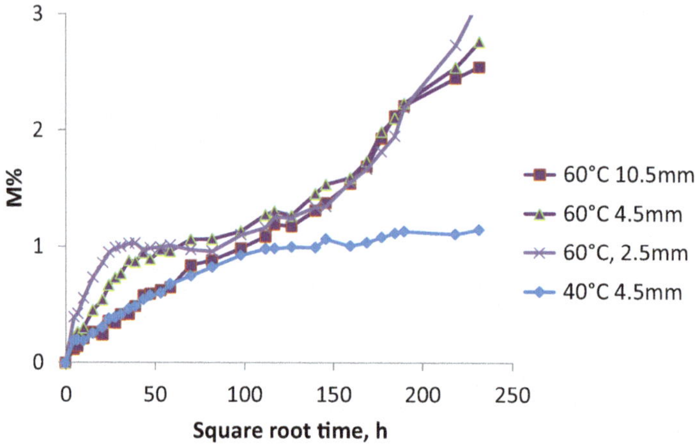

Fig. 13 Weight gain plots, carbon/epoxy, 6 years' immersion 60 °C, three coupon thicknesses

Table 4 Time to reach 1% weight gain, 60 °C immersion

Thickness, mm	Time, days
2.5	32
4.5	144
10.5	886

The plots in Fig. 13 also reveal an increase in weight gain for coupons immersed at 60 °C, which starts after the same immersion time, around 1 year, for all three thicknesses. This suggests that it may be related to a surface degradation mechanism rather than volume degradation, and indeed a surface color change was noted. The increase in weight gain may be caused by the conversion of epoxies to more polar groups; chain scission would be expected to result in a mass loss rather than an increase. At the lower temperature, 40 °C, even after 6 years in water there is no indication of a weight gain increase for any of the coupons tested, but this may simply indicate a strong temperature dependency of this mechanism.

It may be tempting to use these plots to try to estimate the lifetime of these composites. The diffusion kinetics can be used to determine the time necessary to saturate a composite of any thickness, while a criterion based on the time to reach a weight gain increase at 60 °C (about 1 year) might be employed to extrapolate to the time required for the material to start to degrade at a lower temperature. There is no evidence of change after 6 years at 40 °C, and given the accelerating effect of at least 4 in raising the temperature from service immersion conditions, typically 15–20 °C, to 40 °C the lifetime of this composite should exceed 20 years. This provides a very conservative first criterion for lifetime. Clearly in order to examine how the properties of interest change once the resin starts to degrade requires mechanical testing, as the fibers can continue to support loads beyond this first damage initiation. In this study short beam shear tests were performed. Figure 14 shows results for ILSS (interlaminar shear strength) measured on wet 4.5 mm thick specimens after different immersion periods.

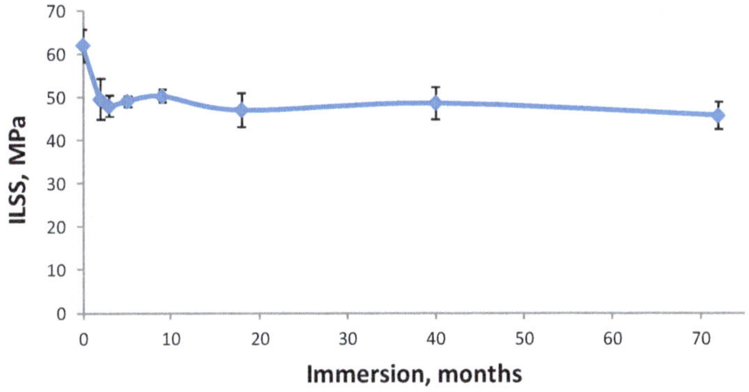

Fig. 14 ILSS values versus immersion time in seawater at 60 °C, carbon/epoxy

Fig. 15 ILSS specimen surfaces
Left: 6 years 60 °C. Right: 6 years 25 °C. Millimeter scale

There is an initial drop in strength that occurs quite quickly, within 2 months, which may be caused by plasticization. However, then the strength remains remarkably constant even after 6 years in seawater at 60 °C. This appears to confirm that the weight gain increase noted at 60 °C is probably a surface effect, and does not affect through-thickness properties significantly. Observation of the specimen surfaces revealed a yellowing of the matrix resin, Fig. 15. Despite the changes on the weight gain plots the ILSS values are still high after 6 years.

8.2 *Glass Fiber Reinforced Composite*

Another long duration aging study was performed on glass/epoxy laminates, lasting nearly 8 years. The material is described in Table 5, and some data from this project have been presented previously [32, 33]. The epoxy resin is a DGEBA based system that has been widely used for filament winding and it is still being used today. The hardener employed in this study was an anhydride, and it is (now) well known that these are susceptible to hydrolysis if exposed to water at elevated temperatures [34, 35]. In the application here the material was to be used in cooling water pipework exposed to temperatures up to 60 °C, so the initial aim was not to accelerate tests but rather to examine the material response at service temperatures; however, a set of samples was aged in tap water at different temperatures, (25, 40 and 60 °C), and it is this database which is of interest here.

Figure 16 shows examples of weight gain plots for this material in distilled water at three immersion temperatures. At 20 °C the diffusion plot stabilized at a weight gain of around 0.5%, and then a small increase was noted after about 7 years. At 40 °C the weight gain exceeded this value after around 2.3 years, while at 60 °C weight gain increased beyond this value after only 1 year. At the highest temperature weight gain increased to over 3%, and then fell to below the initial sample weight (not shown on Fig. 16).

Table 5 Glass/epoxy samples for long term aging study

Fiber	Matrix	Cure temperature	Tg dry	V_f	Thickness, mm	Orientation
E-glass	LY556/HY917	140 °C	115 °C	53%	3	±55°

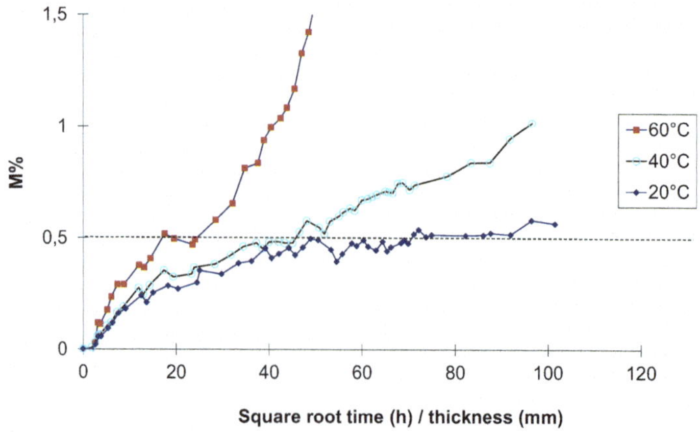

Fig. 16 Examples of weight gain plots, glass/epoxy, immersion in distilled water for 7.7 years

Table 6 Mechanical tests performed wet after 3 years in distilled water at 60 °C

Panel	Flexural strength, MPa		ILSS, MPa	
	35°	55°	35°	55°
Unaged	410	127	44	20
60 °C 1%	302	120	43	19
60 °C 3%	231	87	30	13

Fig. 17 Weight gain panel (200×300 mm^2) after 7.7 years in distilled water at 60 °C, and unaged reference strip

These results underline the risk of raising temperature to accelerate water ingress into this composite formulation. At 60 °C the matrix resin undergoes hydrolysis with chain scission and reduction in cross-link density.

The coupons used for weight gain measurements were large, 200×300 mm^2, and after 3 years two of the five immersed at 60 °C were used for mechanical tests. Three point flexure and short beam shear tests were performed in both directions of the panels. Table 6 shows results. The two panels immersed at 60 °C showed different weight gains after 3 years, 1% and 3%. Specimens from the latter had significantly lower strengths.

Figure 17 shows one of the panels used for weight gain measurements at 60 °C, together with an unaged reference strip. In this case the change in color is observed throughout the composite thickness, in contrast to the carbon/epoxy of the previous section for which it was limited to a very thin surface layer.

If aging tests are only performed for a few months, as is often the case, this hydrolysis mechanism will not be detected, even though it will probably be the limiting factor in long term use of the composite.

9 Other Factors Required to Improve Representativity of Testing

The results from many composite studies show large drops in mechanical properties of composites after quite short accelerated aging times under severe conditions. This creates a quite false impression of composites as materials that are sensitive to water. Interpretation of data from accelerated aging tests must relate sample geometry and test conditions to the form and environment of the structural component, and requires considerable care. The following factors should be considered:

- Many marine composite structures are very thick, and will never saturate with water in service. A difficulty in accelerated testing is to ensure that the accelerating factor (e.g. temperature) does not affect the homogeneity of degradation. For example if temperature is raised the laboratory degradation may be diffusion controlled, mainly affecting surface layers, whereas in service at lower temperatures it may be homogeneous [8]. This will depend on the characteristic times for diffusion and hydrolysis and the relative thicknesses of coupons and structures.
- It is very rare for a composite structure to be immersed at sea without a protective coating. This may be a gel-coat, an anti-fouling paint, or a specific water barrier layer. These will all slow water entry.
- There are few free edges in components, and those that exist will normally be protected, so diffusion times will only depend on the through-thickness diffusion coefficient.
- In many cases, boat hulls for example, water only has access to one composite surface.
- Immersion is often discontinuous, allowing regular drying periods.

The length of this list highlights the limitations of current test practice. Some of these points can be accounted for by modelling. For example, once a diffusion model has been validated by testing thin coupons it can, in theory, be included in a numerical model to allow the water profile through any thicker section to be determined. An example for a 20 mm thick tidal turbine blade can be found in [36]. The model results showed that only 8% of the thickness of the blade would be saturated after 25 years in seawater at 20 °C.

There are very few studies which quantify the influence of coatings on the long term behavior of composite marine structures. This may be justified by assuming a worst case scenario, where the coating is an extra benefit but which may be eroded or otherwise removed during service. Nevertheless, it can significantly reduce water ingress and a realistic lifetime model should include it. Again, if the diffusion properties of the coating are known it can be included in a numerical analysis.

The fact that composite structures are rarely fully immerged suggests that a more representative test would apply water to only one surface of a specimen. Equipment has been developed for uni-facial testing to study coating performance and in particular the kinetics of blistering due to osmosis [37, 38], Fig. 18. It is simply a water-filled, temperature controlled test tank with circular openings, onto which the

Fig. 18 Aging tank for uni-facial exposure of composites to water, without edge effects

samples are clamped. The inner surface of the sample is exposed to water; the outside will be exposed to laboratory temperature and humidity conditions.

This can be applied to study water diffusion kinetics under more realistic conditions, also avoiding coupon edge effects.

Finally, equipment such as oceanographic measurement devices or seismic equipment is immersed during service then stored between campaigns. If drying kinetics are known then they can be used to estimate the effect of drying periods on water profiles. The kinetics of drying may be different from immersion [39] and need to be measured. Again, numerical modelling can allow integration of these data to provide the response to in-service conditions. However, immersion/drying cycles may be more severe than continuous immersion and this is another area requiring more study. Drying cycles also provide valuable information on the reversibility of aging, and are strongly recommended.

10 Conclusions

This paper describes accelerated immersion tests of composites for marine applications. First, the various factors which can be used to reduce test time are discussed. It is shown that reducing sample thickness offers some advantages over the more usual approach of increasing water temperature. Then two examples of long term testing are presented, 6 years' immersion for a carbon/epoxy and 8 years' immersion of a glass/epoxy. These examples show the potential for accelerated evaluation of durability but also the limitations. Finally, a number of factors which need to be included to make test strategies more realistic are discussed.

Acknowledgements Part of the work presented above was supported by the ONR Solid Mechanics Program, with Dr. YS Rajapakse as Program Monitor. The author is also grateful to many colleagues at IFREMER without whose dedicated efforts and expertise over the last 30 years this paper would not have been possible.

References

1. Banks LC, Gentry TR, Barkatt A (1995) Accelerated test methods to determine the long term behavior of FRP composite structures: environmental effects. J Reinf Plast Comp 14:559–587
2. Davies P, Rajapakse YS (eds) (2014) Durability of composites in a marine environment. Springer, Dordrecht
3. Davies P, Rajapakse YS (eds) (2018) Durability of composites in a marine environment 2. Springer, Cham
4. de Souza Rios A, Amorim WF Jr, de Moura EP, de Deus EP, Feitosa JPA (2016) Effects of accelerated aging on mechanical, thermal and morphological behavior of polyurethane/epoxy/fiberglass composites. Polym Test 50:152–163
5. Dey V, Mobasher B (2018) Quantitative characterization of accelerated aging in cement composites using flexural inverse analysis. Cem Concr Compos 89:181–191
6. Barbosa APC, Fulco APP, Guerra ESS, Arakaki FK, Tosatto M, Costa MCB, Melo JDD (2017) Accelerated aging effects on carbon fiber/epoxy composites. Compos Part B 110:298–306
7. Panaitescu I, Koch T, Archodoulaki V-M (2019) Accelerated aging of a glass fiber/polyurethane composite for automotive applications. Polym Test 74:245–256
8. Colin X, Verdu J (2014) Humid ageing of organic matrix composites, chapter 3. In: Davies P, Rajapakse Y (eds) Durability of composites in a marine environment. Springer, Dordrecht
9. Gillen KT, Celina M, Clough RL, Wise J (1997) Extrapolation of accelerated aging data – Arrhenius or erroneous? Trends Polym Sci 5(8):250–257
10. Bellenger V, Verdu J, Morel E (1989) Structure-properties relationships for densely cross-linked epoxide-amine systems based on epoxide or amine mixtures, Part 2. Water absorption and diffusion. J Mat Sci 24:63–68
11. Unpublished results, IFREMER
12. Arhant M, LeGac P-Y, LeGall M, Burtin C, Briancon C, Davies P (2016) Modelling the non Fickian water absorption in polyamide 6. Polym Degrad Stab 133:404–412
13. Wright WW (1981) The effect of diffusion of water into epoxy resins and their carbon fibre reinforced composites, composites, July, 201–205
14. Fahmy AA, Hunt JC (1980) Stress dependence of water diffusion in epoxy resin. Polym Compos 1(2):77–80
15. Weitsman Y (1987) Coupled damage and moisture-transport in fiber-reinforced, polymeric composites. Int J Solids Struct 23(7):1003–1025
16. Perreux D, Suri C (1997) A study of the coupling between the phenomena of water absorption and damage in glass/epoxy composite pipes. Compos Sci Technol 57(9):1403–1413
17. Lundgren JE, Gudmundson P (1999) Moisture absorption in glass-fibre/epoxy laminates with transverse matrix cracks. Compos Sci Technol 59(13):1983–1991
18. Humeau C, Davies P, Jacquemin F (2018) An experimental study of water diffusion in carbon/epoxy composites under static tensile stress. Compos Part A 107:94–104
19. Avena A, Bunsell AR (1988) Effect of hydrostatic pressure on the water absorption of glass fibre-reinforced epoxy resin. Composites 19(5):355–357
20. Pollard A, Baggott R, Wostenholm GH, Yates B, George AP (1989) Influence of hydrostatic pressure on the moisture absorption of glass fibre-reinforced polyester. J Mater Sci 24 (5):1665–1669

21. Hoppel CPR, Bogetti TA, Gillespie JW (1995) Literature review-effects of hydrostatic pressure on the mechanical behavior of composite materials. J Thermoplast Compos Mater 8(4):375–409
22. Davies P, Choqueuse D, Mazéas F (1997) Composites underwater, proceedings progress in Durability Analysis of Composites Systems conference (DURACOSYS). Balkema, Rotterdam, pp 19–24
23. Humeau C, Davies P, Jacquemin F (2016) Moisture diffusion under hydrostatic pressure in composites. Mater Des 96:90–98
24. Wang Y, Zhang X, Cai G, Wan B, Waldmann D, Qu Y (2018) A new thickness-based accelerated aging test methodology for resin materials: theory and preliminary experimental study. Construct Build Mater 186:986–995
25. Shen C-H, Springer GS (1981) Moisture absorption and desorption of composite materials, chapter 3. In: Springer GS (ed) Environmental effects on composite materials. Technomic, Lancaster, pp 15–33
26. Gagani A, Fan Y, Muliana AH, Echtermeyer AT (2018) Micromechanical modeling of anisotropic water diffusion in glass fiber epoxy reinforced composites. J Compos Mater 52 (17):2321–2335
27. Aronhime MT, Neumann S, Marom G (1987) The anisotropic diffusion of water in Kevlar-epoxy composites. J Mater Sci 22:2435–2446
28. Beringhiera M, Simar A, Gigliotti M, Grandidier JC, Ammar-Khodja I (2016) Identification of the orthotropic diffusion properties of RTM textile composites for aircraft applications. Compos Struct 137:33–43
29. Colin X (2018) Nonempirical kinetic modeling of non-Fickian water absorption induced by a chemical reaction in epoxy-amine networks. In: Davies P, Rajapakse YS (eds) Durability of composites in a marine environment 2. Springer publishers, pp 1–18
30. Merdas I, Thominette F, Tcharkhtchi A, Verdu J (2002) Factors governing water absorption by composite matrices. Compos Sci Technol 62:487–492
31. Kootsookos A, Mouritz AP (2004) Seawater durability of glass- and carbon-polymer composites. Compos Sci Technol 64(10–11):1503–1511
32. Davies P, Baizeau R, Choqueuse D, Salmon L, Nagot F, (1999). Aging and long term behaviour of composite tubes, Proc DURACOSYS, Balkema, Rotterdam 1–10
33. Perreux D, Choqueuse D, Davies P (2002) Anomalies in moisture absorption of glass fibre reinforced epoxy tubes. Compos Part A 33(2):147–154
34. Yagoubi JE, Lubineau G, Traidia A, Verdu J (2015) Monitoring and simulations of hydrolysis in epoxy matrix composites during hygrothermal aging. Compos Part A 68:184–192
35. Capiel G, Uicich J, Fasce D, Montemartini PE (2018) Diffusion and hydrolysis effects during water aging on an epoxy-anhydride system. Polym Degrad Stab 153:165–171
36. Tual N, Carrere N, Davies P, (2015). Characterization of sea water ageing effects on mechanical properties of carbon/epoxy composites for tidal turbine blades, Proceeding 20th international conference on composite materials, Copenhagen, Denmark
37. Castaing P, Lemoine L, Gourdenne A (1995) Mechanical modelling of blisters on coated laminates II – experimental analysis. Compos Struct 30:223–228
38. Hewitt JT (2001) Evaluation of the factors that influence the blister resistance of boat hulls and the methods used to test them, AOC Technical paper. Available at: https://www.aoc-resins.com/pdf/technical-blister.pdf
39. Weitsman YK (2012) Fluid effects in polymers and polymeric composites. Springer, New York

Statistical Long-Term Creep Failure Time of Unidirectional CFRP

Yasushi Miyano and Masayuki Nakada

1 Introduction

Carbon fiber reinforced plastics (CFRP) have been used for primary structures of airplanes, ships, automobiles, and other vehicles, for which high reliability must be maintained during long-term operation. Therefore, an accelerated testing methodology is strongly anticipated for the long-term life prediction of CFRP structures exposed to actual environmental temperatures, water, and other influences.

The mechanical behavior of matrix resin of CFRP exhibits time-dependent and temperature-dependent viscoelastic behavior: not only above the glass transition temperature T_g, but also below T_g. Consequently, it can be inferred that the mechanical behavior of CFRP depends strongly on the time and temperature [1–5]. Our earlier reports proposed the formulation of statistical static, creep, and fatigue strengths of CFRP based on the matrix resin viscoelasticity [6, 7].

The tensile strength along the longitudinal direction of unidirectional CFRP constitutes important data for the reliable design of CFRP structures. The authors developed a test method for creep and fatigue strengths as well as the static strength at elevated temperatures for resin-impregnated carbon fiber strands (CFRP strands) combined with T300–3000 and epoxy resin [8]. Furthermore, the authors confirmed that the time-dependent and temperature-dependent tensile static strength is controlled by the viscoelastic behavior of matrix resin based on Rosen's shear lag model [9]. Additionally, the authors developed a test method for a CFRP strand of T800–12000 and epoxy resin with a highly reliable co-cured tab. We evaluated the temperature-dependent tensile strength of this CFRP strand [10]. Our most recent study undertook the prediction of the statistical creep failure time under tension loading along the longitudinal direction of unidirectional CFRP performed using

Y. Miyano (✉) · M. Nakada
Materials System Research Laboratory, Kanazawa Institute of Technology, Hakusan, Japan
e-mail: miyano@neptune.kanazawa-it.ac.jp; nakada@neptune.kanazawa-it.ac.jp

S. W. Lee (ed.), *Advances in Thick Section Composite and Sandwich Structures*,
https://doi.org/10.1007/978-3-030-31065-3_18

CFRP strands of T300–3000 and epoxy resin [11]. The statistical creep failure time of CFRP strands at a constant load and temperature was predicted using statistical results of static tensile strengths of CFRP strands measured at various temperatures and using the viscoelastic behavior of matrix resin. The predicted results agreed well with the experimentally obtained results measured using creep tests for CFRP strands.

In this chapter, the statistical static strengths under tension loading along the longitudinal direction of unidirectional CFRP with carbon fibers of various kinds are measured at various temperatures and are discussed in terms of the role of the matrix resin viscoelasticity. First, the time-dependent and temperature-dependent statistical static strength for CFRP laminates is formulated based on the matrix resin viscoelasticity. Second, the statistical static strengths of carbon fibers of four kinds are measured using many specimens of high-strength and high-modulus PAN-based carbon fibers and low and high modulus pitch based carbon fibers. Third, the temperature-dependent tensile static strengths of unidirectional CFRP with carbon fibers of four kinds are measured statistically at various temperatures using resin-impregnated CFRP strands as unidirectional CFRP specimens. The fractographs of CFRP strands with carbon fibers of four kinds show failure at various temperatures. Finally, the temperature dependences of tensile static strengths are discussed from the role of the matrix resin viscoelasticity based on Christensen's model of viscoelastic crack kinetics [12].

2 Statistical Prediction of Creep Failure Time of CFRP

We have proposed the formulation of a statistical measure of static strength σ_s of CFRP based on the matrix resin viscoelasticity, as presented in the following equation in our previous paper [7] as

$$\log \sigma_s(P_f, t, T) = \log \sigma_0(t_0, T_0) + \frac{1}{\alpha_s} \log\left[-\ln(1 - P_f)\right] - n_R \log\left[\frac{D^*(t, T)}{D_c(t_0, T_0)}\right], \tag{1}$$

where P_f signifies the failure probability, t denotes the failure time, t_0 represents the reference time, T is the temperature, T_0 stands for the reference temperature, σ_0 and α_s respectively denote the scale parameter and the shape parameter on the Weibull distribution of static strength. Also, n_R is the viscoelastic parameter. In addition, D_c and $D*$ respectively represent the creep and viscoelastic compliances of matrix resin. The viscoelastic compliance $D*$ for the static load with a constant strain rate is

$$D*(t, T) = D_c(t/2, T) \tag{2}$$

The statistical static strength σ_s is shown by the following equation by substituting Eq. (2) into Eq. (1).

$$\log \sigma_s(P_f, t, T) = \log \sigma_0(t_0, T_0) + \frac{1}{\alpha_s} \log[-\ln(1 - P_f)] - n_R \log\left[\frac{D_c(t/2, T)}{D_c(t_0, T_0)}\right] \tag{3}$$

Figure 1 shows how to make the creep strength master curve from the static strength master curve. The creep strength master curve is obtainable by horizontally shifting the static strength master curve by the amount of log A. Therefore, the statistical creep strength σ_c is shown as equation [11].

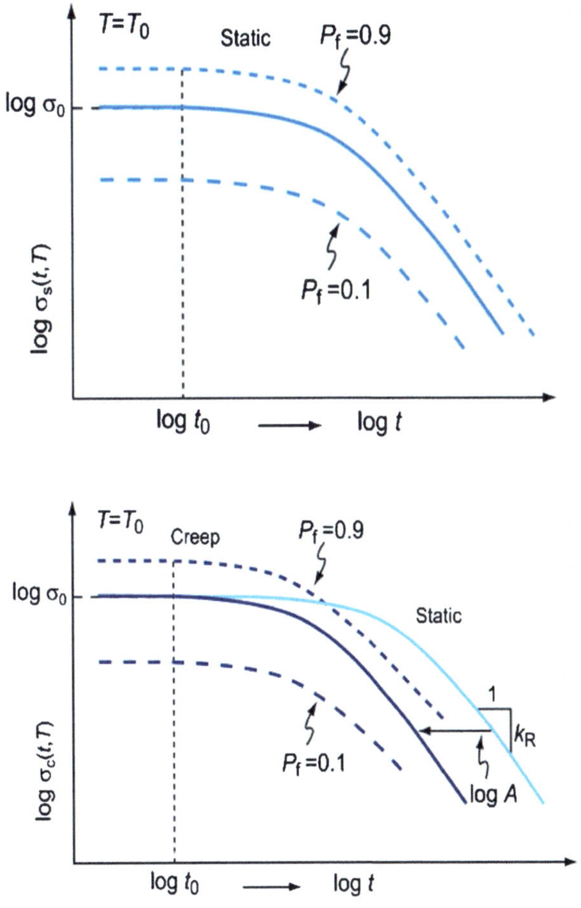

Fig. 1 Creation of the creep strength master curve from the static strength master curve

$$\log \sigma_c(P_f, t, T) = \log \sigma_0(t_0, T_0) + \frac{1}{\alpha_s} \log[-\ln(1 - P_f)] - n_R \log\left[\frac{D_c(At/2, T)}{D_c(t_0, T_0)}\right] \tag{4}$$

The shifting amount log A is determined by slope k_R of the static strength curve shown in Fig. 1 as

$$\log A = \log(1 + 1/k_R),\ k_R = n_R m_R \tag{5}$$

where n_R stands for the viscoelastic parameter in Eqs. (1) and (3), and where m_R is the slope of logarithmic creep compliance of the matrix resin against the logarithmic time [12].

Figure 2 shows drawing of the failure probability against failure time at constant creep strength σ_{c0} from the creep strength master curve. The failure probability of CFRP under a constant creep stress σ_{c0} can be shown by the following equation from Eq. (4).

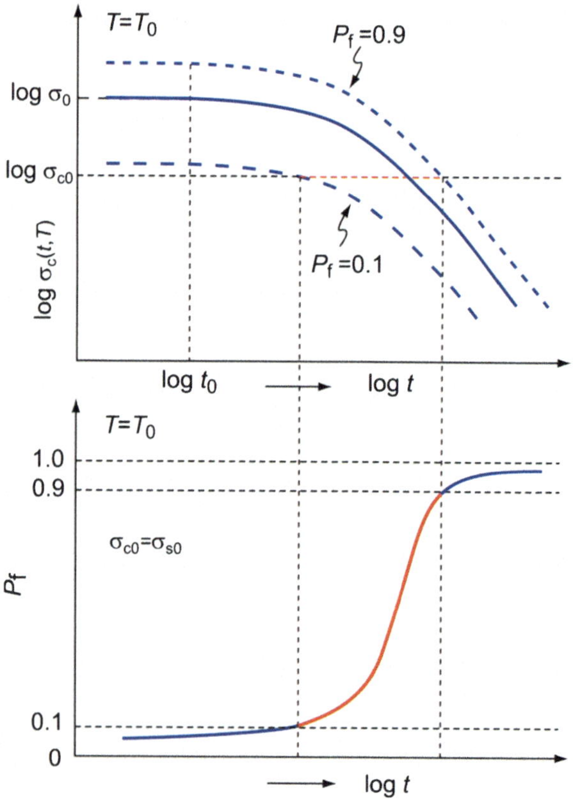

Fig. 2 Failure probability versus failure time at constant creep strength

$$P_f = 1 - \exp(-F), \; \log F = \alpha_s \log\left[\frac{\sigma_{c0}}{\sigma_0}\right] + \alpha_s n_R \log\left[\frac{D_c(At/2, T_0)}{D_c(t_0, T_0)}\right] \tag{6}$$

The creep failure probability of CFRP is a function of the applied creep stress, scale, and shape parameters of the static strength of CFRP and viscoelastic behavior of the matrix resin.

Rosen's model is well known as a failure model for the tensile static strength under the longitudinal direction of unidirectional CFRP. Viscoelastic parameter n_R in Eq. (1) can be represented by the following equation based on Rosen's model [9].

$$n_R = 1/2\alpha_c \tag{7}$$

Therein, α_c represents the tensile strength shape parameter of the single carbon fiber.

3 Molding of CFRP Strands and Testing Methods

Carbon fibers of four kinds were used in this study: high-strength PAN-based carbon fiber T300–3000 and high modulus PAN based carbon fiber M40J-6000 (Toray Industries Inc.); and low modulus pitch based carbon fiber XN05–3000 and high modulus pitch based carbon fiber XN50–3000 (Nippon Graphite Fibers Corp.). The carbon fibers used in this study and their mechanical properties referred from catalogs are shown in Table 1.

We molded the CFRP strands of four kinds combined with the respective fibers and a general purpose epoxy resin (jER828; Mitsubishi Chemical Corp.) using a filament winding system developed by the authors, as shown in Fig. 3 [8]. Actually, 200 specimens for the CFRP strands were molded respectively at one time using this system. The epoxy resin composition and the CFRP strand curing conditions are presented in Table 2. The gage lengths of CFRP strands are approximately 200 mm, as shown in Fig. 4. Mechanisms for grips to the CFRP strands were developed for creep tests and static tests at various temperatures.

The glass transition temperatures $T_g = 160$ °C of the epoxy resin were ascertained from the peak of loss tangent against temperature at 1 Hz using Dynamic Mechanical Analysis (DMA) testing machine. The fiber volume fraction $V_f = 55\%$ of CFRP strand is ascertained from the CFRP strand weight.

Table 1 Carbon fiber of four kinds and their mechanical properties

Carbon fiber	PAN or Pitch	Density [g/cm^3]	Tex [g/km]	Elastic modulus [GPa]	Tensile strength [MPa]
T300–3000	PAN	1.76	198	230	3530
M40 J-6000	PAN	1.75	225	377	4400
XN05–3000	Pitch	1.65	410	54	1100
XN50–3000	Pitch	2.14	450	520	3830

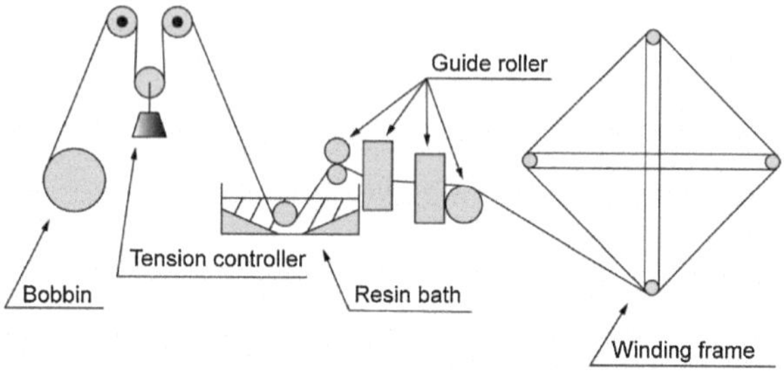

Fig. 3 Molding system for CFRP strands as specimens

Table 2 Composition and cure schedule of CFRP strand of four kinds

CFRP strand	Carbon fiber	Composition of resin (weight ratio)	Cure schedule
T300/EP	T300–3000	Epoxy: jER828 (100)	70 °C × 12 h
M40 J/EP	M40 J-6000	Hardener: MHAC-P (103.6)	+ 150 °C × 4 h
XN05/EP	XN05–3000	Cure accelerator: 2E4MZ (1)	+ 190 °C × 2 h
XN50/EP	XN50–3000		

The original static testing machine with the temperature chamber shown in Fig. 5 can achieve a constant elongation rate in a wide range and a constant temperature in the range of 25 °C and 170 °C. The original creep testing machine with temperature chamber shown in Fig. 6 can realize a constant load in the maximum load 1 kN and a constant temperature in the range of 25 °C to 150 °C.

4 Results and Discussion

4.1 Creep Compliance of Matrix Resin and Static Strength of Carbon Fibers

The dimensionless creep compliance D_c/D_{c0} measured at various temperatures is shown on the left of Fig. 7. Long-term D_c/D_{c0} at T = 120 °C is obtained by

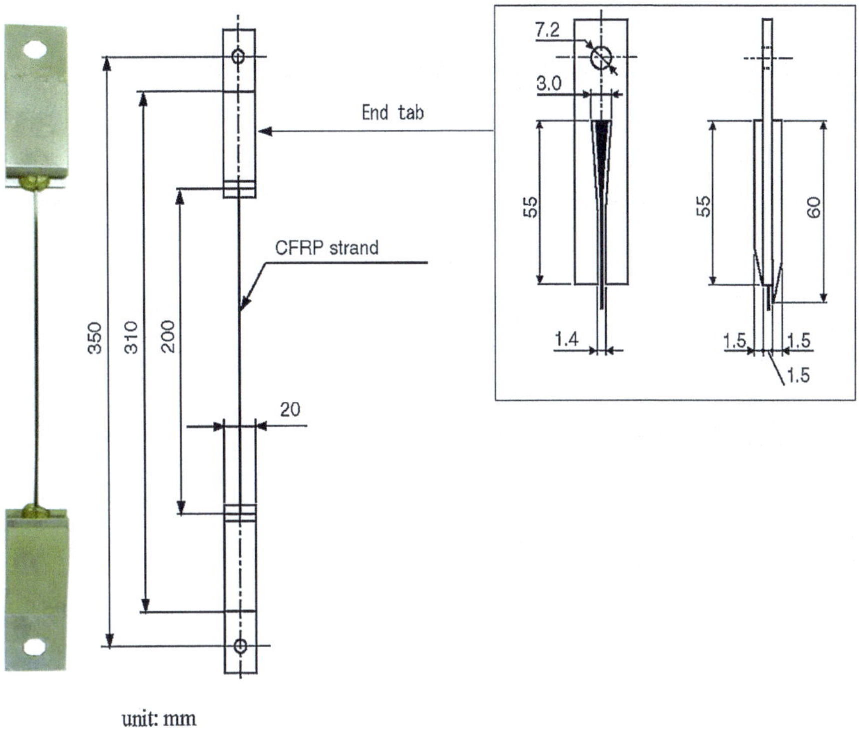

Fig. 4 Configuration of CFRP strand specimens

horizontally shifting those at various temperatures, as shown at the right of Fig. 7 [11]. The reference temperature and time are selected for this study as $T_0 = 25$ °C and $t_0 = 1$ min. The creep compliance at reference temperature and reference time D_{c0} is 0.33 GPa^{-1}. The dashed curve is the dimensionless viscoelastic compliance $D*$ of matrix resin under the constant strain rate at $T = 120$ °C.

The carbon fiber tensile strength σ_t measured statistically at 25 °C is obtained using the following equation. The carbon fiber gage length is 25 mm. The number of specimens for each fiber is 50 (Fig. 8).

$$\sigma_t = \frac{W_{max}N}{t_e}\rho \tag{8}$$

Therein, W_{max} denotes the maximum load [N]. Also, ρ, t_e, and N respectively represent the density of the carbon fiber [kg/m^3], the tex of the carbon fiber strand [g/km], and the numbers of fiber in one strand. Figure 4 shows the Weibull distributions for the tensile strengths of the carbon fibers. The table in this figure presents shape parameters α_c and scale parameters β_c for these carbon fibers.

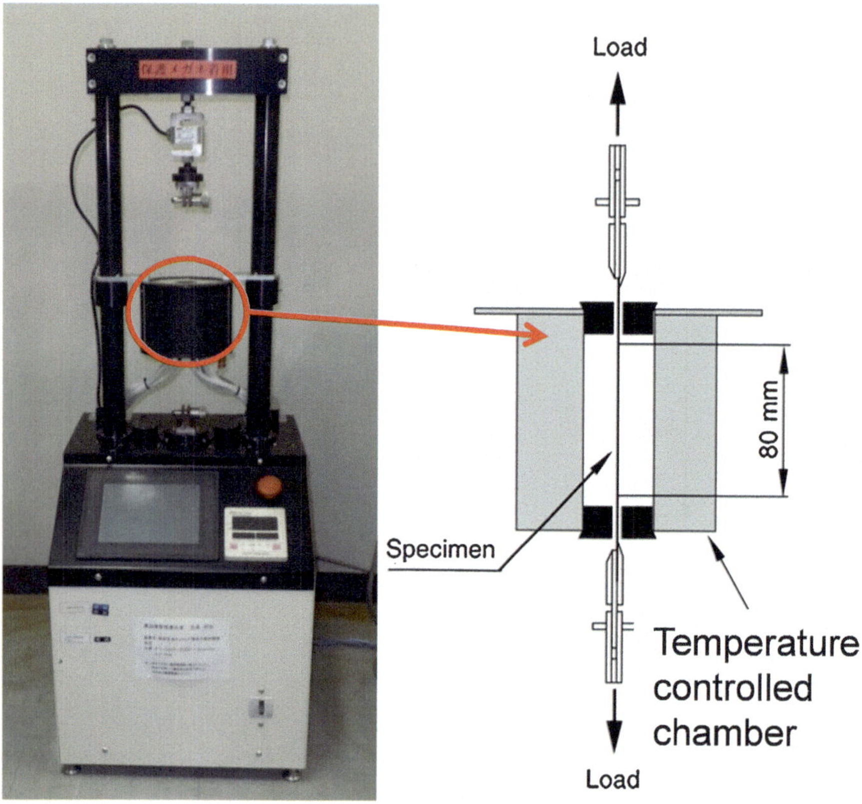

Fig. 5 Static tests for CFRP strands

4.2 Static Tensile Strengths of CFRP Strands at Various Temperatures

Static tension tests for CFRP strands of four kinds were conducted at four or five temperature of 25 °C, (120 °C), 135 °C, 150 °C, and 170 °C with cross-head speed of 2 mm/min. The CFRP strand tensile strength σ_s is obtained using the following equation.

$$\sigma_s = \frac{W_{max}}{t_e}\rho \quad (9)$$

In that equation, W_{max} represents the maximum load [N]. Also, ρ and t_e respectively denote the density of the carbon fiber [kg/m^3] and the tex of the carbon fiber strand [g/1000 m].

Figures 9 and 10 show static strength versus temperature for the CFRP strands of four kinds. The static strengths for T300/EP and XN05/EP decrease markedly with increasing temperature. This strength for M40 J/EP decreases only slightly with increasing temperature. That for XN50/EP remains constant with increasing temperature.

Fig. 6 Creep tests for CFRP strands

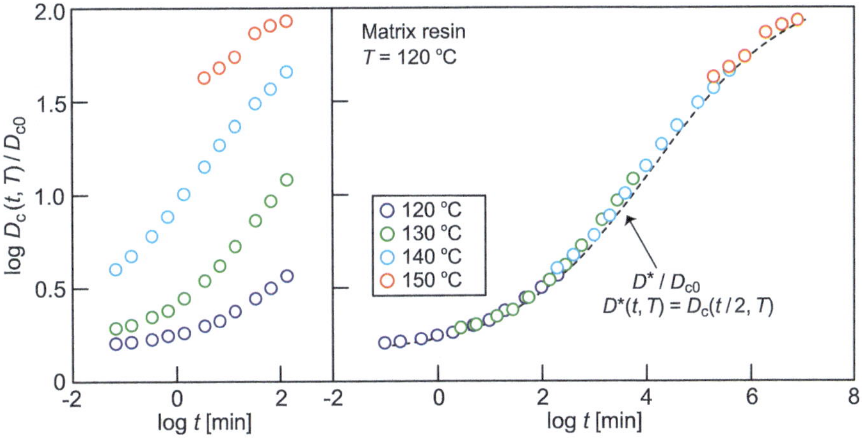

Fig. 7 Dimensionless creep compliance of matrix resin at $T = 120$ °C [11]

The Weibull distributions for static strength at various temperatures are depicted in Figs. 11, 12, 13 and 14 for the respective CFRP strands. In these figures, α_s is the shape parameter and β_s is the scale parameter of CFRP strand. Although the scale parameter decreases or keeps a constant value according to the temperature rise, the shape parameter maintains almost a constant value for all CFRP strands of four kinds

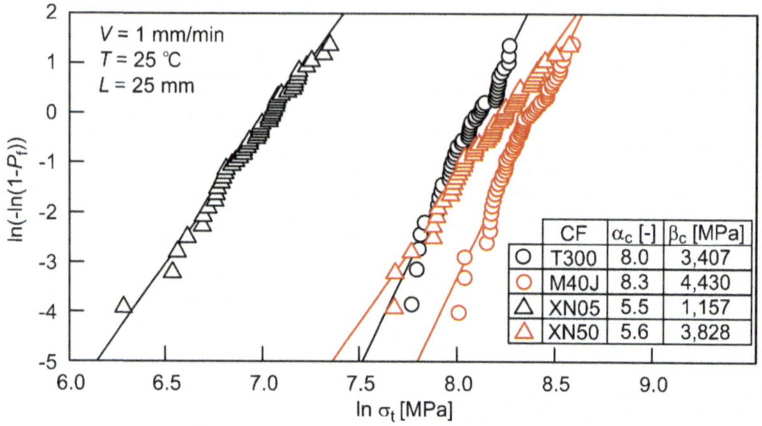

Fig. 8 Weibull Distributions of tensile strengths for carbon fiber of four kinds

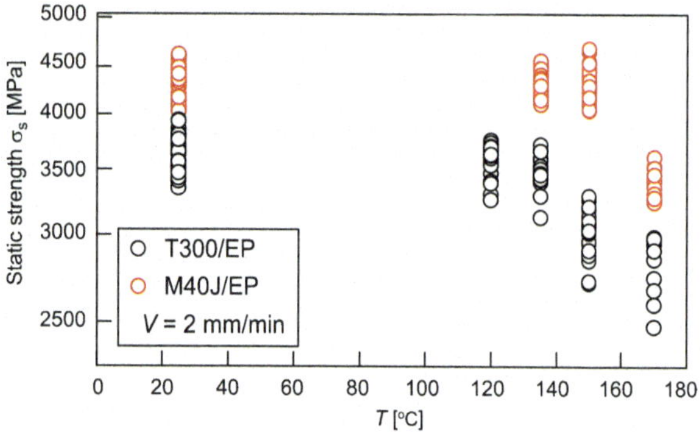

Fig. 9 Static strengths of T300/EP and M40/EP versus temperature

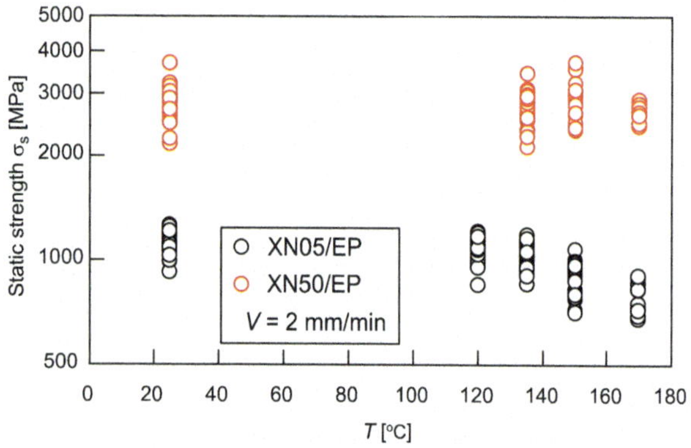

Fig. 10 Static strengths of XN05/EP and XN50/EP versus temperature

to the temperature raise. Shape parameter α_s and the scale parameter β_s at the temperature $T = 25$ °C in these figures can be inferred as shape parameter α_s, scale parameter σ_0 of static strength at the reference temperature $T_0 = 25$ °C and the reference failure time $t_0 = 1$ min used in Eq. (1).

Fig. 11 Weibull distributions of static tensile strength of T300/EP at various temperatures

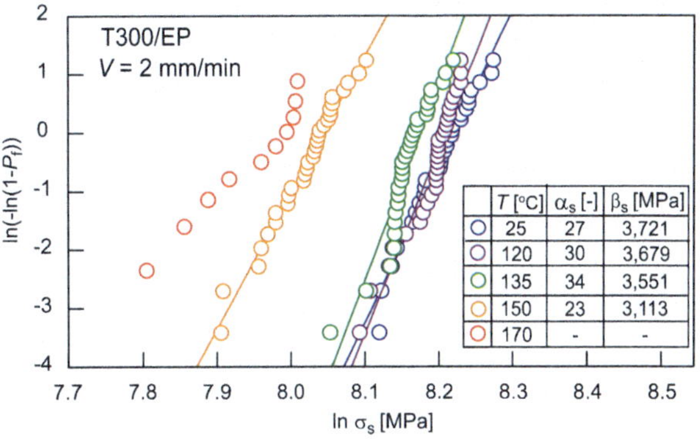

Fig. 12 Weibull distributions of static tensile strength of M40 J/EP at various temperatures

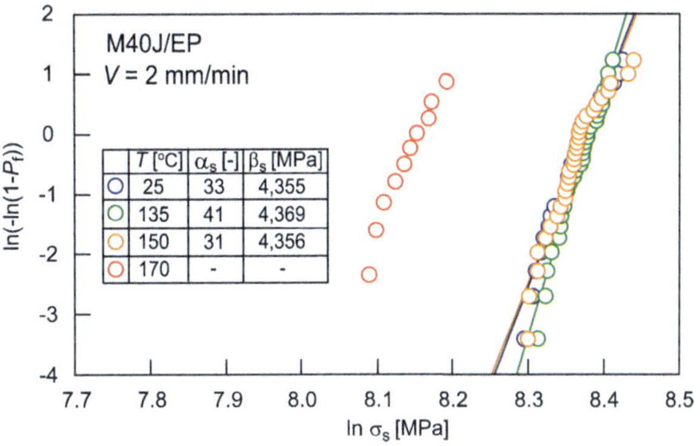

Fig. 13 Weibull distributions of static tensile strength of XN05/EP at various temperatures

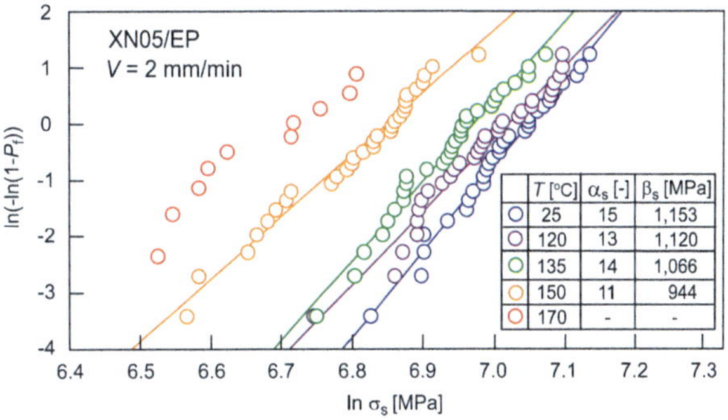

Fig. 14 Weibull distributions of static tensile strength of XN50/EP at various temperatures

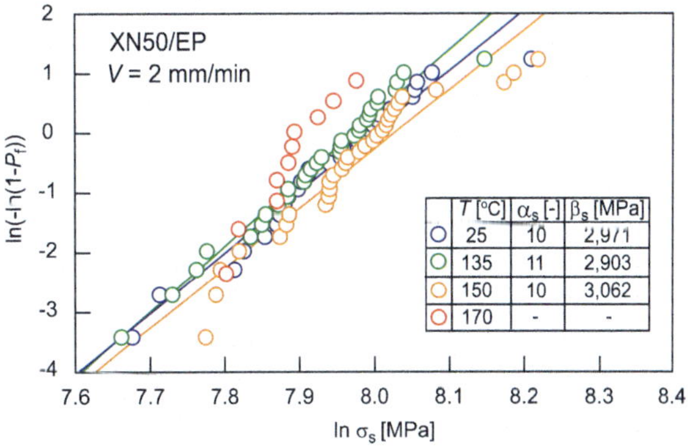

4.3 Static Tensile Strength of CFRP Strand Against Viscoelastic Compliance of Matrix Resin

Figures 15, 16, 17 and 18 show the dimensionless static strength σ_s/σ_0 of a CFRP strand against the dimensionless viscoelastic compliance of matrix resin $D*/D_{c0}$ simultaneously and temperatures for CFRP strands of four kinds. The relation of

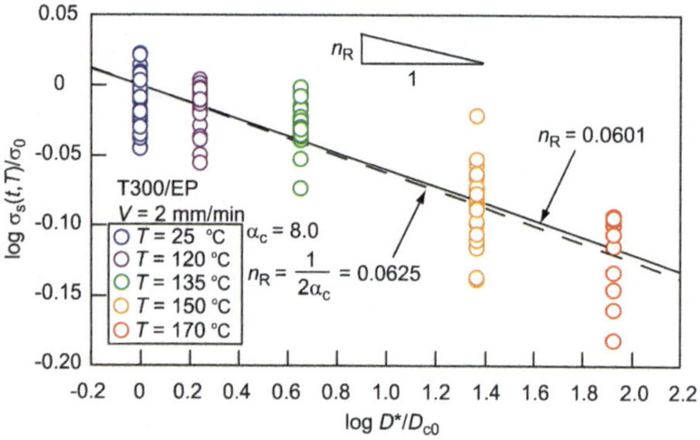

Fig. 15 Static strength of T300/EP against viscoelastic compliance of matrix resin

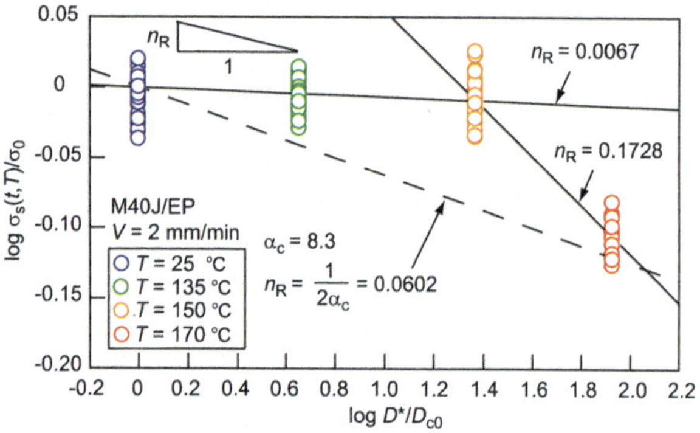

Fig. 16 Static strength of M40 J/EP against viscoelastic compliance of the matrix resin

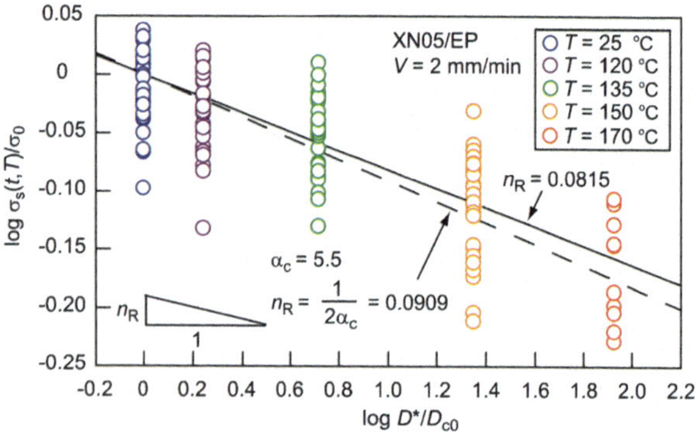

Fig. 17 Static strength of XN05/EP against viscoelastic compliance of the matrix resin

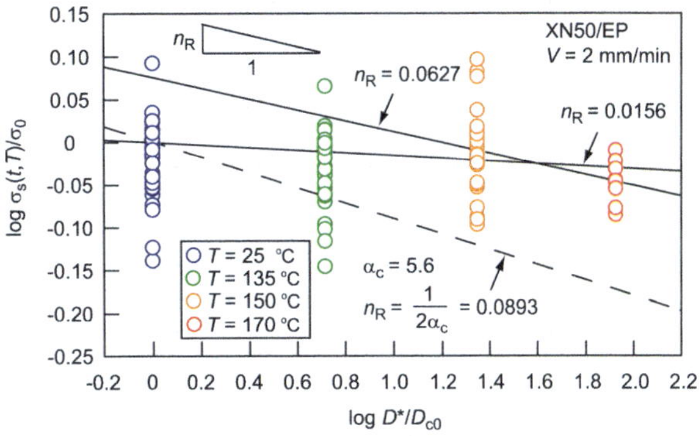

Fig. 18 Static strength of XN50/EP against viscoelastic compliance of the matrix resin

Table 3 Parameters of statistical static strengths for various CFRP strands

CFRP strand	σ_0 [MPa]	α_s [−]	n_R
T300/EP	3721	27	0.0601
M40 J/EP	4355	33	0.0067 (Low temp.) 0.1728 (High temp.)
XN05/EP	1153	15	0.0815
XN50/EP	2971	10	0.0156 (Low temp.) 0.0627 (High temp.)

σ_s/σ_0 against $D*/D_{c0}$ can be represented as one or two solid straight lines with the slope of n_R, which is the viscoelastic parameter in Eq. (1), as obtained from Christensen's model of viscoelastic crack kinetics. The dotted lines in these figures are straight lines with the slope of viscoelastic parameters n_R obtained by substituting shape parameter α_c of single carbon fiber strength into Eq. (4) based on Rosen's model.

The slopes of CFRP strands of four kinds in Figs. 15, 16, 17 and 18 show characteristic behaviors. The slopes of T300/EP and XN05/EP coincide well with those obtained from the shape parameter α_c of single carbon fiber strength based on Rosen's model. The slopes of M40 J/EP and XN50/EP are approximately zero in the wide range of compliance except for high compliance of the matrix resin.

4.4 *Master Curves of Static Tensile Strength for Various CFRP Strands*

All parameters in Eq. (3) for the master curve of static strength are determined for CFRP strands of four kinds and the matrix resin as shown in Table 3: shape parameter α_s and scale parameter σ_0 of static strength at the reference temperature $T_0 = 25$ °C; reference failure time $t_0 = 1$ min; viscoelastic parameter n_R; and the dimensionless viscoelastic compliance of matrix resin $D*/D_{c0}$.

The master curves of statistical static strengths for CFRP strands are shown in Figs. 19, 20, 21 and 22 with experimental data measured at various temperatures. Each master curve in these figures shows characteristic behavior. The master curves of T300/EP and XN05/EP show a smooth decreasing single curve with increasing

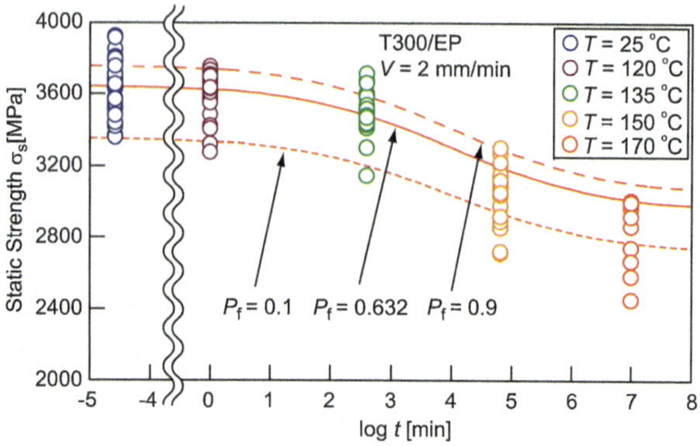

Fig. 19 Master curve of static strength of T300/EP

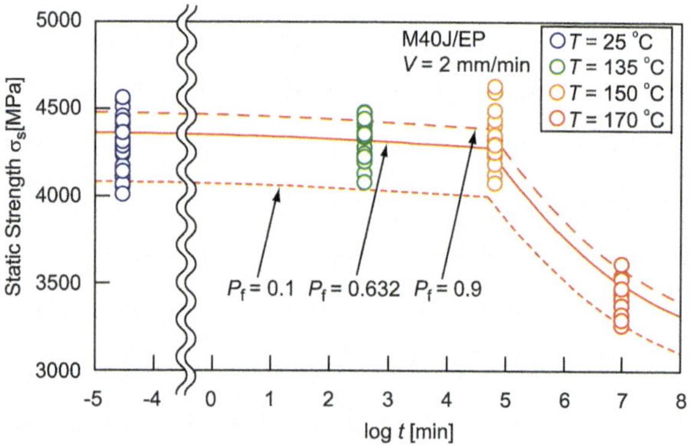

Fig. 20 Master curve of static strength of M40 J/EP

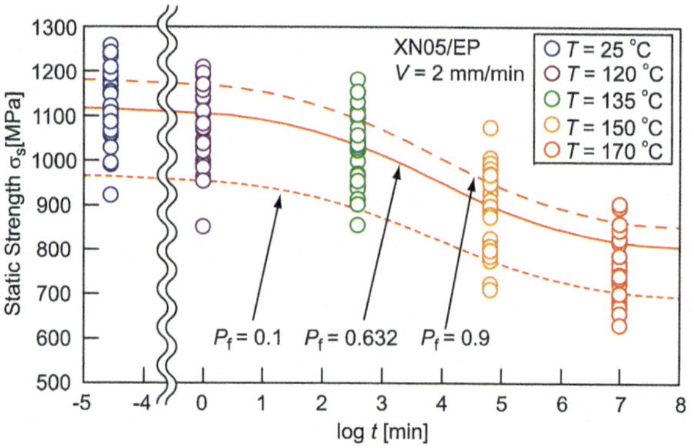

Fig. 21 Master curve of static strength of XN05/EP

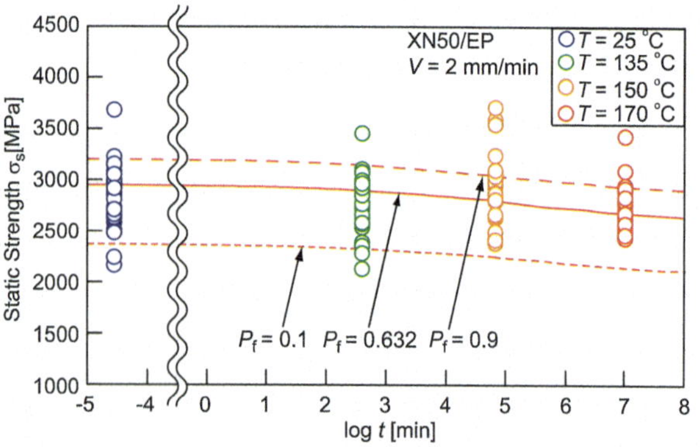

Fig. 22 Master curve of static strength of XN50/EP

failure time. The master curves of M40 J/EP and XN50/EP show approximately zero slope in the wide range of failure time. They decrease remarkably at a high failure times.

4.5 Experimental and Predicted Statistical Creep Failure Times for Various CFRP Strands

The statistical creep strengths for CFRP strands of four kinds at 120 °C are presented in Figs. 23, 24, 25 and 26 with experimental data measured at the same temperature. The upper sides of these figures portray master curves of statistical creep strengths σ_c for the CFRP strands. The lower sides of these figures depict failure probabilities P_f

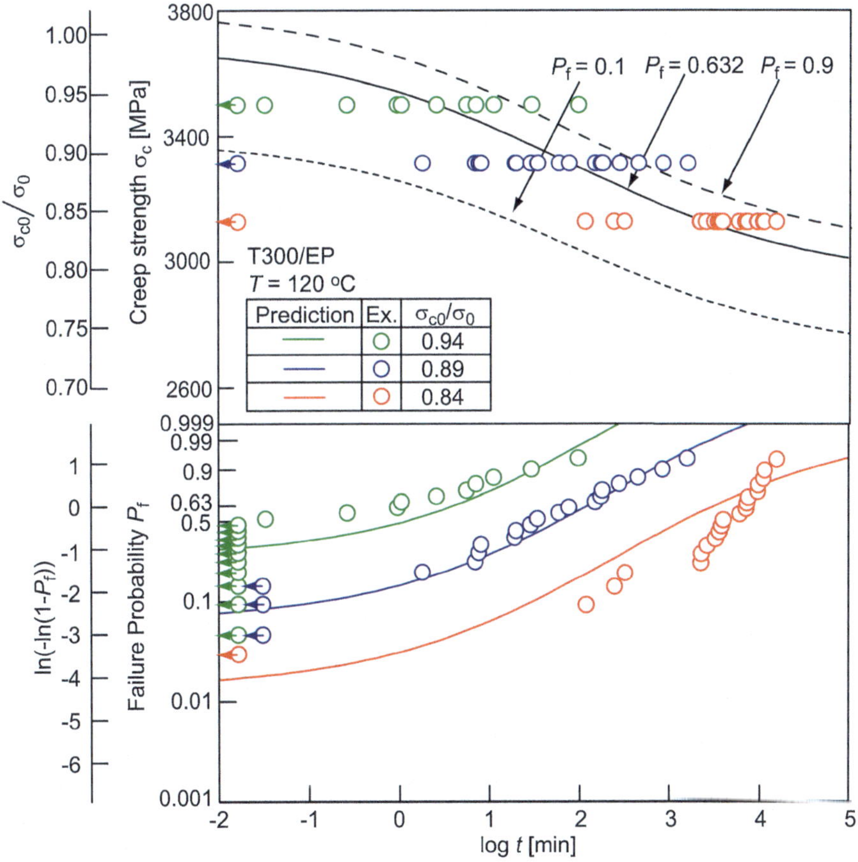

Fig. 23 Experimental and predicted statistical creep failure time for T300/EP

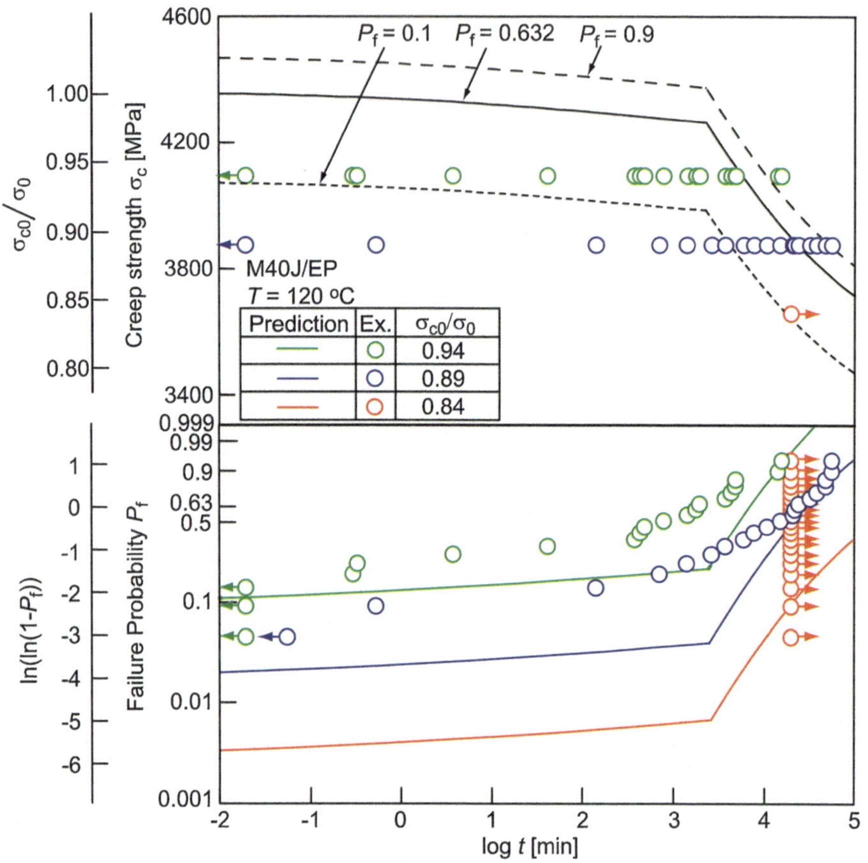

Fig. 24 Experimental and predicted statistical creep failure time for M40 J/EP

against the creep failure time t at several constant creep stress levels σ_{c0} for CFRP strands of four kinds.

These figures clarify that each of the predicted statistical creep strengths for four kinds of CFRP strands agrees with the corresponding experimental ones measured directly using creep tests. Each master and probability curve in these figures shows characteristic behavior. The creep strength master curves of T300/EP and XN05/EP show a smooth decreasing single curve with increasing failure time. The master curves of M40 J/EP and XN50/EP show approximately zero slope for a wide range of failure times. They decrease remarkably in times of high failure. The failure probability curves of T300/EP and XN05/EP show a smooth increasing single curve with increasing failure time for each creep stress level. The failure probability curves of M40 J/EP and XN50/EP show approximately zero slope for a wide range of failure times. They increase remarkably in high failure times for each creep stress level.

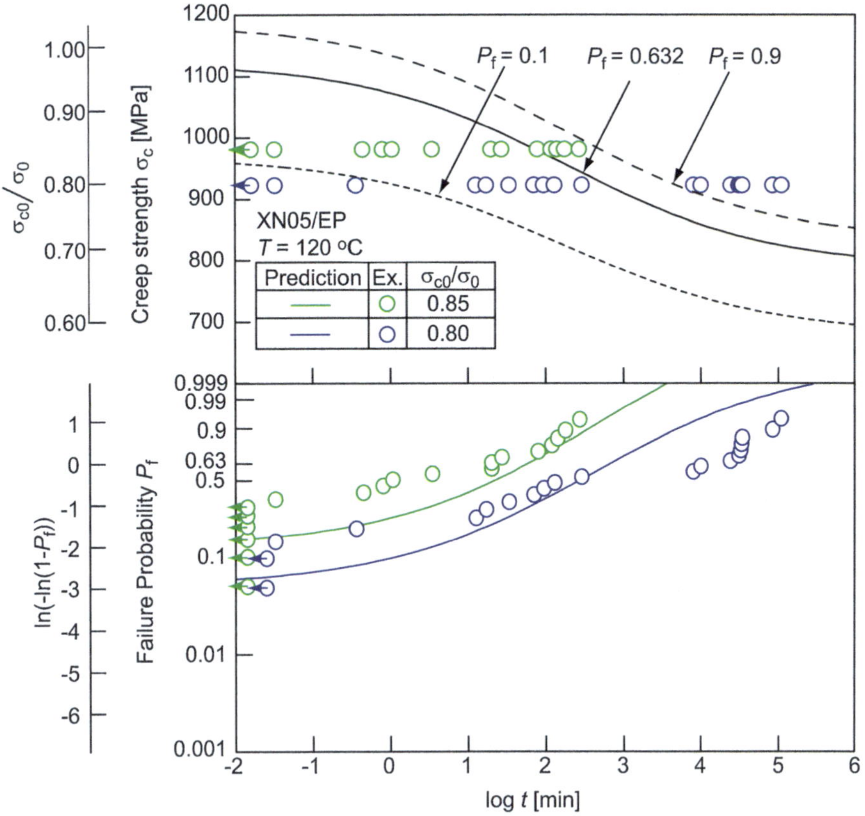

Fig. 25 Experimental and predicted statistical creep failure time for XN05/EP

4.6 Fractographs Obtained After Static and Creep Tests

Two fracture modes exist. Viscoelastic parameter n_R changes drastically with the failure mode. The n_R of T300/EP and XN05/EP with low modulus fibers coincide well with the values predicted from a_c of the carbon fiber strength. The n_R of M40 J/EP and XN50/EP with high modulus fibers are approximately zero in a wide range of compliance of the matrix resin. Figures 27 and 28 show that the former mode is cumulative fracture. The latter is brittle fracture.

Figure 28 shows the definition of depth of tensile fracture surface of CFRP strand. Figures 29 and 30 show the depth of tensile fracture surface for CFRP strands. These behaviors are well correspond respectively to the behaviors of the static strength against the viscoelastic compliance of matrix resin shown in Figs. 15, 16, 17 and 18.

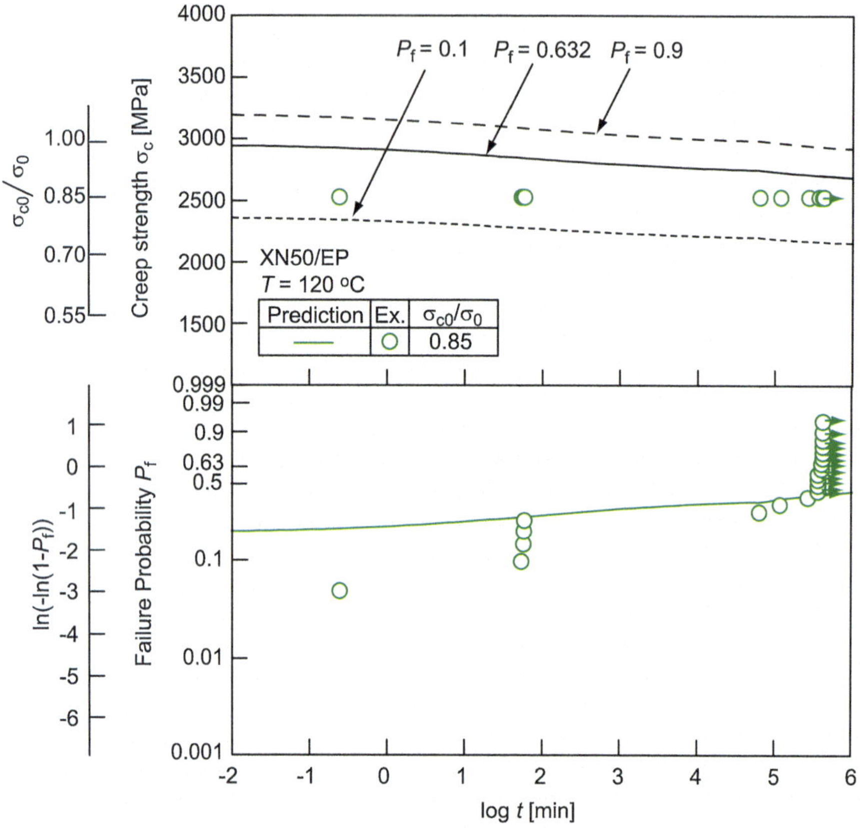

Fig. 26 Experimental and predicted statistical creep failure time for XN50/EP

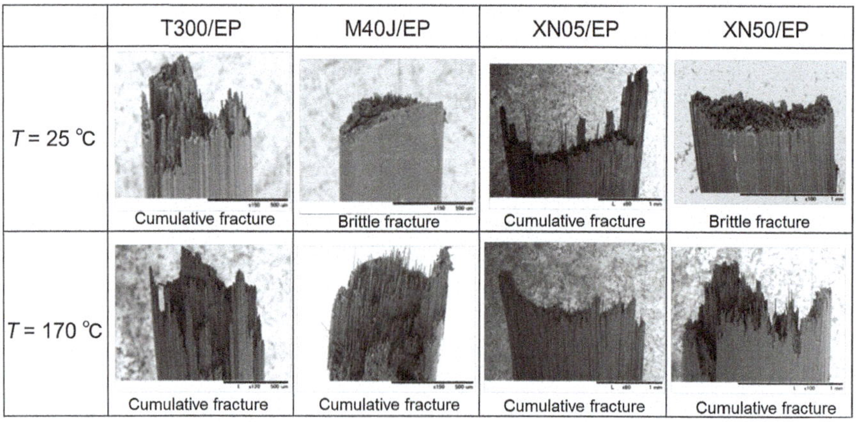

Fig. 27 Fractographs after static tension tests for various CFRP strands

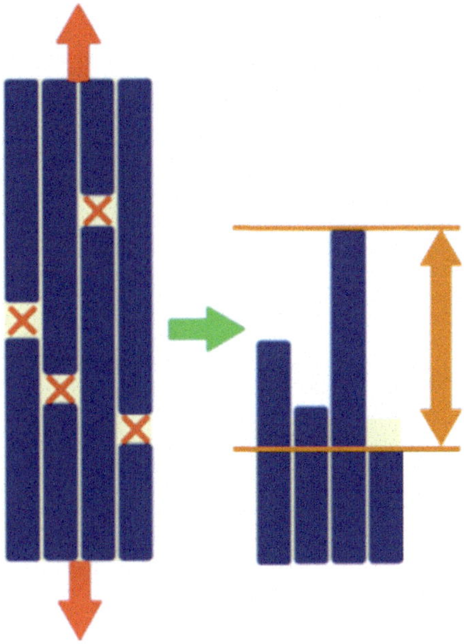

Fig. 28 Definition of depth for tensile fracture surface for CFRP strands

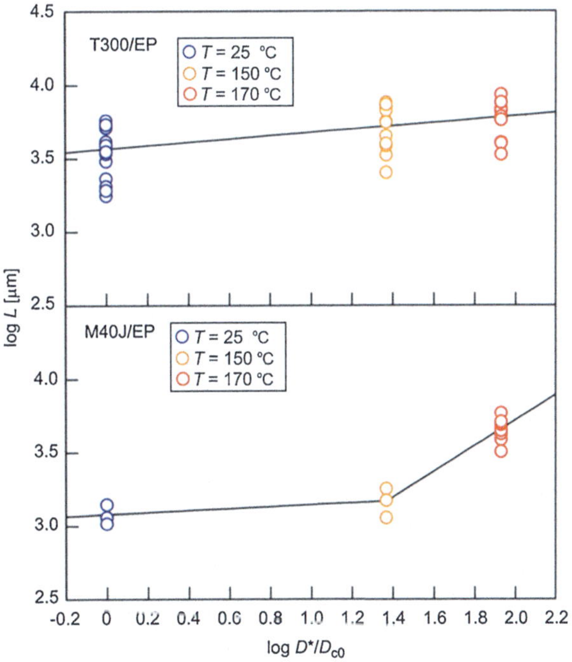

Fig. 29 Depth of tensile fracture surface for T300/EP and M40 J/EP against viscoelastic compliance of matrix resin

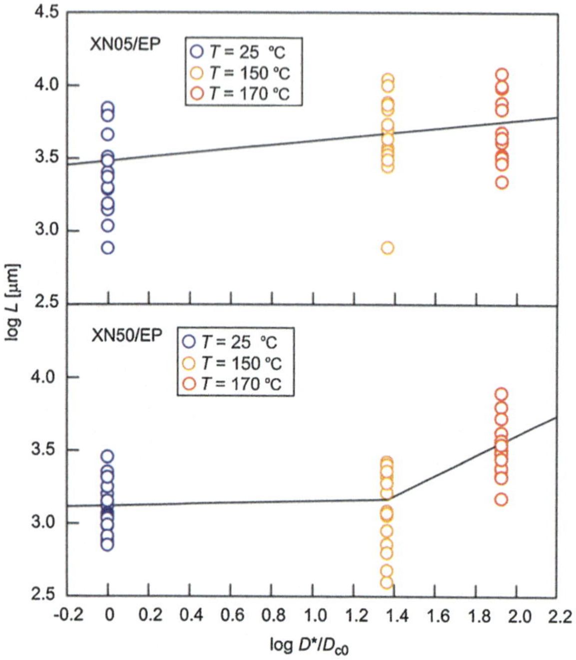

Fig. 30 Depth of tensile fracture surface for XN05/EP and XN50/EP against viscoelastic compliance of matrix resin

5 Conclusion

Formulation for the statistical time-dependent and temperature-dependent static and creep strengths for CFRP based on the matrix resin viscoelasticity was proposed. The validity of formulation was clarified experimentally for tension loading along the longitudinal direction of unidirectional CFRP using our developed CFRP strand system. Results clarified that the long-term creep failure time of unidirectional CFRP under tension loading can be predicted statistically and easily using static tests for CFRP strands at various temperatures. Furthermore, results clarified that two fracture modes exist in CFRP strands under static and creep loadings. Viscoelastic parameter n_R showing sensitivity to the matrix resin viscoelasticity changes drastically with the fracture mode.

Acknowledgments The authors thank Office of Naval Research for supporting this work through ONR award (N62909-16-1-2132) with Dr. Yapa Rajapakse and Dr. Ming-Jen Pan as ONR Program Officers. The authors thank Professor Richard Christensen of Stanford University as one investigator of this project.

References

1. Aboudi J, Cederbaum G (1989) Analysis of viscoelastic laminated composite plates. Compos Struct 12:243–256
2. Sullivan J (1990) Creep and physical aging of composites. Compos Sci Technol 39:207–232
3. Gates T (1992) Experimental characterization of nonlinear, rate dependent behavior in advanced polymer matrix composites. Exp Mech 32:68–73
4. Miyano Y, Nakada M, McMurray MK, Muki R (1997) Prediction of flexural fatigue strength of CFRP composites under arbitrary frequency, stress ratio and temperature. J Compos Mater 31:619–638
5. Kawai M, Yagihashi Y, Hoshi H, Iwahori Y (2013) Anisomorphic constant fatigue life diagrams for quasi-isotropic woven fabric carbon/epoxy laminates under different Hygro-thermal environments. Adv Compos Mater 22:79–98
6. Miyano Y, Nakada M, Cai H (2008) Formulation of long-term creep and fatigue strengths of polymer composites based on accelerated testing methodology. J Compos Mater 42:1897–1919
7. Nakada M, Miyano Y (2015) Advanced accelerated testing methodology for long-term life prediction of CFRP laminates. J Compos Mater 49:163–175
8. Miyano Y, Nakada M, Kudoh H, Muki R (1999) Prediction of tensile fatigue life under temperature environment for unidirectional CFRP. Adv Compos Mater 8:235–246
9. Nakada M, Miyano Y, Kinoshita M, Koga R, Okuya T, Muki R (2002) Time-temperature dependence of tensile strength of unidirectional CFRP. J Compos Mater 36:2567–2581
10. Okuya T, Nakada M, Miyano Y (2013) Reliable test method for tensile strength in longitudinal direction of unidirectional carbon Fiber-reinforced plastics. J Reinf Plast Compos 32:1579–1585
11. Nakada M, Miyano Y (2016) Statistical creep failure time of unidirectional CFRP. Experimental Mechanic 56:653–658
12. Christensen R, Miyano Y (2006) Stress intensity controlled kinetic crack growth and stress history dependent life prediction with statistical variability. Int J Fract 137:77–87

Effect of Seawater on Carbon Fiber Composite Facings and Sandwich Structures With Polymeric Foam Core

Dayakar Penumadu

1 Introduction

An increasing interest in applying polymeric composites and sandwich layups to ship structures requires a fundamental understanding associated with the degradation of static and fatigue behavior of polymer composites due to long term exposure to marine environment. Stitched cross ply carbon fiber and vinyl ester resin based polymeric composites (facings) and sandwich structures consisting of thick PVC foam core materials with very low density are currently being considered by US Navy and the degradation in mechanical properties of these materials due to harsh naval environment is of high interest.

While extensive data are available for polymers and polymeric composites exposed to sea water environment [1, 2], only a small amount of information has been available for carbon fiber vinyl ester composites and sandwich structures manufactured using conditions relevant to the current needs. It is important to have better understanding that marine composites degrade under sea water exposure. The sorption of water in polymeric composites and their effects on composites' performance is a very complex issue requiring polymer science for fluid interactions with polymer and fiber surfaces at a molecular level and applied mechanics for a detailed understanding of the mechanical response. Issues associated with the mechanical performance of composites and sandwich structures in the presence of water concern their dimensional stability, static strength, fatigue response, performance under blast and impact loading, and delamination aspects of facing-core interface. The utilization of polymeric composite based sandwich structures in naval craft is of current interest to US and several European navies. Typically,

D. Penumadu
Department of Civil and Environmental Engineering, 227 Tickle College of Engineering, University of Tennessee, Knoxville, TN, USA
e-mail: dpenumad@utk.edu

S. W. Lee (ed.), *Advances in Thick Section Composite and Sandwich Structures*,
https://doi.org/10.1007/978-3-030-31065-3_19

those lay-ups are made from load carrying efficient materials consisting of closed-cell polymeric foams placed between fiber reinforced polymeric facings. The resulting sandwich components possess an exceedingly light weight, thereby increasing submersibles' buoyancy and, when employed in superstructure designs, may enhance a ship's stability by lowering its center of gravity. In marine applications, such materials and associated structural components are exposed to sea environment over extended period of time. The degradation in mechanical properties from moisture absorption and temperature variations is a major concern for naval structures.

2 Materials

Three individual materials, closed cell PVC foam, carbon fiber reinforced vinyl/ester facings (composite laminates), and foam-facing sandwich layup structure were considered for the study. PVC closed cell H100 foam material (nominal density of 100 kg/m^3), provided by DIAB, was chosen in this study due to its frequent use in naval sandwich structures [3]. The foam panel was obtained from the vendor in the form of panels with 25 mm thickness. Figure 1 shows an example picture of foam panel, its closed cellular micro-structure, and custom developed tensile/torsional specimens with dumbbell shape for suitable mechanism to grip during a tensile/torsional test, and associated dimensions.

The facing material consisted of stitch-bonded fabric of carbon fiber tows resin infused with vinyl ester. Each carbon tow consisted of 12 k Toray's Torayca T700 individual fibers and the facing was laid up into an equibiaxial fabric. The vinyl ester compatible sizing based fabric, designated by LT650-C10-R2VE, was supplied by the Devold AMT AS, Sweden. The aerial weight of the biaxial fabric was 634 g/sq.m

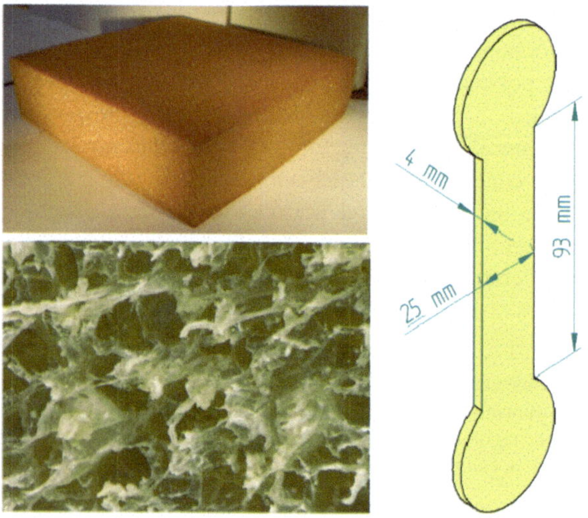

Fig. 1 Polymeric H100 foam

with 315 g/sq.m of fiber in the 0° direction and 305 g/sq.m in the 90° direction. The fibers were stitched together with a 14 g/sq.m polyester knitting thread. Toray's Torayca T700 carbon fiber was chosen because of its cost, stiffness, and strength. From manufacturer's data, the T700 fiber had a tensile strength of 4.9 GPa (711 ksi), a tensile modulus of 230 GPa (33.4 Msi), and an elongation of 2.1%. The matrix material was Dow Chemical's Derakane 510A-40, a brominated vinyl ester, formulated for the VARTM process [4]. The bromination imparts a fire-resistant property to the composite which is essential for ship structures which have a potential to be exposed to onboard fire issue. Vinyl ester has a higher fracture strain than the typical polyesters, and hence produces composites with superior mechanical properties and impact resistance. The fiber volume was found to be 58% by the areal density method and includes 2.2% weight of polyester stitch fiber. Carbon fiber reinforced/vinyl ester laminated composites consisting of $[0/90]_{2s}$ and $[\pm 45]_{2s}$ cross-stitched lay-ups were utilized in this study. Test materials were fabricated following a standard manufacturing protocol using a resin whose viscosity was optimized for Vacuum Assisted Resin Transfer Molding (VARTM) process and all panels were post-cured for consistent quality [4] after infusion and target time of curing at room temperature. The degree of curing was ascertained using differential scanning calorimetry by extracting samples from various locations from a panel.

The sandwich layup consists of a closed cell polymeric foam layer placed between thin carbon fiber reinforced polymeric composite facings that are described above. The composite sandwich panels of size 60 cm × 90 cm (2 ft. × 3 ft) and 2.54 cm (1 in) thickness were fabricated using the VARTM process. Typical test specimens of facing and sandwich beams specimens are shown in Fig. 2.

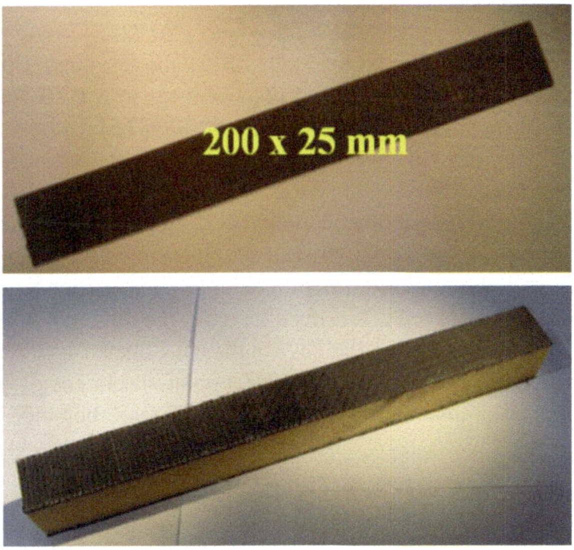

Fig. 2 Carbon fiber reinforced vinyl ester facing (top) and sandwich lay-up made of facing/H100 foam (bottom)

3 Materials Preconditioning to Simulate Marine Environment

To evaluate the effects from sea environment, simulated sea water using coarse sea salt from Frontier with a final density of 1.025 g/cc was used to immerse all samples in controlled water bath at 40 °C. Immersion duration ("Wet" specimens) was decided when they reached a fully saturated state, determined from moisture weight gain data with time. For weight gain measurements, each soaked sample after a given duration was removed from the water bath and patted dry with a paper towel. Micro-scale was used to record the water uptake data prior to determining mechanical properties. Some of the saturated composite coupon specimens were also subjected to low temperature between 0 °C to −15 °C in a freezer or environmental chamber, and subsequently tested at target duration to evaluate cryogenic effect. The test equipment includes an environment chamber that used liquid nitrogen (LN2 and the objective was to understand the material behavior as function of aging condition at cryogenic temperature in sea water environment to mimic the exposure of ship structures in extreme cold temperatures.

4 Seawater and Temperature Effects on Constituent Materials

Polymeric H100 foam

A biaxial servo-hydraulic mechanical testing MTS 858 system with a full-scale capacity of 25 kN axial force and 250 N-m torque was fitted with an external 110 N (25 lbf) axial load cell for tensile tests and a 0.7 N-m torque load cell for torsional tests to conduct mechanical tests on PVC foam core material. To obtain tensile modulus of foam core, extension tests were performed at strain rate of a 0.5 mm/min and up to a peak stress of 0.6 MPa, a state of stress that is well below the non-linear stress-strain relationship. In a view of previous experience, each specimen was subjected to three load/unload cycles at room temperature to establish a repeatable value of Young's modulus, *E*. Subsequently, the same specimens was immersed in sea water roughly 2 months and frozen down to a temperature of about −5 °C in a freezer and tested in tension after exposure duration of 2 and 6 weeks, respectively. It is important to test the environmental effects on the same specimen due to large variation in properties of foam coupon specimens obtained from same panel. The properties also vary through thickness as will be noted below.

Based on previous research, it was concluded the sea water penetrated only the outer cellular layer of the foam. This explains why tensile mechanical properties were insensitive to the presence of sea water as discussed in the next section. Therefore, a torsion test, used to measure shear modulus *G*, was deemed to be more likely to capture the effects of sea water, since in that test the highest shear stresses develop at the outer locations of the prismatic shaped specimen. Ten specimens were used to determine freezing effects on dry specimens and additional 10 specimens were prepared and used to test wet specimens which were soaked in sea water for 8 weeks and tested at room temperature. These wet coupons were subsequently kept in a freezer at −5 °C for 4 weeks and tested again in torsion. A rotation range of $-5^\circ \leq \theta \leq 5^\circ$ at rate of 10 degree/min was used to study the effects of sea water and low temperature on the shear modulus G of foam. These values of rotation correspond to shear strain values well below yield state of stress.

Carbon vinyl/ester facing laminates

Tensile tests were performed on 200 mm long and 25 mm wide specimens and approximately 2 mm thick to determine the effect of freeze/thaw cycling on the Young's modulus, E of $[0/90]_{2s}$ and $[\pm 45]_{2s}$ layups. An MTS 810 test system with 250 kN load cell was employed under strain control at rate of 300 με/min and up to 750 με, which is well within the linear elastic axial strain level. Failure strength test was also performed using the same equipment and procedure. Strains were recorded by an extensometer. Tabs were mounted at the ends of the specimens, in order to direct failure away from the stress concentrations at the gripped portions. Prior to the tensile tests, all the specimens were completely dried at ambient temperature in desiccators.

The facing specimens were enclosed within an environmental chamber connected to liquid nitrogen to evaluate the low temperature effect on water soaked specimens. Note that, as desired, the profile was controlled to vary between the ambient temperature down to 0 °C, −15 °C, and then back up to 0 °C, with a final return to room temperature. Each of the above temperature levels were maintained for at least 2 h. The sample was held at zero load in the servo-hydraulic system throughout the thermal excursions to allow for free thermal expansion (or shrinkage).

Foam/facing interface fracture toughness

The composite sandwich panels of size 60 × 90 cm as mentioned above were cut and machined to form 254 mm long, 25.4 mm wide, and approximately 29 mm thick sandwich specimens. In addition to the foam and facings, the thickness dimension included also the thickness of top and bottom adhesive interfaces. Fifty millimeter long pre-cracks were cut with a sharp edge along the top facing/core interface in order to stimulate the growth of the interfacial delamination [5]. Load application was facilitated by using hinges glued to the facings in a modified double cantilever beam (DCB) configuration which was followed the procedure in previous work [6]. The hinges were mounted 38 mm away from the edge of the loaded, top, facing and the sandwich specimens were centered about the sides of the steel plates. The effect of exposure to sea water on fracture toughness was investigated by pre-soaking the specimens for 3 months in simulated sea water using natural coarse sea salt at a controlled temperature of 40 °C in a water bath prior to the initial delamination testing. Subsequently the specimens were soaked for interim periods of 2 weeks during intermittent crack growth between unloading and reloading cycles. It was necessary to re-soak each "wet" specimen for at least 2–3 weeks, in order to

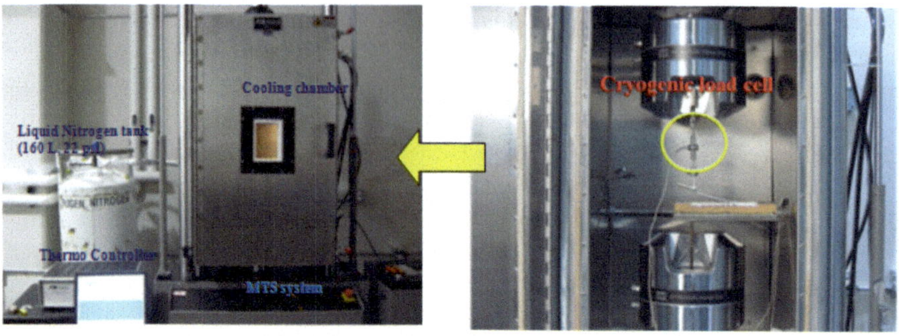

Fig. 3 Delamination test set up including an environmental controlled chamber

maintain a fully saturated crack tip region after each unloading and prior to each reloading.

A 250 kN MTS test system was employed in combination with an MTS 609 alignment fixture and a smaller 50 kN axial force transducer. Loading was introduced by means of the larger, 250 kN, MTS machine under displacement control at a crosshead rate of 2 mm/min. The specimens were placed an inside environmental chamber, within which was encased custom made fixture attached to 1 kN cryogenic tensile load cell perform delamination tests as shown in Fig. 3. Programmed loads were monotonically increased under displacement control until noting an abrupt drop in their amplitudes, at which stage crack extensions were observed and the machine unloaded back to zero displacement. This procedure was repeated until delamination's approached the far edge of a specimen. The digital images of the interfacial cracks were analyzed using digital image analysis software (ImagePro®). Crack morphology was also determined for possible use in calculating values of critical energy release rate. This was done by transferring the gray scale image to binary form using thresholding technique and by focusing on the region of crack-tip. The exact length of crack was obtained by tracking the number of pixels and conversion to length units using the known optical magnification at which digital images were obtained.

Tensile modulus of foam

Figure 4 depicts the typical loading data of PVC foam and a custom made fixture and setup. Based on previous experience, each specimen was subjected to three load/unload cycles to establish a repeatable value of *E*. The modulus was obtained from averaging only the second and third load cycles, since those presented almost identical information. Table 1 shows average tensile modulus value from 20 specimens at up to the same target stress level. The data for the same tests, conducted after 2 and then 6 weeks exposure to −5 °C, exhibit no degradation in the values of *E*. It can be seen that exposure to low temperature over period of time does not significantly affect the value of *E* for specimens. The minute degradation of *E* falls well within the data scatter caused by foam inhomogeneity. Foam specimens were tested before immersing in seawater for 10 weeks and tested at room temperature to obtain

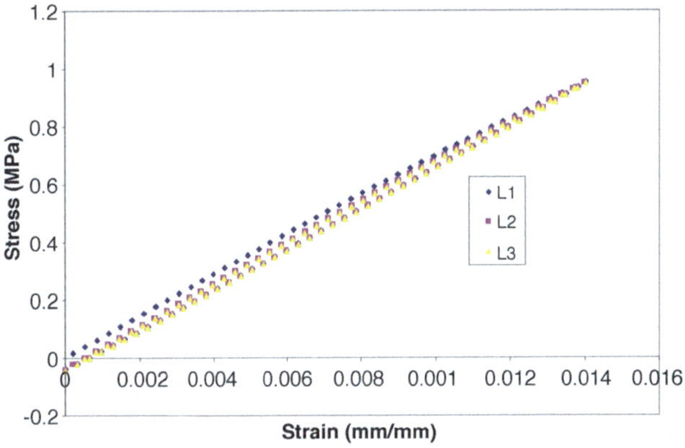

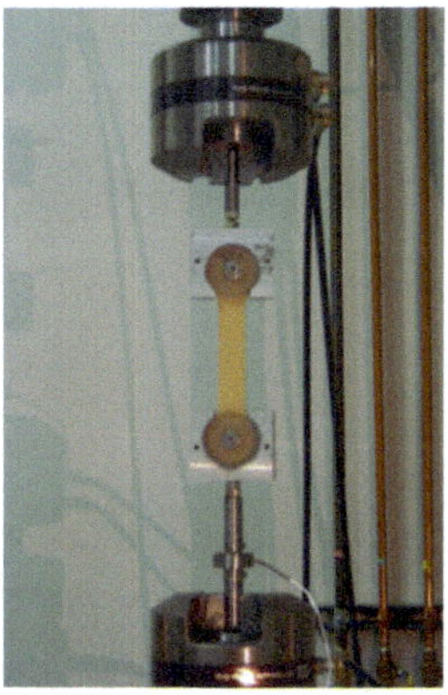

Fig. 4 Typical three cycles used to determine *E* modulus and tensile test set up

Table 1 Low temperature effect on foam modulus in tension

Condition	Room	0 °C (2 weeks)	0 °C (6 weeks)
Dry	70.66 MPa	70.84 MPa	70.42 MPa

Fig. 5 Summary of tensile modulus variation of H100 PVC foam before/after sea water saturation and freezing for 4 weeks

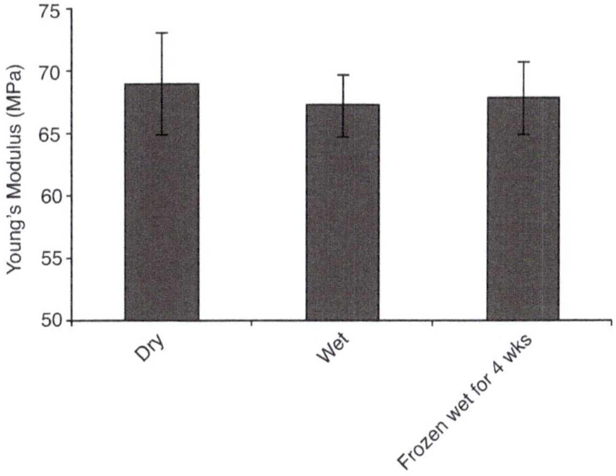

comparative values of *E*. It was noted the sea water causes approximately 5% degradation in *E*. Subsequently, those "wet" specimens were frozen for 4 weeks and tested in tension to obtain the corresponding values of *E*. The data yielded nearly identical results before and after freezing as shown in Fig. 5.

For carbon fiber vinyl/ester facing, tensile tests were performed on 200 mm long and 25 mm wide specimens to determine the effect of freeze/thaw cycling on the Young's modulus, *E* of $[0/90]_{2s}$ and $[\pm 45]_{2s}$ layups. The tensile modulus was determined from the slope of the stress-strain curve in elastic range prior to yield. In the dry state, the facing composites were found to have approximate average

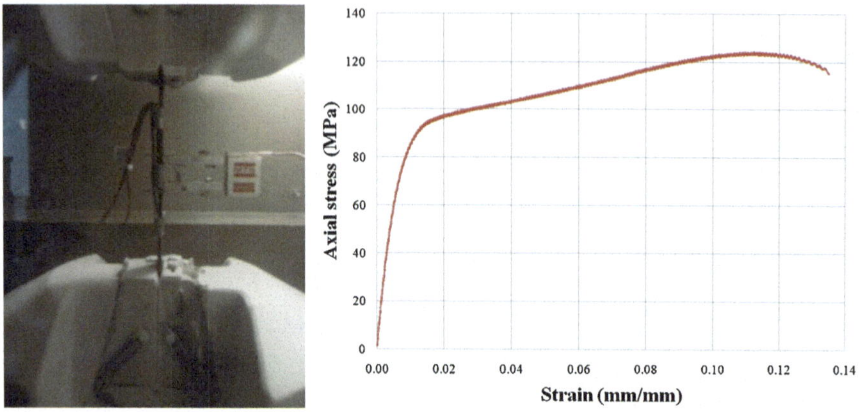

Fig. 6 Low temperature setup and behavior of $[\pm 45]_{2s}$ carbon fiber vinyl ester facing

moduli of elasticity of 80 GPa and 15 GPa, respectively. Figure 6 shows the setup of a tensile test and a typical stress-strain result. In a view of authors' past research to evaluate environmental effects, $[\pm 45]_{2s}$ facing layup show more pronounced degradation; therefore, this study focused only on this specific orientation of facing, where resin behavior also comes into play and thus most vulnerable for environmental effects. The initial values of E were ascertained by subjecting each specimen to three cyclic tests prior to thermal exposure, resulting in 15 GPa $\leq E \leq$ 16.5 GPa.

A similar procedure was employed to record values of E after the aforementioned exposure to cyclic temperature. The sea water effect was evaluated by pre-immersing the sample at 40 °C at least 3 months and recording E immediately upon their removal from the bath. It was noted that immersion caused only small differences in properties of the $[\pm 45]_{2s}$ facings. A similar procedure was employed to record values of E after exposure to sea water and cyclic low temperature. After measuring stiffness immediately upon their removal from the bath, it was followed by freezing the wet specimens at −10 °C for 2 weeks and obtaining the subsequent values of E inside chamber.

The procedure was repeated 6 weeks later. To simulate a cold environment, the target temperature in the chamber was held roughly 2 h before testing. The overall results were close to those of dry facing where the average stiffness at low temperature increased by 3–5% at −10 °C beyond those of the dry specimens as shown in Fig. 7. The slightly increased values are possibly attributed to an improved mechanical factor by the generation of thermally induced internal compressive stresses within the $[\pm 45]_{2s}$ coupons which enhance the friction between the contracting resin and the fibers.

Shear modulus of foam core material

The shear modulus G was recorded by means of torsional tests, twisting the sample over the angular range of $-5° \leq \theta \leq 5°$. The base-line variability ranged between 23.3 MPa $\leq G \leq$ 28.9 MPa tested under torsion at room temperature. It was

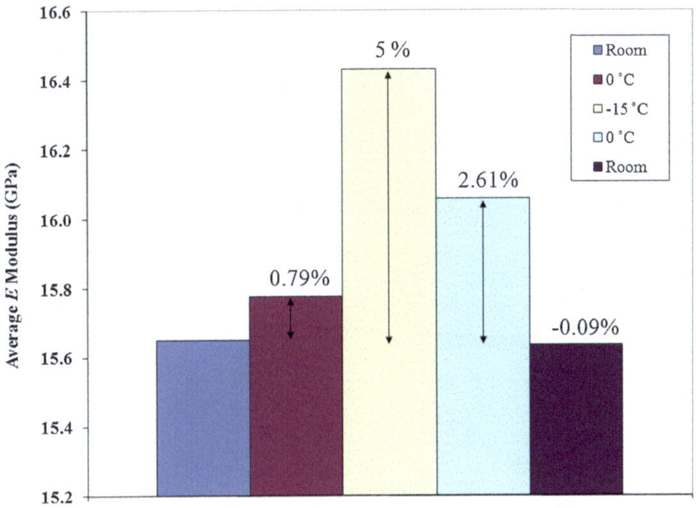

Fig. 7 Average values of dry elastic modulus for composite facings (45 degree orientation) subjected to low temperatures

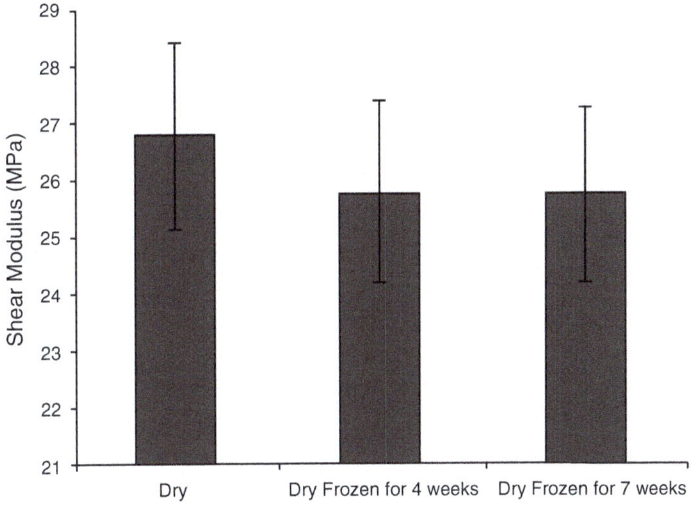

Fig. 8 Shear modulus before/after freezing

necessary to establish their reference values for a comparative purpose. Shear modulus values for 10 specimens tested in the dry state exhibit fairly comparative reductions from room temperature levels down to those due to subsequent freezing for 4 and 7 weeks. Reductions of shear modulus 3–6% were noted after 4 weeks of freezing and no further drop at 7 weeks shown in Fig. 8.

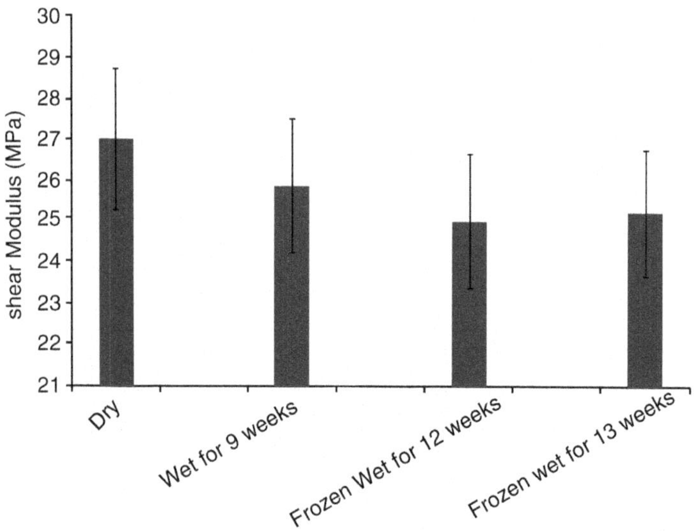

Fig. 9 Shear modulus due to sea water and low temperature

Additional ten specimens were used to study the combined effects of sea water and cryogenic temperature. The results in Fig. 9, show an overall degradation of 5% with a further of slight decrease around 2% after freezing for additional 13 weeks. Immersion in sea water resulted in an overall reduction of up to 5% in G, though it should be borne in mind that this reduction is affected by sea water that is confined to a region that is only 0.4–0.5 mm (δ_T) deep along the outer boundary of the specimen.

To evaluate the wet shear modulus of the foam G_w [7], Fig. 10 shows all known dimensions and materials properties, except G_w. Employing the well-known expression relating the torsional behavior of a solid rectangular sections of height D and width W, and a similar relationship for a hollow rectangular tube of the same outer dimension and thickness δ_T. These are

$$\left[\frac{M_T^{(s)}}{\theta}\right]_{solid} = \frac{1}{3}H_s h_s^3 G_s \text{ and } \left[\frac{M_T^{(o)}}{\theta}\right]_{hollow} = \frac{4A^2 G\delta_T}{L} \tag{1}$$

where $A = (D - \delta_T)(W - \delta_T)$ and $L = 2(D + W - 2\delta_T)$

The saturated sample consist of an inner, essentially dry, core of modulus G and dimensions $H_s = D - 2\delta_T$ and $h_s = W - 2\delta_T$. This core is encased within a thin saturated tube of thickness δ_T. Denote by R the ratio $\left(\frac{M_T}{\theta}\right)_{dry} \Big/ \left(\frac{M_T}{\theta}\right)_{wet}$. One obtains

$$R = \left(\frac{H_s}{D}\right)\left(\frac{h_s}{W}\right)^3 + \frac{12A^2\delta_T}{LDW^3}\frac{G_w}{G} \tag{2}$$

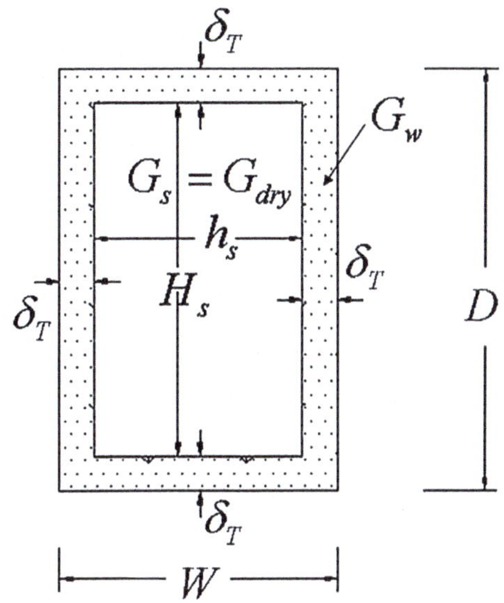

Fig. 10 A sketch of the cross section subjected to torsion. Note that G_{dry} and the overall torsional resistance are experimentally determined. The only unknown is G_w of the saturated region

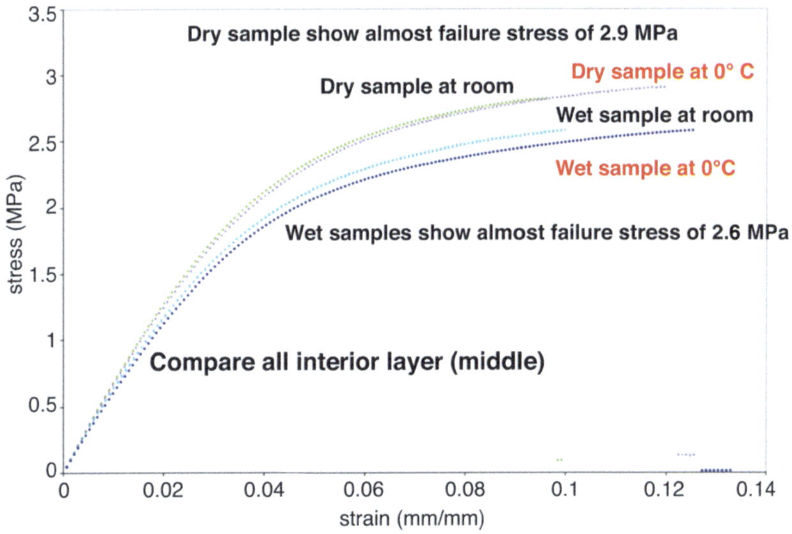

Fig. 11 Comparison of failure stress of PVC foam subjected to sea water effect

In summary, it was shown that low temperature and sea water do not significantly reduce tensile properties but that the overall shear modulus had decreased by 7% at sub-zero temperature.

Failure strength of foam core material and matrix dominated facing
Strength of PVC foam under tensile test indicated an average value of 2.5 MPa in virgin state of the materil. Figure 11 shows a general trend of polymeric foam

Table 2 Critical energy release rate for interface delamination of face sheet and PVC foam core

Condition (No of samples)	G_c (Front) (N/m^2)	G_c (Back) (N/m^2)	% Avg. Degradation
Dry (> 15)	580–952	541–963	
Wet (> 10)	432–619	451–632	~30%
Cyclic wet frozen (>8)	405–690	420–620	No additional degradation

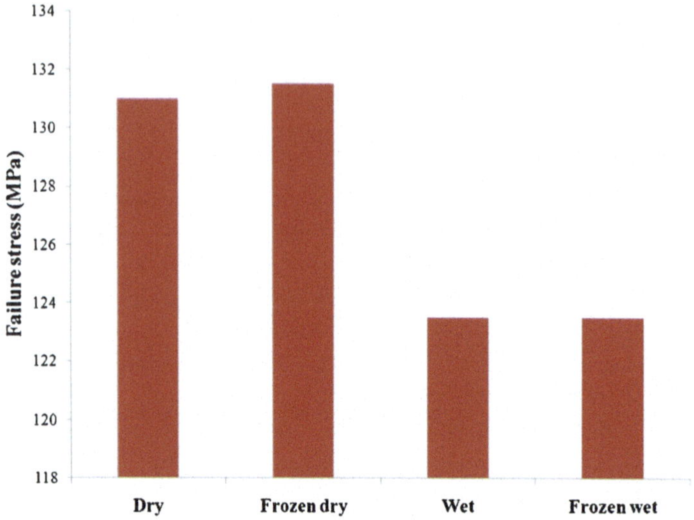

Fig. 12 The $[\pm 45]_{2s}$ degradation of failure stress before/after exposed to sea environment and low temperature

subjected to sea water and low temperature. The results in Table 2 show that the sea environment could degrade the strength up to 10% and combined effect of sea water and low temperature only slightly degrades further.

Comparative study on carbon fiber reinforced vinyl/ester facing shows that strength data of facing for coupons extracted along $[\pm 45]_{2s}$ yielded approximately 130 MPa and no difference resulted in exposure to low temperature, but a further decrease of 5% due to sea water effect. No further significant reduction shows on combined effects of sea water and low temperature. Comparative results are summarized in Fig. 12.

Delamination of face-foam core for sandwich material

The resistance to delamination growth can be characterized by the strain energy release rates (G), where the critical energy release rate (G_c) is used as a measure of the interminar fracture toughness. Assuming a linear elastic response, it was chosen to determine the critical strain energy release rates G_c by means of the "area method". The area under each load/unload cycle was calculated numerically using the trapezoidal rule. Critical energy release rate, which defines the fracture toughness, was obtained using the expression:

$$G_c = \frac{1}{b} \cdot \frac{\Delta U}{\Delta a} \tag{3}$$

Where ΔU is the area under the load-displacement trace as the crack grows; Δa is the extended crack length recorded during the test, and b is the width of specimen. The value calculated is the average energy consumed for the crack extension Δa. The wide range of values listed in Table 2, where values are listed for crack lengths observed on both sides of the specimens, reflects the cycle to cycle variability in the thickness of the facings as well as the random nature of the foam cell structure. Accordingly, exposure to sea water resulted in approximate 30% reduction in G_c and exposure to combination of sea water and cryogenic temperature slightly degrades it. These results indicate that exposure to sea environment significantly reduces the interfacial toughness between facings and foam core material.

5 Effect of Seawater and Microstructure on Composite Laminates (Facings)

CF/VE specimens were prepared using different ply lay-ups, namely $[0/90]_{2S}$ and $[\pm 45]_{2S}$ as described earlier to achieve fiber and matrix dominated failure respectively. Tensile tests were performed under deformation control at a constant cross-head rate of 0.1 mm/min, at room temperature in air. In order to minimize the effect of gripping stress on the mechanical behavior and to avoid failures within tab section, 25 mm long tabs were attached to the ends of the specimens using a suitable adhesive. Where experimentally feasible, an extensometer was used to record strain data for comparison purposes. It showed comparable results with Digital Image Correlation (DIC) technique used to obtain surface strain data.

The DIC system (commercially available, VIC-3D) uses dual-cameras to measure surface displacements, and full-field surface strains in three dimensions (Fig. 13). Mechanical properties of carbon fiber reinforced vinyl ester composites (CF/VE), consisting of fiber dominated samples of $[0/90]_{2S}$ and matrix dominated samples of $[\pm 45]_{2S}$ orientation, using variable specimen sizes were evaluated.

Fiber and matrix dominated specimens, $[0/90]_{2S}$ and $[\pm 45]_{2S}$ repectively, showed similar stress-strain behavior for all sample sizes considered in this study. No damage was found until catastrophic failure as shown in Fig. 14(a) for fiber dominated specimens corresponding to $[0/90]_{2S}$ lay-up. The strain information reported in these stress-strain curves was obtained using DIC technique. Figure 14(b) shows the deformation behavior for a matrix dominated specimen, showing significant damage accumulation beyond 80 MPa axial stress. Typical mechanical properties of fiber and matrix dominated carbon fiber reinforced vinyl ester composites can be found in Table 3.

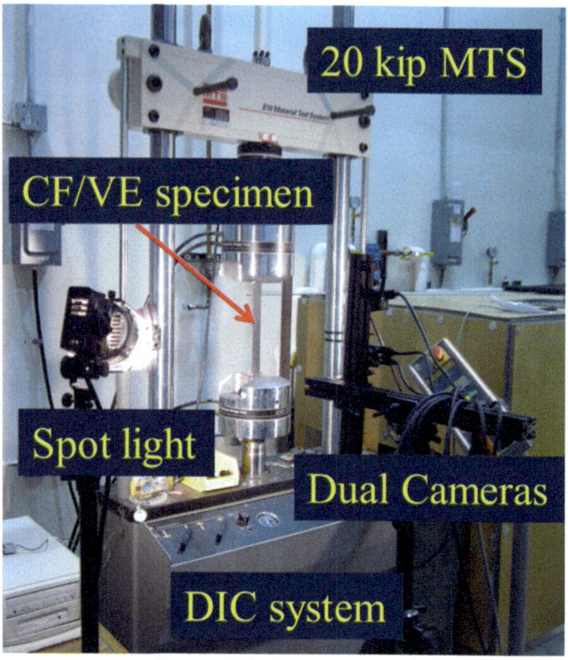

Fig. 13 Experimental set-up for a tensile test using a servo-hydraulic loading system, coupled with DIC system for obtaining strain data on the specimen surface

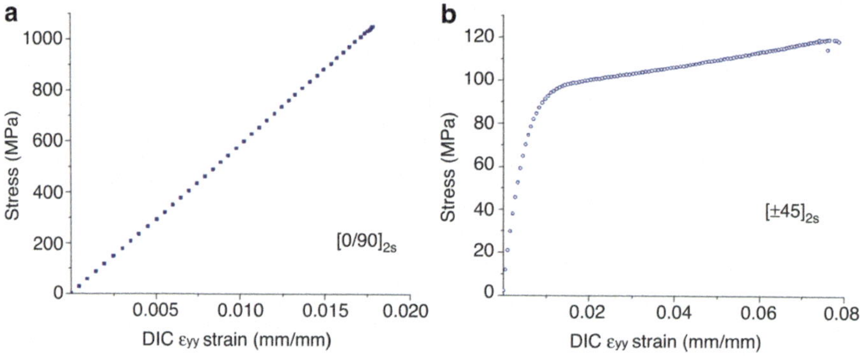

Fig. 14 Typical stress-strain behavior of (**a**) fiber $[0/90]_{2S}$ and (**b**) matrix $[\pm 45]_{2S}$ dominated CF/VE sample dimensioned 300 mm long by 25 mm wide

Table 3 Typical material properties of CF/VE composite facings

Property	Value	Dimension
Longitudinal modulus $[0/90]_{2S}$	60	GPa
Longitudinal modulus $[\pm 45]_{2S}$	15	GPa
Failure strength $[0/90]_{2S}$	1000	MPa
Failure strength $[\pm 45]_{2S}$	120	MPa

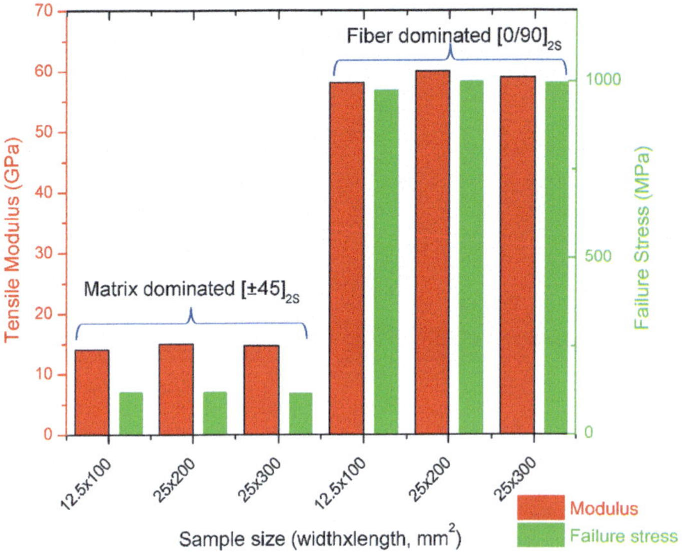

Fig. 15 Effect of specimen size on mechanical properties of laminates

CF/VE specimen sizes were scaled up of 2 and 3 from the baseline case with keeping the constant thickness from same four cross stitched plies (2.8 mm). The large size of both fiber and matrix dominated specimen showed similar strength and moduli as summarized in Fig. 15.

Using the observations obtained from strain variations using 3D-DIC, it is concluded that the failure of the CF/VE samples are dominated by strain localization (Figs. 16 and 17) possibly from manufacturing defects sites expected of VARTM technique and specimen size effects were not significant. Since strength and moduli are controlled by defects, especially voids and manufacturing defects associated with varied amount of cross-link density and fiber interfacial shear strength during specimen preparation, it is crucial to make sure that good manufacturing quality of test specimens is achieved in production. A fundamental question of why certain zones lead to large accumulation of strain localization for the carbon fiber composites was further explored by a using x-ray based tomography in non-invasive mode.

CF/VE samples were also subjected to controlled rate of strain loading with ten tensile load-unload loops (corresponding to unloading at 20, 40, 60, 80 MPa, and 5 loops at 100 MPa) to evaluate accumulated plastic strain as shown in Fig. 18a. Fiber dominated CF/VE $[0/90]_{2S}$ composites under ten cycles of up to 80% failure stress load-unload did not show accumulated permanent strain.

In order to evaluate the damage evolution due to sea water effects compared to time aged sample, a saturated CF/VE $[\pm 45]_{2S}$ that was soaking in sea water for 3 years was also subjected to similar testing. This particular sample had a moisture uptake of 0.8% based on periodic weight gain measurement. Results included in Fig. 18 shows that exposure to sea water substantially increased the damage measured at various strain or stress levels quantifying the relative effect of marine

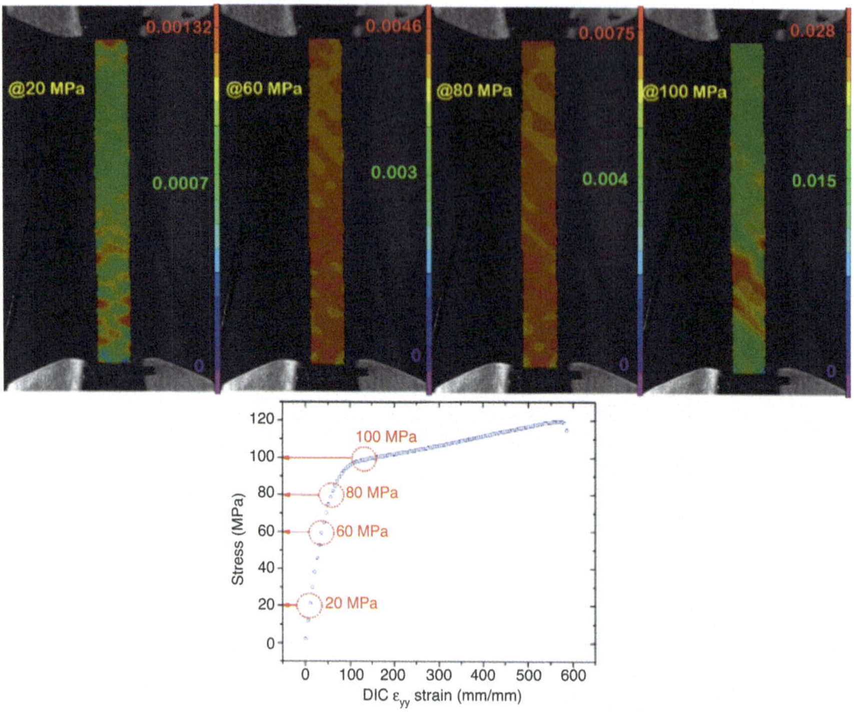

Fig. 16 Deformation variation of matrix dominated CF/VE using major principal strains at various specific stress levels

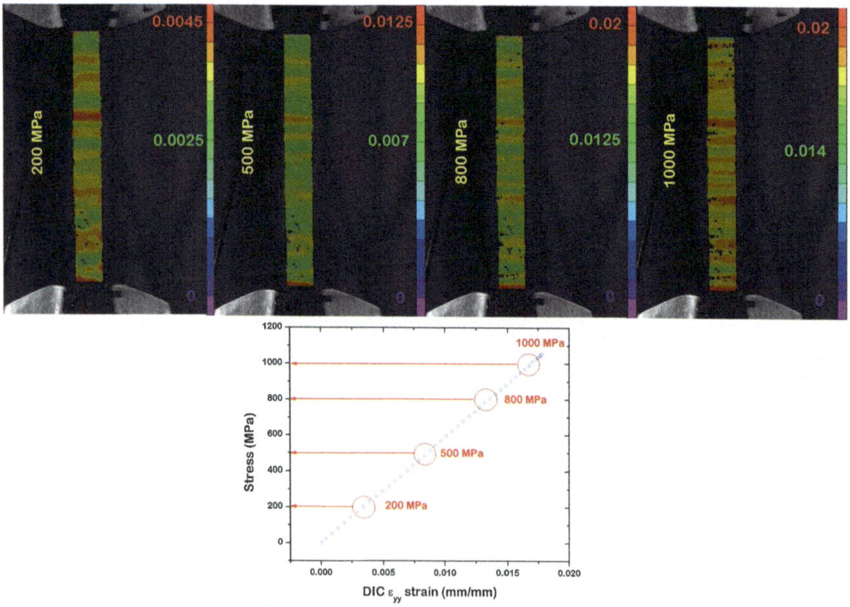

Fig. 17 Deformation variation of fiber dominated CF/VE using major principal strains at various specific stress levels

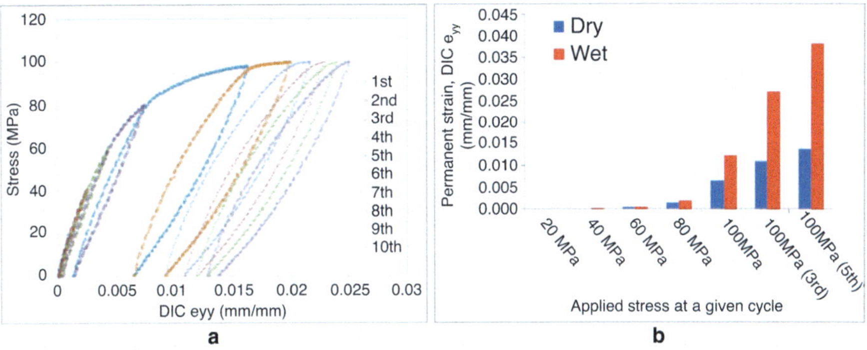

Fig. 18 (**a**) Typical load-unload behavior of undamaged $[\pm 45]_{2S}$ CF/VE, (**b**) Damage evolution of "wet" and dry matrix dominate CF/VE subjected to tensile load-unload varying from first cycle to 10th cycle

environment on damage evolution. After 10 cycles of tensile load-unload, saturated $[\pm 45]_{2S}$ CF/VE sample showed substantial matrix cracking on the sample surface unlike the case of dry and aged specimen of $[\pm 45]_{2S}$ CF/VE composite. Corresponding to the 10th load-unload cycle at the amplitude of axial stress prior to unloading of 100 MPa, large strain variations along the sample length with high strain localization at mid height were observed.

6 Fatigue Behavior of Sandwich Facing Material and Seawater Effect

Under tension-tension fatigue test, orientation dependence of CF/VE exhibit disparate failure mechanisms when fatigued in air or under immersed conditions that lead to earlier failures when submerged in fluid. The secant and tangent modulus, E_S and E_T respectively were used to determine the degradation due to sea water confinement effect at each orientation [8–12]. The fatigue life of composites is denoted by the number of load cycles to failure defined as a target state of strain accumulation or catastrophic failure. It was observed when comparing peak strain subjected to same stress level of $0.5\times\sigma_{\text{failure}}$ between fiber and resin dominated composite laminates, sea water confined effects did not significantly reduce fatigue life on fiber dominated specimens. On the other hand, sea water confinement has a significant effect to reduce fatigue life on resin dominated CF/VE. In the present research, tangent moduli comparison between fiber and matrix dominated CF/VE at the same stress level of $2/3\sigma_{\text{failure}}$ showed the similar observation as mentioned above. Note that a cyclic fatigue tests were suspended when 1 millions of cycles were achieved. This fatigue endurance limit was used in several studies in the past [12, 13]. Based on the results, only matrix dominated samples are considered for detailed studies on degradation in fatigue life due to seawater exposure and confinement.

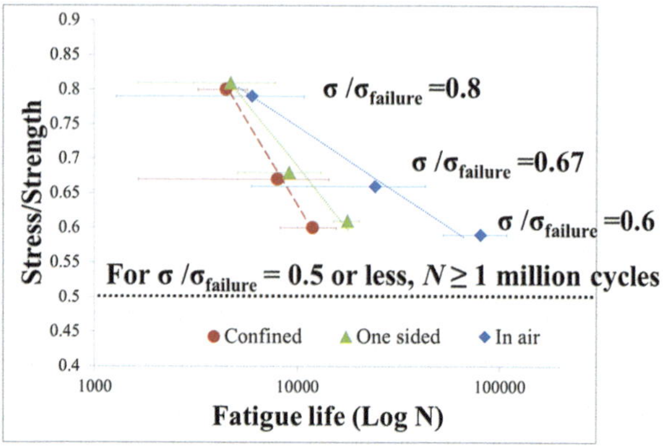

Fig. 19 Fatigue life of $[\pm 45]_{2S}$ CF/VE in air, and water confinement

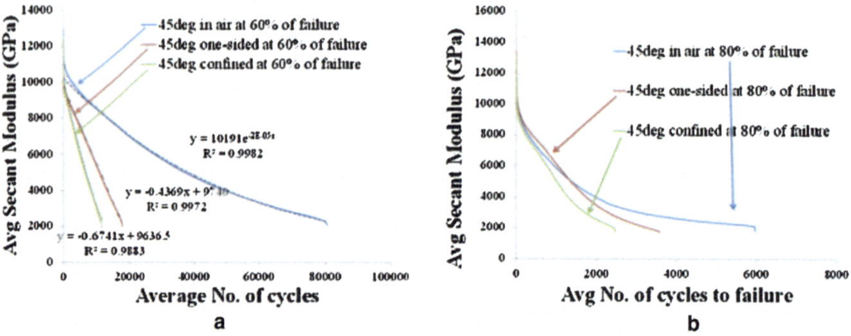

Fig. 20 (**a**) Fatigue life data of $[\pm 45]_{2S}$ CF/VE to in air, water confinement, and one-sided confined, (**b**) Secant modulus degradation model using simple regression analysis

Fatigue life of CF/VE composite samples on $[\pm 45]_{2S}$, which have the average static failure strength at 120 MPa, were obtained at maximum applied stress levels of 0.8, 0.67, 0.6, and 0.5. Fatigue life due to sea water confinement and one-sided confined effects are summarized in Fig. 19 when compared to cyclic loading in air. A significant degradation of number of cycles to failure from sea water confinement effects was observed for all stress levels considered in this study, up to 80% of static strength as the maximum cyclic stress value for a stress ratio value R of 0.1. Prediction of fatigue life in composites is very complicated because of multiple failure mechanisms possible including fiber interface initiated damage, fiber breakage, matrix cracking and associated debonding. With different damage accumulation mechanisms in composites, secant modulus degradation as in Fig. 20a from measured data was found to be a representative and simple parameter to provide tension-tension fatigue life prediction. Due to water confinement effects a simple linear regression based predictions will provide an adequate estimate of number of cyclic failures and the fatigue life available. However, accumulated fatigue damage of CF/VE using secant modulus degradation in air does not appear to show this linear dependency with cycles of loading. These results are important to develop the

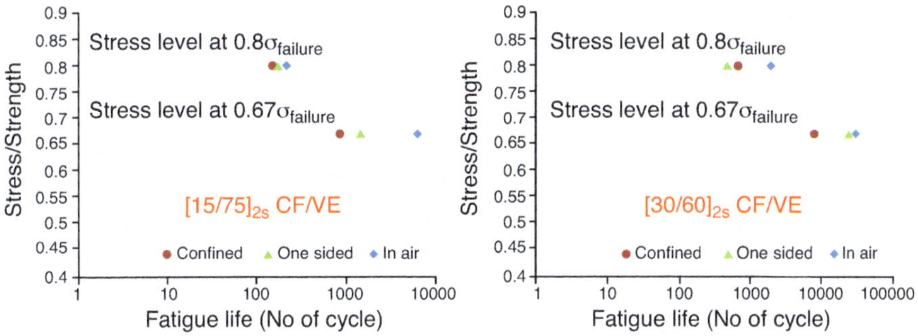

Fig. 21 Fatigue life data of $[15/75]_{2S}$ and $[30/60]_{2S}$ CF/VE due to in air, water confinement, and one-sided confined

endurance fatigue limit for use in marine design. Similar observations were found in $[15/75]_{2S}$ and $[30/60]_{2S}$ from Fig. 20b and the number of cycles to failure reduced considerably when compared to dry condition in air. True strain data prior to failure showed a similar trend when comparing dry sea water confinement or one-sided confined samples to dry in air samples.

Comparison of experimental results for $[15/75]_{2S}$, $[30/60]_{2S}$, and $[\pm 45]_{2S}$ laminates of CF/VE composites yielded failures under much fewer numbers of cycles at $2/3\sigma_{failure}$ when fatigued under completely immersed conditions than in air can be found in a previous year report. Fatigue life of $[15/75]_{2S}$ and $[30/60]_{2S}$ due to sea water confinement is compared in Fig. 21. It was supported by X-ray tomography results that the damage development in matrix dominated samples is the combination of the coalesced matrix cracks parallel to the fiber and delaminations; however, voids between matrix/fiber rapidly grow and lead to the early fatigue failure mode when confined in sea water. Fatigue behaviors of CF/VE at various orientations as function of stress levels and different R-ratio have been considered and master curves developed for naval design applications.

7 CFVE Sandwich Structure Fatigue Behavior Exposed to Sea Water

Past research on CFVE sandwich structure response has been focused largely on delamination fracture testing and response from three/four point bending. Those studies involved fiber dominated CFVE sandwich materials in the experiments. It is important to consider matrix dominated sandwich structures, especially under water confinement conditions due to the large role matrix resin plays in damage evolution under fatigue loading in tension. Recent work of the author showed that water confinement effects are very pronounced for matrix dominated CFVE results. Therefore, similar experiments were conducted, now using actual sandwich structures in matrix dominated mode, and results are being evaluated by comparing

Fig. 22 Experimental set-up with dry matrix dominated CFVE sandwich specimen

fatigue response of dry vs. one sided immersion in terms of cyclic fatigue life (No of cycles) and associated damage evolution.

The experimental setup shown in Fig. 22: Experimental Set-up with Dry Matrix Dominated CFVE Sandwich Specimen setup was utilized for this research to evaluate tensile fatigue behavior using specimens with a total length *L* of 225 mm, a gage length *GL* of 150 mm, and a width *w* of 25 mm. A 3-D DIC system described earlier was used to obtain spatially resolved strain as a function of the state of loading. Figure 23 shows the setup for one-sided exposure to seawater using a latex bag to make water available while the composite facing is undergoing cyclic loading. Due to water available during loading, as matrix cracks develop, water wicks into these cracks and causes crack coalescence and growth quickly, thus causing large loss in fatigue life with one side of the sandwich panel exposed to seawater, a condition very similar to actual ship structure in the ocean.

Four or more samples in each case were evaluated to infer the sea water confinement effects. Figure 24 shows a typical tensile result of 2 ply CFVE sandwich structure studied in unison with spatially resolved tensile strain during monotonic loading. A relative comparison of modulus and failure strength in tension of 2-ply sandwich CFVE composites and typical 4-ply CFVE laminate are summarized in Table 4. The large decrease in modulus and failure strength is partially due to reduction in fiber volume fraction and changes in degree of cross-linking and matrix properties due to change in cure kinetics with PVC foam having some influence on the cross-linked state of vinyl ester resin (Fig. 25).

Tensile fatigue life of 2-ply CFVE sandwich composites while sea water was allowed to interact with composite sandwich specimen on one of its outer faces was found to substantial decrease (56% due to sea water confinement) as shown in Fig. 26. Fatigue life of CFVE sandwich composites with matrix dominated samples due to sea water confinement effects were obtained in this case at a stress level of 0.67 based on previous study of the author in evaluating laminate behavior by in isolation.

Fig. 23 Matrix dominated wet CFVE specimen with one-sided water confinement

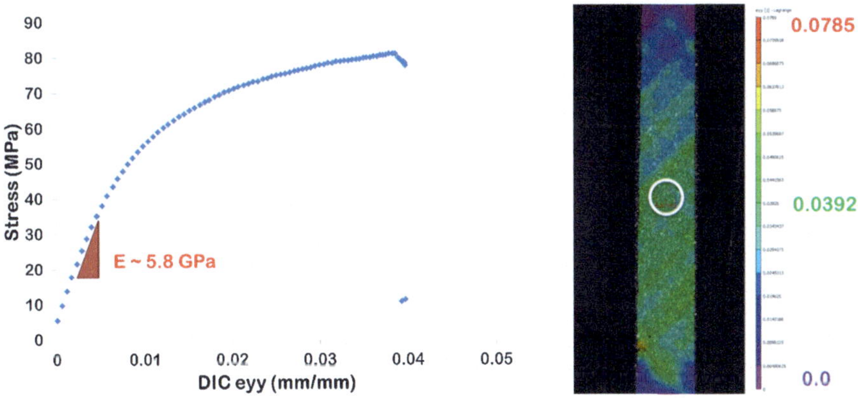

(a) (b)

Fig. 24 CFVE sandwich structure subjected to tensile loading coupling DIC results

Table 4 Typical results of 4-ply and 2-ply matrix dominated $[\pm 45]_{2S}$ CFVE

Conditions	Average modulus (GPa)	Failure stress (MPa)
4-ply dry CFVE $[\pm 45]_{2S}$	15	120
2-ply dry CFVE $[\pm 45]_{2S}$	5.8	84.5

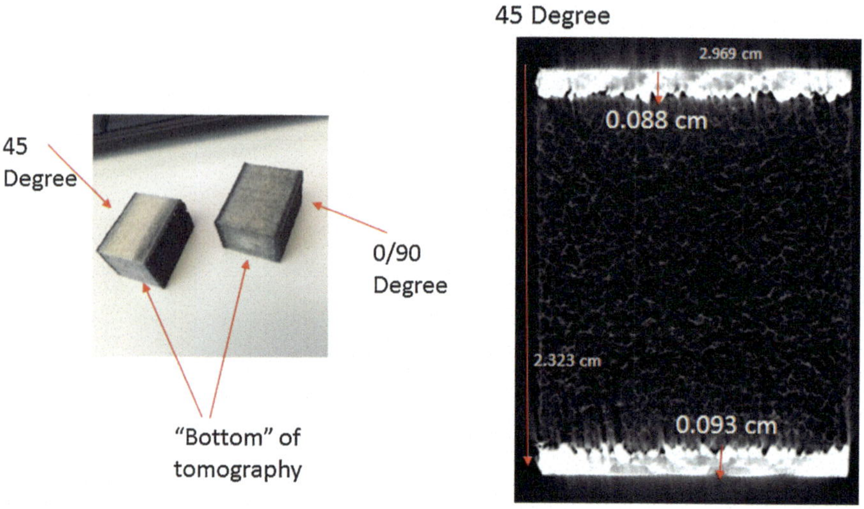

Fig. 25 X-ray computed tomography of sandwich facing and foam core showing resin rich areas with lower fiber volume fraction and saw-tooth microstructure resulting in large stress concentrations during cyclic loading

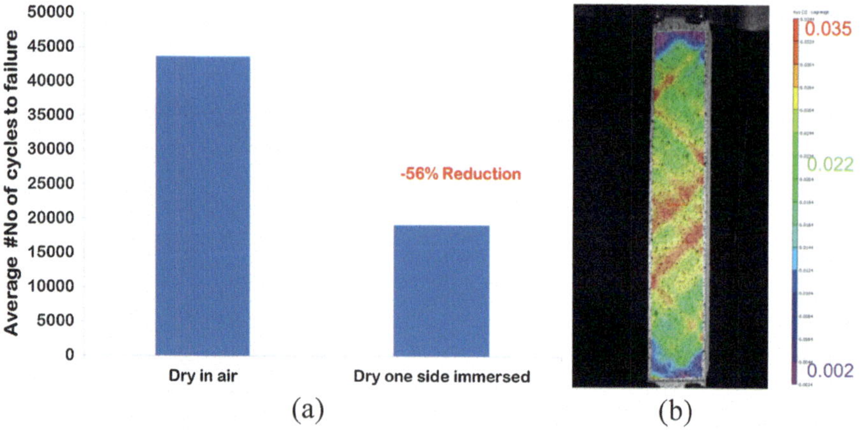

Fig. 26 (**a**) Comparison of matrix dominated CFVE dry in air and dry with one sided water confinement, (**b**) Typical DIC strain results along the sample length

8 Compressive Behavior and Seawater Effects

The complex behavior of CFVE composites subjected to tensile loading in monotonic and cyclic conditions and seawater effects have been addressed in the previous sections of this chapter. The associated degradation from the isolated and coupled effects of sea water, hydrostatic pressure, and cold temperature were also summarized. Since naval designs incorporate vinyl ester resin due to its superior performance in sea environment compared to epoxy based system, due to its low cost, and tunable properties for interface with a given sizing of considered carbon fiber type and intermediate form as stich bonded fabric, and VARTM manufacturing technique is well suited for naval needs, it is important to understand and evaluate this material system for particular manufacturing conditions in compression also. Such data and three dimensional behavior does not exist in the literature. It is also not clear how the sea water diffused CFVE system will degrade under compression. It is to be expected that the mechanical response of the CFVE will be very different in compression compared to tension and matrix will have a larger effect than in tensile loading, with corresponding degradation effects from marine environment potentially more significant. Compression behavior of fiber reinforced composites has proven to be a very important but poorly understood aspect fiber reinforced composites and over the years there has been a considerable lack of understanding on best ways to measure it experimentally. This led to the development of Combined Load Compression (CLC) approach as the test fixture is easier to use, repeatable to manufacture, and less massive than the Illinois Institute of Technology Research Institute (IITRI) developed compression fixture.

Standards exist for both approaches (ASTM D 6641 and D 3410). CLC test fixture (Fig. 27) is more recently widely adopted by the composites industry as the fixture is small, relatively easy to fabricate, and the combined use of end loading and shear loading in the grip region leads to more uniform state of compressive stress in the gage section. A key advantage is also that the use of tabs are avoided for the specimen, thus if a high degree of care was taken in ensuring uniform thickness samples during fabrication, eccentricity in the load train and resulting in bending stresses is minimized further. A specimen with a gage length of 12.7 mm and an overall size of 5.5″ (140 mm) × 0.5″ (12.7 mm) × t" (2.5 mm nominal) is obtained from the VARTM CFVE panel in a target orientation. It can be cut in the direction of warp to get the 0^0 sample, perpendicular to the direction of warp to get the 90^0 sample or at an angle of 45^0 to the warp.

Specimens from two different panels were evaluated as shown in Table 5. Part of the specimens were time aged (dry) while other specimens from a given panel and location and orientation were soaked in sea water at 40 °C for several weeks (wet) till the moisture uptake reached saturation equilibrium monitored by periodic weight gain data. Many tests were conducted prior to obtaining the compression response summarized in Table 5 during the process of perfecting the process to conduct highly reliable compression testing using CLC fixture. Testing improvements included procedures to fabricate and assemble flat specimens in test fixture with minimum misalignment, torque controlled gripping for repeatable lateral restraint in the grip

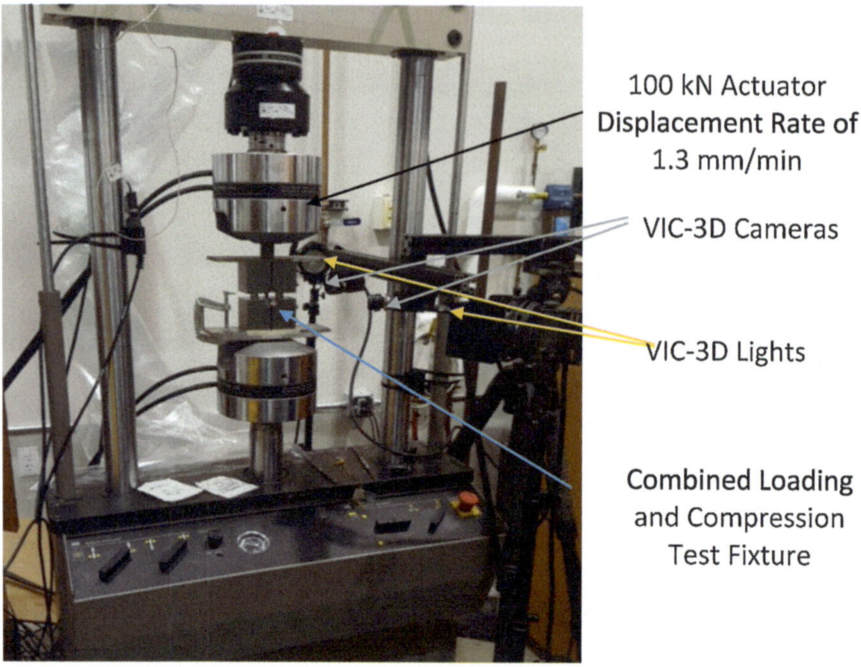

Fig. 27 Compression testing setup with spatially resolved strain mapping

Table 5 Summary results of compression behavior and sea water effects

Panel	Environment	Specimen Orientation (degrees)	Compression Modulus (GPa) Average Value	Compression Modulus from Individual Tests	Compression Strength (MPa) Average Value	Failure Stress from Individual Tests	Compression Strength Degradation Due to Sea Water (%)
Panel 1	Dry	0	49.35	41, 57.7	399	461, 237	
Panel 2	Dry	0	49	52.2, 56.6, 44.5, 44.2, 48	461	464.3, 455.9, 492.52, 389, 502.88	
Panel 1	Wet	0	43.75	44.6, 42.9	269	278, 260	32
Panel 2	Wet	0	57.4	57.4	349	349	24.3
Panel 1	Dry	45	11.6	10.3, 13.4, 10.6, 10.6	127	129, 125, 160, 125	
Panel 2	Dry	45	14.96	15.2, 18.3, 11.2, 14.2, 15.7	124	120.68, 122.7, 126.9, 123, 127.4	
Panel 1	Wet	45	11.55	11, 12.1	118.5	121, 116	7
Panel 2	Wet	45	14.55	15.1, 14	118.1	114.9, 121.3	4.7
Panel 1	Dry	90	46.96	46.1, 48.6, 52.2, 46.6, 41.3	422	431, 450, 462, 376, 389	
Panel 2	Dry	90	55.5	54.2, 55.5, 58.7, 53.5	489	552.35, 487.7, 453.9, 462.5	
Panel 1	Wet	90	47.35	48.1, 46.6	415	435, 395	1.6
Panel 2	Wet	90	56.1	54.9, 57.3	444	488.6, 399.2	9.2

regions, procedures to mount strain gage rosettes on the gage section, and simultaneously obtaining spatially resolved strains during CLC testing, first of its kind in this study. As can be seen from Table 1, there was consistent difference in compression properties for Panels 1 and 2 though they were identically prepared during the manufacturing using vary carefully controlled VARTM process. For a given panel, specimens that were exposed to long-term sea water showed substantial degradation in compressive strength for fiber dominated samples $[0/90]_{2S}$, with a loss in compressive strength in the range of 24% and 32% for Panel 1, and 1.5% & 9.2% for Panel 2. Currently detailed evaluation of the role of microstructure and corresponding failure modes is ongoing to ascertain such large difference for observed average degradation in strength for two panels. For matrix dominated samples $[\pm 45]_{2S}$, the degradation in compressive strength was found to be 4.7% and 7% and is much less pronounced. The tensile response of these composites under identical manufacturing conditions show much less degradation in fiber direction and similar degradation in the matrix direction. The ingress of water is associated with increased separation between the molecular chains, thereby inducing expansion strains and may have significantly more deleterious effect in compression. This is a very important finding and the root cause of the measured degradation will require further evaluation of the effect on matrix and fiber-matrix interface as well. An example result of axial strain measured during two load-unload cycles during a compression test for one of the $[0/90]_{2S}$ specimens is shown in Fig. 28.

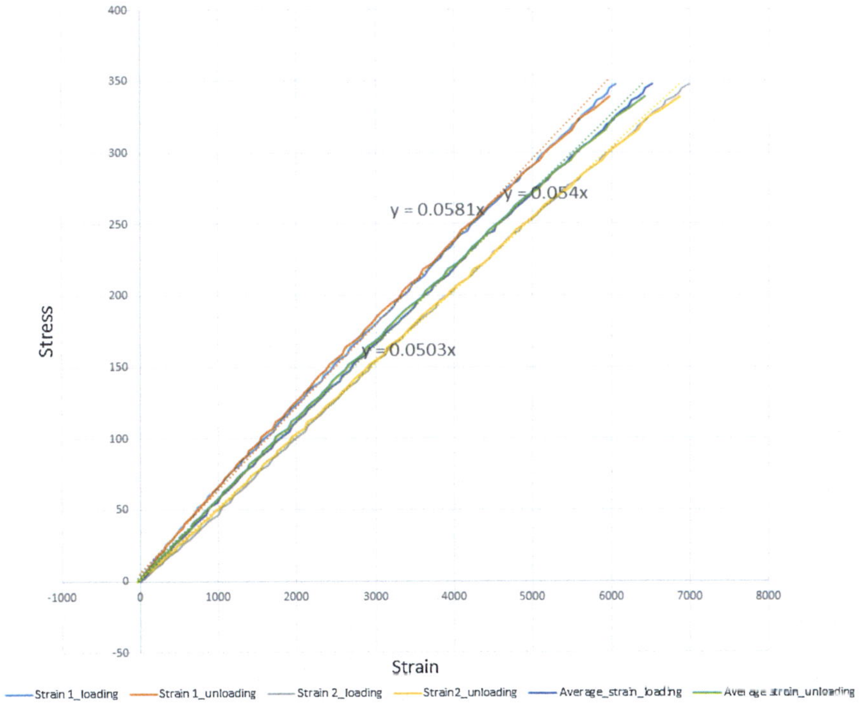

Fig. 28 Compressive stress-strain data from two strain gages mounted front and back sides of the specimen, and average compressive strain showing the effects of small misalignments during a CLC based compression test

Since the tensile strength of approximately 1 GPa was reduced by 60% for fiber dominated $[0/90]_{2S}$, it appears that the resulting microstructure at fiber tow level using VARTM approach with relatively large aerial density biaxial fabric is impacting the loss in strength in compression loading substantially. Currently detailed investigation of interpreting the initial compression data reported above considering strain gage data, spatially resolved 3D-DIC, and microstructure evaluations using high resolution computed X-ray tomography is ongoing. It is expected that the results will lead to important basic understanding of characterizing compression behavior of composites and degradation from harsh marine environment.

9 Summary

Mechanical degradation of polymeric composites was experimentally investigated due to exposure to sea environment and low temperature over extended period of time. Following observations are summarized associated with static, fracture, fatigue loading in tension, and degradation in compression response from seawater exposure.

- Only the outer boundary of PVC foam core was penetrated and degraded by the sea water for long-term exposure (several years) while the inner part still remained dry. The shear and Young's moduli of polymeric foam were tested by means of novel techniques and found to be $G = 25$ MPa and $E = 60$ MPa, respectively. Exposure to sea environment and low temperature resulted in approximate 8% degradation under torsion but no significant reduction could be detected under tensile tests. σ_{failure} of foam core was approximately 2.5 MPa and reduced up to 10% due to sea environment. No significant reduction is found when combined effects of sea and low temperature environment.
- Elastic properties of carbon fiber and resin dominated samples, $[0/90]_{2s}$ and $[\pm 45]_{2s}$, were insensitive to coupling effects. However, it was shown that σ_{failure} of $[\pm 45]_{2s}$ decreased by 5% due to sea water and has slightly reduced further with subsequent freezing after soaking.
- It has been shown that the interfacial fracture toughness G_c of sandwich structure decreased by 30% due to sea environment. The result indicated that exposure to sea water may severely reduce the interfacial toughness.
- Fatigue life of CF/VE composite samples on $[\pm 45]_{2S}$, which have the average static failure strength at 120 MPa, were obtained at maximum applied stress levels of 0.8, 0.67, 0.6, and 0.5 under tension-tension cyclic loading. Fatigue life due to sea water confinement and one-sided confined effects is evaluated against response from cyclic loading in air. A significant degradation of number of cycles to failure from sea water confinement effects was observed for all stress levels considered in this study up to 80% of static strength as the maximum cyclic stress value for a stress ratio of 0.1.

- A combined load compression fixture integrated with most advanced imaging and radiation based measurements systems is established for the study of marine composites under compression.
- The degradation associated with long-term exposure to sea water on the compression behavior of composite facings and tensile fatigue behavior of sandwich structures under one sided exposure was evaluated using very carefully controlled experimental data and advanced mechanics based modeling. The effect of sea water on CF/VE facings was evaluated for elastic and failure behavior in compression at different orientation of layups using an integrated approach of obtaining deformation data using on specimen strain gages on front and back face of the CLC specimens, spatially resolved 3-D digital image correlation, and microstructural changes and associated failure modes using high resolution and non-invasive computed tomography techniques.
- Substantial degradation of compression strength in both fiber and matrix dominated orientations for the carbon fiber reinforced vinyl ester composite laminates has been observed and microstructural investigations are currently underway to ascertain the dominant failure modes associated with each sample failed in compression. Noticeable variation in compression properties were observed for two different panels made using identical VARTM protocols, but for a given panel, there was a consistent and substantial degradation in compression strength due to long term sea water soaking.
- Monotonic tensile loading on composite facings on sandwich structures using a custom developed specimen configuration from a sandwich structure shows monotonic tensile modulus and failure strength considerably smaller than the conditions associated with identical laminate only manufacturing technique. Detailed microstructure from x-ray computed tomography verified the cause for the loss in tensile properties associated with sandwich facings and this aspect needs to be considered carefully for naval design.
- One sided water exposure and corresponding degradation in tensile fatigue life for these facings was also obtained for a constant R ratio and a maximum cyclic stress of 67% of the tensile strength.

Acknowledgments This research was supported by ONR Contracts through Solid Mechanics Program for the past decade, under a program managed by Dr. Yapa Rajapakse and is gratefully acknowledged. Majority of the material presented in this chapter was based on doctoral student's research work of Dr. Akawat Siriruk, Mr. Vivek Chawla, and Mr. Zachary Arwood and is greatly appreciated. Dr. Penumadu also would like to acknowledge the indirect support to this research from US Department of Energy and National Science Foundation for related investments in his fiber reinforced composites research program, resources, and facilities and more recently from the Institute for Advanced Composites Manufacturing Innovation.

References

1. Li X, Weitsman YJ (2004) Sea-water effects on foam-cored composite sandwich lay-ups. Compos Part B 35(6–8):451–459
2. Weitsman YJ (2012) Effects of fluids on mechanical properties and performance fluid effects in polymers and polymeric composites. Springer, Boston
3. DIAB Divinycell, Divinycell International, DeSoto, TX, USA
4. Shivakumar KN, Swaminathan G, Sharpe M (2006) Carbon/vinyl ester composites for enhanced performance in marine applications. J Reinf Plast Compos 25(10):1101–1116
5. Wood CA, Bradley WL (1997) Determination of the effect of seawater on the interfacial strength of an interlayer E-glass/graphite/epoxy composite by in situ observation of transverse cracking in an environmental SEM. Compos Sci Technol 57(8):1033–1043
6. Siriruk A, Penumadu D, Jack Weitsman Y (2009) Effect of sea environment on interfacial delamination behavior of polymeric sandwich structures. Compos Sci Technol 69(6):821–828
7. Siriruk A, Weitsman YJ, Penumadu D (2009) Polymeric foams and sandwich composites: material properties, environmental effects, and shear-lag modeling. Compos Sci Technol 69 (6):814–820
8. Guedes RM (2007) Durability of polymer matrix composites: viscoelastic effect on static and fatigue loading. Compos Sci Technol 67(11–12):2574–2583
9. Hwang W, Han KS (1986) Fatigue of composites—fatigue Modulus concept and life prediction. J Compos Mater 20(2):154–165
10. Lee LJ, Yang JN, Sheu DY (1993) Prediction of fatigue life for matrix-dominated composite laminates. Compos Sci Technol 46(1):21–28
11. Petermann J, Schulte K (2002) The effects of creep and fatigue stress ratio on the long-term behaviour of angle-ply CFRP. Compos Struct 57(1–4):205–210
12. Siriruk A, Penumadu D (2014) Degradation in fatigue behavior of carbon fiber–vinyl ester based composites due to sea environment. Compos Part B 61(0):94–98
13. Lee LJ, Fu KE, Yang JN (1996) Prediction of fatigue damage and life for composite laminates under service loading spectra. Compos Sci Technol 56(6):635–648

Failure Mechanics of Low Velocity Dynamic Impact on Woven Polymeric Composites in Arctic Conditions

Pavana Prabhakar

1 Introduction

The rise and demand in arctic exploration has necessitated an in-depth understanding of the mechanical behavior and failure mechanisms of materials exposed to arctic conditions. Structures in such applications are often subjected to adverse environments like sea water, wave impacts or extreme low temperatures, which can cause surface alterations, internal damage, and degradation of chemical and mechanical properties that may ultimately compromise the safety of the naval structure. Therefore, the materials used in these structures must be able to withstand harsh environmental conditions of extreme low temperatures in addition to mechanical loads. Fiber reinforced polymeric composite (FRPC) have become an attractive option for this type of applications due to their corrosion resistance, high strength-to-weight ratio, ability to absorb noise and vibration damping, ease of fabrication, maintenance and repair [10, 18, 38]. They have been successfully integrated in offshore applications such as offshore vessels, ships hulls, tanks or pipes [16, 39]. Despite several advantages that FRPCs offer, a major drawback is their low resistance to impact damage due to their layered nature. Therefore, the motivation of this paper is to investigate the influence of combined arctic temperature (-50 °C) and low-velocity singles and repeated impact loading on the damage and failure mechanisms of woven carbon/vinyl ester laminates.

Dynamic impacts on structures can occur under several different scenarios, including but not limited to, tool drop during maintenance and repair, hail strikes, iceberg collision, wave slamming [13, 25, 47], etc. These impacts are divided into low- and high-velocity. Low-velocity impacts typically occur at velocities below

P. Prabhakar (✉)
Civil & Environmental Engineering, Engineering Physics (Affiliate),
University of Wisconsin – Madison, Madison, WI, USA
e-mail: pavana.prabhakar@wisc.edu

S. W. Lee (ed.), *Advances in Thick Section Composite and Sandwich Structures*,
https://doi.org/10.1007/978-3-030-31065-3_20

10 m/s [42], which may produce barely visible damage (BVD) on composite surfaces, but with the possibility of significant internal damage. This is deemed very dangerous, as BVD could result in catastrophic failure of the structure without warning. The energy ranges vary between applications, but the velocities are maintained below 10 m/s [13, 25, 47]. The impact energies were chosen in the current paper based on the work presented by previous researchers. Common failure modes observed during low-velocity impacts are matrix cracking, fiber breakage and delamination [1, 9, 35]. Of these, delamination is one of the most common failure mechanisms [44], which often results in the reduction of stiffness, strength, durability and stability of composites [5, 40].

To establish the life and durability of FRPC in arctic conditions, in-depth investigation into the influence of combined impact and low temperature needs to be conducted. In real applications, structures are not impacted once, but are constantly subjected to repeated impacts like in the case of wave impacts, main shutdown of an off-shore platform, drifting supply vessels or ice impacts [19, 24]. Most of the previous impact studies have focused on single low-velocity and repeated impact at room temperature. Naik et al. [32] investigated the damage imparted to woven-fabric and cross-ply E-glass/epoxy and carbon/epoxy laminates under low-velocity (1 m/s and 3 m/s) impact. They reported that woven-fabric laminates were more resistant to in-plane impact damage than cross-ply laminates. Rajkumar et al. [34] studied the effect of repeated low-velocity impacts on glass fiber metal composites, and established that the peak load, impact energy, and failure strain decreased with increasing number of impacts. Sayer et al. [37] investigated the impact response of hybrid composite plates (glass-carbon/epoxy) with different stacking sequence for impact energies ranging between 25–75 J, and concluded that fiber fracture was the dominant failure mode as the impact energy increased.

Murat et al. [31] tested woven carbon/epoxy prepreg laminates at different impact energy levels in the range of 1 J – 10 J, and observed that thicker samples manifested higher resistance to impact damage and the damage area increased with increasing impact energies. Morais et al. [30] investigated the effect of repeated low energy impact response of carbon-epoxy composites with different stacking sequences, and reported that cross-ply and non-symmetric laminates have better endurance against low energy impacts than unidirectional laminates. Li et al. [27] studied the influence of the thickness of carbon fiber composites under low-velocity impact with energies of 17 and 18 J, and reported that the contact force, absorbed energy and bending stiffness decreases with reducing sample thickness. Nguyen et al. [33] investigated the influence of low, medium and high velocity impact on carbon fiber reinforced polymeric composites with impact energies of 10 J, 40 J and 120 J, respectively. They reported that the predominant damage mode was minor delamination, large delamination and fiber fracture and perforation, corresponding to low, medium and high velocity impact, respectively. Sultan et al. [43] studied woven carbon fiber reinforced prepreg laminates with impact energies ranging between 0.4–42 J, and reported that matrix cracking occurred below 21 J and fiber breakage manifested at energies between 21–31 J.

Despite an extensive amount of research conducted at room temperature (RT) under single and repeated impact loading, seldom work has been reported on repeated impact in arctic temperatures (AT). Icten [22] studied the influence of temperature on single and repeated impact of woven glass-epoxy composites at room temperature and −50 °C. They observed that the laminates impacted at −50 °C recorded higher peak forces and absorbed less energy than the samples impacted at room temperature. Ibekwe et al. [21] investigated the impact response of glass fiber reinforced unidirectional and cross-ply laminated composite beams at 20 °C, 10 °C, 0 °C, −10 °C and −20 °C, and observed that more damage was induced in the specimens impacted at lower temperature than those at higher temperatures. Salehi-Khojin et al. [36] investigated three combinations of fiberglass and Kevlar woven composites. Three different impact energies were tested (8 J, 15 J, and 25 J) for temperatures ranging from −50 °C to 120 °C. They reported that the deflection, maximum force and energy absorption increases with increasing temperature (from −50 °C to 120 C) and impact energy (from 8 J to 25 J). Lopez-Puente et al. [29] investigated the influence of low temperatures on the damage imparted in carbon fiber/epoxy laminates (tape and woven) by impact velocities ranging from 60 to 525 m/s and at three temperatures (25, −60 and −150 °C). They reported that higher kinetic energy and low temperature resulted in larger damage in the laminates. In addition, they concluded that as the velocity increased, damage saturation occurs and temperature will not influence the damage extension. Im et al. [23] investigated the effect of different temperatures (−30, 20, 90 and 120 °C) on carbon fiber/epoxy and carbon fiber/peek laminates with lay-up $[0_6/90_6]_s$ and $[0_4/90_4]_s$. They concluded that as the temperature increases, the delamination areas decrease. Gomez-del-Rio et al. [17] recorded the response of carbon fiber reinforced polymeric composites with different stacking sequences (unidirectional, cross-ply, quasi-isotropic and woven laminates) in ambient temperatures ranging from 20 °C to −150 °C. They reported that the extent of damage and absorbed energy increased with the decrease in temperature for all tape laminates, however, woven laminates did not exhibit this trend.

There have been mixed observations with respect to the extent of damage under low-velocity impact loading at low temperatures for composites and seldom work has been reported on low-velocity repeated impact on carbon fiber reinforced polymeric composites. Therefore, in the current study, the response and failure mechanisms of woven carbon/vinyl ester laminates subjected to low-velocity single and repeated impact loading at room (25 °C) and arctic temperature (−50 °C) for a range of impact energies are investigated. Vinyl ester is considered in the current study due to superior UV resistance and low water absorption as compared to polyester resins [28, 41], which makes it attractive for ship and offshore applications. The variations in impact response in terms of force, displacement, energy and damage mechanisms is studied in detail and presented here.

2 Experimental Procedures

2.1 Manufacturing

Carbon fiber laminates were fabricated using vacuum assisted resin transfer molding (VARTM) process. Material system, fabrication process and sample dimensions used in this study are discussed next.

2.1.1 Material System

3 K tow (3000 filaments per tow) plain weave carbon fiber fabrics were used to manufacture the laminates investigated in this paper. These fabrics typically create lightweight and tensile stiffened structural products, and are also compatible with a variety of thermoset and thermoplastic resins. This type of fabric is commonly used in aerospace, marine and automobile applications. Hetron 922 vinyl ester resin, formulated for 1.25% MEKP hardener, was the resin system used for impregnating the dry carbon fabric during the VARTM process. Hetron 922 is a thermoset with low viscosity, which is advantageous for easy infiltration through the dry fabric during the VARTM process. Manufacturer provided mechanical properties of the carbon fibers and vinyl ester resin are given in Table 1.

2.1.2 Laminate Fabrication

Carbon fiber reinforced composite samples tested at 25 °C and −50 °C were manufactured by vacuum assisted resin transfer molding (VARTM) process [11]. Plain weave carbon fabric and vinyl ester resin were purchased from Fibre Glast (www.fibreglast.com). Laminates with 16 layers of dry fabric were manufactured according to the ASTM Standard D7136/D7136M [4]. The layers of fabric were placed between 2 layers of flow-media, 2 layers of breather and 4 layers of nylon peel ply. All layers were cut to dimensions of 305 mm × 305 mm. The arrangement of fabrics was placed between 2 aluminum molds, wrapped with Stretchlon 800 bagging film and sealed with vacuum-sealant tape, ensuring spaces for both inlet and outlet connectors (Fig. 1).

Table 1 Typical constituent material properties

Property	Carbon fiber	Vinyl-ester
Tensile strength	4.2–4.4 GPa	82.7 MPa
Tensile modulus	227.5–240.6 GPa	3.7 GPa
Elongation	1.4–1.95	4.6–7.9
Flexural strength	–	131 MPa
Flexural modulus	–	3.4 GPa
Nom. thickness	0.3048 mm	–

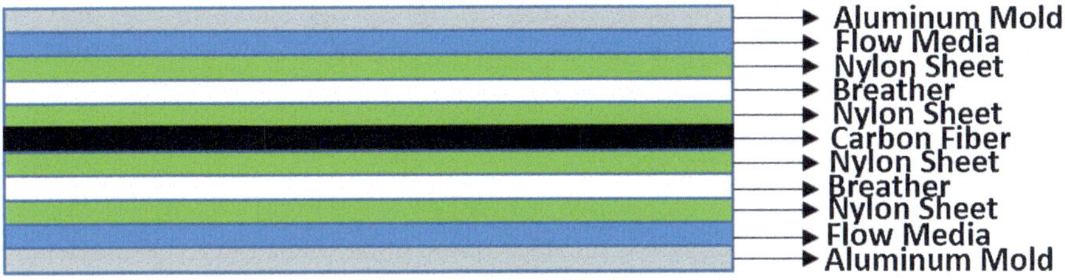

Fig. 1 Layer distribution of materials used in the VARTM process

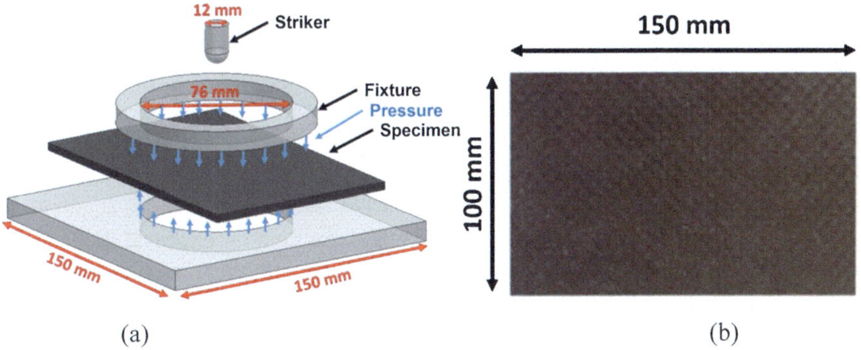

Fig. 2 (**a**) Dynamic impact fixture with striker; (**b**) Dimensions of impact sample

The woven dry fabrics were reinforced with a mixture of vinyl ester resin and Methyl Ethyl Ketone Peroxide (MEKP) hardener. The resin was catalyzed with 1.25% MEKP by weight and mixed thoroughly as recommended by the manufacturer. The resin/hardener mixture was placed in a desiccator first to remove bubbles from the mixture. The outlet was then connected to a vacuum pump until the vacuum bag achieved a pressure of approximately 80 MPa. The inlet of the vacuum bag was then submerged in the resin/hardener mixture for transferring resin through the laminate. Upon completion of the resin transfer process, the laminate was cured at room temperature for 24 h. A total of 6 plates of 305 mm length by 305 mm width were manufactured, and six samples were obtained from each plate. To ensure that the curing conditions were identical for all the samples at room and arctic temperature, half the samples from each plate fabricated were set for testing at 25 °C and the other half for −50 °C for a given impact energy.

2.2 *Impact Tests*

Drop-weight impact tests were performed using a CEAST 9340 Drop Tower Impact System on rectangular laminate samples of 150 mm length × 100 mm width (refer to Fig. 2(b)) with an average thickness of 4.1 mm. The laminates were clamped between two metal fixtures with a test area of 46 cm^2 as shown in Fig. 2(a). A hemispherical striker with a mass of 3.0 kg and diameter of 12.7 mm was used to impact the samples at their centers in the out-of-plane direction [4] with kinetic energies of 20 J, 25 J, 30 J, and 35 J for both temperatures (25 °C and −50 °C). The impact velocity was calculated based on the mass of the striker and kinetic energy, using the equation, $E_k = ½\, mv^2 = mgh$, where, E_k is the impact energy or kinetic energy, v is the impact velocity, m is the mass of the impacting striker, h is the height of the striker measured from the surface of a sample in the impact drop tower, and g is the gravitational acceleration.

For a particular impact energy (20 J, 25 J, 30 J or 35 J), the impact velocity and the striker falling height were adjusted accordingly by the Drop Tower Impact System, where the impact velocities were 3.64 m/s, 4.07 m/s, 4.46 m/s and 4.82 m/s, respectively. All the tests were low-velocity impacts, that is, below 10 m/s [8]. Single impact tests were conducted first. In order to establish the durability under repeated impact loading, each sample was repeatedly impacted until complete perforation of the striker through the sample thickness. Force-time, displacement-time and energy-time responses were recorded by the data acquisition system "CEAST DAS 8000 Junior" of the impact machine for each test. Schematic of an impact test fixture is shown in Fig. 2(a). Four samples were impacted for each combination of impact energy (20 J, 25 J, 30 J and 35 J) and temperature (25 °C and −50 °C). Corresponding force-time, displacement-time, energy-time and force-displacement responses were obtained for each test.

The samples planned for testing under in-situ arctic conditions were placed in a Thermo Scientific™ freezer at −50 °C for a period of 90 days to reach a uniform temperature. A basic heat transfer analysis was performed which showed that a sample at room temperature can −50 °C in 15–20 min when subject to a constant ambient temperature of −50 °C. A 90 days exposure prior to testing was chosen to subject the samples to arctic pre-conditioning. To perform the in-situ arctic tests, the samples were removed from the freezer and placed within a temperature controlled environmental chamber, which was connected to the CEAST 9340 Drop Tower Impact System. Prior to every impact test, the environmental chamber was conditioned for 15 min with Liquid Nitrogen (LN_2) to reach a uniform temperature of −50 °C within the chamber.

2.3 Quasi-Static Tests

To examine the strengthening effect of low temperatures, compression tests were performed on pure vinyl ester samples and tension tests were performed on woven carbon/vinyl ester samples at room and arctic temperatures. Compression Test: Three samples each of vinyl ester were tested at in-situ 25 °C and −50 °C under flat-wise compressive loading. Cylindrical specimens with a diameter of 25.4 mm and a height of 50.8 mm were tested according to ASTM D695 standard [3]. These tests were performed using an ADMET eXpert 1654 testing system with a crosshead displacement rate of 1.3 mm/min. Tension Tests: Five samples each were tested at in-situ 25 °C and −50 °C under tensile loading. Rectangular specimens with a width of 15 mm, thickness of 1 mm and length of 250 mm were tested according to ASTM D3039 standard [2]. These tests were performed using an ADMET eXpert 1654 testing system with a crosshead displacement rate of 2 mm/min.

2.4 Micro Computed Tomography (micro-CT) Scanning

Typically, low velocity impacts produce barely visible damage (BVD) on composite surfaces after a single impact. Hence, the samples were examined under a micro-computed tomography (micro-CT) scanner to evaluate the internal damaged regions in arctic and room temperature. The samples were reduced to a rectangle of 145 mm in length and 90 mm in width (the original dimensions were 150 mm in length by 100 mm in width), such that they can fit in the scanner chamber. The impact damage was centered in this rectangle and cutting around the edges did not alter the damage that occurred predominantly at the center of the samples and far away from the edges. A small hole with a diameter of 1.6 mm was drilled at the center of the impacted region of the laminates for applying a dye-penetrant at these holes, upon which the samples were held in a vacuum chamber for 5 min. This procedure was repeated three times to ensure that all damaged regions were filled with the solution. For the first 2 applications of the dye-penetrant, the solution was completely absorbed. A third application ensured that the sample was saturated with the solution, which resulted in complete solution penetration in all available openings, such as delamination and cracks. Zinc iodide solution was used as the dye-penetrant, which has a high absorption coefficient in comparison to the constituents of the composite materials i.e. carbon fiber and vinyl ester.

The Zinc iodide solution was a mixture of alcohol (10 ml), distilled water (10 ml), Kodak photo solution (1 ml) and zinc iodide powder (60 g). Excess dye-penetrant was evaporated by placing the laminates in an oven at 50 °C prior to x-ray scanning. Excess dye penetrant in its liquid phase is not preferred as its motion inside the crack during a scan adversely affects the quality of the 3-D reconstruction. This was eliminated through drying the dye penetrant, which deposits a saline residue on the crack area and in turn provides greater resolution of the damage. Hence, drying the dye penetrant is beneficial. Also, 50 °C is enough to dry the samples without creating any thermal damage in the composite. All laminates were scanned with a SkyScan 1173 X-ray micro-tomography with the same resolution of 35.9 μm and an angle step of 0.19. The X-ray tube voltage and current were set to 60 kV and 120 microampere, respectively. All the scans were performed using built-in Al filter, and a flat field correction was applied for each scan. The reconstruction was performed using the NRecon commercial software.

3 Results and Discussion

Deformation-time, energy-time and force-time responses recorded for impact energies of 20, 25, 30, and 35 J at 25 °C and −50 °C are discussed in detail in this section. Durability of laminates upon repeated impact is assessed in terms of number of impacts required to perforate a laminate through the thickness and the rate of reduction in the peak force for a combination of impact energies and temperatures. The response of the laminates in terms of visual damage, degree of damage and failure mechanisms is also evaluated and elucidated next.

3.1 Laminate Strengthening

Mechanical properties of woven carbon/vinyl ester composites change when cooled to arctic temperatures (AT). Prior research by Dutta [14] on the compressive response of glass-fiber-reinforced polymer composites at the U.S. Army Cold Regions Research and Engineering Laboratory (CRREL) showed that their strength and stiffness increases with reducing temperatures. But, also become brittle and are susceptible to cracks due to increase in thermal residual stresses caused by mismatch in the coefficient of thermal expansion (CTE) between the fibers and matrix.

3.1.1 Compression Test Results

Compression tests on pure vinyl ester were also conducted in-house as part of the current study to investigate the influence of low temperature on these composites. Table 2 shows the results from the compressive testing of vinyl ester, where the yield strength (coined as the stress value where the response starts to become non-linear), ultimate strength and elastic modulus increased by approximately 55%, 49% and 28% when cooled from 25 °C to −50 °C. Typical compressive stress-strain response (one sample) of vinyl ester is shown in Fig. 3, where the final failure strains reduced with reduction in temperature, which implies that deformation of vinyl ester will be lower at AT as compared to those at RT. The increase in matrix strength is attributed to the binding forces between molecules, which are tightly frozen at AT [12]. Therefore, the compressive strength of vinyl ester increases at low temperatures. Garcia et al. [15] investigated the flexural response of woven carbon/vinyl ester composites in AT, and reported that dry arctic conditioned samples manifested an ≈23% increase in flexural strength with respect to those at room temperature.

Table 2 Results from compressive testing of vinyl ester at RT and AT

Mechanical properties	Temperatures (C)		Percentage change (%)
	25	−50 °C	
Yield strength (MPa)	47 ± 5	72 ± 4	55
Ultimate strength (MPa)	85 ± 4	126 ± 12	49
Elastic modulus (GPa)	2.6 ± 0.2	3.4 ± 0.1	28

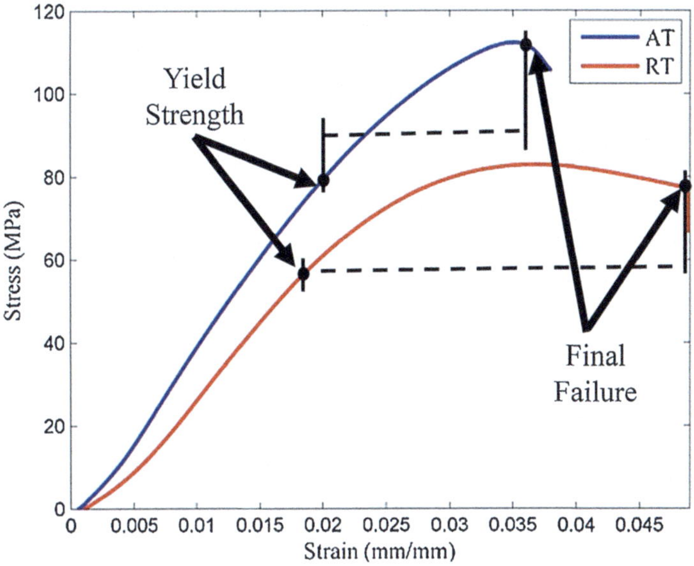

Fig. 3 Typical compressive stress-strain plots of vinyl ester at RT and AT

Table 3 Results from tensile testing of woven carbon/vinyl ester samples at RT and AT

Mechanical properties	Temperatures (C)		Percentage change (%)
	25	−50 °C	
Ultimate strength (MPa)	562 ± 31	623 ± 29	11
Elastic modulus (GPa)	37 ± 4	42 ± 4	15

3.1.2 Tension Test Results

Tension tests on woven/carbon vinyl ester samples were also conducted in-house as part of the current study. Table 3 shows the results from the tensile testing of the woven carbon/vinyl ester composite, where the Young's modulus and ultimate tensile strength increased by approximately 15% and 11% respectively when cooled from 25 °C to −50 °C. Typical tensile stress-strain response of woven carbon/vinyl ester is shown in Fig. 4(a), where the final failure strains reduce with reduction of temperature. This implied that there is a reduction of ductility and increase in brittleness of the composites at low temperature [20]. Kim et al. [26] attributed such increase in brittleness at low temperatures to predominantly fibers, which increased rapidly within a temperature range from RT to −50 °C.

On the other hand, the increase in the laminate strength and stiffness is attributed to the strengthening of the matrix. Therefore, there will be less damage at low temperatures initially, but it continues to increase as the load approaches a critical value where the fibers fail. However, matrix cracking and delamination will be dominant at room temperature [45]. Figure 4(b) shows the failure regions of one

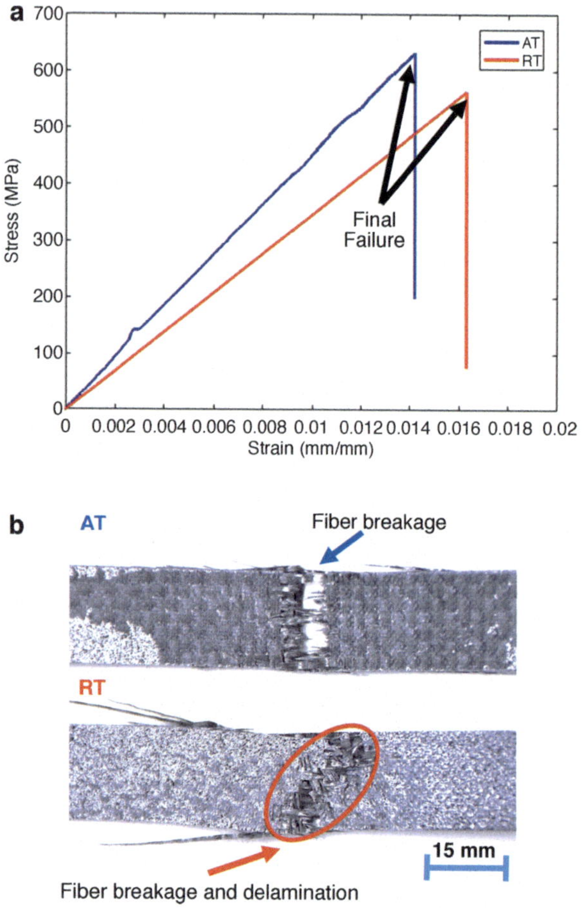

Fig. 4 (**a**) Typical tension stress-strain plots of woven carbon/vinyl ester at RT and AT; (**b**) Failure regions of the composite specimens at RT and AT under tension

set of specimens tested at 25 °C and −50 °C under tension. It can be seen that at AT, the damage was localized in just one region (across the transverse direction of the sample) and fiber breakage was the dominant failure mechanism. On the other hand, the samples at RT experienced a more dispersed damage (across the longitudinal direction of the sample). In addition, matrix cracks and some fiber breakage were the main failure mechanisms. The samples at RT were painted white to show this failure pattern more clearly.

3.2 Contact Force and Deflection

During an impact test, contact force is generated by the contact of the striker with the impacted face of a sample, which is recorded as the force-time response by the data acquisition system of the impact machine.

3.2.1 Single Impact

Figure 5(a) shows a representative force-time graph of an impact event, where the maximum impact force corresponds to the peak value of the graph. Figure 5(b) shows the variation of maximum impact force for different impact energies for the specimens tested at room and arctic temperature. At both temperatures, the

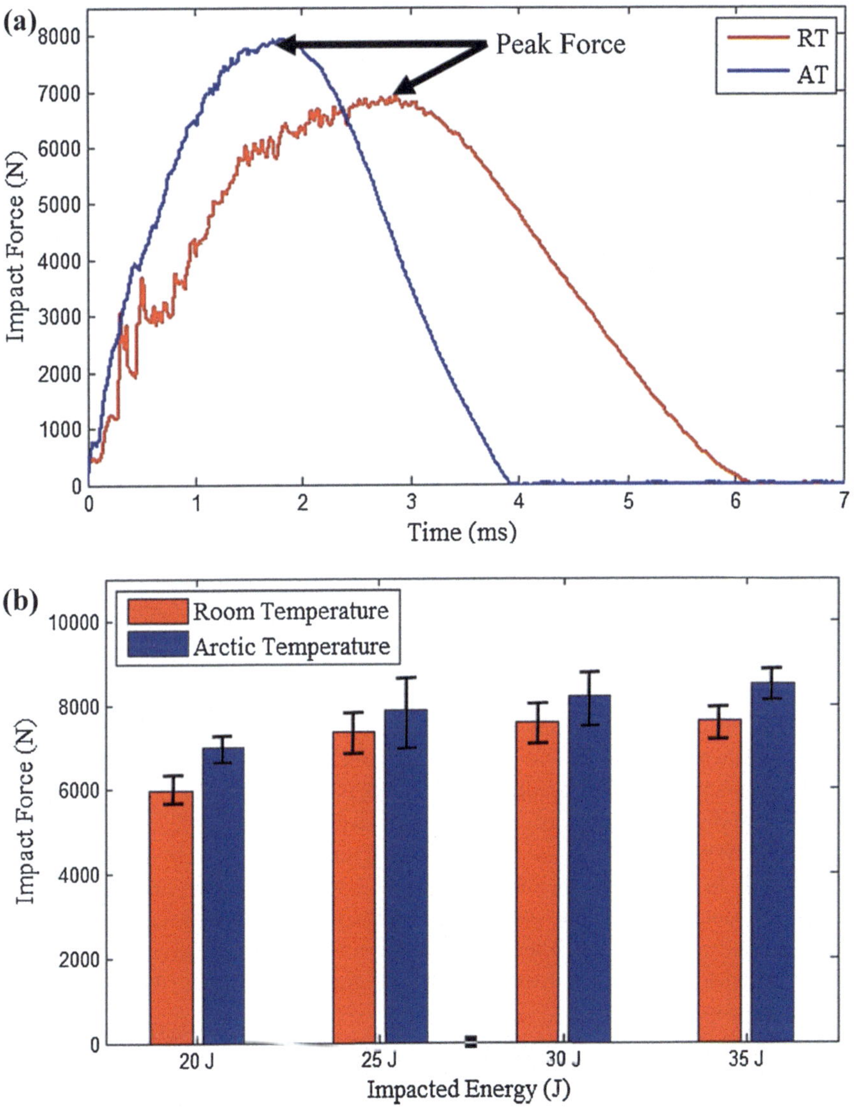

Fig. 5 (**a**) Representative force-time graph of an impact event at 20 J for RT and AT; (**b**) Impact contact force at RT and AT for 20 J, 25 J, 30 J and 35 J

maximum impact force increased with increasing impact energies. In addition, the specimens tested at AT experienced higher impact forces as compared to those tested at RT, which is attributed to the increase in strength of composites when subject to low temperatures as discussed in the section on "Laminate strengthening". Also, laminates became stiffer at low temperatures rendering them less flexible as compared to those at room temperature.

The average initial slopes of the force-displacement responses for the samples tested at different impact energies for both temperatures is shown in Fig. 6, where the samples impacted at arctic temperature manifested higher stiffness values as compared to those at room temperature. This is attributed to the strength increase at low temperatures of the laminates. Due to the increased strength at arctic temperature, lower deflections were observed as compared to the samples tested at room temperature.

Figure 7 shows the deflection with varying impact energies. As expected, the deflections at both temperatures increased with the increasing impact energies.

3.2.2 Repeated Impact

Typical repeated impact responses of laminates impacted at two energies of 20 J and 25 J are shown in Figs. 8(a) and 8(b). In general, it is expected that the peak force recorded reduces with increasing number of impacts due to accumulation of damage. However, the specimens manifested two different responses at 20 J and 25 J. At 20 J, the peak forces increased initially upon repeated impacts, but reduced after several impacts finally resulting in perforation. At 25 J, the trend was as expected, where the peak force gradually reduces with increasing number of impacts.

Bienias et al. [7] categorized the repeated impact response into phases of force change. The first phase is called "stabilization", represented by letter A in Fig. 8(a), which is the very first impact on a laminate where the impact energy is insufficient to cause damage for decreasing the stiffness of the laminate. Icten [22] attributes this to the contact of the impactor with a relatively compliant matrix material. The second phase is known as "force increase", given by letter B in Fig. 8(a), which consists of multiple impacts before the maximum peak force is reached. In this phase, laminates experienced higher contact force after each impact due to the compaction of matrix under the striker. The third phase is "maximum force", represented by letter C, which corresponds to the number of impacts at maximum peak force beyond which force reduction occurs due to the presence of damage, such as matrix cracks, fiber breakage and delamination. The last phase is "force decrease", given by letter D, where the peak force and stiffness recorded gradually reduces with increasing number of impacts.

Specimens repeatedly impacted at 20 J at room and arctic temperatures manifested all four phases of force change (Fig. 8(a)), whereas, those impacted at 25, 30 and 35 J showed only phases C and D (Fig. 8(b)). Lower impact energies are not sufficient to damage the laminate in the first impact, thereby, causing phases A and B, as opposed to higher impact energies that manifest only phases C and D.

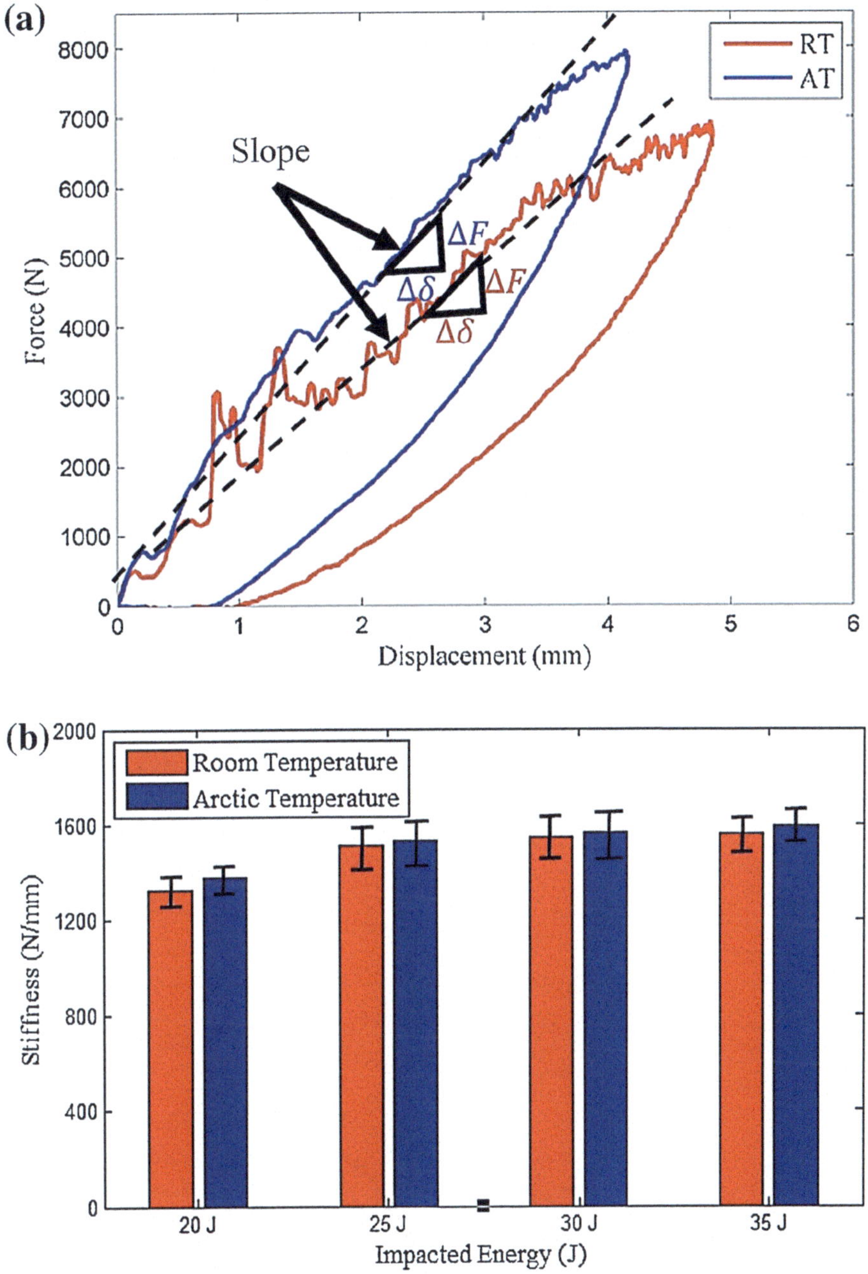

Fig. 6 (**a**) Representative force-displacement graph of an impact event at 20 J for RT and AT; (**b**) Stiffness (initial slope) at RT and AT for 20 J, 25 J, 30 J and 35 J

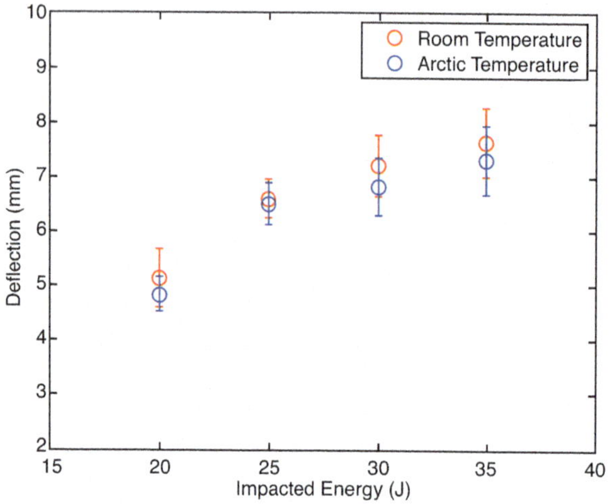

Fig. 7 Deflection for each impacted energy at room and arctic temperature

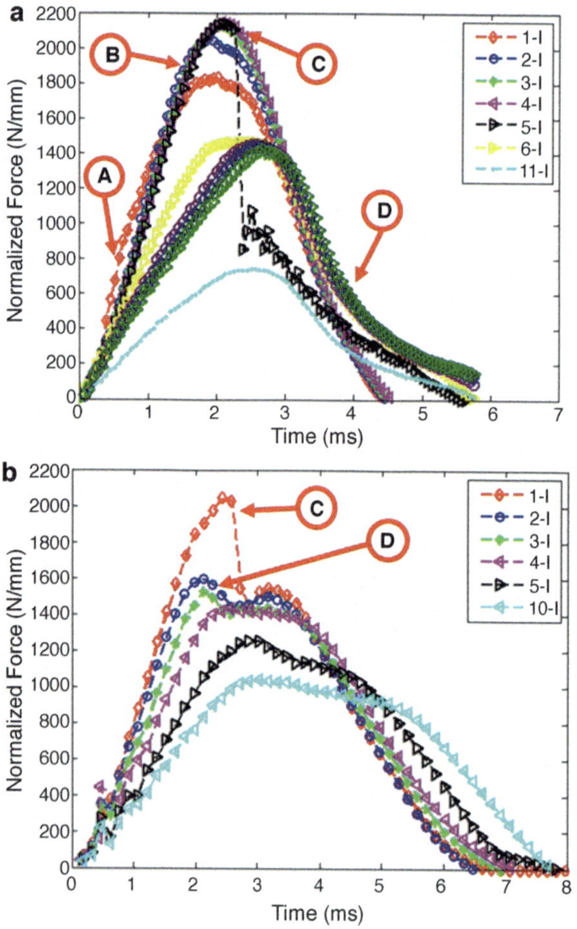

Fig. 8 Representative force-time responses for a single sample subject to repeated impact: (**a**) 20 J at arctic temperature; (**b**) 25 J at arctic temperature

3.2.3 Temperature Effect on Impact Force

Peak impact force is plotted against the number of impacts at room and arctic temperature in Fig. 9. For all energies, the samples impacted at −50 °C experienced higher impact forces and required a greater number of impacts to perforate the laminate as compared to the samples tested at 25 °C. The samples impacted at 20 J at both temperatures experienced the four phases of force change described above, which are, stabilization, force increase, maximum force and force decrease. Figure 9(a) shows the response for 20 J for both temperatures, where an increase in impact force after the first impact is observed corresponding to the force increase stage. Upon reaching a maximum impact force, a decrease in impact force is observed with further impacts. With increasing number of impacts, significant difference between the impact forces is observed at both temperatures. For 25, 30 and 35 J impact energies, the samples experienced only 2 phases of force change: maximum force and force decrease as shown in Figs. 9(b), 9(c) and 9(d).

The slope of the force versus number of impacts plot indicates the rate of reduction in impact force with increasing number of impacts, which is higher at

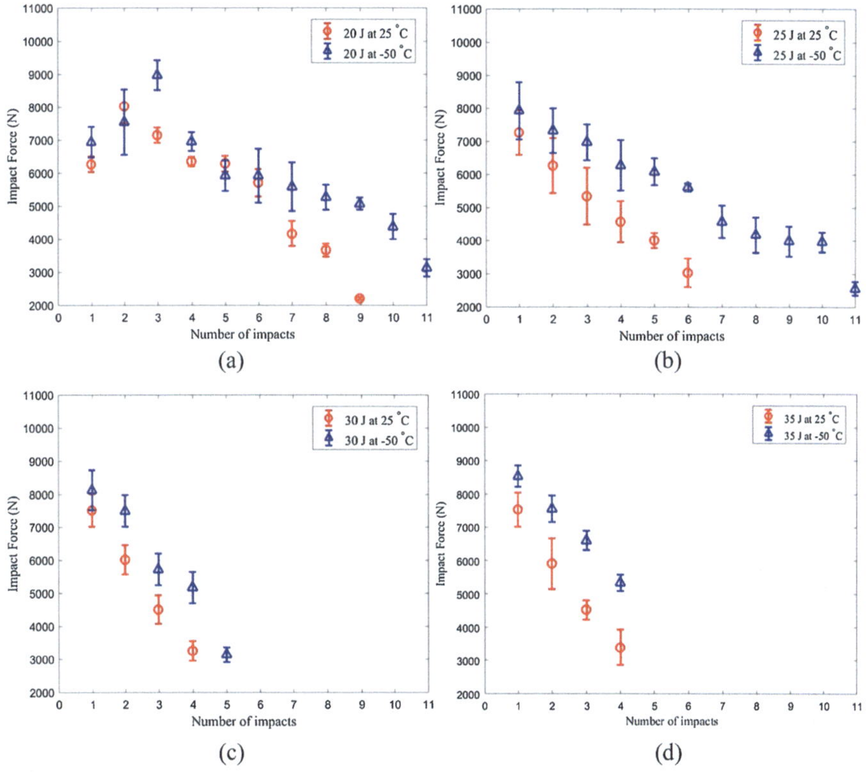

Fig. 9 Contact force with increasing number of impacts at 25 °C and −50 °C: (**a**) 20 J, (**b**) 25 J, (**c**) 30 J and (**d**) 35 J

higher impact energies of 30 J and 35 J, and also similar at room and low temperature. Fiber fracture is the dominant failure mechanism at higher impact energies as opposed to matrix cracking at lower impact energies. This is attributed to two factors: (1) lesser influence of low temperature on carbon fibers and damage saturation. The influence of low temperature on carbon fiber is less significant as compared to matrix, from what is observed in the case of coefficient of thermal expansion (CTE) [46]; (2) damage saturation occurs when the temperature has no influence on damage extension. This corroborates the similar responses at low and room temperature. As expected, the impact force at both temperatures increased with the increasing impact energy.

3.2.4 Temperature Effect on Deflection

Figure 10 shows the deflection versus number of impacts at room and arctic temperatures. Increase in rigidity of the laminates due to matrix strengthening at

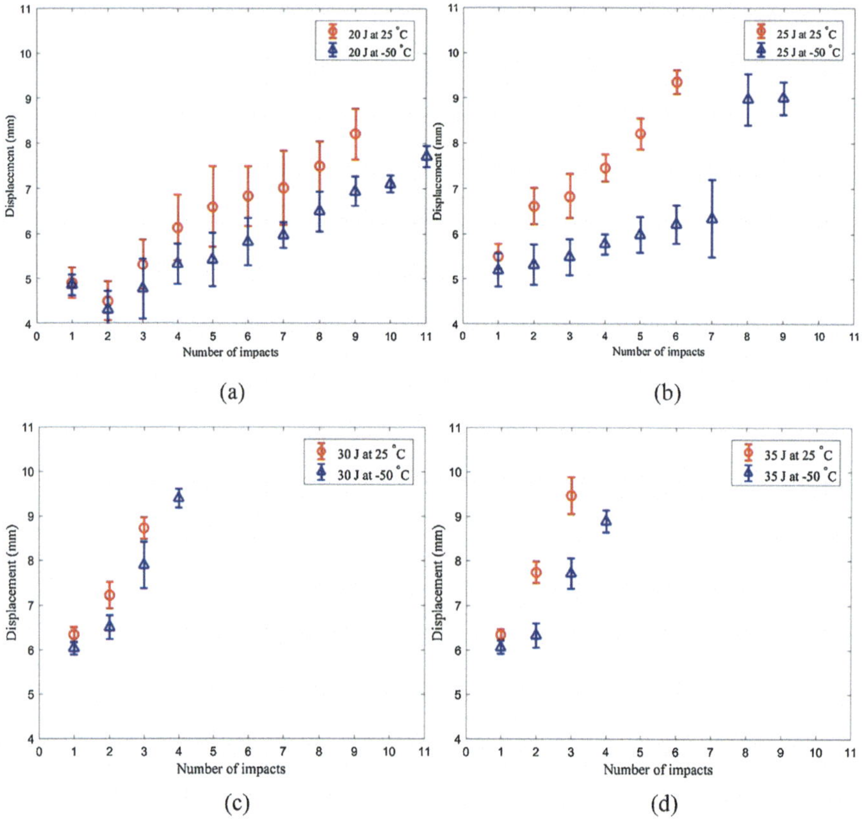

Fig. 10 Deflection with increasing number of impacts at 25 °C and −50 °C: (**a**) 20 J, (**b**) 25 J, (**c**) 30 J and (**d**) 35 J

arctic temperatures manifests lower deflections in samples impacted at –50 °C as compared to those at 25 °C. The deflection at both temperatures increased with the increasing impact energies. For 20 J (Fig. 10(a)) and 25 J (Fig. 10(b)), the difference between the deflections at room and arctic temperatures under repeated impact loading is more prominent, as compared to that observed for 30 J (Fig. 10(c)) and 35 J (Fig. 10(d)).

3.3 Absorbed Energy

Figure 11 shows a typical energy-time response obtained during an impact even on fiber reinforced laminates. The impacted energy is the peak value on the graph and the post peak plateau region is the energy absorbed by the laminate that is manifested as failure mechanisms like matrix cracking, delamination or fiber fracture. If the impact energy is equal to the absorbed energy, the laminate is deemed completely perforated by the strikers. Increasing absorbed energy implies more damage in the laminate. Hence, the degree of damage (D) for a laminate is defined as the ratio of the absorbed energy to the impact energy, which limits the values to be between 0 (no damage) and 1 (complete damage). The impacted laminates tested here were thin with an average thickness of 4 ± 0.1 mm. As a result, if the impact energy was equal to the absorbed energy, based on experimental observations, the laminates were deemed here to be completely perforated by the striker as a consequence of no rebound. However, in general, impact energy equals absorbed energy means that there is no rebound and all the impact energy is absorbed by the specimen typically as inelasticity or damage, which is especially true when the specimens are relatively thick or highly damping, in which case complete penetration is difficult to achieve. Increasing absorbed energy implies more damage in the laminate. Figure 12(a) shows representative energy-time graphs of the samples impacted at room and arctic temperature for 20 J, 25 J, 30 J and 35 J. In general, it is observed that the samples impacted at arctic temperature (blue color) absorbed less energy than the samples impacted at room temperature (red color).

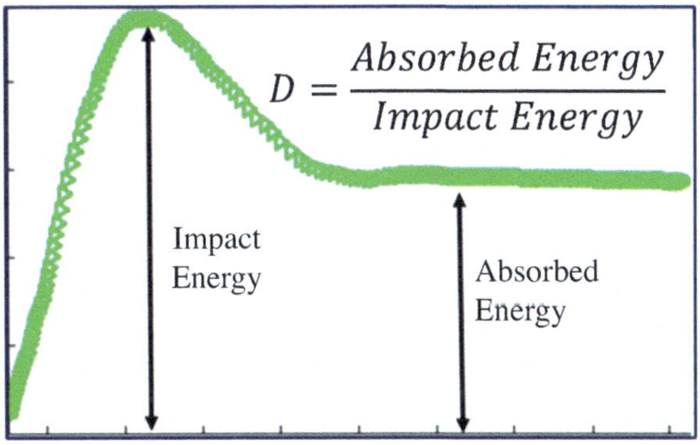

Fig. 11 Typical energy-time response of an impact event

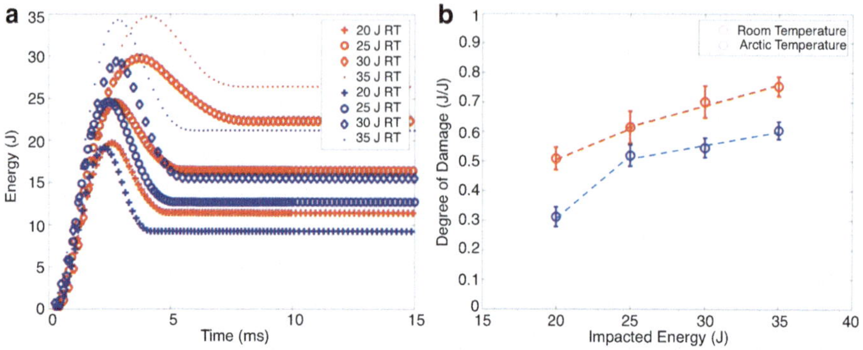

Fig. 12 (**a**) Representative energy-time graph for room (RT: 25 °C) and arctic temperatures (AT: −50 °C) at 20 J, 25 J, 30 J and 35 J, (**b**) Degree of damage for each impacted energy

To quantify the damage in the laminates, a term called degree of damage (D) coined previously by Belingardi and Vadori [6] was calculated, which is defined as the ratio between the absorbed energy and the impact energy. Figure 12(b) shows the degree of damage for different impact energies, where an increasing trend is observed with increasing impact energies. The samples impacted at AT recorded lower degree of damage as compared to the samples tested at RT for a specific impact energy, and is attributed to matrix strengthening at AT.

3.3.1 Single Impact

The trend lines in Fig. 12(b) show a knee formation at 25 J, below which the slope is higher than after. It is known that for fiber reinforced laminates, matrix failure is prevalent at lower impact energies and fiber failure is dominant at higher impact energies. Therefore, the contribution of matrix cracking at lower energies to the degree of damage is significant, whereas, a combination of fiber breakage and matrix cracking contributes at higher energies. However, due to matrix strengthening at AT, the damage in the matrix is lower than damage at RT at low impact energies as explained in the "Laminate strengthening" section. Therefore, the reduction in the degree of damage measured at 20 J between RT and AT is high, about 38% for these laminates. Whereas, at higher impact energies, that is above 25 J, this difference reduces (about 15–22% here) due to lower (but not insignificant) influence of matrix cracking on the degree of damage.

3.3.2 Repeated Impact

Figure 13(a) shows a representative energy-time graph for repeated impacts at 20 J, where the energy absorbed after the first impact (red) decreased first for the next

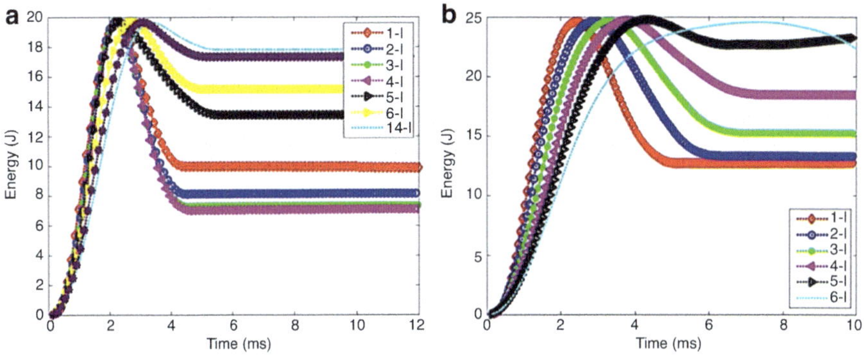

Fig. 13 Representative energy-time responses for repeated impact: (**a**) 20 J at arctic temperature; (**b**) 25 J at arctic temperature

three impacts (blue, green and magenta) due to the compaction of matrix during the "force change" phase as described before. Upon reaching the maximum force (black graph), there was a significant increase on the absorbed energy, which continues to increase gradually after consecutive impacts until laminate perforation. For 25 J, 30 J and 35 J, the energy absorbed increased gradually starting from the very first impact, as seen in Fig. 13(b).

3.3.3 Temperature Effect on the Degree of Damage Under Repeated Impact

Figure 14 shows the degree of damage versus number of impacts for room and arctic temperatures, where the values of D increase with increasing impact energies and also the number of impacts. In general, the samples impacted at −50 °C recorded lower degree of damage as compared to those at 25 °C for a specific impact energy. The damage in matrix is lower at arctic temperature due to matrix strengthening than those at room temperature at low impact energies. It has been previously established that the matrix failure is prevalent at lower impact energies and fiber failure is dominant at higher impact energies. Therefore, the contribution of matrix cracking at lower energies to the degree of damage is significant, whereas, a combination of fiber breakage and matrix cracking contributes at higher energies. Therefore, the difference in the degree of damage measured at 20 J Fig. 14(a) and 25 J Fig. 14(b) between RT and AT is high, about 21–29% for these laminates. Whereas, at higher impact energies (Fig. 14(c) for 30 J and Fig. 14(d) for 35 J), this difference reduces to about 10–15% due to lower (but not insignificant) influence of matrix cracking on the degree of damage.

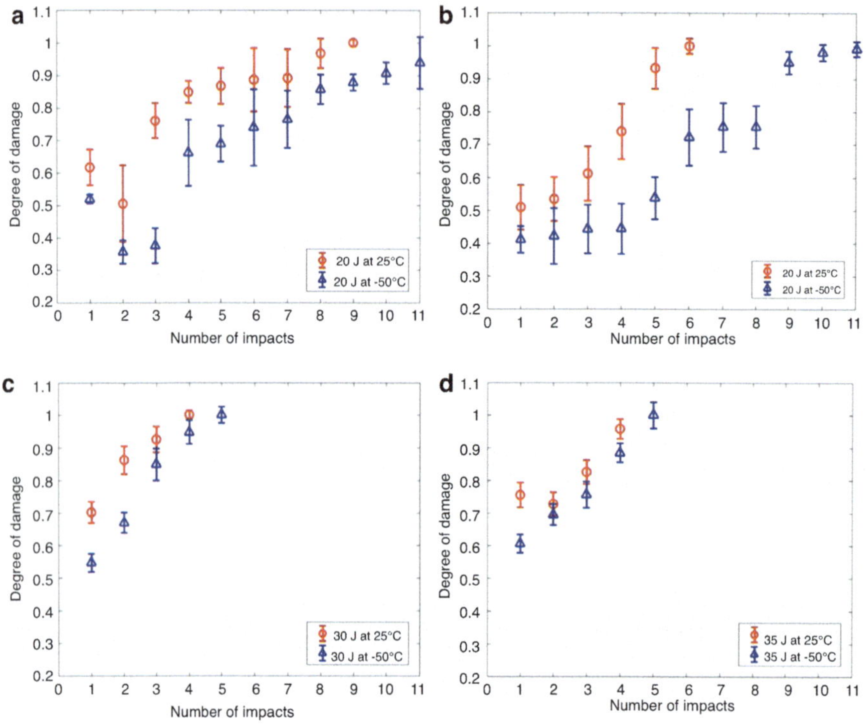

Fig. 14 Degree of damage as a function of number of impacts at 25 °C and −50 °C: (**a**) 20 J, (**b**) 25 J, (**c**) 30 J and (**d**) 35 J

3.4 Damage Mechanisms

3.4.1 Single Impact

Figure 15 shows the micro-CT scan images of samples impacted at 20 J and 30 J in room and arctic temperatures after a single impact. Regions identified as matrix cracking/delamination (smeared areas) and fiber breakage (sharp defined areas) are highlighted in the images. From the images shown in Fig. 16, the samples impacted with 20 J energy at both temperatures manifested small regions of visible damage on the impacted and the back faces. However, the micro-CT scan images show considerable internal damage in terms of matrix cracking/delamination and fiber breakage through the thickness of the samples. Fiber breakage is concentrated on the impacted surface with significant matrix cracking and delamination through the thickness of the laminate impacted at 20 J energy in room temperature. On the other hand, at arctic temperature' the overall spread of damage is confined to a smaller region with higher fiber failure traversing through the thickness of the laminate. The reduction in

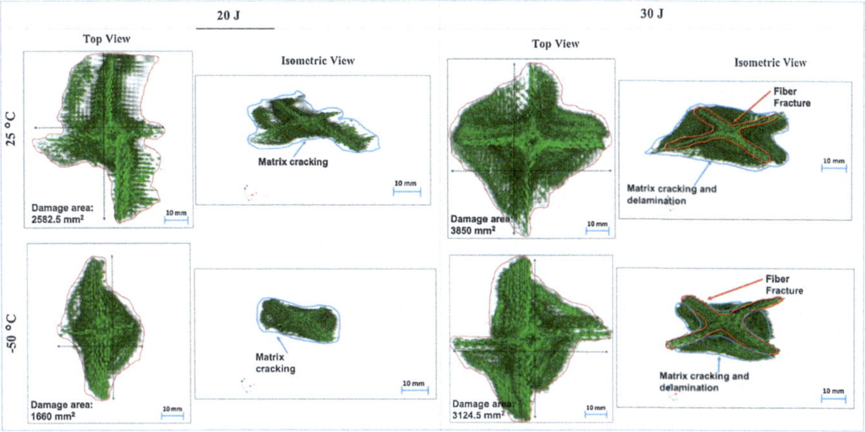

Fig. 15 Micro-CT scan after the first impact for 20 and 30 J at 25 °C and −50 °C

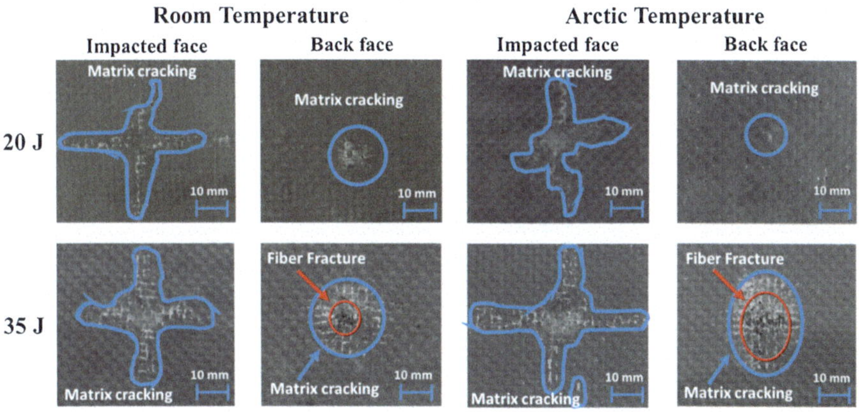

Fig. 16 Impacted face and back face at 20 and 35 J impact energies at 25 °C and −50 °C under single impact

the overall damaged area projected on to the plan of the samples is approximately 36% between RT and AT under 20 J of impact energy.

The micro-CT scan images for 30 J shown in the Fig. 15 exhibit significant fiber failure along with matrix failure, and is representative of the samples impacted at higher energies of 25, 30 and 35 J. Fiber failure through the thickness of the samples is higher at AT as compared to RT when impacted by 30 J impact energy. Also, the reduction in the overall damaged area projected on to the plan of the samples at 30 J is approximately 19% between RT and AT. The observed difference in percentage reduction in damaged area between RT and AT under 20 J and 30 J is due to the increase in strength when composites are subject to low temperatures. The yield strength of the vinyl ester matrix increased up to 55% at AT as compared to RT based on the compression tests conducted in the current study, which indicates a delay in the onset of matrix cracks at AT. Consequently, the composite manifested significantly higher fiber failure than matrix cracking as compared to RT for the same impact energy.

Figure 16 shows the impacted face and back face of two samples after testing. The samples impacted at 20 J in room and arctic temperature did not exhibit fiber breakage, however, exhibited small regions of matrix cracking at the impacted and back face of the laminate as seen by the enclosed blue curves. On the other hand, the samples at 35 J exhibited a combination of fiber breakage (enclosed by the red circles) and matrix cracking at the back face in both temperatures. Even though the samples at 20 J did not exhibit a significant amount of visible damage externally, there was significant internal damage.

3.4.2 Repeated Impact

Figure 17 shows the impacted face and back face of samples repeatedly impacted at 20 J and 35 J to complete perforation. The samples impacted at 20 J at both temperatures experienced a combination of fiber fracture (enclosed by the red curves) and matrix cracking (enclosed by the blue curves). At 25 °C, these samples also manifested small regions of fiber bridging at the back face (enclosed by the green circles). On the other hand, at −50 °C, fiber bridging was minimal and showed predominantly fiber fracture at the back face of the laminate as predicted from the "Laminate Strengthening" section. This was also representative of the samples impacted at 25 J.

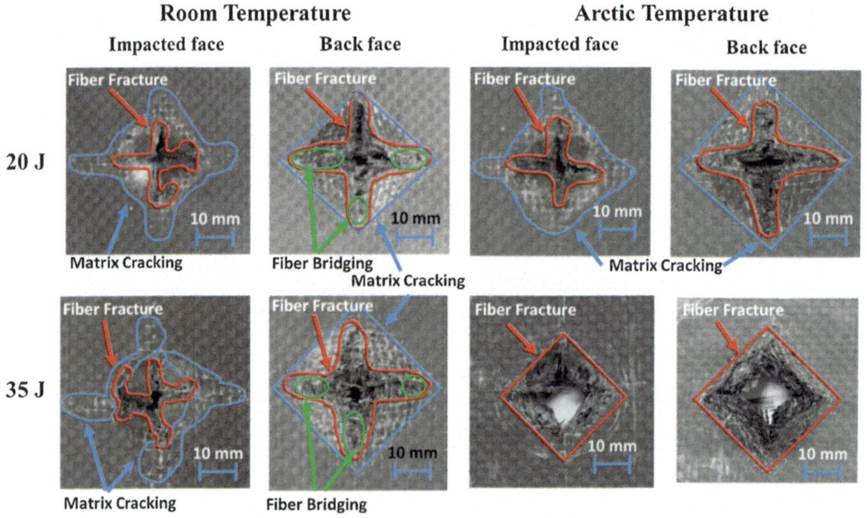

Fig. 17 Impacted face and back face for 20 J and 35 J impact energies at 25 °C and −50 °C under repeated impact

The samples impacted at 35 J at room temperature exhibited a combination of fiber breakage and matrix cracking along with fiber bridging at the back face of the laminate. On the other hand, at arctic temperature they experienced significant fiber fracture at the impacted and back face. The perforated region is sharp and well defined at 30 J in arctic temperature as compared to lower impact energies and at room temperature. The failure mechanisms were similar for the samples impacted at 30 J. Overall, arctic temperature renders the composite brittle thereby promoting more fiber fracture than matrix cracking, which is accentuated at higher impact energies.

4 Conclusion

Dynamic single and repeated impact response and failure mechanisms of woven carbon/vinyl ester composites at room (25 °C) and arctic (−50 °C) temperatures were investigated in this paper in view of increasing interest in arctic explorations and the need to characterize these composites for arctic applications. Four impact energies of 20 J, 25 J, 30 J and 35 J were considered for dynamic impact testing at room and in-situ arctic temperatures, where the samples were repeatedly impacted until perforation. Key observations in terms of the contact force, displacement, energy absorbed and failure mechanisms were re- ported in this paper. Key conclusions are summarized here.

Single Impact

- At −50 °C, the rigidity of the laminates increased due to resin strengthening that resulted in higher initial stiffness as compared to room temperature.
- Higher peak forces manifested in laminates impacted at −50 °C for all energies.
- Increased in strength also resulted in lower displacements in the laminates at −50 °C.
- The laminates absorbed less energy and correspondingly the degree of damage was lower at −50 °C than at 25 °C.
- At RT, the dominant failure mechanism was matrix failure (micro cracks/delamination) at low impact energies (20 J). Whereas, for higher impact energies (25 J, 30 J and 35 J), the dominant failure mechanism was fiber fracture along with considerable matrix failure.
- At AT, the overall damage area projected on to the plane of the composite reduced, however, manifested significantly higher fiber failure as compared to RT for the same impact energy, due to the brittleness of the sample at AT.
- The percentage reduction in damage area and the average degree of damage value between RT and AT exhibited a decreasing trend with increasing impact energies, as the influence of matrix failure is more dominant at lower impact energies and contribute higher to the extent of damage in a laminate.

Repeated Impact

- During a repeated impact event at low impact energies, a laminate experiences four phases of force change: stabilization, force increase, maximum force and force decrease. If the impact energy is sufficiently high to cause significant damage during the first impact, then only two phases of force change will be present: maximum force and force decrease.
- At −50 °C, increase in rigidity of a laminate results in higher initial stiffness, lower defections and higher impact forces than those impacted at 25 °C for all energies.
- The laminates absorb less energy at −50 °C due to matrix strengthening, which results in lower values of degree of damage than at 25 °C. Consequently, the number of impacts needed for complete perforation of laminates increase at low temperature.
- At room temperature, the dominant failure mechanism is matrix cracking at low impact energies (20 J) as compared to higher impact energies (30 J and 35 J), where the dominant failure mechanism is fiber fracture with lesser matrix cracking.
- Significant shift in failure mechanisms occurs at arctic temperature, where fiber fracture is promoted due to matrix strengthening. This manifests as sharp defined perforated regions at low temperature with minimal fiber bridging at the back face of the laminate.
- Overall, the difference in response of laminates at lower energies is more distinct with temperature change from room to arctic, whereas, if velocity increases there will be a damage saturation effect where the temperature will have lower influence on the damage extension.

In conclusion, failure mechanisms shift from matrix failure towards fiber failure at arctic temperature, even though the measured degree of damage and deflection from the impact tests provide lower value at AT as compared to RT. This shift in mechanism can have significant detrimental effect on the tensile residual strength (as fiber fracture will be the main failure mechanism at AT) and durability of the composite. Also, this study is very relevant for developing appropriate repair techniques for composites for use in arctic applications.

Acknowledgments The author would like to acknowledge the partial support by the Department of Defense (DoD) Office of Naval Research – Young Investigator Program (ONR-YIP) Award [N00014-19-1-2206] and DoD HBCU/MI Basic Research Grant [W911NF-15-1-0430] for conducting the research presented here. The author would like to acknowledge the great amount of research work and input from Alejandra Castellanos, who has contributed significantly to the work presented here.

References

1. Abrate S (1994) Impact on laminated composites: recent advances. Appl Mech Rev 47:517–544
2. ASTM D3039M-17 (2017) Standard test method for tensile properties of polymer matrix composite materials. J ASTM Int D3039M-17 1–13

3. ASTM D695-15 (2015) Standard test method for compressive properties of rigid plastics. J ASTM Int D695-15 1–8
4. ASTM D7136/D7136M-15 (2015) Standard test method for measuring the damage resistance of a Fiber-reinforced polymer matrix composite to a drop-weight impact event. J ASTM Int D7136/D7136M-15 1–16
5. Babu S, Shivanand HK (2014) Impact analysis of laminated composite on glass Fiber and carbon Fiber. Int J Emerg Technol Adv Eng 4:824–829
6. Belingardi G, Vadori R (2003) Influence of the laminate thickness in low velocity impact behavior of composite material plate. Compos Struct 61:27–38
7. Bieniaś J, Surowska B, Jakubczak P (2015) Influence of repeated impact on damage growth in fibre reinforced polymer composites. Maint Reliab 17:194–198
8. Cantwell WJ, Morton J (1990) Impact perforation of carbon fibre reinforced plastic. Compos Sci Technol 38:119–141
9. Caprino V, Lopresto V, Scarponi C, Briotti G (1999) Effect of ZnO nanowire morphology on the interfacial strength of nanowire coated carbon Fibers. Compos Sci Technol 59:2279–2286
10. Chalmers DW (1994) The potential for the use of composite materials in marine structures. Mar Struct 7:441–456
11. Chittajallu K (2004) Computational modeling of the Vacuum Assisted Resin Transfer Molding (VARTM) process (master's thesis). Clemson University
12. Chu XX, Wu ZW, Huang RJ, Zhou Y, Li LF (2010) Mechanical and thermal expansion properties of glass fibers reinforced PEEK composites at cryogenic temperatures. Cryogenics (Guildf). 50:84–88
13. Dempsey JP (2000) Research trends in ice mechanics. Int J Solids Struct 37:131–153
14. Dutta PK (1994) Low-temperature compressive strength of glass-fiber-reinforced polymer composites. J Offshore Mech Arct Eng 116:167–172
15. Garcia R, Castellanos AG, Prabhakar P (2019) Influence of arctic seawater exposure on the flexural behavior of woven carbon/vinyl ester composites. J Sandw Struct Mater 21(3):1190–1208. https://doi.org/10.1177/1099636217710821
16. Gibson AG (2003) The cost effective use of fibre reinforced compopsites offshore. Rep Heal Saf Exec 2003
17. Gomez-del-Rio T, Zaera R, Barbero E, Navarro C (2006) Damage in CFRPs due to low velocity impact at low temperature. Compos Part B 36:41–50
18. Greene E (1990) Use of Fiber reinforced plastics in the marine industry. Home-Transport Res. Int Doc **SSC**-360
19. Guedes Soares C, Garbatov Y (2017) Progress in the analysis and Design of Marine Structures. CRC Press/Balkema, London
20. Hartwig G, Knaak S (1984) Fibre-epoxy composites at low temperatures. Cryogenics (Guildf) 24:639–647
21. Ibekwe SI, Mensah PF, Li G, Pang SS, Stubblefield MA (2007) Impact and post impact response of laminated beam beams at low temperatures. Compos Struct 79:12–17
22. Icten BM (2014) Low temperature effect on single and repeated impact behavior of woven glass-epoxy composite plates. J Compos Mater 49:1171–1178
23. Im K, Cha C, Kim S, Yang I (2001) Effects of temperature on impact damages in CFRP composite laminates. Compos Part B 32:669–682
24. Jones N (1993) Recent studies on the response of structures Subected to large impact loads. Soc Nav Archit Mar Eng Sh Struct Comm:1–22
25. Julias A, Murali V (2014) Effect of carbon Fiber position on the impact behavior of glass/carbon Fiber hybrid composite laminates. Int J Appl Eng Res 9
26. Kim M, Kang S, Kim C, Kong C (2007) Tensile response of graphite/epoxy composites at low temperatures. Compos Struct 79:84–89
27. Li XK, Liu PF (2017) Exppcrimental analysis of low-velocity impact beaviors of carbon fibre composite laminates. J Fail Anal Prev 17:1126–1130

28. Lima Sobrinho L, Ferreira M, Bastian FL (2009) The effects of water absorption on an ester vinyl resin system. Mater Res 12:353–361
29. Lopez-Puente J, Zaera R, Navarro C (2002) The effect of low tempertures on the intermediate and high velocity impact response of CFRPs. Compos Part B 33:559–566
30. Morais WA, Monteiro SN, d'Almeida JRM (2009) Evaluation of repeated low energy impact damage in carbon-epoxy composite materials. Compos Struct 91:318–323
31. Murat BIS, Rahman AAA (2017) Study of impact damage beahvior in coven carbon fiber plates. Eng Phys Int Conf Epic 2016:47–54
32. Naik NK, Chandreshakarkhar Y, Mduri S (2000) Polymer matrix woven fabric composites subjected to low velocity impact: part I. Damage initiation studies. J Reinf Plast Compos 19:912–943
33. Nguyen S, James T, L Iannucci (2016) Low, medium and high velocity impact on composites. 16th International Conference Composite Structures. 1–2
34. Rajkumar GR, Krishna M, Narasimha HN, Sharma SC, Vishnu KR (2012) Experimental ivestigation of low-velocity repeated impacts on glass fiber metal composites. J Mater Eng Perform 21:1485–1490
35. Safri SNA, Sultan MTH, Yidris N, Mustapha F (2014) Low velocity and high velocity impact test on composite materials – a review. Int J Eng Sci 3:50–60
36. Salehi-Khojin A, Bashirzadeh R, Mahinfalah M, Nakhaei-Jazar R (2006) The role of temperature on impact properties of Kevlar/fiberglass composite laminates. Compos Part B 37:593–602
37. Sayer M, Bektas NB, Sayman O (2010) An experimental investigation on the impact behavior of hybrid composite plates. Compos Struct 92:1256–1262
38. Selvaraju S, Illaiyavel S (2011) Application of composites in marine industry. J Eng Res Stud 2:89–91
39. Setvati MR, Shafiq N, Mustaffa X and Syed ZI (2014), A review on composite materials for offshhore structures. in ASME. International conference on offshore mechanics and Arctic engineering 5, (ASME)
40. Shyr TW, Pan YH (2003) Impact resistance and damage characteristic of composite laminates. Compos Struct 62:193–203
41. Signor AW, Chin JW (2003) Effects of ultraviolet radiation exposure on vinyl ester matrix resins: chemical and mechanical characterization. Polym Degrad Stab 79:1–12
42. Sjoblom PO, Hartness JT, Cordell TM (1988) On low-velocity impact testing of composite materials. Compos Mater 22:30–52
43. Sultan MT, Worden K, Staszewski WJ, Hodzic A (2012) Impact damage characterization of composite laminates using a statistical approach. Compos Sci Technol 72:1108–1120
44. Szekrényes A (2002) Overview on the experimental investigations of the fracture toughness in composite materials. Hungarian Electron J Sci:1–19
45. Wang XF, Zhao JH (2001) Monte-Carlo simulation to the tensile mechanical behaviors of unidirectional composites at low temperature. Cryogenics (Guildf) 41:683–691
46. Yusriah L, Mariatti M, Bakar AA (2010) The properties of vinyl ester composites reinforced with different types of woven fabric and hollow phenolic micospheres. J Reinf Plast Compos 29:3066–3073
47. Zhu L, Faulkner D (1996) Damage estimate for plating of ships and platforms under repeated impacts. Mar Struct 9:697–720

Behavior of Composite Materials and Structures in Low Temperature Arctic Conditions

K. T. Tan

1 Introduction

1.1 Background and Motivation

The arctic region is severely affected by the effect of global warming. Over the last three decades, the polar ice regions have experienced dramatic shrinkage, resulting in a loss of approximately 50% of ice area [1]. This climatic change leads to the opening of new sailing passages in the polar oceans, which could be utilized for shorter transportation and more efficient shipping routes. However, operation in the arctic region is extremely challenging, due to the harsh climatic conditions. Therefore, marine and naval structures need to be specially designed to endure and be tolerant against the extreme low temperature environment.

Composite sandwich structures have been increasingly used in aerospace, marine, wind-energy and various industrial applications. This is attributed to their high specific flexural stiffness, ultra-low structural weight, and excellent flammability resistance. Composite sandwich structures generally consist of two thin carbon fiber reinforced polymer (CFRP) facesheets adhered to a light-weight core. CFRP facesheet layers are designed to withstand bending and in-plane loading; whilst the core layer is responsible for carrying the transverse shear load. Despite the numerous advantages of using composite sandwich structures, they are extremely susceptible to out-of-plane impact damage. For marine structures in arctic region, impact damage can easily occur when ice particles collide with the outer hull of the marine composite sandwich structures. The resulting damage comprises of complex mixed failure modes involving matrix cracking, delamination, fiber breakages, interlayer debonding, and shearing and crushing of the core layer.

K. T. Tan (✉)
Department of Mechanical Engineering, The University of Akron, Akron, OH, USA
e-mail: ktan@uakron.edu

S. W. Lee (ed.), *Advances in Thick Section Composite and Sandwich Structures*,
https://doi.org/10.1007/978-3-030-31065-3_21

Much research effort have been devoted to understand the impact behavior of composite sandwich structures. In particular, there is considerable interest in using polymeric foam cores due to their low cost, ease of manufacturing, good moisture properties and high chemical and flammability resistance. Several studies have discussed the contribution of the core layer properties on the impact behavior [2–5]. In regards to the effect of low temperature, there exist several studies that have investigated the low temperature effects on impact strength of composite laminates [6–12]. However, there are only limited studies focusing on the impact damage behavior of composite sandwich structures in extreme low temperature environments. Liu et al. [13] investigated the low temperature effect on strength and crushing behavior of carbon fiber composite truss sandwich cores, which is analogous to out-of-plane impact behavior, but without the influence of dynamic impact energy. Erickson et al. [14] studied the effect of temperature on impact and bending strength of glass fiber reinforced sandwich composites with a honeycomb core, with the lowest temperature of −25 °C. Salehi-Khojin et al. [15] also analyzed the impact behavior of honeycomb-core sandwich composites at a lower temperature of −50 °C. They concluded that impact damage is significantly influenced by the test temperature, and the largest damage area is observed at −50 °C. Both Erickson et al. [14] and Salehi-Khojin et al. [15] studied honeycomb-core sandwich composites, but not foam-core sandwich composites. In a more recent study by Yang et al. [16], the authors investigated the effect of temperature on impact behavior of polyvinyl chloride (PVC) foam-core sandwich structures at high temperature of 82 °C and low temperature of −46 °C. It is clear that no current research explores the extreme low temperature impact performance of composite sandwich structures in arctic condition, which could possibly reach down to −70 °C. Therefore, there exists a research gap in understanding the coupling phenomenon between dynamic loading and extreme arctic cold environments in foam-core composite sandwich structures. There also exists a research gap in regards to the residual strength and the associated damage mechanisms of polymeric foam core sandwich structures in low temperature environments.

In this study, low-velocity impact behavior of composite sandwich structures is investigated at four different temperatures. Room temperature of 23 °C is chosen as a benchmark case; 0 °C is a second benchmark case to represent the standard freezing temperature; −30 °C is the average temperature in the arctic region; and −70 °C is the lowest recorded temperature in the arctic region. The main objective of this study is to understand the impact damage response and to elucidate the impact damage mechanisms and characteristics associated with extreme low temperature in the arctic region. Compression-after-impact (CAI) tests and Flexural-after-impact (FAI) will also be performed. The state-of-the-art Instron CEAST impact test machine with environmental chamber is employed to perform the impact study. Force/displacement-time history data will be analyzed to identify critical damage forces, panel deflection, structural stiffness and absorbed energy. X-ray micro-computed tomography technique is further utilized to observe complex damage modes. A longer overarching goal of this work is to provide a fundamental knowledge to design new sandwich construction with enhanced damage tolerance in the arctic region for future naval applications.

1.2 Influence of Low Temperature on Composite Laminates

This section presents a quick summary on the mechanical properties of composite laminates at low temperature. A literature review is performed on published work involving the experimental testing of composites at low temperature conditions. Results from these papers are summarized using normalized parameter approach, whereby the property at a specific temperature is normalized with the same property at room temperature, as given in Eq. (1). A normalized property of less than one means that the property is degraded by the specific temperature change; whilst a normalized property of more than unity means that the property has been enhanced by the temperature change.

$$\text{Property}_{\text{normalized}} = \frac{\text{Property}_{\text{specific temperature}}}{\text{Property}_{\text{room temperature}}} \tag{1}$$

Figure 1 shows the normalized tensile and flexural properties of carbon epoxy composite laminates at various temperature conditions [17]. It is clear that tensile and flexural strength are both degraded by low temperature, while tensile and flexural modulus are generally increased due to low temperature effect. Figure 2 presents the effect of temperature on impact delamination area, as well as the consequent residual fracture bending strength [6]. It is observed that delamination area of impacted specimens are larger at lower temperature, thus leading to reduction in residual fracture bending stress. The opposite is true at higher temperature (more than room temperature).

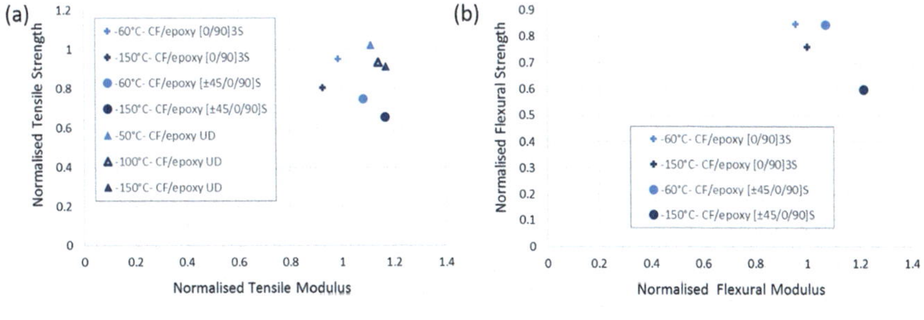

Fig. 1 Effect of low temperature and laminate lay-up on (**a**) static tensile; (**b**) static flexural properties. Data extracted from [17]

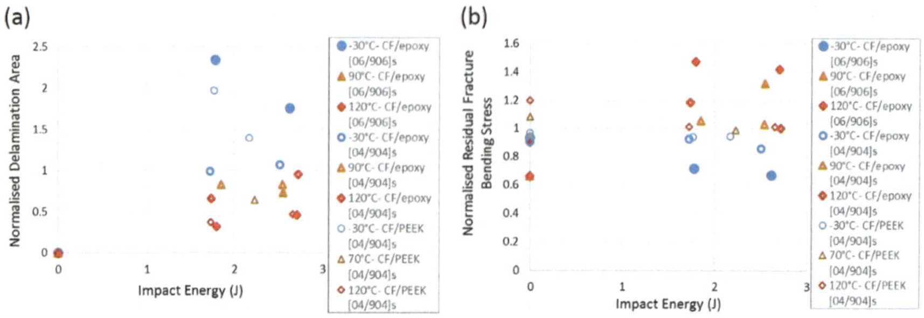

Fig. 2 Effect of temperature and material system on (**a**) impact delamination area; (**b**) residual fracture bending strength. Data extracted from [6]

2 Experimental Details

2.1 Specimen Preparation

Composite sandwich specimens are made of 0°/90° woven carbon fiber/epoxy matrix facesheets attached to a layer of Divinycell polyvinyl chloride (PVC) H100 foam core. Two sets of samples with different facesheet thicknesses of 0.254 mm and 0.762 mm are used in this work. The Divinycell PVC H100 foam core has a relatively larger thickness of 6.35 mm (Fig. 3). The specimens are cut into a standard size of 150 mm × 100 mm from a motherplate using a sharp diamond cutter. For specimens that are to be tested at −30 °C and −70 °C, they are first kept in a low temperature freezer at −22 °C for 24 h, and then conditioned at their test temperature (−30 °C or −70 °C) for 20 min before impact and during impact. For specimens that are tested at 0 °C, they are kept in the freezer at 0 °C for 24 h and then conditioned at 0 °C for 20 min before and during impact. Benchmark room temperature specimens are not conditioned and are impacted at room temperature of 23 °C.

2.2 Impact Testing Setup

The state-of-the-art Instron Ceast 9350 impact drop tower machine, equipped with an environmental chamber, is employed to perform impact testing on composite sandwich panels at low temperature ranges. Liquid nitrogen tank is connected to the environmental chamber to generate low temperature condition. The impact machine is linked to a control system to monitor and record the test data. Test results are acquired by DAS64K data acquisition system. The sampling frequency of the readings is 500 kHz measured at 1500 points. An impactor of 16 mm in diameter and a nominal mass of 1.182 kg is attached to a tub holder of 2.3 kg. The total impact mass of 3.482 kg is used throughout the tests. The impact height is adjusted to reach

Fig. 3 A sample of the foam-core composite sandwich structure. [Reprinted from Composite Structures, Vol. 192, Elamin M, Li B, Tan KT, Impact damage of composite sandwich structures in arctic condition, Pages 422–433, Copyright (2018), with permission from Elsevier]

the desired impact energy level. The correspondent impact velocities to the impact energy levels are 2.4 m/s and 1.7 m/s for 10 J and 5 J, respectively. For flexural-after-impact specimens, they are impacted at 8 J and 4 J instead. The samples are placed on a round support frame with a cut-out window of 76 mm in dia meter, and top clamped with 100 N clamping force. At least two samples are tested for each energy level, facesheet thickness, and test temperature, so as to ensure the accuracy and repeatability of the measurements. The experimental set-up of the impact test is depicted in Fig. 4. More experimental details can be found in [18].

2.3 X-ray Micro-computed Tomography Analysis of Internal Damages

X-ray Micro-computed tomography (Micro-CT) is a non-destructive technique that uses three-dimensional x-ray imaging to inspect internal structural damages. Many projection images are captured, as the sample rotates a complete 360° revolution. Micro-CT is an efficient technique to inspect deep details of internal damages in composites using a single scan and to visualize internal cross-sectional damages [19]. This method of inspection provides accurate and in-depth information of complex internal damage modes. Area measurement software can be further employed to measure the damage areas of the post-processed micro-CT images.

3D renderings and cross sectional images of the samples are acquired using X-ray micro-CT (Phoenix Nanotom MTM 180, GE sensing Inspection Technologies GmbH, Germany; NSF Grant 260145). X-ray emission parameters for the scans are 90 keV and 50 μA. Each single rotational position image is averaged twice, and 1800 scans per acquisition are collected. Voxel size for the collected data is 41.7 μm. After

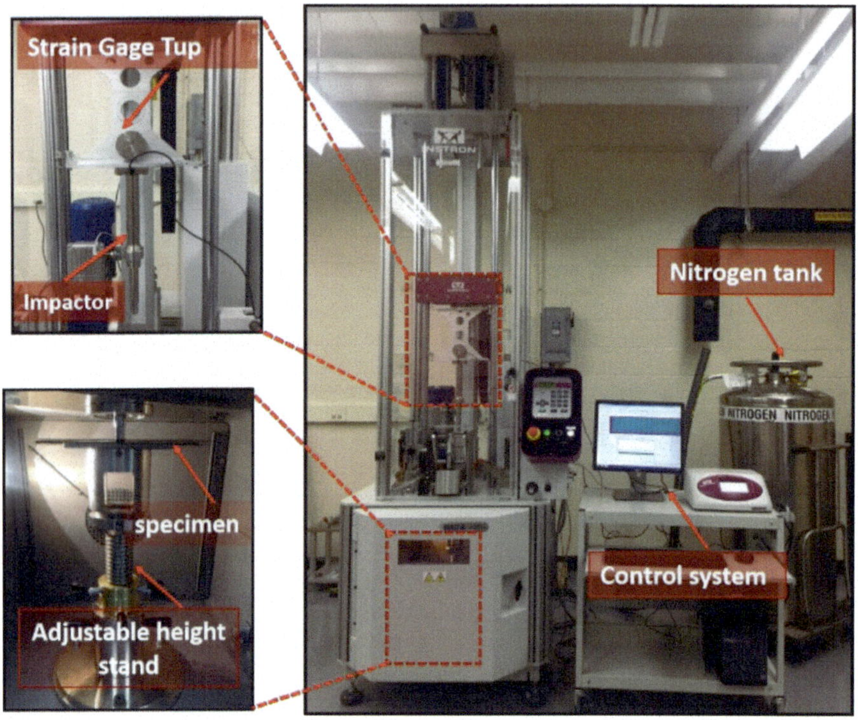

Fig. 4 Impact drop weight test system. [Reprinted from Composite Structures, Vol. 192, Elamin M, Li B, Tan KT, Impact damage of composite sandwich structures in arctic condition, Pages 422–433, Copyright (2018), with permission from Elsevier]

reconstruction of the 3D virtual object utilizing Phoenix software, analysis of the reconstructed volumes is investigated using VG Studio Max software.

2.4 Compression After Impact Test Setup

After completion of impact tests, benchmark specimens are taken directly to perform the CAI test at room temperature. The low-temperature impacted specimens are conditioned inside the freezer at −22 °C for 24 h prior to the CAI test. Unimpacted specimens are also conditioned at the test temperature and subsequently subjected to compression testing, so as to act as benchmark references with impacted CAI test specimens. The specimens are perfectly clamped using ASTM compression after impact fixture. This CAI fixture fully clamps the specimen sides by independent restrains. The lateral side is restrained by four knife-edges plates that prevent the global buckling of the sample, while the bottom and the top sides prevent the axial motion. The fixture is made of stainless steel with hardened coating to avoid corrosion while operating in extreme environments. It is designed to meet the global testing

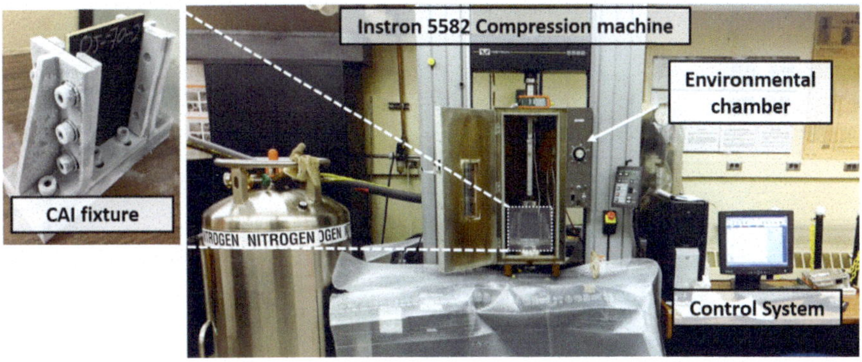

Fig. 5 Compression-after-impact (CAI) test system

standards of ASTMD 7137 M, with a capability to operate in extreme temperatures ranging from −75 °C to 250 °C. Compression loads are applied by Instron 5582 test machine with a 100 kN load cell. The machine is connected to an environmental chamber and a liquid nitrogen tank is used to cool the chamber to low-temperature conditions. The load is delivered from the load cell to the CAI fixture by a stainless steel extension rod that passes through the environmental chamber. The entire system is controlled using Bluehill software to acquire the test readings. The CAI test set-up is shown in Fig. 5. More experimental details can be found in [20].

2.5 *Flexural After Impact Test Setup*

Flexural after impact (FAI) tests are conducted to investigate the residual strength and stiffness of the samples. A three-point bending apparatus is used following ASTM C 393 with a crosshead speed of 0.5 mm/min and a span of 100 mm. Samples are tested in bending conditions at room temperature, −30 °C and −70 °C, using an environmental chamber in Instron 5582 machine, as shown in Fig. 6. The sample is placed such that the first face sheet contacted in the impact tests is under compression. This orientation is chosen because compression is often the critical loading mode after an impact event. Following impact, the fibers may not be extensively damaged, but the matrix damage may be significant. Bending test is also carried out placing distal unimpacted facesheet towards the anvil. In all cases (as shown in Fig. 7), two specimens were tested under each condition. More details can be found in [21].

Bending Test Matrix	
ASTM standard	Test Temperature
C 393	23°C, -30°C, -70°C

Fig. 6 Flexural three-point bending test

3 Results and Discussion

3.1 Impact Damage Response

Impact history curves are analyzed to understand the dynamic response and impact behavior of composite sandwich structures during the impact event. Representative force/energy against time curves for thin specimens (0.254 mm facesheet thickness) impacted with 10 J of impact energy at various temperatures are presented in Fig. 8.

The force-time curves, represented by solid lines in Fig. 8, show an initial linear behavior as the impactor contacts the specimen. The curve then reaches a first peak, followed by a sharp drop indicating the penetration failure of the top carbon fiber/epoxy composite facesheet (impact face). This failure exists in the form of delamination, fiber breakage and matrix cracking of the composite facesheet. Subsequently, impact force value continues to increase with time, inducing crushing and shearing in the foam layer, causing damage to the core layer. For specimens impacted at low temperature (Figs. 8b, c, d), the foam core is severely damaged, and the specimens experience tensile failure at the bottom composite facesheet, evidently indicated by the second peak and sudden drop in the force-time curves.

3.2 Impact Damage Critical Forces

Force-time history graphs can be categorized into two ways. First, for specimens with both top and bottom facesheets damages, the critical forces are characterized by damage forces of top facesheet and back facesheet, denoted by Fskin and Fback, respectively. Second, for specimens with only top facesheet damage, the critical damage forces of the top facesheet and core layer are denoted by Fskin and Fcore, respectively. Figure 9 summarizes the impact damage force values for thin and thick

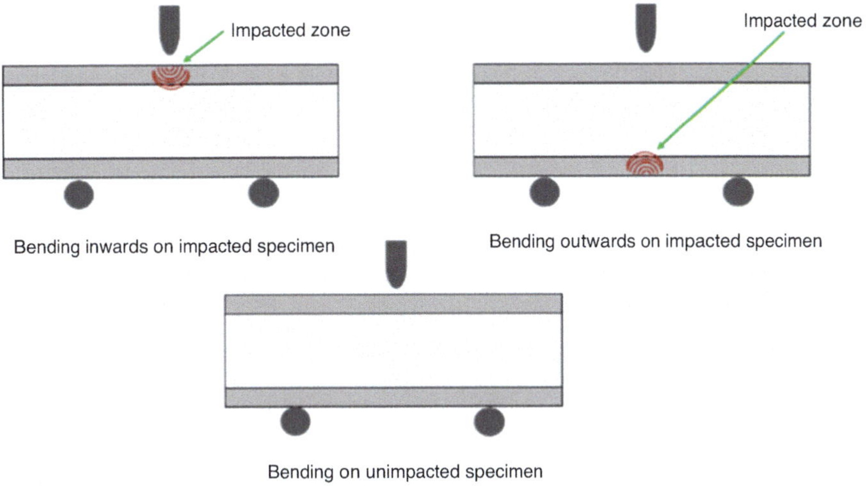

Fig. 7 Schematic view of test configuration for three-point bending test

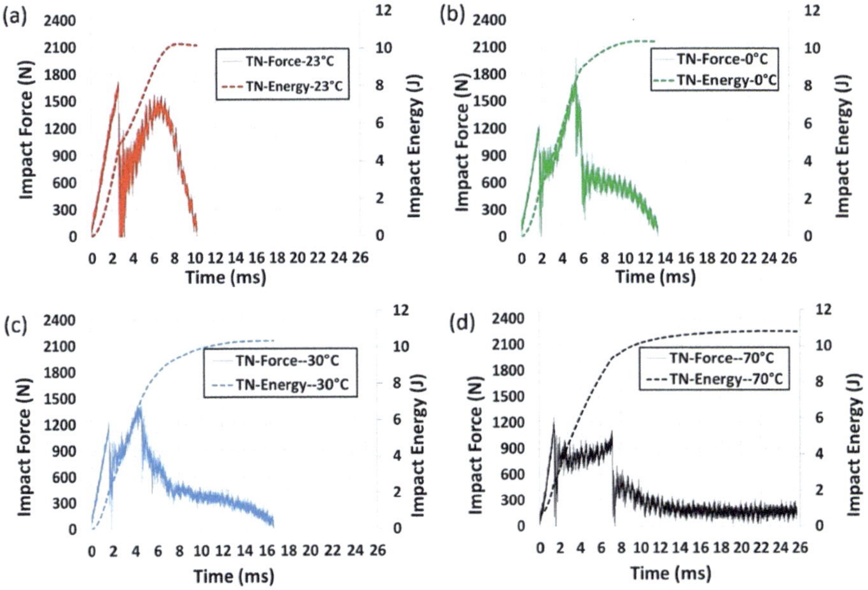

Fig. 8 Thin specimens impacted with 10 J at (**a**) 23 °C; (**b**) 0 °C; (**c**) −30 °C; (**d**) −70 °C. [Reprinted from Composite Structures, Vol. 192, Elamin M, Li B, Tan KT, Impact damage of composite sandwich structures in arctic condition, Pages 422–433, Copyright (2018), with permission from Elsevier]

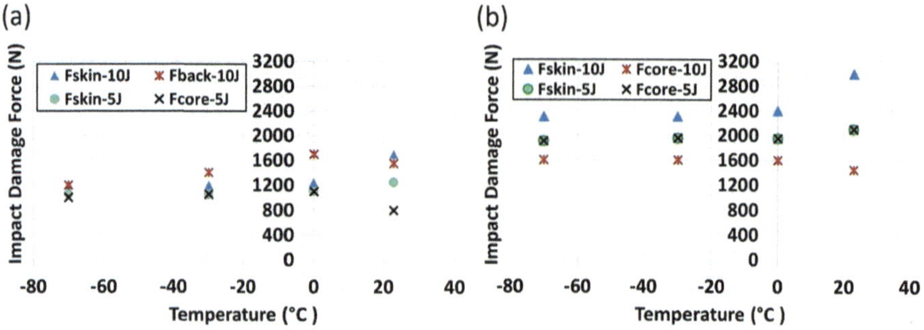

Fig. 9 Critical impact damage forces for (**a**) thin specimens; (**b**) thick specimens. [Reprinted from Composite Structures, Vol. 192, Elamin M, Li B, Tan KT, Impact damage of composite sandwich structures in arctic condition, Pages 422–433, Copyright (2018), with permission from Elsevier]

specimens across all test temperatures. The data points are obtained from the average of at least two test specimens. Error bars are omitted to improve readability of the figure, since the measurements are reasonably accurate with very low marginal errors. By comparing the critical damage forces for the top composite facesheet (Fskin), it is clear that Fskin decreases as temperature decreases. This is true for both thin (TN) and thick (TK) specimens and for both 5 J and 10 J impact energy levels. This implies that the failure strength of the composite facesheet is significantly reduced by extreme low temperature, with the maximum strength reduction of up to 30% in the extreme arctic temperature of −70 °C, making composite facesheets very susceptible to impact damage in low temperature environment. Low temperature influences the properties of the material constituents, resulting in decreased impact toughness, thus leading to a more brittle material behavior. Furthermore, the matrix interface becomes weaker, and this results in easy debonding of carbon fibers from the epoxy matrix.

The values of the back facesheet damage force (Fback) are usually greater than the top facesheet damage force (Fskin) due to the back facesheet membrane resistance and densification of the core layer before back facesheet penetration (Fig. 9a). The same force reduction trend along decreasing temperature is observed for Fback values (Fig. 9a), which similarly indicates that it is easiest to penetrate the back facesheet at low temperature, especially at −70 °C.

The core damage Fcore values are generally lower than the corresponding Fskin forces since the core can typically resist a lower impact (indentation) load than the facesheets. The Fcore values show slight increase at low temperature range (see Fcore-5 J in Fig. 9a and Fcore-10 J in Fig. 9b). This is likely due to the fact that the stiffness of the foam core increases at low temperature, thus resulting in greater force resistance during the impact event. In the case of Fcore-5J for TK specimens (Fig. 9b), Fcore-5 J values are assumed to be the same as Fskin-5 J values, implying only top facesheet damage is considered, since it is difficult to distinguish between the critical damage forces for the top composite facesheet and the foam core.

3.3 *Impact Damage Absorbed Energy*

In regards to the energy-time curves, the impact energy values increase along time during the impact event. The peaks of the energy-time curves are the impact energy given to the system, which in some cases are more than the defined impact energy of 5 J or 10 J, due to the energy added by the global panel deflection. The intersection point between the solid line (force-time) and the dotted line (energy-time), corresponds to the time when top facesheet penetration occurs. The first peak force for the force-time curve is the force required to penetrate the top facesheet; whilst the penetration energy in the energy-time curve is the amount of energy required to penetrate the top facesheet of the composite sandwich structure. Figure 10a portrays the energy-time characterization graph. The top facesheet layer is perforated upon reaching the penetration energy, which is denoted by Ep. The total impact energy is given by Emax, while Eabs indicates the amount of energy absorbed by the panel at the end of the impact event. The rebound energy, referred to as Emax-Eabs, is the

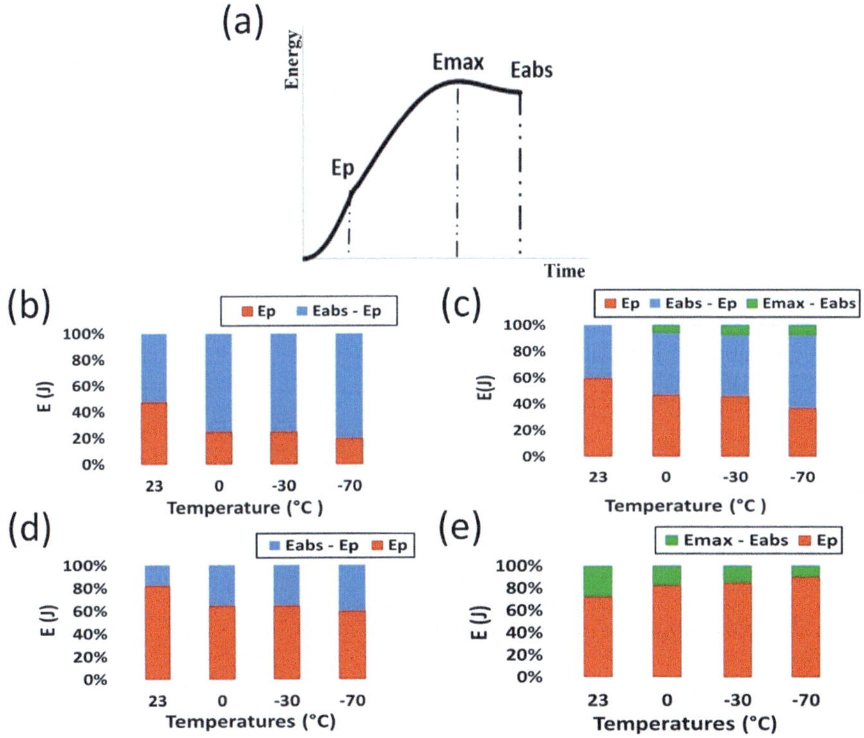

Fig. 10 Impact energy (**a**) characterization graph; (**b**) thin specimens at 10 J; (**c**) thin specimens at 5 J; (**d**) thick specimens at 10 J; (**e**) thick specimens at 5 J. [Reprinted from Composite Structures, Vol. 192, Elamin M, Li B, Tan KT, Impact damage of composite sandwich structures in arctic condition, Pages 422–433, Copyright (2018), with permission from Elsevier]

amount of energy that returns back to the system at the end of the impact event in the form of impactor rebound. In the case when there is complete penetration of the specimen, there is no rebound energy and the impact energy is fully absorbed by the specimen in terms of damage failure.

Based on Figs. 10a, 10b, 10c, 10d and 10e present a summary of the amount of energy absorbed by the composite sandwich specimens. The various layers of facesheets and foam core contributed to the absorption of impact energy depending on the material strength, thickness and failure mechanisms of each layer. A clear general trend shows that as the test temperature decreases, the amount of penetration energy of the top facesheet, Ep, is observably reduced. This observation agrees with earlier force-time data, in which the Fskin critical force value decreases with decreasing temperature (Fig. 9). For thin specimens impacted with 10 J energy (Fig. 10b), the penetration energy (Ep) is reduced by half at extreme low temperature of -70 °C, indicating a higher susceptibility to failure at low-temperature environment. Furthermore, at low temperature, the core layer absorbs approximately 80% of the total impact energy which results in severe crushing and shearing of the foam core. It is worth noting that for Fig. 10b, the Eabs-Ep energy also accounts for energy required to penetrate the bottom facesheet, especially for low test temperature cases. At 5 J impact energy (Fig. 10c), it is similarly observed that the required penetration energy decreases with decreasing test temperature. Interestingly, the amount of rebound energy increases incrementally with decreasing temperature due to an increased stiffness in the specimens at low temperature.

The thicker facesheet of the TK specimens enhances the impact resistance of the composite sandwich panel significantly. At 10 J impact energy (Fig. 10d), across all temperatures, the required penetration energy almost double that of the thin specimens (Fig. 10b). Composite facesheets absorb most of the energy in the form of matrix failure, delamination and fiber breakages. The rest of the energy is absorbed by the partial crushing of the foam core layer. At low-temperature condition, less energy is required to penetrate the facesheet, so the remaining energy is absorbed primarily by the foam core, resulting in greater damage in foam layer. At 5 J impact energy (Fig. 10e), it is difficult to clearly characterize Ep, since post-mortem observation shows that the facesheet skin is only slightly damaged (not penetrated), and the core damage is assumed negligible. As a result, a considerable part of the impact energy is returned to the system as rebound energy of the impactor.

3.4 Impact Damage Mechanisms

Understanding barely visible impact damages (BVID) is crucial since impact energy is dissipated through internal damage mechanisms such as delamination, core crushing and debonding. Figs. 11a, 11b and 11c show the cross-sectional views of the micro-CT images for the thin specimens impacted at 10 J impact energy, at room temperature (23 °C), average arctic temperature (-30 °C) and lowest arctic temperature (-70 °C), respectively. All images are taken at the impact region where

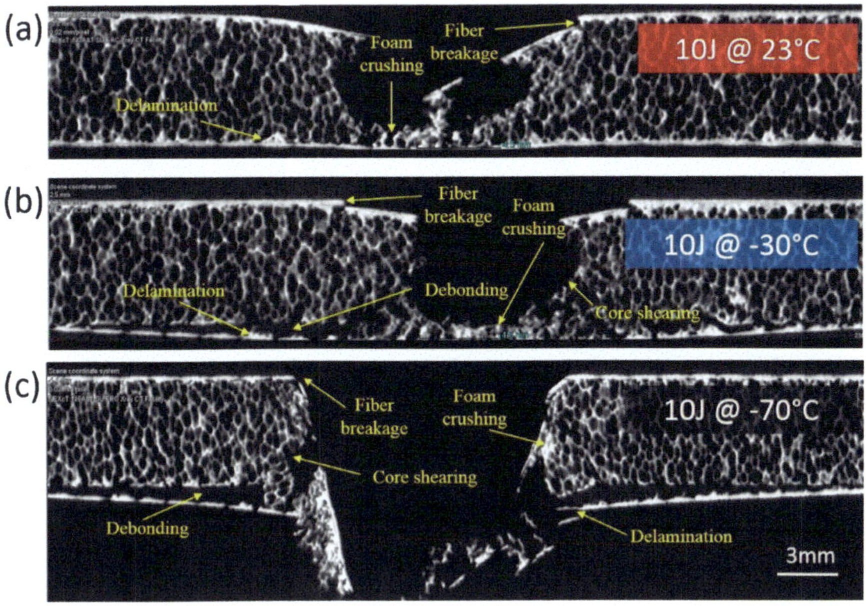

Fig. 11 Micro-computed tomography identification of damage mechanisms at cross-sectional view of impact region with 10 J impact energy at (**a**) 23 °C; (**b**) −30 °C; (**c**) −70 °C. [Reprinted from Composite Structures, Vol. 192, Elamin M, Li B, Tan KT, Impact damage of composite sandwich structures in arctic condition, Pages 422–433, Copyright (2018), with permission from Elsevier]

maximum damage occurs. It is undeniable that at low test temperatures, the specimens experience more severe internal damages than at room temperature. Micro-CT images reveal several damage mechanisms for each layer of the composite sandwich panel. Fiber breakage and delamination of the facesheets are caused as the result of the penetration failure of the top facesheets and the tension of the back facesheet respectively.

The continuous loading of the foam core during the impact event causes the foam core layer to experience severe central crushing and side shearing. The core layer can also be seen to have compressed and crushed towards the bottom facesheet. Since the support of the top facesheet layer is reduced in low-temperature environments, severe foam damage and back interface debonding exist at very low temperatures (Figs. 11b and 11c). Furthermore, the possibly brittle failure of the adhesive layer at extreme low temperature leads to the debonding of core layer from the bottom facesheet. At −70 °C, it is shown that the specimen have more complicated fiber breakage, greater debonding, and severe foam core crushing mechanisms.

The plan views of the micro-CT scanned specimens are illustrated in Fig. 12. The debonded area between the foam core and the bottom facesheet is determined by the white region in the micro-CT images. It is clear that, as the temperature decreases, the debonded area increases. This large debonded area poses serious implication to the reduction of residual bending and compressive strength of the composite

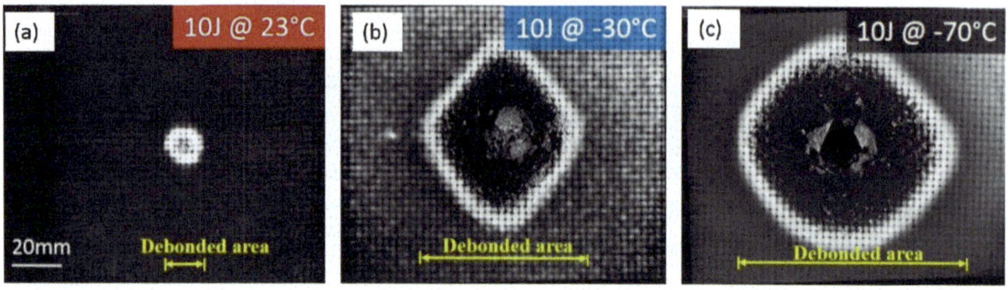

Fig. 12 Micro-computed tomography plan view showing interfacial damage at bottom facesheet and core with 10 J impact energy at (**a**) 23 °C; (**b**) −30 °C; (**c**) −70 °C. [Reprinted from Composite Structures, Vol. 192, Elamin M, Li B, Tan KT, Impact damage of composite sandwich structures in arctic condition, Pages 422–433, Copyright (2018), with permission from Elsevier]

Fig. 13 Facesheet-core debonded area at different temperatures. [Reprinted from Composite Structures, Vol. 192, Elamin M, Li B, Tan KT, Impact damage of composite sandwich structures in arctic condition, Pages 422–433, Copyright (2018), with permission from Elsevier]

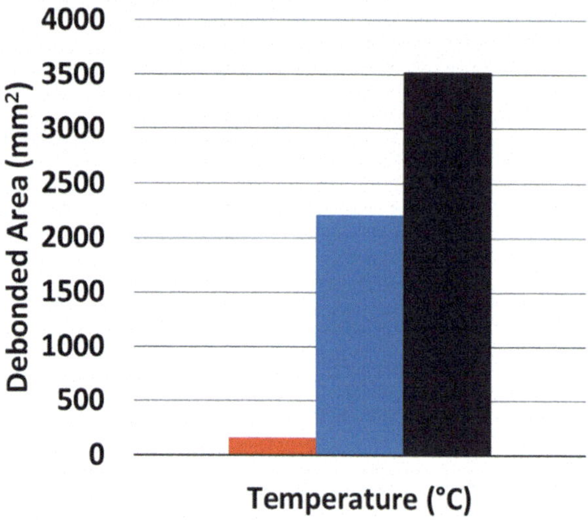

sandwich panel. Post-process measurement of the debonded area size are achieved by using an area measurement tool, and plotted in Fig. 13. It is clear that there is a substantial increase in debonded area when specimens are impacted at extreme arctic low temperatures compare to the debonded area at room temperature.

It is also interesting to note that the effect of low temperature apparently changes the damage mode from a typical low-velocity impact to a typical high-velocity impact. Typical low-velocity impact damage with barely visible damage and small dent is observed at room temperature. However, as the temperature decreases to a low temperature of −70 °C, the damage mode changes to that of a high-velocity impact damage with a clear localized indentation or penetration, taking the shape of the impactor [22].

3.5 CAI Performance

The compressive response of the thinner set of samples at different temperatures is shown in Fig. 14. For all temperature cases, the non-impacted specimens (0 J) are compared with the impacted sets at 5 J and 10 J. The compressive load-displacement curves increase at the beginning of the compression event until it reaches the brittle fracture load. This is indicated by a quick drop in the load value and the sound of the fracture. As the compression continues, each curve shows an approximately constant compressive load, which represents the load taken by the core layer. Non-impacted samples, have higher residual strength than 5 J and 10 J since the impact-induced damages greatly reduce the total strength of the composite structure. Therefore, non-impacted facesheet layers withstand higher compressive loads before they fail, and the rest load is taken by the core layer until the end of the compression event. Hence, non-impacted specimens have higher compressive core damage values compared to the impacted samples. The influence of low temperatures on the CAI strength is summarized in Fig. 15.

To correlate the effect of low temperatures with the maximum compressive load and the maximum compressive energy, Fig. 15a is generated as a quantitative summary of the compressive failure values across different temperatures. It is

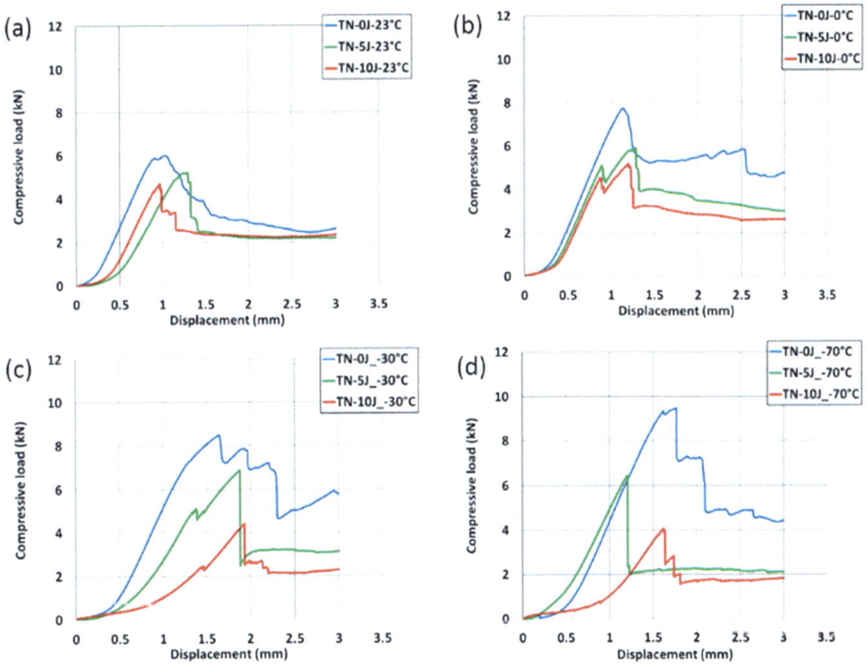

Fig. 14 Compression load-displacement curves for thin specimens (**a**) 23 °C; (**b**) 0 °C; (**c**) −30 °C; (**d**) −70 °C

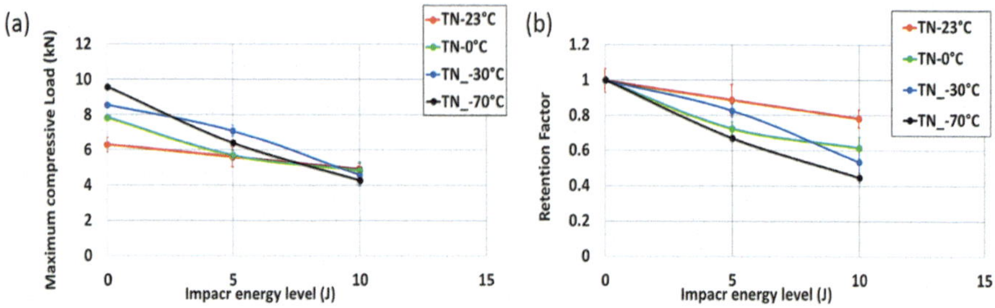

Fig. 15 (**a**) Maximum compressive strength against impact energy; (**b**) retention factor against impact energy

shown that, non-impacted samples (0 J) have higher maximum compressive damage values at low temperatures. For instance, the maximum compressive load in room temperature is increased by almost 34% and 27% for −70 °C and −30 °C respectively. In general, for non-impacted samples, there exists an evident relationship such that, the lower compression test temperature the higher compressive strength. This is attributed to the increase in composite sandwich structure stiffness at low temperatures due to the higher atomic binding force of the polymer matrix. The coefficient of thermal expansion of the fibers is lower than the coefficient of thermal expansion of the matrix. As a result, low temperatures induce internal compressive stresses at the fiber-matrix interface which enhanced the facesheet buckling load and consequently the panel compressive load.

To fully understand the effect of the temperature on the CAI residual properties, it is crucial to compare the compressive failure for the non-impacted and the impacted samples at the same temperature. This comparison is very important as it could be used to show the effect of the environmental conditions on the impacted structures. Figure 15b studies the change of the retention factor with test temperature. Retention factor is the ratio of the strength of the impacted samples to the strength of the non-impacted (benchmark) samples. The retention factor indicates the severity of the impact failure by presenting the reduction on the residual strength due to the pre-compression impact failure. As illustrated in Fig. 15b, when a structure is operating at room temperature (23 °C), its compressive strength is reduced by around 10% and 20% when impacted at 5 J and 10 J impact energy, respectively. However, when the same structure operates in the extreme cold arctic temperature of −70 °C, the residual compressive strength will be drastically reduced by 35% and 50% for 5 J and 10 J, respectively. Generally, Fig. 15b reveals that retention factor is drastically decreased in low-temperature environments.

3.6 Flexural After Impact Performance

Flexural after impact three-point bend test plots for post impacted specimen along with non-impacted specimen are shown in Fig. 16. The force-displacement curves for almost all specimens show a linear elastic regime followed by a load drop.

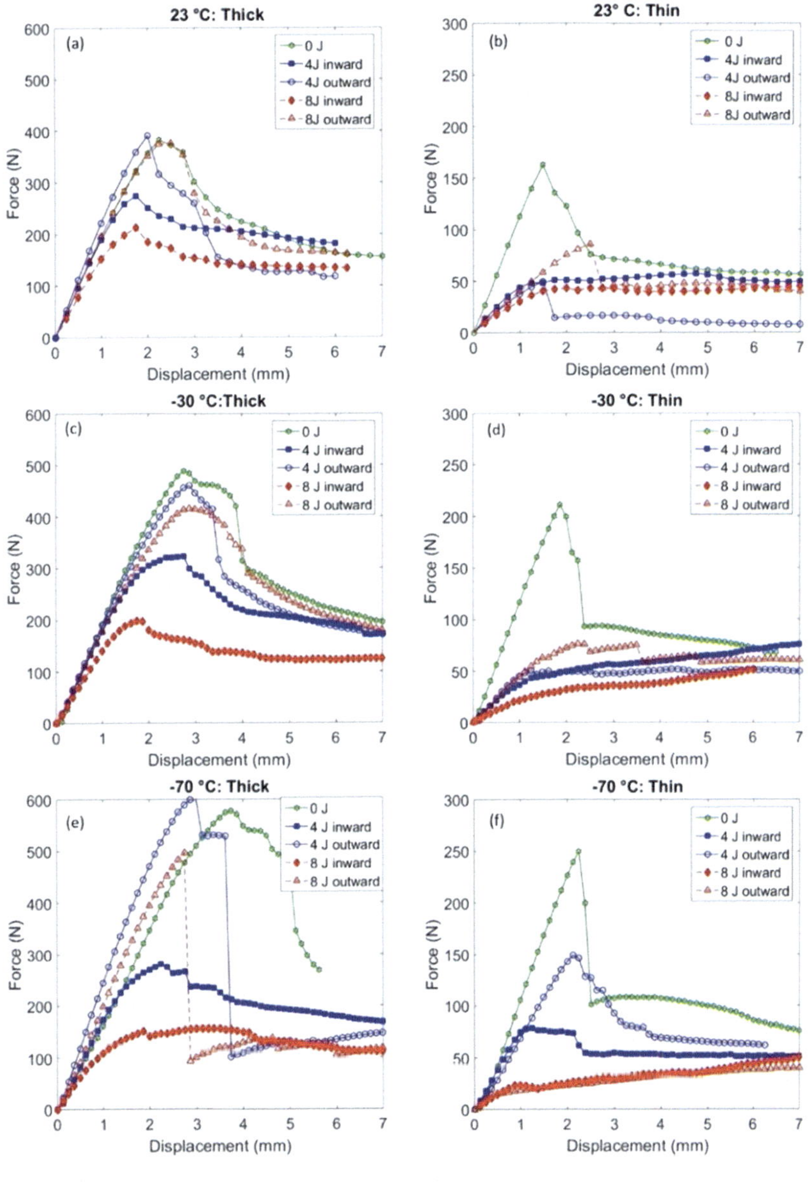

Fig. 16 Force vs displacement plots for (**a**) thick and (**b**) thin at 23 °C; (**c**) thick and (**d**) thin at −30 °C; (**e**) thick and (**f**) thin at −70 °C

Non-impacted specimen shows a trend of increasing peak force or flexural strength for decreasing temperature. The thick specimen breaks catastrophically for bending outward at −70 °C (Fig. 16e), while at other temperature the specimens show a linear portion followed by a decline in flexural force. However, for thin specimen the load-displacement curves do not show clear load drop at −30 °C and −70 °C (Figs. 16d and 16f). This is due to the fact that prior impact front facesheet and the core is severely damaged which results in energy to be released only by crack propagation through the core. There is a trend of superiority of bending outward to inward case. This is because for inward cases, the top facesheet is already reduced in compressive properties due to prior impact and this reduction is more for 8 J impact energy compared to 4 J. However, for bending outward cases, the top face sheet is still undamaged, thereby providing required compressive properties. Moreover, bottom facesheet can still provide tensile properties, since the fibers are not damaged (only matrix is damaged). Therefore, bending inward cases show inferior flexural properties to bending outward cases.

4 Conclusion

This study investigated the impact response and damage mechanisms of composite sandwich structures in arctic condition. Carbon fiber/epoxy matrix composite facesheet with PVC foam core sandwich structures were impacted at two energy levels of 10 J and 5 J, across test temperatures ranging from 23 °C to −70 °C. Results revealed that extreme arctic low temperature had a significant influence on the impact behavior of sandwich composites. Force-time curves showed that specimens exhibited less impact strength at low temperature. Critical impact force values pointed to the fact that arctic low-temperature environment affected the strength of composites severely and induced complicated damage mechanisms. Energy-time curves demonstrated that lower penetration energy was required to perforate the top facesheet of the specimens at low temperature. Force-displacement plots showed that the specimens experienced greater displacement at low temperature, primarily due to impactor penetration. Moreover, the structural stiffness of low-temperature specimens was higher than room temperature specimens.

X-ray micro-computed tomography was employed to analyze the internal damage extent of the impacted specimens. Results from the micro-CT confirmed that the specimens suffered severe damages at arctic low temperature, by various failure mechanisms such as fiber breakage, delamination, interface debonding, and foam core crushing and shearing. Furthermore, micro-CT results elucidated that at low temperatures, the specimens experienced severe debonding of the bottom facesheet from the core layer. The debonded area size was drastically increased with decreasing test temperature. The results from this work provided important understanding of how composite sandwich structures behave in low temperature arctic condition, which would be useful in future numerical modelling and design of impact damage tolerant naval composite structure.

Results from this study also showed that low-temperature environments severely influenced the residual strength of the arctic structures. A significant drop in the CAI strength on the impacted samples is observed with decreasing of test temperature. At −70 °C, which typically represents the lowest temperature in the arctic region, a very significant reduction in the retention factor has been obtained when the specimens are impacted at 10 J. Also, higher reduction in the retention factor was observed at −30 °C compared to 0 °C and 23 °C. Flexural properties for impacted specimens generally decrease with higher impact energy. Flexural properties are superior during bending outward cases to bending inwards cases, due to reduced compressive properties at the impact face. Therefore, operation in arctic environments extremely magnifies the effect of the impact damages on the residual strength of the structure.

Acknowledgments The work was supported by research grant N00014-16-1-3202 provided by the Office of Naval Research – United States, Solid Mechanics Program (ONR Program Manager: Dr. Yapa Rajapakse).

References

1. Vihma T (2014) Effects of arctic sea ice decline on weather and climate: a review. Surv Geophys 35(5):1175–1214
2. Hazizan MA, Cantwell W (2002) The low velocity impact response of foam-based sandwich structures. Compos Part B Eng 33(3):193–204
3. Zhang G, Wang B, Ma L, Wu L, Pan S, Yang J (2014) Energy absorption and low velocity impact response of polyurethane foam filled pyramidal lattice core sandwich panels. Compos Struct 108:304–310
4. Wang J, Waas AM, Wang H (2013) Experimental and numerical study on the low-velocity impact behavior of foam-core sandwich panels. Compos Struct 96:298–311
5. Styles M, Compston P, Kalyanasundaram S (2007) The effect of core thickness on the flexural behaviour of aluminium foam sandwich structures. Compos Struct 80(4):532–538
6. Im KH, Cha CS, Kim SK, Yang IY (2001) Effects of temperature on impact damages in CFRP composite laminates. Compos Part B Eng 32(8):669–682
7. López-Puente J, Zaera R, Navarro C (2002) The effect of low temperatures on the intermediate and high velocity impact response of CFRPs. Compos Part B Eng 33(8):559–566
8. Gómez-del Rıo T, Zaera R, Barbero E, Navarro C (2005) Damage in CFRPs due to low velocity impact at low temperature. Compos Part B Eng 36(1):41–50
9. Lopresto V, Langella A (2014) Composite laminates under dynamic extreme conditions. Procedia Eng 88:173–179
10. Ibekwe SI, Mensah PF, Li G, Pang SS, Stubblefield MA (2007) Impact and post impact response of laminated beams at low temperatures. Compos Struct 79(1):12–17
11. Sánchez-Sáez S, Barbero E, Navarro C (2008) Compressive residual strength at low temperatures of composite laminates subjected to low-velocity impacts. Compos Struct 85(3):226–232
12. Kang KW, Kim JK (2009) Effect of shape memory alloy on impact damage behavior and residual properties of glass/epoxy laminates under low temperature. Compos Struct 88 (3):455–460
13. Liu J, Zhou Z, Ma L, Xiong J, Wu L (2011) Temperature effects on the strength and crushing behavior of carbon fiber composite truss sandwich cores. Compos Part B Eng 42(7):1860–1866
14. Erickson MD, Kallmeyer AR, Kellogg KG (2005) Effect of temperature on the low-velocity impact behavior of composite sandwich panels. J Sandw Struct Mater 7(3):245–264

15. Salehi-Khojin A, Mahinfalah M, Bashirzadeh R, Freeman B (2007) Temperature effects on kevlar/hybrid and carbon fiber composite sandwiches under impact loading. Compos Struct 78 (2):197–206
16. Yang P, Shams SS, Slay A, Brokate B, Elhajjar R (2015) Evaluation of temperature effects on low velocity impact damage in composite sandwich panels with polymeric foam cores. Compos Struct 129:213–223
17. Sanchez-Saez S, Gomez-del Rio T, Barbero E, Zaera R, Navarro C (2002) Static behavior of CFRPs at low temperatures. Compos Part B Eng 33:383–390
18. Elamin M, Li B, Tan KT (2018) Impact damage of composite sandwich structures in arctic condition. Compos Struct 192:422–433
19. Tan KT, Watanabe N, Iwahori Y (2011) X-ray radiography and micro-computed tomography examination of damage characteristics in stitched composites subjected to impact loading. Compos Part B Eng 42(4):874–884
20. Elamin M, Li B, Tan KT (2018) Compression after impact performance of composite sandwich structures in extreme low temperature arctic condition, under review
21. Khan MH, Li B, Tan KT (2018) Impact performance and bending behavior of composite sandwich structures in cold temperature arctic conditions, under review
22. Khan MH, Elamin M, Li B, Tan KT (2018) X-ray micro-computed tomography analysis of impact damage morphology in composite sandwich structures due to cold temperature arctic condition. J Compos Mater 52:350\–3522

Mapping Interior Strain Fields in Thick Composites and Sandwich Plates With Digital Volumetric Speckle Photography Technique

Lingtao Mao and Fu-Pen Chiang

1 Introduction

Optical full-field measurement techniques such as speckle photography [1], speckle interferometry [2], geometric moire [3], moire interferometry [4], digital speckle photography (DSP) [5] and digital image correlation (DIC) [6, 7] have all been applied to determining properties of composite materials [8]. Thanks to their simplicity, DSP and DIC have become ubiquitous in recent years as the preferred tools. However, all these full–field measurements techniques can only measure surface displacement of the specimen. Due to material heterogeneity at different length scales, the deformation of a composite structure is always 3D in nature. Thus, a 2D surface measurement technique can never fully reveal the failure mechanism of composites. As a result whenever a naval composite structure is designed, a high safety factor is often used. This adds weight and cost to the resulting structure. Thus there is a strong need to develop a technique whereby one can probe into the interior deformation of solids. While there already exist several stress/strain analysis techniques that can probe the interior of solids (for example frozen-stress photoelasticity [9, 10], moire [11], and embedded speckle method [12, 13]), they all require that the specimen material be transparent, in one case even birefringent.

In recent years X-ray micro-computed tomography (CT) has become a familiar tool in materials research. Applications of micro-CT to composite materials have mainly been concentrated in two fields. One is for the visualization of internal features, such as void measurement, fiber location and waviness, fiber breakage,

L. Mao
State Key Laboratory of Coal Resources and Safe Mining,
China University of Mining & Technology, Beijing, China

F.-P. Chiang (✉)
Laboratory for Experimental Mechanics Research, Department of Mechanical Engineering,
Stony Brook University, Stony Brook, NY, USA
e-mail: fu-pen.chiang@stonybrook.edu

S. W. Lee (ed.), *Advances in Thick Section Composite and Sandwich Structures*,
https://doi.org/10.1007/978-3-030-31065-3_22

interface delamination, internal damage and crack growth, etc. [14, 15]. The other application deals with internal displacement measurement. Based on the volumetric image capability of CT scanning, Digital Volume Correlation (DVC) technique was proposed to assess the internal displacement fields of solid objects. Roux, et al. [16] used DVC to evaluate the internal displacement of a solid foam. Brault, et al. [17] performed volume kinematic measurements of laminated composite materials with metallic particles imbedded into the specimen for the purpose of contrast enhancement. Lecomte-Grosbras, et al. [18] investigated the free-edge effects in laminate composites by using DVC. Recently we developed an effective 3D experimental strain analysis technique called DVSP (Digital Volumetric Speckle Photography) [19] in which we use the internal features of opaque solids as 3D volumetric speckles. We developed an algorithm to process the CT recorded volumetric speckles in a way similar to the algorithm we developed for the 2D digital speckle photography technique [20, 21]. DVSP can be applied to probing the internal deformation of almost any solid material. We have successfully applied the technique to coal [22], rock [23] and concrete [24]. In this study, we describe in detail the theory and practice of DVSP and its application to composites. Three examples are chosen for the demonstration: a woven composite beam under three-point bending, a woven composite beam with a prepared slot under three-point bending and a foam composite sandwich beam under three-point bending. Displacement and strain distributions of many internal sections were mapped in detail.

2 Theory of Digital Volumetric Speckle Photography (DVSP)

The CT system is used to scan the specimen before and after the application of load. The two reconstructed digital volume images are defined as reference volume image and deformed volume image, respectively. Both digital images are subdivided into volumetric subsets with voxel arrays of 32 voxel × 32 voxel × 32 voxel, for example, and 'compared' with the procedures schematically shown in Fig. 1.

Let $h_1(x, y, z)$ and $h_2(x, y, z)$ be gray distribution functions of a pair of generic volumetric speckle subsets, before and after deformation, respectively, and that

$$\begin{aligned} h_1(x,y,z) &= h(x,y,z) \\ h_2(x,y,z) &= h[x-u(x,y,z), y-v(x,y,z), z-w(x,y,z)] \end{aligned} \tag{1}$$

where u, v and w components of the displacement vector collectively experienced by the speckles within the subset of voxels along the x, y, and z directions, respectively. A first-step 3D FFT (Fast Fourier Transform) is applied to both h_1 and h_2 yielding

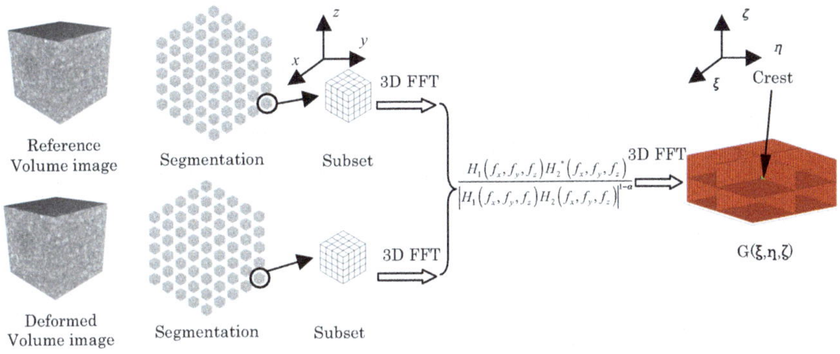

Fig. 1 Schematics demonstrating the processing algorithm of DVSP

$$
\begin{aligned}
H_1(f_x,f_y,f_z) &= \Im\{h_1(x,y,z)\} = |H(f_x,f_y,f_z)| \exp\left[j\phi(f_x,f_y,f_z)\right] \\
H_2(f_x,f_y,f_z) &= \Im\{h_2(x,y,z)\} = |H(f_x,f_y,f_z)| \exp\left\{j\left[\phi(f_x,f_y,f_z) - 2\pi(uf_x + vf_y + wf_z)\right]\right\}
\end{aligned}
\tag{2}
$$

where $H_1(f_x,f_y,f_z)$ is the Fourier transform spectrum of $h_1(x, y, z)$, $H_2(f_x,f_y,f_z)$ is the Fourier transform spectrum of $h_2(x, y, z)$, and $\Im$ stands for Fourier Transform. $|H(f_x,f_y,f_z)|$ and $\phi(f_x,f_y,f_z)$ are the spectral amplitude and phase fields, respectively.

Then, a numerical interference filter between the two 3D speckle patterns is performed at the spectral domain, i.e.

$$
F(f_x,f_y,f_z) = \frac{H_1(f_x,f_y,f_z)H_2{}^*(f_x,f_y,f_z)}{|H_1(f_x,f_y,f_z)H_2(f_x,f_y,f_z)|^{1-\alpha}} \tag{3}
$$

in which * stands for the complex conjugate, and α is an appropriate constant $(0 \le \alpha \le 1)$.

When $\alpha = 0$, Eq. (3) can be expressed as

$$
F(f_x,f_y,f_z) = H_1(f_x,f_y,f_z)\frac{\exp\left\{-j\left[\phi(f_x,f_y,f_z) - 2\pi(uf_x + vf_y + wf_z)\right]\right\}}{|H(f_x,f_y,f_z)|} \tag{4}
$$

where $\frac{\exp\left\{-j\left[\phi(f_x,f_y,f_z) - 2\pi(uf_x + vf_y + wf_z)\right]\right\}}{|H(f_x,f_y,f_z)|}$ is essentially an inverse filter (IF).

When $\alpha = 0.5$, Eq. (3) can be expressed as

$$
F(f_x,f_y,f_z) = H_1(f_x,f_y,f_z)\exp\left\{-j\left[\phi(f_x,f_y,f_z) - 2\pi(uf_x + vf_y + wf_z)\right]\right\} \tag{5}
$$

where $\exp\{-j[\phi(f_x,f_y,f_z) - 2\pi(uf_x + vf_y + wf_z)]\}$ is a so-called phase-only filter (POF).

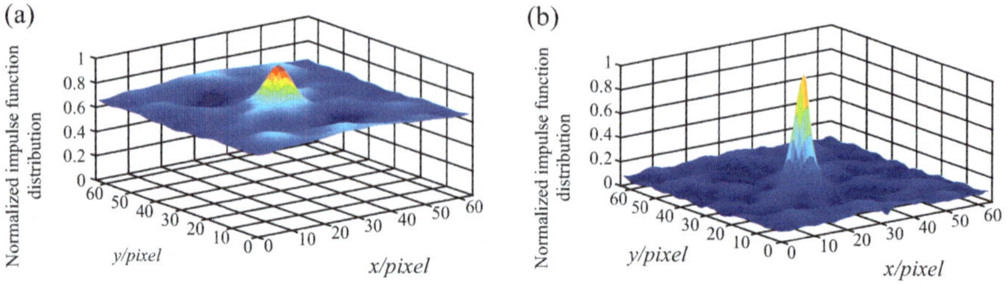

Fig. 2 Normalized impulse function distribution with different filters: (**a**) CMF filter; (**b**) POF filter

When $\alpha = 1$, Eq. (3) can be expressed as

$$F(f_x, f_y, f_z) = H_1(f_x, f_y, f_z) H_2^*(f_x, f_y, f_z) \tag{6}$$

where $H_2^*(f_x, f_y, f_z)$ can be viewed as a classical matched filter (CMF). When a correlation filter is chosen, peak sharpness and noise tolerance are the factors that need to be considered. In the 2D digital speckle photography technique, α is 0.5, and the algorithm is essentially a POF. The influence of CMF, POF and IF filters on the accuracy of 2D speckle photography were analyzed and the results indicate that IF is extremely sensitive to noise, thus cannot be used as a reliable filter. There is no significant difference between CMF and POF filters [25]. But while the POF filter provides somewhat more accurate estimates of the peak position, the reliability of the CMF filter is better. In Fig. 2, normalized impulse function distributions for the two filters are shown. It is noted that the peak impulse with POF is sharper, which can provide a good compromise between peak sharpness and noise tolerance in the correlation theory.

In this paper, $\alpha = 0.5$ is adopted. As a result, Eq. (3) can then be written as

$$\begin{aligned} F(f_x, f_y, f_z) &= \frac{H_1(f_x, f_y, f_z) H_2^*(f_x, f_y, f_z)}{\sqrt{\left|H_1(f_x, f_y, f_z) H_2(f_x, f_y, f_z)\right|}} \\ &= \left|H_1(f_x, f_y, f_z)\right| \exp\left\{j\left[\phi_1(f_x, f_y, f_z) - \phi_2(f_x, f_y, f_z)\right]\right\} \end{aligned} \tag{7}$$

where $\phi_1(f_x, f_y, f_z)$ and $\phi_2(f_x, f_y, f_z)$, are the phases of $H_1(f_x, f_y, f_z)$ and $H_2(f_x, f_y, f_z)$, respectively. It is seen that

$$\phi_1(f_x, f_y, f_z) - \phi_2(f_x, f_y, f_z) = 2\pi(uf_x + vf_y + wf_z) \tag{8}$$

Finally, a function is obtained by performing another 3D FFT resulting in

$$G(\xi, \eta, \zeta) = \Im\{F(f_x, f_y, f_z)\} = \overline{G}(\xi - u, \eta - v, \zeta - w) \tag{9}$$

which is an expanded impulse function located at (u, v, w). This process is carried out for every corresponding pair of subsets. By detecting crests of all these impulse functions, an array of displacement vectors at each and every subset is obtained.

It is well known that in 2D DSP the random error is a function of the subset size, the speckle size, and the amount of decorrelation. It is also true in DVSP. If the displacement between the corresponding subsets is large, the increase of nonoverlapping area would result in an increase of decorrelation giving rise to an enhanced random error [26]. In order to reduce the random errors, a coarse-fine calculation process is adopted. In the coarse calculation, the first integer voxel prediction indicates that the largest size of the subset ($2^p \times 2^p \times 2^p$) should not exceed the size of the region-of-interest in the reference volumetric image. And thus, the subset with the size ($2^{p-1} \times 2^{p-1} \times 2^{p-1}$) is used, and the corresponding subset in the deformed volumetric image is chosen based on the first integer voxel predicted displacement. By repeating this process, the optimal subset size is obtained for the fine calculation. In this study, we selected 32 voxel ×32 voxel ×32 voxel as the optimal size for the final calculation, and the subset shift was 5 voxels.

Because of the discrete nature of digital volume images, the displacement vectors evaluated from the above coarse-fine calculation process are integral multiples of one voxel. In order to obtain more accurate and sensitive characterization, a sub-voxel investigation of the crest position is needed. To achieve this, we selected a cubic subset with $3 \times 3 \times 3$ voxels surrounding an integral voxel of the crest and a cubic spline interpolation was employed to obtain the interpolated values among the integral voxels in each respective dimension. After interpolation, the cubic subset was enlarged and a new three dimensional array was generated with a size depending on the interpolation interval. The smaller the interval and the bigger the array size give rise to a higher interpolation accuracy. The price to pay, however, is the need for more computational time and more memory space. In practical applications there would be a tradeoff between the two competing needs. By detecting the positions of peak values of the new array, displacements of subvoxel accuracy can be obtained. The interpolation procedure is illustrated schematically in Fig. 3.

3 Strain Estimation

The internal strain tensor ε can be derived from displacement fields. Due to the influence of unavoidable noise contained in the CT images, the displacements thus determined contains discontinuities or noises that are not a feature of the material but a consequence of the discrete nature of the analysis performed. The errors in local displacements may be amplified during the strain computation process. By using the PLS (Point Least-Squares) approach, the errors can be substantially reduced during the process of local fitting, and the strains thus estimated will be more precise [27].

The element of PLS approach is as follows. To compute the local strain tensor of each considered point, a regular cubic box with size of $(2N+1) \times (2N+1) \times (2N+1)$ discrete points surrounding the point of interest is selected. If the strain calculation

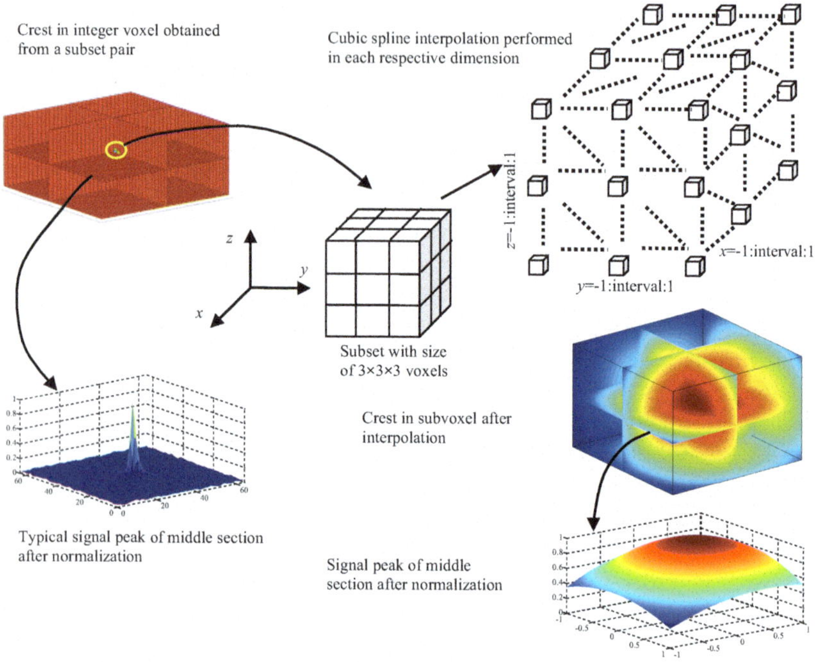

Fig. 3 Schematics showing the interpolation procedure

window is sufficiently small, the displacement in each direction can be reasonably assumed to be linear, and they can be mathematically expressed as

$$\begin{aligned} u(x,y,z) &= a_0 + a_1x + a_2y + a_3z \\ v(x,y,z) &= b_0 + b_1x + b_2y + b_3z \\ w(x,y,z) &= c_0 + c_1x + c_2y + c_3z \end{aligned} \tag{10}$$

where x, y, $z = [-N, N]$ are the local coordinates within the strain calculation box, $u(x, y, z)$, $v(x, y, z)$ and $w(x, y, z)$ are displacements directly obtained by the DVSP method; and $a_{i=0,1,2,3}$, $b_{i=0,1,2,3}$ and $c_{i=0,1,2,3}$ are the unknown polynomial coefficients to be determined. With the Least-squares or Multiple Regression Analysis, the unknown coefficients can be estimated. Then, the six Cauchy strain components ε_x, ε_y, ε_z, ε_{xy}, ε_{xz} and ε_{yz} at the interrogated point can be calculated as

$$\begin{aligned} &\varepsilon_x = \frac{\partial u}{\partial x} = a_1 \quad && \varepsilon_{xy} = \frac{1}{2}\left(\frac{\partial v}{\partial x} + \frac{\partial u}{\partial y}\right) = \frac{1}{2}(b_1 + a_2) \\ &\varepsilon_y = \frac{\partial v}{\partial y} = b_2 \quad && \varepsilon_{yz} = \frac{1}{2}\left(\frac{\partial w}{\partial y} + \frac{\partial v}{\partial z}\right) = \frac{1}{2}(c_2 + b_3) \\ &\varepsilon_z = \frac{\partial w}{\partial z} = c_3 \quad && \varepsilon_{xz} = \frac{1}{2}\left(\frac{\partial u}{\partial z} + \frac{\partial w}{\partial x}\right) = \frac{1}{2}(a_3 + c_1) \end{aligned} \tag{11}$$

4 Experiments & Results

4.1 Experimental Setup

In this study, the main components of the industrial X-Ray computer tomography system consist of a microfocus X-ray source from YXLON (Feinfocus 225 kV), a X-ray detector unit (1024 pixel × 1024 pixel) from PerkinElme (XRD 0822AP 14), and a motorized rotation stage from Newport. The X-ray has a focus with the size of 3 μm × 6 μm, a voltage range of 50–225 kV, and the tube current ranging from 0 to 1440 μA. For each scanning, 720 projections are captured and distributed at equal angles over 360°. The entire process takes 25 min. For high quality image, more projections can be captured. Based on these projections, the Feldkamp algorithm is used to reconstruct a sequence of slice images, and then a volume image of the specimen is obtained by merging these slice images together.

To test a specimen under load, we designed and built a compact loading setup that would allow the operation of micro-tomography of a specimen under load in situ. By using the loading setup, uniaxial compression, uniaxial tension and 3-point bending experiments can been carried out with different rigs. The setup is enclosed within a cell made of PMMA, which is transparent to X-ray. Mechanical loading is provided by an electric motor with an actuator. A load sensor of 20 kN capacity and 0.002 kN resolution, and a grating scale of 20 mm capacity and 0.001 mm resolution are used to record the load and the displacement, respectively. The CT system and loading setup are shown in Fig. 4.

4.2 A Woven Composite Beam Under 3-Point Bending

The specimen is made of a tri-direction fabric composite material with Advantex SE1500 E-glass filaments and 55% epoxy resin by volume. The filament diameter is

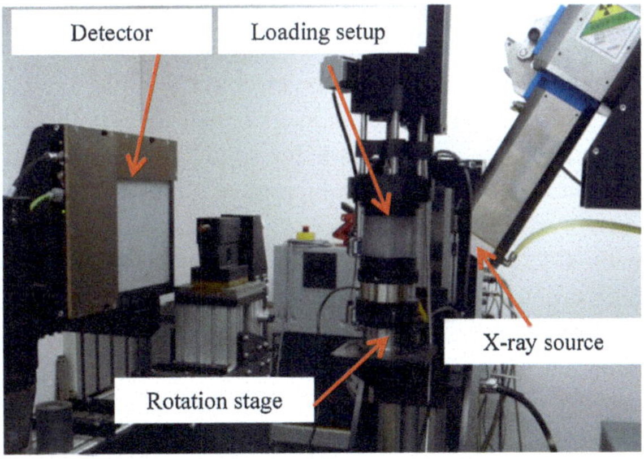

Fig. 4 The Micro-CT system and the loading setup used to record the volumetric image of the specimen at each loading step

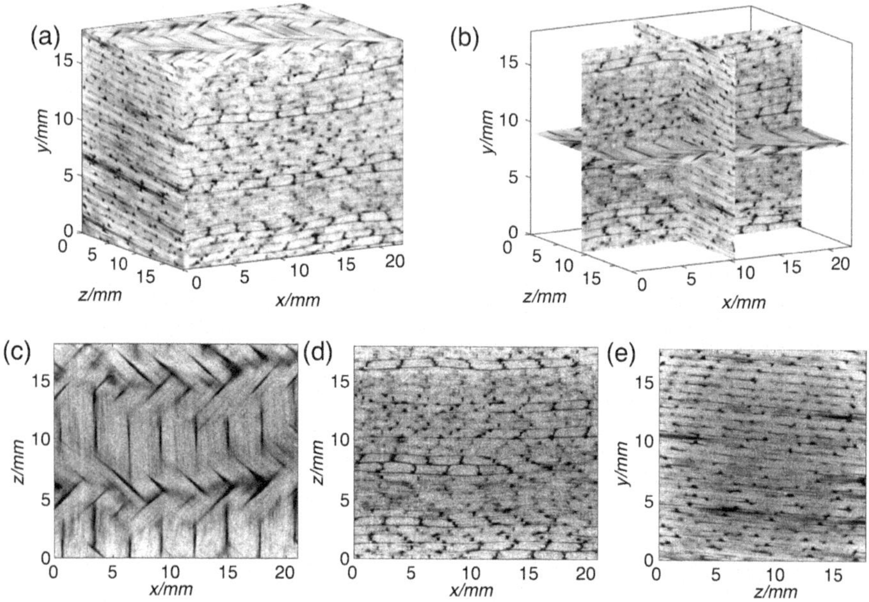

Fig. 5 The woven composite material: (**a**) reconstructed volumetric image based on CT slices; (**b**)- (**e**) Three orthogonal sections of the volumetric image

17 μm, and its density is 2.63 g/cm^3. The fiber orientations are set at +45 °, 0, and −45 ° with respect to the rotation axis. The density of the basic matrix is 1.08 g/cm^3. When applying the DVSP algorithm to composite materials the first issue is whether or not the Micro-CT can capture the minute interior details of the composite material. Figure 5 shows the reconstructed volume image and three orthogonal sections of the woven composite material based on CT slices by using the Micro-CT system. Even though the density of the filament is more than twice that of the matrix, the size of filament is small, and beyond the scanning resolution of the CT system. Therefore, we cannot detect the filament from the slice image. However, there are a lot of pores with low gray value as shown in the images. These internal meso-structures are of sufficient details that they can be treated as 3D speckles. Thus the algorithm of DVSP can be effectively applied.

The dimension of the beam specimen for 3-point bending is 38.8 (L) mm × 18.8 (H) mm × 9.0 (T) mm, while the span between two supports is 30.0 mm. The beam was loaded with step-wise increments until failure appears via visual observation. There were 11 loading steps [28]. After each loading step the specimen was scanned by the CT system. The load-displacement curve is shown in Fig. 6(a). In each step, a reconstructed volume image with size of 900 voxel × 250 voxel × 361 voxel was obtained; and the voxel resolution is 45 μm × 45 μm × 45 μm. Figure 6(b) shows the reconstructed CT image of the specimen at loading Step 1. No damage of the material is discernable from surface inspection of the specimen up to the loading

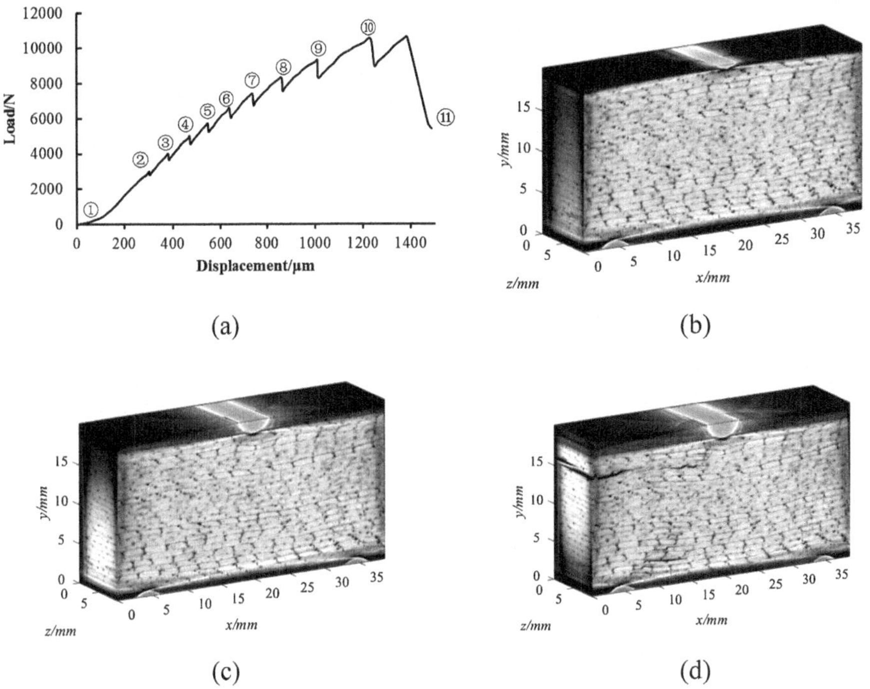

Fig. 6 Loading/displacement curve and reconstructed volume images: (**a**) the load-displacement curve (**b**) the volume image of Step 1 (**c**) the volume image of Step 10 (**d**) the volume image of Step 11

Step 10, as shown in Fig. 6(c). But micro cracks are clearly visible on the surface of the specimen after loading Step 11 as shown in Fig. 6(d).

The volume image of Step 1 is treated as the reference volume image, and volume images of subsequent steps (Step 2-Step 10) are "compared" to the reference volume image via the DVSP algorithm. The coarse-fine calculation process was applied, the final subset size was 32 voxel × 32 voxel × 32 voxel, and the subset shift was 5 voxels.

In Fig. 7, displacement fields of three transverse sections at Step 10 with F = 10,500 kN are depicted. Strain fields were then calculated from the displacement data. The size of the calculation cubic element was 31 × 31 × 31 points, and the shift step was 5 points. In Fig. 8, the strain ε_{xx}, ε_{yy}, ε_{zz} and ε_{xy} of the middle longitudinal section in Step 7, Step 9 and Step 10 are shown, respectively. It is noted that the internal heterogeneity of the woven composite had manifested itself vividly on the strain distributions. The distribution of the normal strain ε_{xx} indicates the periodic distributions of the matrix and fiber, whereas the distribution of normal strain ε_{yy} reflects the layered characteristics of the material. The distribution of the shear strain ε_{xy} as the load increases clearly indicates where the failure would eventually occur, as shown in in Fig. 6(d) at loading Step 11.

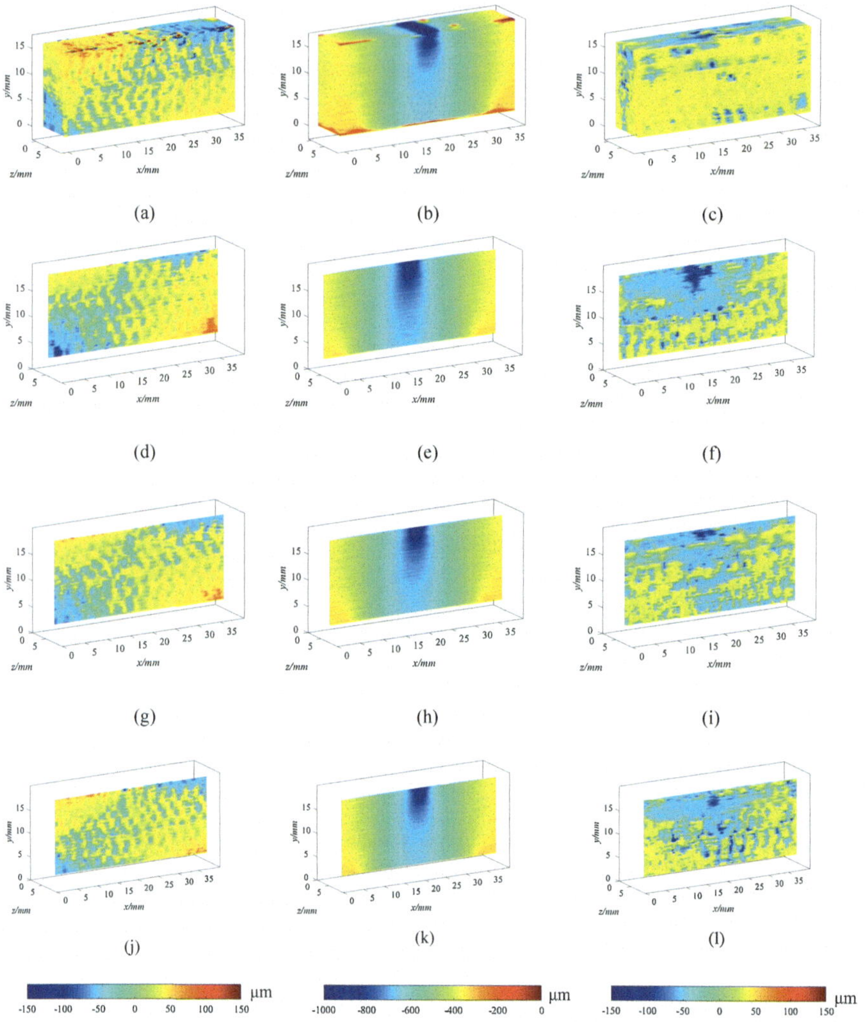

Fig. 7 Displacement fields of the specimen at loading Step 10; (**a**)-(**c**) u, v, w fields of the whole specimen within the cropped region; (**d**)-(**f**) u, v, w fields of the transverse section at $z = 2.25$ mm; (**g**)-(**i**) u, v, w fields of the transverse section at $z = 4.50$ mm; (**j**)-(**l**) u, v, w fields of the transverse section at $z = 6.75$ mm

4.3 A Woven Composite Beam With a Prepared Slot Under 3-Point Bending

The experiment of a woven composite beam with a prepared slot under 3-point bending was conducted. The material of the beam is the same as that in Sect. 4.2. The dimension of the specimen is 39 (L) mm × 18 (H) mm × 8.5 (T) mm, and the size of

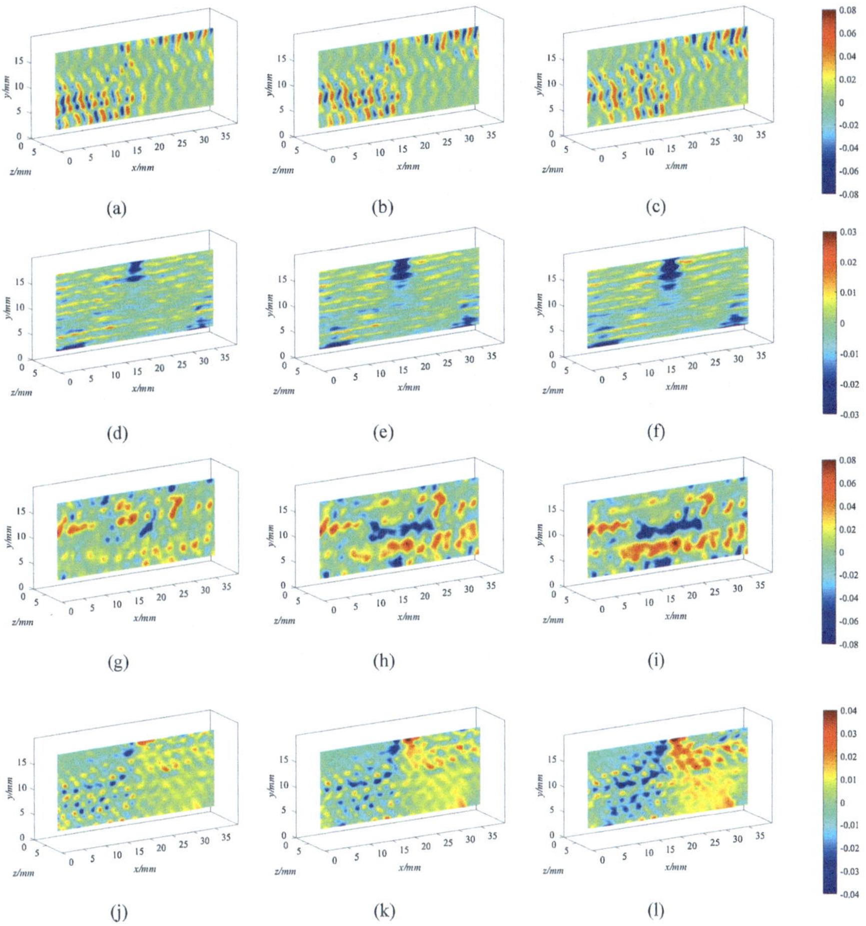

Fig. 8 Strain distribution of the middle longitudinal section at $z = 4.50$ mm under different loadings steps: (**a**)-(**c**) ε_{xx} at Step 7, Step 9 and Step 10, respectively; (**d**)-(**f**) ε_{yy} at Step 7, Step 9 and Step 10, respectively; (**g**)-(**i**) ε_{zz} at Step 7, Step 9 and Step 10, respectively; (**j**)-(**l**) ε_{xy} at Step 7, Step 9 and Step 10, respectively

the slot is 3.60 mm × 0.68 mm. The span between two supports is 30.0 mm. The load was applied incrementally in 10 steps. The load-displacement curve is shown in Fig. 9(a). The size of the reconstructed volume image in each step is 960 voxel × 260 voxel × 424 voxel, and the voxel resolution is also 45 μm × 45 μm × 45 μm. Comparing Fig. 6(a) with Fig. 9(a), the loading capacity of the specimen with a slot is reduced due to the weakening effect of the slot. The maximum loading of the specimen with a slot was 8400 N, whereas that of the specimen without a slot was 11,700 N. From Figs. 6(d) and 9(d), it is noted that there is a delamination crack on the left upper region of those two specimens. In the maximum shear strain region of the specimen without a slot, delamination was the failure mode, whereas for the specimen with a slot, cracking was the failure mode.

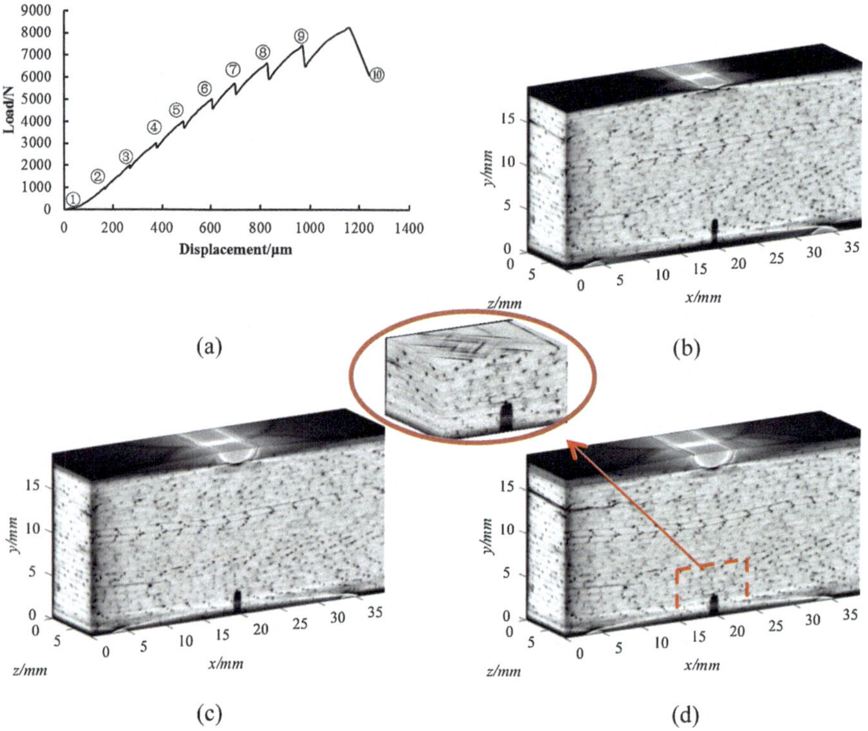

Fig. 9 Loading/displacement curve and reconstructed volume images: (**a**) load-displacement curve; (**b**) volume image of Step 1; (**c**) volume image of Step 9; (**d**) volume image of Step 10

The volume image of Step 1 was treated as the reference volume image, and volume image of subsequent steps (Step 2-Step 9) were "compared" to the reference volume image using the DVSP algorithm. The coarse-fine calculation process was applied, the final subset size was 32 voxel × 32 voxel × 32 voxel, and the subset shift was 5 voxels.

In Fig. 10, 3D displacement fields of the specimen and three transverse sections at Step 9 with F = 7400 N are depicted. Periodic patterns in the displacement field due to the periodic structure of the specimen are also found. Figure 11(a), 11(b) and 11(c) show images of the region near the slot in different sections at Step 10. Corresponding to these regions, the *u* and *v* displacement fields of Step 9 are plotted as shown in Fig. 11(d), 11(e), 11(f), 11(g), 11(h) and 11(i), ε_{xx} and ε_{xy} strain fields of Step 9 are plotted as shown in Fig. 11(j), 11k, 11(l), 11(m), 11(n) and 11(o). Strain concentrations are clearly indicated in various locations. Corresponding to the zone of cracks, there are higher strain values in ε_{xy} fields.

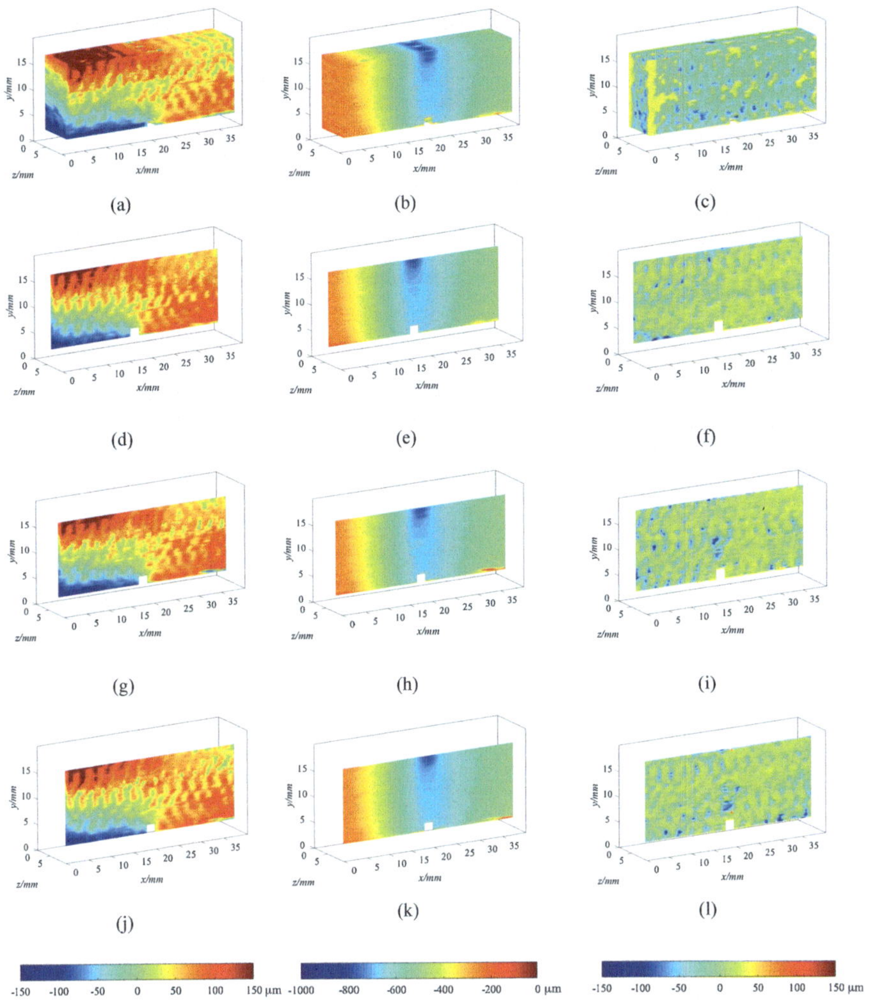

Fig. 10 Displacement fields of the specimen at loading Step 9; (**a**) – (**c**) *u*, *v*, *w* fields of the whole specimen; (**d**) – (**f**) *u*, *v*, *w* fields of the transverse section at $z = 2.00$ mm; (**g**) – (**i**) *u*, *v*, *w* fields of the transverse section at $z = 4.25$ mm; (**j**) – (**l**) *u*, *v*, *w* fields of the transverse section at $z = 6.50$ mm

4.4 A Woven Sandwich Beam Under 3-Point Bending

The size of the sandwich beam is 50.0 mm × 20.0 mm × 33.0 mm, the thickness of the face sheet is 3.8 mm, and the span between two supports is 30.0 mm. The face sheet is made of E-glass vinyl ester (EVE) composite with the fiber woven into a quasi-isotropic layout: $[0/45/90{-}45]_s$. The fibers are made of a 0.61 kg/m^2 areal density plain weave. Ashland Derakane Momentum 8084 resin are used as the matrix material. The two face sheets are of identical layups and materials. The

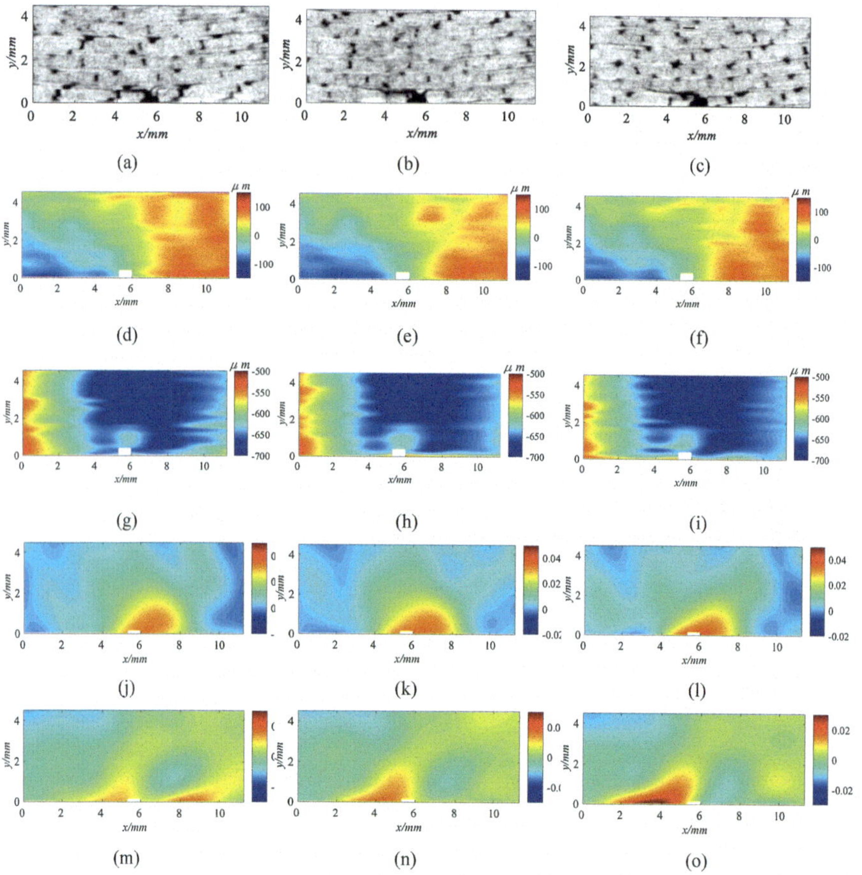

Fig. 11 Displacement and strain distributions of Step 9 near the slot in different sections (**a**)section along $z = 2.00$ mm of Step 10; (**b**) section along $z = 4.25$ mm of Step 10; (**c**) section along $z = 6.25$ mm of Step 10; (**d**) u field of (**a**); (**e**) u field of (**b**); (f) u field of (**c**); (**g**) v field of(**a**); (**h**) v field of (**b**); (**i**) v field of (**c**); (**j**) ε_{xx} field of (**a**); (**k**) ε_{xx} field of (**b**); (**l**) ε_{xx} field of (**c**); (**m**) ε_{xy} field of (**a**); (**n**) ε_{xy} field of (**b**); (**o**) ε_{xy} field of (**c**)

core material is made of Corecell™ P600 styrene foam [29]. The whole loading process is divided into 6 steps [30]. In each step, the specimen was CT scanned and a volumetric image with 890 voxel × 360 voxel × 530 voxel was reconstructed, and the physical size of a voxel is 55 μm × 55 μm × 55 μm. The volume image of the specimen and the loading history are shown in Fig. 12.

Since the length of the beam is rather short, at the end of the loading history we did not observe any shear crack in the foam core or along the interface of face sheet and core. The volume image of Step 1 was designated as the reference image, whereas the subsequent images were the deformed images. By applying the DVSP algorithm, the displacement contours of u, v, w of the core after each incremental loading from Step 2 to Step 5, were calculated. The core region with the size of

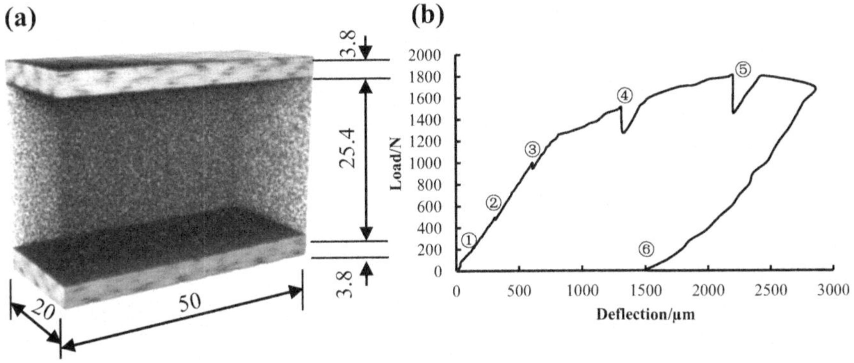

Fig. 12 3D image and loading history: (**a**) Reconstructed 3D image of the specimen; (**b**) loading history

890 voxel × 360 voxel × 430 voxel was boxed in for the calculation. The coarse-fine calculation process was applied, the final subset size was 32 voxel × 32 voxel × 32 voxel, and the subset shift was 5 voxels. The displacement contours of the core under different loadings are depicted in Fig. 13. The first three patterns, Figs. 13(a), 13(b) and 13(c), show clearly the characteristic deformation of a short beam under 3-point bending. The distributions of u and v fields are self-explanatory. The w-field shows a barrel-like bulging of the central part of the beam, a pattern similar to the deformation of a short square column under compression with the movement of both ends constrained. As the loading increases the displacement fields lose some of the symmetry as shown in subsequent pictures in Fig. 13. This is because the loading was not exactly symmetrical, a common occurrence in a typical testing arrangement.

In the process of the strain calculation, the size of the calculation cubic box is 31 × 31 × 31 points, and the shift step is 5 points. Since a sandwich beam under bending tends to fail in the form of shear failure of the core, we only calculated the in-plane shear strain distributions, and only at loading steps 3, 4, and 5 with loads being 1000 N, 1500 N, and 1800 N, respectively. As depicted in Fig. 14, ε_{xy} contours of the entire core within the cropped region for the three loads are shown in Figs.14(a), 14(b), and 14(c), respectively. Figures 14(d), 14(e), and 14(f) depict the in-plane shear strain distributions in the plane of $z = 5.875$ mm; Figs. 14(g), 14(h) and 14(i) depict the shear strains in the plane of $z = 10.00$ mm, and Figs. 14(j), 14(k) and 14(l) depict the shear strains in the plane of $z = 14.12$ mm under the same three different loadings, respectively. It is clearly seen that shear strains are localized along two 45 degree bands fanning out from the point where the concentrated load was applied. If the sandwich beam were longer, it is believed that failure would probably occur along these two shears bands. From the patterns shown in Fig. 14(c), it is reasonable to suspect that failure would probably initiate from the interior of the foam core.

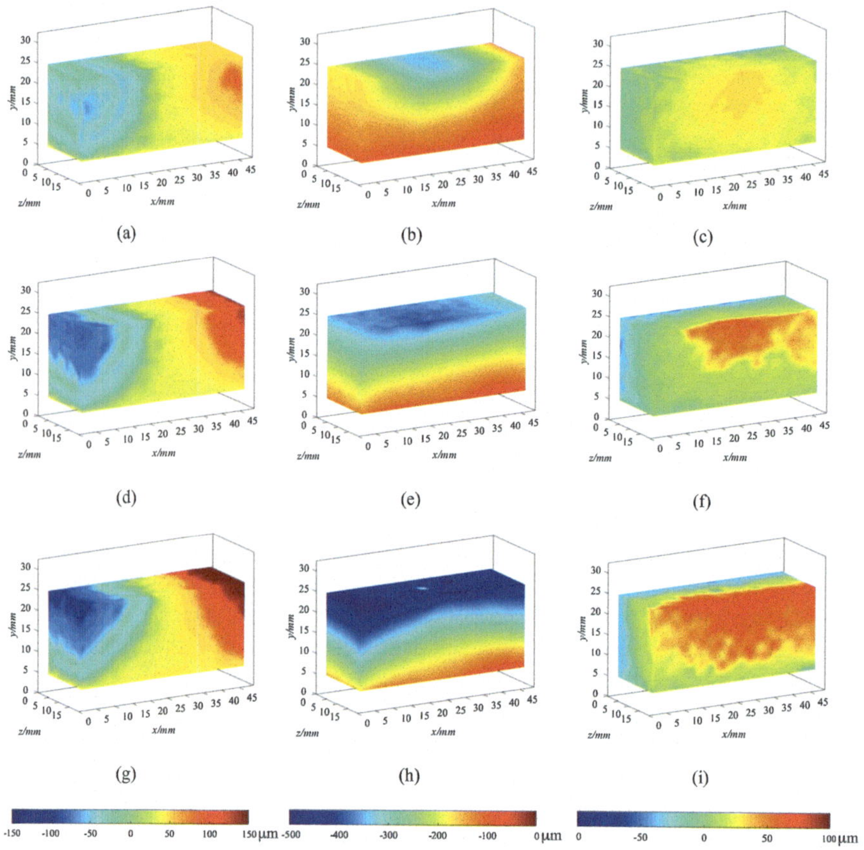

Fig. 13 Displacement distribution of the core under different loadings: (a)-(c) u, v and w fields, respectively, at Step 3 with loading 1000 N; (d)-(f) u, v and w fields, respectively, at Step 4 with loading 1500 N; (g)-(i) u, v and w fields, respectively, at Step 5 with loading 1800 N

5 Discussion

5.1 The Effect of Subset Size

One of the important features of the DVSP method is the fact that the interior meso- or microstructures of the specimen material are treated as volumetric speckles. These structural features tend to vary from material to material and from region to region. They don't form ideal speckle patterns. Thus, it is important that the size of the subset be selected judiciously when applying DVSP. As a demonstration of the size effect of subset, we cropped a cubic block with 200 voxel × 200 voxel × 200 voxel from each of the two volume images of the two tested materials as shown in Figs. 15 (a) and 16(a). These two images are defined as the reference volume images, whereas the "deformed" volume images are obtained by the Fourier shifting method. For

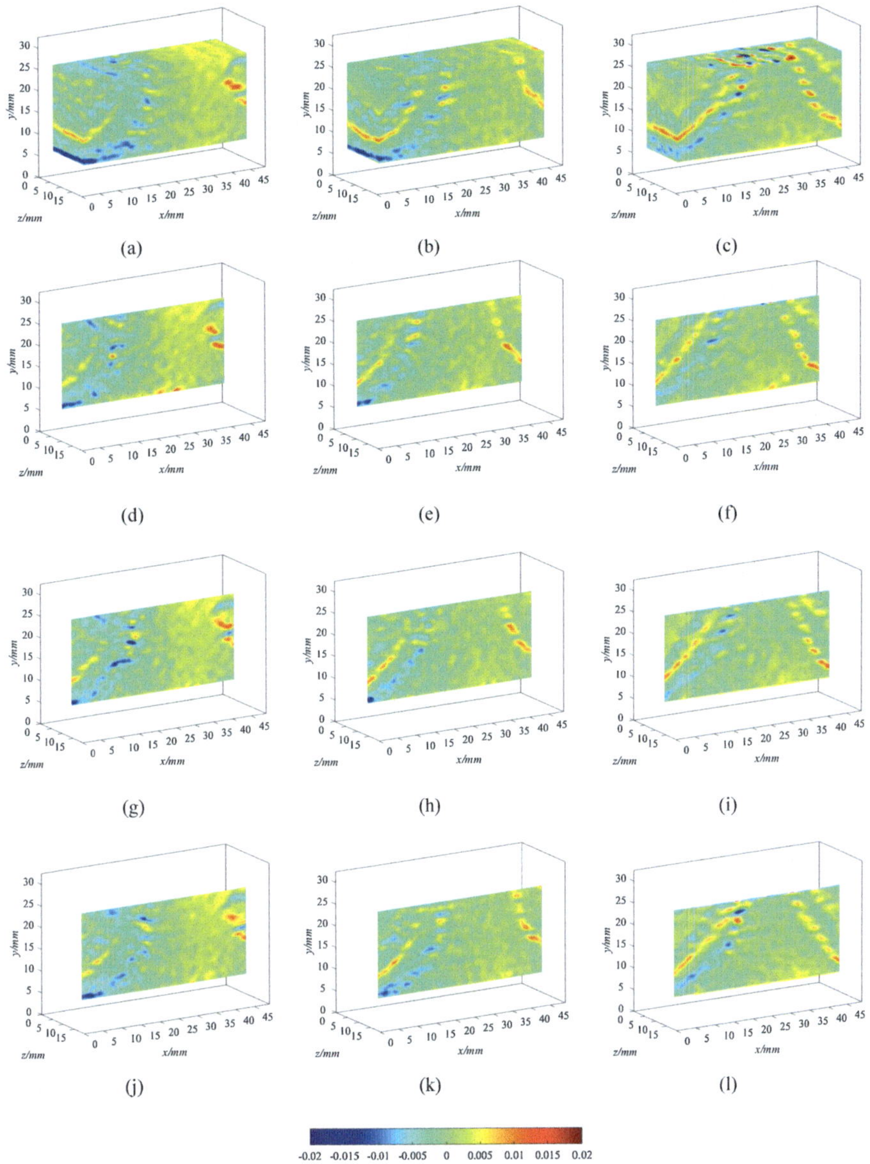

Fig. 14 Shear strain distribution of the core under different loadings: (**a**)-(**c**) ε_{xy} field of the whole core under loadings of 1000 N, 1500 N and 1800 N, respectively; (**d**)-(**f**) ε_{xy} fields of the longitudinal section at $z = 5.875$ mm under loadings of 1000 N, 1500 N and 1800 N, respectively; (**g**)-(**i**) ε_{xy} fields of the longitudinal section at $z = 10$ mm under loadings of 1000 N, 1500 N and 1800 N, respectively; (**j**)-(**l**) ε_{xy} fields of the longitudinal section at $z = 14.125$ mm under loadings of 1000 N, 1500 N and 1800 N, respectively

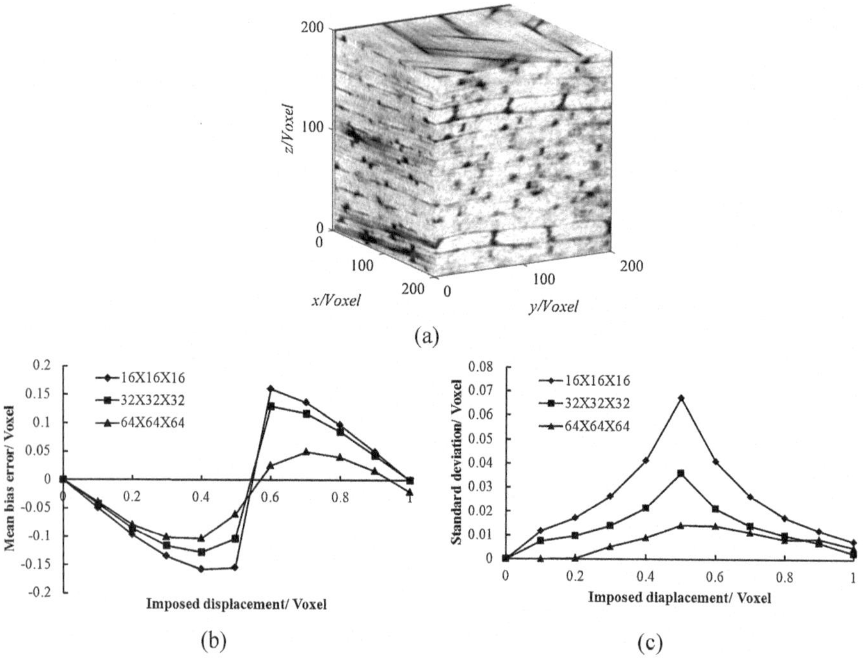

Fig. 15 Accuracy of DVSP on the woven composite specimen: (**a**) a cubic volume image of the woven composite specimen; (**b**) mean bias error of the displacement; (**c**) standard deviation of the displacement

every reference volume image, ten different "deformed" volume images with sub-voxel rigid body translation ranging from 0.1 to 1.0 voxel, respectively, along the z direction were obtained. In the fine calculation, the subset size with 16 voxel $\times$16 voxel $\times$16 voxel, 32 voxel $\times$ 32 voxel $\times$ 32 voxel and 64 voxel $\times$ 64 voxel $\times$ 64 voxel, respectively, were used. And the subset shift is 5 voxels. The resulting mean bias errors and the standard deviation errors of displacements are shown in Figs. 15(b), (c) and 16(b), (c), respectively. It is seen that the larger the subset size, the smaller the resulting error. However, as the subset size increases, fewer independent measurement points can be obtained which will influence the resolution of strain calculation [31]. It is noted that the meso-structure of the foam core is much more uniform than that of the woven composite, thus gives rise to better accuracies.

In the DVSP theory, the subset is assumed to be rigid. In reality there is deformation and rotation of the material within the subset. While by increasing the subset size results in better accuracy of DVSP, the errors caused by ignoring the deformation and rotation of the material within the subset tend to increase too. To estimate the effects of deformation within the subset on DVSP, we used a quadratic displacement model test [32]. The above blocks with 200 voxel $\times$ 200 voxel $\times$

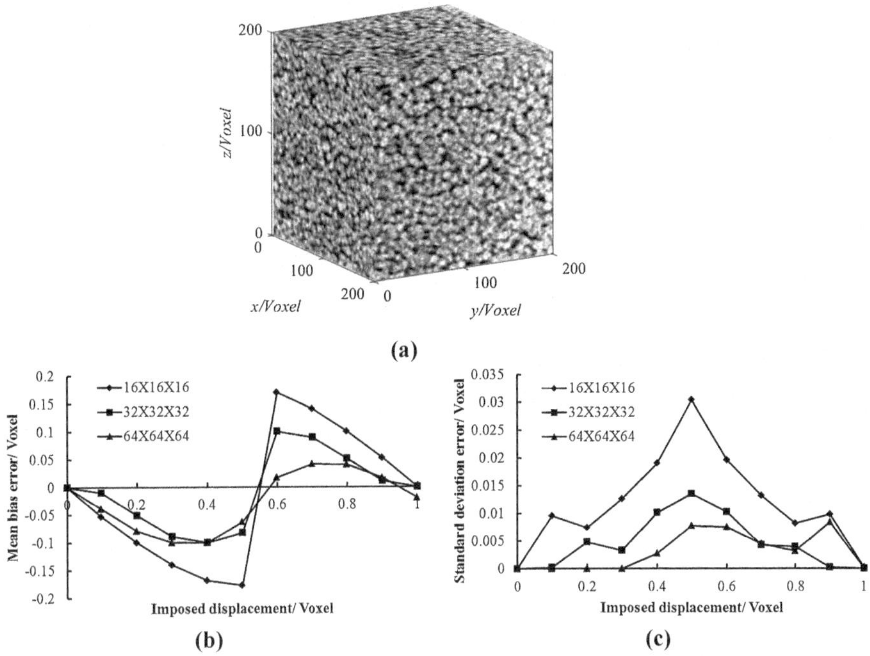

Fig. 16 Accuracy of DVSP on the core of the sandwich specimen: (**a**) a cubic volume image of the woven composite specimen; (**b**) mean bias error of the displacement; (**c**) standard deviation of the displacement

200 voxel was again used as the reference volume image, whereas the "deformed" volume image was obtained as follow: the block was given a rigid displacement of 2 voxels along the z axis, and then divided into three regions along the z axis. Region 1 ($1 \leq z < 68$ voxels) has a linear displacement field with strain $\varepsilon_1 = -0.1\%$; Region 2 ($68 \leq z < 134$ voxels), has a quadratic displacement with strain varying from $\varepsilon_1 = -0.1\%$ to $\varepsilon_2 = 0.5\%$; and Region 3 ($134 \leq z < 200$ voxels) has a linear displacement field with strain $\varepsilon_2 = 0.5\%$. The displacement fields were calculated by using DVSP with different subset sizes having 16 voxel $\times$16 voxel $\times$16 voxel, 32 voxel $\times$ 32 voxel $\times$ 32 voxel and 64 voxel $\times$ 64 voxel$\times$ 64 voxel, respectively, and all with 5 voxels shift. The displacement and strain were calculated and are depicted in Fig. 17. It is noted that the deformation in subset do result in noticeable error. In Region 1 and Region 2, smaller subsets capture the solution more accurately, whereas in Region 3, larger subset has better results. For all these three regions, the optimal subset size is 32 voxel $\times$ 32 voxel $\times$ 32 voxel. In order to reduce the influence of subset deformation, one can use the iterative least-squares method [27], but iterative procedures tend to increase the implementation complexity resulting in more computational expenses.

The deformation between the reference subset and the corresponding deformed subset would cause the correlation coefficient to decrease. We imposed a set of uniform deformation with strains from 0.1% to 1% with the interval 0.1% on the

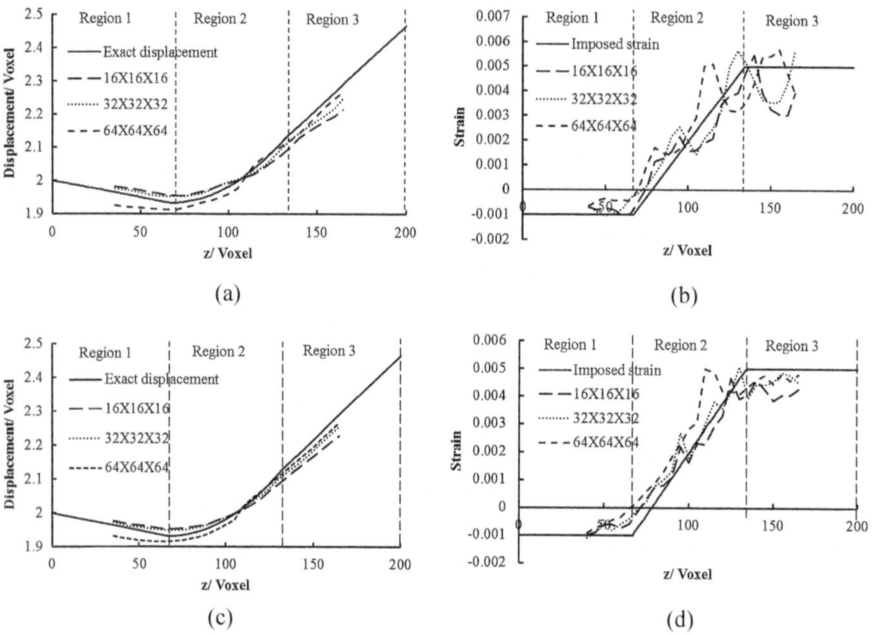

Fig. 17 Calculated results of DVSP in the quadratic displacement test: (**a**) Exact displacement and displacement curves of the woven composite with different subset sizes; (**b**) Imposed strain and calculated strain curves of the woven composite with different subset sizes; (**c**) Exact displacement and displacement curves of the core in the sandwich with different subset sizes; (**d**) Imposed strain and calculated strain curves of of the core in the sandwich with different subset sizes

woven composite block shown in Fig. 15(a) to obtain the deformed volume images, and calculated the correlation coefficients between the reference image and the deformed image. The subset size was 32 voxel × 32 voxel × 32 voxel, and the subset shift was 5 voxels. The correlation coefficient curve with strain is shown in Fig. 18(a). The coefficient decreases in a nearly quadratic function as the strain increases, and the standard deviation rises as the strain increases. When the strain is 1%, the coefficient is 82.2%, and the standard deviation is 0.12. In Fig. 18(b), the curve shows that the influence of the displacement is more than one voxel on the correlation coefficient. When the displacement is 8 voxels, one fourth of subset, the coefficient drops 53.4%, and standard deviation to 0.11. Figures 18(c) and 18(d) are typical 2D normalized impulse function distributions under the influence of deformation and integer voxel displacement, respectively. When the signal-to-noise-ratio decreases, it would affect the detection of the crest of the impulse functions. For the purpose of increasing signal-to-noise-ratio, we carried out a coarse-fine calculation process. In the coarse calculation, integer voxel displacement evaluation is more robust with the large subset size, and strain calculation resolution can be supported in the fine calculation with the small subset size.

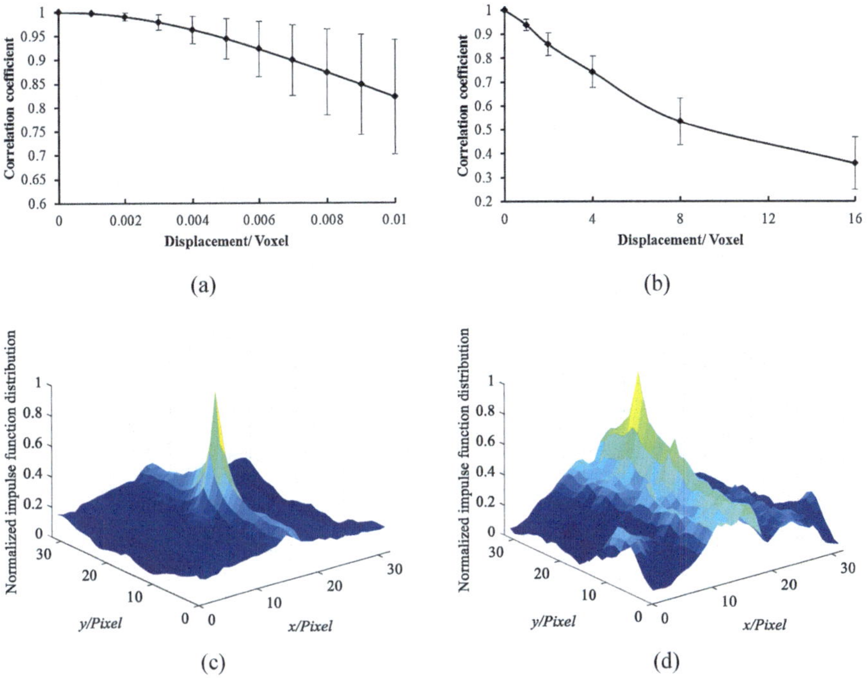

Fig. 18 Influence on correlation coefficient of the deformation and the integer voxel displacement: (**a**) the curve of correlation coefficient with different strain deformation; (**b**) the curve of correlation coefficient with different integer voxel displacement; (**c**) a 2D correlation coefficient distribution with strain deformation influence; (**d**) a 2D correlation coefficient distribution with 8 voxels displacement influence

5.2 *Influence of Artifacts in CT Images*

CT slice images are reconstructed with an appropriate mathematical algorithm from the different angular radiographic projections. The non-uniformity of detector elements, the polychromatic nature of the X-ray, the imperfect motion of the rotation stage, the temperature variation of the X-ray tube, and the possible rigid body motion of the specimen will all give rise to different artifacts, such as streak, ring, and beam hardening [33]. These artifacts will influence the measurement results. To analyze the effects of artifacts of the micro X-ray CT system, we took two consecutive scans of a woven composite specimen with identical settings and without moving (other than the tomographic rotation) or deforming the sample and designated them as Scans 1 and 2, respectively. Noise and artifacts of the system were present in the reconstructed image of both scans. Two volumetric images with a size of 200 voxel × 200 voxel × 200 voxel were cropped from the reconstructed images. The physical size of a voxel was 42 μm × 42 μm × 42 μm.

We defined the volumetric images of Scans 1 and 2 as the reference image and deformed volumetric image, respectively. We then assess the influence of artifacts in

Table 1 Least squares fit of the displacements to the exact solution with rigid body translation and small angle rotation

u_{rigid}	v_{rigid}	w_{rigid}	θ_x	θ_y	θ_z
0.74	0.89	0.60	1.58×10^{-4}	2.22×10^{-4}	1.93×10^{-4}

CT images using the DVSP algorithm with subvoxel shifting. The average measured displacements are $u = 0.75$ voxels, $v = 0.83$ voxels, and $w = 0.69$ voxels, and the standard deviation errors are 0.12 voxels, 0.18 voxels and 0.20 voxels, respectively. All of the measured displacements are greater than the standard deviation errors, indicating that some motion occurred between scans due to physical perturbations in the CT system. Because there are real noise and artifacts in the volumetric images of Scans 1 and 2, the above results can depict the uncertainty of the algorithm in its application.

As a more accurate alternative to fitting just rigid body movements, a least squares fit was used to calculate rigid body translations and rotations, assuming small angles [34], i.e.

$$\begin{aligned} u_{\text{rigid}} + \theta_z Y - \theta_y Z &\cong U \\ v_{\text{rigid}} - \theta_z X + \theta_x Z &\cong V \\ w_{\text{rigid}} + \theta_y X - \theta_x Y &\cong W \end{aligned} \tag{12}$$

where θ_x, θ_y, and θ_z are the rotations about the x, y, and z axes, respectively, and X, Y, Z and U, V, W are the vectors of the x, y, z coordinates and u, v, w displacements, respectively, for all of the correlation points. The results are shown in million. There are rigid body translations similar to the average measured displacements values. Thus, when compared with the volumetric image of Scan 1, the volumetric image of Scan 2 incurred noticeable rigid body translations in all three directions but with small rotations (Table 1).

6 Conclusion

This study describes in detail a new experimental method, called DVSP (Digital Volumetric Speckle Photography), that is capable of probing the interior 3D deformation field of composites. A woven composite beam, a woven composite beam with a prepared slot and a sandwich composite beam are selected to demonstrate the technique's unique capability. DVSP takes advantage of modern industrial X-ray CT's capability to obtain a 3D volume image of a solid object in digital form. The solid's internal meso/micro structures are treated as 3D volumetric speckles. The digitized volume image is subdivided into an array of cubic voxels of certain predetermined size and processed using a 2-step 3D FFT algorithm. The result is a 3D array of displacement vectors, from which the strain field can be calculated using an appropriate strain-displacement relation. Better results can be obtained with a higher power X-ray, which has a better resolution, a more rugged CT system and a more efficient software. Even with the current state of art, we believe DVSP can contribute significantly to the understanding of thick composite's 3D effects and failure mechanism.

Acknowledgments This work was financially supported by the US Office Of Naval Research's Solid Mechanics Program grant No: N0014-14-1-0419, the National Natural Science Foundation of China (51374211) and the Laboratory for Experimental Mechanic Research of the Department of Mechanical Engineering at Stony Brook University. F. P. Chiang wishes to thank Dr. Yapa Rajapakse, Director of the US Office of Naval Research's Solid Mechanics Program, for his support over the years for the advancement of the speckle technique.

References

1. Butters JN, Leendertz JA (1971) Speckle pattern and holographic techniques in engineering metrology. Opt Laser Technol 2:26–30
2. Leendertz JA (1970) Interferometric displacement measurement on scattering surfaces utilizing speckle effect. J Phys 3:214–218
3. Daniel IM, Rowlands RE, Post D (1973) Strain analysis of composites by moiré methods. Exp Mech 13:246–252
4. Post D, Han B, Ifju PG (1994) High sensitivity Moiré. In: Experimental analysis for mechanics and materials. Springer, Berlin
5. Chang S, Hong D, Chiang FP (2004) Macro and micro deformations in a sandwich foam core. Compos Part B 35:503–509
6. Fathi A, Keller JH, Altstaedt V (2015) Full-field shear analyses of sandwich core materials using digital image correlation (DIC). Compos Part B 70:156–166
7. Pollock P, Yu L, Sutton MA, Guo S, Majumdar P, Gresil M (2012) Full-field measurements for determining orthotropic elastic parameters of woven glass-epoxy composites using off-axis tensile specimens. Exp Tech 38:61–71
8. Gre'diac M (2004) The use of full-field measurement methods in composite material characterization: interest and limitations. Compos Part A 35:751–761
9. Ju Y, Wang L, Xie HP, Ma GW, Mao LT, Zheng ZM, Lu JB (2017a) Visualization of the three-dimensional structure and stress field of aggregated concrete materials through 3D printing and frozen-stress techniques. Constr Build Mater 143:121–137
10. Ju Y, Wang L, Xie HP, Ma GW, Zheng ZM, Mao LT (2017b) Visualization and Transparentization of the structure and stress field of aggregated Geomaterials through 3D printing and Photoelastic techniques. Rock Mech Rock Eng 50:1383–1407
11. Sciammarella CA, Chiang FP (1964) The moire method applied to three-dimensional elastic problems. Exp Mech 4(11):313–319
12. Asundi A, Chiang FP (1982) Theory and application of white light speckle methods. Opt Eng 21:570–580
13. Dudderar TD, Simpkins PG (1982) The development of scattered light speckle metrology. Opt Eng 21:396–399
14. Schilling PJ, Karedla BR, Tatiparthi AK, Verges MA, Herrington PD (2005) X-ray computed microtomography of internal damage in fiber reinforced polymer matrix composites. Compos Sci Technol 65:2071–2078
15. Nikishkov Y, Airoldi L, Makeev A (2013) Measurement of voids in composites by X-ray computed tomography. Compos Sci Technol 89:89–97

16. Roux S, Hild F, Viot P, Bernard D (2008) Three-dimensional image correlation from X-ray computed tomography of solid foam. Compos Part A 39:1253–1265
17. Brault R, Germaneau A, Dupré JC, Doumalin P, Mistou S, Fazzini M (2013) In-situ analysis of laminated composite materials by X-ray micro-computed tomography and digital volume correlation. Exp Mech 53:1143–1151
18. Lecomte-Grosbras P. R'ethor'e J. Limodin N. Witz J.-F. Brieu M (2015) Three-dimensional investigation of free-edge effects in laminate composites using X-ray tomography and digital volume correlation. Exp Mech 55:301–311
19. Chiang FP, Mao LT (2015) Development of interior strain measurement techniques using random speckle patterns. Meccanica 50:401–410
20. Chen DJ, Chiang FP, Tan YS, Don HS (1993) Digital speckle-displacement measurement using a complex spectrum method. App Opt 32:1839–1849
21. Chiang FP (2009) Super-resolution digital speckle photography for micro/nano measurements. Opt Lasers Eng 47:274–279
22. Mao LT, Chiang FP (2016) 3D strain mapping in rocks using digital volumetric speckle photography technique. Acta Mech 227:3069–3085
23. Mao LT, Hao N, An LQ, Chiang FP, Liu HB (2015) 3D mapping of carbon dioxide-induced strain in coal using digital volumetric speckle photography technique and X-ray computer tomography. Int J Coal Geol 147-148:115–125
24. Mao LT, Yuan ZX, Yang M, Liu HB, Chiang FP (2019) 3D strain evolution in concrete using in situ X-ray computed tomography testing and digital volumetric speckle photography. Measurement 133:456–467
25. Sjodahl M (1997) Accuracy in electronic speckle photography. App Opt 36:2875–2885
26. Liu JY, Iskander M (2004) Adaptive cross correlation for imaging displacements in soils. J Comput Civ Eng 18:46–57
27. Pan B, Wu DF, Wang ZY (2012) Internal displacement and strain measurement using digital volume correlation: a least-squares framework. Meas Sci Technol 23:45002
28. Mao LT, Chiang FP (2015) Interior strain analysis of a woven composite beam using X-ray computed tomography and digital volumetric speckle photography. Compos Struct 134:782–788
29. Wang E, Shukla A (2012) Blast performance of sandwich composites with in-plane compressive loading. Exp Mech 52:49–58
30. Mao LT, Chiang FP (2017) Mapping interior deformation of a composite sandwich beam using digital volumetric speckle photography with X-ray computed tomography. Compos Struct 179:172–180
31. Bergonnier S, Hild F, Roux S (2005) Digital image correlation used for mechanical tests on crimped glass wool samples. J Strain Analysis 40:185–197
32. Gates MR (2015) Subset refinement for digital volume correlation: numerical and experimental applications. Exp Mech 55:245–259
33. Limodin N, Rethore J, Adrien J, Buffiere JY, Hild F, Roux S (2011) Analysis and artifact correction for volume correlation measurements using tomographic images from a laboratory X-ray source. Exp Mech 51(6):959–970
34. Gates M, Lambros J, Heath MT (2011) Towards high performance digital volume correlation. Exp Mech 51(4):491–507

Index

S. W. Lee (ed.), *Advances in Thick Section Composite and Sandwich Structures*,
https://doi.org/10.1007/978-3-030-31065-3

D

G

H

Q

R

The manufacturer's authorised representative in the EU is Springer Nature Customer Service Centre GmbH, Europaplatz 3, 69115 Heidelberg, Germany. If you have any concerns regarding our products, please contact ProductSafety@springernature.com

Printed and bound by CPI Group (UK) Ltd, Croydon, CR0 4YY
30/09/2026
02225241-0018